MARSHALL B. CLINARD *University of Wisconsin at Madison*
ROBERT F. MEIER *University of Nebraska at Omaha*

Sociology of Deviant Behavior

ELEVENTH EDITION

HARCOURT COLLEGE PUBLISHERS

Fort Worth   Philadelphia   San Diego   New York   Orlando   Austin   San Antonio
Toronto   Montreal   London   Sydney   Tokyo

Publisher	Earl McPeek
Acquisitions Editor	Bryan Leake
Market Strategist	Laura Brennan
Project Editor	Michele Tomiak
Art Director	Brian Salisbury
Production Manager	Diane Gray

Cover credit: *Orange Glow* is a powerful image which we feel conveys a message of optimism, hope, and understanding—an image that can translate well the subject matter of this text.

© Adolph and Esther Gottlieb Foundation/Licensed by VAGA, New York, NY

ISBN: 0-15-506569-6
Library of Congress Catalog Card Number: 00-108414

Copyright © 2001, 1998, 1995, 1992, 1989, 1979, 1974, 1968, 1963 by Harcourt, Inc.

Address for Domestic Orders
Harcourt College Publishers, 6277 Sea Harbor Drive, Orlando, FL 32887-6777
800-782-4479

Address for International Orders
International Customer Service
Harcourt, Inc., 6277 Sea Harbor Drive, Orlando, FL 32887-6777
407-345-3800
(fax) 407-345-4060
(e-mail) hbintl@harcourt.com

Address for Editorial Correspondence
Harcourt College Publishers, 301 Commerce Street, Suite 3700, Fort Worth, TX 76102

Web Site Address
http://www.harcourtcollege.com

Printed in the United States of America

0 1 2 3 4 5 6 7 8 9 076 9 8 7 6 5 4 3 2 1

Harcourt College Publishers

To my children, from whom I have learned a great deal
Marsha
Stephen
Lawrence

Marshall B. Clinard

To my parents, who made my learning possible
Frank
Eileen

Robert F. Meier

Sociology of Deviant Behavior presents a theoretical overview of the nature and meaning of deviance, examining in detail a number of forms of behavior commonly regarded as deviant. Throughout the book, sociological concepts and processes underlie the presentation. We have attempted to identify and explain the leading theories and sociological orientations of deviant behavior: anomie, control, labeling, conflict, and learning. We have also attempted to be sensitive to other perspectives where they apply, both in sociology (such as the rational choice perspective) and in other disciplines, such as biology. The theoretical frame of reference throughout the book is socialization, or learning, theory with a normative perspective. The reader will see that we find the meaning of deviant behavior in the context of the acquisition of all behavior. The central theme of the book is that understanding deviant behavior is no different from understanding any other behavior; deviant behavior is human behavior, understandable within a general context of socialization and role playing. This frame of reference furnishes theoretical continuity throughout the book, although we have taken care to include other viewpoints as well. Where possible, we have also attempted to illustrate sociological ideas from the deviant's own perspective through case histories or personal accounts.

It is not easy to define deviant behavior. Often any consensus that has appeared to exist has been the result of political, social, and economic powers of groups that have succeeded in imposing on others their views of what constitutes deviance. Here, we examine the merits of four definitions of deviance: statistical, absolutist, reactivist, and normative. We have adopted the *normative* definition as best fitting the complex society in which we live and the increasingly complex global community, which is characterized by a high degree of differentiation and, as a result, a high degree of deviance.

As with previous editions, this eleventh edition is a complete revision that incorporates the most recent theoretical developments in the field and the latest research findings. Reviewers of previous editions have suggested changes. In the eleventh edition, we have placed more emphasis on some forms of deviance and issues of social control that are of great contemporary concern—for example, drugs and violence, both personal and family. There is a new chapter on white-collar and corporate crime. And certain chapters have been reorganized in a new part, "Studies in Stigma." Here, we concentrate not only on the nature of deviance but on the reactions toward and consequences of deviant behavior and conditions.

We have augmented the material throughout with first-person accounts to illustrate some of the sociological concepts and theories discussed. We have attempted to devote attention to "newer" forms of deviance, including eating disorders such as anorexia and bulimia. There is more attention throughout on the impact of technology on both deviant behavior and its control. Every chapter has been updated; some, such as the material on heterosexual deviance, has been enlarged.

Chapter 1 deals with the nature and definition of deviance. It introduces the sociological concepts necessary to understand the processes as well as the theories of deviance that follow. Chapter 2 concerns the nature of deviant events and social control. It explores the close link between processes of deviance and its control, and it provides a conceptual background to the discussions of social control in the substantive chapters that follow. Chapter 3 presents a discussion of general sources or contexts of deviance. We have concentrated on a few of the many sources of deviance, with particular attention to the importance of the social context of deviant acts and to current disputes about the role of the media in generating deviance.

Chapter 4 introduces the student to the individual and group processes that shape deviant behavior and deviant careers. A number of select perspectives on individual deviance are also introduced. The next chapters deal with sociological perspectives and theories of deviance. Chapter 5 examines and contrasts two major sociological theories of deviance: anomie and conflict perspective. Chapter 6 examines three other theories: control, labeling, and learning. New to this edition is the use of a short case study to introduce the nature of each theory.

We then shift to an in-depth examination of various forms of deviant behavior. Chapters 7 and 8 identify the processes involved in crimes of interpersonal violence and nonviolent crimes. Chapter 9 is the new chapter on white-collar and corporate criminality. Chapter 10 deals with drug use, while Chapter 11 is devoted to alcohol, the drug most widely used. Chapter 12 focuses on forms of sexual deviance, and Chapter 13 on suicide. Chapter 14 analyzes physical disabilities and eating disorders, and Chapter 15 covers homosexuality and homophobia from a sociological point of view. We conclude in Chapter 16 with a sociological examination of mental disorders and the stigma suffered by persons suffering from these disorders. These last three chapters are examples of conditions often regarded sociologically as deviant, with profound social and personal implications for the self-concept of the individual.

This book first appeared in 1957. It pioneered a major shift from the then characteristic approach to deviance, termed *social disorganization* or *social problems*, to a more basically sociological orientation built around the concept of normative deviance and deviant behavior. Subsequently, the conceptual framework of *deviant behavior* has received wide acceptance and use in sociology. In this eleventh edition, we continue the tradition begun more than 45 years ago of attempting to understand deviance in its social context. This edition also emphasizes that deviance is an inescapable feature of modern, complex societies because such societies are characterized by a system of ranked social differentiation (stratification) that is generally associated with many types of social deviance. We also wish to affirm in this edition the obvious relationship between deviance and social order and the need for a sociological understanding of all aspects of society in order to comprehend the nature and complexity of social deviance.

Over the years, numerous sociologists and friends have contributed the basic data for this book through their theoretical writings and research on deviance. The references in the book acknowledge most, but not all, of them. At various times, other sociologists have critiqued various editions, including the present one, and they have thus contributed valuable ideas and suggestions. We are grateful to all of them. We also wish to acknowledge especially the assistance of June Turner, Geri Murphy, Kris Piessig, Angela Patton, Tim Himberger, and Steve Culver, whose various technical skills help make this edition possible.

A special thanks is due Lin Marshall and Stacy Schoolfield. Stacy's assistance in particular was invaluable and much appreciated.

We wish specifically to thank the following reviewers who provided valuable suggestions for this edition of the book: Christopher F. Armstrong, Bloomsburg University; Korni Swaroop Kumar, SUNY college at Potsdam; Kevin M. Thompson, North Dakota State University; Nathaniel Eugene Terrell, Emporia State University; John Broderick, Stonehill College; and David Struckhoff, Loyola University.

An instructor's manual for this text is available and may be obtained through a local Harcourt representative or by writing to the Sociology Editor, Harcourt College Publishers, 301 Commerce Street, Suite 3700, Fort Worth, Texas 76102. Suggestions are always welcome. Please feel free to contact either of the authors by mail or, if you prefer, by e-mail at rmeier@unomaha.edu.

M. B. C.
R. F. M.

CONTENTS IN BRIEF

CONTENTS

PART I

The Nature of Deviance

DRINKING beer may not be considered deviant for college partygoers, but it would be at a meeting of Alcoholics Anonymous. What is deviant for one person or group may not be for another. But the term *deviance* has a general meaning on which there is substantial agreement. In this first part, we introduce the idea of deviance and identify important elements of its social context. Deviance does not take place in a social vacuum since the concept of deviance is uniquely sociological. It takes place more in some places than in others, at some times more than at others, and in some situations more than in others. This is the basic theoretical problem in the sociology of deviance. Can we account for these times, places, and groups?

As these questions suggest, deviance occurs in a particular context, a combination of time, place, situations, and people. But to understand the structure of deviance, we have to know what deviance is and what it is not. We should also know that deviant acts are seldom ignored and that they often elicit reactions from others, ranging from mild disapproval to scorn to physical punishment. The nature of these reactions, often called social control, tells us much about the nature of deviance and its role in society. One of the contexts for deviance is in the social dynamics found in urban areas. This is not to say that deviance is found only in cities, but that urban forces are often conducive to processes of deviance and social control. ∎

CHAPTER 1

The Nature and Meaning of Deviance

Marsha Lindall is an attractive woman in her late 30s. She has a very successful professional career and recently moved to a new city because of a promotion. In the course of meeting new colleagues and friends, Marsha has increasingly had the sense that she is in some way an outsider. It has become clear to her in conversations that some of the new people she is meeting regard her "family" situation as unusual. Although Marsha was married and subsequently divorced, she has not had children. Some of the people she has been meeting seem to think that this is deviant. And Marsha is beginning to feel excluded from social occasions where there are couples and children. She is beginning to think that she is in some sense deviant.

Before we can see whether this is the case, we have to understand the meaning of the term.

Deviance takes many forms, but agreement remains elusive about which specific behaviors and conditions constitute deviance. This ambiguity becomes especially evident when some people praise the same behavior that others condemn. If the concept were not confusing enough already, many discussions of deviance evoke strong political and moral attitudes, prompting some groups to call on the law to support their views of certain acts (e.g., homosexuality, abortion). Even everyday behavior can have this moral tone. The centuries-long debate on whether mothers should breast- or formula-feed infants contains moral dimensions. A study of first-time mothers in England concluded that whether they intend to breast- or formula-feed, women face considerable challenges from those who disagree with them as the mothers seek to establish that they are not only good mothers but also good partners and good women (Murphy, 1999). Not even the time-honored institution of motherhood is immune from allegations of deviance.

To understand deviance, one must first understand this contradiction: No consensus reliably identifies behavior, people, or conditions that are deviant, although most people would say that they know deviance when they see it. Many lists would include mental disorder, suicide, crime, homosexuality, and alcoholism. Yet disagreement casts shadows over even this basic list of "generally accepted" forms of deviance. Some, for example, deny that homosexuality is deviant in any way. One person may see problem drinking in behavior that may represent no such problem to another. Many segments of the population dispute the harm of certain crimes, such as prostitution and the use of marijuana or cocaine (Meier and Geis, 1997). Attitudes toward deviance may resemble St. Augustine's comment about time: One knows pretty much what it is—until one is asked to define it.

Even scholars who study deviance fail to reach agreement on which people, acts, or conditions fall within that topic. Cohen (1966: 1) says books about deviance address "knavery, skullduggery, cheating, unfairness, crime, sneakiness, betrayal, graft, corruption, wickedness, and sin." Gouldner (1968) complains that the empirical literature on deviance has been limited largely to "the world of the hip, night people, drifters, grifters, and skidders: the 'cool world.'" Howard Becker (1973) limited his influential study on deviance to jazz musicians and marijuana users. A British collection of papers on deviance dealt with drug users, thieves, hooligans, suicides, homosexuals and their blackmailers, and industrial saboteurs (Cohen, 1971). Lemert (1951) illustrates his theoretical position on deviance with references to, among other examples, the blind and stutterers. Dinitz, Dynes, and Clarke (1975) list several types of people as examples of deviance: midgets, dwarfs, giants, sinners, heretics, bums, tramps, hippies, and Bohemians. Becker (1977) finds the "genius" deviant. Liazos (1972) attempts to capture the essence of deviance by with the phrase, "nuts, sluts, and 'preverts.'" Henslin (1972) discusses four types of deviants to illustrate research problems in the field: cabbies, suicides, drug users, and abortion patients. Stafford and Scott (1986: 77) offer the following list of disapproved conditions: "old age, paralysis, cancer, drug addiction, mental illness, shortness, being black, alcoholism, smoking, crime, homosexuality, unemployment, being Jewish, obesity, blindness, epilepsy, receiving welfare, illiteracy, divorce, ugliness, stuttering, being female, poverty, being an amputee, mental retardation, and deafness." Davis (1961) talks about blacks as deviants, and Lemelle (1995) claims that members of the U.S. middle class consider black males as deviant, and both Davis (1961) and Schur (1984) discuss women as examples of deviance. Many people would include witches on their lists of deviants (see Geis and Bunn, 1990), and some others would include people of unusual height (Adler and Adler, 2000: 13).

All of these writers reflect attitudes common in their times. Their lists include behavior (e.g., smoking), physical conditions (e.g., ugliness), and types of people (e.g., bums). Examples cite both voluntary acts (e.g., crime) and involuntary ones (e.g., stuttering). These lists seem to share few common threads, particularly when at least one presumed deviant—the genius—represents a quality often considered a positive value (see Goode, 1993). How can this be? Usually when we think of the idea of deviance, we are thinking about something that is disvalued. But one instructor of a course on deviance assigned her students to do random acts of kindness as a class project (Jones, 1998). Her students discovered that their behavior confused those to whom they were being kind. The recipients of the kindness were suspicious of the students' motives and regarded them as deviant.

Disagreement about whether specific acts are deviant need not prevent full discussion of the important dimensions of deviance: what it is, how to define it, which causes explain it, and what social groups can and should do to reduce it. This book examines a number of forms of deviance, including some mentioned in

FIGURE 1–1
Examples of Deviance

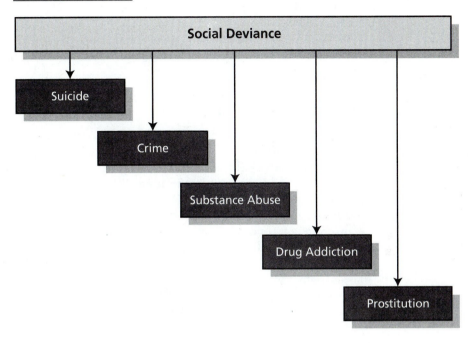

the lists above. Not everyone agrees that each form fits the definition of *deviant*, and the discussion will spark disagreement about the extent to which some acts represent deviance. We feel, however, that many readers will so regard them as deviant.

Clearly, different people would compose different lists of deviant people and acts, sometimes drastically different lists; yet no one should dismiss deviance as an idiosyncratic judgment, because many observers agree that a wide range of acts and people fit within that category. Also, these evaluations represent *group* opinions, not just individual ones. Some social acts and actors do, in fact, share a quality called *deviance*. Observers share some degree of understanding about the deviant nature of *some* acts and conditions. A formal definition identifies the common characteristics among these individual judgments.

What Is Deviance?

Some sociologists conceive of deviance as a collection of conditions, persons, or acts that society disvalues (Sagarin, 1975: 9) or simply finds offensive (Higgins and Butler, 1982: 3). These definitions avoid the critical question of how or why people classify acts or individuals as offensive to them and, hence, on what basis they disvalue those examples. Such a conception also fails to recognize the possibility that deviance might include highly valued differences, that society can encounter "positive" as well as "negative" deviance, as in the cases of the genius (see Dodge, 1985; but also see Sagarin, 1985) and the exceptional child (Zeitlin, Ghassemi, and Mansour, 1990).

But examples give insufficiently precise definitions of *deviance*. Only an explicit definition can fully identify a range of examples that might be considered deviant, in either a positive or a negative sense.

Normative Definition

A normative definition describes deviance as a violation of a *norm*. A norm is a standard about "what human beings should or should not think, say, or do under given circumstances" (Blake and Davis, 1964: 456; see also Birenbaum and Sagarin, 1976, and Gibbs, 1981). Violations of norms often draw reactions or *sanctions* from their social audiences. These sanctions constitute the pressures that most people feel to conform to social norms.

Two common conceptions characterize norms as evaluations of conduct and as expectations or predictions of conduct (Meier, 1981). The conception based on evaluation recognizes that some conduct (behavior or beliefs) ought or ought not to occur, either in specific situations (for example, no smoking in public elevators) or at any time or place (for example, no armed robbery, ever). The conception of norms as expectations or predictions highlights regularities of behavior based on habit or traditional customs. People expect a child to act a certain way in church, another way on the playground.

Norms are not necessarily clear-cut rules. Norms are social properties; they are shared group evaluations or guidelines. Rules come from some authority, which formulates them individually and imposes them on others (such as the laws of a monarch or despot). Rules and norms do share many characteristics, however, including the property of directing behavior; both are necessary components of social order (Bryant, 1990: 5–13).

One virtue of the normative conception comes from its answer to a question that stumps the reactivist conception: On what basis do people react to behavior? In other words, if deviance results only through the reactions of others, how do people know to react to or label a given instance of behavior? Norms supply the only obvious answer to this question. For this reason, the reactivist and normative conceptions may complement one another; norms provide the basis for reacting to deviance, but social reactions express norms and identify deviance.

People are considered deviant because of their behavior or conditions. People risk being labeled deviant by others when they express unaccepted religious beliefs (such as worshiping devils), violate norms pertaining to dress or appearance, or engage in proscribed sexual acts. Certain conditions also frequently lead people to label others as deviant, including physical handicaps and violations of appearance norms (e.g., obesity). People whose identities as deviant result from their beliefs or behavior fall into the category of achieved deviant status, while certain conditions may confer ascribed deviant status (Adler and Adler, 2000: 12–13).

In this book, we adopt a normative definition of deviance. Deviance constitutes departures from norms that draw social disapproval such that the variations elicit, or are likely to elicit if detected, negative sanctions. This definition incorporates both social disapproval of actions and social reactions to the disapproved actions (see also Best and Luckenbill, 1994: 2–4). Perhaps the key element in this conception is the idea of a norm. Norms do not simply operate in society. They are created, maintained, and promoted, sometimes in competition against one another. Society creates norms in much the same sense that the idea of deviance itself results from social construction and negotiation (Pfuhl and Henry, 1993).

Alternative Definitions

One of the most common definitions, often heard in everyday conversation, identifies deviance as a variation or departure from an average. This *statistical definition* emphasizes behavior that differs from average experience; it cites rare or infrequent phenomena. This definition assumes that what most people do determines the correct way to act.

It faces immediate difficulties, however, if it classes any minority as inherently deviant. A statistical definition applies the label *deviants* to those who have never stolen anything or violated the law, never used marijuana, never tasted alcoholic beverages, and never had premarital sexual relations (Rushing, 1975: 3–4). Statistical regularities of behavior do not reveal the *meaning* of deviance. Rather, the definition must connote some difference or departure from a standard of behavior—what "should" or "should not" be rather than "what is."

Another alternative definition of deviance relies on values. Values represent long-term, desired states—such as health, justice, and social equality. A society's values represent a general orientation for behavior without prescribing specific behavior. Values are important parts of an absolutist definition of deviance.

Some observers regard social rules as "absolute, clear and obvious to all members of society in all situations" (Scott and Douglas, 1972: 4; see also Hawkins and Tiedeman, 1975: 20–41). This *absolutist, or arbitrary, conception of deviance* assumes that everyone agrees on obvious, basic rules of a society, leading to general agreement that deviance results from violation of previously defined standards for acceptable behavior. This position takes the definition for granted, as though everyone always agrees that certain violations of rules represent abnormal acts and others do not. It presumes that everyone knows how to act according to universally held values; violations of these values constitute deviance.

History and current practices in some societies today show situations characterized by almost universal acceptance of some set of guiding values, perhaps those of the Catholic church or the Koran. Among other societies, the teachings of the Buddha, the Bible, or other religious leaders or sacred documents have laid out invariable prescriptions for behavior. Western societies have often found general, universal standards in middle-class moral values or in the personal biases of some writers whose rural, traditional, and religious backgrounds have led them to view many forms of behavior related to urban life and industrial society as morally destructive (Ranulf, 1964). Still another version of the absolutist definition asserts that conceptions of what is deviant stem ultimately from preferences and interests of elite segments of society (Schwendinger and Schwendinger, 1977).

The absolutist conception of deviance, however, ignores many facets of social life:

> The absolutist asserts that, regardless of time and social context, certain culture-free standards, such as how fully persons develop their innate potential or how closely they approach the fulfillment of the highest human values, enable one to detect deviance. Thus suicide or alcoholism destroys or inhibits the possibility of the actor's developing his full human potential and is therefore always deviant. . . . The absolutist believes that he knows what behavior *is*, what people *should be*, and what constitutes full and appropriate development. (Lofland, 1969: 23–24)

A final alternative definition is called the *reactivist conception*. The reactivist conception defines deviance as behavior or conditions labeled as deviant by others. As one reactivist puts it: "The deviant is one to whom that label has successfully been applied; deviant behavior is behavior that people so label" (Becker, 1973: 9). The reactivist definition thus identifies acts as deviant *only* according to social reactions to those acts, determined through labels applied by society or agents of social control. Once behavior receives the label *deviant*, an easy extension applies the label to the actor as well. The reactivist conception of deviance has gained strong influence, and reasons for its popularity are easily identifiable. The reactivist conception attempts to concentrate on the truly social identity of deviance—the interaction between the deviant and society (really, its representatives in the form of social control agents, such as the police)—and the consequences of that social relationship. Reactivists reject the notion that deviance results from some innate quality of an act; rather, they claim that this judgment depends exclusively on the reactions of the act's social audience.

 In Brief   **Shorthand Definitions of Deviance**

Statistical definition: Common conditions determine what is normal or nondeviant; anything in the statistical minority represents deviance.
Absolutist definition: Deviance results from a value judgment based on absolute standards. Certain actions and conditions qualify as deviant because they have always defined deviance (through tradition or custom).
Reactivist definition: Deviance is whatever a social audience reacts against (or labels) as deviant. Something that elicits no reaction escapes identification as deviant.
Normative definition: The label *deviant* depends on a group's notion of actions and conditions that should and should not occur. This situational conception can change in different situations.

Critics of the reactivist definition acknowledge the importance of the interactions between deviants and social control agents, but they assert that those interactions do not *define* the term. To grasp the illogical foundation of this view, imagine a man committing burglary and escaping detection; because he evades detection, he experiences no reaction by society and thus escapes classification as a deviant. Furthermore, even those acts that do elicit social reactions do so for some reason. That is, something about the acts must prompt others to react against them in the first place. That quality (such as the innate wrongfulness of the acts or their violation of agreed-upon standard of behaviors) is what really determines deviance.

Now, remember the case of Marsha Lindall at the beginning of the chapter? According to which of the above definitions is she deviant? Statistically, more adults are now refraining from having children than in some earlier times, and divorce is now regrettably common. There are no absolute standards that proclaim that adults should have children. But Marsha is feeling uncomfortable around others; and if she is excluded from social occasions because of her single, childless condition, a reactivist definition would be appropriate. But why would others label her deviant? One answer is that the norms of their groups suggest that couples and parents are statuses that are valued. Marsha may be feeling deviant because her friends who have partners and children think that is the way things *should* be.

Norms and, hence, deviance relate both to small groups and to certain structural features of society. The larger meaning of deviance emerges in the context of social differentiation and stratification. It also grows out of the properties and nature of norms, social groups who subscribe to those norms, and the influence of those norms on behavior. The chapter now turns to an examination of these concepts—differentiation, norm, sanction, and social control.

Deviance and Society

At the simplest level, deviance refers to something different from something else. Deviants are people not like us. "They" behave differently, or so many people think. But deviance extends beyond simple, everyday observations of differences among people and their behavior. Some differences in styles of dress, for example, do not amount to deviance. Persons who wear a common style of clothing may still favor different colors without becoming deviant from one another.

Beyond the idea of differences, deviance implies something evaluated negatively or *disvalued*. Someone's clothing may look different without qualifying as deviant. It may earn that label only if its difference seems in bad taste, such as when colors clash violently or when the clothing is not suitable to the occasion; wearing a bathing suit to a funeral amounts to deviance. Some people would never wear red and orange or black and blue in the same outfit. They would consider these combinations deviant in the context of fashion.

Only some people see problems in colors that clash. Young people who dress in punk-rock or hip-hop fashions may value highly clashing colors and unusual styles in clothing and hair. Others may extend or violate some groups' appearance norms by "going too far," as some older people might regard the current fad of body piercing and tattooing among some young people.

Exceptions like these suggest that *deviance is a relative notion*. It depends upon some audience's definition of something as deviant. These three ideas—differentness, judgment, and relative standards—each have important implications for a sociological understanding of deviance. These ideas support understanding of the meaning of individual deviant conduct, as well as the connection of that conduct to the larger social community.

Norms

Because the definition of *deviance* refers to norms, a fuller explanation must identify the importance of norms to everyday life. Social norms—expectations of conduct in particular situations—regulate human social relations and behavior. Norms vary according to how widely people accept them, how society enforces them, how it transmits them, and how much conformity they require. Some social norms may require considerable force to ensure compliance; others may require little or none. Some norms remain fairly stable in the standards they set; others define more transitory expectations (Gibbs, 1965).

Individuals in a group rarely recognize the often arbitrary origins of the social norms in their group since they have encountered these priorities in the ongoing process of living. Group members learn and transmit norms from generation to generation. In this way, individuals incorporate into their own lives the language, ideas, and beliefs of the groups to which they belong. Human beings thus see the world not through their eyes alone (for then each would see the same thing); rather, they regard the world through the lenses of their cultural and other group experiences.

Even moral judgments generally do not reflect the positions of an individual alone, but those of the group or groups to which the individual belongs. Probably no one has stated the significance of group influence through norms more cogently, even poetically, than Faris did many years ago:

> For we live in a world of "cultural relativity" and the whole furniture of earth and choir of heaven are to be described and discussed as they are conceived by men. Caviar is not a delicacy to the general [population]. Cows are not food to the Hindu. Mohammed is not the prophet of God to me. To an atheist, God is not God at all. (Faris, 1937: 150–151)

Norms make crucial contributions to the process of maintaining order. Some regard them as cultural ideals, while others describe them as expressions of what society expects in certain situations. For example, one may examine sexual behavior as the result of cultural ideals or of specific expectations for certain situations, such as a married couple on their honeymoon. One can infer ideal cultural norms from observations of what people say, sanction, or react against. *Proscriptive norms* tell people what they should not do; *prescriptive norms* tell them what they ought to do. Norms not only set social or group standards for conduct, but they also define categories through which people interpret experiences. Norms establish the basis for interpreting both actions ("He should not have laughed during the funeral") and events ("Funerals are certainly sad").

The social norms and behaviors vary substantially among social classes in the United States, revealing many differences in groups' attitudes and values. The norms of plumbers differ from those of doctors and professors; construction workers display attitudes markedly different from those of college students. Child-rearing patterns differ from one social class to another. Lower-class parents, for example, tend to discipline children by physical punishment more often than middle-class parents do, although the difference is not as large as some expect (Erlanger, 1974). Child-rearing patterns also vary with the intensity of the parents' religious beliefs more than with differences in specific religious affiliations (Alwin, 1986). Members of the lower classes commit the most crimes of violence, such as murder, aggravated assault, and forcible rape; a lower-class "subculture of violence," to be discussed in Chapter 7, may offer a partial explanation. Some norms affect acceptance of others, and these interrelations, in turn, influence the socialization process.

Norms play integral roles in the organization principles of all societies, from small tribal groups to modern industrial societies. In complex modern societies, group norms may differ radically or only slightly; some norms differ only in emphasis between groups. As a result, someone who belongs to a number of groups, each with its own norms or levels of emphasis, may experience personal conflict.

People often feel pressured to act in different ways according to the roles they are performing at the time. A *social role* is a collection of norms that together convey expectations about appropriate conduct for persons in a particular position. Thus, different sets of norms govern the behaviors of husbands and bachelors, the role of a shopper differs from that of a sales clerk, and so on. The norms and roles that a person acquires from the family group do not always agree with the norms and social roles of the play group, age or peer group, work group, or political group. An individual may value membership in certain groups more highly than membership in others, and he or she may, as a result, tend to conform more closely to the norms of the most important groups. Although the family group supplies important guidance, it is only one of several groups that influence a person's behavior, whether deviant or nondeviant. Many other important sources promote norms in modern societies: social classes, occupation groups, neighborhoods, schools, churches, and friends.

Among relatively homogeneous peoples, such as primitive or folk societies, most members perceive common sets of norms and values in similar ways, although differences do emerge (Edgerton, 1976). Members of such societies thus share many common objectives and meanings, in contrast to more modern, complex societies, where social group affiliations reflect race, occupation, ethnic background, religion, political party, residence, and many other attributes. Social class and age or peer group memberships determine particularly important aspects of this differentiation.

Most people are aware of the norms in their everyday lives once their attention is directed toward them. Norms define acts, actors, and conditions as either acceptable or unacceptable examples of deviance. Norms can and do change through the actions of individuals and groups promoting their norms over others, but the continuing influence of norms underlies the meaning and explanation of deviance.

Differentiation and Deviance

People differ from one another in a number of ways, including age, sex, race, educational attainment, and occupational status. *Differentiation* is the sociological term that refers to such variations. At the most general level, *deviance* also refers to differentness. The concept of deviance would have no meaning in an undifferentiated society. Since, however, no group of people could ever share all of the same characteristics, deviance can occur in every society to the extent that some differences will be more highly valued than others. Some sociologists believe that societies can tolerate only limited deviance and that deviance remains relatively constant over time within a society. Emile Durkheim (1895/1982) long ago described deviance as "normal" and asserted that no society could rid itself of deviance. Durkheim argued that by defining what is deviant, societies also define what is not, thereby helping to create shared standards. Some sociologists do not doubt that deviance maintains a constant level, but they assert that the amount of deviance in a society adjusts both upward (Erikson, 1965) and downward according to social conditions. Expansion of standards for deviance in times of scarcity may help to foster social cohesion (Erikson, 1965), while some conditions lose the stigma of deviance when this need subsides (Moynihan, 1993). As a result, while the overall levels are the same, the acts and conditions defined as deviant can change over time.

Durkheim observes that deviance could be found even in a society of saints, where small differences among them would be morally magnified. Some saints, in other words, would still be literally holier than others.

A few examples of religious deviance illustrate the wisdom of Durkheim's observation. In 1987, religious broadcasters approved a code of ethics to prevent scandals like the one that forced evangelist Jim Bakker to resign as head of a popular television ministry. Bakker vacated his ministry after a sex scandal and subsequent questions of financial impropriety ended in a federal prison term. His colleagues, ranked by many among the most moral members of society, felt the need to adopt a special moral code after the Bakker scandal and others raised questions about the ethics of these reputed moral leaders.

The code of ethics also set guidelines for fundraising, a topic that gained importance when the Reverend Oral Roberts declared in March 1987 that God would take his life if he did not raise $8 million by April of that year. (Roberts later said that his life had been spared because he had attained that goal.) Subsequently, another television evangelist, Jimmy Swaggart, faced accusations of sexual improprieties with a prostitute. Disappointed viewers abandoned these and other "televangelists" in significant numbers, cutting the revenues of the ministries in half by the early 1990s. Upon his release from prison in 1996, Jim Bakker indicated that he would not return to televangelism. Still, the revenues of other television ministries, even those not linked to scandals, also declined, including that of Pat Robertson, who laid off 740 employees from his Christian Broadcasting Network (Shupe, 1995: 140).

More recent cases of religious deviance echo these earlier instances—a 1997 scandal in an Indiana diocese; a clergy molestation case prosecuted in another part of Indiana; the publication of the Case Reports on the Mormon Alliance; financial malfeasance on the part of a senior pastor at a Texas church; and molestation and sexual deviance cases in two Florida communities (Shupe, 1998). Such instances of deviance appear to be related not to individual characteristics of clergy, but to organizational features of religious institutions (Shupe, Stacey, and Darnell, 2000).

The conditions that promote social differentiation in society also promote deviance (Meier, 1989). Conditions that increase differentiation also likely boost the degree and range of social stratification by increasing the number of criteria for comparing people. Those comparisons often result in invidious distinctions, or ranks, that identify some characteristics as more highly valued than others. Expansion of the criteria for stratifying individuals also stretches the range of conditions that society disvalues or ranks below others (see also Cohen, 1974).

As people differ from one another in more ways, the likelihood of stratification or at least the degree of stratification increases. Modern, industrial societies differentiate people in extremely complex ways. In addition to such characteristics as age, sex, and race, members of modern societies display greater diversity than those of more traditional, homogeneous societies in behavior, dress, attitudes, and interaction patterns. Within modern societies, differences between urban and rural areas enhance differentiation as well. Sometimes, some people say *deviance* when they mean *diversity*, or behavior that results from social differentiation. A diverse society introduces a number of dimensions for defining deviance judgments: age, sex, ethnicity, heritage, religion, and the like. But the increasingly multicultural nature of a society like that in the United States need not threaten people (Parrillo, 1996).

Beyond this trend toward diversity, however, an increase in stratification clearly seems to raise the chances that some of these rankings will reflect disvalued characteristics. Not only will some individuals fall to lower ranks as a result, but they may also feel disvalued. To the extent that society values education, it disvalues undereducation; to the extent that it values an occupation with high prestige (like Supreme Court justice), it disvalues one with little or no prestige (like ditchdigger). Judgments about "better" or "worse" begin the process of making judgments about deviance.

These kinds of links join deviance to stratification within a society. Status rankings from top to bottom span roughly the same range as negative classifications that make up a structure of deviance. A society with a relatively high level of differentiation generally exhibits a large number of status ranks; a society with relatively little differentiation usually features a small number of ranks. Similarly, a highly stratified society should define a larger number of negative status ranks as compared to a less stratified society. In other words, expanding stratification increases the number of criteria on which to make judgments of deviance. A relatively simpler society should define both a simple structure of social stratification and a simple or narrow structure of deviant status categories.

Age, sex, status, occupation, race, and education differentiate individuals, among many other criteria. A comprehensive definition of deviance would clearly indicate which kinds of differentiation amount to deviance and which just determine differences without any moral connotations. Some sociologists, however, have recommended an alternate strategy: leaving deviance undefined and proceeding with research on "matters dealing with deviance." Lemert (1982: 238), for example, has suggested that "the study of deviance can best proceed by identifying bodies of data through primitive, ontological recognition rather than by formal definition."

Judgments of deviance do not refer to static or constant standards, though. Deviance takes constantly changing forms and elicits varying degrees of disapproval. To understand which conduct or conditions stimulate disapproval, one must first understand social power. *Power* can be defined as the ability to make choices by virtue of control over political, economic, or social resources. People who have money, education, and social influence generally wield more power than those who lack those resources. Powerful people, by virtue of their influence, often define standards for deviance, and they often find more deviance among others with less power than they have themselves. Public opinion often treats white-collar and company crime as less serious than ordinary street crime, even though offenses by these powerful criminals may cause more serious injuries and worse financial losses than street crime.

Lawyers, doctors, and other professionals who commit crimes often escape the criminal label altogether. Reasons for this disparity include the classification of these crimes outside the rubric of criminal law and the habit of most people to conceive of powerful persons not as evil or depraved violators, roles often reserved for lower-class people. White-collar crimes generally draw sanctions defined by administrative law, such as license suspension, or by civil action, such as mandatory restitution. Therefore, many people regard crimes committed with a pen, such as embezzlement, as less serious than those committed with a gun, such as robbery.

The importance of social power can also be expressed in terms of social differentiation. Deviance requires relative judgments not because no trait or act is everywhere and for all time deviant, but because the processes of social differentiation and change alter social opinions. This fact raises the key question of why some acts and actors receive sanctions as deviant while others do not. Sociologists frequently answer this question by referring to power. Powerful groups expand the range of stratified social phenomena through a process of definition and influence (Chambliss, 1976). A generic term for this process, *norm promotion*, indicates an ability to successfully promote particular norms to the exclusion of other, competing norms. Regardless of the process for defining specific acts as deviant, however, social judgments of disvaluement represent a core component of the concept of deviance. This fact presumably explains why some sociologists use the phrase *moral differentiation* to refer to deviance (Lemert, 1982). Deviance judgments are moral judgments.

Subcultures

Norms emerge from groups, and different groups are likely to have different norms. A person encounters varying expectations for behavior depending on the group to which he or she belongs. Acts labeled deviant in one group may be perfectly acceptable behavior in another. Sociologists often refer to such differences as *subcultural differences.*

Sometimes members of a social group share a set of values and meanings not shared by the society of which they are a part. This separation creates a *subculture.* A subculture is a culture within a culture—a collection of norms, values, and beliefs with content distinguishable from those of the dominant culture. This definition implies that members of the subculture participate in and share the larger culture of which the subculture is a part. At the same time, it implies that the subculture observes some norms and meanings peculiar to its members. A subculture need not act in opposition to the larger culture; if it does, the term *counterculture* supplies a more appropriate meaning (Yinger, 1982).

An example of a counterculture is the world of outlaw motorcycle gangs, whose members refer to themselves as "1 percenters." When the American Motorcycle Association condemned the activities of outlaw bikers, it claimed that these cyclists were only 1 percent of the organized motorcycling population. The Hell's Angels and other such groups adopted the term as a symbol of distinction (Thompson, 1966: 13, 18). These bikers live hedonistic lives and often reinforce their image of themselves as social outcasts by engaging in outrageous behavior for the benefit of onlookers. The subculture values mobility, mechanical ability, skill at fighting, adeptness at riding very large Harley-Davidson motorcycles, and ability to manipulate or "con" others (Watson, 1982). Crime is often a part of these cyclists' lives, at least for their street life spans, claimed to last only about 5 years (Quinn, 1987). After that time, the effects of run-ins with the law, brawls, or crashes take their toll and the members move out of the gangs, usually into working-class occupations.

Biker women may lead even more bleak lives. Most often drawn from backgrounds of economic and social deprivation, biker gangs exploit women physically and economically (Hopper and Moore, 1990). Women often participate in various initiation rituals and contribute to the finances of the gang through drug sales or prostitution. For some of the biker women, the gang—as debilitating as it might appear to outsiders—provides a comforting measure of structure and predictability in an otherwise capricious life.

A variety of subcultures and countercultures characterize modern industrial societies. Some of these subcultures gain the status of deviants. Cohen suggests that subcultures arise in highly differentiated, complex societies when a number of persons encounter similar problems living within the prevailing culture. In his view, subcultures represent collective solutions to shared problems posed by the dominant culture (Cohen, 1955: 14). For example, the delinquent subculture represents a response by lower-class boys to the frustration of trying to meet middle-class expectations of him in school. The delinquent subculture provides an alternative status system in which the boy feels better equipped to compete than in the school environment. Lewis (1961) adopts a similar view in his description of a subculture of poverty.

Criminologists have described the same process to explain the origins of subcultures within institutions for deviants, such as prisons (Johnson, 1996). In prison, subcultures represent social alternatives to the prison world. Composed of opposing norms and values, these subcultures may be affiliated with prison gangs that provide support and protection for its inmate members. Prison subcultures differ not only from the larger prison culture but also from one another. Racial and ethnic conflict among inmates now commonly erupts in many maximum security prisons because the contents of these subcultures conflict with one another.

Large city slum areas are more than overcrowded, congested collections of run-down physical facilities. Sociologically, a slum represents a subculture with its own set of norms and values, which favor poor sanitation and health practices, weak interest in formal education, and characteristic attitudes of apathy and isolation from conventional institutions. Inner-city areas also breed subcultural norms conducive to violence, theft, delinquency, vandalism, selling and using illegal drugs, and presence of street addicts. A "slum way of life" emerges from a combination of cultural attitudes toward economic conditions and responses to wider social and economic opportunities. This subculture frequently characterizes high-rise housing projects in major cities (Wilson, 1987).

Subcultures show clear connections to many forms of deviance. The chapters that follow will describe subcultural influences and contexts that affect drug use, homosexuality, skid-row drinking, delinquency and crime, and even suicide. Even chronic psychiatric patients discharged from institutions develop their own subcultures. Deinstitutionalization has resulted in large numbers of chronically mentally disordered people living on the streets in large cities. Subcultures help to solve the problems these people face in meeting the demands of modern urban society by providing social support for members, enhancing self-esteem by suggesting rationales for their conditions, and offering practical suggestions for independent survival. Former mental hospital patients may engage in a broad range of deviant activities, including selling their legally obtained medications, shoplifting, and even prostitution. The subculture's norms set limits on these activities, however, at the same time that it justifies deviance. As one patient phrased it:

> We're not doing anything that's really wrong. We don't murder or rob or things like that. We only take a few groceries once in a while from the A&P store. And we only do that when it's absolutely necessary. Other people who have lots of money do it all the time, and they take things much bigger than we do. We do it for medical reasons—our health, but they just do it for greed. (Herman, 1987: 252)

A few generalizations help to summarize the importance of group norms in modern, complex societies. (1) Groups *within* a society may exhibit differences in the norms of accepted behavior almost as pronounced as the differences between cultures. (2) Any logical explanation of the actions common in certain deviant subgroups must trace the development of the behavior through a process similar to that through which any member of any cultural group learns to act; for example, the process by which Eskimos learn through their culture to act, think, and interpret the world like Eskimos may provide a model for a similar process in a deviant subgroup. (3) A discussion of the norms of any given family probably focuses on those of the social class, occupational group, or some specific subcultural group to which the family belongs.

The Relativity of Deviance

A definition of *deviance* that refers to norms does not identify any particular type of conduct as deviant. This definition also allows for constant changes in standards for and forms of deviance along with the degree of disapproval that each on elicits. In this sense, deviance cites not a unique type of behavior but, rather, common behaviors that happen to offend some group. Because norms imply relative judgments (limited to groups, places, and times), deviance is also a relative phenomenon.

This fact results in an almost endless variety of acts and characteristics qualifying as deviant depending on conditions and circumstances. Debates over prostitution, gambling, nudism, cheating, medical quackery, and marijuana use arise from conflicts between norms about such acts. Just as some people consider some acts as deviant, various kinds of people also become so classified. Social types perceived by some as deviants include reckless drivers, pacifists, racists, "hippie" radicals, "square" conservatives, the very rich, the very poor, old people, drinkers, non-drinkers, and motorcycle gang members. Some liberals, for example, criticize conservatives, considering them deviant, and some conservatives return that criticism.

The identity of deviance as a violation of a norm does not indicate who creates and enforces the norm. Questions about what deviance is and who fits in that category require answers that specify which groups define certain behaviors as deviant. Such questions ask whose norms deviants violate. In this sense, observers view deviance from the perspective of the social audience of the act. Take, for example, the designation of promiscuity.

TABLE 1–1
The Relativity of Deviance

Activity	Probably Not Deviant for	Probably Deviant for
Drinking beer	Fraternity members celebrating a football victory	Baptist deacons celebrating a successful church fundraising campaign
Asking someone of the opposite sex out on a date	Unmarried people	Married people
Setting one's own bedtime	Parents	Young children
Sexual intercourse	Married couples	Catholic priests
Selling drugs	Pharmacists	Illicit drug dealers
Acting "weird"	People who just won the state lottery	Older people who have no reason to act differently

---

❊ *Issue* | **The Example of Nudism**

Nudity, like all voluntary behavior, is governed by norms. These norms vary relative to groups, times, and social situations, so nudity sometimes qualifies as deviant, and sometimes it does not. Many people regard nudity outside one's bathroom or bedroom as deviant, but nudists deny such a conception.

Whispering Pines, a nudist resort in North Carolina, establishes a number of conditions that might seem conducive to sexuality. Like other nudist camps, it is located in a relatively private setting, and members visit voluntarily, presumably bringing with them rather liberal attitudes about public nakedness. Perhaps above all, unclothed people seem to encounter plenty of opportunity to engage in sex. This assumption reflects many people's strong association of nudity with sex. Celebrity advice-giver Joyce Brothers has cautioned parents that children exposed to nudity will develop terrible guilt and frustrations that will lead to an obvious end—incest (Brothers, 1974: C1).

No evidence supports this assertion, and, in fact, nudists construct rigid distinctions between nudism and sexuality (Story, 1993). Like most nudist resorts, Whispering Pines establishes a number of norms that restrict sexual behavior. For example, a club atmosphere legitimates attendance at the resort. Members pay fees of $250 per year plus $32.50 for annual dues to the American Sunbathing Association. Members then pay per-day site rental charges. Whispering Pines allows the assembled members to include no more than 10 percent single males and 10 percent single women at one time, thus restricting the proportion of sexually unattached persons in the group. Children, recognized as incompletely socialized to sexuality, must be supervised at all times in the camp. Someone who behaves rudely, perhaps using binoculars or cameras, may have his or her name and picture placed on a list maintained by the American Sunbathing Association, thus barring visits to nudist resorts around the country. First-timers, called "cottontails," must learn these norms.

In addition to these restrictions, sex in public is explicitly forbidden. Perhaps because of all these norms, the owners of Whispering Pines indicate that open sexuality just doesn't happen. Visitors to Whispering Pines observe expectations of nonsexual conduct. As one owner put it, "There are no bathing beauties. People are just people. Women are always

Suppose that a particular unmarried woman maintains an active and varied sex life. While some people may condemn her as "promiscuous" others may view her and her behavior as "liberated." Not that these highly divergent designations do not stem from differences in the sexual behavior itself. On the contrary, the behavior has been the same; it is only the evaluation of it that has varied. (Schur, 1984: 5)

Those who regard the woman's behavior as promiscuous might not permit her to reenter conventional social sexual roles, even after a long period of conforming behavior.

worrying about fat knees, legs, that sort of thing. You come out here a while, you wouldn't worry about stuff like that. They accept themselves as what they are" (Hill, 1990).

Patrons of nudist camps carefully dissociate nudity from sexuality. They consider nudity as a natural condition, not a "dirty" or deviant one. Such a conception would require considerably more effort to maintain if nudist resorts condoned or encouraged more profuse sexuality.

But visitors to Whispering Pines face charges of deviance not only from critics of public nudity. In fact, another class of nudists finds deviance in the restrictions imposed there and at similar facilities. These *naturists* promote acceptance of nudity in public places such as beaches, not just in nudist camps. They view nudity as a natural condition in virtually all social situations, disdaining visitors to nudist resorts for their willingness to display their nudity only in these limited situations. Naturists also criticize the attempt to divorce nudity from sexuality, which they regard as simply another natural process. Jeannette, a naturist nudist, says:

> We went to a camp where we were the first naturists to visit. The first thing we did was to introduce ourselves by telling everyone our first and last names. They jumped all over us because we gave out our last names. They told us that we should never tell anyone our last name at a nudist camp because we would be giving someone the opportunity to do terrible things with the information that could hurt our reputations. It as pretty obvious to us that they didn't believe that nudity is all right. (Cox, 1989: 123)

Members of the general public may regard patrons of nudist resorts as deviants who violate a commonly held norm that restricts nudity to "private" places or relationships. The naturists also criticize those at nudist resorts for deviance, because they defy the norms of true nudists, who do not fear associating sex with nudity. One naturist reports that people he met at a nudist camp didn't even talk about sex: "I find them to be nauseatingly sterile" (Cox, 1989: 123).

Who is deviant? The nudists in the resort, who dissociate nudity from sex; the naturists who feel comfortable with that association; or nonnudists? Any answer requires a relative judgment.

Sociologists often maintain that no act includes anything inherently deviant; deviance requires a judgment that refers to some norm. In effect, the norms create deviance by creating social differentiation and attaching a moral quality to the act that designates it as something one ought to do or to avoid. This position does not imply complete absence of widespread agreement on the wrongfulness of certain acts, such as deliberately killing a person, physically assaulting an old person, or engaging in sexual intercourse with a child; but it does suggest that moral judgments differ because norms differ.

While norms state relative positions, some receive more attention in society than others, and these differences often depend on the power of certain groups to enforce their norms over members and other people. Criteria for deviance may depend on the relative power of groups to enforce and extend their norms on others. Social power, then, strongly affects an understanding of why deviance is relative. For example, strong negative attitudes toward suicide, prostitution, homosexuality, and drunkenness, among other acts, have stemmed mainly from the actions of certain conservative church groups (see Greenberg, 1988) and from conservative middle-class norms generally. Opposition to marijuana use, nudity, and distribution of pornographic materials originates with other "moral entrepreneurs," who attempt to impose their norms on others (Becker, 1973: 147–163). Some criminologists maintain that actions identified as crimes and the severity of penalties are specified by segments of society with the power to shape criminal policy (Quinney, 1981). Thus, a burglar who nets $200 from her crime may serve a long prison term, while a corporate executive whose actions defraud consumers of millions may suffer only probation, a small fine, or an order to perform some public service for the community.

Creating Deviance

Deviance is often a socially created condition. Society defines an act as deviant through a political process that exerts power within some symbolic and moral context (Ben-Yahuda, 1990). When groups perceive threats to their interests from certain acts or conditions, they may attempt to promote those interests by persuading others of the legitimacy of their priorities.

In this way, successful social promotion creates and maintains attention for some social issues. Such processes create criteria for a number of forms of deviance, including homosexuality, drunk driving, and use of certain drugs. Trebach (1987), for example, relates public attitudes about cocaine to social events, including the deaths of two well-known athletes in 1986, specific television specials about crack and other drugs, and calls by political leaders for a "war on drugs" that would include drug testing and harsher legal penalties. Orcutt and Turner (1993) report on a resulting media "feeding frenzy" in 1986 characterized by attempts to document a serious cocaine problem. Media outlets used graphic depictions to portray modest yearly changes as huge jumps in drug abuse. In fact, the problem was as created as real.

Missing Children   Groups create categories of deviance when they persuade others to acknowledge the legitimacy of their norms. A group determined to promote its norms begins by publicizing a problem with which other groups can relate. The problem of missing children illustrates how one group can enlarge its concerns to interest other groups, as well. A number of concerns have historically focused on children in the United States, including delinquency, poor school performance, and neglect. In recent times, the list has lengthened to include child prostitution, child pornography, negative effects of rock music, Halloween sadism, incest, molestation, involvement with religious cults, and drugs.

The most recent concern highlights missing children, most of whom people incorrectly label as victims of kidnappers (Best, 1990). Fueled by this fear among

✳ *Issue* | **The Real Risk of Child Kidnapping**

Parents who fear that kidnappers will abduct their children have a number of options these days. By fingerprinting and videotaping children, they can prepare tools to aid the police in identifying these young victims in case of tragedy. Nervous parents can also rely on DNA identification kits, wrist alarms, beepers, and child-safety books and tapes to help them protect their kids. Some parents have had identification devices implanted into their children's teeth, while others have bought pagers for their children to allow constant communication with them. In fact, child-protection needs have created a growth industry in the United States in the 1990s. A national club, Safe-T-Child, offers classes and tips for parents who are concerned about strangers abducting their children.

Do real risks warrant such measures? Agent Kenneth Lanning with the FBI's Missing and Exploited Children Task Force explains that all parents worry about their children, but parents have a hard time determining the actual risk of kidnapping of their children. The news media, especially television reports, unintentionally inflate the perception that today's children live under high risk of kidnapping. In the 1950s, by contrast, news coverage centered on primarily local issues, and these reports did not dominate television programming. Many local news programs lasted on 15 minutes. "In most cities today," Agent Lanning said, "news comes on at 4 P.M. and goes off at 7:30 P.M. Proportionately speaking, the same amount [of abductions] may be going on, but when it happens today it is highly unlikely we're not going to hear about it" (Boccella, 1996). Major television networks air weekly news programs most evenings, for example, *60 Minutes, Prime Time, Dateline,* and *48 Hours.*

The National Center for Missing and Exploited Children started keeping statistics in 1982. Since that year, it has reported an increase in the number of missing children from 82,000 to 800,000, but the vast majority of those cases cite runaways or children taken by family members and released soon after. The center also reports that between 3,500 and 4,500 children experience abduction by a nonfamily member, who releases them within 30 minutes. Each year, strangers hold 200 to 300 children for extended periods or never return them, and about 100 of those are eventually killed.

Jan Wagner, an Austin, Texas businesswoman and mother of two, started Safe-T-Child after becoming separated from her son for about 30 minutes at an amusement park. In addition to offering classes for parents, the group has also pressed for legislation requiring schools to offer child-safety classes.

Source: Boccella, Kathy. 1996. "Parents Pushing Panic Button." *Des Moines Register,* November 3, pp. 1–2.

parents, individual parents of missing children have generated a social movement based on their own experiences, supported by national groups formed to help locate missing children, such as Child Find. Activists have promoted this problem by raising the issue with legislators and with the public through television programs and movie adaptations of actual cases. Public-service announcements on television and images of missing children's pictures on grocery bags and milk cartons confirm the success of this movement.

To promote this issue, activists have cited horrific examples, expanded the definition of the problem, and issued unrealistically large estimates of its scope. While the movement's messages imply that strangers have kidnapped the missing children it claims to represent, the actual number of such crimes is small. Unfortunately, even a small number is too large, and this sentiment reinforces public attention to missing children. In late 1993, a stranger abducted a 13-year-old girl from California, Polly Klaas, from her bedroom during a slumber party. She was later found killed. Examples like this one mask the fact that relatives (usually estranged spouses) bear responsibility for most so-called *kidnappings,* and only relatively few children are taken annually even under these circumstances. Most missing children are more properly called *runaways.*

Satanic Cults   Some believe satanic cults worship malevolent beings throughout the United States. Books, articles, and talk show episodes on satanism tend to confirm this belief by highlighting claims that unknown groups increasingly engage in animal slaughter, nocturnal rituals, and even human sacrifices.

Are satanic practices really spreading? One cannot say for certain, although some groups certainly display concern for the occult, witchcraft, and satanism. However, Victor (1990: 288) reports that one claim of ritualistic killing of animals in New Hampshire was "later determined to be only road kills cleaned up by state road workers and deposited in the woods." Hicks (1990) observes that a comprehensive investigation of cult-mutilation claims concluded that the "mutilations" resulted from natural actions of scavengers and predators. A police officer who investigates crimes linked to satanism says:

> I most [often] investigate "occult" crimes that turn out to be false reports. For example, one woman reported a satanic burglary. As it turned out, the symbolism was poorly done and I got her to admit she made it up. We also had a middle-age woman do this to get front page coverage. Her motive was to get support for a teen center in town. I think the greatest danger now is over interpreting "occult" crime. Kids have no idea of the religious significance behind their symbolism. They could not tell you when Walpurgisnacht is, but will happily wear a pentagram because [rock singer] Ozzy [Osborne] does. (Crouch and Damphousse, 1991: 202–203)

Rumor supplies the principal means for propagating such threats to create moral panic. Many people doubt rumors they hear about satanic cults, but some people do believe these stories. Others may justify spreading such rumors with a "better-safe-than-sorry" claim. Rumors spread more easily when authority figures, such as teachers, parents, and ministers, repeat them. Repetition of satanic rumors may encourage people to believe them, because "where there's smoke, there's fire." In this way, even implausible rumors can gain acceptance (Victor, 1993).

Without denying the existence of satanic groups, widespread problems with satanism can originate in socially created phenomena (see also Forsyth and Oliver,

1990). Overemphasis on the activities of a few groups can suggest that larger numbers of these deviants pose a significant threat. Such a view might promote the goals of people and groups who use the idea of the devil to scare others into agreeing with their views.

Determining Norms and the Content of Deviance

Sociologists face a difficult problem determining how strongly various groups within a society oppose certain norms (Sagarin, 1975: 222). They can gauge support for and opposition to some norms more easily than others. Criminal law embodies a set of legal norms that all can see; ambiguous norms regulate sexual behavior, on the other hand, often changing between groups and situations over time.

Normative changes may display predictable patterns. Some observers have argued that some forms of behavior follow a cycle, or sequence of stages, moving from disapproval to greater tolerance (see Winnick, 1990). A behavior that elicits widespread social disapproval may gain legitimacy after some group campaigns for a change in public attitudes. The group may promote its goal by claiming that current sanctions victimize many people. Publication of a major research study (such as a public opinion poll) may also amplify calls for change. Over time, others may come to share the newer norm, in the process changing the social criteria for deviance. This life cycle of deviance may explain changes in public views of cigarette smoking, alcohol consumption, and use of heroin and cocaine, for example.

Changing norms seriously complicate attempts to evaluate standards for deviance. Acceptability of cigarette smoking, for example, has risen and fallen a number of times in the United States since the 1800s (Troyer and Markle, 1983). In the 1870s, many groups and individuals strongly condemned the practice, in part because it was most common among urban immigrants of low social status who were also characterized as heavy drinkers. At that time, smoking by women was considered particularly deviant because of an association with prostitution. In spite of these norms, cigarette smoking increased in the United States, and attitudes began to change following World War I. By the end of World War II, people generally considered smoking acceptable, socially desirable behavior.

Attitudes began to change again in the 1960s as medical evidence linked tobacco smoking with a number of serious physical illnesses. In the 1970s, laws banned cigarette advertising on television and radio. By the 1990s, smoking bans restricted the practice in public places, because many nonsmokers object to smoke and because medical evidence shows heath hazards from inhaling someone else's "secondhand" smoke. Indeed, citizen groups have campaigned for even stronger measures, and several cities enacted ordinances during the 1980s prohibiting smoking in places such as elevators, public meeting rooms, and certain areas in restaurants.

Cigarette smoking has again become deviant through changes in norms regulating the behavior. Anti-smoking attitudes define a norm that advocates have successfully promoted in recent years. Virtually all states now restrict smoking in public buildings, and many states have imposed specific statutes regulating smoking in other public areas. Most of this legislation has flowed from the idea that smoking, regardless of its effects on smokers, adversely affects others (Goodin, 1989). Clearly, the norms on smoking have changed (Mansnerus, 1988).

Normative change results sometimes from very complex reasons (as in attitudes toward cigarette smoking) and sometimes from more identifiable reasons (as in prohibition of alcohol early in the 20th century). Links between cigarette smoking and health hazards and an increasing social emphasis on self-control and physical fitness have helped to change many people's evaluations of the behavior. An even more important link connects smoking with the notion of an individual's right to avoid it. On the other hand, Prohibition originated in a 1920 campaign by some groups to create legislation outlawing production, sale, and consumption of alcoholic beverages. Alcohol offended their sense of morality, and they convinced others to share that evaluation (Gusfield, 1963). Still, relatively clear norms do not force everyone to class cigarette smoking or consumption of alcohol (or even marijuana) as deviance. In fact, serious disputes question how deviant each of these activities is, and positions vary depending on who makes the judgment.

Two Final Observations

It is necessary to make two final observations.

Social Deviance and Social Problems Deviance related to social behavior may differ from deviance related to social problems, even though the two kinds of deviance overlap. Not all social problems represent instances of deviance. For example, many people regard unemployment, population control, and lack of adequate medical care for poor people as social problems; these examples hardly fit the pattern of deviant behavior. The same could be said about other conditions, such as aging and homelessness (see Manis, 1976; and Spector and Kitsuse, 1979).

Consequently, sociologists study forms of deviance that arouse contemporary interest, debate, and concern. In the past, discussions of deviance might have covered different types of behavior. Within the last 300 years, or even less, such topics might have included blasphemy, witchcraft, and heresy, because large numbers of people then regarded these activities as serious forms of deviance often punishable by death. More recently, strong social condemnation of premarital sexual relations would have branded such activity as deviance. In the future, some forms of behavior regarded today as deviant may well lose that identity as new norms arise and new issues replace old ones.

Obviously, space limitations preclude a book such as this from analyzing all forms of social behavior that might possibly represent deviance. Any author must select certain topics to cover. Forms of behavior designated in these pages as deviance reflect the criteria stated earlier, including certain types of crimes (personal and family violence, crimes against property, crimes against the political state, and those committed in connection with an occupation such as white-collar and corporate crime), illegal drug use, deviant alcohol use and problem drinking, prostitution, homosexuality, mental disorders, and suicide. We also discuss severe physical disabilities, such as those experienced by crippled, obese, mentally retarded, and blind people, because these members of society often experience social reactions similar to those targeted at deviants.

Deviance: *A "Charged" Word*   Since sociologists determine deviance relative to groups and their norms, they may judge all manner of acts, thoughts, and conditions as deviant at some times and by the norms of some groups. Some readers will disagree with certain content in this book that discusses acts that spark disagreement about moral qualities. We recognize considerable consensus about the deviant nature of some of these acts and conditions, such as murder and alcoholism.

We also admit less agreement about others, such as homosexual behavior, use of marijuana, and certain types of heterosexual acts. For example, not everyone agrees that homosexuality constitutes deviance, and those who do may not agree about how deviant it is. Those personal judgments exceed the scope of this book and perhaps the scope of sociology. We wish to emphasize that we intend not to make moral judgments but merely to report social reactions.

Even when people share some measure of agreement that something is deviant, they may still strongly disagree about appropriate methods of social control. In the course of our discussions, we will identify divergent ideas and methods of social control with respect to each form of deviance.

Summary

The notion of deviance generally refers to some difference from a social standard in behavior, conditions, and people. Sociologists can define *deviance* in statistical, absolutist, reactivist, or normative terms, although the reactivist and normative conceptions may differ less than some believe. *Deviance* here means deviations from norms that meet with social disapproval and elicit, or would likely elicit if detected, negative sanctions. The amount and kind of deviance in a society is related to the degree of social differentiation in that society.

People judge deviance relative to the norms of a group or society. Just as norms change, so too do criteria for deviance. Observers sometimes encounter difficulty identifying norms before anyone violates them. Further, because not everyone subscribes to a given norm, some may disagree about what constitutes deviance. Deviant acts represent necessary but not sufficient conditions for becoming a deviant. A person does not become a deviant simply by committing deviant acts; if that were true, society would be composed entirely of deviants. Deviance is linked to a society's stratification system. Greater differentiation in society boosts the potential for deviance. Some norms represent properties of groups determined in complex ways. Others represent properties of political units; these legal norms offer opportunities to see the processes by which norms emerge and change.

Deviants are members of society who come to adopt roles identified with deviance. Just as people learn conventional norms and social roles, they also learn deviant roles and patterns of behavior. A complicated relationship links a choice to adopt a deviant role and the commission of deviant acts. A full understanding of a deviant act requires knowledge of the process of committing deviant acts and the role and actions of victims.

Despite some overlap between the notions of deviance and social problems, they are not the same thing. The concept of deviance spans an enormous range of actions and conditions, and this book cannot address every instance of deviance. Therefore, we limit our discussions to instances of deviance about which we recognize strong consensus or which have sparked strong normative dispute. Even within these widely accepted forms of deviance, people disagree about their deviant characteristics. Some, for example, regard homosexuality as unmistakably deviant, while others class it as a biologically natural if statistically rare phenomenon.

Selected References

Becker, Howard S. 1973. *Outsiders: Studies in the Sociology of Deviance*, enlarged ed. New York: Free Press.

A classic statement on the sociology of deviance that proposes an interactionist or labeling perspective (see also Chapter 5). Becker's extremely influential observations on the nature of deviance make this one of the most widely cited contemporary books on the subject.

Rubington, Earl, and Martin S. Weinberg, eds. 1999. *Deviance: The Interactionist Perspective*, 7th ed. Boston: Allyn and Bacon.

A collection of papers reflecting different aspects of the interactionist approach to deviance. Topics include being a deviant, interacting with agencies and agents of social control, and means of dealing with the stigma of being deviant.

Sagarin, Edward. 1975. *Deviants and Deviance: An Introduction to the Study of Disvalued People and Behavior.* New York: Holt, Rinehart and Winston.

An older but not dated discussion of the concept of deviance and the processes of becoming deviant. Written in textbook form, this book offers many sociological insights on deviance and social control.

Tittle, Charles R., and Raymond Paternoster. 2000. *Social Deviance and Crime: An Organizational and Theoretical Approach.* Los Angeles, CA: Roxbury.

An extended discussion of the utility of dealing with crime and other forms of deviance within a common conceptual framework. The authors devote much attention to the processes by which individuals acquire and manage a deviant identity and escape social stigma.

CHAPTER 2

Deviant Events and Social Control

Imagine a teacher, married and a parent of four children, who has sex more than once with one of the teacher's 13-year-old students. The teacher is arrested, convicted of two counts of child rape and given probation with a suspended 7-year prison term. Does it sound like a lenient sentence? It did to some, but the twist here is that the teacher was a woman and her sixth-grade student was a boy. Mary Kay LeTourneau taught in a Seattle suburb where she and the boy met and, according to her, fell in love. After the conviction LeTourneau's marriage ended, and she began a treatment program for sex offenders while in the community.

But things took a different twist when she and the boy were seen together although expressly forbidden to meet. LeTourneau's probation was revoked, and she is presently serving the prison sentence. In the meantime, she and the boy have had two children together.

The notion of deviance is connected closely to that of social control. Often, deviant behaviors represent such undesirable acts that people want to "do something" about them. What they do often results in sanctions or other overt reactions to the behavior or condition. For the purposes of this book, these reactions can be collectively called *social control*. The nature and strength of the reactions vary with the deviant conduct. In the case mentioned above, what if the sex of the parties were reversed? Male teacher, female student. Does this make a difference in the kinds of reactions produced?

This chapter explores the relationship between deviance and social control by examining characteristics of deviant events and processes of social control intended to eliminate those acts or reduce their frequency.

Deviant Events

Deviant events take many forms, but all such events violate some norm. Potentially, therefore, people could commit as many deviant events as they can find norms to violate. Some of these acts involve physical behavior, as in crime, while others may involve verbal behavior, such as children inappropriately scolding their parents.

The term *event* refers to some behavior, but also to the context in which the behavior occurs. That context may involve a deviant, a victim, the circumstances that brought them together, and, depending on the act, a history between the deviant and victim. The understanding of deviant events begins with antecedents or history and encompasses the immediate situation in which the event takes place and its aftermath or consequences (Sacco and Kennedy, 1996). An offender causes an act of simple assault, for example, but the victim and the interaction between the offender and victim also frequently constitute "causes."

Clearly, the word *cause* means something different than *blame*, and analysis of deviance focused on events should consider all of the elements that came together to produce the deviant act. The offender and victim may have continued a dispute over a period of time, or a short argument may have preceded the assault. The assault may have followed an interaction in which one of the parties challenged the honor of the other or said something that the other considered disrespectful (Oliver, 1994). Event analysis requires attention to these and all contributing factors to the deviant act.

Focusing attention on the deviant act itself necessarily neglects the context in which it takes place. For example, some women bare their breasts at Mardi Gras in exchange for beads and other trinkets. They do so not simply because they are exhibitionists who take advantage of many opportunities to take off their clothing (Forsyth, 1992). Rather, this temporary exhibitionism depends heavily on situational variables like alcohol, a party atmosphere, the desire to engage in momentary risk taking, and a physical setting dissociated from sexual activity. Studies have described similar motivations for women who become strippers (Skipper and McCaghy, 1970) and topless dancers (Thompson and Harred, 1996). Many people enter these occupations not because they are exhibitionists, but because specific financial and social circumstances permit the women to undress in public. In this sense, an instance of exhibitionism may be physically isolated in time and space, but the social context defines and shapes the deviant act.

Deviant Roles

Everyone performs a number of social roles in everyday life. At different times, people may act as students, sons or daughters, consumers, friends, and, sometimes, deviants. No one is deviant all the time; the role of deviant, like all roles, only sometimes emerges in the acts that people perform. Some people play roles as deviants more than others, but even those who make their livings from deviance do not commit deviant acts all the time. This description clearly fits people who engage in deviant acts only occasionally, such as a person who has too much to drink on New Year's Eve, but even people who are strongly committed to deviant roles only sometimes perform those roles. Organized criminals, for example, in addition to their criminal behavior, also act as spouses, parents, shoppers, football or soccer fans, and the like.

Most deviant acts do not just happen. Such an act culminates a process or series of stages that develops over a period of time—it has a history. In other words, most deviant acts occur in particular social contexts (Bryant, 1990: 23). Some deviant acts, such as instances of domestic violence, often begin without specific intentions to commit the acts; the acts follow development of interactions with others. "Each action of each party is in some measure dependent upon the previous action of the other party. The outcome of such an interaction process is a joint product of both" (Lofland, 1969: 146).

One can interpret behavior more easily after identifying the roles that the participants are performing. Male patrons of a pornography store, for example, fulfill a number of roles when not in the store (laborer, father, insurance executive, etc.) but another set of roles in the store. Tewksbury (1996) divided such patrons into five types based on their roles in the store: (1) porno watchers (who were interested only in the pornography the store sold), (2) masturbators (who sought sexual gratification via masturbating), (3) sex seekers (who sought other men for homosexual encounters), (4) sex doers (those sought by the sex seekers), and (5) naïve (curious visitors who did not interact with others in the store).

Deviant Places

A deviant act may begin with an interpretation of a situation as an opportunity to commit the act. If a teenager sees a set of keys left in a car, for example, he or she may interpret the situation as an opportunity to steal the car. Another teenager might pay no attention to the same situation (Karmen, 1981). A drug addict may view the presence of drugs in a pharmacy or a doctor's office as a possible supply and burglarize the premises. A difficult or stressful situation may elicit one kind of perception in a person contemplating suicide and a completely different perception in someone else. However it occurs, an analyst must evaluate the act in its social context as the outcome of a particular social process that includes a physical dimension.

Subsequent chapters will show that deviant acts are not random events; they occur more in some places, at some times, and among some groups than in others. The expression *deviant place* describes a physical location typically connected to deviant acts. Conventional crime is more frequent in cities than in small towns and in some neighborhoods more than others within cities. In the 1920s, two researchers at the University of Chicago, Clifford Shaw and Henry McKay, found a relationship between delinquency and certain areas of that city. More recently, Stark (1987) has theorized a relationship between deviant acts and certain types of communities with high population densities and crowded housing conditions. Substantial poverty in these communities along with extensive physical deterioration can affect the social morale and outlook of residents. In these neighborhoods, people tend to spend a lot of time outside, where they encounter strong temptations and opportunities to deviate. These neighborhoods also feature low parental supervision, since the children spend much time out of their homes, decreasing opportunities for oversight. All of these conditions may contribute to deviant acts.

Deviant places are locations likely to host deviant conduct. Neighborhoods often become places where deviance is likely to occur as they decay into disorder (Kelling and Coles, 1996). Small instances of disorder, such as graffiti, panhandling, and gatherings of street people, can lead to larger instances and even crime. A broken window in an abandoned building supports the perception that no one cares about or owns the building. Similar developments often follow such an initial instance of disorder. These may, in turn, lead to the perception of absence of supervision on the street where the building sits. Eventually, those who are not bothered by the unsupervised atmosphere—or who actually like it—may take over the street.

The Deviant Act Over Time

Analysts cannot effectively study deviant acts in isolation from their social contexts, including temporal relations between separate acts. Deviants may learn to commit these acts over long periods of time through a process of realizing pleasure and adventure from committing successive acts. The adult robber, for example, may have begun his or her career in adolescence with minor youth gang delinquencies and other risk-taking activities (Blumstein, Cohen, Roth, and Visher, 1986). A member of the gay community may have engaged in homosexual activities only sporadically as a youngster, acquiring a homosexual identity only through later participation in the gay social environment (Troiden, 1989).

Risk-taking behavior like drug experimentation or low-stakes gambling may seem both financially and socially rewarding for some people. A study of gamblers has suggested that a lower-class, regular gambler may begin this career by pursuing a reputation for "seeking action." The person gambles because the activity offers excitement and confirms a self-image of a lively, interesting person (Lesieur, 1977). Someone who gambles regularly, in other words, acts consistently with the social role of a gambler. As their gambling activities increase, some participants appear to fall into continuing spirals of gambling involvement. As debts mount, the compulsive gambler increasingly views gambling as the only way out of a predicament. After using up other, legitimate options (such as cutting expenses, loans from family, friends, or financial institutions), this person relies on gambling to provide financial relief. Some horse players interviewed by Rosecrance (1990) indicated that they had stayed away from betting for periods up to 3 years but eventually returned to the "action." What began as socially condoned activities thus became a way of life for these individuals. In this way, a penny-ante poker game may eventually escalate into contacts with established gamblers, high-stakes games, and a long list of creditors.

Many deviant acts form part of long chronicles. An apparently simple act of criminal assault, for example, may in fact result from a number of events and interactions. Oliver (1994) describes the importance of social context in understanding the assaultive behavior of urban men. While precipitating conditions, such as an argument between two men, may seem to dominate the situation, the argument clearly takes place in a particular context. One party may have insulted the other or provoked him to a fight. More frequently, one of the parties may have challenged the manliness of the other or insulted his wife, girlfriend, or family. Assaults do not just happen; they are part of a sequence of action, reaction, and interpretation.

The history of deviant acts suggests the possibility of transitory events; that is, some deviant acts occur at some times more than others. Some deviant acts, for example, are tied into particular situations. Nudity at Mardi Gras, as mentioned

✳ *Issue* | **Parade Strippers**

In certain parts of New Orleans during Mardi Gras, some women participate by exposing their breasts in exchange for beads and trinkets thrown from floats in parades. Unlike mooning and streaking, fads that occurred in certain parts of the country, parade stripping has grown in popularity to the point where a widely known term—*beadwhore*—labels women who participate in this activity.

Parade strippers often attribute their participation to either dares from friends or the effects of alcohol. Parade strippers gain some satisfaction from the experience, although like most people at nude beaches, they do not participate for sexual satisfaction. Receiving beads and the excitement of the moment apparently provide sufficient inducements for the strippers. Most parade strippers deny exposing themselves publicly in other situations or at other times. By limiting their participation to Mardi Gras in public areas with friends present, the strippers control the circumstances and ensure safety in the activity. Because Mardi Gras often involves the suspension of many norms and conventions, parade strippers experience less condemnation than those who perform other displays of public nudity. As a result, parade strippers engage in a mild form of public exhibitionism that seems to offend no one and for which the strippers suffer no disapproval.

Source: Forsyth, Craig J. 1992. "Parade Strippers: Being Naked in Public." *Deviant Behavior* 13: 391–403.

earlier, occurs under circumstances that are artificial to the lives of the participants. The party atmosphere, the effects of alcohol, and common expectations that some women will remove their tops—these conditions contribute to a feeling of a moral holiday. Those women who bare their breasts seldom appear publicly nude apart from Mardi Gras, and the time and place of the celebration provides a strong facilitating context for this form of deviance (Forsyth, 1992).

Deviant Acts and Victims

The nature of a deviant act depends not only on the past experiences of the actor but also on the responses of others in the immediate situation. The individual considers these responses in formulating a definition of the situation. The reactions of the social audience help to organize and shape the deviant act.

Unanticipated consequences often arise from events not expected in the early stages of the deviant act. Cases of criminal homicide often result from such surprises. For example, an offender may start out to burglarize a house and end up killing the resident. A number of murders occur in connection with other crimes, such as when a drug transaction goes sour and someone is killed. In crimes of violence, such as homicide and assault, perpetrators and victims frequently know one another (Reiss and Roth, 1993), as in family violence. Research has identified a number of factors usually associated with family violence, including low socioeconomic status, social stress, social isolation, and low self-concept (Gelles, 1985). In addition, a family assault frequently reflects a cycle of violence in which perpetrators often report past family violence by their parents. One cannot understand family violence, in this very real sense, outside the context of the victims of this offense, since victims frequently become offenders later in life.

Not all deviant acts target victims, however, at least in the form of specific people or items of property. People with mental disorders, for example, generally do not inflict harm on victims, although their disorders may severely disrupt marital and family relationships. Similarly, homosexuality, drug addiction, prostitution, and alcoholism are not acts directed toward harming other people, although they too may significantly affect others associated with the deviant (Meier and Geis, 1997). Similarly, many deviants commit their acts outside the presence of any audience. Addicts often take drugs without anyone else present, and even some forms of crime can take place without audiences, such as burglary.

Deviant events gain their significance because they draw attention to conditions that define *deviance* rather than the *deviants* who commit the acts. Deviants are only one part of this social equation. Sociologists must broaden their perspective to examine the nature of the social events associated with deviance (see also Miethe and Meier, 1994; Sacco and Kennedy, 1996). The social context of deviance includes social forces that bring deviants together with potential victims, as well as the times and places of those interactions. But that context also includes efforts designed to reduce deviance, a process that sociologists refer to as *social control*.

Social Control

Many scholars regard the problem of social order as perhaps the fundamental question for all social sciences (Rule, 1988: 224). Why do people conform to rules and norms, even when obedience contradicts their own interests? Why do some people violate laws and others violate deeply held social understandings about appropriate conduct? Most sociologists respond to such questions by talking about social control.

All social groups have means of dealing with behavior that violates social norms. These methods, taken together, are called *social control* (Meier, 1982). A definition might narrow the broad notion of control to a statement such as "overt behavior by a human in the belief that (1) the behavior increases or decreases the probability of some subsequent condition and (2) the increase or decrease is desirable" (Gibbs, 1989: 23). Social control implies deliberate attempts to change behavior. Social control measures serve the social purpose of ensuring, or at least attempting to ensure, conformity to norms. In some situations, people conform to norms because they know of no alternative. In other situations, they conform to gain some inducement to do so. These inducements may represent informal social control mechanisms, such as ridicule, or actions of formal agencies such as the church or government. Like a deviant event it seeks to limit, social control is a process.

Processes of Social Control

Sociologists can distinguish between two basic processes of social control. (1) *Internalization of group norms* encourages conformity through socialization, so that people both know what society expects and desire to conform to that expectation (Scott, 1971). (2) Social reaction influences conformity through *external pressures* in the form of *sanctions from others* in the event of anticipated or actual nonconformity to norms. These possibilities do not define mutually exclusive processes; they can and do occur together.

Internalization Processes   Internalization of group norms achieves social control when a person learns and accepts the norms of her or his group. This process is a result of the overall socialization process that motivates members to conform to group expectations regardless of other external pressures. Society need not exert conscious effort to secure compliance with such norms, for they define the spontaneous and unconscious ways of acting that characterize the bulk of any culture's customs.

People generally learn mechanisms of social control, like customs, traditions, beliefs, attitudes, and values, through prolonged interactions with others. Most wives do not murder their husbands, a fact due not entirely, or even mostly, to the severe legal penalties for criminal homicide; most North American drivers stay on the right side of road not entirely because they worry that other drivers will regard their driving as deviant; not everyone who drinks alcoholic beverages avoids becoming drunk simply through fear that the neighbors will gossip. Rather, most people conform to most norms most of the time because, first, they have learned the content of those norms and, second, because they have accepted the norms as their own and take those standards for granted in choosing their behavior.

A great deal of conformity to norms results from socialization that convinces people that they *should* conform, regardless and independent of anticipated reactions from others. In this sense, socialization deserves the label *self-control* because this conformity often results from the socialization process. Social control consists, in a sense, of processes that teach the person to avoid processes of deviance. Social control processes teach how not to engage rather than how to engage in deviant behavior (Gottfredson and Hirschi, 1990).

Sanctioning Processes   Sanctions are social reactions to behavior. Sociologists sometimes classify them according to their content. Social controls through external pressures include both negative and positive sanctions. A *negative sanction* is a punishment meant to discourage deviant conduct. A *positive sanction* is a reward meant to encourage conduct that conforms to a norm. Sociologists also classify sanctions according to their sources, that is, who supplies the reactions. *Informal sanctions*, such as gossip and ostracism, are unofficial actions of groups or individuals, while *formal sanctions*, such as criminal penalties, are official group expressions meant to convey collective sentiments.

Formal and informal sanctions do not act independently from one another (Williams and Hawkins, 1986). Formal sanctions can reinforce informal sanctions,

TABLE 2-1
Different Kinds of Sanctions

Nature of Sanction	Source of Sanction	
	Formal	**Informal**
Positive	Raise in job salary	Praise
	Medal in the army	Encouragement
	Certificate	Smile
	Promotion	Handshake
Negative	Imprisonment	Criticism
	Dismissal from a job	Spanking a child
	Excommunication from a church	Withholding affection
		Negative gossip

and vice versa. One study found, for example, that a sample of 800 teenage boys expressed more concern for what their families would think of them than about formal penalties associated with arrest by the police (Willcock and Stokes, 1968). Yet, the fear of formal penalties, such as arrest and incarceration, exerted important influence, too. This finding suggests that a combination of both informal and formal sanctions powerfully influences behavior.

Informal Social Controls

Informal social sanctions come from reactions to behavior by people who personally know one another. Informal sanctions act to enforce informal norms, often in small groups.

Informal sanctions such as gossip and ridicule may work especially effectively in relatively small social groups where everyone knows everyone else and the same people spend time in continuing face-to-face contact. An author has reported one example of an informal sanction, in this case gossip, from such a society:

> Early this morning, when everyone was still around the village, Fokanti began loudly complaining to an affine (who was several huts away and was probably chosen for that reason) that someone was "killing her with broken promises." Who? Asibi. He promised to help Fokanti with her rice planting today and now he's reneged. At this point, Asibi appeared and tried to explain how something else had come up which required his attention. This cut no ice with the woman, who proceeded—her voice still at a high volume—to attribute Asibi's unreliability to his "just wanting to go to dances all the time, like last night!" None of this public broadcasting was helping Asibi's reputation any, so he promised to change his plans and make good his original promise. (Green, 1977: 42)

This example clearly illustrates the extremely powerful effect of informal sanctions. Asibi kept his original promise because he cared about what others thought of him. He regarded his reputation in the group as important and he did not wish it damaged. He wished, in other words, to avoid shame and embarrassment in the community for not keeping his word.

In another tribal society, Brison (1992) found that the Kwanga of Papua New Guinea used gossip to attempt to control the behavior of other tribal members. If gossip succeeds, it allows people to avoid other, more confrontational methods of resolving conflicts. Gossip provides an effective means of social control because the group can readily control it, and members can easily defend themselves against accusations. Charges may draw counter charges. Verbal salvos handled in this manner can preempt other forms of interpersonal conflict resolution, such as violence.

Informal sanction reactions inspire a range of specific behaviors: ridicule, reprimands, criticism, praise, gestural cues, glances and other mechanisms of body language meant to convey approval or disapproval, denial or bestowal of affection, and verbal rationalizations and expressions of opinion. "Frequently, the penalty consists of verbal expressions of displeasure; even a glance of annoyance on the face of a friend is often enough to inhibit deviant acts or to arouse feelings of guilt or shame" (Shibutani, 1986: 218). Gossip, or the fear of gossip, is a very effective sanction among people who have close personal relationships.

Arthur (1998) studied a conservative Mennonite community in California where a strict dress code existed for women. The dress code was a symbol of group belonging, and adherence to it was considered a sign of religiosity. It was required for all female church members. Women who deviated from the code were subject to a number of constraints from gossip to expulsion and shunning. Nevertheless, some women managed to create minute changes in the dress code to express some individuality.

Braithwaite (1989) sees an important general crime control tool in informal social controls. Informal sanctions such as shame may help to prevent criminal acts and to reintegrate offenders into the community. He notes that most members of society refrain from crime, not because they fear legal sanctions, but because their consciences do not permit legal violations. Most people do not contemplate "bad" actions because they share society's characterization of those actions as bad behavior that people should avoid. People who do violate the law might respond to appeals to shame, in addition to the formal sanctions of fines, jail, and prison, arousing their consciences and inducing them to control themselves. This use of shame, a common sanction in many countries such as Japan, joins the informal power of the individual conscience with the formal power of the state and its criminal sanctions.

> When a young constable raped a woman in Tokyo several years ago, his station chief resigned. In this way, junior and senior ranks express a shared commitment to blameless performance. This view of responsibility is part of the Japanese culture more largely. When a fighter aircraft struck a commercial airliner, causing it to crash, the Minister of Defense resigned. Parents occasionally commit suicide

when their children are arrested for heinous crimes. . . . Japanese policemen are accountable, then, because they fear to bring shame on their police "family," and thus run the risk of losing the regard of colleagues they think of as brothers and fathers. (Bayley, 1983: 156)

The interplay of formal and informal controls also affects the operation of law enforcement in rural areas. The discretion of local law enforcement officials often helps to keep some suspects out of the system. A local police officer may bring greater information about the suspect or a longer association to an encounter than an urban officer would experience. This relationship might lead the officer to forgo an arrest otherwise justified by a person's offense (Weisheit, Falcone, and Wells, 1996: 81–82). Informal relationships with citizens also aid rural police in solving crimes, and rural police officers are likely to feel stronger appreciation of their communities and local traditions than urban police officers. These informal relationships and community identification result in a different kind of peacekeeping in rural districts than in cities, where the law alone, not personal relationships, provides the main means of social control.

Formal Social Controls

Formal controls involve organized systems of reactions from specialized agencies and organizations. The main distinction between these controls separates those instituted by the political state from those imposed by agencies other than the state. These agencies include churches, business and labor groups, educational institutions, clubs, and other organizations.

The development of formal systems of control may be related to conditions that weaken informal systems of control (see Horwitz, 1990: 142–149). When family, church, clan, or community do not apply controls, as occurs in the process of urbanization, society needs alternative forms of control. These alternative forms involve actions by third parties—such as the state in the form of police, courts, and correctional system—to enforce various norms and regulations.

Because the institutional systems of society incorporate formal sanctions, these sanctions are administered by people who occupy particular positions or roles within those institutions. These people are commonly known as *agents of social control* since their duties include administration of controlling sanctions. In the most general sense, the label fits anyone who attempts to manipulate the behavior of others by imposing formal sanctions. The police, prosecutors, and judges in the criminal justice system obviously qualify as agents of social control, but so too do employers, psychiatrists, teachers, and religious leaders who promise heaven and threaten hell to believers. In each instance, society charges the people who occupy these roles with making authority to determine reactions to (sanctions for) the behavior of others. The behavior of social control agents in effect forms a system of social control that intertwines control efforts from different sources—police, judiciary, corrections, juvenile justice, psychiatry, welfare, the family, and other agencies of the state and civil society—to form a network of control (see Lowman, Menzies, and Plays, 1987).

TABLE 2–2
Some Common Institutions of Social Control

Institution	Agent	Deviance	Sanction
Religion	Minister, priest	Sin	Penance, withholding rites, excommunication
Business	Employer	Absence, laziness, violation of work rules	Dismissal, suspension, fine
Labor union	Shop steward	Failure to obey union rules	Expulsion from union, fine
Professional group	Officer	Ethical violations	License revocation, expulsion from group
Political state	Police, prosecutor, judge	Violations of administrative, civil, or criminal law	Fine, probation, imprisonment, civil suit
Club or social organization	Officers	Violations of club rules	Fines, suspension of privileges, expulsion
Family	Parents	Youthful disobedience	Spanking, "grounding," withholding privileges

Nonpolitical groups impose penalties, some more severe than those imposed by the political state for crimes. A business concern may fire an employee, even after long years of employment, for an act of deviance, such as theft or embezzlement. A professional group or union may suspend or even expel an individual member, an act that may cost the offender his or her livelihood (see Shapiro, 1984: 135–166).

Professional athletes may face fines of several hundred or even thousands of dollars for infractions of the rules of their clubs or leagues, such as insulting spectators, violating club rules, or losing the club's playbook. Violations of such norms may result in fines or even suspensions without pay. Religious organizations may demand penance or withhold certain religious services, such as the wedding privilege or a religious service at death. They may even impose the most drastic punishment of all, especially to members of a particular faith—excommunication from the church. Clubs and similar groups generally define scales of fines, temporary suspension of membership privileges, or even expulsion as means of controlling their members' behavior.

Institutions of social control establish series of specific actions not only to punish transgressors but also to reward those whose compliance with norms equals or exceeds the expectations of the group. Curiously, nonpolitical agencies such as businesses and professional, religious, and social groups, probably use rewards more than punishments to mold the behavior of their members (Santee and Jackson, 1977). Through promotions, bonuses, or tangible tokens of merit, business organizations frequently reward those who make outstanding contributions.

Professional groups reward faithful members with election to honored offices or special citations. Religious groups reward their members with promises of future lives filled with euphoria, by positions of leadership within the church organizations, and by pins and scrolls that recognize exemplary service and commitment. Clubs, lodges, fraternities, and sororities likewise offer diverse prestige symbols for those who walk the path from initiate to full-fledged member without reflecting dishonor on the group. In recognition similar to military awards, a small number of U.S. civilians each year earn Carnegie awards for outstanding heroism.

Unlike many other kinds of organizations, the political state seldom distributes positive sanctions, or rewards, as a way to maintain social control. Citizens seldom receive rewards or commendations for systematically obeying most requirements demanded by the law for such behavior. The state of, say, Nebraska cannot practically award certificates to all those who did not commit burglaries in the past year, even if it could identify them. Some states and cities occasionally publicize the long-term safety records or courtesy of certain drivers, but this is one of the few exceptions. This limitation on positive sanctions from the state has important implications for citizens' expectations regarding the effectiveness of its social control efforts. One might expect only partial effectiveness of official, state-imposed social control, because state sanctions are limited to those that are negative in content. Some jurisdictions occasionally experiment with rewarding citizens who conform to the law, such as cases where drivers are recognized for being accident- or ticket-free, but such programs are rare and do not extend to serious crimes.

The political state can impose a variety of penalties upon those who violate state or legal norms, and some observers have noted increases in the power of the power of the political state over time (Lowman et al., 1987). Law violators below the legal age of adulthood come under the jurisdiction of juvenile courts; those who have attained adulthood are subject to punishment under criminal law. Offenders can face fines, imprisonment, requirements for probation supervision, or in some states even execution. The state also imposes sanctions beyond those of the criminal law to control law violations by business organizations. Administrative sanctions and civil actions may subject firms to monetary payments, court injunctions, and license revocations. States can revoke the licenses of professionals, such as physicians and lawyers, to practice.

Law as an Example of Formal Control

Law interests sociologists because it represents a formal system of social control. Chapter 1 introduced the discussion of the origins of social norms, and this section extends that coverage by examining the origins of legal norms. While it does not provide a comprehensive analysis of the sociological origins of law, it does illustrate the origins of legal, as opposed to social, norms.

The content of law reflects the conditions of its society. In the United States, the content of law developed around central issues of states' rights, slavery, economic development tempered by government regulation of monopolies, and the role of the Constitution in protecting individual civil rights (Abadinsky, 1988: 25–51).

Even the most detailed accounts of legal development, however, do not explain any theories of the origin of law.

Scholars promote two major views of the origins of law. One asserts that law emerges to embody and reflect the strong, majority sentiment of the population; the other asserts that law reflects successful actions by certain groups with enough power to legislate according to their own interests. These *consensus* and *conflict* models, respectively, compete to establish the general orientations of lawmaking.

Clearly, laws originate in actions of government or its agencies. Before any government or state articulated laws, however, society recognized certain acts as wrong, justifying punishment by a central authority, such as a monarch. Such acts as murder, robbery, and assault have long been considered illegal and violations of what is called *common law*, an Anglo-Saxon legal tradition defining law as judicial precedent rather than statutory definition. Emerging state law simply incorporated these common law crimes into the legal system in a formal way by codifying the prohibitions (Thomas and Bishop, 1987). One could interpret these laws as products of the strong social consensus regarding the wrongfulness of the prohibited acts.

Many other criminal laws, however, stir more disagreement about the wrongfulness of specific acts. Laws related to these acts develop from conflict among groups in society (Chambliss and Seidman, 1982). In any society, conflicts inevitably pit the interests of states, groups, and cultural units against one another. Conflict is a normal feature of social life, and it exerts a well-documented effect on the lawmaking process.

To illustrate, historical analyses indicate that statutes against embezzlement and vagrancy evolved through conflict processes motivated by competing economic, political, or social interests. Initial laws against vagrancy represented devices to protect the development of industrial interests in English society at the time by forcing people into the cities to work (Chambliss, 1964). Early legal responses to embezzlement emerged from a requirement for strict measures to protect foreign trade and commerce against the acts of people retained by others as agents (Hall, 1952: Chapters 1 and 2). Under previous social understandings, these agents who legally gained possession of property were not guilty of crimes if they then turned this property into their own use. This arrangement, however, made foreign trade extremely risky at a time when England was expanding its economic influence to other countries. Clearly, a new law was needed to protect merchants and their trading relationships with other countries. The first embezzlement statute overturned this older idea and made the agents responsible for that property. Without such a law, bank tellers, bookkeepers, and others trusted with other people's money could take that money for their own use without legal risk.

Within this general framework of consensus about and conflict over legal requirements and prohibitions, some criminologists have regarded virtually all crime as behavior that conflicts with the interests of segments of society that have the power to shape social policy (Bierne and Quinney, 1982). Although this definition seems appropriate for much crime related to political behavior and personal morality, it states too broad a case to explain the origins of all criminal sanctions. All social strata regard acts such as burglary, larceny, and robbery as crimes; these behaviors would remain crimes no matter who wielded power in the social structure, so legal sanctions seem more properly to originate from general consensus in society. Moreover, Hagan (1980) has concluded, after an extensive review of historical analyses, that many interest groups influence the provisions and passage of most legislation. Hagan denies the accuracy of assertions that laws have benefited only vested business or political interests. Still, a full understanding of the origins of laws clearly requires sensitivity to the roles of various interest groups, both in the formulation of new legislation and in changes in penalties under existing legislation (see Berk, Brackman, and Lesser, 1977).

What Kinds of Problems Can the Law Solve?

Criminal law is a political product, and there are disagreements about many aspects of law, including which acts to prohibit, how severely to punish violators, and which powers the police should exercise under what circumstances. The disagreements spark political debates, arguments before courts and legislatures, and even conversations among neighbors. Because a government enacts laws within a political context, controversy about criminal law is virtually guaranteed.

Despite some disagreement over most laws and their handling in the criminal justice system, society has become increasingly sensitive to the complex relationship between criminal law and the problems it addresses. A full explanation must consider two questions: What kinds of problems can the law solve? What kinds of problems can the law create? The questions themselves suggest that the law can both benefit and harm society, although most conceptions of law give little consideration to its potential to make matters worse. To enforce laws against prostitution, for example, the police must often engage in aggressive tactics that border on unethical practices and even entrapment. Posing as clients, police may go beyond acceptable legal limits to precipitate the very action they wish to prevent. Many people see no legitimate role for the law in regulating drug use; some even advocate complete decriminalization of certain drugs, like marijuana.

The most effective laws reflect social consensus that deems certain problems appropriate for legal intervention (Meier and Geis, 1997). In the absence of such agreement, the law is often an ineffective tool for social control. In 1996, for example, the district attorney in Gem County, Idaho, decided to mount a legal attack on the problem of teen pregnancies. He began to prosecute unmarried teenaged mothers for violations of a long-dormant statute prohibiting fornication, that is, sexual intercourse between unmarried people. He justified his action by noting, "It's a sad thing for a child to only know his or her natural father as someone who had a good time with his mother in the back seat of a car." When identified, the fathers face a similar charge. Typically, convicted offenders suffer court sanctions, including 3-year probation terms and requirements to attend parenting classes together, to complete high school, and to avoid drugs, alcohol, or cigarettes. Civil rights advocates have objected to what they see as discriminatory enforcement of the fornication law only against teenagers (Brooke, 1996).

Clearly, the law cannot solve every behavior problem, even if everyone agrees that specific behavior represents a problem. Some behaviors lie outside the

Issue | **Legal Punishment for Teenagers Who Have Sex**

An Idaho district attorney generated controversy when he decided to prosecute an unmarried teenaged mother for fornication. Some people expressed concern about teenage pregnancies and cited a need for control, so they backed the prosecutor. Others believed that the law cannot solve this problem, and it should not try.

An NBC poll, taken for the *Dateline* show broadcast on November 15, 1996, asked a sample of respondents about the case. The results of this unscientific survey showed that 3 percent agreed with the decision to apply the law in such cases, 17 percent believed religious counseling represented the best reaction for the teenagers, 21 percent cited sex education in the schools as the best solution, and 53 percent favored sex education in the home.

Clearly, people disagreed widely on the appropriate solution. In such instances, the law will have considerable difficulty because it lacks the support of widespread social agreement.

authority of law, and they should remain that way. Failure to brush their teeth causes problems for some people, but the problem may not call for a legal response.

What kinds of problems should the law address? People disagree, but scholars have suggested several criteria to guide these decisions. First, the law should target behavior that represents harm to others. This principle, first articulated by John Stuart Mill in the 18th century, has come to serve as an important social guideline. Mill argued that the state can legitimately exercise power over citizens in a free society, against their will, only to prevent harm to others. Such a criterion recognizes that most crimes pose danger to people, whatever the moral qualities of the acts. Criminal law should focus on restricting the physical, financial, and social costs of crime to members of society.

In contrast, some legal scholars have suggested that the law should highlight behavior that violates the moral beliefs of a large number of people (Packer, 1968). Few concepts of criminal laws state criteria divorced from such moral judgments, since most people's conception of crime includes behavior that violates norms, that is, behavior that should not occur. The law should not try to prohibit every immoral act, however, if only because people's versions of morality differ, and because many violations of moral beliefs do not produce sufficiently serious effects to merit legal prohibitions. Breaking promises, forgetting a friend's birthday, breaches of etiquette and manners are all immoral acts in the sense that they ought not to occur. But these breeches do not by themselves represent good candidates for measures in criminal law. Rather, other forms of social control provide more appropriate responses to these acts of deviance. Instead, acts that generate widespread and strong condemnation are more likely candidates for criminalization.

A third standard suggests that legal prohibitions should target acts for which the state can enforce its laws. A law against cancer would do little good, and it could do a good deal of harm, because legal measures can do nothing about this problem. Police can enforce laws against prostitution and other sexual acts between consenting adults only by engaging in undesirable activities of their own, such as spying on people, paying informers, and listening at wiretaps. In truth, many people pay high prices so that all may live in a free society. The law can do little about most drive-by shootings, random robberies, and residential burglaries. While police apprehend some gang assassins, robbers, and burglars and send them to court for punishment, these offenders often return to crime, and other criminals often take the places of those caught by police. In a very practical sense, democratic principles limit legal measures to reactions after the fact to most crimes. While threats of legal sanctions may deter some criminals from committing offenses, not all think ahead, and even those who do may not be deterred forever. A speeding driver immediately slows down when a police car appears in the rearview mirror (the threat of a ticket deters the offense), but he or she may speed up again once the police car disappears.

While laws do not change circumstances that contribute to crime, society probably benefits from establishing such laws. Even though the police cannot control the conditions that give rise, for example, to most murders—arguments, alcohol consumption, and easy availability of firearms—society should not decriminalize violent crime. If the law does nothing else, it conveys a very powerful message about the value of life and the extreme social abhorrence toward illegitimate violence. In less clear-cut cases, however, society must carry out a delicate and contentious process to decide which acts should be illegal, and the resulting legal controls require careful application.

Legal Sanctions

Legal sanctions represent some of the most visible tools of formal social control. Legal sanctions are penalties imposed for violating laws. Many of these punishments come from criminal law, but other bodies of law specify legal sanctions as well. The regulatory agencies of the federal government, such as the Federal Trade Commission (FTC), Securities and Exchange Commission (SEC), and Occupational Health and Safety Administration (OSHA), impose a variety of sanctions for violations of their rules. An *injunction*, an order to terminate some activity, is one of the most common regulatory sanctions. Regulatory agencies can also administer fines that can become substantial sanctions, depending on the circumstances of the case, sometimes as much as three times the damage caused by the violation.

Sanctions associated with violations of criminal law include court-ordered community service, fines, incarceration, and even execution. Imprisonment creates a particularly visible form of legal control. Increasing use of imprisonment in the United States in recent years has raised the country's rate of imprisonment to the highest in the world. Overcrowding plagues almost all state prison systems, as American prisons and jails housed more than $1^1/_2$ million people in 1997. Still, prison populations continue to climb.

Other legal penalties leave offenders in their communities. Courts sometimes sentence offenders to participate in community supervision programs (probation) instead of sending them to prison or jail. After a period of incarceration, an offender may live under supervision in the community (on parole) instead of serving the remaining sentence in prison. Probation handles more people in the United States than any other correctional program.

In recent years, legislators have developed a number of sanctions for specific offenders and for use under special circumstances. Since prison overcrowding limits flexibility in many states, these sanctions provide alternatives to institutionalization. Certain offenders remain at home under house arrest, enforced by *electronic monitoring* devices, in many communities. A court sentence may order an offender to wear a special wrist or ankle bracelet that sends electronic signals indicating the offender's whereabouts to correctional authorities. Drug offenders in some communities are sentenced through special *drug courts* to participate in treatment programs as a condition of probation. Some communities are experimenting with programs based on *restorative justice*, which makes a strenuous effort to include the victims of crimes in legal proceedings. Courts may order offenders to pay restitution and/or medical bills for their victims. Such sentences may also require offenders to meet with their victims to attempt to reconcile, or they may assume some special obligations to the victims' families as a result of their crimes.

The Irony of Social Control

Deviants interact with associates, victims, and others in committing deviant acts. They may also interact with agents of social control after the acts are committed. Agents of social control represent the community or society in those interactions, with important consequences for the deviants. Under certain circumstances, applications of social control measures may intensify or reinforce deviant acts in unintended ways. Participation in a drug treatment program, for example, may intensify a drug user's self-conception or identity as an addict. The person may come to accept that continued association with other addicts and participation in the addict subculture as a necessary or even "natural" situation in light of this self-conception.

Whether contact with an agent of social control directs a person toward or away from deviance depends on many factors. Contact with social control agents can certainly enhance the feelings of differentness and apartness that most deviants experience (Becker, 1973). This contact even influences some deviants to continue their associations with other deviants and their deviant conduct. In this sense, rather than solving deviance problems, social control agents and agencies can contribute to them.

Some people engage in deviant acts while they continue to occupy conventional status ranks and roles. Some sociologists call this activity *primary deviance* (Lemert, 1951: 75–76). Primary deviants do not form deviant self-concepts, and they tend not to identify with deviant roles. Thus, physicians who work in fee-for-service settings may unethically persuade uninformed patients that they need

✳ *Issue* | The Snake Handlers

> And these signs shall follow them that believe. . . . They shall take up serpents; and if they drink any deadly thing, it shall not hurt them. Mark 16: 17–18

The Holiness snake handlers of eastern Kentucky are a controversial religious group. They sincerely believe that God will protect them from the poisons of the snakes they handle during their religious services. The serpent handlers originated in 1910 in the Appalachian counties of Kentucky (Kimbrough, 1995). Their practices placed them in opposition to other religious groups and a number of times to police authorities, as well. By involving poisonous snakes in their services, these people sought to test their faith. These sincere believers rejected the more intellectual faiths associated with traditional eastern U.S. ministers. The fundamentalist belief system of the serpent handlers emphasized simple good and evil distinctions in determining right and wrong.

Snake handling arose among people struggling to survive a transition from subsistence agriculture to a free-market economy based on mining. Many hesitated to give up subsistence farming, a family enterprise that closely linked the occupation of farming with family life and activities. As this simple life was breaking down, the rural, poor, uneducated people of Harlan County, Kentucky, fought to maintain their traditional ways of life.

As they experienced these attacks on their livelihood and way of life, these people felt that religion was not helping to sustain them. Established Protestant religions came into the area along with the mining operations, but Appalachian folk did not relate to these highbrow, relatively unemotional ministers and their sermons. Over time, local preachers arose to meet the residents' need for a more involved kind of religion. Drawing from members of the Church of God, Pentecostal and Holiness churches, and Free Will Baptist churches, the local preachers stressed not Bible knowledge (which was in short supply in Appalachia in any case), but personal identification by the residents with the messages of the preachers. Those messages stressed empathy with the congregations' problems, something lacking in the services of the eastern ministers.

Mountain evangelists stressed simple messages (right versus wrong) and allowed freedom for self-expression. Participatory services invited members of the congregation to contribute, often in direct and vocal ways. The early evangelists stressed the Bible lessons implied by Mark 16, and soon reports of miracles were circulating the hill

more medical services, such as surgery, than their conditions actually require (Coleman, 1989: 113). These physicians may not view themselves as far outside the norms of their conventional professional roles and acceptable medical practices. If deviant acts do not materially affect a person's self-concept or cast that person in a deviant role, they remain examples of primary deviance. Someone who takes illegal drugs with friends a few times may not sacrifice a self-image as a non-addict as a result.

On the other hand, *secondary deviance* may develop when the deviant role is reinforced through further participation in a deviant subculture that brings country—levitations, fire handling, serpent handling, drinking poisons, and resurrection of the dead.

As the popularity of the lively, involving services spread, another attraction lured people to the snake handlers' services: the use of symbols. The snake handlers became important as much for the rituals they performed as for their beliefs. They conveyed a sense of order, a sense of good, in an increasingly disorderly, seemingly evil world. As social change engulfed the area, as mining and a credit economy increasingly replaced kinship and barter, and as new cultural forms threatened the old hill ways, the snake handlers offered stability and harmony. They helped believers to cope with the humiliation of poverty by offering a moral alternative to the new values. The snake handlers embodied the response of these hill folk to their feelings of dislocation as isolation and the old ways surrendered to the pressures of the modern world.

The snake handlers promised protection for the poor hill people, changing them from individual victims of enormous social changes that threatened their entire way of life to a group capable of a powerful collective response. "We are all in this together" was the message. The church services emphasized common responses and prayers. The congregations assumed that if the Holy Ghost influenced one member, He was likely to influence all. Group prayers and songs lifted sagging spirits, snakes symbolized traditional values and faith, and talk of the millennium evoked images of the thousand years of peace promised in Revelation. These elements all buttressed the people against the social collapse they saw around them.

In pursuing their religious vision, of course, the snake handlers became deviants. Although their services created a unique sense of order and control, they were seen as threats to the order of others. Often disparaged as crazy or psychotic, the snake handlers appeared to be nothing of the sort. They seemed like deviants from the perspective of outsiders, whose norms rejected the apparently extreme use of poisonous snakes in worship services. But from the viewpoint of the people they served, the serpent handlers were important people: They provided harmony in an increasingly disorderly, threatening world. They left believers with a sense, after all the change, of the rightness of old-fashioned beliefs, in spite of what outsiders thought. The snake handlers were thus an important source of social control in their eastern Kentucky counties, but a source of deviance for those outside the local groups.

associations with more pronounced deviants (Lemert, 1951: 75–76). A blind person may begin as a primary deviant, for example, and then develop a self-concept that results in secondary deviance through association with other blind persons or participate in agencies for the blind: A person who engages in occasional homosexual acts may start to frequent gay bars and practice a gay or lesbian lifestyle; a relatively casual drug user may become immersed in an addict subculture for purposes of social support and access to a supply of drugs; an official who accepts one bribe may do additional favors for money and become further involved in a corrupt political machine. These secondary deviants acquire deviant roles that increase their participation in deviant subcultures, promote acquisition of knowledge and rationalizations for the behavior, and boost their skills at avoiding detection and sanctions.

The process of self-evaluation in developing secondary deviance has several additional effects, including a tendency to minimize the stigma of deviation. "Experiences at one time evaluated as degrading may shift full arc to become rewarding. The alcoholic is an example; deeply ashamed by his first stay in jail, he may as years go by come to look upon arrest as a means of getting food, shelter, and a chance to sober up" (Lemert, 1972: 84). The secondary deviant becomes committed to deviance and performing deviant roles. Sometimes, the deviant does not perform this role by choice. Some deviants become trapped in deviant roles by the force of penalties they encounter when they try to establish themselves in nondeviant contexts.

In this sense, social control efforts sometimes backfire and complicate the deviance problems they ostensibly work to correct. Observers sometimes struggle to identify the conditions under which control defeats its own goals in this way, and conditions when social control efforts do counteract deviance. For present purposes, it is sufficient to recognize that regardless of the intent, social control may not effectively inhibit deviance.

As we noted in the previous chapter, sociologists judge deviance relative to applicable standards. It involves a norm violation, but who sets the norm? Behavior that qualifies as deviant in one group is conforming or "normal" behavior in another. Some forms of deviance can arise from changes in a group that set it apart from others. In this way, acts regarded as deviant may actually be responses to establish social control within the group. Such is the case with religious snake handlers.

Summary

Deviant events take place in specific social contexts. They develop their own histories and can evolve over time. Awareness of this context is necessary, not only to understand the events, but also to allow effective social control. Social control efforts usually influence people's actions through sanctions, or specific reactions to behavior. Internalization of norms establishes probably the most effective form of social control because it eliminates the need for sanctions. Social control can operate through either formal or informal methods, and different types of sanctions exert varying effectiveness. The social control process is part of the deviance definition process. Some people control others by defining their conduct as deviant. In this way, the definition of deviance serves the same function as specific sanctions—keeping people "in line" or in their "places."

Law is an example of a formal mechanism of social control. By nature, law applies to everyone in a political jurisdiction, and violations often provoke severe sanctions. Members of society sometimes disagree about what should be against the law. In those cases, the law usually cannot supply effective social control. In other cases, the law creates worse drawbacks than ineffectiveness; it sometimes actually magnifies social problems by amplifying deviance.

Sociologists judge both deviance and social control relative to specific standards. Since behavior that constitutes deviance varies from group to group according to changes in their normative structures, measures of social control also vary.

Selected References

Gibbs, Jack P. 1994. *A Theory about Control.* Boulder, CO: Westview Press.

This is a recent statement from an influential thinker on social control. Tightly reasoned and written, the challenging reading can lead to substantial rewards in the form of a greater understanding of the important concept of control.

Horwitz, Allan V. 1990. *The Logic of Social Control.* New York: Plenum.

A systematic examination of the nature and meaning of social control. In this well-written and very informative book, the author examines social control as something itself to be explained.

Kelling, George, and Catherine M. Coles. 1996. *Fixing Broken Windows: Restoring Order and Reducing Crime in Our Communities.* New York: Free Press.

An extended explanation and application of the broken windows philosophy that small, unattended disorders (e.g., graffiti) can lead to larger, more serious problems (e.g., crime). This book supplies a strong reminder that social control works on many levels and in many ways.

Kimbrough, David L. 1995. *Taking Up Serpents: Snake Handlers in Eastern Kentucky.* Chapel Hill: University of North Carolina Press.

Processes of deviance and social control play out in concrete circumstances with real people. This interesting story highlights the interrelationships of the nature of deviance, social control, and social change.

Meier, Robert F., Leslie W. Kennedy, and Vincent F. Sacco, eds. 2000. *The Process and Structure of Crime: Criminal Events and Crime Analysis.* New Brunswick, NJ: Transaction.

A collection of papers dealing with the criminal events perspective, which highlights the interaction among offenders (and other deviants), victims, and the social situations immediate to the deviant act.

CHAPTER 3

The Social Context of Deviance

Is the availability of drugs related to their use? If so, one might expect physicians to have high rates of drug use. And they do. Drug use and drug dependence are very common among physicians, much more so than for the population as a whole (Goode, 1999: 403–404). Consider the following. One study found that more than 40 percent of a sample of physicians and medical students had taken a psychoactive drug recreationally one or more times, and 25 percent reported they had used psychoactive drugs in the last year. Estimates show that 3 percent of all currently practicing physicians, some 12,000 persons, are currently dependent on a narcotic. Job stress might be related to the high rates of use and dependence among physicians, but the physician's knowledge of the drug, how it should be administered without ill effects, and how to counteract any negative effects of the drug, for example, diminished appetite and nutritional deficiencies, might actually encourage use. The context of medicine is strongly related to drug use among physicians.

In many instances, we see that deviance appears in a social context that sets essential conditions for a complete understanding of deviance. Neighborhood influence sometimes determines the context for deviance, and so do structural forces, such as the spread of urbanization or characteristics of larger societies. For example, in every country, cities show higher rates than rural areas for juvenile delinquency, nearly all types of criminal behavior, prostitution, drug use and addiction, alcoholism, mental disorders, and suicide.

This chapter examines the nature of the social context in which deviance manifests itself and the major sources of deviant attitudes. It discusses the importance of urbanization, the influence of a person's associates, community or neighborhood effects, and other contributions by mass media, occupation groups, and the family. While explanations of such contexts do not in themselves constitute theories of deviance, these reviews represent important elements of a theory of deviance.

The Process of Urbanization

Urban life provides an important context in which to understand deviant behavior. Life in cities has greatly accelerated social differentiation, the clash between different sets of norms and values, and the breakdown of interpersonal relations (see Spates and Macionis, 1987).

The general process of urbanization resulted in today's large cities. Such a population center acts as a *convergent* city, because it gathers in one place all of the components needed to support social, economic, and political activity. This traditional model identifies a city—also often called a *metropolitan area* (Hawley, 1971)—by its central business district, entertainment district, industrial district, transportation corridors, households, and urban villages. A high population density concentrates people in apartments, condominiums, or houses on small lots. Communities form around ethnic identities of people from similar socioeconomic backgrounds. People commute to jobs and shopping centralized in certain areas of the city, but they spend much of their time in their local neighborhoods and communities. Certainly, the growth of urban areas in the United States until the 1950s reflects this trend toward convergent cities.

Some have pointed out a trend away from convergent cities and toward *divergent* cities, a shift with implications for the origins and rates of deviance (Felson, 1998). A divergent city encompasses people spread out from the central part of the metropolitan area to the suburbs. The development of mass housing programs in the 1950s and the general suburbanization of American society reflect this shift.

During the 1970s, rapid growth in suburban areas of large cities contrasted with declining core-city populations caused by an exodus from central areas. The growth of suburbs meant not only that people had moved from the city to suburban areas, but also that manufacturing and service occupations had left the central city. Factories, office buildings, banks, and retail establishments relocated near these workers. Cities and suburbs began to function as separate social and economic entities. The build-up of malls and other shopping services has widened the economic split between suburbs and central cities. Consumers no longer have to commute to the city for their purchases, and urban retail establishments have suffered.

As jobs leave downtown areas, the people follow, seeking housing closer to their places of employment. Suburbanization moved masses of metropolitan populations into single-family dwellings on large lots. These people reached their homes more often by private automobile than by mass transit. The dispersal of economic activities left jobs no longer concentrated in certain areas, but spread throughout the suburban areas. A similar dispersal of shopping took people out of their home communities or downtown areas to shopping malls or strip malls. Dispersal of social activities scattered people as they spent more time outside their homes and in varying parts of the metropolis, since no single physical area in the suburbs concentrated entertainment opportunities.

As a result of these changes, many people developed new lifestyles. They relied more on media reports than on input from their neighbors for news and general information. The media also assumed a growing role in entertainment. Private transportation gained importance because people needed to travel comparatively long distances, and mass transit systems in most suburban areas had not developed as comprehensively as it had in the convergent city. Suburbs featured extensive open spaces, most of them outside the control of residents; parks, vacant lots, and large school grounds surrounded suburban tracts, and highways and sizable parking lots accommodated ubiquitous cars. Changes in time schedules and transportation complicated the process of developing a sense of community with others.

Social Consequences of Urbanization

The shift from cities to suburbs has produced social consequences as powerful as the changes in population and lifestyles. The service sector or blue-collar job opportunities that once employed inner-city residents have diminished and in some places largely disappeared, moving to the suburbs. Therefore, very high unemployment rates characterize many cities, especially in their slum or inner-city areas and among teenagers and young adults. Residents also contend with substantial instability in employment, since available jobs may offer only short-term or temporary employment.

These changes have had serious impacts on African-American youth. In the early 1950s, African-American young people had an unemployment rate almost identical to that of comparable whites (Murray, 1984: 72). Beginning in the late 1950s, the unemployment rate among young African Americans began to increase. Much of this initial increase reflected the loss of agricultural jobs in the South (Cogan, 1982), but subsequent increases in unemployment resulted from the loss of manufacturing jobs in the North. Unemployment for this group stabilized during the 1960s, but at a very high rate; nearly one-quarter of the potential African-American teenage labor force was unemployed. The 1970s brought even worse conditions, and by 1980, nearly 40 percent of 16- and 17-year-old African-American youth were out of work. Many members of this age group found some employment, but the transitory jobs paid poorly and failed to nurture the skills, work habits, and regular work records that would support later moves into the adult world of work.

By the 1980s, jobs were leaving the city bound for the suburbs, where many people had moved. Many well-educated and occupationally trained African Americans moved, too, leaving mainly unskilled and untrained workers behind in the city with few prospects for steady employment. This *underclass* has largely subsisted on support from government programs and work in the few jobs they could find and keep (Wilson, 1987). The loss of many unskilled, entry-level jobs has forced many African-American youths to find employment where they can and when they can.

The migration of middle-class African Americans—members of the group most likely to benefit from shifting economic opportunities—has had marked effects in inner-city communities. Among the more pronounced consequences is a leadership vacuum in these communities that affects most institutions, including politics, education, welfare, and community services. This problem has become all the more acute as city services, such as transportation, have shifted to the suburbs following population shifts. Many inner-city areas have become physically as well as economically isolated. All individuals and groups who remain in the central city experience the debilitating effects of these urban processes. Moore (1987: 7), for example, describes a typical outcome for Chicano gang members in Los Angeles: "Youngsters

Issue | **What Has Happened During the 1900s?**

The U.S. Census Bureau has chronicled the changes that occurred from 1900 to 2000. Here are some of the changes.

	1900	2000
Population	76 million	270 million
Divorces	200,000	19.4 million
Married women in the workforce	800,000	34 million
Life expectancy for men	46	74
Life expectancy for women	48	79
Death rate	17.2 per 1,000 people	8.6 per 1,000 people
Households with telephones	55 percent in 1920	94 percent
Households with televisions	9 percent in 1950	98 percent
Daily newspapers	2,042	1,489
Percentage of population in rural areas	60 percent	25 percent
Average household size	4.8	2.6
Percent of all 14- to 17-year-olds enrolled in high school	11 percent	93 percent

Some of these figures imply a number of changes in what is considered deviant and the extent to which deviance might occur. The increase in population, the increasing education level of the population, the aging of the population, and the growth of cities are all social contexts associated with deviance.

Source: Bureau of the Census. 1999. *Statistical Abstract of the United States*, 1999. Washington, DC: Government Printing Office.

look ahead to a dreary round of welfare payments, living off relatives, street hustling (big-time and small-time), dehumanizing jail terms and short-lived dead-end jobs." Economic well-being is only one indicator of quality of life in urban communities. Rates of deviance suggest another. Inner-city areas are known for high rates of crime, delinquency, and alcohol and drug use. For example, Barr, Farrell, Barnes, and Welte (1993) found little difference between black and white male college graduates, but as educational accomplishment declines, alcohol use increases. African-American males who have not completed high school drink more than three times as much as comparable white males. Not only do lower-class African-American males drink more than lower-class white males, but they also experience more problems as a result of their drinking (Herd, 1994).

The shift of economic opportunities due to urbanization produced a change in the composition of inner-city populations. This shift shows up in the characteristics of the underclass, such as high rates of unemployment, crime, drug and alcohol use, and family disruption. Economic dislocations are reflected in other dimensions as well. For example, large numbers of homeless people live in big, urban areas. Some of these street people are former mental patients whose conditions do not warrant continued mental hospitalization (Herman, 1987). Others have been inmates from correctional institutions and residents of institutional drug rehabilitation programs. This population also includes individuals unable to find and keep steady employment for a number of reasons, including poor job skills or work habits. Most homeless people describe long histories of evolving personal problems featuring illegal drug use, alcoholism, crime, and psychiatric symptoms (Benda, 1987). For many of these individuals, homelessness and dependence on the urban welfare system simply represents the next step in a downward spiral of social debilitation. For others, life on the streets is manageable and offers both economic and social hope. A study of street vendors, for example, reported that many of the vendors, rather than reflecting a degree of urban disorder, actually contribute to the social order of the communities in which they work (Duneier, 1999).

Rates of deviance are high in inner-city areas and slum communities. Deviance there takes the form of a variety of criminal activities, drug use and trafficking, prostitution and commercial vice, and extensive alcoholism and problem drinking. Suicide and various mental disorders also characterize life in these areas. For some, deviance becomes a way of life, much as city activities become a way of life. To see how, sociologists need to examine the nature of city life and the lifestyles found there.

City Life and Deviant Behavior

A close relationship ties certain types of deviance to city life. As compared to rural areas, modern cities appear to experience higher rates of nearly all types of crime, illegal drug use, heavy drinking and alcoholism, homosexual behavior, mental disorder, and suicide. These relationships appear in the United States as well as other countries. Trends for crime show the most rapid increases, almost without exception, in the world's developing countries. This unhappy change results almost entirely from the acceleration of urbanization associated with industrialization (Clinard, 1976; Shelley, 1981). In fact, statistics from these developing countries reveal concentration of most crime in larger cities that account for relatively small proportions of their total populations. The relationship between crime and urbanization spans European, South American, African, and Asian nations, as well (Johnson, 1983). Observers cite similar relationships in such diverse cities as Bogotá, Tokyo, Bangkok, Singapore, and Warsaw (Gomez Buendia, 1990). They have also documented relationships between city life and specific crimes, such as rape (Baron and Straus, 1989).

Statistics may distort this picture somewhat, because rural residents handle some criminal acts through informal means rather than officially reporting them. Also, cities undoubtedly offer more opportunities for crime than rural areas. Still, differential reporting and opportunity account for only part of the enormous differences between rural and urban crime rates (Weisheit, Falcone, and Wells, 1996). Moreover, mixed evidence leaves no clear conclusion about whether the city attracts deviants from rural areas. Studies of the relationship between geographic mobility and crime show no direct effect (Tittle and Paternoster, 1988). Instead, certain conditions and processes in the city appear to induce its residents to commit crimes.

As with crime, other forms of deviance occur disproportionately in urban settings, as compared to suburban and rural communities. As the size of the community increases, for example, the proportion of alcohol users among its members also grows. This relationship relates to growth in the proportion of middle-class and upper-class, white, Protestant alcohol users in relatively urban areas, as well as to a value system conducive to alcohol consumption (Peek and Lowe, 1977). While some of these people moved to the suburbs as part of the general process of social mobility, drinking patterns have remained more established in cities than suburbs or rural areas. Suicide rates in cities also eclipse those in suburban and rural areas (Stack, 1982), and so do the rates of many types of mental disorders (Dohrenwend and Dohrenwend, 1969). City residents use more opiates and cocaine than people who live elsewhere, and drug subcultures are almost exclusively urban phenomena (Stephens, 1987). Deviant street networks also function primarily in cities (Miller, 1986). These groups carry out organized activities for illicit purposes, including crime, prostitution, and drug selling; they survive in an urban environment that provides a market for their services. Homosexuals often establish gay communities in large cities throughout the world where they find environments more tolerant than those in small towns and rural areas (Miller, 1992).

The deviant effects of urbanization often concentrate in the inner-city areas of large metropolitan area. These neighborhoods often suffer from the highest rates of such deviant acts as crime, illicit drug use, problem drinking, and suicide. Not all forms of deviance, of course, appear disproportionately in inner-city areas. Upper-class forms of deviance like white-collar and corporate crime, for example, generally occur elsewhere. Also, the upper and middle classes also share many forms of deviance with the lower classes, although these problems generally do not characterize better-off communities as strongly as slums. Similarly, rural residents also practice many forms of deviance, although they do not engage in those behaviors as actively as people in urban areas do.

Urbanization is not always positively related to deviance. After analyzing historical data, for example, Gillis (1996) found a negative relationship between urbanization and serious crimes in France from 1853 to 1913. Over history, the growth and development of cities may have followed different processes from those common today. In the United States, the rates of most serious crimes have declined throughout the 1990s. At the same time, cities continue to change as residents disperse along with employment and shopping opportunities in the divergent city. Rates of deviance, including crime, drug use, alcoholism, and mental disorders, remain comparatively high in regions of the city that reflect previous patterns of population concentration and economic despair.

Urbanism as a Way of Life

While urbanization provides an important context for much deviance, the growth and change of cities alone does not account for deviance. Cities do not "cause" deviance; they merely represent social situations that facilitate deviance and in which deviance often takes place. The concept of *urbanism* defines a more direct cause of much deviance.

Urbanism differs from urbanization, which refers to the process through which people become concentrated in cities. In contrast, *urbanism* refers the cluster of qualities and characteristics that distinguish the city from rural areas. As such, *urbanism* is not synonymous with *city*. Whereas *city* refers to an area distinguished principally by population size, density, and heterogeneity, *urbanism* implies a complex of social relations embodied in a particular way of life or way of perceiving the world.

Rural areas in an urban-industrial society may become urbanized as the way of life there comes to reflect urban characteristics. This development illustrates why the relationship between deviance and city life requires consideration of more than sheer size alone. The terms *urban* and *rural* do not refer exclusively to the relative size of a community but also to the social characteristics that accompany the development process. The absolute number of people in a community does not predict the extent of deviance; rather, the social dynamics among those people have the essential effects.

Urbanism is an important concept because this process brings about pronounced changes in people's way of life, the patterns of population distribution, contacts with ever-wider circles of people. In turn, these changes contribute to subsequent increases in impersonal relationships, differences in leisure-time pursuits, and the city's great diversity of subcultures and countercultures. Urbanism results in increased social differentiation, clashes between norms and social roles, and a breakdown in informal, traditional controls. Among other effects, urbanism complicates the activities of life, reduces personal links in relationships, encourages proliferation of many different subcultures, and weakens direct behavior controls.

Wirth (1938) provided the first systematic description of urbanism. His work described an interaction between aspects of the physical environments in cities and aspects of their social environments. The size of a city supports a wide range of human interactions, but it may also limit those contacts to relatively superficial exchanges. The high density of people in cities leads to the creation of subgroups and subareas, and, inevitably, some people do not belong anywhere.

Life in the city reflects its heterogeneous population. People often play segmented social roles, and fluid membership often changes the characters of groups over time. Some observers believe that such factors have led to a weakening of a sense of community or belonging, particularly among urban residents (Putnam, 2000). As a way of life, urbanism often features extensive conflicts of norms and values, rapid social change, easy mobility of the population, emphasis on material goods and individualism, and escape from informal social controls.

Deviance relates intimately to the diversity of urban life and to the conditions that create that characteristic diversity and tolerance of alternative lifestyles. Consider the following scene:

Imagine sitting in Union Square in downtown San Francisco noting social differences in people who walk by. A well-dressed young woman carrying an attaché case hurries past, presumably on her way to her next appointment. An old man, unshaven and in shabby dress, reclines on the grass enjoying the sun; occasionally he takes a brown bag from his coat pocket, unscrews the top of a bottle inside, and has a drink. Four long-haired young men in bright floral shirts and dungarees are engaged in a serious discussion about the state of the nation. A short distance away, a black woman plays a guitar as she offers Christian messages to anyone who will listen. Few do. A group of Chinese children, 10 or 11 years old, playfully skip their way through the square. A middle-class couple emerges from the City of Paris department store carrying a large assortment of packages. They buy an ice cream from a street vendor, stroll into the square, and sit wearily but happily on the grass. (Spates and Macionis, 1987: 286)

Similar scenes play out in many large cities, and they all illustrate great social differences between city residents. This diversity results from the process of concentrating people with similar social characteristics: age, sex, income, education, ethnic or racial background, and political and religious beliefs. In rural areas there may be only a few people who like alternative rock music, poetry, Zen Buddhism, or exotic food, but in cities it is more likely that many will share these interests. Similarly, one is more likely to find people who are interested in deviant pursuits in cities. People from like backgrounds who share similar characteristics often choose similar lifestyles and interests. Also, the city itself often shapes and determines these lifestyles by bringing people into contact with one another. The process that describes and explains the social diversity in cities is called *urbanism*.

City living does not, of course, directly cause deviant behavior; the relationship persists because many of the conditions associated with urban life allow and support deviance. The following sections discuss six characteristics of urban life that are conducive to deviance.

1. Norm Conflicts

Diversity of people's interests and backgrounds is a major characteristic of urbanism. Urban residents vary in age, race, ethnic background, and occupation, as well as interests, attitudes, and values. Large cities really collect varying cities within cities, each community with its own subculture, religious affiliation, history, or racial characteristic. Often, groups practice different customs and speak separate languages (Hunter, 1974). Divergent group norms and values, as well as conflicting social roles, result from the city's heterogeneous population, complex division of labor, class structure, and, apparently, the simple physical dimension of its population size (Mayhew and Levinger, 1976).

Urbanites have many opportunities to meet people and make friends. They also involve themselves in many varied activities. For this reason, urban environments reduce residents' involvement with the traditional complex of kin and close neighbors. The importance of urban environments comes from the kinds of social opportunities they provide, including opportunities for deviance as well as conformity. In urban areas, people easily find others like themselves with whom they can build networks and subcultures (Fischer, 1984). The city's large population alone matters less than its social diversity, which really differentiates residents from one another.

As different groups in urban areas pursue widely differentiated and conflicting social goals, individuals often lose sight of conventional ways of behaving and suitable social roles in many areas of their lives. Many, widely differentiated people formulate their own social expectations, and this simple fact results in difficulties in fulfilling role obligations. Thus, urban residents tend to disagree about norms, and they show a tendency to ignore traditional norms. A city does not develop one central value system, but many such systems, and they may conflict at various points, leading to role confusion or conflict. This diversity may change even important values that people strongly maintain or bring with them to the city (Fischer, 1975). As one consequence of this heterogeneity, urban residents tend to tolerate behavioral diversity more easily than nonurban residents do (T. C. Wilson, 1991).

The decline in normative consensus leaves opportunities for alternative norms to influence residents, and some of these norms may permit deviance. Subcultures frequently emerge and gain strength within the urban environment, and the norms from these subcultures often support deviance. In this way, the city creates social groups differentiated from one another by distinctive sets of beliefs and ways of living. Many of them observe conventional norms, but others practice deviance because they lack commitment to the dominant social order, adopting their own normative structures unconventional behaviors. Not all urban deviance results from the activities of subcultures (Tittle, 1989), but sociologists find a strong relationship between overall rates of deviance and the conditions that create and maintain subcultures.

Proliferation of different groups and subcultures creates opportunities to learn norms that others consider deviant. Alternative normative systems also create opportunities to associate with and receive support from others in the group. A large city population may provide the so-called *critical mass* of deviants along with opportunities to commit deviant acts.

> Large population size provides a "critical mass" of criminals and customers for crime in the same way it provides a critical mass of customers for other services. The aggregation of population promotes "markets" of clients—people interested in purchasing drugs or the services of prostitutes, for example; and it provides a sufficient concentration of potential victims—for example, affluent persons and their property. (Fischer, 1984: 93)

2. Rapid Cultural Change

Rapid social and cultural change, both of which are features of urbanization, effectively reduce the importance of long-established traditions and customs. Social changes occur more easily in cities than in rural areas because the diversity of urban life creates new possibilities. The city concentrates sources of change—varying ideas, groups, and values—and this variability gives the city a certain unpredictable quality. The effects can produce political changes in the form of new leadership; social changes in the form of new fashions, music, or lifestyles; and technological changes. Sometimes the practical demands of urban life produce these changes. At other times, the changes seem to grow out of failures by informal controls to uphold and maintain older values and ideologies. Family structure may change as a result of high illegitimacy rates in urban areas. The structure of urban society often resolves itself into fairly distinct peer groups, magnifying age differences and widening communication and status gaps between teenagers and adults. Such a situation promotes delinquency, crime, drug use, and innovative sexual behavior (Friday and Hage, 1976; Glassner and Loughlin, 1987).

3. Mobility

Some claim that less than a century ago, a person might live a lifetime without ever going far from home and or seeing more than a handful of strangers; descriptions of today's mobile society portray quite a different picture. Modern transportation, particularly in urban areas, moves people about rapidly, stimulating frequent contacts with many different people. Urban societies generally tend to regard social mobility favorably, but the frequent physical moves that often requires may have unsatisfactory effects. They tend to weaken attachments to local communities, eroding interest in maintaining certain community standards. Also, contact with secondary groups with diverse patterns may weaken bonds that reinforce social control among members of local groups. Mobile people come into contact with many different norms, and they begin to understand how other codes of behavior differ from their own. Their moves necessitate changes in friendships and social roles, along with adjustments of old norms to new ones.

Some forms of deviance may grow directly out of mobility. For example, suicide (Stack, 1982) and crime (Reiss and Tonry, 1986) become relatively common in areas where people feel few ties to the community and other residents. The highest crime rates (at least for most crimes) occur in areas of a city with the highest transient populations (Brantingham and Brantingham, 1984). Tittle and Paternoster (1988) found that mobility appears to reduce moral commitment, and this change, in turn, increases the chances of criminality. They found little relationship between mobility and marijuana use, however. Some forms of deviance, therefore, bear stronger relationships to mobility than others.

4. Materialism

External appearances and material possessions have ascended to roles of primary importance in modern urban society, where people are known more often for the gadgets they own than for themselves. People increasingly judge others by how well they display their wealth, a display Veblen termed *conspicuous consumption.* Under urban conditions, people can often judge one another solely by the clothes they wear, the automobiles they drive, the costliness of their homes and furnishings, the exclusiveness of the social clubs to which they belong, and the their apparent salaries. If you encounter too many people to know each one personally, you can at least know them by their possessions.

By the 1980s, the emphasis on possessions had become associated with a type of urban resident, the "Yuppie," or *young urban professional*. Partly because of the high cost of living in many urban communities, some have identified another urban type called a "Dink," or *double income no kids*. Both of these social types are characterized by concern over possessions and lifestyles that emphasize high-paying employment to support accumulation of possessions. While the decade of the 1980s was not associated with a specific lifestyle, some described it as the "greed decade," acknowledging the high cultural emphasis on materialism. At the other end of the income range, cities house many individuals and groups who seem trapped in a cycle of poverty. While economic opportunity may abound in urban areas, not everyone can enjoy that opportunity.

Through the 1990s, the economy in the United States provided substantial opportunities for some to obtain wealth. A strong stock market, diversified industrial base, and the growth of a technology infrastructure provided many with investment opportunities that created many new millionaires. By the turn of the century online computer investing was opening these opportunities to many from the comforts of their homes. With the new wealth came new buying and a resurgence in the value of materialism.

Large U.S. cities make room for very wealthy people along with very poor people, particularly in inner-city areas. As the earlier discussion pointed out, Wilson (1996) has observed a group of city residents permanently unable to break from the chains of poverty in spite of sustained governmental and personal efforts to improve their lives. Shifts in the national economy have left many African-American males without employment, as manufacturing has shifted to suburban and foreign facilities. Financial and other professional services have displaced manufacturing in many cities, and this change has decreased employment opportunities for unskilled workers. Loss of these jobs may disenfranchise such people from the larger culture, caused as much by their feelings of alienation as by the lack of economic resources to enjoy the material benefits of urban life. What is worse, these "ghetto poor" live in social circumstances that reinforce their poverty (Wilson, 1991).

5. Individualism

Individualism often plays a more important role in city life than in other community practices. Urban residents may regard their own interests and self-expression as paramount criteria in their social relations, as feelings about self replace feelings of cooperation more characteristic of rural life. Individual competition may also intensify in urban settings, which present not only more opportunities to achieve goals, but also competitors for those same goals. This process may inhibit close social contacts between competitors, such as fellow employees.

This force within urbanism may have intensified in recent years with the rising popularity of various psychological ideas that stress individual self-satisfaction over social relationships and obligations. These therapies also look for solutions to problems within individuals rather than in the social contexts of the problems. Treating an individual's alcoholism in isolation, for example, fails to address the pressures or cultural traditions that contribute to the urge to drink. An Irish proverb states the point: "They talk of my drink, but not of my thirst." When people apply 12-step therapies to a variety of behavior problems, they inherently assume that the problems result only from bad individual decision making or moral choices. Someone primarily interested in personal happiness, self-concept, and other interests risks losing sight of existing and potential social relationships and responsibilities. As Richard Sennett (1977: 259) has observed:

> The reigning belief today is that closeness between persons is a moral good. The reigning aspiration today is to develop individual personality through experiences of closeness and warmth with others. The reigning myth today is that the evils of society can all be understood as evils of impersonality, alienation, and coldness. The sum of these three is an ideology of intimacy: social relationships of all kinds are real, believable, and authentic the closer they approach the inner psychological concerns of each person.

Of course, pursuit of individual desires may cause no ill effects, except perhaps if someone satisfies personal wants at the expense of the larger culture—that is, to the detriment of other people. The intersection of self-interest and community interest sometimes proves elusive, but individualism asserts that individual interests take precedence over social obligations and responsibilities.

6. Increase in Formal Social Controls

As urbanism spreads and informal group controls lose power to induce conformity to social norms, people encounter expanded opportunities and inducements for behavior that deviates from the standard set by others and their norms. As impersonality increases and intimate communication declines, normative violations produce progressively weaker informal censures of the kind known (and feared) by rural residents. Conflicting normative experiences in an urban setting tend to weaken parental authority and other traditional informal controls, not only over youth but also over all individuals.

Alternative normative systems may replace traditional authority with standards more appealing to some people. As a result, responsibility for controlling behavior in cities shifts more and more to the police, courts, and other agencies of government that tend to enforce the norms of certain groups (Meier, 1982). Systems of social control associated with interpersonal relations have given way to systems of control by third parties (see the general discussion in Gibbs, 1989). Marriage counselors, police officers, judges, psychiatrists and other therapists, and mediators have exerted increasing influence on interpersonal relationships as social life has moved to cities. Whereas people once found ways to solve interpersonal problems on their own, they now rely heavily on others for expert assistance.

The city's diverse collection of groups invites people to belong to many groups. As a result, as Simmel pointed out, each individual accepts a large network of group affiliations, but each group may exert little influence over the individual. People may supplement membership in occupational, civic, religious, political, and

✳ *In Brief*   **Six Reasons That Cities Are Conducive to Deviance**

1. Norm conflicts
2. Rapid cultural change
3. Mobility
4. Materialism
5. Individualism
6. Increase in formal social controls

Source: Wirth, Louis. 1938. "Urbanism as a Way of Life." *American Journal of Sociology* 44: 1–24.

recreational groups, as well as other voluntary associations, with participation in other groups composed of particular sets of acquaintances from daily life. Many people belong to and associate with a wide variety of groups, but no single group maintains powerful control over their behavior. As the informal controls associated with group membership dissipate, more formal controls arise to replace them.

Sources of Deviant Attitudes

Urbanism is not, of course, the only source of deviant attitudes. Other sources, such as the family, associates, and mass media, also instill such attitudes. The city often defines the context in which these sources of deviance operate, but a full understanding requires exploration of these other sources of deviant attitudes.

Some specific theories, to be discussed in the next chapter, address the origins of deviance, but certain general observations may establish a preliminary outline of the genesis of deviant attitudes in society. So far, this book has stressed the central theme that deviance represents behavior—much of it learned—or conditions that violate norms. In a pluralistic society like that in the United States, of course, different groups promote varying norms that call for a variety of behavior and conditions, and adherence to one set of norms may violate someone else's norms.

People acquire norms through socialization or learning processes based on interactions with other members of one's own group. Intimate, personal associations exert probably the strongest influence on people's acquisition of deviant norms and attitudes. Most of these associations occur within group settings, as individuals interact with certain companions, gangs, families, occupational colleagues, or neighborhood residents. Some norms that people learn in such interactions tend to push them away from deviance ("It is wrong to steal"), while others pull them toward it ("Everyone cheats on their income taxes; the government expects it"). Interactions teach that marijuana use may violate a norm for appropriate behavior by a member of a church group at a picnic, but not by a member of a college fraternity at a private picnic. Individuals come to learn the norms of the groups in which they maintain membership; some of those norms prescribe and some pro-scribe conduct that another group might evaluated as deviance. When members learn the norms of Group A, behavior based upon them may violate the norms of Group B.

People "naturally" learn both deviance and nondeviance in the sense that these standards grow out of socially determined definitions and other social processes. Conflict between standards emerges naturally, as well, because these processes vary between groups. A number of sources of deviant norms and attitudes affect these learning processes. The following discussion explores the relationship between deviant attitudes and associates, neighborhood, family, mass media, and occupational groups.

Associates

People acquire attitudes favorable to deviance primarily through interactions with companions and by participation in small, intimate groups, such as gangs or fami-lies. Thus, they learn deviant attitudes in much the same manner as they adopt conforming norms. One of criminology's most consistent findings stresses the importance of companions in the commission of illegal acts by both juveniles and adults. People commit most kinds of crimes, and virtually all acts of juvenile delin-quency, in groups or in the context of group associations. (See the review on delinquency in Short, 1990.) Studies confirm this relationship not only in the United States but also in all other countries.

The number of companions may vary from two or three close friends, who join in acts of adolescent shoplifting or vandalism, to a larger number of gang mem-bers or other associates. Even when offenders commit crimes alone, they acquire attitudes and values toward criminality through relations with others.

An individual finds such companions in the places where social life gathers people together: schools, workplaces, and recreational facilities. In these social settings, individuals learn both deviant norms and deviant roles. Deviant compan-ions actively teach others attitudes, norms, and skills conducive to deviant acts (e.g., Akers, 1985). Some observers, however, argue for a less direct process of influence by deviant companions; they assert that peers merely fail to control deviant acts by withholding negative sanctions for others who commit deviant acts for other reasons (Hirschi and Gottfredson, 1980). Since most delinquents com-mit their offenses in groups, it is important to understand the role of peer influ-ence in gatherings of adolescents (Dornbusch, 1989).

In a different venue, business-related violations of law occur more frequently in some industries than others, suggesting that interactions in these industries con-stitute more effective learning environments for law violations than those in other industries with few violations (Clinard, 1990). Even within companies, the num-bers of employees, organizational decision-making structures, and degrees of organizational control over employee actions all help to explain the occurrence of these crimes (Shover and Bryant, 1993). A company can create an effective crime-learning environment when it couples pressure for profits or market stability with examples of unethical behavior by management.

Companions play especially critical roles in developing many forms of deviance, becoming the single most important source of deviant attitudes. Virtually all illicit drug use requires a learning context for users, along with access to the drug. Users must learn who can supply the drug, and participants in that world necessarily limit access to such information. Users learn what they need to know only by participating in a drug using community; for that reason alone, drug dealers typically come from the ranks of users (Adler, 1993). In fact, one cannot become a regular marijuana user (Becker, 1973) or a heroin user (Stephens, 1991) without the help, encouragement, and support of others.

Various sexual acts, whether considered deviant or not, obviously involve people with others, and some acts even require support from others who may compose well-defined communities, such as those found among homosexuals. In the United States and around the world, homosexual communities vary in size and activities. Members create networks, not only for interpersonal support and sociable interactions, but also, in some cities, for political activism (Miller, 1992).

Similarly, gambling, both legal and illegal, requires active participation by other gamblers. An individual may bet against another person, a gaming casino, or a group of undetermined members who organize a particular form of gambling, such as policy or numbers games. While many regard compulsive gambling as a form of addiction, problem gamblers (those who repeatedly lose a lot of money gambling) may just as plausibly suffer more from defective gambling strategies than from a psychological illness (Rosecrance, 1988).

Many people cooperate in the production of pornography, and in the process, they may reinforce in each other the feeling that they merely provide a demanded service (Attorney General's Commission on Pornography, 1986: 284–291). The production of sexually oriented movies and videotapes requires a crew of several people and a cast, although the actual filming may take only a few days. Most sexually oriented books are written not by single authors, but by teams of writers who guarantee quick completion of books on specific topics; pornographic magazines also require production staffs. All pornographic materials move through an extensive and sophisticated distribution system to reach retail sales outlets.

Neighborhoods

The neighborhood or local community is primarily a forum for personal relationships, where people's lives and local institutions are situated. It creates a world for meaningful experiences and intense social relationships with others. At the same time, this local environment often reflects differences in individuals' normative standards that may lead to deviant behavior. Neighborhoods reflect differences between social communities, giving people who live there a sense of belonging. Neighborhoods also provide the physical contexts for much deviance, and differences among neighborhoods contribute to this deviance.

Neighborhoods differ in many ways: the social classes of residents, racial, ethnic, and religious composition, and population stability. Even more important differences reflect pronounced variations in norms, for example, those supporting or restraining criminal activities (Bursik and Grasmick, 1993). Residents of areas with high crime rates may encounter many opportunities for close associations with people who engage in crime or even encourage them to do so. This sense of support may create a self-perpetuating cycle of crime in certain areas of the city, especially those characterized by low income, run-down housing, poverty, and the presence of certain ethnic groups (Brantingham and Brantingham, 1984: 297–331). Such physical conditions do not encourage crime on their own, though; crime results, instead, from the norms and values common in some areas, sometimes persisting for generations. Similarly, residents of middle-class or upper-class neighborhoods may tolerate or even approve of violations of laws governing their own businesses while condemning crimes such as burglary and theft.

Such observations have led sociologists to recognize differences between communities in their social structures and social control mechanisms. Further, these differences are associated with variations in rates of crime and other forms of deviance (Reiss, 1986). Communities characterized by racial income inequality, poverty, and low occupational status—inner-city or slum areas—also experience relatively high rates of many property crimes, such as robbery and burglary, as compared to communities without these characteristics (Sampson, 1986). Such conditions are also related to violent crime. The mechanism by which neighborhood characteristics are related to behavior may be through a "code of the streets" that emphasizes an individual's ability to command respect from others (Anderson, 1999). This unwritten set of norms may be a response to a lack of jobs that pay a living wage, to the stigma of race, to widespread drug use, and to alienation and a lack of hope.

A neighborhood gathers together members of the same social class, so their standards show susceptibility to class influences. Relatively high rates of conventional crime (as opposed to white-collar or business crime) appear associated with relatively low-class neighborhoods, as measured by official police statistics, reports from victims, or self-reports of illegal acts (Elliott and Huizinga, 1983; but compare Tittle, Villemez, and Smith, 1978, and Tittle and Meier, 1990). This observation suggests that rates of deviance, in general, vary by neighborhood, depending on the social classes of the residents. Communities with high rates of crime and delinquency, for example, tend to feature many opportunities for residents to acquire criminal attitudes. Even in areas with high crime rates, however, residents espouse noncriminal norms and values, exerting powerful influence over the everyday conduct of residents (Kobrin, 1951).

Neighborhood influences also show up in measures of other forms of deviance, such as teenage pregnancy. African-American teenagers living in U.S. metropolitan areas tend to begin sexual intercourse at earlier ages than other teenagers, and girls experience higher rates of premarital pregnancy. Nearly 70 percent of African-American girls have had sexual intercourse by age 15, compared to about 25 percent for white girls (Hacker, 1992: 76). Similarly, 41 percent of African-American girls have become pregnant by age 18, compared to 21 percent of their white counterparts. African-American girls use contraceptives less often than white girls do, and, once pregnant, African-American girls carry their pregnancies to term more often than white girls do (Reiss and Lee, 1988: 147).

The risk of teenage pregnancy seems higher in some neighborhoods than in others. Hogan and Kitagawa (1985: 852) cited several characteristics of high-risk environments—lower-class, ghetto neighborhoods, nonintact families, five or more siblings, sisters who became teenage mothers, and lax parental control of dating behavior. Girls from such environments experience rates of pregnancy 8.3 times higher than those of girls from low-risk neighborhoods. Clearly, high rates of teenage pregnancy result, at least in part, from the unfavorable social and economic circumstances in the neighborhoods where these African-American teenagers grow up.

Further, one alternative for the pregnant girl, marriage, is less available in African-American neighborhoods than in white ones. The pool of "marriageable" African-American males is smaller than the comparable selection of potential white husbands, and that pool of candidates continues to shrink (Hacker, 1992: 74). A large proportion of young, African-American males is unemployed, uneducated, and/or in trouble with the law. Many already live under correctional authority or carry arrest records, both conditions that diminish access to good jobs. In fact, nearly 1 in 4 black males and 1 in 10 Hispanic males between the ages of 20 and 29 live in penal institutions or under some form of correctional supervision (e.g., probation); in contrast, correctional control affects only 1 in 16 white males in this age category (Mann, 1993). Another segment of this population may not make suitable husbands due to debilitation by drugs, alcohol, or mental disorders. Also, the death rate for young, African-American males is increasing, mostly as a result of murder by other young, African-American males.

These challenging conditions for African-American girls are also related to neighborhood norms pertaining to sexual activity and pregnancy. The prevalence of early sexual activity in a neighborhood legitimates the acceptability of this practice, in spite of the costs associated with teenage pregnancy. A sister's teenage pregnancy and weak parental control of a teenage girl's dating behavior further contribute to and reinforce a normative climate permissive toward early sex. If neighborhood norms do not reward such behavior, neither do they strongly condemn it. As Hacker (1992: 75) notes, "In far fewer white neighborhoods are out-of-wedlock pregnancies and motherhood seen as customary and ordinary."

The Family

The social institution of the family has undergone great changes that have eroded its importance in general social life (Reiss and Lee, 1988). Increasingly, other groups, such as school associates and even street gangs have participated increasingly in the socialization of young children. As kinship ties weaken and mothers leave for employment outside the home, urban children may spend progressively less time with immediate family members.

Although some argue against overstating the importance of family socialization experiences (Rowe, 1994), many people cite unsatisfactory family social influences as a chief source of deviant behavior. Observers relate a number of family influences to deviant conduct, particularly crime, including single-parent households (or so-called *broken homes*); the family itself may act as a source of deviant norms. For example, one extensive review identified socialization variables such as a lack of parental supervision and weak parent–child involvement among the most powerful predictors of juvenile misbehavior (Loeber and Stouthamer-Loeber, 1986). Furthermore, the delinquency or other behavior problems by one child in a household increases the probability that other children in the family will exhibit these problems.

One general theory of deviance, self control theory, attributes crime and delinquency, as well as other forms of deviant behavior, to low self-control. Further, "the major 'cause' of low self-control . . . appears to be ineffective child-rearing" (Gottfredson and Hirschi, 1990: 97). Building self-control abilities among children requires concerted efforts by parents, whose success is reflected in such indicators as measures of attachment of child to parent, degree of parental supervision, and parental punishment for deviant acts.

The social consequences of shifts in population are reflected in the changing family patterns of urban minority populations. Just 38 percent of African-American children lived with both parents in 1989, less than half of the proportion of white children who lived in traditional homes (Pins, 1990). In the city of Chicago in 1996, for example, the National Center for Health Statistics reported that 79 percent of all African-American babies were born to unwed mothers, up from 55 percent in 1980. Because of economic hardship and the absence of adult males in their homes, many African-American children grow up under the care of grandparents: 1.2 million, or 13 percent, of African-American children under the age of 18 live with their grandparents; in more than half of those cases, the mother also lives with the grandparents. In Mississippi, a rural state, 74 percent of births to African-American mothers occur outside marriage (Hacker, 1996), a figure remarkably similar to that of Chicago. This correspondence demonstrates that urbanism—which can occur in rural areas, too—rather than city size accounts for the link with deviance.

Researchers have persistently tried to link youth crime to homes with only one parent (usually the mother), on the assumption that such a break in family ties may lead a young person to commit delinquent or other deviant acts. Most studies do show high percentages of delinquents (between 30 and 60 percent) with backgrounds in one-parent households, but these figures can mislead. A selection process confounds the relationship between delinquency and broken homes in the juvenile justice system; it appears that juveniles from broken homes are more likely than others to be arrested, convicted, and sentenced to custody in juvenile institutions (Schur, 1973: 121). The way that broken homes influence the acquisition of deviant attitudes seems indirect, however, because reduced supervision over young people permits adolescents to come into contact with deviant norms in the community (Grinnell and Chambers, 1979; Rankin, 1983).

Research does suggest, however, that family and marital disruption powerfully influences crime rates. Juvenile offense rates by both African-American and white youths remain low in cities with a high percentages of both African-American and white households headed by married couples, while high divorce rates appear to have significant effects on adult criminality (Sampson, 1986). Since most one-parent households result from divorces, marital disruption can, indeed, have important consequences for children. Among the obvious discontinuities in child rearing during a divorce, separating parents often fail to cooperate in raising their children (Maccoby and Mnookin, 1992). Boys appear to take longer than girls to adjust to divorce—a process that is most intense in the first year after the divorce (see Hodges, Buchsbaum, and Tierney, 1983; Lowry and Settle, 1985). This difference may reflect the fact that most children of divorced parents live with their mothers, and the absence of the fathers may present greater adjustment problems for boys than for their sisters.

The Mass Media

Controversy still surrounds the question of whether people learn deviant attitudes from the mass media, particularly television and motion pictures. At an earlier time, some worried that reading crime-related comic books would lead youngsters toward delinquency. A more contemporary concern questions the influence of video games on violent behavior by youths who play them.

Since the 1950s, however, television and motion pictures have exerted influence far surpassing any presumed effects from comic books and video games. Without question, television and movies have emerged as dominant media in the United States. Their images present a version of the country's culture that emphasizes wealth, materialism, and conflict, in the process furnishing juveniles with role models conducive to many forms of deviance. Some accounts of deviant acts suggest that offenders encountered deviant ideas for the first time in the media and then acted upon those ideas.

For example, one set of crimes has been attributed to ideas presented on a videotape. Terry Lee Corron viewed a video that described in vivid detail the strangulation of six women by a killer who wore camouflage clothing, lifted weights, and maintained a gallery of pornography (*Spokane Spokesman Review*, November 8, 1987, p. A2). Shortly after 3 A.M. on May 20, 1987, Corron, a security guard, crept up behind a co-worker at the switchboard of the hospital where he worked. He wrapped his hands around her neck and threw her to the ground, breaking her nose and shoulder. He then choked her for a while before fleeing. That attempted murder paralleled the first crime in the video. Both victims wore white nurse uniforms, and both were attacked from behind, picked up by the neck, and thrown down. Many felt that Corron had learned to assault from the video.

How Media Influence People's Perceptions of Deviance   Media make information about deviance more readily available today than ever before. Television, radio, newspapers, and news magazines provide a daily flood of information about deviant acts and actors, some serial reports giving graphic detail as the reporters eagerly follow the unfolding cases. These reports indirectly reach even people who do not receive them directly, for example, those who do not subscribe to newspapers or watch television news, because others who encounter media reports of deviance repeat them in local conversation. Media outlets also provide extremely timely information on deviant events, particularly television news, with its emphasis on live reporting. Highly competitive local news organizations gain high prestige and viewer approval when they report stories before rival outlets. Further, because national organizations or networks back most local news organizations, local media can report on dramatic events—such as spectacular crimes—in distant locations. Rural residents regularly hear about crimes that occurred thousands of miles away on the other side of the world.

✳ *Issue*   **An Example of Media Influence**

The role of the media is powerful, but can it contribute to murder? In one case, the issue was not whether the offender learned to commit his crime through watching television, but whether he was motivated to so do as a result of participating on a television program.

In March 1995, producers of the talk television show *Jenny Jones* decided to have a show on which guests would be invited either because they had a secret crush on another or because someone had a crush on them. But there was a twist: these were to be same-sex secret crushes. One of the guests, 24-year-old Jonathan Schmitz, was told that someone had a crush on him and that he would find out who it was on the air. Schmitz was suspicious and told the producers that he was heterosexual and that he hoped that the person who had the crush was not another man.

But it was. On the air, Schmitz was surprised to discover that the person who had a crush on him was a man, Scott Amedure, who had met Schmitz before. Schmitz expressed a great deal of surprise and embarrassment during the taping, and although he was gracious while on the air, he was clearly upset at the experience.

Within three days of the taping of the show, on March 9, 1995, trouble was reported. Schmitz had been accused of murdering Amedure. The talk show itself also became a defendant in the case since Schmitz pleaded that his appearance on the show was what sent him over the edge. Plagued by depression and drug dependency, Schmitz said that his appearance on the show put so much pressure on him that his killing Amedure was inevitable. A jury in November 1996 disagreed and convicted Schmitz for first-degree murder. He was sentenced in December 1996 to serve 25 to 50 years in prison. Several jurors were interviewed after the trial and indicated that they did think that the show contributed to the murder, but that Schmitz was ultimately responsible for the crime.

Did the *Jenny Jones Show* contribute to Amedure's death? And, what of the *Jenny Jones Show* itself? Have things changed? Since the trial and conviction of Schmitz, the show has had the following themes: "I Let My Lover Have Affairs While I Was Pregnant," "My Mom Had An Affair With My Man," "Living Large" (on breast augmentation), and "She Could Easily Seduce Her Mother's Husband."

Source: See also Scott, Gini Graham. 1996. *Can We Talk? The Power and Influence of Talk Shows*. New York: Plenum Press.

Even when audience members miss news reports about crime, media deliver similar ideas in other forms. Surveys reveal that most people spend most of their leisure time watching television. Extremely popular television shows other than news programs present powerful images of crimes. In addition to dramas about police officers fighting crime, a number of other crime-related television shows blur the distinction between real and fictional crime. Many popular programs show reenactments of crimes, advertise the cases against wanted criminals, and dramatize the work of real police officers. Information about both real and imagined crime help to form people's images of the frequency and scope of criminality in modern life (see Sparks, 1992). Even television programs that are oriented toward "entertainment" have an extremely high violent content (National Television Violence Study, 1999).

Because the media concentrate on unusual events (which, by definition, are also newsworthy events), only the most spectacular instances of deviance reach the public's attention. Not surprisingly, many people share the mistaken impression that violent crime, activities of satanic cults, and child kidnapping have increased in frequency. For example, figures from the U.S. Department of Justice show that rates of violent crime, including murder, rape, assault, and robbery, peaked in the early 1980s and have fallen since then (Reiss and Roth, 1993). Substantial public concern over kidnapping of children has no basis in fact (Boccella, 1996); yet, the perception, not the reality, still motivates parents to worry.

Mass media portrayals of crime illustrate these processes. Media accounts concentrate on violent crimes, such as homicide, rape, and aggravated assault. At the same time, they ignore the crimes most often experienced by the public, such as theft, burglary, and public drunkenness. In the process, they give the distorted impression that violent crimes are increasing throughout the community. In fact, crime, including violent crime, has been declining in the United States for some time. At the same time, the media tend to ignore the harmful acts of white-collar and corporate offenders who are protected by their power and privilege (Chambliss, 1999). Similarly, media attention to cases of child kidnapping by strangers, for example, leaves the impression that the frequency of such acts far exceed their actual number: roughly 200 cases each year for the entire country (Finkelhor, Hotaling, and Sedlak, 1990).

Most crime, in fact, involves unspectacular, everyday violations, usually minor thefts with small monetary values. These generally unsophisticated acts require little planning. Most crime, in other words, is very ordinary (Felson, 1998). For this reason, media outlets do not find most crime worthy of news reports. Statistically rare crimes like homicide, on the other hand, are newsworthy precisely because of their rarity. Yet, coverage of such crimes cannot avoid the impression that such crime pervades the community.

Television news, with its apparent fixation on infrequent and startling crime stories, is a natural target for charges of distorting reality. Other programs help to convey incorrect impressions, as well, however. A TV viewer can watch crime stories at almost any time because of the rise of so-called *tabloid* television networks, news programs, docudramas, and crime-watch programs, such as *Cops*, *911*, Court TV, *Unsolved Mysteries*, and *America's Most Wanted*. These networks and programs deliberately blur fact and fiction, and in the process, they cannot avoid conveying a distorted picture of crime. For example, a few seconds of broadcast video for a show like *Cops* distill down the result of many hours of taping.

Media Impact   According to a realistic appraisal, the direct effects of crime portrayals by both television and motion pictures serve only to aggravate or reinforce current deviant attitudes and subcultural roles rather than creating new ones (Lowery and DeFleur, 1988). Quieting all the media in a society such as the United States would not reduce delinquency and crime. Certainly, extensive delinquency and crime carried on before the present mass media gained influence of any social consequence. Children both learn from and react to influences from the various media of mass communications, but all messages that they receive interact with another set of influences, such as the family, school, relatives, church, and companions before they become real guides to action. Television viewers can no doubt learn different ways or techniques of acting on their current attitudes, but without TV, they may have acted upon those attitudes in any case, although in different ways. Further, some writers have claimed that media influences are mediated by certain psychological characteristics that viewers bring to their media exposures (Feldman, 1977: 86).

The media's role in promoting violence has received special attention, resulting in three different views about whether or not television promotes aggressive acts by viewers: (1) television teaches aggressiveness; (2) it reduces aggressiveness by serving as an outlet; (3) it has no demonstrable influence on viewers' tendencies toward aggression. Some researchers have found a direct role through which television media teach and model aggressive behavior (Bandura, 1973; Liebert, Neale, and Davidson, 1973); others have failed to document such a role (Feshbach and Singer, 1971). An extensive review attempted to summaries the findings of studies examining the relationship between television viewing and violent conduct. The authors concluded that research had provided no evidence of any significant media influence on aggressive behavior (Kaplan and Singer, 1976).

Media effects can last for either the short term or the long term. One media task force evaluated available scientific evidence regarding media influences on violence. Its discussion of major short-term effects concluded that audience members exposed to violence learn how to perform violent acts, and that audience members become increasingly likely to exhibit that learning if they expect rewards for the behavior (Baker and Ball, 1969). This is supported in a major review of the media violence literature: "The strongest supported immediate effect is the following: Exposure to violent portrayals in the media increases subsequent viewer aggression" (Potter, 1999: 42). There are also long-term effects. "Exposure to mass media portrayals of violence over a long period of time socializes audiences into the norms, attitudes, and values for violence contained in those portrayals" (Baker and Ball, 1969: 376). Long-term effects also include a fearful worldview and desensitization to violence (Potter, 1999: 42).

Clearly, mass media images do present the United States as a crime-ridden, violent society. The newspapers and news broadcasts in major cities maintain a daily stream of graphic reports on such incidents, in the process giving the impression that deviants conform to certain stereotypes; the typical television image of a criminal shows a lower-class citizen rather than someone from a white-collar occupational group. Although many people who work in the media may think that they simply tell people what they need to know, public opinion polls indicate a perception that the media spend too much time on crime (Harris, 1987: 184–187). In any case, sociologists need to review much more theoretical and empirical work before they draw firm conclusions about media influences on deviance (DeFleur and Ball-Rokeach, 1982).

Occupation

Sociologists recognize the importance of jobs not for their economic rewards, but because of their contributions to people's personal identities and conceptions about the meanings of their lives. Some workers regard jobs as inconvenient nuisances tolerable only through necessity. To others, jobs form nurturing contexts for formation of employee's self-concepts.

Sociologists commonly expect occupations to reduce deviance. By maintaining a legitimate occupation, a worker can develop a sense of commitment to the job and a sense that deviant acts could jeopardize it (Hirschi, 1969). Workers may commit themselves in this way more readily for jobs with the status of professions rather than those regarded simply as occupations. Sometimes, however, workers in low-paying, menial positions need their jobs more than white-collar workers need theirs, since the better-qualified people can more easily find other alternatives. Still, a commitment to one's occupation and a continuing need for its economic benefits can blunt tendencies toward deviance.

Occupations also bring people into contact with other, often like-minded people. Co-workers may share similar world outlooks, aspirations, and ways of defining social situations. Plumbers resemble other plumbers more than they resemble lawyers. Lawyers resemble other lawyers more than they resemble auto mechanics. As people develop stronger commitments to their jobs, they tend to associate more often with people like themselves, such as co-workers. Such contacts allow them to share thoughts and perceptions of behavior, in the process reinforcing and amplifying common conventions.

But occupations may also provide opportunities both to learn and to commit deviant acts. White-collar workers may pick up techniques for carrying out undetected law violations in business, for example, from conversations with business associates. Training for sales specialists may perpetuate unethical tactics, such as bait-and-switch strategies, in which they will lure customers into stores with promises of low prices only to switch their interest to more expensive alternatives. Sometimes, workers become occupational criminals merely by learning to meet the expectations of good employees, as an antitrust offender explained:

> It certainly wasn't a premeditated type of thing in our cases as far as I can see. . . . To me it's different than _____ and I sitting down and we plan, well, we're going to rob this bank tomorrow and premeditatedly go in there. . . . That wasn't the case at all. . . . It was just a common everyday way of doing business and surviving. (Benson, 1996: 68–69)

Advertising specialists may deviate from the law if they learn how to prepare misleading copy. The line between ads that are false and "just" misleading may depend on nothing more than the size of the fine print. Illegal practices often become diffused among companies and the industries in which they participate. A systematic examination of violations and sanctions by the largest U.S. companies found particularly high violation rates in three industries—pharmaceuticals, the auto industry, and the oil industry—as compared to other industries (Clinard and Yeager, 1980).

As another example, physicians and nurses have access to many kinds of drugs denied to other people. It is not surprising, therefore, that addiction rates are relatively high among physicians. A study of 98 physicians, all of them either practicing or former opiate addicts, found that their drug-using careers began later than those of other opiate addicts, and they almost never associated with other physician addicts (Winick, 1961). Their addictions usually came to light through indirect evidence from reviews of prescription records. Another author has confirmed a very high level of general drug use, excluding tobacco and alcohol, among physicians as compared to other occupational groups (Vaillant, Bright, and MacArthur, 1970). Many physician addicts become users as medical students. One study found that 59 percent of a sample of physicians and 78 percent of a sample of medical students reported use of an illegal psychoactive drug sometime during their lives (McAuliffe, 1986).

Similarly, performers in the entertainment business, such as jazz musicians and performers in rock bands, are more likely than the general population to use marijuana and/or cocaine largely because this kind of drug use appears to be a part of the music subculture. Drug use may build an individual's prestige among other musicians, to fight the fatigue of one-nighters and as a recreational and stimulating pastime.

Cultural Diversity

Immigrant groups have traditionally come to the United States with the goal of assimilating themselves into the dominant culture. Immigrants desired to learn English, send their children to public schools, and to compete in the marketplace. They wished to become "Americans."

Many pursued this goal, not at the expense of their native cultures, some of which they retained. Immigrant groups gathered in ethnic neighborhoods, shopped at stores catering to ethnic tastes in food and clothing, and maintained the religious traditions they brought to the new world. Ethnic neighborhoods grew up in the large cities of the East Coast following waves of immigration from European countries, and many still retain identifiable characteristics. Identifiable communities harbor people of mainly Italian, Irish, Polish, and German ancestry. Similarly, other neighborhoods, especially on the West Coast, concentrate populations of immigrants and their descendants from China, Korea, Japan, and Mexico. Assimilation required new arrivals to gain skills (mainly in language and education) to participate in the marketplace, but it did not force them to abandon their cultures.

More recent immigrant groups, such as Hispanics from Cuba, have fought to maintain more of their original cultures. In fact, several ethnic and racial groups express an increasing desire to maintain ethnic distinctions. Groups throughout the world struggle to maintain ethnic solidarity (Calhoun, 1993), as do those in the United States. These efforts—marked by protective attitudes toward traditions, language, and customs—can lead to various conflicts with other ethnic groups and with the larger majority culture.

In an increasingly multicultural society, many laws fail to reflect shared values, and they can become focal points for disagreement. For example, acceptance varies between communities for laws restricting gambling, drinking, fortune telling, illegal sales of sporting-event tickets, extramarital sex, sexual behavior among married couples, and use of fireworks. Value conflicts over the use of marijuana and alcohol might be expected between, say, college fraternity members and a collection of Baptist ministers. When laws lack the support of a general value consensus, however, they build some potential for violations into their own structures.

Most citizens violate one law or another (Adler and Lambert, 1993). With so many laws on the books and their number increasing each day, even people motivated by conformance may commit violations. Specific laws represent the values of some groups, but not all, which guarantees some disagreement and creates a potential for violations. People violate some laws more often than others. Reports of tax evasion date to biblical times. People commonly consumed alcohol during Prohibition, and today's laws do not prevent use of marijuana and cocaine. Avoiding jury duty, driving under the influence of alcohol, and illegal gambling are all common crimes. Every act listed in Table 3-1 is a crime in at least one state, punishable by criminal sanctions.

Multiculturalism implies a lack of value consensus, not only about crime, but about other forms of deviance, as well. Normative differences among groups play

TABLE 3-1
Commonly Committed Crimes and Their Punishment

Crime	Customary Penalty*
Exceeding the speed limit	Up to $200 fine and possible loss of license
Appearing nude in public (e.g., nude sunbathing)	Up to one year in jail and/or $500 fine
Lying on applications for government job or benefits	Up to one year in jail and/or up to $1,000 fine
Illegal parking	Up to $500 fine
Illegal gambling (e.g., betting on a card game or sporting event)	Up to six months in jail and/or $1,000 fine
Engaging in prohibited sex acts (e.g., sodomy)	Up to six months in jail and up to $500 fine
Failing to recycle (e.g., failure to separate newspaper, bottles, and cans)	Up to $100 fine for repeat offender; $25 fine for first offense
Failure to control one's pet	Up to $100 fine for repeat offender; $25 fine for first offense
Disregarding a jury summons	Up to one year in jail and/or up to $750 fine

*Laws and penalties vary from state to state.

SOURCE: Authors' files and Adler, Stephen J. and Wade Lambert. 1993. "Just About Everyone Violates Some Laws, Even Model Citizens." *Wall Street Journal*, March 12: A1, A4.

central roles in both deviance and multiculturalism; norms define both group membership and the content of deviance. As Chapter 1 explained, one group may label an act as deviant that another group tolerates as normative behavior. These differences can create occasions for culture conflict and even deviance as groups compete with one another for resources, such as political power (see Sutherland, Cressey, and Luckenbill, 1992: Chapter 6). The growth of multiculturalism virtually ensures that some behavior valued in one group will not fit the values of another; in fact, one group might condemn behavior that another welcomes.

Summary

All deviance occurs within a social context, often defined by urban settings. Sociologists have noted a strong relationship between urbanization and deviance for virtually every form of deviance, including crime, delinquency, alcoholism, drug addiction, mental disorders, and suicide. Not surprisingly, the development and growth of deviance accompanies the general process of urbanization found in all parts of the world. In the United States, urbanization initially shifted the population toward urban areas; more recently, it has carried both population and economic opportunities toward suburban parts of metropolitan areas. In some cities, an underclass has developed and remains because its members have been unable to follow those opportunities. These poor urban residents, often people of color, experience very high rates of deviance in their daily lives.

The urban context of deviance reflects certain social characteristics of cities (summed up by the term *urbanism*) more strongly than the simple effects of living among many people. Urbanism is the way of life that develops in cities, reflected in a number of social features of city life, such as norm conflicts, rapid cultural change, mobility, materialism, individualism, and growing reliance on formal social controls. When people come to cities, they must learn new ways of thinking and doing things, and they must manage this change in a setting of widely diverse people, ideas, and events. This differentiation characteristic of cities bears a strong relationship to deviance.

The city creates an atmosphere conducive to deviance in another way by harboring deviant attitudes. These attitudes form part of the structure of a city, permitting a number of different subcultures and patterns of behavior to coexist. City dwellers' attitudes reflect the number and kind of associates they regularly encounter, the neighborhoods where they live, characteristics of their family lives, their exposure to mass media, and their employment situations. City dwellers encounter deviant attitudes from such sources more often than rural and suburban residents do.

An understanding of deviant behavior must explain the roles of associates, neighborhoods, families, mass media, and occupations to describe how people and groups come to commit deviant acts. These broad influences produce effects as important as those of urban influences, and they are closely related to many forms of deviance. Subsequent chapters will review the importance of these sources of deviant attitudes.

Peers and neighborhood influences are among the most important sources for many forms of deviant attitudes. The family and occupation may not directly supply deviant attitudes, but both define social contexts in which individuals can learn and act upon standards for deviance. Widely experienced mass media images of deviance condition the audience members' understanding of deviance and may influence public policy about it. Multiculturalism, a modern reflection of long-established culture conflicts, may spread deviant attitudes and behavior through conflict between different groups' normative evaluations.

Although the sources of deviant attitudes discussed in this chapter define important conditions, they do not in themselves explain the emergence of deviance. They do not, in other words, constitute sociological theories of deviance, even though such theories may incorporate these sources of deviant attitudes. Theories elaborate specific explanations of the deviance process, the topic of Chapter 5.

Selected References

Brantingham, Paul, and Patricia Brantingham. 1984. *Patterns in Crime.* New York: Macmillan.

This collection reports on research findings concerning the relationship between crime and city characteristics. Well-written and informative, this book summarizes a great deal of relevant literature.

Bursik, Robert J., Jr., and Harold Grasmick. 1993. *Neighborhoods and Crime: The Dimensions of Effective Community Control.* New York: Lexington.

This empirical analysis explores neighborhood influences on crime. Derived from a social disorganization perspective (discussed further in Chapter 5), the book presents a reasonable and interesting account of community sources of deviance.

Felson, Marcus. 1998. *Crime in Everyday Life,* 2nd ed. Beverly Hills, CA: Pine Forge Press.

This short, readable account details how groups and structures influence crime. Written from a human ecology perspective, it captures much important influence on crime and other forms of deviance by urban and local communities.

Reiss, Albert J., Jr., and Michael Tonry, eds. 1986. *Communities and Crime.* Chicago: University of Chicago Press.

This recent collection reproduces papers from sociologists interested in a communities orientation that characterized work done within the "Chicago School" in the 1920s. Some sociologists have revived this theoretical perspective in recent years.

Wilson, William Julius. 1996. *When Work Disappears: The World of the New Urban Poor.* New York: Alfred A. Knopf.

Debilitated by shifting economic opportunities and the departure to the suburbs of relatively highly skilled blacks, remaining slum dwellers suffer from a number of intense problems, including poverty, teenage pregnancy, and high rates of deviance. Wilson proposes new federal government initiatives to combat this problem.

PART II

Explaining Deviance

IN 1996, Timothy McVeigh was convicted and imprisoned for bombing the federal building in Oklahoma City. Why did he do it? Was it "bad" genes? Inadequate upbringing? Frustration? His reaction to an oppressive social system? The work of the devil? A deliberate choice on McVeigh's part to get back at someone or something? Some combination of these? McVeigh's own explanation was provided in a national news interview in March 2000, where he suggested that the federal government was a leading teacher of violence and that he was just a good student.

There have always been explanations of deviance. As we shall see in the first chapter of this part, some explanations emphasize the nature of society, some the nature of individuals. Some emphasize the conditions of everyday life, some the structure of life in capitalistic economic systems. Moralistic or spiritual, psychological or biological, there appears to be no end of theories.

Our intent in this part is to emphasize some of the major sociological theories of deviance. Some of these look for the cause in the structure of entire societies (structural theories) while others look for causes in the processes by which individuals come to commit deviant acts (processual theories). Our exploration of these theories is necessarily brief, and we have confined ourselves to those perspectives that have generated substantial interest among students of deviance. Given that sociology as a discipline is little more than 100 years old, it is not surprising that some of the perspectives extend back to the beginnings of scientific social thought while others are more recent. ■

CHAPTER 4

Becoming Deviant

People do not become deviants simply by committing deviant acts. If the sociological criterion for deviance extended no further than commission of a deviant act, society would be full of deviants and the term would have little meaning. A sociological conception of deviance identifies a person who plays a social role that exhibits this behavior.

This chapter addresses the way in which people come to fulfill deviant roles, that is, the process of becoming deviant. To understand these processes of acquiring a deviant role, one must also examine the social nature of human beings, including the self and identity of the deviant, and the process of socialization into a deviant role. This analysis also requires some empathy with deviants, an ability to see the world the way they do. Observers gain important insights into the processes of becoming deviant when they understand the meanings to those people of their deviant acts. Therefore, this chapter critically evaluates various ideas about how individuals come to commit deviant acts, including biological, psychiatric, and psychoanalytic explanations, along with a generalized perspective called the *medical model*.

At the outset, however, one fundamental point requires attention, and the chapter will return to it from time to time: The belief in an inherent difference between deviants and nondeviants relies upon a series of false assumptions. In fact, *all deviant behavior is human behavior,* and the same basic processes produce social behavior for both deviants and nondeviants. Certain subprocesses affect deviants, but they operate within the general framework of a theory of human behavior. The focus of analysis, as well as the fundamental social processes, remains the same for all human behavior, whether one studies inmates in correctional institutions or wardens, mental patients or psychiatrists, corrupt business owners or ministers.

Evaluation of deviance cannot cite unique or clear-cut criteria. No clear-cut distinction separates deviance from nondeviance without reference to norms, and, of course, norms change. Relative judgments of deviance determine class behavior as deviant at one time or in one situation, but the same acts may not seem deviant at another time or in another situation. For much the same reason, *deviant* is an ambiguous label. Human beings must live with changing norms, navigating constant shifts in expectations or norms that govern behavior and continuously reassessing their applicability.

As children mature, for example, their parents' expectations change. Even the rules of daily life change. Children renegotiate bedtimes, permission to travel to certain parts of their neighborhoods, telephone calls, and other privileges as they grow older, and standards taken as given at one time become unclear with passing time.

Similarly, human beings must often resolve conflicting expectations and demands. Teenagers who follow the dictates of their parents, for example, may violate the expectations of their peers. Similarly, employees who honor the expectations of their employers may violate the expectations of their fellow workers.

Since deviant behavior is human behavior, the general explanations of one should apply to the other. One must discuss the social nature of humans to show how deviant and nondeviant conduct stem from the same basic social processes and how deviance becomes the role behavior of an individual. Just as society creates conventional roles, deviant roles also emerge, and people become socialized to accept and fulfill them.

Socialization and Social Roles

In a sense, deviants are hypocrites. They violate some norms but conform to and defend others. They do not appear to define any general behavior pattern of conformity and nonconformity with all social norms. A certain person may deviate from certain norms and comply with others. A criminal may break the law by extorting money from people, but he might avoid an opportunity to cheat on his wife, explaining that marriage is a sacred commitment. Those who deviate from sexual norms may not steal, for example, while many white-collar criminals observe rigid codes of sexual conduct and maintain largely ethical dealings with neighbors. Top managers of criminal corporations may fulfill roles as highly dedicated citizens of their local communities. Even a strongly disapproved deviation may represent only a small proportion of a person's total life activities. Even where deviations constitute a more organized subculture, as among heroin addicts, accepted conduct may coincide in many ways with norms and values of the larger community (Levison, Gerstein, and Maloff, 1983). No one is deviant all of the time, and even the most committed deviant engages in deviant acts only at some times.

Social behavior is an acquired activity. People do not naturally begin interacting socially at birth; this activity develops through socialization. People modify their behavior in response to the demands and expectations (norms) of others, so that practically all behavior is a product of social interaction. Words like *honesty, friendliness,* and *shyness* have meaning only in relation to other people. Even expressions of emotion, such as anger or depression, despite strong physiological components, mostly express social reactions. Individual emotions are social products, too (Scheff, 1983).

Through group experiences, a human being becomes dependent upon others for human associations, conversation, and social interactions. The importance of this dependence on groups becomes apparent in situations that deny group contacts. Admiral Richard Byrd, the first person to fly over the North and South Poles, voluntarily isolated himself for several months in barely habitable polar regions more than 100 miles from the nearest human being of his expedition. Byrd's diary reveals an interest in his own reactions to such isolation, describing his experiences and vividly showing how much an individual depends on social groups.

> Solitude is an excellent laboratory in which to observe the extent to which manners and habits are conditions by others. My table manners are atrocious — in this respect I've slipped back hundreds of years; in fact, I have no manners whatsoever. If I feel like it, I eat with my fingers, or out of a can, or standing up — in other words, whichever is easiest. What's left over, I just heave into the slop pail, close to my feet. Come to think of it, no reason why I shouldn't. It's rather a convenient way to eat. I seem to remember reading in Epicurus that a man living alone lives the life of a wolf.
>
> A life alone makes the need for external demonstration almost disappear. Now I seldom cuss, although at first I was quick to open fire at everything that tried my patience. Attending to the electrical circuit on the anemometer pole is no less cold than it was in the beginning, but I work in soundless torment, knowing that the night is vast and profanity can shock no one but myself.
>
> My sense of [being] human remains, but the only sources of it are my book and myself and, after all, my time to read is limited. Earlier today, when I came into the hut with my water bucket in one hand and the lantern in the other, I put the lantern on the stove and hung up the bucket. I laughed at this; but, now when I laugh, I laugh inside; for I seem to have forgotten how to do it aloud. This leads me to think that audible laughter is principally a mechanism for sharing pleasure. . . . My hair hasn't been cut in months. I've let it grow because it comes down around my neck and keeps it warm. I still shave once a week — and that only because I found that a beard is an infernal nuisance outside on account of its tendency to ice up from the breath and freeze on the face. Looking in the mirror this morning, I decided that a man without women around him is a man without vanity; my cheeks are blistered, and my nose is red and bulbous from a hundred frostbites. How I look is no longer of the least importance; all that matters is how I feel. However, I have kept clean, as clean as I would keep myself at home. But cleanliness has nothing to do with etiquette or coquetry. It is comfort. My senses enjoy the evening bath and are uncomfortable at the touch of underwear that is too dirty. (Byrd, 1966: 139–140)

Deviants and nondeviants perform a variety of social roles that represent the behavior society expects of a person in a given position or with a certain status within a particular group (Heiss, 1981). The daily activities of a human being contribute to performance of a series of roles that the person has learned and that others expect the person to fulfill. People learn to play such roles as son, daughter, man, woman, father, mother, husband, wife, old person, doctor, lawyer, or police officer. Similarly, people learn to perform the deviant roles of gang member, professional thief, drug addict, alcoholic, or mental patient.

Although people experience a great deal of socialization in role playing and role taking during childhood, this social guidance also continues in later life.

Individuals learn new roles and abandon old ones as they pass through their life cycles and encounter new situations. Adolescence represents a period of adjustment to new roles (Hogan and Astone, 1986). Marriage brings a need to acquire new roles, as does entrance into the world of work when someone begins a profession or occupation. Old age also often requires a major role adjustment, as people must leave behind old roles and assume other ones (such as that of a "retired person" or "senior citizen").

Social behavior develops not only through responses to the expectations of others, which force one to confronting their norms, but also through social interactions, which lead one to anticipate others' responses and incorporate them into one's own conduct. When two or more people interact, for example, all are more or less aware of their mutual evaluation of behavior; in this process, each individual also evaluates his or her own behavior in relation to that of others.

The act of orienting one's own behavior to a set of expectations defined by a role is called *role playing*. A *role set* is a complement or collection of role relations that a person acquires by occupying a particular social status. A teacher acquires a role set that specifies relationships to students and to all the others connected with the school. Put another way, a role "is a set of expectations attached to a particular combination of actor–other identities (for example, father–son, father–daughter), and all the roles associated with one of the actor's identities is that identity's role set" (Heiss, 1981: 95).

The effectiveness, or even the possibility, of social control depends on people developing the ability to behave in a manner consistent with the expectations of others. Even self-control—an individual decision to engage in some behavior—is social control in that a person develops a self-concept in reaction to group expectations (Gecas, 1982).

Socialization as Role Taking

Socialization focuses largely on learning norms and roles. Put another way, *socialization* refers to the process by which members of society acquire the skills, knowledge, attitudes, values, and motives necessary to perform social roles. This learning process prepares an individual to meet the requirements of society in a variety of social situations. The required behavior (habits, beliefs, attitudes, motives, and actual conduct) represent an individual's *prescribed roles;* the requirements themselves are *role prescriptions*. People learn role prescriptions or norm requirements through interactions with others. The social structure or society itself largely dictates which roles a child learns in the family, such as a male or female sex role. Groups, then, are multidimensional systems of roles; a group is what its role relations make it. The individual members of a group may change as the group continues. In a delinquent gang, for example, the role of leader and other required roles in the gang may continue despite changes in gang membership. In fact, much deviant behavior directly expresses roles:

A tough, bellicose posture, the use of obscene language, participation in illicit sexual activity, the immoderate consumption of alcohol, the deliberate flouting of legality and authority, a generalized disrespect for the sacred symbols of the "square" world, a taste for marijuana, even suicide—all of these may have the primary function of affirming, in the language of gesture and deed, that one is a certain kind of person. (Cohen, 1965: 13)

Professional thieves, for example, perform a variety of roles. They must punctually keep appointments with partners and honor prohibitions against "squealing" on other thieves (Sutherland, 1937). Social status or position among thieves comes from an individual's technical skill, connections, financial standing, influence, dress, manners, and general knowledge. The professional criminal may play different roles toward victim, friend, spouse, children, father, mother, grocer, or minister (Inciardi, 1984).

Actual role behavior may differ somewhat from specific role prescriptions, because it responds to a variety of influences, such as the behavior of others in the situation, membership in groups with different and confusing role prescriptions, and so forth. *Role strain* may arise in situations with complex role demands and

TABLE 4-1
Keeping Role Terms Straight

Role Term	Meaning
Role	A set of expectations for a person occupying a particular social position. (Social positions are called *statuses*). Also, the behavior expected of a person in a given status within a particular group.
Prescribed role	Required behavior (habits, beliefs, attitudes, motives, and actual conduct) of a status.
Proscribed role	A role not permitted an individual because of other roles the person occupies (e.g., bachelor is a proscribed role to a husband).
Role playing	Orienting one's behavior to a set of expectations bound up in a role.
Role taking	The decision to adopt a particular social role.
Role set	A set of expectations attached to a particular combination of actor–other identities (for example, father–son, father–daughter), and all the roles associated with one of the actors' identities.
Master role	A role so important to the individual that he or she organizes other roles around it.

where a single person must fulfill multiple roles (Heiss, 1981; Parsons, 1951: 280–283). Many of these problems arise in systems of roles because (1) unclear role prescriptions cloud understanding of what is expected, (2) an individual plays too many roles to fulfill all of them adequately, resulting in role overload, and (3) an individual may play conflicting or mutually contradictory roles, forcing that person to perform a role without necessary preparation. The diversity of social roles in modern, urban society is an important determinant of the extent of social deviation in society.

Deviant Role Taking

Sociologists can speak of deviant roles in the same way they can speak of any other social roles. Some members of society perform criminal roles; many people with physical disabilities, such as the obese, the crippled, the blind, the retarded, come to occupy the roles expected of them based on their physical conditions. In fact, such people's social roles often require explanations beyond the physical disabilities themselves. Much behavior attributed to mental disorder makes sense in relation to social roles, as will be shown later, as does the behavior of the homosexual and the organized criminal offender. Even suicide often reflects enactment of a social role to its ultimate conclusion. Sociologists interpret much deviance that appears irrational or senseless as efforts to proclaim or test certain kinds of identity or self.

Role-expressive behavior can include the use of marijuana and heroin, especially in the early, experimental stages; driving at dangerous speeds and "playing chicken" on the highway; illegal consumption of alcoholic beverages; participation in illegal forms of social protest and civil disobedience; and gang fighting. In order to recognize this motivation, however, one must recognize the roles at stake and the "role-messages" carried by specific behaviors in the actor's social world (Cohen, 1966: 99).

A number of compelling reasons support viewing deviant behavior in terms of roles (Turner, 1972). For one reason, this kind of analysis brings diverse actions together into a particular category or style of life, such as the "homosexual," "drug addict," or "criminal." By examining each type of behavior as part of a deviant role, an observer can identify common dimensions. Many people have, at one time or another, engaged in homosexual acts. But an identity as a homosexual requires more than engaging in homosexual acts. It is a role performed to some degree by people who identify with homosexuality. This role may involve a style of dress, gestures, certain language, knowledge of homosexual meeting places, and how to react to heterosexuals. Similarly, many adults have been drunk at some time in their lives, but only a few come to perform the role of alcoholic or problem drinker. People who drink differ in many ways, but far fewer differences separate alcoholics. Once a person assumes a deviant role, deviants become more like one another.

Deviant roles exert powerful effects, both for the people performing the roles and for others. Once a person acquires an identity as an "alcoholic," a "homosexual," a "criminal," or a "mentally disordered" person, other social roles become organized around the deviant role. The deviant role thus becomes a *master role* for the individual. Master roles determine characteristics so important for the individual that he or she begins to identify with the role and to organize other roles around it. The individual may eventually develop a deviant self-conception through selective identification with the deviant role out of the many roles that he or she plays.

When other people stress a person's performance of a particular deviant role, the deviance often plays a central part in that person's identity; physically disabled people frequently accept certain roles in this way. Other deviant roles, too, frequently act as master roles, largely determining the reactions of the people with whom the deviants interact. Because substantial stigmas accompany deviant roles, others tend to reject deviants in society and to cast them outside normal interaction patterns. Some exclude deviants because the deviant roles dominate their opinions. Notice how powerful some names of deviant roles sound: sex offender, drug addict, suicidal. Once that part of a person is known, it becomes a central feature of other people's interactions with the deviant.

Not all deviant roles dominate people's lives, however. Some people engage only occasionally in deviant activities, keeping these acts separate from their "straight" lives. Some exotic dancers, for example, compartmentalize their lives into deviant and conforming parts by justifying their deviant behavior as a reaction to necessity (Reid, Epstein, and Benson, 1996). Some problem drinkers similarly maintain physical separation of their drinking from their employment by drinking only at times that do not conflict with their work. Similarly, some prostitutes attempt to live separate, nondeviant lives outside work.

 Issue   **Exotic Dancing as an Occupation**

Exotic dancing is a euphemism for the job of undressing in public before a paying audience. Also known as *strippers* or *adult entertainers,* exotic dancers work in clubs that specialize in this form of entertainment. The attraction of the job is not hard to see: It offers quick money—sometimes in large amounts—for part-time work that doesn't require training.

A survey of 41 dancers drawn from 12 clubs inquired about the influence of stripping on the women's identities. Does stripping influence their self-concepts? Largely, it does not. Most of the respondents indicated that they worked in the strip clubs for the money, but their dancing did not powerfully influence their personal identities. They viewed themselves not only as students, mothers, consumers, and in other conventional roles, but most also went to some length to distance their identities from their dancing. Few agreed that their dancing reflected their personal values. Justifying their employment by citing their need for money, most dancers indicated that they only "played at" their deviant occupation without permitting it to contaminate their personal lives.

Source: Reid, Scott A., Jonathon S. Epstein, and D. E. Benson. 1996. "Does Exotic Dancing Pay Well But Cost Dearly?" Pp. 284–288 in *Readings in Deviant Behavior.* Edited by Alex Thio and Thomas Calhoun. New York: HarperCollins.

People cannot easily change previously ascribed roles when they desire. Whether or not a person continues to play a role that society has assigned, others often interpret the person's behavior as part of this role and its corresponding status. For example, former prison inmates who return to their home communities may spark rumors interpreting their behavior in a manner consistent with real or imagined "criminal tendencies," despite their determined efforts to go straight. The deviant may encounter barriers that prevent reentry into conventional social roles while, at the same time, having to deal with social rejection and exploitation. As one former mental patient put it:

> Since I was let out, I've had nothing but heartache. Having mental illness is like having the Black Plague. People who know me have abandoned me—my family and friends. And the people who find out that I was in the mental [hospital] . . . treat me the same way. . . . And, at the boarding home where I was placed, I hardly get enough to eat. For lunch and supper today, all we got was a half a sardine sandwich and a cup of coffee, and they take three-hundred and fifty dollars a month for that kind of meals and lousy, overcrowded, bug-infested rooms to live in. (quoted in Herman, 1987: 241)

Several consequences flow from the power of community interpretations to perpetuate a person's identity with a criminal status and role. Sometimes such a person quits resisting and "gives in" to society's definition, actively playing the part that others seem to expect. If others treat a person as generally a deviant rather than as one who commits specific deviant acts, this response may produce a self-fulfilling prophecy, setting in motion several mechanisms that "conspire to shape the person in the image people have of him" (Becker, 1973: 34).

Deviant behavior can also affect the deviant's selection of other roles in life. For example, family life often conflicts with deviant behavior that results from the performance of deviant roles. Marijuana use, for example, seems to be associated with a postponement of motherhood among women and with an increase in the propensity toward marital dissolution among both men and women (Yamaguchi and Kandel, 1985).

Deviant Acts and Deviant Roles: The Example of Heroin Addiction   People assume deviant roles over time. Opiate addiction illustrates this point. Some people think of behavior related to heroin addition as a simple result of physical dependence, actions over which the addict has no control once addicted. Bennett (1986) studied 135 English addicts between 1982 and 1984 to identify the stages of their drug careers. The majority of addicts began their drug taking with marijuana or amphetamines. They later turned to heroin after considerable drug experience, usually when friends offered opportunities.

Most people became addicts in the course of slow developments. The majority of Bennett's addicts took more than 1 year to become addicted. A number of heroin users progressed over many months only occasionally taking the drug. Once addicted, some discontinued their use—sometimes for as long as a year or two. As one of the addicts phrased it: "I usually use every day for a couple of months and then I start cutting down. I have occasions when I dry myself out for 3 or 4 months. I don't want my habit to get too big" (Bennett, 1986: 96). In other words, these addicts performed the role of addict more at some times than others, and they managed to perform other, conventional roles, as well.

Addiction careers vary both in the total amounts of heroin consumed and addicts' socialization to the drug subculture. Although users continued to maintain addict self-conceptions without daily heroin use, they required contact with the drug subculture to ensure future supplies and support. Contact with that subculture greatly increased the chances that a particular drug user would develop a deviant self-concept and begin to adopt the addict role. In this sense, adoption of a deviant role varied by degree. Most persons neither conform totally or completely submerge themselves in deviant roles; most live somewhere between these extremes.

Bennett's research dealt with current heroin users. When asked whether they would abstain from heroin in the future, about one half reported they would like to continue to use heroin. These addicts felt comfortable with their addictions, and they thought they would experience better lives with than without heroin. The other half indicated they would like to quit using within the next decade, usually citing other people's expectations (e.g., spouse, another family member, friend, employer) as the main reason. Thus, the expectations of others supply important motivations for both occupying and leaving deviant roles.

Seeing the Deviant's Perspective

People often easily condemn the norms and values of others because they lack experience of priorities different from their own. Ultimately, understanding requires comprehending the world of the deviant as that individual experiences it, at the same time remaining sufficiently detached to analyze the interrelationships of the deviant world and the larger social order. All too frequently, observers evaluate others only from the perspective of their own worlds. Descriptions of actions as "senseless," "immoral," "debauched," "brutal," and so on often fall on deviants, scattered by outside observers with no awareness that deviant actions might have different meanings to those actors. A sociologist seeks to develop an "appreciation" for deviance not as a form of approval, but as a way to understand actions as the deviant does. Social scientists try to see the world and the meaning of deviance from the perspective of the deviant (Matza, 1969).

Much research on deviance has begun with a motivation to correct it rather than to understand it. On the other hand, an excessive emphasis on appreciating the deviant's world can lead to an overly romanticized view that obscures a meaningful, honest appraisal of deviant lifestyles. Clearly, observers must mix correctional and appreciative perspectives to provide a balanced view of deviance. Indeed, this balance may pose the greatest challenge to the observer: to see much and maintain the authenticity of what is observed.

Understanding Deviant Worlds

To share the deviant's perspective and definition of the situation does not mean that the scientific observer "always concurs with the subject's definition of the situation; rather, that his aim is to comprehend and to illuminate the subject's view and to interpret the world as it *appears to him*" (Matza, 1969: 25). The problem begins, therefore, with gaining access to deviant worlds.

Sociologists and others gather much material on deviant perspectives through in-depth interviewing, by soliciting "insider" reports from deviants themselves, and by participant observation (Cromwell, 1996; Douglas, 1970, 1972). Such sources have generated a great deal of information about visitors to nudist camps, drug users, call girls, homosexuals, youth gangs, pool hall hustlers, Hell's Angels, and topless barmaids. Researchers became insiders to collect some of the information by participant observation, becoming or posing as members of the study groups. This research usually maintains secrecy; the researcher's targets do not know they are under study. The members of the group treat the researcher as one of their own and share their lives.

Obvious ethical and practical considerations limit the use of insider participant observation, as well as what Douglas (1970: 6–8) has called "fictitious membership." In this technique the group members know the researcher's identity, but they also know that the researcher will not report their actions to the police or other officials. This method raises ethical questions of its own, however; the researcher may feel obligated to contact authorities about serious acts of deviance, such as major crimes.

Other firsthand material comes from life histories, diaries, and letters of deviant persons. The chapters that follow frequently feature such material to aid understanding from the perspective of the deviant. For example, analysis of suicide notes helps later analysts to understand the meaning of this act to the participant.

A sociologist need not become a deviant in order to comprehend that person's world, or even to gain access to those who are deviant. In fact, a researcher must evaluate disadvantages of adopting the deviant lifestyle and becoming one of the study's subjects. Deviants have no exclusive claim to knowledge of the subject. Drug addicts do not necessarily become experts on the addiction process as they personally experience it. Homosexuals are not experts on the social dynamics of homosexuality, although they have completed the process of becoming homosexual, and they must continually *manage* (a term to be explained shortly) their identities. No one would claim that only mentally ill persons could understand that condition, although they have intimate knowledge of that life.

Thus, an insider does not invariably acquire knowledge that reliably guides interpretation for others who have undergone roughly the same experience. In fact, insiders may not even supply a valuable type of information about deviance. No one but a heroin addict can know personally the experience of severe pains resulting from withdrawal. Still, the important questions about heroin addiction and the withdrawal process focus not on what it feels like, but rather on the role of the withdrawal process in continued addiction and the importance of the drug subculture in defining that experience.

Deviants, naturally, see the world differently from those who lack inside identities. A researcher must balance sensitivity to the deviant's unique perspective with a concern for objectivity. Drug addicts know where to obtain their supplies of drugs and from whom; this insight does not explain the process through which they came to be addicts. They know which of their acquaintances support and oppose them; they do not necessarily know the extent and types of influence each exerts on them or their deviance. In other words, insiders do not know all that a researcher wants to know about deviance; in fact, what they know often does not provide reliable, generalized knowledge about deviance or even their own deviant behavior (Merton, 1972). Deviants cannot be expected to provide all worthwhile information about deviance merely because they are deviants.

An insider does, however, gain access to certain kinds of information. The insider's knowledge supports fruitful insight only within narrowly prescribed boundaries. An alcoholic may offer valuable information about her own experiences, but even minute questioning of skid-row residents would give little help in studying the nature and extent of alcoholism in the United States, the social processes that generate and inhibit alcoholism, or the most effective means of treatment for the widest variety of alcoholics. As one observer put it, "Just as a boxing commentator does not need to slug it out over 12 rounds to bring a fight to life, so the [researcher] must remain content to 'talk a good fight'" (Pearson, 1993: xviii).

In spite of these limitations, firsthand observation and deviants' own accounts provide important information that fills out a researcher's objective understanding of the phenomena. While a sociologist does not need personal experience as a deviant to formulate valuable questions, deviance works through interactive processes; excluding information from the deviant would ignore one side of that interaction (Adler and Adler, 2000). For example, distant analysis may suggest that topless dancers are attracted to their work only for the money. In fact, Thompson and Harred (1996: 272) show a more complex picture. These dancers reveal combinations of three attractions to their work: (1) a tendency toward exhibitionism for gain, (2) an opportunity to dance topless as an alternative to other occupations, and (3) an awareness of the easy economic rewards for topless dancing. As one dancer explained:

> I had won a couple of bikini contests. One night I was in a club competing in a "best legs" contest and one of the girls took her top off! The crowd all went wild and the MC made a big deal out of it and it was obvious she was gonna win. Well, almost every girl after her took off their tops. By the time it was my turn to go out on stage, I'd had several drinks, and I thought "what the hell?" So, I pulled off my top, strutted my stuff, and it was no big deal. I didn't win the contest, but it made applying for this job easy. I thought, "Why show your tits for free?" So now I do it for about $400 to $500 a night—you can't beat it.

The example of illicit drug users also shows the benefits of a rich appreciation or understanding of deviance through personal input from the deviants. For example, an apparently simple exchange of crack for sex can actually hide a multiplicity of meanings for the participants. Distant observation might lead a researcher to two easy, but mistaken, conclusions regarding sex-for-crack trades: (1) Crack users typically exchange sex for their drug supplies. (2) Sex-for-crack trades result from the moral failings, poor judgment, or depravity of the participants (Ratner, 1993). Seen from the participants' perspective, however, life in the streets and life with crack requires more complex explanations.

The case of heroin addicts also illustrates the value of looking at behavior from the deviant's perspective. Many interpretations view addicts only as emotionally ill retreatists who have dropped out of many social relationships and resorted to the use of heroin. Ethnographic research by urban anthropologists and sociologists portrays addict life differently when it approaches the lives of addicts in a more empathetic manner (Agar, 1973; Hanson, Beschner, Walters, and Bovelle, 1985). Addicts see themselves engaged in a meaningful way of life, in spite of its deviant nature. An urban heroin addict adopts more or less completely a master social role as a street addict that dominates this person's relationships and life activities (Stephens, 1987: 77–79). This role comes to form a personal identity for the addict as he or she learns it in association with other participants in the drug subculture. The subculture also provides access to drugs and support from other addicts. Even an occasional, recreational heroin user adopts a set of attitudes and norms from others that support the process of becoming addicted (Zinberg, 1984). Without firsthand information about these processes, a researcher would not understand the dynamics of heroin addiction, restricting the effectiveness of efforts to deal with it.

Managing Deviance

One of the most valuable benefits of analysis from the deviant's perspective comes from enhanced sensitivity to some of this person's problems. Society's negative sanctions pose obvious difficulties that deviants would like to avoid. In addition to specific negative consequences, the deviant must also deal with the general *stigma* of an identity as a deviant.

Social groups understandably feel compelled to stigmatize some members. Stigma functions to defend the group; it "reaffirms the rule, reaffirms the conformists as conformists, and separates off the wrongdoer who has broken the rule" (Harding and Ireland, 1989: 105). But if stigma benefits the punishing group, it creates a problem for the deviant, who must learn to live with criticism by others as "odd" or "strange" compared to "normal" people. Deviants practice a number of techniques to manage or cope with this kind of stigma, prevent the stigma altogether, or reduce the harm of the stigma. By such techniques, in other words, the deviant tries to save face and ward off social rejection.

Management techniques suit the particular form of rejection that the deviant encounters, but a number of techniques commonly protect many forms of deviance (Elliott, Ziegler, Altman, and Scott, 1982). These techniques might function separately, or in combination with one another.

1. Secrecy   If others never learn about an act of deviance or a person's activities in a deviant role, that person will escape any negative sanction. A homosexual who fears the reactions of others may hide his or her sexuality from family and employer; an obese person may avoid social gatherings and maintain an isolated existence; a heroin addict may wear clothing that hides needle marks on arms and legs; criminals attempt to elude the police through planning and careful execution of their crimes. "Secrecy is [often] urged upon deviants by their in-the-know friends and family among normals: 'That is what you want to do, okay, but why advertise it?'" (Sagarin, 1975: 268).

A number of the topless dancers mentioned earlier reported hiding their occupation from boyfriends, husbands, and fathers to avoid their disapproval. "I told my mom right away," one dancer said, "because we don't keep any secrets, but we both agreed it would be a lot better if my dad didn't find out" (Thompson and Harred, 1996: 274). Sex workers who engage in telephone sex often described themselves as "telemarketers" to others, and one confessed that she was more honest with her credit card company than her family and friends about what she did (Rich and Guidroz, 2000: 41).

Sexually transmitted diseases (STDs) carry with them the potential for powerful stigma to people who disclose this information. Most people wish to have and present to others a sexual identity that is clean, healthy, and attractive. STDs are none of these. One of the most important devices for those with STDs is to therefore attempt to "pass" as someone without such a disease (Nack, 2000). To pass as sexually healthy often involves lying to others or simply not disclosing to others that one has a disease. "I guess I wanted to come across like really innocent and everything," admitted one woman with an STD, "just so people wouldn't think that I was promiscuous . . . " (Nack, 2000: 104).

2. Manipulating the Physical Setting   A deviant can often avoid negative sanctions by creating the appearance of legitimacy for the act or situation, regardless of its true nature. A bookkeeper who embezzles an employer's funds attempts to maintain the appearance of an honest, trustworthy employee. Problem drinkers may turn down a drink when with friends to divert suspicion. An obese person may avoid social gatherings. Prostitutes sometimes operate under the guise of masseuses or escorts (Prus and Irini, 1980: 65–68). The deviant seeks not necessarily to completely conceal the activity, but to maintain the most legitimate possible outward appearance of the setting for the deviant acts. This management technique sometimes works, because legitimate massage parlors and escort services do exist and therefore create doubt about the extent of a person's deviance. Some of the topless dancers told their friends and neighbors that they worked in clubs as waitresses. A study of telephone sex workers reports that the workers would often decorate their workspaces in an attempt to personalize the space and make it more human with recipes, family photos, and cartoons (Rich and Guidroz, 2000: 38). Such efforts at manipulating physical space may be as much motivated to reduce the stigma of sex work to the women as to try to convince outsiders that it is telemarketing, not phone sex, that goes on there.

3. Rationalizations   A deviant may try to avoid sanctions by explaining and justifying the deviance in terms of the situation, the victim (if the act produces one), or some other cause usually beyond the deviant's control. A tax cheat may justify the offense by complaining about paying already excessive taxes. A shoplifter may depict this crime as acceptable behavior because "the store can afford the loss, and insurance will cover it, anyway." An obese person may falsely attribute the results of an eating disorder to a physiological or glandular condition. In the study of women with STDs mentioned earlier, many of the women eventually came to a point where they began to blame others for their medical condition. Speaking of a previous partner, one woman was able to transfer her stigma to him by suggesting that she caught her STD from him even though she had no proof: "I don't know how sexually promiscuous he was, but I'm sure he had had a lot of partners" (Nack, 2000: 107).

If a deviant tries to justify an act after committing it, the term *rationalization* is appropriate; if the justification precedes the act, the term *neutralization* more appropriately describes the management method. Neutralization weakens the strength of the norm by placing the deviance in a more acceptable framework or by convincing the deviant that the norm does not apply for some reason. This technique also provides an effective way to save face when confronted with a troublesome or embarrassing situation. For example, a person who works in a position of financial trust may try to justify embezzling money by citing unique financial difficulties, such the cost of special care for a medical condition or impending foreclosure. The topless dancers denied that their activities hurt anyone, and some indicated that they were really dancing for other, more important reasons: "I'm not proud of what I do—but I do it for my daughter. I figure if I can make enough money doin' this and raise her right, she won't ever have to stoop to doin' the same thing" (Thompson and Harred, 1996: 276).

4. Change to Nondeviance   In another deviance management technique, a person tries to move from deviant to nondeviant status. Criminals usually talk about this technique as "going straight" or becoming "rehabilitated." An obese person may lose weight, a prostitute may marry and settle down to raise a family, and a problem drinker may shun alcohol. Observers often have difficulty determining whether someone has abandoned deviance, since this judgment is often a social one. A heroin addict may no longer inject heroin but may take methadone, itself an addicting drug, though a more socially acceptable one. Another addict may turn to heavy use of alcohol.

The change to nondeviance causes trouble for some stigmatized people. Some heavy cocaine users manage their drug use without developing addictions, ingesting cocaine only under controlled conditions (Waldorf, Reinarman, and Murphy, 1991). Some of the topless dancers explained that they saw their work as a temporary stopgap until they could find something else. Some deviants, such as the physically disabled, cannot practice this technique. Some deviants simply lack motivation to change, even if they could, such as a homosexual who wishes to remain homosexual.

The change to nondeviance also operates on a group level when militants try to affirm their deviance and eliminate sanctions for it. For example, homosexuals in some communities have publicly proclaimed their status, pressuring legislators to

| *Issue* | **Techniques of Managing Homosexuality** |

Some stigma-management techniques work only for certain kinds of deviance. Homosexuals and lesbians, for example, are exposed to considerable social stigma and have used a number of management devices geared toward reducing the stigma they experience as a result of their sexual orientations. Troiden describes four such techniques:

1. *Capitulation.* Those who capitulate to homosexuality refrain from engaging in it and from openly expressing homosexual attitudes.
2. *Minstrelization.* This term, derived from acting like a minstrel in an old-time show, refers to behavior in accord with popular stereotypes of homosexuality, such as dressing and walking in certain ways to affirm a homosexual lifestyle.
3. *Passing.* Probably the most common adaptation for homosexuals, this method requires leading a double life by limiting the flow of information between the straight and gay worlds. Gays who pass do not deny their sexual preference, but neither do they publicize it.
4. *Group alignment.* By belonging to and participating in a homosexual subculture, gays openly acknowledge to themselves and others their identities as homosexuals.

Source: Adapted from Troiden, Richard. 1989. "The Formation of Homosexual Identities." *Journal of Homosexuality* 17: 43–73.

change laws concerning this behavior and urging greater public tolerance. Kitsuse (1980) has suggested the term *tertiary deviant* (in contrast with primary and secondary deviant) to describe someone who presses for redefinition of deviant conduct to change standards for acceptable behavior. Militant prostitutes have taken similar public stands advocating decriminalization of this offense. In each case, deviants try to change to nondeviance, not by altering their behavior, but by redefining standards for the behavior itself. In 1993, gay organizing exerted pressure for full acceptance of homosexuals in the military forces, possibly motivated by a desire to achieve dramatic acceptance by the public in general more than a desire for acceptance specifically in the military. By the year 2000, it was clear that acceptance of gays in the military was still problematic and that gays were still subject to verbal and behavioral discrimination.

5. Joining Deviant Subcultures   Participating in a subculture helps deviants to manage their deviance by reducing contact with "normals" and therefore the chances of suffering negative sanctions (Troiden, 1989). The subculture may also facilitate deviant acts by providing a necessary condition, for example, a supply of drugs, and by reinforcing deviant attitudes. By frequenting gay bars and maintaining interactions only with other homosexuals, at least during those times, homosexuals decrease the chances that outsiders will stigmatize them. Gay bars can also help someone to maintain a homosexual identity by managing interactions with nongays in a situation that gays control; in the process, the subculture reinforces and perpetuates gay life (Reitzes and Diver, 1982). A subculture offers sympathy and support to a deviant, along with association with other deviants. It helps the deviant to cope with social rejection while at the same time providing opportunities to commit deviant acts (Herman, 1987).

Individualistic Theories of Deviance

A sociological theory of deviance explores the social conditions that underlie deviance—how society defines it, how group and subcultural influences relate to it, how deviants come to occupy their roles, why deviance is distributed in time and space, and how others react to deviations from norms. Individualistic theories, on the other hand, seek to explain deviance by evaluating conditions or circumstances uniquely affecting the individual, such as biological inheritance or early family experiences. Theories based on individual choices largely disregard both the process through which people learn deviant norms and group and cultural forces in deviance. This section critically examines several individually oriented explanations, including those based on biological determinants, psychiatric or medical models, psychoanalytic principles, and psychological principles.

Biological Explanations for Deviance

A human being embodies a biological nature and a social nature; obviously, without a biological nature, no human nature would emerge. A person's identity reflects interplay rather than opposition between the two aspects. Humans are animals who must breathe, eat, rest, and eliminate wastes. Like any other animals, they require calories, salt and other chemicals, and a particular temperature range and oxygen balance. Human animals depend on their environments, and certain biological capacities limit their activities.

Some scientists and practitioners claim to trace certain forms of deviant and antisocial behavior to specific physical anomalies, body chemistry compositions, or hereditary characteristics (Fishbein, 1990). These beliefs, in turn, have important consequences for suggestions about prevention and treatment programs. Some observers, for example, advocate sterilizing certain types of deviants, in the process expressing a biological view of human nature.

Biological perspectives usually define positions antithetical to those of psychological and sociological theories of human behavior. A more moderate view might look for interactions between biological and environmental factors to produce particular behavioral outcomes. A biological explanation might, for example, account for the widespread belief among social scientists in the importance of family socialization to determine subsequent behavior. Consider childhood misbehavior. Observers have identified a variety of parenting styles, and no single model always corresponds to specific disciplinary emphases and misbehaviors. One observer points out that children from virtually any kind of family may misbehave: "Many problem youths . . . come from the range of normal parenting variation, from families that are working- or middle-class," (Rowe, 1994: 223), as well as families with ample financial resources. Some might suggest important effects, not from family socialization experiences only, but from combinations of genetic and environmental influences on behavior.

Some biologists believe that specific, biologically inherited traits account for alcoholism, crime, drug addiction, certain types of mental disorders, and certain sexual deviations. Only limited and mixed evidence supports conclusions about such a view, although researchers actively studied such questions throughout the early 1990s. Work in the fields of alcoholism and crime illustrate the larger positions.

Biology and Alcoholism   Interest continues to swirl around the relationship of vulnerability to alcoholism and a complex interplay of genetic and environmental factors (Secretary of Health and Human Services. 1993: Chapter 3). Findings from family, adoption, and twin studies suggest that genetic factors may affect behavior by some chronic drinkers but not others. Researchers have been looking for genetic markers of alcoholism, that is, genes or parts of genes that transmit alcoholism vulnerability. Of the estimated 100,000 genes in the human genome, 20,000 appear to be expressed in the central nervous system. An unknown number of these genes may affect the development of alcoholism. The sheer number of possible genes seriously complicates the search for the appropriate one. In an additional complication, the biology of alcoholism could act not through genetic information, but through such factors as brain chemistry, individual variations in susceptibility to alcohol, and interactions between genetic and environmental factors.

While studies identify a possibility of inherited tendency toward alcoholism, research to date has failed to identify a specific alcoholism gene that predisposes individuals to heavy drinking. A study by the National Institute of Alcohol Abuse and Alcoholism (Secretary of Health and Human Services, 1993) has reported finding no more frequent incidence of a so-called *alcoholism gene* among 40 alcoholics than among 127 nonalcoholics. Nevertheless, subsequent research may eventually isolate such a gene or pool of genetic information. Even if particular genes increase the risk of alcoholism, social and psychological elements still may interact with genetic factors to determine drinking behavior. For example, certain genetic markers may exert important effects only when coupled with particular personality dimensions or in certain social contexts, such as social class or community situations.

A later chapter on alcohol use and heavy drinking will present evidence to suggest that alcoholism and problem drinking represent learned behavior rather than merely biologically determined certainties. One's associates, occupation, racial and ethnic group, religion, and other social factors prove more predictive of both drinking and problem drinking.

Biology and Crime   Several lines of investigation study links between biological characteristics and crime. The earliest scientific analyses of crime, in the 19th century, focused on biological variables (e.g., the work of Cesare Lombroso), and modern biological research has continued this tradition. Biology could influence the origins of crime in many ways, including genetically inherited traits, hormones, body type, neuropsychological (brain) factors, chemical composition of body tissues, and a variety of other physical dimensions. Observers have offered explanations for criminality based on body type, glandular disorders, brain pathologies, and, in the 1970s, chromosome anomalies (XYY) (Brennan, Mednick, and Volavka, 1995). Also, studies have traced many specific crimes, including rape, to biological causes (Ellis, 1989).

The possibility of a genetic basis for some crime has led investigators to explore a number of specific hypotheses. Some have looked at variations in frequency of crime in twins as compared to other siblings. In an analysis of 10 studies that focused on twins to evaluate the genetics of adult crime, Raine (1993: 55–57) found considerable evidence for the heritable determinants of crime. Identical twins showed a much higher tendency toward both committing criminal acts than did fraternal twins. Furthermore, these findings spread across studies conducted in the United States and several different European counties, including Holland, Germany, Finland, Norway, and Denmark.

A similar analysis evaluated 15 studies of crime in biological and adopted families. The researchers found that almost all of the studies reported some genetic predisposition to crime (Raine, 1993: 63). Adoption studies try to separate genetic influences from environmental ones by documenting the lives of adopted children of criminal biological parents. The researchers found higher crime rates among such children than among adopted children without criminals as biological parents. Several independent research teams working in several different countries have confirmed this conclusion.

Still, some believe that crime results from interaction of a number of factors rather than from a single biological system or component (see Knoblich and King, 1992). Evaluation of such interactions would also examine outside factors, such as the effect of consumption of alcohol and other drugs on biological systems. Researchers could also examine the effects of different psychological and sociological contexts for crime.

The notion of inherited tendency or biological predisposition figures prominently in the theory of crime developed by Wilson and Herrnstein (1985). These authors attribute criminality to an individual's acquisition of criminal attitudes and to biological tendencies in some offenders to violate the law. Wilson and Herrnstein assert that crime results from a choice that people make, but biological constraints limit their ability to determine these choices. Biological makeup, for example, may influence a person's range of social interactions and, therefore, her or his learning experiences. While many sociologists would agree with this conception, it stops far short of any claim that offenders simply act out biological predispositions to crime.

Evaluating Biological Approaches   The heterogeneous focuses of crime and alcoholism studies inhibit overall judgments about biological research. Clearly, theories must account for more than one type of alcoholic behavior and more than one type of criminal behavior. As a result, the influence of genetic factors may vary as much among individual alcoholics as between alcoholics and nonalcoholics. Also, genetic factors may influence the behavior of some criminals but not others.

Over a century of research has not yet identified precise biological mechanisms for deviance or means for transmitting them. Recent research supports interesting conclusions that suggest roles for biology, but these studies still have not dispelled the widely accepted notion that biology contributes little to explanations of social or symbolic behavior of humans or of deviant behavior, in general. Uneven quality has left doubts about research exploring biological causes of crime. Overall, however, more recent, better designed and executed studies seem to find no relationships or weak ones, contradicting earlier, less well-designed studies that found relationships (Walters, 1992).

There are no physical functions or structures, no combination of genes, and no glandular secretions contain within themselves the power to direct, guide, or determine the type, form, and course of human social behavior (Fishbein, 1990). Biological structures or properties certainly set physical limits on the activities of people, but any social limits result from the way in which cultures or subcultures symbolize or interpret these physical properties.

Inheritance cannot determine deviant behavior, as a general characteristic, since people cannot inherit knowledge of the social norms that define deviance. While an individual can inherit a particular way of looking or, sometimes, acting, the identity of that appearance or behavior as deviant depends on social, not biological, events. This conclusion holds for crime, since it is "obviously impossible for criminality to be inherited as such, for crime is defined by acts of legislatures and these vary independently of the biological inheritance of the violators of law" (Sutherland and Cressey, 1978: 123). This fact undermines complete support for direct inheritance of deviance. Instead, some have argued for inherited tendencies for such behavior. In many ways, this idea establishes an even more unscientific and vague position, since it usually fails to specify the nature and physiological location of this tendency.

Research may also incorrectly ascribe a role for inheritance in behavior when actions really result from social transmission of somewhat similar ways of behaving from one generation to another in a culture or from one family to another. Actually, heredity plays no role in this perpetuation of deviance, because genes cannot possibly detail so-called *family behavioral traits* or culture. Transmission of cultural or family attitudes and values would require an inconceivably complex gene structure and biological heritage. On the other hand, families easily pass on behavioral traits by sharing common experiences and attitudes. In this fashion, and not through biology, people who know one another, or who share family relationships, may come to carry out similar actions.

From time to time, various explanations account for certain forms of deviance by citing biological characteristics, but sooner or later all of them disappear as individually valid theories. While particular offenders may indeed possess abnormal chromosome patterns, this characteristic, like other biological characteristics, ultimately fails to explain deviance because it ignores the relativity of deviance and the essentially social process for determining and judging human behavior. Some recent theories have combined biological and nonbiological explanations, but they also have failed through their inability to explain how physical and, say, social dimensions come together to form single, unitary explanations.

Psychiatric Model of Deviance

Psychiatrists regard deviants as patients with psychological illnesses. They view deviant behavior as a product of some fault within the individual, such as personal disorganization or a "maladjusted" personality. These theorists treat culture, not as a determinant of deviant and conforming behavior, but rather as a mere context within which individuals express inappropriate tendencies.

Psychiatric explanations for deviance commonly emphasize that every person at birth feels certain inherent, basic needs, in particular the need for emotional security. Furthermore, deprivations of these universal needs during early childhood lead individuals to form abnormal personality patterns. Psychiatrists assert that childhood experiences, such as emotional conflicts, largely but not exclusively determine personality structures and thus patterns of behavior in later life. They see a direct relationship between the degree of conflict, disorder, retardation, or injury to the personality and the degree of deprivation. By affecting personality structures, psychiatrists claim, children's family experiences largely determine their behavior, deviant or nondeviant, in later life. They particularly stress the need for maternal affection in developing a healthy personality structure.

According to this theory, extreme cases of so-called *general personality traits* characterize deviants but not nondeviants. These personality traits are said to include emotional insecurity, immaturity, feelings of inadequacy, inability to display affection, and aggression. These traits result from early childhood experiences in the family. Psychiatrists point out that a child's first experiences with others occur within the family group, so traits arising from these experiences form the basis for the entire structure of the individual's personality. Deviant behavior often reflects a way of dealing successfully with such personality traits; for example, an immature person may commit crimes, or emotional insecurity may lead a person to drink excessively and become an alcoholic.

The psychiatric position implies that *certain childhood experiences produce effects that transcend those of all other social and cultural experiences.* Its proponents suggest that certain childhood incidents or types of family relationships lead individuals to form certain types of personalities that contain within themselves the seeds of either deviant or conforming behavior, irrespective of culture. Thus, childhood determines the development of personality traits that encourage or inhibit deviance, and a person's behavior after the childhood years fundamentally represents efforts to act out tendencies formed at that time.

This theory views deviance as merely a symptom of some underlying psychological sickness that afflicts an individual unless professionals detect and treat it. Those who take this view regard most deviance as some form of mental illness or psychological disorder. The significance of a criminal act, they claim, comes not from the behavior itself, as serious as it may be, but from the underlying, "real" problem deep within the criminal's personality structure. They evaluate the crime as a symptom of such a hidden problem.

The psychiatric approach to deviance has moved toward an increasingly medical point if view over time. Until about 1960, psychoanalytic concepts and theories commonly dominated diagnosis and treatment of deviants (MacFarquhar, 1994). Over the past three decades, however, biochemical interventions and drug therapy have almost supplanted purely psychoanalytic treatments.

The Psychoanalytic Explanation

Psychoanalysis addresses most issues in ways closely related to general psychiatric methods, but it promotes its own explanation of deviant behavior. The orientation of psychiatry as a medical specialty differs somewhat from that of psychoanalysis, which deemphasizes the medical model both in its orientation and in the backgrounds of its practitioners.

Psychoanalysis was founded by Sigmund Freud, a Viennese physician who died in 1939 (see Gay, 1988). Psychoanalytic writers look for their chief explanation of behavior disorders in analysis of the individual's *unconscious mind*, which they regard as a world of inner feelings unlikely to express themselves in obvious ways through behavior or to respond to attempts at recall. Antisocial conduct, according to psychoanalysts, results from the dynamics of the unconscious mind rather than from conscious mental activities. Much of an adult's behavior, whether deviant or nondeviant, owes its form and intensity to certain instinctive drives, particularly sexual ones, and to early childhood reactions to parents and siblings.

Psychoanalysis assumes that the conscious self overlies a great reservoir of biological drives. A psychoanalyst defines personality as an amalgamation of three parts: the *id*, the *ego*, and the *superego*. The id represents a buried reservoir of unconscious, instinctual animal tendencies or drives. The ego, on the other hand, represents the conscious part of the mind. Thus, Freud postulated a dualistic conception of mind in which the id, or internal, unconscious world of native or biological impulses and repressed ideas, competes and often conflicts with the ego, the self, which operates consciously to control behavior. The superego operates partly consciously to mediate this conflict; the conscious part corresponds to the individual's conscience (see Lilly, Cullen, and Ball, 1989: 38). Within the mind, superego defines a human's social self, following principles derived from cultural definitions of appropriate conduct.

Psychoanalysts also assert that a normal personality *develops through a series of stages*. The development of personality proceeds with shifting interests and changes in the nature of sexual pleasure from the so-called *oral* and *anal* preoccupations of infant life to love of self, love of a parent of the opposite sex, and, finally, love of a person of the opposite sex other than a parent. Some of these stages overlap, and an individual may advance simultaneously through more than one stage. Some people do not progress satisfactorily through all of them, however, and they experience conflicts and personality difficulties as a consequence.

Thus, psychoanalysts attribute activities of deviants to unconscious attempts to satisfy unresolved infantile desires. Some believe, for example, that the type of crime a person commits and the types of objects involved in the crime often indicate specific types of infantile regressions. Others characterize the etiology of schizophrenia, a form of mental disorder, as a retreat to a form of infantilism. Psychoanalysts often describe alcoholics as passive, insecure, dependent, "oral" stage personalities whose latent hostility has been obscured. Some have compared drug usage to infantile masturbation (Rado, 1963).

Evaluating the Psychiatric and Psychoanalytic Perspectives

Criticisms of the psychiatric or medical model largely cite confusion about illness and norms, the lack of objective criteria for assessing mental health, overemphasis on early childhood experiences, and a lack of scientific verification for these claims.

Sociological critics explain that psychiatric explanations of deviant behavior blur the line between illness and relatively simple behavioral deviations from norms. Deviant behavior thus becomes a criterion for a diagnosis of mental abnormality. In this sense, deviations from norms, such as illegal behavior like delinquency and crime, infer some illness or mental aberration. Yet, the commission of deviant acts does not necessarily imply a mental "problem" anymore than the commission of nondeviant acts implies the absence of a mental problem. Another criticism cites the unreliability of psychiatric diagnoses and the failure of psychiatrists to agree among themselves about objective criteria for assessing degrees of mental well-being or aberration. This absence of objective criteria for either mental disorder or mental health allows psychiatrists to equate illness with examples of deviance like delinquency and crime (Hakeem, 1984). Even within broad diagnostic categories, practitioners have reached little agreement on the nature of psychiatric disorders. This deficiency leads some critics, such as the psychiatrist Thomas Szasz (1987), to describe psychiatry as more religion than science and to assert that psychiatrists have too much power.

Critics also discourage application of the medical model to the study of deviance, because that model is particularly prone to the logical fault of *tautology*. This logical fallacy results, in essence, from circular reasoning. A tautology is a needless repetition of the same sense in different words; it is a redundancy. Consider an example. Jeffrey Dahmer was one of the best-known mass murderers in U.S. history. Dahmer also practiced sadism and cannibalism. Upon his arrest in Milwaukee, Wisconsin, police found many human body parts in his apartment, and Dahmer confessed to crimes so horrible that many people could scarcely comprehend a person behaving as he had. Virtually the only explanation described Dahmer as "crazy" because of the particularly shocking nature of his crimes. To support this claim, reports cited the behavior itself as evidence, since someone would *have* to be crazy to commit such atrocious acts.

The psychiatric model, therefore, uses the term *illness* in two ways. One takes the deviant act as evidence of the illness, while the other cites the concept of illness to explain the deviant act. Dahmer's actions provided evidence of illness that some described as the cause of the behavior. This circular reasoning commonly limits the value of the psychiatric model, since it interprets deviance as evidence of some underlying problem and then presumes that the problem caused the behavior that occurred. This is circular reasoning. Obviously, the psychiatric theory's defenders can break the tautology only by presenting evidence of the presumed but hidden problem that is independent of the deviant act, which they describe simultaneously as the cause and effect of the problem.

Sociologists criticize psychoanalytical explanations of deviance by asserting that human behavior follows from social experience rather than from any innate reservoir of animal impulses (the psychoanalyst's concept of the id). Depending upon social and cultural experiences, a person can act either cruelly or gently, aggressively or pacifically, sadistically or lovingly. A single individual can be a savage Nazi or a compassionate and tender human being like Albert Schweitzer or Mohandas Gandhi.

Finally, sociologists deny that psychoanalysis provides a scientific explanation of human behavior. Most psychoanalytic claims lack scientific verification. No one has yet devised a way to measure or otherwise verify the actions of the id, ego, and superego; instead, psychoanalysis asks others to accept this and other claims only on faith.

Psychological Explanations of Deviance

Many researchers, primarily psychologists, have tried to develop various tests to identify personality traits that distinguish deviants from nondeviants. Such an effort assumes that the basic components of any personality are individual personality traits or generalized ways of behaving. Psychologists have identified many personality traits and ascribed behavior patterns to them, such as aggressive or submissive, intensely emotional or inappropriately unresponsive, suspicious or credulous, self-centered or solicitous for the welfare of others, withdrawing or eager for contact with others, and expecting affection or dislike from others. At one time, some applied the term *temperament* to encompass all such personality traits.

Many researchers once believed that heredity determined an individual's personality traits and that some people "naturally" acted in aggressive or shy ways. Substantial research has shown that such behavior patterns develop primarily out of social experiences. Other research has attempted to link crime with such psychological characteristics as feeblemindedness, insanity, and stupidity (measured by IQ tests). These studies have yielded disappointing results, since they have failed to find strong relationships (Lilly et al., 1989: 39).

Nevertheless, psychologists have applied dozens of personality tests, rating scales, and other devices to try to distinguish deviants from nondeviants. Some tests have remained popular for many years. Therapists and others try to ascertain and measure traits by a variety of pencil-and-paper tests, such as the MMPI (Minnesota Multiphasic Personality Inventory) and the CPI (California Personality Inventory). Projective tests seek to evoke responses for analysis. For example, the TAT (Thematic Apperception Test) confronts subjects with a series of pictures about which they comment; the Rorschach test displays cards containing standardized inkblots, and subjects respond by telling what the shapes mean to them.

Psychologists often seek to explain nearly all forms of delinquent and criminal behavior as products of abnormalities in the psychological structures of individual deviants. They believe that inadequacies in personality traits interfere with such an individual's adjustment to the demands of society. Eysenck (1977) proposed interactionist theory, one of the broadest personality test theories. Eysenck claims that criminal behavior results from a combination of certain environmental conditions and inherited personality traits. Some people, born with genetic predispositions toward crime, encounter adverse environmental conditions such as poverty, poor education, and unemployment, creating criminal deviance.

Unfortunately, psychological evaluation methods experience several major difficulties in distinguishing the personality traits of offenders from those of nonoffenders. In fact, tests like those described earlier have not effectively distinguished the personality traits of criminal offenders from those of nonoffenders. Psychologists have not yet identified a set of personality traits that consistently differentiates deviants from nondeviants. This fact does not eliminate any meaning for individual personality variables (Andrews and Bonta, 1994: 62–63). It does suggest, however, that an exclusive focus on individual personality will not support full understanding of the context of deviance, its process and history. Like some biological factors, personality factors may cause some risk of deviance, but they seldom add predictive value outside the larger social context of deviance.

Psychologists widely believe that differences in personality traits or attempts to escape explain addiction to opiates. They can cite no evidence of anything approaching an "addict personality," however, or any cluster of personality traits that are consistently associated with addiction. Some observers regard alcoholism as the result of personality maladjustment. In this view, early childhood experiences produce feelings of insecurity; together with difficulties in adult interpersonal relations, these feelings produce tensions and anxieties. The use of alcohol reduces anxiety, and drinkers may come to depend on it for this purpose. However, efforts to document such an "alcoholic personality" have not succeeded. Moreover, supporters of this position offer no reason to believe that people with one type of personality are more likely to become alcoholics than people with another type. The view that alcoholism results from particular personality traits often fails to take into account the effect of prolonged use of alcohol on aspects of the drinker's personality. Efforts to identify the personality traits that would distinguish homosexuals from heterosexuals have also ended without success.

The psychological literature devoted to many types of deviant behavior refers to a deviant personality type termed a *criminal psychopath* or a *psychopathic personality*. The more modern term for this kind of offender attributes an *antisocial personality disorder* to him or her. Journal accounts often describe this habitual criminal as without guilt or remorse for offending behavior. Despite considerable dispute over the meaning of the term *psychopath*, some of the characteristics of this person include demonstration of poor judgment and inability to learn from experience, shown by pathological *lying*, repeated crime, delinquencies, and other antisocial acts. Descriptions of psychopathic traits often lack precision, however, as demonstrated by wide differences in diagnoses of psychopathic criminals in various institutions and by research on the associated personality traits. Furthermore, some psychologists identify subjects as psychopaths merely because they have repeated or persisted in offending behavior, committing the same error of circular reasoning described for psychoanalysts. Writing on the characteristic of persistent antisocial behavior as a criterion for designation as a sexual psychopath, Sutherland (1950: 549) stated: "This identification of a habitual sexual offender as a sexual psychopath has no more justification than the identification of any other habitual offender as a psychopath, such as one who repeatedly steals, violates the antitrust law, or lies about his golf scores."

Evaluation of the Psychological Explanation　　Sociologists often criticize the psychological explanation for deviant behavior on the following grounds:

1. Human behavior results primarily from variable, socially determined roles that rather than static conditions like so-called *personality traits.* Psychological theory also fails to explain how deviants acquire specific behavior, such as techniques of stealing.

2. Psychology gives almost no tools for isolating the effects of societal reactions on the behavior of deviants. A psychologist can never say for sure whether given personality traits manifested themselves before development of the deviant behavior or whether experiences encountered as a result of the deviation produced the traits. An alcoholic or a drug addict may develop certain personality traits as a result of a long period of alcoholism or drug addiction, in reaction to consequent rejection and stigma, rather than the trait preceding and perhaps causing the deviance. An accurate scenario might portray interactions of psychological and social factors over time resulting in behavior.

3. Finally, psychology has produced no evidence of associations between so-called *personality traits* and deviations from disapproved norms. Comparisons with control groups have found no series of traits that can distinguish deviants from nondeviants in general. The studies do not show particular traits that all deviants share and that do not occur among nondeviants. Some deviants, for example, display emotional insecurity, but so do some nondeviants. On the other hand, some deviants seem like emotionally secure people. Psychology has difficulty interpreting such mixed results and accounting for the presence, though in varying proportions, of the same characteristics in both deviant and nondeviant groups. For example, an analysis of aggression has pointed out, "Aggressive deviant acts share so much in common with nonaggressive deviant acts that individuals prone to commit aggressive criminal acts are prone to commit nonaggressive criminal acts as well. Thus, no individual-level trait of aggression is consistent with the results of behavioral research" (Gottfredson and Hirschi, 1993: 65).

Rational Choice Theories

Many people explain a great deal of deviance simply as purposeful behavior; such an action represents a choice made by the actor to behave in a certain way, to think in a certain way, or to live a certain kind of lifestyle. In this view, analysis requires little attention to such ideas as self-concept, socialization, role, status, or identity, except of course as they influence individual decision making. Rather, one can evaluate deviance simply by understanding that criminals choose to commit crimes, alcoholics choose to drink as much as they do, and cocaine users choose to consume their drug. The deviant might decide over a long period of time or at the moment to engage in a particular activity, depending on the circumstances.

The idea that deviants may choose some of their situations is not new (see also Akers, 2000: 24–26). Remember that deviant behavior is human behavior, and people exercise considerable choice in all behaviors. Some 200 years ago, the English philosopher Jeremy Bentham and the Italian jurist Cesare Beccaria each explained crime as the result of choices. Offenders weigh the consequences of committing their crimes and the alternatives, these observers explain, and then make their decisions. Both Bentham and Beccaria referred to this process as *hedonism*, or the choice of behavior that would maximize an individual's pleasure and minimize personal pain. This behavioral restatement of a simple economic ratio of costs to benefits determines whether the individual chooses to commit a crime, drink excessive amounts of alcohol, or use drugs.

Contemporary rational choice theorists begin to explain crime with "an assumption that offenders seek to benefit themselves by their criminal behavior" (Cornish and Clarke, 1986: 1). Through this process, a person makes specific decisions or choices about whether or not to engage in crime. These choices exhibit some rationality—although based on the offender's situation—under constraints determined by available time and incomplete information about the choices. A rational choice theory of crime need not view criminals as highly rational, fully informed individuals. Rather, it describes how an individual makes decisions within a context defined by social, economic, and political factors. Some of these forces set conditions for the choice, and some do not.

All people make choices, but all do not agree on the wisdom of others' choices. What seems rational to one person may not appear the same way to another. The decision to take drugs, commit suicide, or steal, for example, may appear rational to the actor but irrational to an audience of that action. Floyd, a convicted offender, was asked about his choices regarding shoplifting:

Q: So how often did you commit them [shoplifting offenses]?

A: Anytime I could. Anytime that I only stood a 50 percent chance of making it. Sometimes if I only stood a 25 percent chance of doing it, I would do it, because I enjoyed it. I'm the type of person, man, if I could steal something from way in the back row or if the store manager is standing here and I could take something right under his nose, that's what I'd get.

Q: Why would you prefer that?

A: Because it's more of an accomplishment. (Tunnell, 1992: 122)

An audience to this crime—police, store security officers, other shoppers— might not act as Floyd did, but this behavior was freely chosen and rational from his perceptive.

Like other rational choice theorists, Wilson and Herrnstein (1985) argue for evaluating behavior by its consequences. The consequences of committing a crime include both rewards (or "reinforcers") and punishments; the consequences of not committing a crime also entail costs and benefits. While offenders gain immediate rewards for their crimes, nonoffenders gain rewards for refraining from crime only in the future. Standards for rationality vary, for one reason, because some people seem better able to anticipate future events than others (Pallone and Hennessy, 1992).

Rational choice theorists have applied their analysis to a variety of settings and forms of deviance, including alternative theoretical perspectives (Clarke and Felson, 1993). Even heavy cocaine users can and do decide to quit the drug (Waldorf et al., 1991). That choice, which evidently proves surprisingly easy for some users, reflects a history and certain conditions. Most of these former users, in fact, terminated their use without any outside therapy. Understanding the context of the decision and the factors that brought users to that decision provide important guidance in predicting the decision itself. To say that users exercised rational choice does not mean that their behavior was nonrandom.

Evaluation of Rational Choice Analysis   Some persons society considers deviant do not choose their deviance. Homosexuals do not choose to acquire their sexual orientation, for example, any more than heterosexuals do. Some dispute about methods for acquiring such orientations divide observers who emphasize the importance of inherited characteristics from others who favor socialization processes. In either case, the individual does not choose, in the normal understanding of that term, his or her sexual orientation. Once the orientation is acquired, however, people do choose in specific instances whether or not to act upon their sexual orientations to form relations with other people. But people do not develop sexual identities and orientations from specific behavioral choices. Similarly people with mental disorders are often considered deviant but do not obviously choose to behave oddly.

One may easily agree with the truism that some criminals, chronic drinkers, drug users, and other deviants choose—in some meaningful sense of that term— to commit specific deviant acts. Nevertheless, the notion of choice offers limited help in achieving a full understanding of many kinds of deviants, such as opium addicts, drug-addicted criminals, and people with mental disorders. A chronic user who has developed a physical dependency chooses to take a drug for reasons quite different from those of a recreational or experimental user, in spite of apparent similarities. While some addicts clearly plan to participate in the drug scene prior to their first involvement with heroin (Bennett, 1986), continued use after addiction seems like a much different kind of choice than the one to begin initial use. While a teenager might choose to shoplift from a department store, the same behavior has a different meaning for an addict who wants money to support a drug habit. Individuals who exhibit aberrant behavior or thought patterns, such as those who are hospitalized in a mental facility, do not choose independently to behave or think as they do; other factors affect such choices.

Rationality makes perhaps the most sense not as a yes-or-no condition but as a matter of degree. At present, there is little consideration in rational choice models of emotions, although there is reason to believe that immediate conduct reflects a combination of rational and emotional factors (Bouffard, Exum, and Patnernoster, 2000). Everyone makes choices, but some people choose among more and better options than others. These statements may have resolved the free will/determinism debate about as completely as the current argument will resolve it, but one may expect some progress by viewing rationality along a continuum rather than a categorical variable. Some may reason that all offenders choose whether or not to commit crimes, but it seems that business executives choose among noncriminal options not available to homeless people. The individuals of each group can decide whether or not to commit crimes, but a different context and circumstances surround each decision.

McCord (1992: 126) offers some guidance for evaluating crime:

> Criminal behavior ought to be studied with recognition that crime is a consequence of motives to injure others or to benefit oneself without a proper regard for the welfare of others. Practices that foster these motives are likely to promote crime. Claims that all behavior is egoistic, that crime requires no explanation and that beliefs are irrelevant to criminal action have been a disservice to criminological theory.

Surely, such a temperate view of rational choice can apply to all forms of deviant behavior. Perhaps the major questions deal, not with whether or not to commit a deviant act, but the conditions under which people make those decisions.

Summary

Deviant behavior is human behavior, and one may expect to understand it only within the larger framework of other human actions and thought. People become deviant just as they become anything else—by learning the values and norms of their groups and in their performance of social roles. Some values support conventional behavior, and some support deviance; some norms call for conventional behavior, and some expect deviance; some roles encompass conventional behavior, and some define deviant actions. Behavior differs according to the content of the values, norms, and roles that influence it.

Viewing deviance in the context of roles enables a sociologist to interpret the meaning of deviance both for the deviant and for others. Deviant roles often exert powerful force, because they tend to overshadow other roles that people may play. To the extent that deviance at least partially expresses role behavior, it conforms to certain expectations about behavior in particular situations. A drug addict may conform to the demands of an addict role, just as crime may reflect an individual's conformance to those of a criminal role. But most deviants perform deviant roles infrequently and usually only for short times.

Evaluation of deviance usually views it from the perspective of a nondeviant. But a full understanding requires an attempt to understand the meaning of the deviance for the deviant. Observational studies can achieve insights into deviance that other methodologies miss. To appreciate deviance means to understand, but not necessarily to agree with, the deviant's view of the world. The methods by which deviants handle rejection or stigma from nondeviants are called *management techniques*. No one technique allows deviants to manage living in a world that rejects them, but not all techniques work for each deviant. These techniques include secrecy, manipulating aspects of the physical environment, rationalization, participation in deviant subcultures, and changing to nondeviance.

Individualistic explanations of deviance attribute the process of becoming deviant to some biological or psychological cause within a person. Substantial and continuing research addresses the role of biological causes in deviant behavior, including alcoholism and crime. This research has not yet identified specific physical structures invariably linked with deviance and never linked with nondeviance. Also, sociologists point out that a single person may act in a deviant manner at one time and a conforming manner at another, suggesting a complex role for biological determinants of deviance.

Individualistic theories reflect a medical model that likens deviance to illness in need of treatment and correction. The psychiatric and psychoanalytic viewpoints share a premise that the roots of deviance lie in early childhood experiences. The psychoanalytic perspective puts more stress on inadequate personality development, however, along with sexual conflicts and the influence of the unconscious mind. Neither view scientifically establishes the accuracy of its position, and years of psychological testing have not yet revealed a method for consistently distinguishing deviants from nondeviants based on their personality traits.

Rational choice models of human behavior reflect an economic framework, but they rely on principles consistent with a number of sociological theories (to be discussed further in Chapters 5 and 6). In a strict sense, all behavior reflects a choice, but specific cases demand varying interpretations of the term *chosen*, the degree of voluntary selection among viable alternatives in the choice, and the extent of rational decision making in such processes. Some people choose among behavioral alternatives not available to others, and even apparently completely unrestrained behavior occurs in a social context that sets conditions for choice.

Selected References

Adler, Patricia A., and Peter Adler, eds. 2000. *Constructions of Deviance: Social Power, Context, and Interaction*, 3rd ed. Belmont, CA: Wadsworth.

This impressive collection of papers on various aspects of deviance includes coverage of constructing deviant categories, processes of becoming deviant, and the nature of deviant identities. Some excellent ethnographic accounts of deviance contain many firsthand reports.

Brennan, Patricia A., Sarnoff A. Mednick, and Jan Volavka. 1995. "Biomedical Factors in Crime." Pp. 65–90 in *Crime*. Edited by James Q. Wilson and Joan Petersilia. San Francisco: ICS Press.

This review evaluates a large body of literature dealing with biological and medical factors associated with crime. It contains much current information.

Fishbein, Diane H. 1990. "Biological Perspectives in Criminology." *Criminology* 28: 27–72.

This extensive review documents biological approaches to the study of criminality. Someone interested in this subject should start with this article.

Raine, Adrian. 1993. *The Psychopathology of Crime: Criminal Behavior as a Clinical Disorder.* New York: Academic Press.

This intelligent and careful examination of the arguments for evaluating crime from a biological perspective is supported by much empirical literature. While the book focuses on the question of whether or not crime is a clinical disorder, it produces a wide-ranging and interesting discussion.

Secretary of Health and Human Services. 2000. *Alcohol and Health: Tenth Special Report to the U.S. Congress.* Rockville, MD: National Institute of Alcohol Abuse and Alcoholism, Government Printing Office.

This report provides an excellent source for information about the application of biological ideas, methods, and studies to alcohol use and heavy drinking. This report is updated and reissued periodically.

CHAPTER 5

General Perspectives on Deviance: Anomie and Conflict Theories

- Is it possible to agree on what creates a healthy society?
- Is deviance related to social change and conflict?
- Is deviance a reflection of an imbalance of values and norms in society?
- Do rules, norms, and laws better define deviance than simply observing the reactions of others to behavior?

Sociologists have not arrived at any single, generally accepted theory that explains all deviance. Some have developed several important theories, however, and one can understand deviance only by comprehending the various theoretical frameworks for studying it and the contexts in which these theories were developed.

Two of the earliest perspectives remain important, although one (social pathology) no longer has a sizable sociological following, and another (social disorganization) is undergoing revision and extension. Sociologists should study these two perspectives, however, because they set the intellectual stage for subsequent thinking and theorizing about deviance, particularly for more contemporary theories—the anomie, labeling, conflict, control, and socialization perspectives.

Sociologists began addressing deviance in the earliest stages of the discipline's development in the United States in the late 1800s. Since that time, sociological perspectives have emphasized a common theme: Society prepares the crime, and the criminal commits it. Different theories promote varying explanations of the nature of that social preparation and the context in which the criminal commits the act. The social pathology perspective emerged first, and the social disorganization viewpoint followed.

Social Pathology

Sociologists developed the concept of *social pathology* in the late 1800s, and they continued to apply it until the end of the 1930s. The social pathology theory linked deviance to pathological conditions in cities and the larger society. Its explanation cited an *organic analogy* that likened society to a biological organism. Theorists described a pathology in society as the direct counterpart to physical illness in the individual. In this way, the social pathology perspective represented a sociological counterpart to the medical model that underlies today's individualistic theories of deviance.

Social pathologists believed that they had identified some universal criteria for any healthy society, and they also pointed out pathologies or abnormalities, such as crime, that could develop in a society. They described deviant conditions as situations that interfered with the normal or desirable workings of society. This theory labeled conditions such as crime, suicide, drunkenness, poverty, mental illness, prostitution, and so forth as *deviance* because they caused known, socially pathological effects (Higgins and Butler, 1982: 20). Even today, reports on social problems commonly evoke the concept of a "sick" society.

Within the social pathology framework, deviance represented a universal disease, an unhealthy deviation from some assumed, universally applicable norm of conduct. People or situations that diverged from such widely formulated expectations represented social illnesses. The social pathologists interchanged and confused the concepts of sin, sickness, and deviance; for them, a social problem or an instance of deviant behavior was, in the end, a violation of *moral* expectations.

The social pathologists defined a universal moral standard—rather than a normative one—against which they judged the deviance or acceptability of all behavior and members of society (Meier, 1976). These social reformers wished to remove the perceived evils of society and to salvage it for the middle class (Davis, 1980: Chapter 3). Their reform efforts dealt principally with poverty, child labor, divorce, ordinary crime, drinking, and prostitution. They treated these vices as city behavior in need of change.

The social pathology theory largely coincided with the personal ideologies and social backgrounds of its advocates. Many early U.S. sociologists interested in social pathology grew up in small, rural, midwestern communities surrounded by a sense of the importance of traditional religion and a distrust of social change and city life (Mills, 1943; Schwendinger and Schwendinger, 1974). As a result, they brought to sociology an attitude perhaps best described as *sacred provincialism*—a moralistic view that stigmatized deviance and social problems and sought to eradicate them. The social pathologists saw no need to understand relative perspectives on the behavior they condemned; they branded it obviously wrong and in need of change. One observer wryly noted, "Like General Custer's, their [the social pathologists'] tactics were simple; they 'rode to the sound of the guns'" (Lemert, 1951: 3).

Evaluation

In spite of their good intentions, the views of the social pathologists do not receive serious attention today for a number of reasons.

Deviance and Illness   Rather than an individual or social illness, deviance represents a departure from norms. Today's sociologists deny that "deviance and crime are reflections of fundamental disorders in society or in the individual or both" (Aday, 1990: 64). The social pathology model cannot effectively divide individuals between normal and pathological categories (Matza, 1969: 41–46). Instead, norms determine socially pathological effects of behavior and situations. No universal and unchanging standards define this distinction, as they do for physiological pathologies.

The Universality of Norms Norms set changeable standards, and changes in norms produce corresponding changes in forms of deviance. Social pathologists failed to recognize variations in standards for deviance over time and from group to group, determined by changes in norms. This characteristic prevents effective comparisons between deviance and diseases, such as cancer, which are universally designated as illnesses. People who agree on standards for a healthy organism do not similarly agree on what constitutes a healthy society. In fact, "it is impossible to find [a criterion] that people generally accept as they accept criteria for health for the organism" (Becker, 1973: 5).

Social Disorganization

The social pathology perspective declined with the growth of the cultural relativity principle (the view that judges cultures not as better or worse than one another, but only as different), which led many to question the validity of supposedly universal values. As Matza (1969: 43) has pointed out, "the idea of diversity contested pathology." The move away from social pathology theory accelerated because of pronounced social changes following World War I and the Great Depression, along with extensive immigration, urbanization, and industrialization in the United States. As newly arrived immigrants from diverse cultures crowded into urban areas, conditions conducive to deviance rapidly developed. The change required a different explanation for deviance based on new concepts.

The need for a new framework was satisfied with the evolution of social disorganization concept, elaborated originally by Thomas and Znaniecki (1918) and by Cooley (1918). The term *social disorganization* refers to both a theoretical explanation of deviance and a state of society that produces deviance. The concept is associated with the Chicago School of sociology because most of its original supporters came from there (Bulmer, 1984).

Social disorganization theory viewed deviance as a product of uneven social change and conflict that affect the behavior of individuals. The theory emphasized the origins of society in presumed agreement forged among individuals about fundamental values and norms, leading to substantial behavioral regularity. Social organization (also called *social order*) emerges from strong internal cohesion that binds individuals and institutions in a society. This cohesion consists largely of consensus about goals worth striving for (values) and about how people should and should not behave. Disruption of consensus concerning values and norms reduces the apparent force of traditional rules, and social disorganization results.

In contrast, successful social organization, the theory held, promoted integration of customs, effective teamwork, and high social morale; these conditions led, in turn, to harmonious social relationships. A well-integrated social group would demonstrate solidarity and homogeneous, traditional standards for behavior. This image allowed little unconventional behavior and deviance, and informal controls usually sufficed to regulate behavior (Meier, 1982).

When the theory developed, the United States, particularly the urban areas, did not correspond well with this idyllic set of characteristics. Theorists interpreted mismatches as clear signs of social disorganization in much of city life. The social patterns of the urban environment appeared to nurture social disorganization, which then led to the deviant behavior. Social disorganization theorists highlighted the concentration of widely heterogeneous people (and their conflicting values and norms) in the same geographic areas; in addition, they noted that people commonly moved into and out of the areas without ever grasping the principles of neighborhood social organization or developing feelings of community (Davis, 1980: 67–69).

Lacking a real social organization, such a neighborhood could become socially disorganized, and this condition itself resulted in high rates of deviance, particularly in comparison with better organized suburban and rural neighborhoods. Within these disorganized parts of the city, certain areas, particularly inner-city slum neighborhoods, appeared to foster even more deviant behavior. According to the theory, the inner-city core featured a high rate of deviance, particularly delinquency and crime, because it lacked basic social solidarity. These theorists explained the development of deviance as a collapse of original group norms that immigrants carried with them to the United States coupled with a failure to solidify effective, new standards; as a result, social disorganization developed and spread (Kornhauser, 1978: Chapter 3).

More contemporary statements derived from this perspective continue its emphasis on community control and neighborhood patterns of interaction. Bursik and Grasmick (1993), for example, cite neighborhood instability and racial and ethnic heterogeneity as key determinants of community control over deviance. Community change and diversity limit the range of relationships in the community and reduce social cooperation, a necessary condition for social control. Instability and heterogeneity also challenge the community's supervisory and control capabilities. When coupled with rapid change, such conditions, even if they do not directly increase actual rates of deviant acts, seem to promote fear of deviance (Taylor and Covington, 1993). These situations may cause even greater consequences if people do not anticipate them. Furthermore, there is some evidence that these dynamics are equally relevant to crime and violence in rural areas (Osgood and Chambers, 2000).

One way to counter these tendencies is to develop networks of friends and neighbors to regulate activities in the community (Bursik, 2000). These networks serve to provide good information to residents and help to coordinate community cohesion efforts.

Evaluation

Sociologists came to recognize a number of problems with the social disorganization concept, and these problems contributed to a decline in its popularity.

Confusing Cause and Effect   The social disorganization perspective meaningfully describes community characteristics that may bear relationships to deviance, but it sometimes fails to distinguish the consequences of social disorganization (e.g., crime) from disorganization itself. Particularly troubling confusion has resulted from a tendency to equate social disorganization with the phenomenon it was intended to explain: deviance (Bursik and Grasmick, 1993: 34). For example, when theorists define delinquency or alcoholism as instances of social disorganization, they mix together elements of cause and those of effect. Early theorists did not clarify the concept of disorganization with particular care.

Subjectivity   A fundamental problem with social disorganization resulted from its subjective and judgmental standards, which theorists presented as an objective conceptual framework (Gibbons and Jones, 1975: 19). Social disorganization applies standards that seem almost as subjective as those of social pathology. In effect, social disorganization simply applies the familiar concept of pathology to groups instead of individuals. While it no longer branded individual members of society as pathological, it applied a similar label—*disorganized*—to communities. Observers designated phenomena as deviant and equated deviance with disorganization, based on their own interpretations rather than on findings from studies designed to determine what members of society would term *social disorganization*. Theorists usually treated social disorganization as a bad condition, judging goodness and badness by their own values and those of their own social classes or other social groups.

Deviants and the Lower Class   Concentrating on inner-city areas, social disorganization theory tried to explain deviant acts almost entirely as lower-class phenomena, effectively ignoring deviance in the middle and upper classes. This focus reflected a bias in favor of middle-class values and standards. The theory's supporters assumed higher deviance rates among the lower class because its members lived in the most disorganized areas of the city. This argument fails through circular reasoning: The lower class includes the most deviants because it is the most disorganized, and it is the most disorganized because it includes the most deviants (Traub and Little, 1994: 56). If the lower class does encompass a higher proportion of deviants than other classes do, reasons other than social disorganization may explain this predominance.

Criteria for Disorganization   Events and conditions that seem like disorganization may at times reflect competition between highly organized systems of norms and values. Subcultures often create highly organized systems of deviant behavior, as occurs in youth gangs, organized criminal syndicates, homosexual communities, prostitution rings, political corruption systems, and business crime networks. Even the norms and values of the slums display strong organization, as Whyte (1943) showed in his classic study of a slum area, *Street Corner Society*. What social disorganization theorists often described as *disorganization* actually represented diversity (Scull, 1988). Such confusion reflects the basic problem of this theory—the ambiguity of the concept itself (Liska, 1987: 88).

General Sociological Theories: Anomie and Conflict Theory

General sociological theories of deviance can be divided into two main types: structural and processual theories. Structural theories emphasize the relationship of deviance to certain structural conditions within a society, while processual theories describe the processes by which individuals come to commit deviant acts.

These types of theories also differ in scope. Structural theories address the *epidemiology* of deviance—that is, its distribution in time and space—while processual theories reflect more interest in *etiology*—that is, the specific causes of deviant acts. For these reasons, structural theories often attempt to explain such phenomena as concentration of certain forms of deviance in the lower classes (an aspect of epidemiology), while processual theories attempt to explain the conditions that lead specific people to commit deviant acts.

Because they analyze deviant phenomena at different levels of aggregation, structural theories are sometimes called *sociological theories*, while processual theories are called *social psychological theories*. Actually, both classes of theory share a good deal of overlap, since many theoretical principles have implications both for epidemiology and etiology. Still, the distinction usefully characterizes a given theory's effectiveness in accounting for these two dimensions of deviance.

The remaining sections of this chapter discuss two structural theories of deviance. Anomie theory attempts to explain differences in rates of deviance between groups. Conflict theory offers a general perspective on both the origins of laws and norms and the behavior of people who violate them.

Anomie Theory

Kenneth was from a lower-class family. His father had been a laborer, and his mother worked for a janitorial service. The family lived in a very poor urban community characterized by substantial unemployment and physical deterioration. Kenneth was a pretty good student at school, but that just seemed to feed his ambitions. He began to dream of a high-paying job and moving away from his neighborhood. He wanted to do better than his father and the rest of the adult men he saw in the neighborhood. He wanted to become a physician.

Kenneth's grades and the quality of the high school from which he was going to graduate were not good enough to qualify for anything other than community college. Kenneth began to get the picture: He was never going to be a physician, and he ought to be looking for work more in line with what he could reasonably expect. But work was scarce in his neighborhood. All the good jobs had left the central city for the suburbs. Kenneth had seen drug dealers in the neighborhood making a lot of money, driving expensive cars, and wearing good clothes. Kenneth didn't want to make money that way, but what else was there? When his girlfriend got pregnant, Kenneth realized that he wasn't going to escape his neighborhood.

Kenneth came to realize that although he might be destined to remain in his old neighborhood, his dream of high pay wouldn't have to come from a conventional job. Dealing drugs was almost commonplace in the neighborhood. He would work hard in that job to get what he wanted.

Kenneth's case is consistent with the perspective known as *anomie theory*. The anomie perspective explains deviance in a way related to the principles of social disorganization. It offers a general explanation of a number of forms of deviance, including crime, alcoholism, drug addiction, suicide, and mental disorders. Anomie theory accounts for both social organization conducive to deviance and the origins of deviant motivations, although its implications for social organization have received less attention than the other aspects (Messner, 1988). While Shaw and McKay emphasized how deviance results from social disorganization, Merton claimed that deviance results from a particular kind of social organization.

Anomie theory advances the core idea that elements in society's structure promote deviance by making deviant behavior a viable adaptation to living in the society (see Aday, 1990: 63–64). The theory describes deviance as a result of certain social structural *strains* that pressure individuals to become deviant. Sociologist Robert Merton originally proposed this view as a general theory in the 1930s (Merton, 1968: 185–248; also see Clinard, 1964: 1–56).

Modern industrial societies create strains by emphasizing status goals like material success, in the form of wealth and education, while simultaneously limiting institutional access to certain segments of society. Important status goals remain inaccessible to many groups, including the poor, the lower class, and certain racial and ethnic groups who suffer discrimination, such as blacks and Chicanos. *Anomie* develops as a result of an acute disjunctive between culturally valued goals and the legitimate means through which society allows certain groups to achieve those goals.

Cultural assumptions generally expect members to achieve success goals by *legitimate means*—through regular employment, relatively well-paid occupations, and completion of education. These channels, however, exclude certain members of society. Thus, while everyone learns to aspire to the "American dream" of financial success, in reality the social structure can provide opportunities for only a small number, so it reserves this dream for a few favored members.

Anomie is the social condition that results from emphasizing success goals much more strongly than the acceptable means by which people might achieve them. Consequently, some persons feel compelled to achieve them through *illegitimate means*, including such forms of deviance as crime, prostitution, and illicit drug selling. Others turn to alcoholism or addiction, and some fall victim to mental disorders when they fail to achieve general social goals. In attempting to explain these forms of deviant behavior, anomie theory has pointed out that official rates of deviance peak among poor people and members of the lower class, who encounter the greatest pressure for deviation and only limited opportunities to acquire material goods and higher education (Clinard, 1964).

Adapting to Strain

The anomie perspective highlights several adaptations that help members of an anomic society to cope. According to Merton, the most common adaptation leads people to conform to society's norms and avoid becoming deviant. Some individuals adapt by becoming ritualists, conforming to society's norms without any expectation of achieving its goals.

People may also choose among several illegitimate adaptations when they cannot reach valued goals through legitimate means. The adaptations relevant to the study of deviance include rebellion, innovation, and retreatism. The choice of an adaptation depends on the individual's acceptance or rejection of cultural goals and willingness to adhere to or violate accepted norms.

Some people adapt to anomie by rebelling against the conventional cultural goals that they feel unable to achieve. Through this *rebellion*, they may seek to establish a new or greatly modified social structure. They often try to set up new goals and procedures that would change the social structure instead of trying to achieve the goals established by society. Political radicals and revolutionaries practice this type of deviant adaptation.

Innovation is an adaptation to anomie that works toward culturally prescribed goals of success by illegitimate means such as theft, burglary, robbery, organized crime, or prostitution. Anomie theory describes this response as "normal" where society limits access to success through conventional means (Merton, 1968: 199). As evidence, Merton has cited the prevalence of crime and delinquency in the lower strata of society. The poor find their opportunities largely restricted to manual labor, which often carries a social stigma. Low status and income prevent them from competing for goals measured by established standards of worth. Therefore, they may likely engage in crime as an alternative way to achieve those goals.

Retreatism, according to Merton, represents an adaptation to anomie that substantially abandons the cultural goals that society esteems and the institutionalized means for achieving them (Merton, 1968: 203–204). An individual may move toward retreatism after fully internalizing the cultural goals of success but finding them unavailable through established, institutional means. Internalized pressures prevent the person from adapting through innovation, so, frustrated and handicapped, he or she adopts a defeated and even withdrawn role. The person retreats by becoming addicted to drugs or alcohol or escaping through a mental disorder or suicide. Retreatism represents a private rather than a group or subcultural form of adaptation, even though the person may have contact with others who adapt to the same conflict in a similar fashion. The retreatist compounds emotional withdrawal by also withdrawing from social life or even life itself.

Extensions of the Anomie Perspective

Cloward and Ohlin (1960) extended Merton's ideas by pointing out varying access to *illegitimate* means of achieving goals. Opportunities for illegitimate adaptations and legitimate ones vary by social strata for many of the same reasons. Lower-class and poor people encounter more opportunities than other members of society do to acquire deviant roles. They gain access to these roles largely through inner-city deviant subcultures, which also support implementation of such deviant social roles, once members acquire them.

Anomie theory explains delinquency as a result of the disparity between goals that society leads lower-class youths to want and their available opportunities. As much as they desire to reach such conventional goals as economic and educational success, many find barriers in legitimate avenues to success. Unable or unwilling to revise their goals downward, they become frustrated and turn to delinquency, if they find such norms and opportunities. A similar argument by Agnew (1992) explains that delinquency may result from an inability to avoid negative or painful situations in life. Limited opportunities may lead adolescents to feel trapped with few prospects for the future. As applied to school crime, for example, personal strain may result from a variety of negative school and interpersonal experiences by the student (Agnew, 2000).

Simon and Gagnon (1976) propose another reformulation of anomie theory. They have pointed out that Merton formulated his theory in the 1930s, a period of chronic economic depression, while the economic affluence of the 1970s appeared to produce a substantially different impact on deviance. Although those least able to gain access to success goals (e.g., lower-class groups) may feel the greatest strain, anomie theory also explains deviance among better-off people who want even more resources. Thus, anomie results, not just from absolute economic position but also from relative position. Well-paid executives may commit white-collar crimes to expand their companies' market shares or to keep their jobs.

Agnew (1995) argues that anomie theories have concentrated only on the type of strain discussed by Merton: failure to achieve positively valued goals. In addition, however, social structure may create other types of strain through actions such as removing positive stimuli (e.g., the end of a relationship, dismissal from a job) or applying negative stimuli (e.g., unpleasant school experiences, poor peer relationships). These other types of strain may accumulate over time until their compounded effects produce deviance.

✳ **In Brief**   **Crime and the American Dream**

Does crime come from the same values that have motivated Americans since the Revolution? Messner and Rosenfeld (1997) argue that the power of the "American dream" set the stage for offending through a set of shared values that include:

1. An achievement orientation that is felt in a pressure to make something of oneself. The kind of achievement most valued is found in material success.
2. Individualism in choices and personal goals. The pursuit of private gain often leads to competition with others for rewards and status.
3. Universalism holds that the promise of success is open to everyone in society regardless of background.
4. The obsession with money is prominent in the United States because this is often how people "keep score" of how they are doing.

The pressure to achieve materially is pronounced when people are pursuing individual goals of success often at the expense of others who are likely to be competitors in the process.

Source: Steven F. Messner and Richard Rosenfeld. 1997. *Crime and the American Dream*, 2nd ed. Belmont, CA: Wadsworth.

Social structure can contribute to the explanation of serious crimes in the United States. Messner and Rosenfeld (1997) fault Merton's concentration on unequal access to legitimate means of success and resulting deviant motivation. They observe additional values promoted by American society, including achievement, individualism, and universal access to success. Together, they assert, these elements of economic and social motivation define the American dream.

While such motivation systems influence behavior in other societies as well, the United States is characterized by "the exaggerated emphasis on monetary success and the unrestrained receptivity to innovation" in reaching it (Messner and Rosenfeld, 1997: 69). This combination has tended to devalue social institutions, such as the family, as compared to economic activity. Family functions, such as regulating sexual behavior and raising children, lose importance in comparison to making money. Similarly, many Americans view the country's political institutions as important only for their functions to facilitate making money (or limiting taxes). Even everyday conversation reflects the dominant position of the economic institution (e.g., everyone knows what the business term *bottom line* means). The strong economic message in the American dream and the country's preoccupation with economic activity create social conditions conducive to much serious crime.

Evaluating Anomie Theory

Explanations of deviance in terms of anomie tend to oversimplify extremely complex problems. This reason probably explains why anomie theory is almost ritually constructed and demolished every school term in courses on deviance and crime. This section points out only a few of its more important inadequacies (Clinard, 1964; Liska, 1987: 54–55).

The Assumption of Universality   Anomie theory assumes universal standards that distinguish legitimate means of pursuing social goals from illegitimate ones. This assumption is invalid, for definitions of delinquent and criminal acts vary in time and place. Deviance is a relative concept; it differs for different groups. For example, the use of marijuana, cocaine, and opium does not constitute deviance in many parts of the world today. In fact, Western societies established laws prohibiting opiate use less than a century ago. Even within the United States, criteria for acts considered deviant depend on the norms that those acts violate, which groups subscribe to those norms, and with what intensity.

Class Bias   Anomie theory also assumes that deviant behavior concentrates disproportionately in the lower class. The theory justifies this assumption by reasoning that members of the lower class experience the greatest gap between pressures to succeed and the reality of low achievement. Considerable evidence certainly suggests a disproportionate likelihood that members of the lower class and minority groups will become *detected and labeled* as delinquents, criminals, alcoholics, drug addicts, and mental patients, as compared to members of the middle and upper classes who may engage in the same behavior.

Studies of occupational, white-collar, and business crime confirm, however, that deviance occurs in the highest social strata (Clinard and Yeager, 1980), despite comparatively light pressures from an anomic society. Even conventional offenders come from groups outside the lower class. For example, middle-class people and college students apprehended for shoplifting do not fit the poverty-based explanation that describes crime as a means of breaking free from severe material deprivation.

Simplicity of Explanation   While some individuals may feel pressure that resembles the strain of anomie, many other factors clearly influence deviant acts, as well. Although some deviants undoubtedly experience frustration when they cannot legitimately achieve success goals, most deviant acts arise out of interaction with others. These audiences may serve as reference groups for deviants and provide advice that the individual values. Many deviant acts, in fact, result from efforts to fulfill role expectations rather than adaptations to disjunctives between goals and means. Anomie theory disregards deviant subcultures, deviant groups, the characteristics of urban life, and processes of interpersonal influence and control. Many forms of deviance, such as drug addiction, professional theft, prostitution, and white-collar crime, actually represent collective acts explained by association with group-maintained norms.

The Trouble with Retreatism   Anomie theory states that some deviants adapt means to goals through retreatism. This explanation lacks precision and oversimplifies a much more complex process through which alcoholism, drug addiction, mental disorder, and suicide develop. As a later section will show, people become alcoholics or mentally ill for much more involved reasons than simply to retreat from success goals. In fact, this process involves normative actions and role-playing.

Drug addicts are not retreatists in any conventional sense. Rather, they participate actively in their deviant social worlds (Hanson, Beschner, Walters, and Bovelle, 1985). Further, few physician-addicts fit the label *retreatists* or suffer from any general inability to achieve culturally prescribed goals (Vaillant, Bright, and MacArthur, 1970). The explanation of retreatism also fails to distinguish the origins of deviance from its effects. Long periods of excessive drinking or drug use may impair a person's social relations and ability to achieve certain goals in society; in this way, anomie theory may confuse cause and effect.

Alternative Perspectives   The broad, social structural system laid out by anomie theory allows only one meaning for an act of deviance. Thus, while anomie theory describes drug use as an escape from economic failure, users may cite different purposes. They may take drugs as a form of innovative behavior, such as risk taking or "getting kicks," a ritual act (such as American Indian use of peyote), an expression of rebellion, an act of peer conformity, or an act of social consciousness (as reflected in instances of medical experimentation) (Davis, 1980: 139).

Merton's theory of anomie, one of the most famous general theories of deviance, attempts to explain a variety of different deviant acts.

Bare Bones Summary of Theory

Society is composed of two structures: a value structure and a normative structure. The value structure determines culturally identified desired end-states (or goals), while the normative structure defines culturally prescribed means to achieve those goals. Socialization initially prepares every person in society to accept each of these two structures. In some societies, however, an imbalance results when certain values carry more weight than the standards for acceptable means to attain them. This imbalance creates social strain, which affects some groups more than others, and the members of those groups must adapt to the social circumstances. *Anomie* is the condition in society that results when the normative structure does not let individuals achieve valued goals. Therefore, people are not anomic, whole societies are.

Adaptations

Living in an anomic society sometimes requires adaptations to its strain. Individuals may exhibit five such adaptations:

Modes of Adaptation to an Anomic Society

	Values (Goals)	Norms (Means)
Conformity	+	+
Innovation	+	−
Ritualism	−	+
Retreatism	−	−
Rebellion	±	±

Legend:

+ = Acceptance
− = Rejection
± = Substitution

Source: From Merton, Robert K. 1968. *Social Theory and Social Structure*. New York: Free Press.

The adaptations of innovation, retreatism, and rebellion constitute deviance. Ritualism, while typically seen as odd, seldom draws social sanctions since it appears on the surface to be a form of conformity.

Policy

The policy implications of anomie include nothing less than drastic social change. In this sense, anomie theory resembles Marxist theories, which also call for social change to solve deviance problems. Altering values or norms will balance the society's goals and opportunities, reducing deviance by freeing individuals from adapting to the new conditions.

Conflict Theories

The conflict theories focus their explanations more on deviance than on deviant behavior. That is, these theories address the *origins of rules or norms* rather than about the origins of behavior that violates established standards. Most writings about deviance within the conflict perspective have related to criminality, but this set of theories appear to cite explanations relevant to a number of other forms of deviance as well (Spitzer, 1975).

The conflict view stresses the pluralistic nature of society and the differential distribution of power among groups. Some groups wield social power, according to this body of theory, so they can create rules, particularly laws that serve their own interests. In the process, they often exclude the interests of others from consideration. In this respect, the conflict perspective conceives of society as a collection of groups with competing interests in conflict with one another; those with sufficient power create laws and rules that protect and promote their interests (Quinney, 1979: 115–160).

Conflict writers display considerable interest in the origins of norms that define certain acts as examples of deviance. Some groups promote their own ideas by trying to persuade other groups of the special importance of certain norms, advocating strong sanctions for violations in these areas (Becker, 1973). Religious groups, driven by abhorrence for acts they regard as immoral, have successfully established norms expressing their strong negative attitudes toward suicide, prostitution, homosexuality, drunkenness, and other behavior (Davies, 1982; Greenberg, 1988; McWilliams, 1993). Other "moral entrepreneurs" have aligned society's norms with their own opposition to marijuana use, public nudity, and distribution of pornographic materials (Attorney General's Commission on Pornography, 1986). According to the conflict view, deviance represents behavior that conflicts with the standards of segments of society with the power to shape public opinion and social policy. This perspective regards crime, along with other forms of deviance, as a socially constructed category (Hester and Eglin, 1992).

Deviance and Marxism

Many contemporary ideas on the importance of general social conflict derive from the work of past sociological theorists such as Marx, Simmel, and more recently Coser and Dahrendorf. These authors describe society, not as a product of consensus about shared values, but as the outcome of a continuing struggle between social classes. Definitions of deviance, then, emerge from class conflict between powerful and less powerful groups. In fact, most writers who apply conflict theories to deviance and crime issues identify themselves as Marxists, although Marxists disagree on the extent to which crime should form the basis of their common view of society (see O'Malley, 1987).

Marx himself viewed society primarily as an uneasy relationship between two groups with incompatible economic interests: the bourgeoisie and the proletariat. The bourgeoisie act as the society's ruling class. These wealthy members of society control the means of economic production and exert inordinate influence over society's political and economic institutions to serve their own interests. The proletariat, on the other hand, fills the ranks of the ruled members of society—workers whose labor the bourgeoisie exploit.

The state acts, not as a neutral party to balance the inevitable conflicts between the two groups, but mainly as a shield to protect the ruling class against threats from the ruled masses. It works primarily to foster the interests of the rulers. Marx believed that developing capitalism would force proliferation of criminal laws to act as important mechanisms by which the rulers could maintain order (Beirne and Quinney, 1982; Cain and Hunt, 1979). First, he explained, laws prohibit certain conduct, particularly conduct that might threaten the rulers' interests. Second, laws legitimize intervention by society's social control apparatus, including the police, courts, and correctional systems; these forces operate against the ruled masses, whose behavior is most likely to violate laws established by their antagonists. In this way, Marx explained, criminal law comes to side with the upper classes against the lower classes. His conception of social conflict is ultimately tied to the economic relationships of capitalism. Marx described an inevitable trend toward alienation between workers and owners, culminating in a major division based on control of the means of production and fed by differing economic interests between those own productive resources and those who work for the owners (Inverarity, Lauderdale, and Feld, 1983: 54–99).

Other Conflict Theorists

Conflict-based explanations of crime and deviance in general have come from Vold (1958); Quinney (1980); Turk (1969); Taylor, Walton, and Young (1973); Platt (1974); Takagi (1974); Chambliss (1976); and others. Despite significant theoretical differences among these writers, they generally share a view of criminal behavior as a reflection of social power differentials: Society defines crime as a function of social class position. Since the elite and the powerless have different interests, measures that benefit the elite work against the powerless. Conflict theorists see nothing surprising, therefore, in official statistics that show substantially higher crime rates in the lower classes than in the more privileged, elite segments of society. Since the elite control the law-making and law-enforcement processes, the goals and provisions of criminal laws coincide with elite interests (Krisberg, 1975). Laws relating to theft, enacted by people in positions of power, protect the interests of who stand to lose the most from theft. Conflict theorists see no social accident, moreover, when offenders who violate these particular laws invariably come from the lower, less powerful classes, who face the greatest temptation toward theft.

Conflict theorists often regard crime as a rational act (Taylor, Walton, and Young, 1973: 221). They explain that thieves steal because social conditions created by an inequitable distribution of wealth force them to do so. Conflict theory views business and white-collar crime as activity designed to protect and augment the capital of owners (Simon and Eitzen, 1987). From this perspective, organized crime represents a rational way of supplying illegal needs in a capitalist society (Block and Chambliss, 1981). Noting that a relatively weak commitment to the dominant social order often accompanies membership in the lower classes, one analysis combined a conflict orientation with a version of control theory (detailed in Chapter 6) to explain persistent crime and delinquency among working-class youth. Colvin and Pauly (1983) have argued that economic repression of workers creates alienation from society; in turn, this alienation, apparent in weakened bonds to the dominant social order, produces criminality.

Conflict theory ascribes a wider role to law than simply protecting the property amassed by the elite. It gives the ruling class a tool by which it can exercise many kinds of control over the ruled. In addition to protecting property, law represses other political threats to the elite through the coercive response of the criminal justice system. The elite establish this important protection for their position, recognizing the inevitability of conflict over opportunity and power and the potential of law to serve the interests of some groups to the detriment of others. In many respects, political criminality amounts to membership in a class in which the lack of opportunities invite them to challenge the authority of the powerful group (Turk, 1984). Under such circumstances, some persons in the lower classes may feel they have little to lose in striking against the system that has denied them greater opportunities. Conflict theorists perceive crime as an unchangeable feature of capitalist society. The United States, one of the world's most advanced capitalist societies, suffers from crime rates among the highest in the world. The state, organized to promote capitalism, also serves the interests of the dominant economic class, the capitalist ruling class. Conflict theorists describe recent developments in criminal justice, including rising imprisonment rates, stiffer penalties, and growing interest in retribution, as reflections of the influence of the bourgeoisie (Horton, 1981). Also, access to criminal opportunities varies by class; unable to engage in embezzlement or business crime, the poor burglarize and mug instead.

Conflict theory also provides an explanation for crime control policies. Beckett and Sasson (2000), for example, argue that the war on crime and war on drugs represented an attempt by conservative legislators and other economic elites to oppose the expansion of the welfare state in the 1960s and 1970s in the United States. By emphasizing a "get tough" approach to crime, these forces paved the way for dramatic cuts in government spending for poverty relief and a massive expansion of the criminal justice system. Crime had become politicized (see also Chambliss, 1999).

Social Threat

By applying a variation of conflict theory, Liska (1992) has offered an explanation of social control efforts. Conventional wisdom holds that crime control efforts respond to crime; that is, as crime increases, so do crime control efforts. The social threat hypothesis denies this direct relationship, explaining enthusiasm for crime control as a function of perceived social threats in society. These threats may come from behavior defined as undesirable or from people or groups defined as inherently dangerous, regardless of their behavior.

Certain social control efforts represent direct measures against perceived threats. Some authors trace the origins and growth of mental hospitals, for example, to the desire for systems of confinement that would control lower-class urban residents (Foucault, 1965) and immigrant groups (Davis and Anderson, 1983). Many residents of mental hospitals were confined there without clinical diagnoses of mental disorders, although they may qualify as deviants in the sense that their indigent behavior violated dominant norms that demanded they maintain positions as employed and productive members of society.

The social threat hypothesis is also consistent with the continuing increase in U.S. correctional populations at a time of relatively stable or even declining crime rates. Recent concerns over drug use have led to a significant increase in the proportion of inmates incarcerated on drug charges. In 1980, 57 percent of all state and federal prisoners were incarcerated for violent crimes. Among the rest, 30 percent had been convicted of property offenses, 8 percent for drug violations, and 5 percent for public-order crimes (Beck and Gilliard. 1995). By 1993, offenders convicted of violent crimes accounted for only 45 percent of state and federal prisoners, while 26 percent of the inmates entered prison for drug crimes, 22 percent for property offenses, and 7 percent for public-order crimes. In 13 years, the composition of inmate populations changed drastically because changes in sentencing practices increased the priority of punishing drug offenders. Today's prisons incarcerate more inmates for violent crimes than for any other type of offense, but their proportion has dropped since 1980, and the number of drug offenders imprisoned has more than tripled during this time. The new drug laws contributed substantially to prison overcrowding throughout the United States.

Left Realism

Another group of theorists have modified the traditional conflict perspective to present what they describe as a more realistic view of the factors involved in crime. Many conflict-based explanations view criminals as members of repressed classes reacting against an unjust social, economic, and political order; in this way, they support images of criminals as romantic figures. Yet, conventional offenders often do not fit the mold of Robin Hood rebelling against an oppressive status quo. Further, these theorists observe, the Marxist explanation of crime recognizes political change as the only major agent of crime reduction. But crime derives from everyday behavioral motivations as well as political ones. Conventional crimes, committed largely by members of the lower and working classes, often victimize their peers. The grim reality of street crime in impoverished neighborhoods does not coincide with an image of the criminal as a social rebel.

Left realism advocates an approach to crime control that recognizes this reality (see Beirne and Messerschmidt, 2000). In this view, crime emerges when four components intersect: offenders, victims, the state, and public opinion. This idea is called "the square of crime" (Matthews and Young, 1992: 17–19). Like other conflict-based perspectives, left realism acknowledges the state's active role in the process of criminalizing certain behaviors. The state enacts laws that reflect established political and economic inequalities and interests. Public attitudes toward crime can shift legal and correctional priorities, making them another part of the general crime process. The ability to influence public opinion provides an important determinant of the legal order, as does the ability to influence actions of the state crime control apparatus.

In developing its realistic approach to crime, left realism recognizes both the danger of crime and its origins in conditions of social and economic inequality (Young and Matthews, 1992). This position implies that reductions in this inequality, and more general promotion of social justice, will substantially reduce crime. Recognizing the reality of crime leads to realistic solutions, according to this theory, some of which operate in local jurisdictions. Crime control depends not only on changes to a society's political structure but also on immediate and short-term measures, including judicious action by the criminal justice system.

Left realists recognize that police cannot control the conditions that bring about crime, but they perform essential actions within any systematic crime control program. The police should remain accountable to the community, however, and involve the public in their activities as much as possible (Matthews, 1987). Citizen patrols could augment policing efforts in high-crime areas, for example, but they would have to cooperate actively with the police.

Left realists advocate a social action program with options other than marginalizing offenders and warehousing them in prisons (Lea and Young, 1986). Instead, they assert, prisons should house only the most serious offenders—people who have shown they cannot live in society without harming others—playing an otherwise minimal role (Lowman, 1992: 156). Offenders should remain in the community as long as possible, controlled by sanctions such as probation and emphasizing community service and restitution.

Evaluating the Conflict Model

The conflict model has made an important contribution to the study of deviance. It has focused attention on the role of the political, economic, and social structure, particularly laws enacted by the political state, in defining deviance. Conflict theorists point out some basic problems and contradictions of contemporary capitalism. They note that definitions of crime reflect society's values and not merely violations of those values. Conflict theories highlight the basic issue in deviance: society's translation of values into laws and other rules. The left realist perspective has offered concrete suggestions for crime control within existing political structures. Despite these strengths, however, several problems still limit the conflict view.

Explanation of Rules or Behavior? Conflict theories offer little information about the process by which an individual comes to commit crimes or becomes a deviant. These theorists raise pertinent questions about the origins of laws and norms, but they work primarily to explain how society forms and enforces certain rules and laws. When conflict theories do address individual actions, they assume that deviance is a rational and purposive activity, for example, an expression of group conflict such as a hate crime or political protest (Akers, 2000: 184). Conflict theories ignore the socialization process and assume that political considerations alone motivate deviants.

Who Benefits?   One may doubt the conflict premise that one particular group devises and enforces all laws for its sole advantage. The conflict approach may offer its best insights in areas where people disagree about the deviant characters of certain acts, such as political crime, prostitution, use of certain drugs, and homosexuality. It may give weaker explanations of acts that spark no such disagreement.

In fact, however, U.S. residents generally agree about the illegal character and seriousness of most acts presently defined as conventional or ordinary crimes (see Hamilton and Rytina, 1980). Laws against homicide, robbery, burglary, and assault benefit all members of society, regardless of their economic positions. Any statement that such laws disproportionately benefit the elite neglects the fact that most of these offenses victimize other poor, lower-class urban residents, not members of any elite, however broadly defined. Although the elite have more property to lose from theft or robbery, most losses from these crimes afflict those least able to afford the cost.

On the other hand, one aspect of the criminal justice system's operations reveals considerable validity in the conflict perspective. Clearly, offenders who commit conventional crimes (generally members of the lower classes) are much more likely to be arrested and convicted, and they serve longer prison terms, than white-collar and business criminals (Reiman, 1984).

Powerful Groups and Social Rules   Conflict theories rely on an overly broad assumption that powerful groups dictate the contents of the criminal laws, as well as other rule-making processes, and their enforcement solely to promote their own interests. A variety of groups, each with its specific interests and concerns, contribute to lawmaking. Powerful groups provide substantial input, but they strongly influence legal structures in all social systems, capitalist, socialist, or communist. By penalizing violators, the criminal law always defends the established order and those who hold power within it. Conflict theories provide little help when they say that those who have something to gain from the rules help to make them. This position fails to answer important questions about the characteristics of a society's powerful groups, the process that translates some norms but not others into law, the selective enforcement of those laws, and differences in lawmaking and enforcement processes in different economic and political systems.

Law and the Causes of Behavior   Conflict perspectives hold that criminal law, supported by certain interest groups, ultimately causes criminal behavior by defining specific acts as crimes. One cannot logically say, however, that the law's standards for deviance induce people to commit deviant acts. In a comparable criticism of the labeling perspective, which generates similar confusion with its emphasis on rule making and deviance by interest groups, Sagarin (1975: 143–144) observes that "without schools, there would be no truancy; without marriage, there would be no divorce; without art, there would be no art forgeries; without death, there would be neither body-snatching nor necrophilia. Those are not causes; they are necessary conditions."

Society could free itself from crime simply by eliminating laws that prohibit certain behaviors. This fact does not imply, however, that the existence of such a law accounts for the prohibited behavior.

Theory as Ideology The collapse of the former Soviet Union and its satellite countries has increased skepticism toward the Marxist approach to social order. Acceptance of conflict theories, particularly the Marxist perspective, depends ultimately on acceptance of its ideological base. Some believe that reconstruction of Marxism would effectively address and solve contemporary problems (see Wright, Levine, and Sober, 1992), but they have not yet convinced everyone. Other sociological perspectives are not completely free from ideology, but conflict theorists emphasize measures to combine theory with practice in a socialist framework. This effort explicitly promotes the political principles that underlie their explanatory scheme.

Full development of conflict theories would require movement toward a socialist society. These theories lose much of their appeal if members of society decline to dissolve capitalism and carry out a transition to socialism in order to eliminate deviance and crime. Conflict theory refuses merely to analyze the conditions under which deviance develops; it also demands a willingness to implement political changes to revise social conditions. Not content with social science as a means to discover the characteristics of the real world, a conflict theorist expresses a commitment to a political ideology that claims the power to eradicate deviance. Appeals to scientific evidence alone fail to reach the ideological component of conflict theories (Gouldner, 1980: 58–60). Finally, the potential for an intellectual and political revival of Marxism remains to be seen.

Summary

General theories of deviance attempt to explain virtually all instances of deviance. They provide frameworks for understanding deviance regardless of its frequency or form (e.g., crime, mental disorders, suicide). Early theoretical perspectives led to two major theories of general deviance: social pathology and social disorganization theories. The social pathology perspective likened society to a biological organism and deviance to some illness or pathology afflicting that organism. It represented a sociological counterpart to the medical model that some psychologists and psychiatrists advanced to explain deviant acts, as discussed in the previous chapter. The social disorganization perspective sought the meaning of deviance in malfunctions of local community institutions. Each of these views made important contributions to the development of subsequent theoretical insights on deviance.

Anomie theory emerged as a major structural theory of deviance more than 50 years ago. This perspective locates the cause of deviance in an imbalance of values and norms in society that emphasizes the desirability of culturally determined more strongly than the availability of socially approved means to achieve those goals. Individuals and groups in such a society must adapt to this mismatch, and some of those adaptations may lead to deviance. Groups that experience unusually high strain from this social imbalance (e.g., members of the lower class) are more likely than others to make deviant adaptations.

Conflict theories, another group of explanations for deviance, have developed their most detailed applications in explanations of criminality, although the same principles can address other forms of deviance as well. These theories concentrate more on the origins of norms, rules, and laws than on the origins of specific rule-breaking behaviors. Socially powerful individuals and groups influence and shape public policy by establishing laws. Elite groups define the contents of law and the responses of the criminal justice system to offenders. Other social norms may originate in the same way.

Some groups may develop sufficient power to raise their own norms to dominate the society's standards for behavior. This process accounts for such norms as those that proscribe homosexual relations, overindulgence in alcoholic beverages, and suicide, usually citing moral or religious reasons. Recent conflict theories have sought to explain crime control. One of them, social threat hypothesis, suggests that social control comes about in reaction to perceived threats; another, left realism, concentrates on politically feasible measures to reduce crime within a conflict perspective.

Selected References

Chambliss, William J. 1976. "Functional and Conflict Theories of Crime: The Heritage of Emile Durkheim and Karl Marx." Pp. 1–28 in *Whose Law? Whose Order? A Conflict Approach to Criminology*. Edited by William J. Chambliss and Milton Mankoff. New York: John Wiley & Sons.

This paper offers a concise comparison between functional and conflict approaches to crime and law. It represents essential reading on applications of these general perspectives to deviant behavior.

Clinard, Marshall B., ed. 1964. *Anomie and Deviant Behavior*. New York: Free Press.

This collection of papers explores the theoretical and empirical status of anomie theory and its applications to forms of deviance including alcoholism, crime, gang delinquency, mental disorder, and drug addiction.

Merton, Robert K. 1997. "On the Evolving Synthesis of Differential Association and Anomie Theory: A Perspective from the Sociology of Science." *Criminology*, 35: 517–525.

This paper represents a recent statement by Merton on the concept of anomie. Written from a sociology of science approach, it addresses how and why macro and micro theories might be integrated.

Messner, Steven F., and Richard Rosenfeld. 1997. *Crime and the American Dream*, 2nd ed. Belmont, CA: Wadsworth.

This short, readable book examines crime from the anomie-strain perspective. The authors extend Merton's argument on the importance of anomie. Their many interesting observations about crime and American society introduce ideas beyond the specific theory discussed in the chapter text.

Quinney, Richard. 1980. *Class, State and Crime*, 2nd ed. New York: Longman.

This book applies the conflict perspective to the problems of crime and crime control. Many of Quinney's observations on crime could be extended to represent the conflict position on other forms of deviance as well.

Taylor, Ian, Paul Walton, and Jock Young. 1973. *The New Criminology: For a Social Theory of Deviance*. London: Routledge and Kegan Paul.

This venerable but still valuable book discusses and evaluates major theoretical arguments within a conflict perspective. The social theory of deviance promised in the subtitle never quite materializes, but the reader takes a very rewarding journey through and around other theories.

CHAPTER 6

Labeling, Control, and Learning Theories of Deviance

The anomie and conflict perspectives establish general theories that attempt to explain all instances of deviance, but other sociological perspectives limit their attention to more specific behaviors. This chapter discusses and evaluates attempts to understand particular kinds of deviance through labeling, control, and learning theories. Each of these theories seeks to explain only certain types of deviance, or even one particular form. For example, labeling theory attempts to explain secondary deviation, which represents role behavior for the deviant, and the society's reactions to deviance. The theory does not attempt to explain primary deviance.

Similarly, control theory gives a systematic explanation for only one form of deviance (crime). Subsequent chapters will identify other theories for specific forms of deviance, such as alcoholism, drug addiction, and suicide.

A single theory can incorporate both structural and processual elements. Learning theory, for example, purports to explain not the processes through which specific individuals come to commit deviant acts, but also the structural causes of differences in rates of deviance among groups. This chapter discusses one example of learning theory, Sutherland's theory of differential association.

To say that a theory is "limited" to certain instances of deviance does not mean that it might not apply to other forms, as well. Some observers describe deviance as a general tendency rather than a specific behavior pattern, so an explanation for one form might offer insight about other forms, as well (Hirschi and Gottfredson, 1993; Osgood, Johnston, O'Malley, and Bachman, 1988). The truth of this assertion remains to be seen. In any case, instructors may prefer to limit the scope of their attention to certain theoretical perspectives as a useful way to target their teaching.

Labeling Theory

Twelve-year-old Jason loved sports cards. His favorite team was the Green Bay Packers and his favorite player was Brett Favre, the quarterback. He had been successful in getting all of Favre's cards, except his rookie year card. Then one day, he experienced good news and bad. The good news was that he saw the rookie card in a store; the bad news was that it cost much more money than he had. Although he knew it was wrong, Jason took the card, put it in his pocket, and tried to leave the store. The clerk saw what he did and stopped him. The clerk called the police, and Jason was charged with the crime of shoplifting.

Jason was deeply embarrassed. He had never done anything like this before. He was taken to juvenile court, where the judge declared him delinquent and placed him on probation. Jason was resentful. He knew he had done something wrong and was sorry about that, but people began to treat him differently. His parents were more suspicious of him and demanded to know more about his whereabouts. It seemed as if his teachers were not as pleased about his performance in school as they had been. His friends became a little more distant and sometimes called him "juvvy" because he had been in juvenile court. He felt he was less welcome in groups at school and among fellow band members.

Jason's case is consistent with what labeling theory would predict. Labeling theory offers a *processual* explanation for deviance. Recall from Chapter 5 that processual theories concentrate on the social psychology of deviance, that is, the conditions that bring about deviant acts by individuals and small groups. Labeling theory, also called the *interactionist* perspective, focuses on the consequences of deviants' interactions with conventional society, particularly with official agents of social control. The major models within this perspective are based on the writings of Lemert (1951; 1972), the most recent of them now over 20 years old. Additional contributions came from similar ideas expressed by others, many of them before Lemert's work, particularly W. I. Thomas (Thomas and Znaniecki, 1918), Mead (see Blumer, 1969: 62, 65–66), Tannenbaum (1938), and Schutz (1967). Over the years, many theorists have contributed to the literature on labeling, including Becker (1973), Garfinkel (1967), Goffman (1963), Scheff (1984), Erikson (1962), Kitsuse (1962), Schur (1979), and others (e.g., Plummer, 1979).

The labeling perspective devotes little effort to explaining why certain individuals begin to engage in deviance. Rather, it stresses the importance of the process through which society defines acts as deviant and the role of negative social sanctions in influencing individuals to engage in subsequent deviant acts. These theorists shift their attention away from individuals and their actions and toward the dynamics of social definitions that label particular activities or persons as deviant. They also focus on the consequences of committing deviant acts. The theory's emphasis on the developmental process leading to deviance seeks to detail a sequence with "varying stages of initiation, acceptance, commitment, and imprisonment in a deviant role are primarily due to the actions of others" (Traub and Little, 1999: 376). This analysis of the process highlights the reactions to individuals or their actions by others (termed *definers* or *labelers*) or on acts perceived negatively by these *evaluating others.*

Therefore, labeling theory incorporates two important components: a particular conception or definition of deviance (the reactivist conception) and a concern with the consequences of social control efforts (the theory of secondary deviation).

Deviance as Reaction

Labeling theorists claim that one can understand the relative and ambiguous concept of deviance only by examining the reactions of others to the behavior.

Becker's definition of deviance may have become the best-known labeling definition. Becker (1973: 9) has described deviance as a "consequence of the application by others of rules and sanctions to an 'offender.' The deviant is one to whom the label has successfully been applied; deviant behavior is behavior that people so label." The crucial element of this definition is society's reaction to an act, not the act itself. Labeling theorists determine deviance, not by any reference to norms, but by reference to the reactions (notably sanctions) of the act's social audience. In this view, deviance does not evoke social control efforts, but the reverse: Social control efforts "create" deviance by defining acts this way and making these standards known to others (see Rubington and Weinberg, 1996).

Labels That Create Types of Deviants

In emphasizing the label society places on deviants, these theorists shift their interest from the origin of the deviant behavior to (1) characteristics of the societal reactions experienced by labeled individuals and (2) consequences of this label for further deviation by those individuals. An official label that tags a person as delinquent, criminal, homosexual, drug addict, prostitute, or insane may have serious consequences for further deviation. Schur (1971: 27) has asserted that the emphasis on labeling decreases the importance of efforts to distinguish the causes that induce individuals to offend, instead calling for a more intensive study of the processes that have produced the deviant outcomes. Lemert (1972: ix) particularly stressed this viewpoint and its consequences for deviance, in the process moving labeling theory a big step away from older methods of sociology, with their reliance on the idea that deviance triggers social control. "I have come to believe that the reverse idea [that social control triggers deviance] is equally tenable and the potentially richer premise for studying deviance in modern society."

According to labeling theorists, the deviant label may produce a basic change in the labeled individual's perception of the deviance. They distinguish *primary deviance*—behavior that arises for a number of reasons, including risk-taking, chance, and situational factors—from *secondary deviance*, which Lemert has described as behavior, or a role based on behavior, intended as a defense against or adjustment to the problems caused by the label (Lemert, 1951: 76). The label produces a deviant social role and confers a social status on the deviant. Two observers explain:

> The idea of a master status is the end result of the entire model. It refers to a dominant status either socially conferred by rule enforcers and the audience or individually by the deviant actor. The emphasis here is on the development of a deviant self-concept and the consequent probability of a deviant career. (Dotter and Roebuck, 1988: 28)

This subtle process produces its effects over an extended period. For example, an individual who acts eccentrically may gain an identity as mentally disturbed if he or she receives formal treatment from a psychiatrist or enters a mental hospital; a drinker labeled a drunk by his or her family may drink further to cope with this rejection. In each case, the person develops the master status of deviant (that is, carries the label *mental patient* or *alcoholic*), and others react toward the individual in a manner consistent with that status. Any other status that characterizes the person becomes secondary, less important than this central, identifying trait.

Labeling theorists cite a reinforcing sequence of further deviance for a person labeled by an arrest, confinement in a mental hospital, or other action by an official agency. A spiral of events and reactions lead to further deviance as a response to the label's stigmatizing effect. In a sense, labeling someone as a deviant may result in a self-fulfilling prophecy. The persons may continue to commit acts of deviance associated with the label, perhaps even developing a deviant career through this secondary deviance. The label tends to exclude the person from participation in conventional groups, moving him or her toward an organized deviant group instead.

Kitsuse (1980) has suggested further that some deviants rebel against their labels and attempt to reaffirm their self-worth and lost social status. These *tertiary deviants* may join social movements to combat negative images associated with their behavior, in effect denying that their actions make them deviants. Kitsuse distinguishes these people from secondary deviants, because tertiary deviants actively protest their labels, whereas secondary deviants passively receive their status. While secondary deviants adapt to the labeling process, tertiary deviants "reject the rejection" and attempt to neutralize their labels. Recent activities by groups of homosexuals, prostitutes, and physically handicapped people reflect this movement toward tertiary deviance.

The transition from primary to secondary deviance may require a lengthy process, during which many labelers apply varying labels to the same person or behavior. Family members may defend behavior considered odd by some, describing it as simple eccentricity rather than evidence of a mental disorder. Reactions over a period of time by school officials, employers, and psychiatrists may move such a person to the status of secondary deviant. Similarly, a person adopts a homosexual role through a complex process that involves the acquisition of a homosexual identity early in life and perhaps reactions by family, friends, and others over time.

The Power to Label

The labeling perspective has also promoted a useful focus on the significant role played by social power in determining standards for deviance. Certain groups may influence the criteria that guide administration of criminal law, for example, through agents such as the police and courts. Observers have documented similar processes in other areas of deviance. For example, the purposeful actions of certain agents of social control have designated the kinds of people considered mentally ill and society's reactions to them, in either institutional or community settings.

By emphasizing the importance of rules, social control efforts, and the effects of stigma on deviants, labeling theorists explore the nature of deviant labels— who creates the rules that define deviance and how society singles out certain individuals and groups for labeling. In short, they deal with the power and politics of deviance. As part of this effort, Schur (1980) has advocated a concept of deviance in terms of "stigma contests" between different groups who promote competing rules and definitions of deviance; society's determination of behavior as deviant always reflects the relative power of these groups. Thus, the least powerful groups most often carries labels of deviants, including drug addicts, alcoholics, mental patients, and, Schur (1984) has argued, even women.

Labeling theorists claim that powerless groups appear disproportionately in official statistics on deviance because class bias influences the actions of social control agents. In addition, relatively powerful people and groups define the behavior of other, less powerful, groups as deviant, further inflating their numbers in counts of deviants. Furthermore, since such statistics underreport the distribution of deviance in powerful segments of society, characteristics other than deviant acts appear to elicit official deviance labels; these traits may include the deviant's age, sex, race, or social class along with characteristics of the social control agency (Box, 1981). This may be one reason for the formation of the juvenile court: to reduce the stigma of delinquency among young offenders. Whether the juvenile court does in fact reduce stigma and labeling is another matter (Triplett, 2000).

The ability to attach labels has important implications for the ability to remove or challenge labels. Some extremely powerful labels completely overshadow everything else known about the people who carry them. When learn of the label "mentally ill" attached to someone, they naturally interpret all behavior from that person in light of the label. "Child abuser" is another powerful label.

> . . . [T]he personal, social, and legal stigma resulting from designating this label is enormous. Once the impression has been formed that a person is a child abuser, the expectation exists that he or she will continue to be abusive. Moreover, there is little a person can do to remove this label. It exists as part of a permanent record that can be recalled whenever a person's childcare capacity or moral standing are questioned. (Margolin, 1992: 67)

Evaluating Labeling Theory

The popularity of the labeling perspective and the intuitive appeal of many of its ideas have not protected it from a number of criticisms. Many critics cite problems with imprecise statements in labeling theory. Ambiguity still obscures key points in the theory.

Where Is the Behavior?   A deviance label does not create the initial behavior it stigmatizes. By ignoring this fact, labeling theory denies the reality of the first deviant act and the basis for society's reaction to it. Ultimately, that reaction results from a violation of some normative standard or expectation. "People often commit acts that violate the law or social norms for reasons that have nothing to do with labels that others apply to them" (Akers, 2000: 126). In fact, most people who commit deviant acts do not carry the stigma of official labels, despite behavior such as stealing, homosexuality, marijuana use, drunken driving, or business or political crimes.

Who or What Labels?   Three groups could conceivably label a deviant: official agents of social control, society at large, and the immediate group to which she or he belongs, along with the significant others who supply cues about role performance. A full explanation must define specifically which group applies a label, since labels from one group may differ substantially in significance from those of others. In general, labeling theorists have focused almost exclusively on the labels applied by formal agencies, according only minor importance to informal sanctions imposed by family, friends, employers, and others. While this concentration may reflect an assumption that formal sanctions bring comparatively significant problems for labeled deviants, the theory ignores evidence that informal sanctions powerfully stigmatize deviants, too. To offer truly useful insights, labeling theorists "must provide more information on the process by which informal and/or formal labels actually effect delinquent or criminal behavior" (Wellford and Triplett, 1993: 18).

Similarly, the labeling perspective neglects the context in which labels are defined and administered. The sanctions implied by a deviant label depend on the nature of the act, the perceived seriousness of its consequences, and the extent of the actor's responsibility for it, among other factors (see Felson and Tedeschi, 1993). As a result, the labeling process is likely to produce variable and uneven effects as perceptions of these matters vary. Labeling theory does not explain this complexity.

How Much of a Label?   Writers promoting reactivist conceptions of deviance have only vaguely explained how much societal reaction constitutes effective labeling. In other words, they have not clearly stated how harshly society must react to label a person as deviant. One might ask whether labels come only from formal social control agencies and, if so, how severely these officials must penalize a deviant act to successfully label a deviant person. Do arrest, conviction, imprisonment, mental hospitalization, and so on constitute effective labels, and how do informal social sanctions, such as those exercised by family and neighbors, influence these labels? Those who define deviance by reactivist criteria have not specifically detailed the kinds of reactions that identify behavior as deviant (Gibbs, 1996: 66).

Who Is Deviant?   As a major consequence of its principles, a reactivist conception of deviance largely restricts the concept of deviance to the lower classes, since the behaviors that trigger labels occur far more frequently in this group than in other classes. Labeling theory allows influential people to engage in disruptive and destructive acts, while largely escaping the deviant label through the protection of their social and economic positions.

> Because of these biases, there is an implicit but very clear acceptance by [labeling theorists] of the current definitions of "deviance." It comes about because they concentrate their attention on those who have been successfully labeled as

"deviant" and not those who break laws, fix laws, violate ethical standards, harm individuals and groups, etc., but who either are able to hide their actions or, when known, can deflect criticism, labeling, and punishment. (Liazos, 1972: 109)

What Are the Effects of Labeling?   Labeling theory contradicts expectations derived from deterrence or rational choice perspectives. Available evidence denies the truth of labeling theory claims that formal sanctions, even when severely and frequently applied, always strengthen deviant conduct patterns (Akers, 2000: 127–128). Labeling theorists argue that people assume deviant roles primarily because others have labeled them as such and because associated sanctions prevent them from resuming nondeviant roles in the community. Such claims fail to acknowledge other possibilities; in fact, labeling is not a necessary and sufficient condition for all secondary or career deviance, although they show a strong association in certain instances.

This argument calls for a distinction between *achieved* and *ascribed* rule breaking (Mankoff, 1971). In ascribed deviance, social rule breaking takes the form of particular, visible physical disabilities, such as mental retardation; achieved deviance involves activities of the rule breaker, such as those of a professional criminal. As Mankoff (1971: 207) says, "Ascribed deviance is based upon rule-breaking phenomena that fulfill all the requirements of the labeling paradigm: highly 'visible' rule-breaking that is totally dependent upon the societal reaction of community members while being totally independent of the actions and intentions of rule-breakers."

Severely crippled, blind, obese, spastic, mentally retarded, and facially disfigured people may become targets of labeling because others recognize undesirable differences from what they regard as normal or appropriate characteristics. The stigma implied by such a label affects the social identity of the labeled person. The physical condition constitutes a necessary condition for labeling and career deviance (see generally Stafford and Scott, 1986).

On the other hand, many forms of achieved deviance may develop without any labeling. Achieved deviants may embark on deviant careers without prodding by agents of social control (Mankoff, 1971: 211). Many choose deviance as a way of life, not because any stigma forces this choice; they simply do not wish to conform.

A deviant career, or secondary deviation, can develop in the absence of arrest or other sanctions, a conclusion amply supported in studies, for example, of embezzlement (Cressey, 1971) and homosexuality. Most people who develop homosexual identities do so independently of contact with police officers or psychiatrists (Langevin, 1985). Delinquent gang behavior may occasionally become highly sophisticated in a person who experiences minimal or no contact with the law. Offenders who maintain legitimate occupations, such as white-collar criminals, may pursue careers in deviance without ever experiencing sanctions and often without fear of future sanctions (Coleman, 1985). Women alcoholics often keep others unaware of their problem drinking, and many of these women perpetuate entire drinking careers without receiving labels (Wilsnack, Wilsnack, and Klassen, 1987). Despite physicians' high rate of narcotic addiction, they seldom face detection and labels as drug addicts (Vaillant, Bright, and MacArthur, 1970). Thus, experience

◆ *Quick Summary* | **Labeling Theory**

Labeling theory, also called the *interactionist perspective*, combines two distinct components: a definition of deviance and the theory of secondary deviation.

Bare Bones of the Theory

The reactivist conception of deviance, championed by Howard Becker, determines deviance as any behavior labeled or sanctioned by others. The theory of secondary deviation, described most elegantly by Edwin Lemert, holds that patterned deviance (or sustained or career deviance) arises in response to sanctioning efforts by agents of social control. Labeling theory identifies two kinds of deviance: primary deviation, or casual and occasional acts not supported by the individual's self-concept, and secondary deviation, acts committed frequently that are reinforced by the actor's self-concept. The secondary deviant—someone labeled, perhaps repeatedly, as a deviant—comes to use deviance as a defense mechanism or an expression of role behavior.

Supporters have applied the theory of secondary deviation to explain a number of different forms of deviance, including homosexuality, drug use, crime, alcoholism, radicalism, and mental disorders. In each instance, the theory proposes the same general process of labeling.

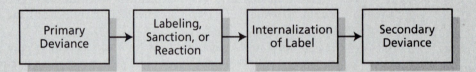

The theory relies unmistakably on interactions between deviants and others in society. Individuals become deviant through interactions with social audiences to their deviance,

surely does not justify a statement that "most deviantness is ascribed, not 'achieved'" (Schur, 1979: 261).

Labeling theorists have so far failed to specify the conditions under which deviant labels produce further deviance. Negative sanctions sometimes produce secondary deviance, and sometimes they produce deterrence. The effects of labeling also probably work in concert with other conditions, such as the individual's self-concept. Kaplan and Damphousse (1997), for example, found that students with low self-esteem (high self-rejection) relatively frequently engaged in subsequent deviance after experiencing sanctions. A clear picture of the relationship between sanctions and behavior requires more attention to the conditions under which labels result in particular outcomes.

In sum, the empirical literature lends little support to the sweeping generalizations of the labeling perspective. Evidence does not confirm the theory's predicted represented largely by the actions of social control agents. Agents such as police officers, psychiatrists, alcoholism counselors, priests or ministers, and teachers may apply society's sanctions for deviant acts. Anyone who accepts responsibility for monitoring compliance with rules and applying sanctions to others is an agent of social control. These people administer societies' sanctions and label violators. While they technically react only to behavior ("using drugs is wrong"), in practice, they frequently label people who behave in deviant ways (e.g., as a "drug addict").

Secondary deviation is only one of the negative consequences of labeling. Labels sometimes force people to continue to occupy deviant roles. Some deviant labels exert very powerful effects; once someone acquires a label (e.g., as a homosexual, criminal, or addict), that label comes to represent a master status that organizes the person's social identity around the deviant label. Others interact with the deviant largely on the basis of the label, as though the label states virtually all anyone needs to know about the person. Deviants find that obstacles hinder their reentry into normal or conventional roles. Forced by continuing sanctions to assume deviant roles, secondary deviants learn to function in a deviant manner to survive.

Policy

The theory of secondary deviation implies benefits from limits on deviant labeling. As a response to delinquency, for example, labeling theorists have advocated such measures as decriminalization, deinstitutionalization, and diversion. Not all deviants are appropriate candidates for such "hands-off" treatment, but labeling theorists argue that such policies would represent effective reactions to many. Both deviants and society would benefit by limiting sanctions and avoiding labels.

effects of sanctions, in particular correctional institutionalization, on deviance (Gibbons, 1994: 36–37). Needed refinements to the perspective should identify specific sanctions, specific deviants, and the social conditions that bring them all together to form deviant identities.

Nevertheless, the labeling perspective identifies some important components in a sociological understanding of deviance. (1) Social control agents play an important role in creating and reinforcing deviance, rather than limiting it as they intend, under certain conditions. (2) Sociologists must understand the important ways in which deviants interact with the social audiences to their acts. Remember that much deviance is a process or event. Theories sensitive to the contexts of deviant acts provide more satisfying explanations than those that ignore the larger situations, and the interactionist perspective attempts to do just this (see Rubington and Weinberg, 1996).

Control Theory

Derek had always been a hard-working student. His good grades got him into a local college, where he was majoring in sociology. With graduation coming up, he was starting to think about life after school. Derek had some good leads on jobs, although nothing was certain, and he was excited about leaving school. But he needed some new clothes, shoes in particular. The best shoes were at the university bookstore's clothing section. The major problem was price. The brand and style of shoes he liked were over $100. He couldn't afford them, and this depressed him.

He knew it was wrong to steal, but Derek began to think that this might be the only way he could get the shoes. It was a big store, he told himself. They would never miss one lousy pair of shoes. They were probably overpriced, so the store wouldn't be out that much money anyway. In fact, if they were properly insured, the store wouldn't be out any money. They would simply make a claim and be reimbursed.

To top it off, Derek realized that the bookstore had been making money on him for the past four years while he was in college. He had purchased his books and supplies there, as well as a good deal of clothing. He had been a very good customer over the years. In some inexplicable way, Derek began to define his relationship with the store in a different way. Because he had given them so much money in the past, he was in some sense *entitled* to those shoes. The restraint he had felt in the past from the law against shoplifting and his feelings of correct behavior were weakened.

Derek stole the shoes. After his apprehension, he was remorseful, but it seemed a little too late.

Derek's case is consistent with control theory, a widely established general perspective on deviance. Like anomie theory, control theory expresses some of the main ideas of the social disorganization perspective. In fact, one author treats some sociologists associated with social disorganization (such as Frederick Thrasher, Clifford Shaw, and Henry McKay) as early control theorists (Kornhauser, 1978). Control theory, however, has focused mainly on explaining crime and delinquency rather than other forms of deviance.

Control theory bases its arguments on the central principle that deviance results from an absence of social control or restraint. Control theorists propose different causes for this lack of control, but they agree that a reduction in control—for whatever reason—will generate more deviance by freeing people to follow their "natural" inclinations. Reckless (1973) calls his version a *containment theory*. He argues that controls over behavior can come from interpersonal, political, and legal sources. Two basic types of containment are *inner containment*—restraints that act within the person—and *outer containment*—that arise from forces in the individual's environment. These sources of controls combine to keep most people from deviating from social norms most of the time.

Control theory does not advance especially new ideas. It originated in the emphasis on social integration in the pioneering work of Emile Durkheim. Durkheim pursued an interest in methods for maintaining social order in a complex society with a sophisticated division of labor and substantial social differentiation, both of which appear conducive to social disorder. Durkheim sought the answer in the notion of *integration* and bonds of commitment that develop between individuals and their larger social groups (Durkheim, 1933; see also Fenton, 1984). Durkheim ambitiously studied an apparently highly individualistic behavior, suicide, and he revealed this form of deviance as a social phenomenon related to the degree of integration in social groups. His analysis confirmed the prediction that suicide rates varied inversely with the degree of social integration; for example, Catholics committed suicide at lower rates than Protestants because, Durkheim explained, the Catholic church provides its members with a greater sense of group belonging and participation.

Similarly, contemporary control theories predict the greatest deviance among the groups and individuals least integrated with conventional society. They reverse the assumption of most other theories, such as the theory of anomie, that conformity is a nonproblematic state. That is, other theories describe conformity as the natural order of things, requiring no explanation. Control theory, on the other hand, asserts that *conformity, not deviance, requires an explanation.* "The important question is not 'Why do men *not* obey the rules of society?'" (Hirschi, 1969: 10). Rather, control theorists focus on questions about why they conform.

Control theory attempts to combine theories of conformity with theories of deviance, finding the causes of deviance not so much in forces that motivate people to deviate as in simple failure to prevent deviance (Nye, 1958: 3–9). Anomie theory posits a motivation derived from a disjuncture between goals and means, inducing individuals to choose deviant adaptations, such as innovation. Control theory, on the other hand, assumes that *everyone* feels urges to commit criminal acts, so theorists need not cite special motivations to explain deviant behavior. Rather, these motivations emerge simply from human nature and situations. Some people act on such motivations because they feel temporarily released from restraints that hold others in check.

Assumptions and Structure of Control Theory

Underlying their view of deviance as the absence of controls, control theorists make certain assumptions about human nature. A recent application of control theory to juvenile delinquency explicitly states two of these assumptions:

> (1) That human nature is on the "bad" side of a neutral position; that is, humans are naturally egocentric and seek to satisfy their wants and needs by the easiest means available, even if those means are illegal. (2) That decreases in prolegal controls (internal and external) allow delinquent behavior. (Arnold and Brungardt, 1983: 398)

Control theorists share certain assumptions with psychoanalysts regarding human nature and the importance of controlling supposedly innate tendencies.

This control, they say, leads to a person's commitment to conform (Reckless, 1973: 55–57).

Hirschi, a leading exponent of control theory, has provided the clearest statement of its principles, identifying four components of a person's bond with society that tend to prevent deviance (Hirschi, 1969: 16–26):

- *Attachment* refers to the extent to which a person feels bound to specific groups through affection, respect, and socialization to group norms.

- *Commitment* describes the degree to the importance of a person's stake in conforming behavior, so that acts of deviance jeopardize other, more valued conditions and activities (Toby, 1957). Concerns about one's reputation or losing one's job are examples of commitment.

- *Involvement* refers to nondeviant physical activities. At the simplest level, an adolescent can spend little time in delinquent acts if he or she spends much time playing basketball, for example. Continued involvement in conventional activities strengthens commitment.

- *Belief* refers to personal allegiance to the dominant value system in a group. These values may assume the importance of moral imperatives for the individual, rendering violations unthinkable acts.

Hirschi (1984: 51) has suggested applying the general logic of control theory to other forms of deviance as well. Delinquency results, according to this argument, from "the tendency or propensity of the individual to seek short term, immediate pleasure," a tendency that may relate to other forms of deviance.

Other than Durkheim, most control theorists stress social processes rather than structures. Hirschi, Reckless, and Nye all talk about the process by which certain individuals escape controls and come to commit acts of deviance. Similarly, Sykes and Matza (1957) describe delinquency as the result of a process that weakens controls over deviant impulses, and they note that the deviant may aid this process by "neutralizing" the restraining effect of norms and laws. Shoplifters may persuade themselves that "the store will never miss the item," or "they really owe me this item for overcharging me on other items all these years" (Meier, 1983). This thinking neutralizes normal restraints, so that they no longer hold the individual's behavior in check.

Control theory also cites termination of criminal careers as reestablishment of bonds and enhancement of integration between deviants and conventional society. Many drug addicts reject treatment, even if they wish to escape their addictions, because they feel they can make the change themselves or that treatment will not help them (Biernacki, 1986: 74). But that rejection of deviant behavior does not lead automatically to recovery unless the addict also removes herself or himself from the drug-using world and reestablishes ties with conventional society. Similarly, many criminals report leaving lives of crime, resolved to support themselves in less risky undertakings (Shover, 1985: 94–97). Different life contingencies may account for this resolve, including development of ties to a noncriminal of the opposite sex and to noncriminal activities.

Crime and Low Self-Control

Gottfredson and Hirschi (1990) have expanded the general argument that crime results from inadequate social controls by emphasizing the importance of an individual's *lack of personal self-control.* They point out that crime immediately and often easily gratifies desires, but it offers few long-term benefits. This immediate gratification can take the form of money, sex, drug-induced euphoria, or stolen property. Criminals also experience excitement while committing their crimes, most of which require little skill or planning. This argument does not imply that the choice to commit a crime is an irrational decision. Rather, crime represents a utilitarian activity—a means to an end. The end is immediate satisfaction, and the benefits last only for the short term rather than the long term. Crimes are committed by people who value immediate gratification and base decisions on only short-term views of life and their own personal goals. These criminals lack self-control. Conversely, people with strong self-control should demonstrate low rates of criminality (see also Hirschi and Gottfredson, 2000).

Gottfredson and Hirschi (1990: 90) describe criminals by saying that "people who lack self-control will tend to be impulsive, insensitive, physical (as opposed to mental), risk-taking, short-sighted, and nonverbal, and they will tend therefore to engage in criminal and analogous acts." Low self-control contributes not only to criminal behavior; it also shows an association with related activities, such as accidents, smoking, and alcohol use. Gottfredson and Hirschi seek the origins of low self-control in the functioning of the family and early childhood development. Indeed, "the major 'cause' of low self-control . . . appears to be ineffective child-rearing" (p. 97). Parents do not raise their children to become criminals, but they may not give proper care either, perhaps failing to provide nurturing family atmospheres or to supervise children's behavior. Parents also may not define some socially disvalued conduct as wrong or deliver appropriate corrective punishment. The socialization process can go wrong at many different points.

Control theory associates crime with social and interpersonal conditions that lead to low self-control. Social conditions that inhibit proper child socialization and supervision, for example, contribute to both crime and low self-control. Control theorists cite these reasons for the association between crime and economic deprivation, working parents, one-parent households, family disruption, poor school adjustment, and other conditions that interrupt effective socialization.

Evaluating Control Theory

Control theory suffers from a number of inadequacies that limit its effectiveness as a complete explanation of deviance.

Where Is Deviant Motivation?   Control theorists assume that everyone feels equal motivation to commit deviant acts, but some do not because something exerts control over their conduct. But observation suggests that some people feel more strongly motivated than others to engage in deviant acts. Because control theorists assume that everyone would practice deviance given the chance, they, like psychoanalysts, must also assume that humans carry these invariant tendencies from birth. Thus, they describe any natural state or human nature as self-seeking, harmful to others, and generally evil. This perception does not conform to the view shared by many sociologists and anthropologists, that diversity is the most striking characteristic of humans; individual intentions vary widely according to cultural background and normative circumstances and show no common tendency toward wrongdoing.

Disagreement Among Control Theorists   Control theory predicts the highest rates of deviance in groups with the least effective social controls. In Hirschi's language, rates of deviance should soar highest among groups with the weakest attachment to, commitment to, involvement with, and belief in the values of conventional society; familiar examples include the lower class and particular racial and ethnic groups. These groups have participated least in the American dream, and control theory expects them to develop the weakest bonds with a society that has not treated them with much kindness.

Yet, Hirschi's (1969) own data find no strong relationship between the probability of delinquency and a person's social class. Probably for this reason, Hirschi and other social psychological control theorists avoid the direct predictions that come from control theorists such as Durkheim (1933) who stress the structural relationships between groups and society.

Even when control theorists hazard specific predictions of high deviance among, say, members of the lower class, their assertion does not explain the extensive deviance among other social classes. As one possible reason for this confusion, note that the theory, like many others in criminology, is stated loosely and open to varying interpretations. Efforts to systematize or formalize the theory into more precise language or symbols have succeeded (LeBanc and Caplan, 1993).

The Assumption of a Central Value System   Most versions of control theory assume that some central value system explicitly informs all members of society what constitutes deviance. Without such an assumption, control theory would describe deviance as the result of people learning different moralities. But control theory is not a learning theory. It relies on the assumption of one central value system so that it can attribute variations in deviance differences in controls alone, and not variations in learned beliefs or values. Sociologists recognize no single, central value system in a pluralistic society like that of the United States, however, and most modern industrial societies seem to feature the same kind of variation. Rather, many value systems affect people's behavior, some of which condone certain acts considered deviant in others.

Lack of Sufficient Empirical Evidence   Applications of empirical evidence give particularly important evaluations of control theory, because competing theories may present stronger theoretical arguments than others. As Hirschi has put it:

> The primary virtue of control theory is not that it relies on conditions that make delinquency possible while other theories rely on conditions that make

delinquency necessary. On the contrary, with respect to their logical framework, these [other] theories are superior to control theory, and, if they were as adequate *empirically* as control theory, we should not hesitate to advocate their adoption in preference to control theory. (Hirschi, 1969: 29; emphasis added)

Researchers have, however, reported mixed empirical evidence regarding the adequacy of control theory. Hirschi's (1969) own study supports the basic ideas of control theory, with some exceptions, as do some other studies (see Wiatrowski, Griswold, and Roberts, 1981). However, Matsueda (1982) and Matsueda and Heimer (1987) found that data supported attributing crime more to acquisition of criminal norms than to weakening of social controls. In perhaps the most extensive analysis to date, Kempf (1993) has evaluated a large body of empirical literature, testing various components of social control with a variety of illegal behaviors. She has described an empirical literature composed mainly of disparate studies with different conceptualizations of control theory, different measures, different samples, and different analytic methods. The studies do not build on one another, and, as a result, "the research reveals little about the viability of social control as a scientific theory" (Kempf, 1993: 173).

Research in areas of deviance other than crime also reveals mixed empirical evidence. Seeman and Anderson (1983) found that social integration offered no effective buffer against heavy drinking. In fact, this study has reported a relationship between high social, conventional involvement and heavier drinking, not the reverse, as predicted by control theory.

Crime and Low Self-Control   As evidence for the effect of low self-control on deviance, Gottfredson and Hirschi cite the behavior that Gottfredson and Hirschi try to explain. A control theory explanation includes a logical fallacy when it says that crime results from low self-control, presenting the behavior of criminals as evidence for low self-control. Control theorists appear to offer the expression *low self-control* as a synonym for crime or other forms of deviance. Consider the following statement from Gottfredson and Hirschi (1990: 119): "Our perspective asserts that crime can be predicted from evidence of low self-control at any earlier stage of life. . . . Our perspective also asserts that low self-control can be predicted from crime at any earlier stage of life." Or the following: ". . . we ended up with 'criminal and analogous behaviors' as the conceptual equivalent of 'low self-control'" (Hirschi and Gottfredson, 2000: 64).

The problem is that crime and low self-control are so closely connected or defined that they form parts of the same thing (Geis, 2000: 39). This problem resembles the one that Bernard (1987: 417) identified with control theory: "If conformity is defined as acts controlled by attachments, involvements, commitments, and beliefs, conformity cannot be explained by the same factors." Research on this perspective has to date produced mixed results, with some studies finding results consistent with low self-control theory and others finding inconsistent results (Akers, 2000; 115–116; Moffitt, Krueger, Caspi, and Fagan, 2000).

This aspect of control theory raises an additional question: Even if one grants that people with low self-control commit crimes, one must then identify the

 **Quick Summary** | **Control Theory**

Control theory offers perhaps the most popular current explanation of crime, judging from professional journal reports intended to "test" it and from mentions in the literature. It is also the oldest sociological perspective on crime, dating back to Durkheim.

Bare Bones of the Theory

Control theory finds the cause of crime in the lack of restraint, or control, over individual conduct. It accounts for deviance, not by showing the development of deviant motivation, but by laying out the consequences of weak social controls. Such controls emerge from particular kinds of relationships with groups, particularly those that form part of conventional society. Durkheim used the term *integration* to denote this restraining relationship, while more recent theorists, such as Hirschi, prefer the term *bond*. A strong relationship— or bond—prevents deviance; a weak or broken bond permits and ensures deviance.

Effective integration into relevant groups reduces the likelihood that people will commit deviant acts. Hirschi discusses four properties that make up this relationship or bond:

Element of Bond	Definition
Attachment	Feelings of respect and affection for relevant groups; concern for what others think
Commitment	Feelings that conventional activities offer rewards; development of a "stake in conformity" and the feeling that deviance jeopardizes valued benefits
Involvement	The amount of time spent in conventional activities; insufficient idle time for possible deviance
Belief	The extent of internalization of conventional norms; development of "inner controls" over deviance

SOURCE: Hirschi, Travis. 1969. *Causes of Delinquency.* Berkeley: University of California Press.

These elements of the social bond determine the nature of an individual's relationship, or the degree of her or his integration, with the group. Deviance can thus be explained by accounting, not for the genesis of motivation, but for the absence of controls.

Policy

If control theory is correct, society should respond to deviance by trying to strengthen people's social relationships and bonds. Improvement comes from programs and activities that promote conventional activities (e.g., staying in school, getting conventional jobs) and development of conventional career activities, such as employment, typical relationships (e.g., marriage), and promotion of the benefits of staying out of trouble.

conditions under which that low self-control is expressed in crime. At some times, these impulses produce crime, while at other times, alternative conduct might result (such as smoking or alcohol use). Control theory should explain why. A theory of criminality should differentiate the conditions conducive to crime from those that lead to noncriminal behavior.

Learning or Socialization Theory

The final general theory of deviant behavior discussed in this chapter can be called *socialization*, or *learning theory*. The remaining chapters adopt this perspective as the central frame of reference for their analysis. The discussion will point out certain weaknesses in the socialization perspective from time to time, but it seems best suited of the available theories to account for many facts about deviance that require explanation.

This perspective treats deviant actions as learned behaviors developed according to the same basic processes through which nondeviants learn conformity (Akers, 1998). Deviance results from learned acquisition of deviant norms and values, particularly those learned within subcultures and among peers. Although the same basic processes create deviant and conforming norms and behavior, the direction and content of the learning may differ. Previous chapters have discussed the processes of acquiring norms, social roles, and self-conceptions, so this one will omit those details. It will also bypass the central contexts of learning deviant behavior, such as urbanism, also discussed in earlier chapters.

Sutherland's Theory of Differential Association

Jeanette was an adventuresome adolescent who seemed to crave new experiences. She always seemed to be the first one to try different things. She became known as a trendsetter for her clothing. She liked many rings in her pierced ears, and she was always trying different colors of polish for her fingernails. Her clothing at times was different because she experimented with different colors and fabrics. She would put together items of clothing that didn't seem to go together. She liked to be different.

Jeanette also liked different people. She had met Anne at school. Like Jeanette, Anne was adventuresome, an average (or worse) student, and attracted to different things. Jeanette had never tried drugs, but Anne had some marijuana she got from her brother. Anne had tried some before and had persuaded Jeanette to try it, too. Jeanette was reluctant because her parents were so strongly against drugs, but Anne told her that drugs were fun and harmless. Anne also told Jeanette that most of the kids in her high school used drugs and that she would not be popular unless she at least tried them. "It's harmless," Anne said. "Everyone's doing it, and who listens to their parents anyway?"

One weekend night, after telling her parents she was going to the mall, Jeanette met Anne in a public park, and they tried the marijuana. Anne taught Jeanette how to smoke it to get the right effects and how to identify the physical effects of the drug as pleasurable. She also told Jeanette where she could get more, if she wanted some.

Jeanette had learned to become a marijuana smoker. Jeanette had learned to become a deviant.

The best-known general learning theory is Edwin H. Sutherland's theory of *differential association*. It has become one of the most widely known theories in sociology since it first appeared in his *Principles of Criminology* in 1947. It has been further discussed, without changes, in subsequent editions of that book (Sutherland, Cressey, and Luckenbill, 1992: Chapter 5).

Sutherland's theory, developed to explain criminal behavior, accounts for both the etiology of deviance, or the cause of an individual's deviant act, and the epidemiology, or distribution of deviant behavior as reflected in various rates. This combination requires analysis of conflicting deviant and nondeviant social organizations or subcultures (differential organizations) and social psychological analysis of individual deviation by comparing conflicting deviant and nondeviant associations of deviant norms (differential associations).

Sutherland argued that deviant group behavior resulted from normative conflict. Conflict among norms affects deviance through differential social organization, determined by neighborhood structures, peer group relationships, and family organization. An individual's normative conflict results in criminal behavior through differential association in which the deviant learns criminal definitions of behavior from personal associates.

The formal proposition statements that express Sutherland's theory apply only to delinquent and criminal behavior, but this chapter's discussion modifies the concept to apply to other forms of deviant behavior as well, such as prostitution, drug addiction, and alcoholism. A relatively simple list summarizes the propositions of the theory of differential association (from Sutherland, Cressey, and Luckenbill, 1992: 88–90), supplemented by amplification on each proposition adapted from McCaghy and Capron (1994: 76–77):

1. Deviant acts represent learned behavior. Deviance is not inherited, nor does it result from low intelligence, brain damage, or the like.

2. Deviants learn this behavior through interactions with others in a process of communication.

3. The primary learning of deviant behavior occurs within intimate personal groups. Communications from sources like the mass media of television, magazines, and newspapers play at most a secondary role.

4. The behavior that deviants learn includes (a) techniques of deviance, which range from very complicated to quite simple, and (b) the specific direction of motives, drives, rationalizations, and attitudes that characterizes the particular form of deviance.

5. The deviant learns this specific direction of motives and drives from definitions of norms as favorable or unfavorable standards. This proposition acknowledges the potential for conflicts between norms, since an individual may learn reasons for both adhering to and violating a given rule. For

example, a person might argue that stealing violates a norm—that is, unless the thief takes fully insured goods, in which case it really hurts no one.

6. A person becomes deviant because definitions that favor violating norms exceed definitions that favor conforming to norms. This key proposition ties up several elements of the theory. An individual's behavior reveals the effects of contradictory learning experiences, but a predominance of deviant definitions leads to deviant behavior. Note that the associations reflect both deviant persons and also deviant definitions, norms, or patterns of behavior. Furthermore, the notion of a learning theory implies a different phrasing for the proposition: A person becomes nondeviant because definitions unfavorable to violating norms exceed those that favor violations.

7. Differential associations may vary in frequency, duration, priority, and intensity. *Frequency* and *duration* refer to the length of time over which a deviant is exposed to particular definitions and when the exposure began. *Priority* refers to the time in the deviant's life when he or she encountered the association. *Intensity* concerns the prestige of the source of the behavior pattern.

8. A person learns deviant behavior by association with deviant and nondeviant patterns involving all of the mechanisms involved in any other learning. No unique learning process leads people to acquire deviant ways of behaving.

9. While deviant behavior expresses general needs and values, those general needs and values do not fully explain it, since nondeviant behavior expresses the same needs and values. Someone might cite a need for recognition to explain actions as diverse as mass murder, a presidential campaign, or a .320 batting average; in fact, this principle explains nothing, since it apparently accounts for both deviant and nondeviant actions.

While all of these propositions may not apply to every form of deviance, Sutherland did intend to apply all of them to all forms of criminal behavior.

Differential Association-Reinforcement Theory

Akers (1985, 1998) has attempted to explain deviance on the basis of learning principles in a *differential association-reinforcement theory* of deviance. Like Sutherland, Akers (1985: 51) claims that deviance results when a person learns definitions that portray some conduct as a desirable, even though deviant, action. "Definitions are normative meanings which are given to behavior—that is, they define an action as right or not right." This sentence states what is learned; these meanings motivate the deviant and create the willingness to violate norms. Over time, individuals come to learn that some behavior and attitudes lead to reinforcement. This learning, in turn, increases the probability of the behavior.

Sociologists have applied this theory to many different forms of deviance, including drug and alcohol use, mental disorders, and suicide. Observers have noted, for example, how some individuals learn that expressing suicidal thoughts and even actual suicide attempts may evoke certain reactions from others, including sympathy, concern, and attention. This ability to generate desired behavior from others reinforces the suicidal behavior, ultimately leading in some cases to successful suicides (Lester, 1987).

Similarly, Wilson and Herrnstein (1985) present a theory of criminality with relevance for other forms of deviance. The theory asserts that criminality is essentially learned behavior within certain biological constraints, some of which may predispose individuals to crime. Wilson and Herrnstein, like Akers, argue that the benefits of a successful crime can reinforce the definitions that motivated it: money, sex, drugs, status. Without a countervailing penalty, crime, like all other behavior, can become a self-reinforcing behavior pattern.

Evaluation of Learning Theory

The notion of socialization makes a central contribution to virtually every theoretical perspective discussed so far, although this element does not qualify them as learning theories. Anomie theory requires that individuals learn success goals and agree upon general social values; control theory works only if people become socialized into a conventional value system through which they develop bonds; conflict theory supposes that socialization develops members' commitments to the interests of their groups; labeling theory describes socialization promoted by society's reactions to deviant roles and statuses. While these perspectives stress socialization with varying intensity, the concept of socialization clearly plays an indispensable role in fully understanding deviance.

Learning or socialization theories, such as Sutherland's process of differential association, have established wide acceptance among sociologists. Studies have found support for socialization theory in applications to many forms of deviance, from crime (Matsueda and Heimer, 1987) to adolescent drug and alcohol use (Akers, 1998), as well as in diverse settings, such as counseling groups in correctional programs (Andrews, 1980). These and other studies suggest the importance of socialization principles to any full understanding of deviance processes, regardless of other causal forces at work. The concept of differential association appeals to theorists for its flexibility in simultaneously resolving both the sociological and social psychological aspects of deviance.

The theory explains variations in behavior between individuals and groups with equal clarity. In particular, it helps analysts to account for the differences among groups in rates of deviance. Arrest and conviction statistics reveal, for example, disproportionately high deviance among young males, urban residents, people of low socioeconomic status, and some minorities as compared with the distribution of these groups in the general population (Brantingham and Brantingham, 1984; Federal Bureau of Investigation, 1999). Sutherland's theory explains these patterns by noting differences in exposure to deviant norms; this exposure varies the probabilities that individuals will learn, internalize, and act on these norms. Thus, comparatively high crime rates among young people largely reflects the importance of peer influence. Changes in peer relations over the human life span substantially mirror changes in criminal behavior, a result consistent with the general expectations of differential association (Warr, 1993).

The official rates of deviance for another group, females, also tend to support the differential association theory. Official statistics have typically reflected low rates of crime committed by females, except for a short period of increased rates during World War II. Rates of female criminality generally declined again after the war, only to increase once more in recent years (Simon, 1975). Differential association theory would explain these variations as results of general increases in the number and range of opportunities for women to participate in society during World War II, leading to increased exposure to deviant norms. The theory would describe the more recent jumps in crime rates as part of increased learning opportunities for women, together with changes in traditional sex roles, away from standards that emphasized submissiveness and stay-at-home isolation. Some observers also believe that early family learning experiences are particularly important for females who later engage in deviance (Giordano and Rockwell, 2000).

Socialization theory offers similar general descriptions for any learning process that ultimately leads to deviance. Forcible rape, for example, may result from the separate and unequal socialization processes for males and females that translate traditional masculine qualities (for example, aggressiveness, power, strength, dominance, and competitiveness) into aggressive sexual behavior over females (Randall and Rose, 1984). Similarly, peers and drinking companions provide important socialization that helps to define situations as appropriate for drinking and to influence attitudes toward and behavior with alcohol (Downs, 1987; Orcutt, 1991).

Critics have attributed some shortcomings to socialization theory, both as a general term for processes of learning deviant norms and values and in its more specific applications, as in the theory of differential association. The most common criticism maintains that the theory tends to present an oversocialized conception of human beings, with insufficient attention to differential responses in the form of individual motivations and rational actions. (For other criticisms, as well as responses to them, see Sutherland, Cressey, and Luckenbill, 1992: 93–99.) Wrong (1961) has argued, for example, that socialization theory commonly claims that people internalize social norms and seek favorable self-images by conforming to the expectations of others. Clearly, there are other considerations in determining behavior that may not stem solely from social learning.

Continuing research on delinquency reveals enduring but complex effects of interactions with peers. For example, a recent study reported that associating with delinquent peers does indeed lead to increases in delinquency, as socialization theory expects, but also that increases in delinquency also increase individuals' associations with delinquent peers (Thornberry, Lizotte, Krohn, Farnworth, and Jang, 1994). Therefore, delinquency emerges as part of a dynamic process of mutual influence rather than as a simple outcome of a process of influence. Most socialization theories, like Sutherland's, do not properly account for the mutually reinforcing effects of peer influence and delinquency.

Other criticisms of socialization theory cite problems of logic, such as a tendency to commit the logical error of tautology. For example, socialization theories attribute deviance to a person's learning of deviant norms, and then they often present deviant acts as evidence of those deviant norms. This circular reasoning

Quick Summary   **Learning Theory**

Learning theory takes many different forms, from the strict behaviorist applications in psychology to some more symbolic applications in sociology. The most popular learning perspectives in the theory of deviance have derived from Sutherland and, more recently, Akers.

Bare Bones of the Theory

Learning theorists claim that deviant behavior results when people learn deviant norms, values, and/or attitudes. Different theories propose variations in the learning process, but they agree that each person is born with an essentially blank slate and that people become fully social human beings through socialization into groups. Environment is everything, and individuals acquire their identities as humans.

The most frequent behavior depends on what receives reinforcement; the least frequent behavior depends on what receives punishment. Someone who acquires deviant norms (i.e., norms that permit or condone deviant conduct) will likely behave in a deviant manner. Sutherland has expressed his version of learning theory, called *differential association theory*, in nine propositions that refer specifically to delinquency, but the processes are clearly general enough to apply to other forms of deviance as well.

Sutherland's Theory of Differential Association

Proposition	Commentary
1. Criminal behavior is learned.	Criminal behavior is not an inherited trait.
2. Learning takes place through interaction with others in a process of communication.	Humans communicate both verbally and symbolically, with each style important to the learning process.
3. The main learning for criminality occurs in intimate groups.	People learn deviance from those they like and respect the most—primary group members.

requires independent evidence of deviant norms aside from the deviant conduct (Kornhauser, 1978). Other observers suggest that Sutherland's long-standing theory could benefit from a more specific discussion of its principal components—definitions favorable to crime, differential social organization, and normative conflict (Matsueda, 1988).

The processes for learning deviant norms and behavior patterns parallel those for learning nondeviant norms and behavior patterns. Only the content of the

Proposition	Commentary
4. The learning includes (a) techniques of committing crime and (b) the specific associated directions of motives, drives, attitudes, and rationalizations.	The content of the learning includes both specific techniques and the cognitive sets necessary to employ those techniques.
5. People learn the content of attitudes from norms, especially laws, as favorable or not.	People learn to accept the permissibility or even desirability of violating some norms.
6. A person becomes criminal when definitions favorable to violating the law exceed those unfavorable to violating the law.	When deviant norms outnumber conventional norms or exceed them in importance, deviance results.
7. Differential associations vary in duration, priority, frequency, and intensity.	These are the "variables" of the theory.
8. People learn criminality through the same process as they complete any other learning.	The process of learning deviance is no different from that for becoming an athlete or a nurse.
9. Criminality cannot be explained by general needs and values.	Such general conditions as "greed" do not fully explain crime, because they also explain noncriminal behavior.

SOURCE: Paraphrased from Sutherland, Edwin H., and Donald R. Cressey. 1978. *Criminology*, 10th ed. Philadelphia: Lippincott.

Policy

Learning theories suggest that deviants can learn conforming behavior instead of deviance. This possibility highlights the importance of programs to bring offenders into contact with law-abiding people (so that the offenders can learn law-abiding norms, rather than allowing criminal norms to corrupt the law-abiding people). Drug and sex education programs in schools and suicide-prevention efforts are examples.

learning differs. The criticism for learning theory's oversocialized conception of human beings highlights real dangers of claiming that all deviant acts result from learning, or that learning *alone* fully explains deviant acts. This is not the case. This book, for example, discusses one form of deviance that does not require learning: physical disabilities. (See Chapter 14.) In spite of criticisms, socialization theory offers the most adequate perspective to explain deviance and make sense of the facts of these behaviors and conditions.

Summary

Whereas general theories of deviance attempt to explain all deviant acts, limited theories narrow their scope. Some theories might address only certain types of deviance, particular substantive forms of deviance (such as alcoholism or homicide), or the origins of deviant acts, ignoring the structures that more generally support deviant behavior. This chapter discussed three perspectives: labeling, control, and socialization theories.

The *labeling* perspective emphasizes interactions between deviants and agents of social control and the consequences of those interactions. According to labeling theory, social control efforts sometimes cause deviance instead of restraining it by pushing people toward deviant roles. Closed off from conventional roles by negative or stigmatizing labels, people may become secondary deviants, partly in self-defense. Society may resist an individual's movement back to conventional, nondeviant roles, leaving that person feeling like an outsider. In this sense, labeling theory claims, sanctioning or labeling efforts designed to control deviance amplify it instead.

Control theory usually limits its explanations to the phenomena of delinquency and crime. It locates the cause of crime in weak bonds or ties with society, that is, a general lack of integration. Groups that do not become integrated into conventional society (such as members of the lower class) may violate the law because they feel little commitment to the conventional order that establishes it. People who feel close to conventional groups, on the other hand, hesitate to deviate from established rules. Social distance that results from broken bonds tends to free people to deviate. Another version of control theory describes crime and other forms of deviance as the results of low self-control.

Socialization or *learning theory* asserts that deviant behavior arises from normative conflict sparked when individuals and groups learn norms that permit or condone deviance under some circumstances. This learning may convey subtle content, for example, that deviance sometimes goes unpunished. Such socialization can also lead people to acquire seriously deviant norms and values that define deviant acts as either necessary or desirable under certain circumstances, such as the company of certain people. Sutherland's theory of differential association is one of the best-known learning theories of deviance; although it focuses on establishing a general explanation of criminality, it fits other forms of deviance as well. Virtually every sociological theory of deviance assumes that socialization influences individuals to become members of groups or the general society. Some theories emphasize this learning process more than others.

Selected References

Akers, Ronald L. 1998. *Social Learning and Social Structure: A General Theory of Crime and Deviance*. Boston: Northeastern University Press.

A systematic exposition of learning theory. The book brings together the development, refinement, and tests of learning theory and offers information on both the content of socialization to deviance (i.e., what is learned) and the process (i.e., how it is learned).

Hirschi, Travis. 1969. *Causes of Delinquency*. Berkeley: University of California Press.

This book presents the most influential version of traditional control theory. It elaborates the theory and then tests it with respect to delinquency. It is required reading for anyone who desires a complete understanding of this perspective.

Lemert, Edwin M. 1951. *Social Pathology*. New York: McGraw-Hill.

After nearly 50 years, this book still retains a freshness not found in many books on deviance. It lays out Lemert's theory of secondary deviation. Despite its misleading title, the book bears no relationship to the social pathology perspective discussed in Chapter 5.

Sutherland, Edwin H., Donald R. Cressey, and David F. Luckenbill. 1992. *Principles of Criminology*, 11th ed. Chicago: General Hall.

The most recent edition of this classic text discusses Sutherland's theory of differential association. Luckenbill was a student of Cressey and Cressey, a student of Sutherland. The book systematically applies the theory to all topics in criminology, making it one of the most theoretically coherent textbooks of crime available.

Traub, Stuart H., and Craig B. Little, eds. 1999. *Theories of Deviance*, 5th ed. Itasca, IL: F. E. Peacock.

A collection of papers on the major theoretical traditions in the sociology of deviance. This volume contains most of the classic statements as well as a few more contemporary commentaries on them. Useful for both classroom use and research.

PART III

Forms of Deviance

CRIME, drug addiction, alcoholism, illicit sex, and suicide. These are powerful terms. Few people are indifferent to their existence, and many have strong and sometimes contradictory ideas of how best to deal with them.

In this part, we discuss some of the more common forms of deviant behavior. Our intent is not to cover all aspects of the behavior or people who engage in the behavior, nor is it to identify every current controversy that has arisen over the nature and control of these acts. Rather, we present a sociological perspective on deviance that is tied to the social nature of these acts. We recognize that there are different approaches to learning about deviance. Some observers employ a moral perspective that condemns the acts and actors as deliberately sinful. Others prefer to look at aspects of the deviant's psychology or biology to understand these acts.

The approach used here is that each of these forms of deviance has a social reality. Deviance is structured behavior; it is not random. Rather, deviance is related to the social conditions people find themselves in and the social positions (roles) they employ. Deviance is related to some of the most fundamental features of social life: age, sex, social class, and residence. It is the relationships between deviance and social life that occupy our attention here. ■

CHAPTER 7

Crimes of Interpersonal Violence

Betty Lou Beets was convicted of the 1983 killing of her fifth husband, Jimmy Don Beets. She had also been charged earlier, but not tried, for the shooting of her fourth husband and had been tried and convicted of shooting and wounding her second husband. According to her daughter, Betty Lou Beets experienced considerable emotional and physical abuse during her marriages. Prosecutors argued that there was no evidence of physical abuse in Betty Lou's marriage with Beets, a point agreed to by Betty Lou's lawyer, who countered that there was plenty of emotional abuse and that Beets's insurance and death benefits appeared to be the motive for the crime. Appeals failed, and on February 24, 2000, Betty Lou Beets was executed, only the second woman in Texas to be so punished since the Civil War.

Murder. Spouse abuse. These and the circumstances under which they occur are among the subjects of this chapter. Substantial consensus affirms the deviant nature of some crimes, such as murder, forcible rape, burglary, and assault. People disagree, however, about the association between deviance and other crimes, such as prostitution, pornography, and the use of marijuana and certain other drugs. This uncertainty confirms that crime is a highly diverse form of deviance. A full understanding of the diversity of criminality begins with a discussion of the nature of crime as a form of deviant behavior. Crime is one of the most widespread forms of deviance, and this chapter introduces the subject by dealing with major forms of illegal interpersonal violence: murder, assault, and rape. Chapter 8 then covers crimes mainly targeted to thefts of property.

Crime as Deviance

Criminal behavior is deviant behavior. A crime results from an act that violates a law, which is a particular kind of norm. Actually, one may examine a crime in two ways: as a violation of the criminal law or as a violation of any law that triggers punishment by the state. Sociologists regard crime as any act considered to cause socially injurious effects and subject to punishment by the state, regardless of the type of punishment. Certainly, behavior that violates a specific criminal statute, such as legislative prohibitions against robbery or fraud, fits the criteria for crime and merits study as such. But the broader, sociological conception of crime expands the topic to study violations of other bodies of law as well. These additional categories include regulatory law created by agencies of the federal government, such as the

Federal Trade Commission (FTC), Securities and Exchange Commission (SEC), and Occupational Safety and Health Administration (OSHA). These kinds of violations, usually called *white-collar* or *corporate crimes*, prominently affect the everyday behavior of certain individuals and groups, and they bring serious social impacts.

Types of Crimes

Criminologists often break down criminal activity into three categories: common-law crimes (conventional or street crimes), white-collar crimes, and adolescent violations (delinquency). Common-law crimes include offenses that virtually everyone would regard as criminal, such as murder, rape, robbery, burglary, and assault. Lawyers often refer to these violations as *mala in se*, meaning they are bad in themselves. Societies judged these acts as illicit behavior before any had developed written, state-enacted laws; in those times, formal standards for behavior came only from common law, a term that refers to legal traditions in the form of judges' decisions. At some time or another, common law has set standards for a variety of behavior, including recreational activities on the Sabbath, the practice of witchcraft, cigarette smoking, selling of alcoholic beverages, and women's wearing of one-piece bathing suits, among many others.

Legal prohibitions of certain other types of behavior come from no such principles in common law. Lawyers refer to these crimes as *mala prohibita*, or bad simply because the law prohibits them. Most of these acts became offenses as reactions to technological and social changes in society. Many are associated with the automobile, building codes, activities to manufacture and sell impure food and drug products, and sales of fraudulent securities.

Sociological analysis must separate conventional crimes from white-collar crimes, perhaps more accurately termed *occupational crimes*. Criminal law deals with conventional crimes, but its provisions seldom apply to occupational crimes. These violations include illegal acts by employees and others associated with business organizations, from small firms to leading corporations, along with politicians, government workers, labor union leaders, doctors, and lawyers in connection with their occupations (Friedrichs, 1996). Because criminologists wish to encompass these violations in their studies, many determine crime not only by the standards of the criminal law but in broader terms as any acts punishable by the state through criminal, administrative, or civil penalties. Administrative law gives the state many ways of compelling individuals, business concerns, and labor unions to obey its regulations. It may withdraw the license that confers a doctor's, lawyer's, or druggist's right to practice; it may suspend business by a tavern or restaurant for a few days or even permanently shut the doors. Clearly, these steps, though technically outside criminal law, can impose very severe sanctions.

The law generally treats a person below the age of 18 who commits a crime as a delinquent rather than a criminal, but this difference does not imply that *juvenile delinquency* is comparable to adult criminality in all respects but age. Many offenses committed by juveniles also represent crimes when committed by adults. Juveniles seldom commit acts of white-collar or occupational crime, however, and some jurisdictions punish delinquency that results from offenses only juveniles can commit. These violations, called *status offenses*, include unmanageable behavior at home, running away from home, and truancy. For this reason, no study of crime should assume that only "junior criminals" face charges of delinquency and incarceration in state training schools. The great majority of adolescents who commit acts of delinquency never "graduate" to adult criminality, and the circumstances of many delinquency offenses suggest experimentation rather than enduring patterns of behavior.

Legally, only behaviors (as opposed to thoughts or beliefs) deserve criminal punishment. Moreover, the law can punish only culpable people for crimes. These legal principles, called *actus reus* and *mens rea*, form the basis of the criminal law. Any crime must incorporate these elements. Behavior that violates the law *(actus reus)* constitutes a crime only if the actor pursues some criminal intent *(mens rea)*. Behavior alone does not create crime; the actor's mental attitude must also contribute.

Further, the criminal law prohibits a multitude of different kinds of behavior, ranging from very minor, petty acts to major acts with enormous social, political, and economic implications. This varied scope of activities guarantees that everyone who reaches adulthood must at some time violate some law. If the term *criminal* refers strictly to anyone who has ever violated the law, then everyone is a criminal.

Criminals obviously do not form some homogeneous group within society. Instead of treating criminals as a class of people, sociologists can more meaningfully refer to *types* of criminals. Distinct types of criminals often differ more from other types of criminals than from noncriminals.

Types of Criminals

The notion of classifying criminals introduces nothing new. In everyday discourse, people commonly refer to robbers, burglars, and rapists, as well as other criminals. Such conversation classifies offenders according to the categories of legal offenses that describe their behavior. This classification scheme gives little useful information, however, since individual offenders frequently commit different kinds of crimes, complicating the classification of any one offender.

A more useful distinction separates criminal offenders according to a typology based on *behavior systems*. This method distinguishes among offenders based on the extent to which they pursue long-term careers in crime. The term *career* may imply a financial occupation, but the term here refers to something else. The notion of a criminal career implies an individual's commitment to crime as a continuing activity. In a sociological sense, *career* can refer to an action or activity that defines a pattern for an individual.

A criminal career differs from a noncriminal career in the acquisition of criminal norms that lead to criminal acts and the individual's view of the criminal behavior. A criminal career involves a life organization of roles built around criminal activities, such as:

1. *Identification* with crime
2. *Commitment* to crime as a social role and characteristic activity

3. *Progression* in crime through development of increasingly complex criminal techniques and increasingly sophisticated criminal attitudes

As offenders identify themselves progressively more completely with crime, they become more committed to criminal careers. As they commit progressively more serious crimes, they develop criminal self-concepts. They also associate more often with other criminals. A career criminal is someone who identifies with crime and has developed a self-concept as a criminal, someone who demonstrates a commitment to criminality through frequent offenses over a period of time, and someone who shows progressive acquisition of criminal skills and attitudes. Over time, such individuals come to organize their life activities and interests around criminal behavior. A noncareer criminal displays no such identification or commitment, has developed no criminal self-concept, and has not progressed in techniques or attitudes.

Sociologists can arrange offenders in behavior system types along a continuum from those without criminal careers at one end to career criminals at the other end. The distinction between career and noncareer offenders does not define a precise separation, but it does capture a major difference between types of offenders. Most people who commit acts of interpersonal violence are noncareer offenders, or primary criminal deviants; property crimes are more often the work of career criminals, or secondary criminal deviants.

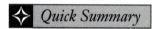 *Quick Summary*   **A Behavior System Approach to Deviance**

A behavior system perspective views deviance on a continuum. It differentiates deviants on the extent to which they perform the role of deviant. Some persons perform this role infrequently; they can be considered noncareer or primary deviants. Other persons perform this role frequently; they can be considered career or secondary deviants. Still others fit somewhere between these two extremes. This view is consistent with a number of theoretical perspectives, including virtually all discussed in this book.

1	2	3	4	5	6	7	8

Noncareer   Career

No role behavior   Role behavior

Primary deviance   Secondary deviance

Criminals can be distinguished from one another by their:

1. *Commitment*   The extent to which an individual is committed to deviance as a behavior pattern.

2. *Identification*   The extent to which an individual identifies with deviance or other deviants.

3. *Progression*   The extent to which an individual progresses in the acquisition of deviant skills or attitudes.

This chapter discusses the crimes of noncareer offenders, while Chapter 8 deals more with career offenders. At one end, the continuum shows violent offenders and occasional property offenders; at the other end, it collects organized and professional offenders. In between, political offenders, occupational and business violators, and conventional criminal offenders appear (Clinard, Quinney, and Wildeman, 1994). (See the "Quick Summary: A Behavior System Approach to Deviance.") Later chapters analyze violations of public-order standards that also often represent crimes—illegal drug use, drunkenness, prostitution, and sexual deviance.

Assault and Murder

Results from the most recent National Crime Victimization Survey released in August 2000 point to a dramatic drop in crime, including violent crime. A reduction of 10 percent was recorded for violent crime and a reduction of 9 percent for property crime. Yet, fear of crime continuously clouds life in the United States, and most of this concern reflects anxiety about violent crime. In every violent, personal crime, the offender tries to accomplish some objective through violence. Common objectives involve closing an argument, winning a personal dispute, or forcing sexual intercourse on an unwilling partner. Offenders generally do not pursue criminal careers in such crimes. In fact, most murderers and assaulters do not conceive of themselves as criminals. They seldom identify with criminal motives and acts, and criminal behavior, as such, plays no meaningful part in their lives. Most murderers do not progressively acquire new criminal techniques or attitudes. As later discussion will explain, murderers have very low recidivism rates (the rate of repeating crimes) compared with the rates of some other types of offenders.

The word *homicide* means to cause a person's death. Homicides can be either criminal (unlawful) or noncriminal acts such as court-authorized executions or accidents. For the purposes of this chapter, *murder* means a criminal homicide, a killing without legal excuse. Technically, a criminal court determines whether a homicide qualifies as murder through a legal process. A related term, *manslaughter*, refers to accidental killing. This chapter's discussion of criminal homicide covers both murder and nonnegligent homicide, but not justifiable homicide, accidental death, or negligent manslaughter.

Another violent crime, *aggravated assault*, represents an application of physical force with the intent to do severe bodily harm as a way to settle a dispute or argument. Nearly all criminal homicide represents some form of aggravated assault in which the victim dies. Most criminal homicides and assaults grow out of personal disputes and altercations, some resulting from immediate conflicts and some from long-standing ones. A few of these offenses occur during commission of other crimes, such as robbery or drug trafficking.

Frequency of Assault and Homicide

Society never learns about every crime committed, if only because victims and witnesses do not report some crimes to the police. Therefore, researchers have developed techniques other than reviewing official statistics to estimate crime. Methods include surveys that ask respondents whether they have committed crimes known to the police or not. Other surveys ask whether respondents have been victims of crimes. Victimization surveys usually identify more crime than police reports reveal. Such techniques suffer from some problems, however, including selective memory, sampling error, and other methodological problems that may limit the value of information from these sources.

Official information about homicide appears to give a reasonably accurate impression, however, because police records sooner or later gather input about

---

❋ *Issue* | **Common Questions on Violent Crime* in the United States**

Is the United States more violent than other countries?	Yes and no. The homicide rate in the United States is higher than that of any other *industrialized* nation but is lower than those of many Central and South American countries. The United States is also a leader among industrialized countries in other forms of violence, such as assaultive behavior and sexual assaults.
Have rates of violence been increasing?	No. The national homicide rate peaked twice in this century, with each peak followed by a decline. The first was during the early 1930s, but then the rate declined for more than 30 years. The second peak was about 1980, declining until 1985 or so. Since that time, the rate has been relatively stable. There is some evidence, however, that the rates for certain subgroups have been increasing. Rates of violence among teenagers, for example, have been increasing in recent years.
Who commits violent crimes?	Offenders and victims share similar demographic characteristics. They are overwhelmingly male and disproportionately drawn from certain racial and ethnic minorities. Men in the 25–29 age group are more likely to commit violent crimes than any other age group. Youth gang violence is common in some cities.
Are most violent crimes the work of "violent career criminals?"	No. While a few individuals commit violent crimes frequently, they account for a small share of the total number of violent crimes in the United States. Serial murderers account for only 1 to 2 percent of all murders per year. Most violent crimes are committed in the course of a long criminal career marked mainly by property crimes.

*Violent crimes include murder, aggravated assault, rape, and robbery.

Source: Reiss, Albert J., Jr., and Jeffrey A. Roth, eds. 1993. *Understanding and Preventing Violence.* Washington, DC: National Academy Press. This is the report of the Panel On The Understanding and Control of Violent Behavior of the National Research Council.

most murders. Police agencies compile more limited information about assault, since many assaults, like property crimes, go unreported.

Incidence of Homicide   The *Uniform Crime Reports* (UCR), the major program of crime statistics for U.S. police organizations, recorded 16,910 murder and non-negligent manslaughters in the United States during calendar year 1998 (Federal Bureau of Investigation, 1999). This figure represents about 1 percent of all violent crimes committed during that year. Although the number varies annually, the statistics show a downward trend in murder. The 1998 figure represents a reduction of 7 percent from the total the year before and a 27 percent drop below the number for 1994.

Incidence of Aggravated Assault   Statistics reveal many more assaults than homicides. In 1998, the police recorded 974,400 aggravated assaults—unlawful attacks by one person upon another for the purpose of inflicting severe or aggravated bodily injury (Federal Bureau of Investigation, 1999). This number represents a 5 percent decrease from the previous year and a decrease of more than 12 percent since 1994. Also, while researchers can expect that the police eventually learn of most murders, they have no such reliable information on assaults, even the aggravated kind. Police receive no reports on many assaults because victims often regard the incidents as private matters between themselves and their assailants.

Group Variations in Homicide

Homicide is not just individualistic behavior. Rather, this class of offenses shows certain structures or patterns in particular societies. The rates of homicide rise in some groups, at some times, and in some situations relative to others. Social acceptance of murder as a method of solving interpersonal conflicts varies a great deal over time, from country to country, from region to region, and by local area, race, social class, and age. These variations offer clues to guide potential explanations of this type of crime.

Variations by Country   People commonly settled disputes through personal violence, despite the risks of assault and murder, throughout nearly all of Europe a few centuries ago, even among the upper classes. Such offenses have become rare in most European countries, particularly most Scandinavian countries, the United Kingdom, and Ireland, but they occur at high rates in Latin American and African countries (Reiss and Roth, 1993). The high homicide rates in Latin American countries seem related to prevailing attitudes of masculinity or *machismo*, which call for recourse to violent responses to personal insults or challenges to one's honor. The rates of violence in Turkey and Finland are higher than those in other European countries, and the rate in Sri Lanka has been traditionally high for Asia (Ferracuti and Newman, 1974: 194–195). The United States ranks near the middle of the world's nations listed in order by rates of violence, although it leads all industrialized countries. In a common variation documented by Archer and Gartner (1984), homicide rates increase in most countries after wartime.

Regional Variations   Within the United States, the southern states have homicide rates considerably higher than those in other regions, although the rates in the West show rapid increases (Federal Bureau of Investigation, 1999). The differences may result largely from regionally important cultural definitions that demand personal violence in certain situations and comparatively easy availability of weapons in some areas.

Homicide rates show variations within many other countries as well (Clinard and Abbott, 1973). For example, the rate in Sardinia exceeds that in any other part of Italy; there, a set of norms forms a "code" that regulates violence, particularly vendetta (homicides), effectively superseding Italian criminal law. Residents learn this code and maintain it through social reinforcement by others (Ferracuti, Lazzari, and Wolfgang, 1970).

Local Differences   Within regions, rates of interpersonal violence peak in the inner-city areas of large metropolitan areas (Brantingham and Brantingham, 1984). These high rates are related to the slum way of life, which approves of force to settle disputes. Studies of homicides in Houston found concentrations of these crimes in a relatively small area inhabited largely by lower-class residents (Bullock, 1955; Lundsgaarde, 1977: 47–50, 105–106). Similarly, another study found that two-thirds of the homicides in Cleveland occurred in 12 percent of the city, primarily in black, inner-city areas (Bensing and Schroeder, 1960). Researchers have documented similar patterns in other large cities in the United States (Wolfgang and Zahn, 1983) as well as other countries (see Reiss and Roth, 1993).

Racial Variations   Reports consistently characterize blacks as more likely to be involved in homicide and assault, both as offenders and victims, than their proportion in the population would suggest. Racial disparities in arrest rates are greatest for crimes of violence, due largely to inner-city living. In a now-classic study of homicide in Philadelphia, Wolfgang reported a homicide rate for blacks four times that for whites (Wolfgang, 1958). Other authors have detailed a similar finding for other cities (Reiss and Roth, 1993).

Nearly all crimes of violence in the United States involve offenders and victims of the same race. This fact remains true for homicide, aggravated assault, and rape. Crimes of violence result from *intraracial* rather than interracial attacks. In most of these crimes, blacks murder or assault other blacks; in most of the rest, whites victimize other whites. The homicide information for 1995 collected by the Federal Bureau of Investigation (1999: 17) shows that black offenders killed 94 percent of black murder victims, and white offenders killed 87 percent of white victims.

Variations by Social Class   Crimes of violence are heavily concentrated in the lower class (Luckenbill, 1984). Wolfgang's study (1958: 37) attributed 90 percent of the homicides in Philadelphia to killers from lower-class occupations. Most of the victims were also members of the lower class. Another study found that most homicide and assault offenders in London came from the lower class (McClintock, 1963: 131–132). Similar results have been reported for other locations (Reiss and Roth, 1993: 129–130).

Further, the specific circumstances of murder seem to vary by social class. A study of middle-class and upper-class homicides found differences between the patterns for these crimes and those for the vast majority of all homicides among members of the lower class. Premeditated planning contributed to almost three-fourths of the middle-class and upper-class murders, the offenders killed largely for personal gain, and alcohol played no part in the great majority of these crimes (Green and Wakefield, 1979). By contrast, Wolfgang (1958) found that alcohol was an element in—but not necessarily a cause of—almost two-thirds of lower-class murders in Philadelphia. In fact, alcohol frequently accompanies both homicides and assaults, both in the United States and in other countries (Parker, 1993).

Variations by Age and Sex   Personal violence in urban areas is highest among young age groups and males. Specifically, in the United States males aged 18 to 24 show much higher homicide rates than any other group (Wolfgang and Zahn, 1983; Luckenbill, 1984). The 15-to-24 age group also has the highest rate of aggravated assault. In fact, many studies of criminal violence reveal the heaviest incidences of assaultive behavior, including murder, during late adolescence and early adulthood (Weiner, 1989: 118). Most victims of murder and assault are also

---

✳ *Issue* | **Trends in Juvenile Violence**

Recent reports of declining overall rates of violence, including murder, rape, and serious assault, mask to some a potentially serious problem. There are really two trends, one for adults, the other for juveniles, and the juvenile trend is disturbing, for it is moving in the opposite direction of that for adults.

While the national murder rate has declined slightly since 1990, the rate of murder committed by teenagers 14 to 17 increased by 22 percent. Furthermore, this recent surge in the murder rate among teenagers occurred while this segment of the population was actually in slight numerical decline. But this will soon change since the "baby boomerang" (the offspring of the baby boomers who were born right after World War II) will soon catch up. There are currently 39 million children under the age of 10, many more than we've had for decades. These children are now reaching the high crime-committing age categories, and some criminologists feared that juvenile violence rates would increase even more.

But the most recent official statistics actually show that the arrest rate for juvenile violent offenders declined in recent years, and some of the criminologists who had earlier predicted the increase were forced to retract their predictions in an article in *USA Today* (December 13–15, 1996). The concern over rising juvenile violence rates has thus far been unfounded.

Sources: Fox, James Alan. 1996. "The Calm Before the Juvenile Crime Storm." *Population Today*, 24: 4–5; and Federal Bureau of Investigation. 1999. *Crime in the United States, 1998*. Washington, DC: Department of Justice, Government Printing Office.

young people. Offenders tend to murder and assault victims similar to themselves in age, sex, race, and social circumstance.

Some worried that rates of criminal violence might escalate among females during the 1980s and 1990s, driven by large social changes concerning the status and role of women. Research has not confirmed such a trend, however. Women contribute a relatively small part of the overall rates of violent crime, not only in the United States but in other countries as well (Simon and Baxter, 1989). In fact, some observers have the idea that once women are freed from the constraints of patriarchy and experience more freedom in the world of work, they will become more criminal (Brownstein, 2000: 107).

On the other hand, studies have increasingly implicated juvenile gangs in violence committed in many U.S. cities. Some gangs have become extremely large organizations with members in a number of states. Two such gangs, the Crips and the Bloods, are said to pay for extensive arsenals of weapons by selling and distributing illegal drugs. In Los Angeles, Bing (1991: 154–155) has reported identifying 43 known Blood sets (groups) and 56 Crip sets. The largest Crip set includes more than 1,000 members, although some of the smaller sets number fewer than 20 each. Gang members direct their violence mainly at other gang members, and it is hard to know precisely how much violence to attribute to them (Klein, 1995). Their violence may take the form of individual assaults or "drive-by" shootings using automatic weapons. These attacks sometimes injure or kill innocent bystanders along with or instead of the intended victims.

Interaction Between Offender and Victim

Most murders and aggravated assaults represent violent responses to social interactions between one or more parties. These crimes result when the situations acquire definitions that call for violence. Such an act of violence may result from a single argument or dispute, or it may complete a long series of disputes between intimates, such as husband and wife, lovers, close friends, or fellow employees. Many murders grow out of conflicts between intimates, such as lovers or family members. Still, more than half of all murderers know their victims, but not in the context of intimate relationships. Rather, killers and their victims share some kind of relationship: friends, neighbors, casual acquaintances, workplace associates, associates in illegal activities, or members of a single gang or two rival gangs (Reiss and Roth, 1993: 79).

The victims of homicides and assaults sometimes precipitate the attacks against them. In such a homicide, the victim may draw a weapon first, strike the first blow in a fight, or in some other way precipitate the victimization. One study described such conditions in more than one in four homicides (Wolfgang, 1958: 252). Another study found one out of every three victims initiating the violent confrontations (Voss and Hepburn, 1968: 506).

Assaultive crime, whether or not it leads to homicide, is intragroup behavior. Males attack males with traits similar to their own, and people over 25 attack others of the same age group. In other words, offenders kill and assault others with whom they likely share group interactions. A study of 8,000 murder cases involving 10,000 offenders in the 75 largest counties of the United States found that 80 percent of murder victims were killed by relatives or acquaintances (Dawson and Boland, 1993).

Most cases of violence develop from disputes that may seem trivial to outsiders. A person's judgment of a trivial dispute reflects age, social class, and other background characteristics. People may commit homicides over nonpayment of a small debt, petty jealousy, or a small neighborhood quarrel. These apparently insignificant events may become extremely important to the people involved.

As these discussions suggest, conditions conducive to social interaction in general also create opportunities for interpersonal violence. The Federal Bureau of Investigation (1999) reports higher rates of homicides and assaults on weekends than during the week and generally higher rates during the summer than during the winter months. Such times encourage social interactions, and such socializing

 Issue **Computer Confessions**

For many months before his crime, Liam Youens, age 21, used the Internet to describe his loneliness, his suicidal fantasies, and his love-hate relationship with a former classmate. His Web site was filled with reminiscences, tirades, and confessions in which he planned his crime, explained his motives, and sought understanding for what he was going to do. No one listened.

On October 15, 1999, the Nashua, New Hampshire, teenager drove his mother's car to where Amy Boyer worked as a dental assistant. Youens waited until Boyer got into her car, shot her, and then killed himself.

Youens was a troubled, lonely youth. He was friendless in high school, often teased by others; he never dated, and he ate lunch alone, standing in a corner of the cafeteria. He fell in love with Boyer in eighth grade but never had the courage to tell her. He became jealous when Boyer called to another boy on the school bus one day and decided she had to die. But he confided his thoughts and feelings only in his computer.

"The NPD [Nashua Police Department] believed it could prevent me from getting guns. HA!" he wrote. "Some people thought that me working at 7-11 was hilarious. Idiots! The only reason I could get that job would be to spend every cent I earned on powerful assault rifles to execute my vengeance. . . . I have always lusted for the death of Amy."

The Web site opens with a picture of Youens wearing sunglasses and holding a rifle. His Web site had a number of sections that contained mainly his writings, potential targets for his anger, and rough plans for carrying out various killings. In one such section, Youens detailed plans for a mass murder at his high school:

"I'm trying to remember when lunch starts. 10:05, I think. I believe 10:20 would be a good time for the attack. I plan to start shooting people in the courtyard as fast as I can. Hopefully I'll get to the second clip. If so, I'll go for head shots, head shots, head shots! They are a MUST for a high body count."

Source: *Omaha World Herald*, December 12, 1999, p. 15-A.

sometimes leads to violence. The involvement of alcohol in many homicide and assault cases also reflects the link between these crimes and particular social situations. Violence is a kind of interaction, just as loving is a form of interaction. Therefore, no one should express surprise when conditions that bring people together for social purposes also encourage violence.

Understanding and Explaining Violence

The relationship with social situations helps to explain homicide and assault crimes, but one must also understand another characteristic of offenders and victims: differential power. Violence frequently results from an attempt by one party to establish or reestablish a position of power over the other party (Hepburn and Voss, 1973). Criminologists build on this idea when they cite the importance of *asymmetrical power relationships* in crimes of violence.

Acts of violence can flare up between husbands and wives, business partners, parents and children, and siblings. Each instance may result in a different form of the violence (homicide, aggravated assault, child abuse, spousal abuse), but power differences always separate the participants. These crimes occur most often when a powerful person feels some threat to or shift in power that would favor the less powerful person. Under these circumstances, violence may help the powerful person to reestablish control of the relationship.

This view resembles that offered by Daly and Wilson (1988), who conceive of homicide as behavior that grows out of particular competitive relationships among people, usually males. "[T]wo individuals will perceive themselves to be in conflict when the promotion of one's expected fitness entails the diminution of the other's" (Daly and Wilson, 1988: 293). Many such conflicts erupt between young males as they compete for women and status. The principle also supports the idea that violence can result when one party attempts to correct a perceived imbalance of control over another (Tittle, 1995).

Many people, of course, do not resort to violence to correct power imbalances in interpersonal relationships. Most acts of interpersonal violence appear to grow out of situation definitions that identify those acts as expected or required of offenders. Crimes of violence do occur more often in some groups and in some places than in others, suggesting a relationship between those acts and social characteristics.

Some criminologists have tried to explain this relationship by suggesting that subcultural patterns determine the frequency of violent crimes. They note variations by neighborhood, social class, occupation, sex, race, and age in the use of violence as a method of social interaction. Explanations of these variations describe *subcultures of violence*, or group normative systems that condone assaults and even homicides (see Table 7-1). Wolfgang and Ferracuti (1982) found differences in attitudes toward the use of violence between specific populations, for example, social classes and ethnic groups. These groups organize attitudes favorable to violence into sets of culturally transmitted norms. Such norms define expected conduct in specific situations, set the value of human life in the group's overall scale of values, and define shared perceptions and interpretations of situations.

TABLE 7-1
What Is Deviant Violence?

We begin to answer this question in the same manner as determining whether anything is deviant—with reference to norms. What are the norms in the following situations?

The Situation	Meaning 1 Nondeviant Interpretation	Meaning 2 Deviant Interpretation
A man deliberately and not in self-defense runs into another man he does not know and injures the second man so severely that the man is hospitalized for several weeks.	Football	Street fight
A man picks up a stick and proceeds to hit another man he does not know with the stick.	Hockey	Aggravated assault
A mother deliberately and not in self-defense strikes her daughter, causing pain.	Parent spanking own child for disciplinary reasons	Child abuse

Proponents of this explanation and the idea of a subculture of violence have based their conclusions on differences among groups in *rates* of violence. Aggregate figures do not, of course, imply that all members of a group share or act on the values that favor violent behavior.

After learning these violent norms, group members interpret some situations, such as disrespect for their honor or reputations, as occasions for violence. Sometimes failure to act violently may cause an immediate and perhaps irrevocable loss of status in a group, say, when a male declines to fight to defend his girlfriend's sexual reputation. Under such conditions, the subculture defines violence not only as appropriate, but as an action worthy of rewards. The loss of status from failure to act violently may even lead to banishment from the group. One assaulter described the circumstances of his assault this way:

> I can only take so much. If I tell a person to leave me alone, leave me alone right then and there. But he pushed me to a point where I could just take so much. He's making me feel less than a man, by not leaving me alone when I ask him to. When he pushed me, he moushed me in my face. I don't go for that—nobody touches me. (Oliver, 1994: 106)

Similar circumstances can lead to group violence. One gang member tells an observer about the process of challenges and reactions (Bing, 1991: 40–41):

> "See them two dudes?" Faro's voice, unaccountably, has dropped to a whisper. I nod my head. "I'm gonna look crazy at 'em. You watch what they do." He turns

away from me, and I lean forward over the wheel so that I can watch the faces on these two guys. . . . His eyes connect with Faro's, widen for an instant. Then he breaks the contact, looks down, looks away. And there is no mistaking what I saw there in his eyes: it was fear. . . .

I ask Faro what would have happened if the guy had looked crazy back.

"Then we woulda got into it."

"With me sitting here next to you? Are you kidding?" I can hear an edge of shrillness in my voice.

He laughs softly. "Never would have happened. That was just some damn preppy out on his lunch hour."

But if he had returned the challenge. What then?

"Then I woulda killed him."

The distribution of violent crimes suggests looking for the subculture in the inner-city regions of urban areas. The theory describes how some residents of these places, primarily lower-class young males from minority groups, come to subscribe to the subculture's violent norms or values. Some sociologists, however, doubt that subcultures of violence can explain patterns of homicide and assault. They cite significant regional differences in homicide, for example, as phenomena not easily attributable to a subculture of violence (Dixon and Lizotte, 1987).

Other studies offer more direct challenges to the theory. One looked for empirical evidence for the subcultural theory based on measures of self-esteem, violent behavior in the form of fighting, and esteem conferred by others for using violence. It found no convincing evidence of these elements of a subculture, casting doubt on the idea that lower-class groups actually reward violence (Erlanger, 1974). Another study found no expected value differences among people who reported differences in violent behavior at different times of their lives (Ball-Rokeach, 1973). A third study reported at best only partial support for the subcultural theory in data from adolescents (Hartnagel, 1980). Finally, a study of the general population found no support for the existence of a separate black subculture of violence (Cao, Adams, and Jensen, 1997). Despite these results, violent values may remain important, but only at certain times and under certain conditions; until further research identifies those times and conditions, the theory needs more evaluation (Erlanger, 1979).

At least one researcher has reported substantial support for the subculture of violence thesis. In interviews with men who had committed assaults, Oliver (1994) reported that most of the assaultive episodes resulted from a series of precipitating conditions he termed "autonomy transgressions." These included perceived threats in the form of name calling, insults, and unacceptable accounts, or unbelievable stories, from others. The importance of these factors was something shared among the assaultists and seemed to form the basis on which violent behavior could be initiated. While Oliver does not conclude that there is a distinctive subculture in which such values are learned, his findings are consistent with that idea.

In addition to subcultures, people can learn violence in other, less direct ways. Archer and Gartner (1984) have claimed that soldiers can learn to accept violence more easily after their war experiences than they did before fighting in combat. Phillips (1983) has found that homicide rates increase after widely publicized championship heavyweight fights. Phillips cites this relationship as support for the notion that violent acts can represent "imitative" effects of other forms of violence. These findings suggest that offenders may model some instances of criminal violence after other, more socially acceptable forms of violence.

Homicide rates vary in relation to certain features of cities, particularly degrees of income inequality. High income inequality can inspire hostility and frustration, as people resent their economic and social positions relative to those of others. Research has confirmed this idea by reporting a relationship between rates of interpersonal violence and degrees of racial income inequality in cities (Blau and Blau, 1982). Other tests of this idea have found support by comparing homicide rates and rates of income discrimination in a number of countries (Lee and Bankston, 1999). Empirical research also provides reason to believe that racial inequality affects homicide rates in cities (Messner and Golden, 1992).

These studies evaluate a structural theory, that is, one that attempts to explain the violence rates of aggregates rather than individuals. How do individuals act on these ideas? For one possibility, resentment of income inequality may lead, not to revolutionary behavior such as intentional violent acts against rich persons, but to opportunistic attacks on convenient targets. Instead of the real focus of his or her aggression, a violent person may harm others who are physically and emotionally close within a neighborhood, peer group, or family.

Another theory invokes an association between violence and conditions that generate social strain. This theory (Ogle, Maier-Katkin, and Bernard, 1995) has explored homicidal behavior among women. It has explained violent acts as results of strain generated by structural conditions that tend to place women in inferior jobs. Comparatively low status and salaries lead to dependence on men and, therefore, to resentment. Women also tend to internalize stress as guilt and hurt rather than releasing it by directing anger outward. This characteristic contributes to substantial social control over women and lower overall rates of deviance compared with those of men. These lower rates are punctuated, however, by occasional cases of extreme violence, particularly in long-term abusive relationships.

Development of Dangerous Violent Offenders

Along with violence tied to specific social situations and power differentials, some violent crimes result when certain people characteristically resort to violence in varied situations and circumstances. These people often have compiled extensive criminal records and lengthy histories of antisocial conduct in other areas, such as school maladjustment and family problems. One study of 50 such people concluded that they experience increasing acceptance of violence as a solution to many kinds of problems (for example, arguments or feeling pushed around) and thus come to affirm their self-images as violent people (Athens, 1992).

A person becomes a dangerous violent offender through a relatively unique series of four stages (see also Rhodes, 1999). In the first stage during their formative years, *brutalization*, these offenders experience physical coercion to submit to authority. They witness the brutalization of others and learn from others in their primary groups to accomplish objectives through violent means. Such experiences move these offenders to a stage of *belligerency*, in which they conclude that they must typically resort to violence in future relations with people. In the third stage, they set out on sequences of *violent performances*, in which they intentionally and gravely injure others. In the final stage, *virulency*, these violent performances convince others in the deviants' primary groups to see them as violent people instead of merely as people capable of violence. This reaction of others confers a sense of power on these offenders that reinforces their preference for violence. The development of virulency is illustrated by a quote from a teenager recently convicted of aggravated assault (Athens, 1992: 76–77):

> After the stabbing, my friends told me, "Hey man, we heard about what you did to Joe. It's all over school. Everybody's talking about it. You must really be one crazy ass motherfucker." My girlfriend said, "Wow, you stabbed that dude." Finally, things came together and hit right for me. My girlfriend and all my other friends were impressed with what I had done. I didn't really care what my parents thought. Everybody acted like nobody better piss me off any more unless they wanted to risk getting fucked up bad. People were plain scared to fuck with me. My reputation was now made.
>
> I was on cloud nine. I felt like I climbed the mountain and reached the top. I had proven to my friends and myself that I could really fuck somebody up. If something came up again, I knew I could hurt somebody bad. If I did it once, I could do it again. . . . I knew I could fuck somebody's world around, send them sideways, upside down and then six feet under. there was no doubt at all in my mind now that I was a bad son of a bitch, a crazy motherfucker. I could do anything, kill or murder somebody.
>
> Now that I had reached the top of the mountain, I was not coming down for anybody or anything. The real bad dudes who wouldn't associate with me before because they thought I was a nobody, now thought I was a somebody and accepted me as another crazy bad ass.

Some dangerous offenders may go on to perform additional violent acts and learn from the reactions of others to affirm their self-conceptions. As these people assume increasingly violent roles, they relish the reactions of others, who regard them as powerful and dangerous, perhaps even unpredictable, renegades. In this way, other people confer a sense of power that reinforces the offender's use of violence. It is possible that the offender will embark on a career in violence that will further escalate (Athens, 1997: chapter 8).

Domestic Violence

Domestic or family violence has recently gained increasing recognition as a serious social problem in the United States (Wallace, 1999). The family structure tends to hide patterns of assaultive behavior from official view, but increased sensitivity in recent years has encouraged a greater awareness of the problem and its pervasive effects in the American family structure. Any family member can become the victim of family violence, although children and wives suffer these assaults more often than fathers and husbands do.

Child Abuse

One general definition describes child abuse as "nonaccidental physical injury" (Helfer and Kempe, 1974), but such a broad conception may complicate focused analysis. Sociologists have trouble defining physical child abuse more precisely, however, since many parents discipline children through physical punishments. No one disputes the label *child abuse* in cases with injuries recognized as clearly excessive by any reasonably strict person. Children with broken bones, bruises, cuts, and burns would, by most people's definition, qualify as victims of abuse. People agree less easily on a definition of psychological or emotional child abuse, except in very extreme instances. Some also disagree about the abusive character of less serious injuries that might have resulted from more or less reasonable parental punishments. Without any universal standard for judging the best or most desired child-rearing practices, sociologists can apply no universal standard for judging cases of child abuse and neglect.

Yet, parents and policy makers can look for guidance to some established definitions. The official definition of child abuse in federal law, stated in the Federal Child Abuse Prevention and Treatment Act of 1974 (PL 93-237) considers child abuse as:

> the physical or mental injury, sexual abuse, negligent treatment, or maltreatment of a child under the age of 18 by a person who is responsible for the child's welfare under circumstances which would indicate that the child's health or welfare is harmed or threatened thereby. (quoted in Gelles, 1985: 351)

The clearest example of child abuse, child battering, results from physical assaults on children. Children suffer a range of injuries as a result of battering from relatively minor scratches and scrapes to life-taking injuries. Cases of child battering record virtually every kind of assault that an adult can suffer. In addition to child battering, child abuse also includes exploitation of children through pornography and sexual assault, malnutrition, educational neglect, medical neglect, and medical abuse.

No one can give a precise figure on how many children experience child abuse or assault. Clearly, many cases go undetected and unreported. Estimates state that more than 2 million children a year are subject to abusive behavior from family members or others in the United States (Gelles and Straus, 1979b). Further, one observer estimated that more than 300,000 children a year suffer sexual abuse (Sarafino, 1979). A national estimate figured that as many as 1 child in 100 may have experienced physical maltreatment, and even that startling number may underestimate the problem (Garbarino, 1989: 224). A Gallup poll estimated that 25 million Americans, or 15 percent of the adult population, suspected physical or sexual abuse of children they knew (Sagatun and Edwards, 1995: 4), and one observer estimates that 250,000 children annually become victims of sexual molesters (Lloyd, 1991).

Ongoing patterns of abuse complicate efforts to count such episodes, since battering and sexual abuse only rarely represent a one-time event. Such a victim frequently endures assaults many times during a single year. The continuing

---

✳ *Case Study*   **The Cycle of Violence**

Tattoos, faded from decades of wear, adorned Angelo's forearms. "Ink," he said, "tattoos from when I was a kid on the street." I'd seen hundreds of tattoos on the arms of active criminals hanging around street corners and on prison inmates filling cellblocks, but these tattoos seemed out of place poking out from under the turned-up cuffs of Angelo's long-sleeve dress shirt. Angelo, now a veteran motorcycle cop in southern California, was reporting eyewitness accounts about youngsters he's known whose parents are criminals, fathers like Maniac.

"We busted Maniac, me and my partner did," said Angelo with anger in his voice. "This mother . . . was a slime ball. We went to his house, a . . . shack. They live in these awful, filthy places, but every one of 'em got a shrine, a . . . altar to the Hell's Angels. It's the only clean spot in the house. They put their plaques on the walls neatly, sweep the floor, set out their trophies.

"I got inside; my partner was outside watching for any more of 'em comin' up. I cuffed Maniac, told him to put his fat, ugly . . . ass down in the living room. His ol' lady was there. Sleazy . . . !

"They had a kid, a son, maybe about three years old. The kid was dirty and had a [soiled] diaper. Looked like it wasn't changed in days. Kid had . . . dried on the back of his legs. Cute little guy, too. Ace, they called him.

"That . . . Maniac sat there and cursed and yelled at the kid, 'Come 'ere you little mother. . . . I'll kick your . . . ass if you don't get over here, now.' Ace stood there. . . . 'Get over here. . . . I'll kick your . . . ass. Come here, you . . . !'

"It was brutal. There was nothing we could do. The little guy walked over to that . . . , stood in front of him, looked up at him with his big eyes, and put his head on [his father's] leg, and stood there like that, didn't move.

"I knew the kid'd get the . . . knocked out of 'im. There was nothing we could do. Nothing!"

Source: Fleisher, Mark. 1995. *Beggars and Thieves: Lives of Urban Street Criminals.* Madison: University of Wisconsin Press, pp. 3–4.

pattern of assault most likely reflects the victim's relative powerlessness to terminate the abuse. Some observers express concern about apparent increases in the rate of sexual abuse of children, not only in the United States but in other countries as well (Finkelhor, 1982). They cannot say, however, whether the increase in reports of child sexual abuse during the past decade reflect increasing frequency of this crime or increased awareness that brings reports of offenses that earlier would have escaped attention.

Concern over child abuse extends beyond the immediate physical effects of this behavior on young victims. Some evidence suggests that people who experience physical abuse as children become increasingly likely to abuse their own children later in life (Straus, Gelles, and Steinmetz, 1980). Indeed, a number of studies of family violence have documented an intergenerational *cycle of violence* (Gelles, 1985). Nevertheless, a rather exhaustive review of the relevant research literature has found little evidence that "abuse leads [directly] to abuse" later in life (Widom, 1989). Early experiences of violence do appear, however, to constitute a risk factor in later violent behavior. Also, experience of abuse increases the odds of future criminality and delinquency by nearly 40 percent (Widom, 1992). These relationships represent an important research topic for the new century, and sociology needs more carefully conducted studies that examine the possibilities of intergenerational transmission of violent values.

Violence may come to define a normative pattern in some families, part of an accepted manner of child rearing. Even in relatively nonviolent homes, parents may regard violence as an acceptable last resort to obtain compliance from children. The effects of abuse may conceivably contribute to subsequent violence by the victim through some process, such as learning violent values, alienation from authority figures, or a weakened parent–child relationship. The violence that some children receive may actually establish a model for their own behavior later in life. Consider the following report from a young man who was beaten as a child:

> The beatings my stepfather laid on me, the terrible beatings he laid on my mother, and all the violent rhetoric took their toll on my mind. It inflamed me and made me want to go to bad. I was tired of always being messed with by people. I was ashamed of being weak and lame and letting people mess with me all the time. I didn't want to be messed with by people any more. People had messed with me long enough. If anybody ever messed with me again, I was going to go up against them. I was going to stop them from messing bad with me. If I had to, I would use a gun, knife, or anything. I didn't mess with other people, and I wasn't letting them mess with me any more. My days of being a chump who was too frightened and scared to hurt people for messing with him were over. (Athens, 1992: 60–61)

Research has identified a number of characteristics of home life associated with child abuse. Stress, low incomes, low levels of parental education, and family problems (such as divorce or emotional conflict) all show associations with child abuse (Gelles, 1985). Likewise, patterns of child discipline within neighborhoods and ethnic groups provide important influences. While child abuse can take place among all social classes, it appears—along with other forms of violence—to occur most frequently in lower-class homes. Also, abusing parents frequently live in social isolation from others, who might provide support, particularly in times of emotional crisis. Without close friends or isolated from family and neighbors, such a parent must handle family, economic, and social stress alone. Parent support groups develop in some communities in response to the awareness that some parents must cope without essential resources unless other parents who face similar problems provide that interpersonal support.

In related, family-centered crimes, some children suffer sexual abuse or incest. Fathers usually commit these crimes against their daughters, but mothers sometimes offend against sons as well (Finkelhor, 1984). Sexual abuse occurs when an older person initiates contact with a child to gain sexual stimulation. Two conditions distinguish child sexual abuse: the "abuser is older than the child and in a position of authority over the child" (Sagatun and Edwards, 1995: 21). Child sexual abuse involves behaviors such as exhibitionism, fondling of genitals, and mutual masturbation, as well as intercourse. Until recently, sociologists regarded incest as an extraordinarily rare phenomenon, but a recent renewal of interest in family violence has encouraged studies of incest which suggest that it constitutes a more prevalent situation than previously thought. Conflicting estimates probably do not reflect a spread of this behavior but, rather, greater attention and sensitivity to this problem.

Child sexual abuse is an international problem. A review of 24 surveys of people outside the United States found child sexual abuse rates comparable to those in the United States (Finklehor, 1994). All studies have estimated child sexual abuse rates in line with comparable North American research, ranging from 7 percent to 36 percent for women who have been abused and 3 percent to 29 percent for men. Most studies found that females experience abuse at $1\frac{1}{2}$ to 3 times the rate for males. While the surveys differ in many ways, their consistent results clearly indicate an international problem with child sexual abuse.

Spouse Abuse

Like estimates of child abuse, analysis of the extent of spouse abuse in this country must resolve problems with varying definitions of this form of violence. This discussion defines *spouse* to include someone related by marriage to the offender, as well as any unrelated adult living with the offender who performs spousal roles, such as unmarried adults who live together. In recent years, the term *violence among intimates* has been used increasingly. While early definitions of spouse abuse concentrated on damaging physical violence, conceptions broadened as the situation gained recognition as a national problem. Reports of spouse abuse may now include sexual abuse, marital rape, and even use of pornography (Gelles, 1985). The typical understanding of spouse abuse evokes images of husbands offending against their wives, but wives sometimes assault their husbands as well. A national incidence survey (Straus et al., 1980) found victimization of men in one-fourth of the homes where couple violence had occurred. In another one-fourth, women were victims but not offenders, and in one-half of the violent homes, both men and women committed offenses. In fact, in this survey, the percentage of husbands victimized exceeded the percentage of wives victimized.

Schwartz (1987) has analyzed National Crime Survey data and concluded that women strike men more often than men hit women, and men more often call the police after such attacks. These findings contradict the expectations of most people. However, about 95 percent of the victims of serious spouse abuse are women, because larger and stronger men commit more dangerous abuse. As a result, women face substantially higher risk than men of serious injury due to domestic violence. Thus, the real problem of spouse abuse results from physical victimization of women, not of men.

National surveys concerning violence toward women have revealed that more than 2.5 million women annually experience violence (Bachman, 1994), but this figure has been declining. Women are about equally likely to experience violence perpetrated by relatives or intimates, acquaintances, or strangers. Thus, nearly two-thirds of female victims of violence are related to or know their attackers. About one-fourth of attacks on females involve weapons; about one-third of these involve firearms. About three-fourths of the victims have reported resisting the offenders' actions through either physical or verbal reactions. About one-third suffered injury during these crimes. About one-half of the victims reported the crimes to the police. Among those who did not report the crimes, about 60 percent said they considered the attacks as private or personal matters or they described them as minor offenses. Nearly one-half of rape victims perceived influence of alcohol, other drugs, or both in the offenders' behavior.

Estimates from national victimization surveys suggest that rates of intimate partner violence have been declining. Between 1976 and 1998, the number of female victims of intimate violence declined by 1 percent, while the number of male victims of intimate violence declined by 4 percent. In 1998, women experienced about 900,000 violent offenses in the United States at the hands of an intimate, down from 1.1 million in 1993 (Rennison and Welchans, 2000). The number of men who were victims remained at 160,000 in both 1993 and 1998. The highest rates of violence were experienced by women ages 16 to 24.

Recognition of the seriousness of spouse abuse has increased throughout the world. In the United States, this concern has inspired the Federal Violence Against Women Act, which took effect in 1994. The act provides $1.6 billion over the following 6 years to fund improved training for police and prosecutors to help them deal effectively with these cases. Awareness of spouse abuse has grown in other cultures as well. One study targeted family violence among Chinese residents of Hong Kong, for example. It found that 75 percent of a sample of students reported verbal or symbolic aggression between their parents, and 14 percent reported instances of physical violence (Tang, 1994). In general, fathers engaged in more verbal aggression against their spouses than did mothers. However, mothers used physical force toward their spouses as often as fathers did. According to a national survey of violence against women in Canada, 3 in 10 women currently or previously married had experienced at least one incident of physical or sexual violence at the hands of their spouses (Rodgers, 1994).

Historically, some societies have not considered husbands deviant for beating their wives on occasion and within certain limits (stopping short, for example, of death or disfigurement). Prevalent ideologies permitted husbands wide latitude as heads of their households with authority to manage their wives' affairs, and this authority extended to physical actions. In some historical sense, husbands felt obligations to monitor their wives' behavior and exercise certain physical control over them. The so-called rule of thumb permitted husbands to beat their wives with sticks no thicker than their own thumbs. Society has come a long way, of course, from these historical roots to modern attitudes toward physical assaults on spouses, usually wives.

Research associates certain conditions with spouse abuse, such as consumption of alcohol (Secretary of Health and Human Services, 1990: 172–174), economic stress on husbands, and interpersonal conflict within families. The proximity of male family members may also affect violence. Baumgartner (1993) has argued that wives who lived shorter distances away from their extended families experienced less abuse. On the other hand, when the wife's kin live far away, the husband feels weaker deterrence from threats of retaliation.

Family violence can establish a behavior pattern in some homes. Spouse abuse, like other forms of violence, occurs largely, but not exclusively, among members of the lower class. Middle-class and upper-class families also experience spouse abuse, although they report fewer incidents. However, research has not determined clearly whether the variation on recognized spouse abuse by class reflects genuine differences in behavior or comparatively low visibility of spouse abuse in middle-class and upper-class homes. Clearly, spouses from these homes can choose alternatives not available to lower-class spouses. For example, a poor wife may well lack enough money to flee to a motel room for the night, forcing her to maintain physical contact with her abusing husband. Other differences—access to counselors, access to friends with resources, and awareness of community services—might also distinguish lower-class victims from those in the middle and upper classes.

Analyses of information from the federal Bureau of Justice Statistics and the Federal Bureau of Investigation on violence between people who have an intimate relationship—spouses, ex-spouses, boyfriends, girlfriends, and former boyfriends and girlfriends show that there has been a decline in various categories of violence. This decline parallels that for other crimes and is an encouraging development in spite of the high levels of violence that remain. For example, in 1996 just over 1,800 murders were attributable to intimates; nearly 3 out of 4 of these had a female victim. In 1976 there were nearly 3,000 victims of intimate murder (Greenfeld et al., 1998). The decline in lethal violence was greater for spouse killings, compared with the killings of other intimates. The number of female victims of intimate violence has been declining. In 1996 women experienced an estimated 840,000 rape, sexual assault, robbery, aggravated assault, and simple assault victimizations at the hands of an intimate, down from 1.1 million in 1993. Women aged 16–24 experience the highest per capita rate of intimate violence. Over the past two decades or so, intimate murder rates dropped far more rapidly among blacks than among whites.

Criminal justice officials handle spouse abuse cases as instances of assault or, if the situation warrants the more serious charge, aggravated assault. Offenders convicted in court of these charges may receive sentences of incarceration, probation, and monetary fines. Police have commonly tried informally to resolve domestic disputes without criminal charges, sometimes physically separating the conflicting parties for periods of time or referring them to social service agencies. Critics charge that these measures fail to protect victims and that such a response amounts to no response at all. As a result, some states have now implemented mandatory arrest laws, *requiring* police to make an arrest every time they respond to a domestic dispute.

One evaluation has suggested that such a law in Minnesota has deterred offenders from subsequent violations (Sherman and Berk, 1984). That result may have encouraged mandatory-arrest policies in other jurisdictions, but a replication of the same research methods in Nebraska failed to detect any deterrent effects from mandatory arrests (Dunford, Huizinga, and Elliott, 1990). The serious policy implications of this question have encouraged a number of other evaluations of mandatory-arrest laws (Berk, Campbell, Klap, and Western, 1992; Pate and Hamilton, 1992; Sherman and Smith, 1992). These reports seem to justify a belief that mandatory arrest may result in a short-term deterrent effect for some offenders, according to both official- and victim-reported measures. Mandatory arrest also shortens the time that victims of domestic abuse must wait for help, and it communicates to the community a definition of spouse abuse as unacceptable behavior. On the other hand, a mandatory-arrest law requires a substantial and long-term commitment of resources (Zorza and Woods, 1994).

In other cases, victims are able to seek legal protection from court restraining orders that prohibit contact between the batterer and victim. Whether such orders are effective or merely "a piece of paper" depends on the case. In one study the women who obtained the orders believed they were effective in reducing or stopping the violence (Ptacek, 1999: 171). Courts usually impose criminal sanctions on offenders convicted of spouse abuse. As with most assault cases, however, prosecution relies on the assistance of the victim. Officials can rarely complete successful prosecution of these cases without victim cooperation; this is so of most criminal cases. Unfortunately, prosecutors sometimes lose this assistance. A wife may resist prosecuting her husband, particularly if she fears retaliation or plans on reuniting with him. As a result, criminal justice officials often have trouble ensuring justice to all parties in spouse abuse cases.

Elder Abuse

In a problem similar to spouse abuse, elderly people sometimes experience abuse in nursing homes or while living with younger family members (Pillemer and Wolf, 1986). Research has so far provided no reliable estimates of the extent of this problem, although some evidence projects a growing problem as the number of elderly people increases (Pagelow, 1989). Victims sometimes withhold information about abuse because they fear retaliation or transfer to nursing homes. Even when reports reach officials, they often have difficulty proving abuse of an elderly person. Because aging skin bruises easily and elderly people fall rather often, evidence of some physical injury does not provide sufficient proof of abuse, and accused abusers can always propose alternative explanations for physical injuries. Consequently, many instances of elder abuse never reach the attention of authorities. Even when prosecutors can validate reports, the victims, especially if limited by senility or mental confusion associated with aging, may not provide credible testimony. Clearly, children and elderly people share many characteristics as potential victims for abuse, including their general lack of power over their own lives.

Forcible Rape

Forcible rape, another category of interpersonal violence, results when one person unlawfully compels another to engage in sexual intercourse against that victim's will. State laws distinguish this offense from statutory rape, or sexual intercourse with consent of a partner under the legally allowed age of consent, which varies from state to state (Posner and Silbaugh, 1996). Most rape statistics also include sexual assaults without actual intercourse or attempts to commit rape by force or threat of force. Those crime figures omit statutory rape, however, assuming that it involves no force or threat of force.

The Federal Bureau of Investigation (1999: 23) recorded 93,103 rapes in 1998, but criminologists warn against judging the actual incidence of the offense only from official statistics. As explained shortly, forcible rape is notoriously underreported to the police, so these official numbers underestimate the total number of rapes. Nevertheless, the figures do reveal an interesting change: The number of reported rapes in 1998 is the lowest since 1992.

Victimization surveys detect more rapes than reach the attention of the police. One study has revealed, for example, that only one-third of all rapes were reported to the police in 1994 (Perkins and Klaus, 1996). For comparison, consider the reporting percentages of aggravated assault and robbery: 52 percent and 55 percent, respectively.

Patterns of Forcible Rape

The threat of rape seriously concerns a large number of women. Many women recognize the possibly of sexual assault in many social situations, and this awareness leads them to take precautions at times when men would not think about crime prevention. College women, for example, often plan explicitly for times that call for precautions, such as walking on campus at night or from class to class, returning to a dormitory from the library, and attending a social event alone. Fear of rape generates a very real and powerful motivation in women's behavior (Warr, 1985), and this fear seems related to their fear of other crime (Ferraro, 1995). Although a number of situations create possibilities for rape, the threat of an attack by a stranger often elicits the most fear.

One such case illustrates this threat. Jack Doe (not his real name) was arrested in Spokane, Washington for several rapes in 1987 (Spokane *Spokesman-Review*, November 13, 1987: 1, 6). He admitted to raping at least eight females who ranged in age from 5 to 60 years old. This offender described his methods for raping victims in their homes: "It's like doing a burglary. There is nothing to it. All you have to do is go down a street, see a house, see a light on and go up to the house and look in." He prepared for a crime by keeping a periodic watch on the house, monitoring the comings and goings of the residents, and waiting to catch the intended victim home alone. When he was ready, the rapist would approach the house by bicycle, cut the phone lines, and enter the house, usually through a window. After the assault, he would leave the house and pedal away. He carried no weapon and used force only once, when a victim actively resisted. After the crime, the rapist would not immediately return home. Instead, he would ride around, sometimes for hours. Why did he do it? He describes his offenses as a cry for help. "The only way I thought I could get help was to do the rapes," he said. The man's previous record included instances of sexual abuse and considerable institutionalization in foster homes and detention centers.

No one should evaluate the case of Jack Doe as an example of a typical rapist, because no woman suffers an attack by a "typical" rapist. Different men may commit rape in many different situations. The Bureau of Justice Statistics has compiled information from rape victims' reports of the circumstances surrounding those crimes. The latest available victim survey results (from 1994) show that rape often involves an offender known to the victim, and such crimes often occur in places supposedly controlled by people known to the victims. Many rapes, for example, occurred either at or near the victims' homes (37 percent) or the homes of relatives, neighbors, or friends (21 percent) (Perkins and Klaus, 1996: 7). The rapes were almost equally distributed over the 24-hour day, with about one-third committed during the daytime, one-third in the evening, and one-third at night.

Rape victims frequently know their attackers—most casually and for short periods of time—but strangers commit some rapes. In 1994, offenders and victims knew each other in 67 percent of rapes, either as relatives, people well-known to each other, or casual acquaintances (Perkins and Klaus, 1996: 7). In the other 33 percent of rapes, offenders and victims were strangers to each other. Research offers little support for the notion that victims often seduce rapists, although some offenders make such claims to rationalize their behavior (Scully and Marolla, 1984). Young men (between 15 and 25 years of age) commit most rapes, and alcohol seems to play no major role in most of these crimes (Amir, 1971). In any case, statistical evidence characterizes most rapes as planned crimes rather than unexpected, explosive events (although some spontaneous rapes do occur). Most convicted rapists have no prior records for sex offenses, although a substantial number have had previous convictions for other crimes (Deming and Eppy, 1981).

Contrary to a common assumption, forcible rape resembles homicide as a predominantly *intraracial* crime, that is, a crime with both offender and victim of the same race (Randall and Rose, 1984). However, black offenders rape white victims more often than white offenders rape black victims.

Date Rape

One form of rape, *date rape*, has received increasing attention in recent years. The victims of these sexual assaults know their attackers, who usually rape them after dates or other social occasions that the couples attend. During the course of the evening, or even before, the male may have come to expect that the evening will include sexual relations, an expectation that the female does not share. Such a rape seldom involves extreme force, but the male attempts to fulfill his sexual expectations and fails to seriously consider resistance by his victim. The male may feel that he "deserves" sex because he paid for his victim's dinner, a movie, or some other entertainment. The male refuses to accept the female's refusal, sometimes interpreting her resistance as consent. ("She said no, but she really meant yes.") Many date rapes never result in police reports for the same reasons that other rapes go unreported: further embarrassment to the victim.

Estimates of the incidence of date rape remain elusive, in part because of the lack of agreement on a definition. Some observers have reported that more than 25 percent of high-school girls have experienced some kind of sexual assault, and nearly 10 percent of them report they have been raped (Maine, 2000: 152). Perhaps the most widely cited figures come from a 1989 survey, the Stanford Rape Education Project, which asked students whether they had experienced sex "when you did not want it because you were overwhelmed by continual arguments and pres-

✳ *Issue*   **Is Date Rape a Widespread Problem on College Campuses?**

Yes, and Maybe More

Andrea Parrot believes that date rape is a serious crime. "She [Ms. Roiphe] is doing a disservice to the movement that is trying to educate young people about date rape by making it look so rigid, so stupid, so out of touch with reality." Parrot has published a book with Carol Bohmer on this subject, *Sexual Assault on Campus* (Lexington, 1993).

Teresa Nichols, coordinator for women's programs and services at the University of Cincinnati, thinks that date rape is preventable: "In no way are we trying to create a population of victims. We are trying to empower women. We talk to them about communication, about their sexuality, and we try to give them options if they find themselves in a difficult situation. Most often we see freshmen who are experiencing things for the first time. They really want to fit in, and they put themselves in risky situations."

Not as Much as People Think

Katherine Roiphe is the author of a controversial book titled *The Morning After: Sex, Fear, and Feminism on Campus* (Boston: Little, Brown, 1993). She views data such as that generated by the Stanford Rape Education Project as suspect. Women, Roiphe claims, know when they have been raped. "We all agree that rape is a terrible thing, but we no longer agree on what rape is. Today's definition has stretched beyond bruises and knives, threats of death or violence to include emotional pressure and the influence of alcohol. The lines between rape and sex are beginning to blur."

Ms. Roiphe sees few passive women on campus who need help defining or dealing with rape. "Most women feel strong and capable of taking care of themselves. But the rhetoric you hear is that women don't have free will, that they are not in control. Who are these women who are so gullible, so naive that men can get them drunk? Where are these legions of passive, pathetic women?"

Source: Collison, Michele. 1993. "Article's Attack on 'Hype' Surrounding Date Rape Stirs Debate among Researchers, Campus Counselors." *The Chronicle of Higher Education*, July 7: A41.

sure." The investigators concluded that nearly 1 in 3 female students and 1 in 10 male students had been raped, almost always by someone they knew (Jacobs, 1990). Only 10 percent of the women and 25 percent of the men ever mentioned their victimization to anyone, much less reported it to officials. In fact, many of the women did not even regard these incidents as rape. Only 5 percent of the women indicated that they had been coerced into unwanted sex at any time in the past by some degree of physical force; a similar number blamed alcohol or other drugs.

The debate over the meaning of date rape began when a 3-year study commissioned by *Ms.* magazine appeared in 1985. In its most controversial finding, the study concluded that 1 in 4 women had been the victim of rape or attempted rape. The finding drew immediate attention and reaction, since the study described most of those incidents as date rape (see the summary in Hoff Sommers, 1994: 209–226). Critics examined the questions by which the study categorized behavior and found them wanting. But perhaps the most serious challenges to the study's conclusions cited the fact that only 27 percent of those it categorized as rape victims themselves believed that they had been raped. This means, of course, that 73 percent of the women said that their experiences did not constitute rape. The lack of agreement on a definition of date rape and corresponding disagreement on the seriousness of that act have so far impeded attempts to measure this behavior.

Male and Prison Rape

The conventional view depicts rape as a crime committed by men against women, but men can be raped as well. In either instance, the offender is usually male. Like rape against women, male rape often remains hidden from the police by a reluctance to report the crime. The only known data on sexual assault in adult jails comes from a study done in the late 1960s in the Philadelphia correctional system (Scacco, 1982). That study found that almost 4 percent of all males who passed through the facilities became victims of sexual assault. Extrapolating these figures for jails in the United States as a whole, 14,300 inmates incarcerated in 1994 were victims of sexual assault. The Philadelphia study characterized likely victims as young, physically small, nonviolent first-offenders, and the risk of sexual assault increased if the inmate was not "streetwise," was obviously homosexual, or was without a gang affiliation.

One study of adult prisons found that 22 percent of a sample of 452 inmates in Nebraska prisons reported that other inmates had pressured or forced them to have sexual contact (see also Knowles, 1999). One-third of the victims reported only one incident, while the others experienced more than one victimization. The victims were disproportionately white, and most were heterosexuals. All but 13 percent reported negative psychological consequences of these attacks, including depression, flashbacks, nightmares, and suicidal thoughts. Most of the victims did not report the crimes to prison authorities. Extrapolating these figures to the prison population of the United States, nearly 200,000 adult male inmates experienced sexual assault in 1994. If such figures accurately represent the true situation, more males than females become victims of rape in any given year.

It does appear that males who are raped are more reluctant to report the crime than are females (Pino and Meier, 1999). Male victims may feel that rape is especially humiliating and that they should have been able to control the situation. That they were victimized anyway may serve as a powerful attack on their sense of manhood and sexual identity.

The Political Context of Rape

Police statistics and victim surveys do not tell the whole story of rape. A full understanding of this offense, both as a crime and as a form of interpersonal violence, requires an awareness of the political dimensions of this crime. The traditional image of forcible rape described it as a sexual crime motivated by the offender's desire for sexual relations, but an adequate description must refer explicitly to the use of violence. The changing conception of rape as a crime of violence rather than a crime of sex has helped dramatically to reorient thinking about the offense and to challenge long-standing myths about it. Perhaps one of the most persistent myths claims that rape is impossible if the victim puts up sufficient resistance. In truth, however, rape resembles other predatory crimes, leaving the victim no choice but to submit under the threat or the application of physical force (Randall and Rose, 1984).

Within the past decade or so, people have come to recognize rape as a violent crime rather than a sexual one. The women's movement has effectively promoted this conception of rape as violence, leading to a number of legal reforms and alternative theories of rape that recognize its linkage with other violent crimes. Many observers regard forcible rape as a political act, because it reflects an exercise of power by one group (males) over another (females). Brownmiller (1975: 254) has said that "rape is to women what lynching is to blacks: the ultimate physical threat." Many conclude that rape represents an overt act of control that ensures the continued oppression of women and the perpetuation of a male-dominated society. While recognizing the violent origins of forcible rape, the offender's expectation of sexual gratification seems to require further explanation. This element distinguishes rape from ordinary assault or beating of women.

Another political consequence of rape has resulted because, critics say, the law adopts a "paternalistic" attitude toward women. Some observers claim that society has typically viewed females as weak creatures in need of protection and shelter from the harsh realities of life. They criticize rape laws for treating females as the property of men. Such an attitude would punish rape as a crime against men (husbands and fathers) as much as against women. However, one cannot attribute the development of rape laws solely to conceiving of women as property (Schwendinger and Schwendinger, 1981).

Forcible rape shares a number of characteristics with other forms of interpersonal violence, such as homicide, assault, and spouse abuse, including common traits of offenders and victims. These similarities suggest that the subculture of violence may contribute usefully to an explanation of rape as well as other crimes of violence. When young people learn sexual roles, an important part of this socialization covers expectations about the roles of the other sex. Rape statistics report offenders disproportionately from groups that share especially strong conceptions of females as objects of sexual gratification.

In addition, offenders often develop images of the male role that highlight occasional physical aggression and assertion of masculinity, perhaps through competitive sports or other displays of physical prowess; forcible rape may represent such an opportunity for male self-assertion. Such activities constitute important elements of the everyday lives of some males in inner-city neighborhoods and other lower-class areas. Members of the lower class often have difficulty constructing their identities in terms that make sense according to middle-class and upper-class values—materialism, occupational success, and social mobility. As a result, lower-class males may feel forced to develop their identities by emphasizing differences between themselves and women. Physical force and strength suit this purpose; these readily available traits represent a biological difference from women that conveys a sense of power (Hills, 1980: chapter 3).

Still, any understanding of rape must emphasize its identity as a crime of violence and force. The interactions between rapists and their victims, even in conversation, reveal the rapists' interest in manipulating and exercising power over their victims (Holmstrom and Burgess, 1990). An important step toward understanding rape requires an image of the offense as more than merely a convenient substitute for other forms of violence, such as assault or murder. Sex is an important component of rape, and the concept of such an offense makes little sense without a sexual element. On the other hand, some feminists have destructively characterized *all* sexual intercourse as a form of sexual assault (Dworkin, 1987), broadening the concept of rape to the point that it loses any meaning.

Rape Reporting

Surveys indicate that police never learn about many cases of forcible rape. Victims hesitate to file reports for several understandable reasons. (1) Rape is an emotionally upsetting and deeply humiliating experience for the victim. (2) A victim often encounters a strong stigma, even within her own family, although attitudes are now slowly changing. (3) Victims sometimes must deny implications that they consented to the sex acts by either resisting too weakly or by leading on their assailants. (4) Some victims who have reported rapes have then confronted officials in the criminal justice system who doubted their stories and treated them as parties to the crime. (5) Rape victims have faced embarrassing questioning by the police and prosecutors to verify the details of the crimes. (6) Courtroom treatment of rape victims has sometimes allowed very unethical public questioning about the victims' previous sexual experiences with insulting implications about supposed provocative circumstances contributing to the rapes and indecisive physical resistance.

Some evidence suggests that the number of officially recorded rapes has risen, in part, because of increased reporting. Until the mid-1990s, the FBI's *Uniform Crime Reports* indicated increases in both the number of rapes and the rate of commission for that crime (that is, numbers increasing faster than population changes). Victimization surveys of the general population, however, have found a very constant rate of forcible rape during that same period of time (Bureau of Justice Statistics, 1987). This result means that the increase in police rape statistics most likely reflects an increase in victims' willingness to report the crime rather than a jump in the number of crimes. Undoubtedly, encouragement by rape crisis centers and other, usually private, organizations has aided in this increased reporting.

Reports of rape reach the police for varied reasons. In 1985, female rape victims cited two principal reasons for filing reports: to stop or prevent a repetition of this kind of incident and because they identified reporting as the right thing to do (Perkins and Klaus, 1996: 4). Rape victims report the crimes for other reasons, as well, including the need for help after the rape and the desire to obtain evidence or proof. Victims often cited multiple reasons for reporting rapes. Reasons for withholding reports included the feeling that nothing would happen as a result of contacting the police. Male rape victims are less likely to report the crime than are female victims (Pino and Meier, 1999).

Legal Reforms for Rape Cases

A number of changes in the content of laws about rape and the criminal justice system's handling of these cases reflect public awareness of the special needs of such cases and their victims. Most jurisdictions have reexamined rape laws during the past two decades (Spohn and Horney, 1992), and some have introduced changes in their definitions of and official responses to rape. While the traditional legal conception limited rape charges to acts of penile–vaginal penetration, newer statutes set broader standards. For example, Michigan's law (Michigan Comp. Laws Ann. Section 750.520h) defines *sexual penetration* as intercourse, oral sex, anal intercourse, and "any intrusion, however slight, of any part of a person's body or any object into the genital or anal opening" (see also Posner and Silbaugh, 1996: 18–19). Other states have dropped the term *rape* and substituted *sexual assault*, a more general term defined in several degrees that reflect levels of injury or severity. Such reforms are largely unrelated to rape rates, although reforms in many states have expanded rape limits to include any nonconsensual sexual behavior, boosting the number of apprehensions for rape (Berger, Neuman, and Searles, 1994).

Consent reflects an important element of rape, and reviews of related statutes have also focused on standards for evidence of nonconsent. Earlier statutes required physical resistance, while newer ones have dropped that requirement, recognizing the probability of injuries resulting from such resistance.

Regardless of the legal relevance of consent, perceptions of victims' resistance have influenced the outcomes of rape cases in the criminal justice system, as have their actions being perceived as actually precipitating the crimes. A study of rape in Philadelphia described 1 in 5 forcible rapes as "victim-precipitated" crimes, in which the victims had agreed to sexual relations, at least in the judgment of the offenders, but then either retracted their consent before the acts or failed to resist with sufficient strength (Amir, 1971: 266–270). This result does not, of course, condone forcible rape when the victim somehow "invites" sexual intercourse and then refuses. Rape occurs whenever the victim must submit to sex acts without consenting, regardless of the surrounding circumstances. The idea of victim precipitation of rape lacks enough precision to adequately describe some aspects of the relationship between the victim and offender.

Further, other studies have not found the same role for victim precipitation. Curtis (1974), for example, compared degrees of victim precipitation in four crimes against the person: murder, aggravated assault, robbery, and forcible rape. The study found some provocation by the victims of many murders and aggravated assaults, less frequent but still noteworthy provocation by robbery victims, and the least effect of provocation in rapes. Greater explication of conditions of victim precipitation would have tremendous influence in reforms assigning some rapists' responsibility for their actions.

In other changes, many states have relaxed requirements for corroboration by others of rape victims' testimony, and some have defined victims' previous sexual history as irrelevant to courtroom proceedings. Still, credibility of testimony powerfully affects the outcomes of rape trials, and some prosecutors have warned that female jurors may doubt the claims of rape victims as a method of psychological self-protection (Wright, 1995). These jurors may feel reassurance of their own safety when they distinguish themselves from the victim and blame her for her victimization.

In perhaps the most controversial legal change, so-called *shield laws* have prohibited publication of rape victims' names. These laws seek to balance the victim's desire for privacy in view of a humiliating crime and the defendant's constitutional right to confront his accusers and gather evidence for his defense. When challenged, courts have upheld most shield laws (Spohn and Horney, 1992: 28–29).

The criminal justice system has made a number of changes in its procedures for handling rape cases, and these changes may have influenced the number of cases reported to the police. Police departments in most large cities, for example, set up special units to handle rape complaints. Some evidence indicates that specialists in these units tend to drop fewer cases as unfounded charges, which increases the number of cases recorded by the police even with the same frequency of rape reporting (Jensen and Karpos, 1993). In this sense, the police themselves can contribute to high official rape rates.

In a related issue, rape victims have sometimes complained that police grilled them or interrogated them harshly in efforts to substantiate charges of rape. Investigators seemed to presume that victims were lying. Needless to say, the process of police questioning could and did produce additional trauma for victims. Today, police often question rape victims under less confrontive circumstances. A woman officer might handle this part of the investigation, and treatment conveys an empathetic attitude.

Theories of Rape

A theory of rape is not the same as a catalog of motives for rape, although offender motives have a relevant place in a theory explaining offender behavior. Psychiatric and psychological approaches to rape stress causes like rapists' hidden aggression and the classification schemes with categories like power rapists (Groth, 1979) and sexual rapists (Cohen, Garofalo, Boucher, and Seghorn, 1975). Some element of aggression clearly contributes to rape, although theorists have not resolved the psychological meaning of sexual assault to rapists. Also, sexual motivation definitely contributes to most rapes (see Felson, 1993). Some theorists have attempted to combine notions of learned attitudes toward violence with situational inducements to explain incidents of rape (Gibson, Linden, and Johnson, 1980). These theories have not undergone systematic tests, however, and theorists will continue to debate until research supplies empirical verification.

Rape theories must resolve vital questions about the importance of changing sexual roles, the effects of family background, and expectations regarding sexual relationships. If rape serves as a means for men to control women (Brownmiller, 1975), then offense rates should remain low in situations with high ratios of men to women, because these conditions allow men to use their structural position to exercise that control. Preliminary evidence finds some support for this idea (O'Brien, 1991), although the theory requires clarification of the linkage between the intentional act of rape and unintentional structural male control.

Theorists also have not established a clear connection between using sexual means and other means to communicate aggression and control. Tests of such theories in offender populations raise questions, however, since incarcerated rapists may differ substantially from noninstitutionalized rapists. Thus, results from tests of the institutionalized sample may not generalize accurately to noninstitutionalized offenders (Deming and Eppy, 1981: 365–366).

Sociological theories advance several explanations for rape:

1. Rape represents an extension of legitimate violence in society.
2. Rape varies with the degree of gender inequality; increasing equality between the sexes reduces the likelihood of rape.
3. Rape results in part from depictions of women as sexual objects in pornography.
4. Rape results from value conflicts in the larger society.

A test of these theories found support for the last three, but it did not confirm the theory that rape represents some spillover from accepted or legitimate violence in society (Baron and Straus, 1989). Baron and Straus suggest an exceedingly complicated network of causes contributing to rape. Such complexity interferes with efforts to develop unambiguous interpretations of empirical tests. The relationship between rape and use of pornography, for example, might suggest that audiences for pornographic materials learn rape, but some third cause, such as a masculine ethic or culture, may contribute to both.

Another sociological theory of rape invokes the idea of a subculture of violence (Amir, 1971). This theory, introduced primarily to explain murder and assault, also suggests applications to rape. This connection becomes particularly evident in an examination of the characteristics of rapists. The general profile of a rape offender resembles that of a murderer and an assaulter: a young, black, lower-class male living in an inner-city area of a large metropolitan community. Wolfgang and Ferracuti (1982) identify the same group as participants in the subculture of violence. Those who subscribe to the subculture view women as sexual objects and possibly as accepted targets for other forms of aggression as well. As with other theories of rape, however, the subculture of violence theory requires more complete testing before sociologists will widely accept it. Further, many cases of rape and many rape offenders do not fit the profile implied by the subculture of violence view.

Society's Reaction to Crimes of Personal Violence

Society expresses an extremely severe reaction to murder, aggravated assault, and forcible rape by enacting strict laws. Legal penalties include lengthy prison terms and, under some circumstances, execution. As part of the strong societal reaction against these offenses, many people work to develop preventive measures intended to reduce these crimes.

Reactions to Murder

Most murderers display the weakest identification of all offenders with sociological standards for criminality. People who commit murder in the course of personal disputes do not conceive of themselves as criminals, and rarely are they recidivists. Such offenders do not engage in criminal careers, as defined in this book, nor do they progress to more serious criminal offenses. Their criminal careers usually terminate with their apprehension for murder.

Life imprisonment is society's most common legal sanction for murder. While capital punishment laws in a number of states provide for execution of some murderers, courts do not order this penalty extensively in the United States. As of 1994, more than 2,500 inmates lived on death rows in U.S. prisons. Officials executed only 38 people during 1993; at that rate, it would take more than 65 years to execute all current death row inmates, and that time will stretch as additional murderers receive death sentences. Most murder sentences deny eligibility for probation, and inmates convicted of this offense tend to spend more time in prison than other types of offenders. In large part, the severity of their punishment relates to the seriousness of the offense. While some murderers have records of assaultive behavior, not all do. As a result, many murder inmates never cause problems in correctional programs.

Statistical evidence shows a relatively constant rate of homicide over time, but this fact masks an increase in youth violence, particularly offenses associated with gang activities. This trend has led to calls for both neighborhood control of violence and effective gun-control measures. The precise techniques of neighborhood intervention should probably vary from community to community, and some observers have even suggested using gangs themselves in the process (Bursik and Grasmick, 1993). For example, in some communities, gangs sponsor local athletic leagues and events that bring together rival gangs to compete on playing fields and in gymnasiums rather than on inner-city streets. Gangs may well dismiss such activities, however, especially when they see opportunities to make money instead in illegal drug sales.

While such proposals often generate controversy in local communities, only national efforts can meet the need for gun control. The availability of automatic weapons and the financial appeal of the drug trade have increased the threat of well-armed gangs. Still, organized pressure from such groups as the National Rifle Association has so far hindered progress toward far-reaching gun-control legislation. In 1994, the U.S. Congress passed the Violent Crime Control and Law Enforcement Act, raising over $30 billion to assist local law enforcement and prevention programs. President Bill Clinton signed the bill into law in September 1994. Part of that legislation, the Brady bill, mandates a 5-day waiting period between application and actual purchase of a handgun. The bill also banned 19 types of assault weapons.

Effective programs to control violent crime probably also require close cooperation between the police and other criminal justice officials, especially probation and parole officers. Ultimately, however, local leaders, parents, school officials, and other neighborhood citizens must bear increasing responsibility for preventing murder. Local actions might include local gun-control ordinances and perhaps legalization of certain kinds of drugs.

Reactions to Assault

People who commit assault typically compile longer records of offenses than other criminals, and most of these offenses represent crimes against the person. Those who commit aggravated assault do not usually progress to homicide, however, although some aggravated assaults differ from homicides only because the victims live. Many violent acts risk killing the victims, and someone with a history of assaultive behavior may likely repeat this kind of crime, perhaps eventually murdering a victim. Because assaults can result in a range of injuries from minor to major, these offenders may experience punishments ranging from periods of probation and other alternatives to incarceration.

Society should focus on preventing violence, however, rather than reacting once it occurs. As for murder, the risk of assault suggests a need for more community efforts to control violent crime. These efforts begin with socialization of children to define violence as an unacceptable solution to interpersonal problems. Parents must work hard to counteract violent images children see in the mass media and in their everyday lives, especially those who live in inner-city areas. Community activities may promote this goal by helping to integrate younger and older residents through neighborhood festivals, sporting events, and other celebrations. School-based programs have successfully addressed bullying, suggesting an effective way for parents and schools to work together to combat this behavior (Farrington, 1993). For success, school personnel must show concern about this behavior and set a tone that does not tolerate bullying. Such an attitude may also determine the success of community control over other forms of violence. Effective programs to control minor interpersonal violence may continue to work when extrapolated, with modifications, throughout the community.

Reactions to Rape

Official statistics show an increase in reports of rapes to the police. Sociologists may debate whether this trend reflects an increase in actual rape behavior or a growing willingness to involve the police, but evidence does confirm improvement in the criminal justice system's definition and processing of these cases. This change, along with the development of rape crisis centers, may have increased reporting by victims of rape. Still, victimization survey figures disclose considerably more rapes than ever reach the attention of the police.

Rape cases have received much attention, both in law and in the criminal justice system. Considerable legal reform during the past two decades has revised the definitions of sexual assaults and procedures for handling such cases. Rape law reform has followed a pattern in most states of increasing penalties for a growing range of behaviors branded as violations.

Within the criminal justice system, handling of rape cases has improved through greater sensitivity to the potential to magnify the victim's ordeal through callous management. Police departments now most often assign women officers to obtain information from victims, and investigators establish atmospheres of concern and avoid the appearance of grilling victims. Concern over the sensitive nature of this crime continues as a case passes through the criminal justice system, from interviews with prosecuting attorneys to trials and testimony in criminal court. Despite growing concern over such matters in recent years, however, victims still endure a hard process that questions their credibility and publicly examines personal, intimate details of their lives.

Offenders convicted of forcible rape often face long prison sentences, and those who inflict serious physical injuries on their victims serve even longer terms. The full extent of such a crime's victimization, however, extends beyond physical injuries. Rape victims also experience considerable psychological injury. These considerations all justify a strong societal reaction against rapists.

In recent years, concern over other types of rape, particularly date rape, has spread. Clearly, widespread education on this kind of behavior would contribute to an effective response. Rape-prevention programs in high schools, colleges, and universities have attempted to educate students about the dimensions of rape, including date rape. Public gestures affirming the seriousness of this crime, including reminders of the severe penalties awaiting someone convicted of it, would also form important parts of such a program. Educational campaigns should also devote much effort to ensuring that everyone knows the definition and meaning of this kind of crime.

Summary

Crimes of violence inspired more fear among the general public than most other crimes. These offenses, including murder, aggravated assault, and forcible rape, involve attacks, or threatened attacks, on a victim's person. The legal system regards such acts as extremely serious crimes requiring severe responses.

Official police statistics have generally reported increases for crimes in these categories throughout the past decade, although victimization surveys suggest relatively stable rates of homicide and rape. Similar patterns of offending behavior, although not identical ones, characterize these crimes. Offenders likely come from young, lower-class, minority male populations of inner-city communities. Victims likely come from similar groups. Statistics depict murder and aggravated assault as typically unplanned crimes tied to particularly emotional circumstances and use of chemical substances such as alcohol and drugs.

In contrast, rapes more often result from planning and careful execution. More than other crimes, rape continues to undergo a social redefinition process. People now generally regard it as a crime of violence, but many people still react with confusion to some rape situations, such as date rape and spouse rape. Ambiguous standards for consent cloud some perceptions of particular incidents as instances of rape.

Family relationships may also lead to instances of interpersonal violence. Family violence has generated much national attention in recent years. Although these episodes can victimize husbands, most concern and policy efforts have worked to prevent wives, children, and elderly people from becoming victims.

Observers have arrived at no generally accepted theory of interpersonal violence, although the similar patterns in offending suggest the possibility that one general theory might explain murder, rape, instances of family violence, and assault. Sociological explanations, such as the subculture of violence theory, have emphasized that offenders learn violent values (leading to homicide and aggravated assault) or values that identify women as sexual objects and targets for domination. Most crimes of interpersonal violence result from unequal power relations among the participants, and offenders commit violence as an attempt to restore previous power relationships. In asymmetrical and unequal relationships, violence may help offenders to maintain or reestablish power relationships.

Violent offenders receive severe penalties from the criminal justice system. While police eventually learn of most instances of homicide, they never receive reports of some violent crimes such as forcible rape, since many victims hesitate to reveal these crimes. Increased awareness encouraged by the women's movement has promoted more reporting, thereby increasing official figures for incidences of rape and inducing more active official involvement from criminal justice officials.

Selected References

Blumstein, Alfred, and Joel Wallman, eds. 2000. *The Crime Drop in America*. New York: Cambridge University Press.

A collection of papers addressing possible reasons for the decline in violent crime throughout the 1990s. Such factors as gun availability, drugs, the use of imprisonment, changes in policing, and changing demographics are addressed.

Brownstein, Henry H. 2000. *The Social Reality of Violence and Violent Crime*. Boston: Allyn & Bacon.

Violence has a social reality that affects how we think about it. This book analyzes a number of violence topics, including violence among intimates, murder and assault, workplace violence, and the relationship between drugs and violence.

Brownmiller, Susan. 1975. *Against Our Will: Men, Women, and Rape.* New York: Simon and Schuster.

This book offers a feminist perspective on rape. Brownmiller conceives of rape as only one aspect—but perhaps the most violent one—of the "battle of the sexes." She discusses rape as a means by which men control women and maintain their own privileged positions.

Rhodes, Richard. 1999. *Why They Kill.* New York: Alfred A. Knopf.

Rhodes, an accomplished journalist, reviews a good deal of research on interpersonal violence both in the United States and elsewhere. He concludes that violence is learned initially and reinforced early in life in a process of "violentization" where violence comes to be an accepted means of solving problems.

Reiss, Albert J., Jr., and Jeffrey A. Roth, eds. 1993. *Understanding and Preventing Violence.* Washington, DC: National Academy Press.

This book presents the report of the Panel on the Understanding and Control of Violent Behavior of the National Research Council. It collects an extensive compendium of research and thinking about the nature and extent of violence.

Weiner, Neil Alan, Margaret A. Zahn, and Rita J. Sagi, eds. 1990. *Violence: Patterns, Causes and Public Policy.* New York: Harcourt Brace.

This useful collection gathers papers dealing with, among other topics, individual violence, rape, violence in the workplace, mass media effects, causes, and treatment/policy approaches.

Weiner, Neil Alan, and Marvin E. Wolfgang, eds. 1989. *Violent Crime, Violent Criminals and Pathways to Criminal Violence.* Beverly Hills, CA: Sage.

These two volumes assemble papers on a variety of aspects of violence, such as the relationship between violence and drug use, mental disorder, race, and gender.

CHAPTER 8

Nonviolent Crime

Credit card theft is a concern that prevents many people from purchasing products over the Internet. In early 2000 a hacker called Maxus broke into a large database of user credit cards at a site named CD Universe (Walker, 2000). Many thought it was the combination of a security fluke and brilliant code breaker. But when MSNBC, operating on a tip, tried the same thing with seven other e-commerce Web sites, it succeeded in obtaining a wide selection of personal data, including billing addresses, phone numbers, and, in some cases, employee Social Security numbers. Most disturbing was the ease with which MSNBC was able to obtain this information: It took just a few minutes per site. The network discovered that about 20 Web sites either had no password protection at all on their database servers—in each case, they were running Microsoft's SQL Server software—or had password information exposed on their Web site. Connecting to each site was as simple as starting SQL Server and opening a connection to the Web site. Security risks to the nation's computer networks are growing so fast that government and private industry are scrambling to address them. President Bill Clinton proposed $91 million in new federal spending to protect computer networks and create a Federal Cyber Service that would enlist college students in the anti-hacker wars. Attorney General Janet Reno also proposed a national anti-cybercrime network that would function around the clock.

Most people have personally experienced crime sometime in their lives. Many have also committed crimes in their lifetimes, probably minor ones like traffic violations, and many have learned from personal experience the unpleasant feelings of crime victims. Although the rates of most crimes have declined in recent years, in 1998 U.S. residents aged 12 or older experienced more than 31 million crimes according to the *Bureau of Justice Statistics*. Most of these victims suffered property offenses, and most of these crimes involved small dollar amounts of loss or damage. Still, by any indication, crime is a major form of deviance.

Criminality reflects diverse behavior by diverse people. Some criminal acts, such as stealing something with little value, produce only limited consequences in themselves, while other criminal acts, such as spying for a foreign government, can have enormous consequences for many people. Criminals differ in the extent to which they identify with crime and other criminals, the strength of their commitment to crime as a behavior, and the extent to which they progress in acquiring ever more sophisticated criminal norms and techniques.

An important theme of this chapter is an exploration of this diversity. It discusses a variety of criminal behavior systems, including those of occasional

---

✳ *Issue*   **Deviance on the 'Net**

The development of the Internet has broadened the opportunity for many people for communication, commerce, and education. It has also broadened the opportunities for new forms of deviance and a computer-based subculture. This subculture, or underground, is composed of a number of different kinds of computer users who can find gain and adventure behind their computer monitors.

There are four kinds of actors in this subculture:

1. *Hackers* are people who want to know more about computer programming and programs. They sometimes consider themselves to be wizards or computing wizards. Some hackers are interested in obtaining access to unauthorized computer systems and programs, but often hackers are merely computer enthusiasts who enjoy extending their knowledge about computer systems.
2. *Crackers* are computer users who break the security of computer systems to browse through information, to damage files on those systems, or to alter information in those files. Some crackers develop computer viruses to corrupt computer files.
3. *Phreakers* are computer users who find ways to use computer technology for their own gain by obtaining free telephone services without being billed. Phreakers attempt to crack the computer systems of telephone companies or steal phone-card numbers that they in turn use.
4. *Warez d00dz* (singular "warez d00d") are computer users who pirate software by copying and distributing unauthorized copyrighted material. "Warez" is an abbreviated form of the word used in the computer subculture for software. Once copied, the unauthorized software is then distributed via modem or, more usually, on a Web page. The greatest feat for a warez d00d is to emit 0-day warez, a term for pirated commercial software that is cracked on the first day of its release for retail sale.

Source: McCaghy, Charles H., Timothy A. Capron, and J. D. Jamieson. 2000. *Deviant Behavior: Crime, Conflict, and Interest Groups,* 5th ed., pp. 369–373. Boston: Allyn & Bacon.

property offenders, conventional criminals, political offenders, organized crime figures, and professional criminals.

Occasional Property Offenders

Many offenders compile relatively tame criminal records consisting of little more than infrequent property offenses, such as illegal auto joyriding, simple check forgery, misuse of credit cards, shoplifting, employee theft, or vandalism. Such crimes remain largely incidental to the way of life of an occasional offender, who does not make a living from crime or play a criminal role. This type of criminal behavior usually occurs when a situation creates favorable conditions; the offender often acts alone and seldom brings experience from prior criminal contacts

(Miethe and McCorkle, 1998). With some exceptions, such an offender acts with little group support. Such crimes usually demand few skills. For example, inadequate supervision of mass-displayed merchandise in stores presents almost limitless opportunities for shoplifting, so thieves need no training in sophisticated shoplifting techniques.

Occasional offenders do not conceive of themselves as criminals, and most rationalize their offenses and convince themselves that they have not committed criminal acts (Clinard, Quinney, and Wildeman, 1994). A shoplifter, for example, might justify the behavior by arguing that a large store can afford shoplifting; a joyrider may profess no intention to steal the car, only to "borrow" it. No evidence implies that occasional offenders make any effort to progress to types of crime requiring greater knowledge and skills. This section discusses only a few kinds of occasional crime: auto theft, check forgeries, shoplifting and employee theft, and vandalism.

Auto Theft

Joyriding is not a career offense. Strictly speaking, auto theft involves stealing a car with the intent to keep it or its parts; joyriding involves taking a car without the owner's permission but with no intent to keep it. Joyriding is mainly a crime of youth, and offenders usually commit only infrequent violations (Wattenberg and Balistrieri, 1952). They take a car, drive it for a time, and then abandon it. For many youthful offenders, this activity is often a part of an adventuresome evening, and the event is not often planned (Fleming, 1999). This activity involves no technique associated with the conventional auto-theft career, in which a criminal learns to strip stolen cars, select the right kinds of cars, and work through fences to sell the cars or their parts (Steffensmeier, 1986).

Not all auto theft results from joyriding, and not all offenders are young people. Some offenders steal automobiles not only for short-term transportation or joyriding, but for long-term transportation, to support commission of another crime, or to sell the car or its parts (Miethe and McCorkle, 1998: 170). Career auto thieves can dismantle cars quickly and sell the parts. Age and sophistication in stealing techniques help to differentiate these types of auto theft.

Check Forgeries

Estimates suggest that three-fourths of all check forgeries are committed by offenders with no previous pattern in such behavior. Lemert (1972) studied a sample of nonprofessional forgers and concluded that they generally do not come from areas of high delinquency, that they have clean criminal records, and that they have had no contact with delinquents and criminals. They do not conceive of themselves as criminals, nor do they typically progress to more serious forms of criminality. Lemert describes this offense as a product of certain difficult social situations, some social isolation, and a process of "closure" or "constriction of behavioral alternatives subjectively held as available to the forger" (Lemert, 1972: 139). While many people still pay by check for purchases, the declining popularity of checks reflects the increasing importance of credit-card payments for all types of purchases. Many shoppers prefer these cards because they may not require sufficient funds on deposit at banks, as checks do. When cards are stolen, however, their owners may pay for fraudulent transactions (Greenberg, 1982).

Shoplifting and Employee Theft

Shoplifting and employee theft are closely related crimes, sometimes combined into a broader category labeled *inventory shrinkage*, a term that denotes a business's loss of merchandise from illegal activities (such as shoplifting and employee theft) as well as honest, unintentional mistakes (such as bookkeeping errors). This blanket definition prevents any accurate estimate of the total amount lost annually by merchants to illegal activities (Meier, 1983). A check in March 2000 of the Internet Web site for Shoplifters Alternative estimates that annual loss from shoplifting alone is $10 billion and that 10 million shoplifters have been apprehended in the past five years.

The most extensive study of employee theft reported in criminology literature involved an examination of nearly 50 businesses in three metropolitan areas, including retail stores, general hospitals, and electronics manufacturers (Clark and Hollinger, 1983). In retail stores, the most common form of theft was abuse of employee discount privileges for purchases by others. Hospitals reported loss of medical supplies as their most common theft activity. In the manufacturing firms, employees most frequently stole raw materials. Generally, young, new, never-married employees were more likely than others to steal from their employers; these employees also expressed the most dissatisfaction with their jobs. These violations compound losses from other counterproductive but not illegal behavior of employees, such as taking excessively long lunch and coffee breaks, purposely slow or sloppy workmanship, and misuse of sick leave. Businesses and corporations are increasing attempts to reduce employee theft (Traub, 1997).

 In Brief **Anti-Shoplifting Statutes**

By the end of 1990, 39 states had enacted legislation to help prevent—not just control after it occurs—shoplifting. The legislation permits retailers to send "civil demand" letters to people accused of shoplifting. The letters ask the person to pay a penalty of $100 to $200 in addition to returning the merchandise. In exchange, the retailer agrees not to sue for civil damages.

One security officer credits the law with a reduction in shoplifting: "It's an extremely significant loss prevention issue. We saw that tacking on an additional $200 penalty impacted the parents' wallets."

The reduction in juvenile shoplifting alone may be huge. To date, there have been no systematic studies of the law's effectiveness.

Source: *Wall Street Journal*, October 15, 1990, p. 21.

Shoplifters come from all groups in society. Generally, however, most shoplifters appear to fit two categories: youths and "respectable," employed members of the middle class, even housewives (Cameron, 1964: 110; Klemke, 1992). Some of these offenders can afford to buy the things they steal, but they sometimes also take products for which they have no need. Rates of shoplifting are also high among drug users, especially heroin addicts, and homeless persons (Klemke, 1992). The variety of kinds of shoplifters undoubtedly reflects the relatively unsophisticated nature of this crime; a shoplifter does not require much training, and there are many opportunities wherever there are retail stores. Most shoplifters are motivated by the simple desire to obtain the item, but a substantial number of juvenile shoplifters appear to enjoy the risk of doing something forbidden (Cromwell, Parker, and Mobley, 1999).

Most nonprofessional shoplifters take small, inexpensive items. Current methods of large-scale merchandising in supermarkets, discount stores, and merchandise marts seem to invite shoplifting by leaving goods where customers can handle them without supervision. Researchers demonstrated the tiny risk of apprehension in a study in which they deliberately stole products, known to management but not to employees, from department and grocery stores (Blankenburg, 1976). Store workers detected less than 10 percent of all shoplifting activities; further, most customers were unwilling to report even the most flagrant cases they observed.

Occasional shoplifters, as opposed to professionals, generally do not conceive of themselves as criminals, and many cease their crimes after one apprehension. Most offenders rationalize their offenses by expressing mitigating beliefs: "The store can afford the loss," "The store owes me this item because I have bought so much here in the past," and so on. Employees rationalize theft from their employers in a similar manner (Robin, 1974). Research gives reason to believe that employer sanctions can diminish employee theft, although informal sanctions from fellow workers are more effective deterrents (Hollinger and Clark, 1982).

Vandalism

It is virtually impossible to estimate the amount of property lost through vandalism, although an estimate in 1975 reported costs of around $500 million for school vandalism alone (Bayh, 1975). Vandalism is almost exclusively a crime of juvenile offenders, although some young adults may commit this crime. Moreover, vandals worldwide target similar property: schools and their contents; public property like park equipment, road signs, and fountains; cars; vacant houses and other buildings; and public necessities such as toilets and telephones. Likewise, graffiti painted and scrawled on walls and public places require expensive removal measures.

Most vandals have no criminal orientation, conceiving of their acts more as "pranks" or "raising hell" (Wade, 1967). Often stealing nothing, a vandal's limited actions reinforce this self-conception as prankster, not delinquent. In spite of this conception, one study has identified peer relationships and adult–child conflict as the best predictors of vandalism among a sample of middle-class adolescents (Richards, 1979). Groups commit acts of vandalism, but not as expressions of any subculture. Acts of vandalism seldom apply or even require prior sophisticated knowledge; they grow out of collective interactions. Few are planned in advance; they represent essentially spontaneous behavior.

Society's Reaction to Occasional Offenders

Occasional offenders seldom experience severe reactions to their crimes unless they commit particularly large thefts or cause extremely serious damage. Occasional offenders frequently lack any prior offense records, and courts often dismiss their cases or sentence them to probation. Many occasional offenders, nonprofessional shoplifters and others, terminate law-breaking behavior upon apprehension (Cameron, 1964: 165). Some offenders continue to commit crimes, however, particularly in the absence of social controls, as in the case of employee thieves (Hollinger and Clark, 1982).

Also, virtually all career offenders begin with occasional property crimes. Still, the great majority of people who engage in occasional property crimes do not progress to subsequent criminality.

Because occasional offenders seldom establish commitments to delinquency and criminality, they often make good candidates for sanction programs that operate outside the justice system. In the Netherlands, for example, juvenile vandals might be required to repair or replace the property they damaged (Kruissink, 1990). This obligation might take the form of painting over graffiti, paying for repairs, or replacing property destroyed. In the United States, juveniles apprehended for occasional property offenses may receive assignments to diversion programs or probation.

Conventional Criminal Careers

Conventional criminal offenders fit the stereotype of a serious criminal shared by most people. Their careers progress from violence and theft in youth gangs to more serious and frequent adult criminal behavior, chiefly drug crimes and property offenses of burglary and robbery. This experience with crime does not, however, lead them to commit only one kind of offense; on the contrary, these offenders rarely specialize in this way, although property offenders tend to commit subsequent property offenses and violent offenders tend to commit subsequent violent offenses (Kempf, 1987). One study of "street kids" found little specialization in particular offenses. Inciardi, Horowitz, and Pottieger (1993) interviewed 611 youths in the greater Miami area, all of whom had committed a variety of at least 10 serious crimes during the 12 months prior to the interviews.

A large proportion of conventional offenders come from the inner-city areas of large cities. These rundown, congested neighborhoods provide homes for economically deprived residents. Physical conditions do not account for crime, though; rather, norms and values guide and justify offenses, spreading crime through transmission from generation to generation and from one ethnic or racial group to another. In their inner-city neighborhoods, potential criminals experience high incidences of crime committed by others, youth gang activity, drunkenness, illegal drug use, prostitution, illegitimacy, and dysfunctional family life. Residents frequently encounter criminal behavior patterns and role models.

Typically, conventional offenders begin their careers with membership in youth gangs. These groups represent predictable outcomes of family life featuring child neglect and abuse and inattentive supervision, perhaps because parents must work. Gangs may come to perform the social roles of families for some members, although these groups develop only weak social ties (Sanchez Jankowski, 1991). Gangs indoctrinate new members into the techniques of theft, selling illegal drugs, and robbery. They also furnish needed support and rationalizations for such behavior. Often, they weave criminal behavior into activities that seem exciting to novice members, featuring conflicts with other gangs and contests for status through acts of courage and bravery. Gangs often develop into organized entities with names, leaders, and longevity that may continue over generations (Klein, 1995). Gangs help to establish continuity between the criminal behavior of youth and that of later adulthood.

Conventional offenders continuously acquire increasingly sophisticated criminal techniques and develop rationalizations to explain their crimes as they move from petty offenses to more serious ones. As they progress in their careers, they have many contacts with official agencies, such as the police, courts, juvenile authorities, institutions, reformatories, and, finally, prisons. Their behavior continues to progress, but they never become as sophisticated as professional offenders, and their careers in crime usually terminate in middle age.

The career patterns of conventional offenders show clearly in a comparison of people convicted of armed robbery with a group of other property offenders (Roebuck and Cadwallader, 1961). As juvenile delinquents, the armed robbers frequently carried and used weapons, and their arrest histories averaged 18 arrests each. The armed robbers showed early patterns of stealing from their parents and schools that continued on the streets; they also engaged in truancy, street fighting, associations with older offenders, and membership in gangs. Compared with other offenders, the armed robbers also had more extensive records of previous acts of violence, basing their claims to leadership in gangs on superior strength and skill.

Studies of people arrested by police offer further insights into the careers of conventional criminals. One such study evaluated over 4,000 persons who faced robbery or burglary charges in Washington, D.C., between 1972 and 1975 (Williams and Lucianovic, 1979). These persons differed from other property offenders; younger, more often male, and more often black, they were also less likely to be employed than other defendants. The robbers had a median age of 22, compared to 24 for the burglars. Two-thirds of the offenders had been arrested previously as adults (researchers could not study confidential juvenile records), and most were recidivists.

Another study examined about 30,000 incarcerated property offenders (people imprisoned for robbery, burglary, or both) and analyzed the relationships between their crimes and their participation in conventional occupations (Holzman, 1982). This research discovered that about 80 percent of the offenders had held jobs prior to their offenses, suggesting that people who work in legitimate occupations, at least part-time, commit much repeated property crime. In interpreting these results, one must remember that institutionalized offenders do not represent all offenders. Institutionalized robbers and burglars may have approached the end of their criminal careers and the transition to conventional jobs and society.

In any case, sociologists may need to revise the stereotype of the conventional criminal as an unemployed person or, at least, as someone less likely than other offenders to work. In fact, participation in a conventional occupation can sometimes support crimes, as one burglar observes:

> I went in [the house] to work, to install cable . . . [and] the first thing I seen was this chandelier sittin' up in the living room and, like I say, I been doing burglaries and messin' with crystal and jewelry so much that I knew that was an expensive chandelier. The person that was lettin' me in to install the cable had about three rings on they finger . . . and I know the difference between [fake] and real diamonds. Fake stones and fake gold, I know the difference. . . . So that made up my mind [to commit the burglary] right then. When I put the cable in, I seen how easy it was to get in; they had a patio door with no security system on it. (Wright and Decker, 1996: 37)

Gibbons distinguishes between conventional and professional property offenders but suggests that they differ only in degree (Gibbons, 1965: 102–106). Professional, "heavy" criminals (that is, those who commit crimes with the threat and willingness to act violently) show more sophistication than semiprofessional property offenders. These relatively unskilled criminals make less money from their crimes than professionals make, and they work part-time at crime while maintaining some employment in legitimate occupations. While professional property offenders view their relationship to the police as a sort of occupational contest, the semiprofessional offenders express more negative attitudes. A study found that property offenders and addicts ranked the occupational prestige of a police officer lower than that of a cleaning person or a house painter (Matsueda, Gartner, Piliavin, and Polakowski, 1992). Semiprofessional offenders tend to develop lengthy records of arrests and imprisonment for their crimes.

Offenders with active criminal careers often accumulate evidence of other personal failure as well. In a study of 60 repeat offenders interviewed in Tennessee prisons, Tunnell (1992) characterized the subjects as underclass individuals with histories as losers in a number of life pursuits, including family, school, and occupation. Most of them excused their behavior as the result of alcohol or drug use, and few reported that they thought about the consequences of their acts. In other words, they did not mentally calculate the probabilities of apprehension and punishment. Clearly, this attitude makes an important contribution to continued criminal behavior and apprehension.

Self-Conception of Conventional Offenders

As conventional offenders continue to associate with other youths of similar backgrounds in juvenile gangs, and as they progress in their offense careers, they develop self-conceptions as criminals. Growing personal identification with crimes encourages progressively more concrete criminal self-concepts. Sporadic offenders tend to vacillate in their images of themselves as criminals, but such a self-conception becomes almost inescapable for regular offenders who live in continuous isolation from law-abiding society. The law also deals more and more severely with continuing offenses, further cementing their regard of themselves as criminals.

The development of a criminal self-conception can lead to a fatalistic attitude about crime and a cycle of defeatism, as illustrated in the comments of this property offender:

> It . . . gets to the point that you get into such a desperation. You're not work-ing, you can't work. You're drunk as hell, been that way 2 or 3 weeks. You're no good to yourself, and you're no good to anybody else. Self-esteem is gone [and you are] spiritually, mentally, physically, financially bankrupt. You ain't got noth-ing to lose. (Shover and Honaker, 1996: 19)

Offenders who maintain conventional occupations can more easily maintain noncriminal self-concepts, although these criminals must struggle to reconcile their legitimate and illegitimate incomes. By committing crimes as a form of moonlighting in addition to a regular job, an offender may well acquire a substan-tial record by her or his mid-20s. The study reported earlier found an average of four incarcerations—not convictions—for subjects with an average age of about 27 (Holzman, 1982: 1, 791). Any offender would have a hard time maintaining a noncriminal self-concept under those circumstances, although some moonlighters undoubtedly do so.

Society's Reaction to Conventional Offenders

Growing concern about conventional criminality has led some jurisdictions to in-crease criminal penalties. At the same time, some have experimented with alterna-tive modes of control.

Increasing Penalties   Many members of society believe that the community can best protect itself by imprisoning conventional offenders for long periods of time. The severe punishment of a lengthy prison term for a robbery or burglary commu-nicates the probability of a strong societal reaction to conventional offenders. In part, such penalties reflect society's desire to protect property and to punish harshly anyone who tries to obtain it by violence. They also reflect a difference between attitudes toward this type of lower-class crime and occupational or business crime. The criminal justice system often handles a conventional property offender in a way that compiles a long arrest record and a series of incarcerations. The risk of apprehension affects the decision-making processes for these offenders, as do the gains they anticipate from their crimes (Decker, Wright, and Logie, 1993).

The offenses that conventional criminals commit, often with little skill, may place them at considerable risk of apprehension and high risk of conviction and imprisonment. As a result, studies indicate strong chances of recidivism for conventional offenders (see Vera Institute of Justice, 1981). Many spend long periods of time in correctional institutions, and they carry heavy stigmas, since society generally disapproves of the fact of imprisonment as much as what the person did.

The United States imprisons more of its citizens than any other country in the world, including previous governments in South Africa, the former Soviet Union, and the People's Republic of China (Irwin and Austin, 1997). As of 1999, prisons confined more than 1.3 million inmates, and that figure does not reflect local jail populations or count past offenders subject to some other form of correctional supervision, such as probation or parole. Many prison systems are now exceeding their capacity, and there is no end in sight. U.S. prison populations have risen since the 1980s for a number of reasons: (1) Legislators have imposed mandatory prison sentences for increasing numbers of crimes upon conviction. (2) Laws have also increasingly set determinate sentencing requirements (for example, 3 years in prison) rather than indeterminate sentences (for example, 2 to 5 years in prison), in the process increasing the lengths of many prison terms. (3) Many jurisdictions have also increased the severity of penalties for many crimes, especially those subject to strong current concern (lately, drug crimes and violence). In fact, prisons now hold many offenders who do not represent direct threats to persons and property in order to satisfy mandatory penalties. For example, the law might prescribe a prison term for possession of a small amount of cocaine.

Some offenders' actions justify prison sentences, but such harsh steps cannot represent the first line of defense against crime. Less than 3 percent of all people arrested for felonies eventually end up in prison. Almost all prison inmates eventually return to the community and the same conditions that nurtured their offenses in the first place. Offenders can associate with other criminals in prison, perhaps becoming reacquainted with former members of the gangs they joined on the streets.

Boot Camps   Not all conventional offenders represent acts of career criminals. Courts increasingly sentence relatively young offenders who commit less serious crimes to nontraditional correctional programs such as *shock incarceration*. These programs, often called *boot camps*, plunge inmates into highly structured environments patterned after military training. Inmates wear only military-style uniforms and duplicate military base routines. Days begin early with physical exercise, and inmates spend most of the day occupied by some sort of physical labor. Such programs reserve evenings for education and counseling sessions. Inmates stay between 90 and 180 days, perhaps reforming their behavior through structure and self-discipline. The program seeks to shock offenders out of careers in crime via inflexible discipline, rigor, and order (MacKenzie and Souryal, 1991). Supporters claim that these elements challenge the inmates' chaotic and otherwise purposeless lives (see Inciardi, 1990: 595).

Evaluations suggest that such programs often lead inmates to develop some positive evaluations of themselves and their imprisonment terms. Still, boot camps have not reduced recidivism more effectively than either conventional incarceration programs or community supervision (e.g., MacKenzie and Shaw, 1993). And a number of states have reexamined their boot camp operations in recent years. The state of Maryland, for example, temporarily suspended its boot camps in 1999 after complaints of brutality by guards toward inmates. Nevertheless, it appears that properly supervised boot camp programs do provide correctional authorities with alternatives to regular prison sentences for certain kinds of offenders.

Political Criminal Offenders

Political crimes fall into two categories: crimes against the government and crimes by the government. Each type involves different offenses, and society reacts differently to each.

Crimes Against Governments

Governments may define many acts as crimes against the state, including attempts to protest, expressions of beliefs contrary to accepted standards, or attempts to alter current social and political structures. Perpetrators of these acts may act as agents of foreign governments or as citizens expressing deeply felt personal convictions. Examples of specific offenses in this category include treason, sedition, sabotage, assassination, hijacking, violation of draft laws, illegal civil rights protests, and actions based on conflicts between state-imposed duties and religious tenets. Most such acts reflect the offenders' desire to improve the world or their country's political system, and some regard such a political motivation as one of the main defining criteria of political crime (Minor, 1975).

Characteristics such as age, sex, ethnicity, and social class do not differentiate political offenders as a group from the general population. Despite this diversity, Turk (1982) has found some common characteristics of people who express dissent and push for government change. These activists more likely come from middle-class backgrounds than from other classes, and they express stronger political consciousness. Few political offenders conceive of themselves as criminals; in fact, many may contend that their actions violate no legitimate criminal laws. For example, draft resistors during the Vietnam war frequently saw their behavior as morally superior to that of the government that labeled and punished them as criminals.

Some political offenders perceive themselves as revolutionaries. They claim to pursue ideological goals rather than personal ones. A typical political offender expresses a commitment to some form of political and social order, usually not the current one. Schafer (1974: 146) has referred to the political offender as a *convictional criminal* "because he is convinced of the truth and justification of his own beliefs, and this conviction in him is strong enough to cause him to give up egoistic aspirations as well as peaceful efforts to attain his altruistic goals." Such offenders may well recognize that their actions violate legal prohibitions or requirements, although they may not feel bound by the laws they violate.

Groups that resist a political system, whether through dissent, evasion, disobedience, or violence, often recruit members from the more politically sensitive

---

✳ *Issue* | **Cyberhackers Strike a Political Blow**

The Associated Press reported on October 20, 1999, that the day after presidential candidate George W. Bush redesigned his campaign's Web site, hackers broke in and vandalized it by replacing his photograph with a hammer and sickle and calling for "a new October revolution." The hackers replaced a news story about Bush on his Web site with a note that "the success or failure of the working class to achieve victory depends upon a revolution (of) leadership."

The embarrassing security lapse came the day after the Bush campaign launched what it described as its "innovative new design" for its Internet site. The campaign's more sensitive computer operations such as its e-mail system and contribution records were protected on other machines and were not believed to have been compromised. A Bush spokesperson indicated that steps were taken to increase security on the site.

The Web site runs software from Microsoft Corporation, called Internet Information Server, which had suffered several serious security problems during 1998. Microsoft had distributed patches in each case but relied on local computer administrators to install them correctly.

---

groups in the community. Turk (1982: chapter 3) suggests this fact as a reason that representatives of both the upper and lower classes may come to resist the government and why upper-class people may be even more likely than others to become political resistors. Stereotypes about conventional or street criminals simply do not apply to political criminals. In a general sense, however, political offenders learn values that contribute to their crimes in the same way as many conventional offenders learn norms that permit or condone their criminal activities.

In the 1990s, distrust of government has led to the formation of underground militia groups, which actively express not only misgivings about the government, but physical opposition to it as well. Some groups, like the Freemen in Montana, believed that U.S. laws did not govern them. They claimed that they had, in fact, set up their own government. Other individuals not affiliated with anti-government groups may operate alone, committing acts of sabotage that symbolize their opposition to the government. In April 1995, such a person with sympathies for the positions of anti-government militias bombed the Muragh federal building in Oklahoma City. Timothy McVeigh was found guilty after a lengthy trial. Other acts the following year, such as the bombing at the Olympic games in Atlanta, suggested that such groups would continue to challenge the government.

Political criminals may receive group support from others who share their views. Such support may come from sympathetic political groups or simply from interested individuals. Sometimes these offenders also receive both social and material support from segments of society less strongly committed to overt social action. Such groups serve the same purpose as subcultures do for other kinds of deviants; they weaken the stigma imposed by the larger society, offering solidarity and forums in which to interact with others who share similar ideas. Political offenders also benefit from contacts in social networks committed to political resistance.

Other political criminals commit acts against the government for their own gain rather than to promote desired changes. An example is a spy under the control of a foreign government. Spy scandals have affected the CIA and the FBI in the 1990s; both agencies reacted with surprise to disclosures that longtime, trusted employees had supplied sensitive information in exchange for money. The motives of these criminals differ little from those of conventional offenders, whose activities serve their own self-interest rather than social change (Hagan, 1997).

Crimes by Governments

Governments have extensively violated their own laws in many countries over a long period of time. Just in the United States, corruption charges affected close advisers to Presidents Dwight Eisenhower and Lyndon Johnson, forcing some officials to resign. In the 1970s, violations by President Richard Nixon and a variety of his associates resulted in the imprisonment of 25 high-ranking officials, including the U.S. attorney general and two top presidential aides (Douglas, 1974; Jaworski, 1977). A pardon spared the former president himself from possible prosecution after he resigned. Violations of law during the Watergate scandal included obstruction of justice, conspiracy to obstruct justice, perjury, accepting contributions or bribes from business concerns, bribing individuals to prevent testimony, illegal tactics in election campaigns, and misuse for personal purposes of government agencies.

In the summer of 1987, during the presidency of Ronald Reagan, the nation witnessed televised hearings concerning similar White House efforts to sell arms to Iran to gain freedom for political hostages. The same high government officials then tried to funnel the profits from the Iranian arms sales to aid a guerrilla army opposing the government in Nicaragua. While the arms sales may have represented only a misjudgment (since no hostages were returned in exchange for the arms shipments), Congress had expressly forbidden Americans from funding the Nicaraguan rebels (called *Contras*) at the time.

Political scandals continued to swirl around the federal government in the 1990s. The long investigations that such allegations require have not yet supplied complete information on some of them. Perhaps the best-known of these scandals involved Bill and Hillary Rodham Clinton. Critics have accused the Clintons of trying to cover up illegal activities associated with a land development project called "Whitewater" while Bill Clinton was governor of Arkansas. Though highly publicized and the basis for an impeachment trial, subsequent misconduct by President Clinton involving Monica Lewinsky was more personal than political in nature. Also during the Clinton administration, former Secretary of Agriculture Ron Espy resigned over accusations that he accepted gratuities from a large chicken-processing company in his home state. During the presidential election of 1996, the Democratic party and some White House officials were accused of soliciting and accepting illegal campaign contributions from foreign nationals, in direct violation of national law.

Local and state governments may also commit crimes. Police officers may commit acts of misconduct and brutality as well as corruption, illegal use of force, harassment, illegal entry and seizures, and violating citizens' civil rights. The police departments in New York City and Los Angeles were involved in a number of acts of misconduct. In New York City in 1999, police officers beat and sodomized a suspect in a police station, while in Los Angeles officers were accused in the year 2000 of beating and framing suspects of crimes they did not commit. Other public officials may deliberately neglect duties or abuse their privileges. Agents of regulatory agencies, such as building inspectors, may permit contractors to build without necessary permits in exchange for bribes (Knapp Commission, 1977). Politicians may do illegal favors for those who make substantial campaign contributions. Police officials may suppress or manufacture evidence to convict accused criminals, and prosecutors may engage in conduct that is prohibited by procedural law.

In other countries, governments have extensively practiced imprisonment without due process, torture, and murder. Perpetrators of these acts include Uganda, South Africa, Chile, Cuba, Argentina, China, and El Salvador (see Ramirez Amaya, Amaya, Avilez, Ramirez, and Reyes, 1987).

Societal Reaction to Political Offenders

The strength of society's reaction to political offenses by groups and individuals against government depends on public acceptance of the government's authority. Public officials usually do not recognize the moral justifications of political offenses; rather, they react severely toward such offenders, who threaten the current political structure in which the officials participate. Official reaction may, therefore, involve severe sanctions.

Offenses committed by the government, on the other hand, seldom draw strong reactions, except in particularly flagrant abuses, like those associated with the Watergate scandal. However, a change in government may lead to dramatic sanctions for abuses by displaced officials. Governments more frequently punish individual acts of politicians and officials, but these sanctions do not match the severity of those imposed on conventional criminal offenders. The public reacts differently to political offenses and conventional offenders, undoubtedly because political figures wield substantial power and influence. Nevertheless, crimes by government result in progressively stronger social reactions, and participants in political corruption face increasingly powerful negative sanctions.

Organized Crime and Criminals

Members of organized criminal syndicates earn their livings from criminal activities such as controlling prostitution, selling pornography, making loans with usurious terms, running illegal gambling, selling illegal narcotics, racketeering, and reselling stolen goods. Some of the profits from these illegal operations fund legitimate business concerns (Ianni and Reuss-Ianni, 1983), and these relationships of illegitimate businesses to legitimate businesses complicate analysis and responses to organized crime. Additional complexity results from public demand for the services that crime organizations provide and the links between those organizations and local political structures.

Some observers describe a feudal structure for organized crime that organizes activities in "families" cooperating in a larger system known as the *Mafia* or *Cosa Nostra* (see Figure 8-1). Criminologists and crime control specialists commonly

FIGURE 8-1
An Organized Crime Family

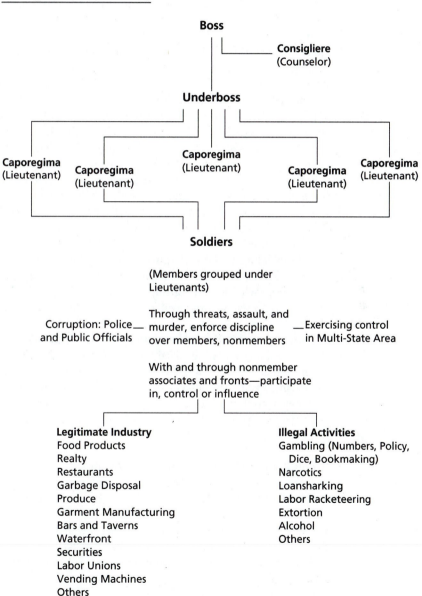

believe that U.S. criminal syndicates arose from earlier criminal structures, especially the Sicilian Mafia (Catanzaro, 1992). These Mafia groups evolved throughout the 20th century to dominate all syndicated crime. Investigations in the 1950s and 1960s identified about two dozen Cosa Nostra families operating in large cities under the direction of a national governing council (Cressey, 1969). This council designates operational territories and settles jurisdictional disputes. Only one such family, if any at all, operates in a typical city, coordinating the criminal activities of as few as 12 or as many as 700 persons, but reports detail five families operating in New York City. The wealthiest and most influential families are said to reside in New York, New Jersey, Illinois, Florida, Louisiana, Nevada, Michigan, and Rhode Island, and less prosperous syndicates operate in other places as well (Rowan, 1986).

Some experts question descriptions of such groups and organizations. They characterize the belief in a U.S. Mafia as the result of distortion driven by ulterior motives and sensational reporting (Smith, 1975). Other critics deny that anyone has identified a national organization of criminal syndicates (Morris and Hawkins, 1971; Reuter, 1983). Yet others interpret the weight of historical evidence as support for a scenario in which organized crime initially represented an alternative ladder of social mobility for immigrants, after which legitimate economic avenues opened up, inducing particular ethnic groups to participate less actively in organized crime (O'Kane, 1992). A resurgence of crime syndicate activities, particularly surrounding the drug trade, may reflect broad economic trends in the United States. As unskilled and semiskilled manufacturing jobs have declined in the U.S. economy, for example, newcomers have found fewer visible points of entry to the legitimate economic system.

Today, sociologists do not argue about whether criminal syndicates operate in the United States; no one now disputes the fact of their presence. Some disagree, however, about whether such groups coordinate their activities through a national organization. A review of the evidence suggests the following conclusion:

> Is there organized crime? Unquestionably. Are Italian-Americans and Sicilian-Americans involved in organized crime in the United States? Plenty of evidence suggests that these groups have been extensively involved. But are racketeers of Italian and Sicilian descent the only ones in organized crime? Clearly not. Are they as well organized, as bureaucratically structured, as nationally powerful as [some] believe? In our opinion, no, no, no. (Kenney and Finckenauer, 1995: 255)

To resolve conflicting conceptions of the organizational structure of syndicated crime (see Abadinsky, 1981), sociologists need working definitions that avoid reference to that structure. Rhodes (1984: 4) offers an illuminating definition:

> Organized crime consists of a series of illegal transactions between multiple offenders, some of whom employ specialized skills, over a continuous period of time, for purposes of economic advantage, and political power when necessary to gain economic advantage.

The most accurate image of the structure of organized crime might depict an organization that facilitates business dealings (which involve illicit activities) among members (Haller, 1990). The Rotary Club serves the economic interests of legitimate businesspeople by facilitating business contacts; illicit enterprises, like organized criminal syndicates, perform similar functions for some criminals. These syndicates also provide forums in which members can settle disputes that arise from these illicit dealings in smuggling bootleg liquor, illegal gambling, prostitution, or other enterprises.

In addition to facilitating illegal activities, organized syndicates also help to keep members out of legal entanglements. Connections with political machines or branches of the legal system, such as the police or courts, confer some immunity from arrest alternatives to minimize the harm that members suffer as a result of criminal charges. Syndicates maintain close relationships with members of conventional society through politicians, police officers, and civil servants. They maintain these links through direct payments or by delivering votes, through either honest or fraudulent means. One study of the relationship between an organized criminal syndicate and city officials has indicated that organized crime probably could not flourish without active support from those in legitimate positions of power in city government or public affairs (Chambliss, 1978).

Organized Criminals

The literature reports practically no firsthand research on the characteristics of organized criminals, and the large-scale organization of criminal syndicates obscures generalizations about members' backgrounds (Cohen, 1977). Some reports suggest, however, that most syndicate members came originally from inner-city areas, and most have compiled records of youth crimes (Anderson, 1965; Ianni, 1975). Many such histories resemble conventional criminal careers, featuring progressions through long series of delinquent acts and crimes as well as association with tough gangs of young offenders. Instead of ending their criminal careers as delinquents, however, these offenders continue similar activities in association with organized criminal syndicates (Miethe and McCorkle, 1998). The choice depends largely on opportunity, and not all youthful offenders who wish to join criminal organizations manage to do so.

Organized Criminal Activities

For the most part, seven areas of illicit enterprise predominate in the activities of crime organizations: (1) illegal gambling, (2) racketeering, (3) distributing illegal drugs, (4) usury or loan sharking, (5) illicit sex, (6) reselling stolen or hijacked goods, and (7) controlling legitimate businesses. While some syndicates may participate in other, often related illegal activities, these categories appear to dominate the interests of syndicates. One estimate has calculated an annual gross income from the activities of organized crime in excess of $50 billion, a figure that exceeds the combined sales of all U.S. steel, iron, copper, and aluminum manufacturers—amounting to about 1.1 percent of the U.S. gross national product

(GNP) (Rowan, 1986). Such an estimate is conservative for, as we shall see, others have placed the figure at a much greater cost.

At the close of the 20th century, organized crime groups had found new territory to mine: Medicare and Medicaid. According to the Government Accounting Office, criminal groups have formed scores of "medical entities"—phony labs, clinics, doctor groups, equipment suppliers, and diagnostic laboratories—in order to bilk the government health programs of hundreds of millions of dollars (Omaha World Herald, November 5, 1999: 7). The companies exist only on paper.

These criminal groups are quite transitory and should not be confused with traditional organized crime families. Members from the criminal groups would recruit Medicare beneficiaries, who would be sent to clinics and offices for unnecessary tests or services. The recruiters would receive a fee for each patient they found, and the fees would be shared with the patients. Medicare was then billed for services and equipment never provided. In some instances licensed physicians participated in the fraud by exchanging their signatures on medical records for cash without actually providing or overseeing any medical procedures. In some instances, the patients actually had died prior to the date when the medical service was supposedly provided.

As mentioned earlier, traditional organized crime activities include illegal gambling, racketeering, illegal drug activity, usury, illicit sex businesses, the resale of stolen or hijacked goods, and the infiltration of legitimate businesses. Following are some of the ways in which those activities work.

Illegal Gambling   The lucrative returns from illegal gambling attract organized crime to this activity. In most cities, illegal bookmakers associated with crime syndicates take bets on sporting events, such as horse races and football, basketball, and baseball games. At one time, nationwide betting organizations concentrated on professional events, but college football and basketball have also drawn a great deal of attention.

"Numbers" games offer another very popular form of gambling, especially in East Coast cities. A player places a bet on some random sequence of three numbers, such as the last three digits of the daily U.S. Treasury balance (Light, 1977). This game works through a complicated organization to secure the bets, record them, and pay winners. Numbers runners (offenders who record the bets and collect money) are usually small-time criminals within larger syndicates. They generally keep between 15 and 25 percent of all bets they collect, although this amount may rise to between one-third and one-half of the losses of a new customer, and a customer who wins pays the runner a 10 percent tip (Plate, 1975: 75). The success of numbers games has attracted competition from state governments, who have set up legal lotteries that appropriate the basic idea of numbers gambling.

Racketeering   Criminal syndicates draw substantial income from racketeering, or programs of systematic extortion that demand money from individuals and organizations in forced purchases of services, loans, or simply permission to continue controlling their own activities (Block and Chambliss, 1981: chapter 4). Federal law defined racketeering as a crime in 1970 as part of the Organized Crime Control Act. The *Racketeer Influenced and Corrupt Organization* (RICO) section of that act is the major federal legislation on this crime. The RICO section prohibits actions that form a pattern of racketeering activity, which it defines as two or more such felonies committed within a 10-year period (Albanese, 1996: 50).

Most racketeering targets organizations that deal with services or commodities, such as wholesalers of perishable goods and laundry and cleaning businesses. Criminals force legitimate business concerns to pay tributes to buy "protection" against violence and damage. Businesses that deal extensively in cash (such as vending machine and pinball and video game operators) are favorite targets because investigators cannot easily trace the extortion payments. Racketeering has particularly affected the motion-picture industry, the building trades, liquor stores, laundry and cleaning establishments, and freight handling and trucking businesses (Ianni and Reuss-Ianni, 1983).

Illegal Drugs   Organized crime plays a major role in importing and distributing illegal drugs. Chapter 10 discusses this activity in more detail.

Usury   Usurious lending, or loan sharking, is another important source of organized crime profits. Such lenders offer money at interest rates far above the legal limits to borrowers in desperate need of cash but lacking the required collateral and financial reputations to secure it through legitimate sources. While banks and other legitimate lending institutions calculate interest annually, a loan shark may compound accrued interest on a monthly, weekly, or even daily basis. Nonpayment usually draws both physical and financial penalties; loan sharks actively try to collect what they believe borrowers owe them, and they also intend to make public examples of people who do not repay their loans.

Illicit Sex   Organized crime groups control prostitutes, operate nude bars, and distribute pornographic literature and films. Recent concern over pornography has attracted public attention to questions about the extent to which criminal syndicates control the production and distribution of pornographic materials (Attorney General's Commission on Pornography, 1986; Reuter, 1983: 95–96).

Reselling Stolen or Hijacked Goods   Organized criminals have engaged extensively in the sale of stolen valuables, often through interstate commerce. Sometimes these groups organize the actual thefts as well, and sometimes they buy the goods from others and resell them at a profit. Some of these goods are stolen from airports and similar loading or storage facilities. More recently, crime syndicates have begun to trade in stolen credit cards (Tyler, 1981).

The Infiltration of Legitimate Business   Besides controlling illegal activities, organized crime has also infiltrated legitimate businesses (Jacobs and Gouldin, 1999). These criminals have gained control through illegal threats and actions or by investing large sums of money from illegitimate sources. Two accounts have identified close relationships between legitimate sources of income and illegitimate sources for criminal syndicates. Ianni (1972) has reported that the "Lupollo" family (the fictitious name for the crime family he studied) operated several legitimate businesses, including a realty company, food-processing companies, an ice and coal delivery business, and garbage-disposal enterprises. Anderson (1979) has reported involvement by the "Benguerra" family (another fictitious name for a real crime family) in 144 legitimate businesses, including eating and drinking establishments, retail trade businesses, firms that manufactured and processed food, construction and building services, casinos and travel services, and vending machine operators. While organized criminals may participate in legitimate businesses to hide their illegal activities, these firms also provide major sources of revenue, some of which may finance illegal operations.

Societal Reaction to Organized Crime

Although society reacts strongly against organized crime, research has revealed a good deal of public ambivalence about these groups (Morash, 1984). Many outsiders express fascination with the Mafia and its purported power. Organized crime has inspired relatively romantic treatments in several films (from *Scarface* in 1931 to *Casino* in 1996), and other media also often portray organized criminals as sympathetic and even kindly people. In any case, most observers feel that strong legal measures have so far failed to exert much control over organized crime. Many reasons may explain this failure.

First, the very nature of organized crime creates a major enforcement problem. An illegitimate business usually encompasses different types of individual crimes; the "organization" comes from interactions between criminals rather than from the crimes themselves:

> It is not against the criminal law for an individual or group of individuals rationally to plan, establish, develop, and administer an organization designed for the perpetration of crime. Neither is it against the law for a person to participate in such an organization. What is against the law is bet taking, usury, smuggling, and selling narcotics and untaxed liquor, extortion, murder, and conspiracy to commit these and other specific crimes. (Cressey, 1969: 299; also see Homer, 1974: 139–168)

Another problem results from public appetites for many of the services provided only by criminal syndicates, such as gambling and usury. Also, criminal investigators struggle to obtain proof of criminal organizations' activities, particularly after these groups intimidate witnesses. Further, the organizations often corrupt public officials to prevent prosecution. Even willing enforcement suffers from a lack of resources to deal effectively with potentially nationwide operations of some crime syndicates. Poor coordination hampers the efforts of local, state, and even federal organized crime agencies. Ineffective application inhibits the effectiveness of sanctions that are available, since prison sentences seldom relate to the enormous financial rewards that crime organizations reap from their illegal activities. Finally, certain slum dwellers regard organized criminals as success stories and even heroes, so enforcement efforts often lack community support (Ianni and Reuss-Ianni, 1983).

Unfortunately, organized crime has become an integral part of U.S. society. These groups have firmly planted their priorities in the values and desires of large segments of the population, offering goods and services demanded by many. People rely on organized criminals for gambling, illicit sex, otherwise unavailable loans, illegal drugs, and, during the time of Prohibition, alcohol. Organized crime could not operate without public demand for these goods and services, although it must sometimes compete with legal alternatives to satisfy some desires, such as gambling trips to Las Vegas or Atlantic City. Even with the expansion of casino gambling on federal land in many states, bettors still support a strong demand for illegal gambling, particularly sports betting. As they work to satisfy current demands, criminal syndicates also attempt to ensure continued high demand by promoting their services.

Legal reforms offer the most logical method for controlling organized criminal activities, in the form of far-reaching new statutes that set more severe penalties for conviction. Indeed, some observers assert that such legal strategies have seriously harmed traditional Italian and Sicilian organized crime groups and that their virtual survival is at stake (Jacobs and Gouldin, 1999). Other solutions include legal changes that decriminalize some activities of organized crime. If government were to legalize gambling and presently prohibited drugs, it might anticipate two possible consequences. First, organized criminals would lose lucrative opportunities in these areas, squeezed out by competition with legitimate businesses to supply these services; this change would weaken crime syndicates, depriving them of important sources of revenue. Second, the state could benefit by collecting some of the revenue that would otherwise flow to organized crime groups. A number of states now offer some forms of gambling themselves in the form of lotteries, although some wonder whether these games have created much competition for the illegal gambling operations of crime groups. Also, legalization of gambling might not eliminate the incomes of members of criminal syndicates, who might well shift to other criminal activities, just as they turned to gambling after the repeal of Prohibition decriminalized trade in alcohol.

In any case, initiatives to deal with organized crime seem to offer more fruitful alternatives than programs targeted at organized criminals. Police apprehension of any individual offender does little to alter the basic structure of public demands for currently illegal goods and services. The enormous demand for such goods and services would attract someone else to provide them. One estimate projected over $100 billion in potential revenue in 1974 from the "hidden economy" (a term that denotes economic exchanges, such as trades for drugs and sex, not recorded through legitimate business reporting systems based on taxes and other normal accounting routines). This figure requires dramatic adjustment upward to account for inflation since that time (Simon and Witte, 1982: xiv). Organized crime groups control most of these exchanges, and no one has developed a way to alter consumer preferences for illegal goods.

Professional Offenders

Of all criminal deviants, professional offenders develop the most extensive criminal careers, the highest social status among criminals, and the most effective skills (Roebuck, 1983). Among all forms of deviance, sociological concepts may provide the best understanding of this behavior. This criminal elite has probably never numbered very many, and research indicates that their numbers have declined enough to refer to them as "old-fashioned criminals" (Cressey, 1972: 45). Various preventive devices have created effective obstacles to professional forgery, for example, possibly contributing to an overall decline in this and other types of professional crime. The term *professional criminal* does not necessarily refer to someone who commits many crimes or someone who makes a living, altogether or in part, from crime. Such broad conception would have little meaning.

Professional offenders distinguish themselves by sharp criminal skills, since they apply relatively elaborate sets of techniques in committing their crimes. They also enjoy high status and admiration in the world of crime. Professional criminals share substantial consensus about values, attitudes, and beliefs with other professional criminals. Their lives revolve around criminal associations, as they interact primarily with other criminals. Finally, professional criminals act within established organizations, frequently committing their crimes as members of larger groups from which they also derive social support and assistance (Sutherland, 1937).

Skill

Professional criminals as a group develop advance abilities to commit a variety of highly specialized crimes. The type of crime a criminal commits does not necessarily class her or him as either a professional or an amateur. Both amateurs and professionals can commit such crimes as picking pockets, shoplifting, and forgery. Instead, the skills that offenders apply to their crimes usually separate professionals from others. Professional offenders acquire substantial skills in committing particular crimes, such as pickpocketing, shoplifting, confidence games, stealing from hotel rooms, passing forged checks and securities, counterfeiting, and various forms of sneak thievery from offices, stores, and banks. The specialized skills of a professional check forger differ substantially from the unrefined techniques of the amateur or naive forger (Inciardi, 1975; Lemert, 1958).

Professional pickpockets, or *class cannons* as they call themselves, ply their trade at airports, race tracks, amusement parks, and other areas frequented by tourists (Maurer, 1964). They work in "mobs" of from three or four up to 10, each member playing a specific role. They select a "mark" (victim) on the basis of guesses, as well as clues from dress and demeanor, about how much money the person may carry. The actual snatch happens quickly, sometimes following a jostle or bump, and the cannon quickly passes the loot to another member of the mob moving in the opposite direction. This team member, in turn, passes the loot quickly to another, and such handoffs may continue until the mob leaves the area. Although the amounts taken by pickpockets vary, estimates from their own reports describe common average incomes of six figures a year for class cannons; of course, this amount represents untaxed income. A hardworking mob may victimize 10 to 15 marks a day with an average take of $50 to $100 per crime. When victims detect the thefts, and they do so only infrequently, pickpockets frequently talk their way out of trouble by offering to return the money, often with a bonus (Inciardi, 1984).

A professional shoplifter steals merchandise for resale rather than for personal use, the sign of an amateur. Professional shoplifters usually work in small groups, often touring the country and staying in one place only long enough to "clout" (shoplift) and dispose of some merchandise. Professionals apply various skillful techniques, usually concealing merchandise in devices such as booster bags (purses or other bags with false bottoms), specially prepared coats, and boxes like those supplied with purchases from other stores (Cameron, 1964: 42–50; Meier, 1983).

Some professional criminals run confidence games, divided into "short cons" and "big cons." In the former, an offender illegally obtains money from an individual victim in a brief, direct encounter such as a sale of false jewelry (Gasser, 1963; Roebuck and Johnson, 1963). Some short cons set up fixed gambling events, such as card or dice games (Prus and Sharper, 1979). A big con requires more time and targets a larger sum of money by "putting the mark on the send" (inducing the victim to go somewhere—a bank or place of business—to get the money). Most confidence games benefit from victims' self-concern by inducing them to agree to violate the law themselves for money, either by accepting illegal propositions or by engaging in crime to raise the money for the confidence game. Clearly, the confidence swindler must be extremely persausive to influence a mark to come forth with extra money. A special kind of confidence criminal, the professional hustler, makes a living betting—often with support from a backer—against opponents in various types of pool or billiard games. The hustler initially hides well-developed skills to induce the victim to place large bets. His conning involves an "extraordinary manipulation of other people's impressions of reality and especially of one's self, creating 'false impressions'" (Polsky, 1964: 14; see also Hayno, 1977; Walker, 1981).

A type of swindle perhaps more known to the public is telephone scams. Run by "boilers," these scams attempt to sell fictitious products or services by calling unsuspecting persons (Stevenson, 1998). The setting for the swindle is a "boiler room" where several salespeople (swindlers) attempt to convince those they have called to make a purchase based on a smooth sales pitch. It is critical for the boiler to construct and maintain an air of respectability and trustworthiness. Most boilers come to refine their skills through the use of tested, highly manipulative techniques to overcome resistance or objections to sales.

Status

The high status of professional criminals is reflected in the attitudes of other criminals and by the special treatment accorded them by police, court officials, and others. Nonprofessionals tend to admire the professionals and aspire to their status. Desire alone, however, does not achieve the position. To become a professional thief, one must encounter an opportunity to learn the techniques of the specific crime and develop the attitudes needed to apply them. The status of the professional criminal results partly from their small number; as with gold, rarity creates value.

Criminal Associations

As in other professions, criminals become professionals through training in their occupations. Practicing professionals teach newcomers techniques that offenders have applied for generations to commit crimes (Inciardi, 1975: 5–13; Inciardi, 1984). Recruiters may draw a promising amateur into professional crime in his or her 20s, or they may target someone who shows some specialized skill, as when professional counterfeiters recruit an engraver. Initially, the developing criminal will take on small jobs to improve his or her skills. During this apprenticeship, the novice progresses in skills and criminal attitudes, learning the code of the profession and methods for disposing of stolen goods (Klockars, 1974; Steffensmeier, 1986).

Associations with other professional criminals also provide important sources for information about criminal opportunities and for news in the profession. Professional criminals may, for example, share information such as "the heat is on in Milwaukee" or "Des Moines is a good town at the moment." Some crimes require associations with other professional criminals because such crimes require more than one person.

These associations also provide background in the specialized language, or argot, of professional crime. The language is not employed to hide anything, and group members avoid using it in public, fearing that they might attract attention among laypersons. Instead, this jargon provides a link between members that they hand down from generation to generation; hence, many terms used by professional criminals, like some terms used by doctors, can be traced back several hundred years (Maurer, 1949: 282–283; Maurer, 1964: 200–216; Sutherland, 1937: 235–243). This common language gives unity to professional criminals, nurturing a sense of history and some measure of pride as part of such a tradition.

Organization

Professional criminals generally work within loosely organized groups, although these allegiances depend somewhat on the nature of the crimes the group commits. For example, pickpockets work in stable groups over long periods of time, although they usually experience some turnover in personnel. Professional "heavy" criminals (robbers and burglars) often form groups for specific jobs and then disband (Chambliss, 1972; Eisenstadter, 1969). Relatively few of the crimes that professional offenders commit suit the requirements of an individual working alone; most require planning, input, and participation by others. As a result, many offenders are required to associate with other criminals.

Societal Reaction to Professional Crime

Generally, professional criminals achieve less public awareness than organized crime or conventional criminality. Occasionally, a well-publicized case or a popular motion picture, such as *The Sting* (1973), brings this type of crime to the public's notice. Even then, however, reports sometimes romanticize this form of crime and give it an innocuous image.

Even when police apprehend a professional criminal, he or she may well escape conviction or punishment. Professional criminals often receive preferential treatment by the police, who often see themselves as craftsmen in the same profession, and they recognize and respect the professional criminal's ability and skills. Professionals who face prosecution may manage to "fix" their cases by bribing police officers, judges, or prosecuting attorneys. In some jurisdictions, however, courts consciously impose severe sanctions on shoplifters who use booster devices in their offenses. Still, professional criminals usually do not compile lengthy arrest records because their skills help them to escape detection, and this characteristic may make a given professional resemble a noncriminal worthy of only a light sentence.

Summary

Criminals engage in heterogeneous types of behavior. Groups sometimes act in concert, and sometimes individual offenders work alone. Some crimes involve violence, while others, such as instances of theft, do not. Members of the middle and upper classes sometimes commit crimes in the course of their occupations. Business organizations and other groups of individuals may also perpetrate offenses to achieve organizational goals. Criminals differ in the extent to which they identify with crime and other criminals, the strength of their commitments to crime as a behavior, and their progressive acquisition of increasingly sophisticated criminal norms and techniques. This chapter stresses such diversity as an important theme in understanding criminal deviance.

Occasional property crime requires specific situations and conditions. Occasional property offenders do not usually conceive of themselves as criminals, and they rationalize their criminal acts to themselves. These offenders usually conform to the general standards of society and find little support for their behavior in subcultural norms. Most of these offenders do not progress to increasingly serious forms of criminality, and they usually experience only mild societal and legal reactions, having compiled no records of previous crimes.

Conventional criminal behavior is sometimes called *street crime*. Offenders begin their careers early in life, often in associations with youth gangs. They frequently vacillate between the values of conventional society and those of criminal subcultures. Some continue their careers with progressively more serious crimes, while others abandon crime after adolescence. These offenders usually accumulate many arrests and convictions for their crimes, and they often face relatively severe legal penalties.

Governments create laws to protect their own interests and continuity. Behavior that violates these laws amounts to political criminality. Governments may create specific criminal laws, such as conspiracy laws, and apply traditional laws to control and punish those who threaten the state. Political offenders usually act out of conscience, and they seldom conceive of themselves as criminals; indeed, they often regard the government they want to change as the real criminal. A government can act illegally itself through its agents. Crimes by the government usually receive weaker societal condemnation than crimes against the government.

Organized crime groups, or syndicates, pursue crime as a livelihood. At the lower levels of these syndicates, offenders conceive of themselves as criminals and live in isolation from the larger society. At the upper levels, syndicate members often associate with leaders in society, such as politicians and lawyers. Syndicates provide illegal goods and services to meet demand of legitimate members of society. Opinion remains divided about the extent of coordination between syndicates at the national level. The public tolerates this form of crime, partly because people want the services the syndicates provide and partly because it requires a complex response.

Professional criminals pursue crime to make a living and as a way of life. They develop committed self-concepts as criminals, and many take pride in their skills and criminal accomplishments. They associate with other criminals and enjoy high status among other kinds of offenders. Professionals and amateurs may commit the same kinds of crimes, but professionals apply more sophisticated methods. Few accumulate lengthy criminal records, not only because they become good at what they do and often successfully elude police, but also because many manage to "fix" their cases in the course of legal processing.

Selected References

Klein, Malcolm W. 1995. *The American Street Gang: Its Nature, Prevalence and Control.* New York: Oxford University Press.

This examination of gangs in the United States discusses different aspects of gang life and different methods for studying them, along with various policy alternatives.

Miethe, Terance D., and Richard McCorkle. 2001. *Crime Profiles: The Anatomy of Dangerous Persons, Places, and Situations,* 2nd ed. Los Angeles: Roxbury.

A readable, concise introduction to various types of offenders. The authors provide information on the nature of the crimes, trends in offending, and profiles of both offenders and victims.

Neubauer, David W., ed. 2001. *Debating Crime: Rhetoric and Reality.* Belmont, CA: Wadsworth.

A recent, nicely conceived collection of papers on many topics dealing with criminal behavior and crime policy. Papers were selected for their ability to stimulate debate and to identify controversial issues.

Steffensmeier, Darrell J. 1986. *The Fence: In the Shadow of Two Worlds.* Totowa, NJ: Rowan and Littlefield.

This readable account considers a type of property offender that facilitates other kinds of offenses and offenders. Without fences, many property offenders could not dispose of their illegally obtained goods.

Turk, Austin T. 1982. *Political Criminality: The Defiance and Defense of Authority.* Beverly Hills, CA: Sage.

This valuable discussion of the concept of political crime considers its definition and reviews research on the topic. The book goes beyond a mere review, however, by suggesting new directions for research and theory in political criminality.

CHAPTER 9

White-Collar and Corporate Crime

What do the following have in common?

- In September 1998, on President Bill Clinton's visit to Russia's then-President Boris Yeltsin to offer support for that country's growing economic problems, Moscow police arrested First Deputy Finance Minister Vladimir Petrov for accepting a bribe and taking an illegal action on behalf of a bank.

- Archer Daniels Midland in the fall of 1996 pleaded guilty to criminal price-fixing and paid a record $100 million fine. A federal jury in late September 1998 convicted three past and present executives of the company of conspiring with competitors to fix the price and worldwide sales volumes of the livestock feed additive lysine. The three—Michael Andreas, on leave as executive vice-president of ADM, Terrance Wilson, retired head of ADM's corn-processing unit, and former ADM biochemist Mark Whitacre—face a maximum 3-year prison sentence and at least a $350,000 fine. The evidence against the three executives was supported by audio- and videotapes of conversations among the executives with their counterparts in other countries. The company now faces the possibility of paying close to $200 million or more to settle civil suits from customers and shareholders.

- Televangelist Jim Bakker was indicted on 24 counts of fraud and conspiracy for taking huge bonuses out of his company's general fund and overselling his Lifetime Partnership Plan, under which some 9,700 hapless "partners" were offered the right to stay regularly in what turned out to be a single bunkhouse with 48 beds.

- John Watts, Jr., owner of a medical home care company, pleaded guilty in 1995 to defrauding Medicare of $1.5 million. Watts was a convicted cocaine dealer who started the business after release from prison. He submitted bills for services never provided for people, most of whom were dead.

There is much dissimilarity about these cases—the backgrounds of the offenders, the geographic location, the amount of money involved (although large in each case), and the specific criminal offense. What is the same in each case is the fact that each crime was committed in the context of the person's legitimate occupation. The term *white-collar crime* refers to offenses committed by workers in the course of their commercial activities (Sutherland, 1949, 1983).

White-collar crime is crime that occurs in the context of a legitimate occupation, but it is not confined only to the actions of individuals. Companies, corporations and organizations of all types, including political parties, church organizations, and government, can also violate the law in the pursuit of profit, power, and position within their organizational environment. For offenses involving entire firms or industries, sociologists apply the terms *organizational* or *corporate crime.*

Undeniably, there is much white-collar crime, and its consequences are widespread and serious. People in business may engage in income tax violations, illegal financial manipulations such as embezzlement and various types of fraud, misrepresentation in advertising, expense-account misuse, illegal campaign contributions, and bribery of public officials, to name only a few examples. In the investment business, fraudulent securities may be sold, asset statements may be misrepresented, insider information may be abused, and customer assets may be used illegally by brokers. In retail trade, "copied" goods are now common. One can get fake beanie babies (or whatever the current toy of the year is) and Barbie dolls, sophisticated electronic equipment, facsimile brand-name clothing, and computer software. Virtually all goods can be copied. One step beyond copied goods is counterfeit goods using well-known brand names, such as Gucci, Rolex, and Calvin Klein.

Practically no occupation is immune from white-collar crime. Even occupations that are designed to restrain and exercise social control, such as the police and mental health professionals, have experienced white-collar crime by practitioners who violate the ethics and laws of their professions. Professors have committed plagiarism to bolster their reputations and salaries, and clergy have defrauded believers out of millions of dollars. Physicians and dentists have charged private and public insurance programs for services not rendered or have used their positions to obtain drugs. Therapists have sexually abused patients, violating the very nature of their trusted positions. White-collar crime is common crime.

The Example of Computer Crime

Recently, computer technology has been used to commit crimes within the context of legitimate businesses and occupations. As computers play an increasingly large role in almost every aspect of the lives of people and businesses, there is much room for the misuse of computers for personal gain. Financial and academic records may be changed via computer, assets may be transferred from one account to another, and unauthorized services and products may be initiated. Occupational violations may be committed by professionals as well, for example, physicians who defraud government programs like Medicare and lawyers who misappropriate funds in receivership or secure perjured testimony from witnesses.

It is possible that white-collar crime is increasing and that this increase is due to the changing ways of conducting business, but it is hard to know whether this is the case. Much business is now done on computer and over phone lines, and there is much more room for abuse of these tools than when business was conducted wholly on paper. While there is a greater degree of security in dealing with hard copies of documents and personal communication, modern business depends on the computer and the electronic transmittal of information. Much of the modern world is computerized. All kinds of bank transactions, coordinating air traffic, and administering a 911 emergency telephone system, among other tasks, all require the extensive use of computers, as people learned during the Y2K "crisis" that was anticipated at the turn of the century. Even if the overall amount of white-collar crime is increasing, there is no dispute over the fact that we are becoming much more aware of this kind of crime. It is common now to learn of such crimes in the media with special television shows, newspaper accounts, and an increase in the number of books and magazine articles devoted to the subject.

Computers offer increasingly obvious capabilities as weapons. For example, they can help offenders perform unauthorized financial functions, such as monetary transactions and transfers. In a large bank, a criminal might divert very small amounts of funds from all account balances into another account that he or she controls. One business may break into another firm's computer system and alter

 Issue | **To Recall or Not Recall?**

In August 2000 Bridgestone/Firestone Inc. announced a recall of some 20 million tires for light trucks and sport utility vehicles that have been implicated in more than 65 deaths. The Firestone ATX, ATX II, and Wilderness AT tires are used mostly on Ford Explorers—the industry's top-selling SUV—but the recall included tires on other brands of vehicles as well. The complaints alleged that Firestone tires appeared to experience weakened tread bonding when traveling at high speeds. The tread would separate from the tire casing, especially in warm climates, causing blowouts and a loss of control. Firestone had been contacted by hundreds of consumers as well as State Farm Insurance Company after it became apparent that the tires were failing in large numbers. The problem was either that the tires were being manufactured in a defective manner or Firestone's quality control was inadequate.

Firestone and Ford Motor Companies maintained that the tires are safe, but both companies began a recall of the same tires for free on vehicles sold in Venezuela, Ecuador, Thailand, Malaysia, Colombia, and Saudi Arabia after tires failed in those countries in the previous few years. Firestone resisted pressure to do so in the United States, saying the matter was under investigation.

Ironically, in 1978, Firestone recalled 14 million of its 500-series tires that had tread separations and blowouts. Federal regulators fined the former Firestone Tire & Rubber Co. $500,000 then for concealing the safety problems. Firestone nearly went bankrupt after that recall and was subsequently bought by Bridgestone Corp. in 1988.

Also in August 2000, it was discovered that employees at Mitsubishi Motors had been hiding customer complaints in a secret place to avoid disclosing them. The complaints had to do with defective automotive parts, including failing brakes, fuel leaks, malfunctioning clutches, and fuel tanks prone to falling off. Each complaint was handled on an individual basis, but the company took care not to advertise the problems to avoid the humiliation of a recall. The company admitted that the practice dated back to 1977 and was carried out with full knowledge of workers, managers, and even one current board member.

data to reduce its account balance (Tien, Rich, and Cahn, 1986). Other computer crimes include gathering information without authorization or falsifying stored data, such as grades in a university registrar's office or account balances in a bank. And these offenses can be committed anywhere there is a computer, a modem, and access to an Internet service provider (ISP). These crimes can be committed at any time during the day or any day of the week.

The federal government began to take serious notice of computer security in year 2000, when a new program was introduced to help train people in computer security. One reason for this attention is the extent of loss businesses and other organizations can sustain once their computer systems are compromised. Studies have provided only sketchy estimates of losses from computer crime, but Friedrichs (1996: 177) establishes a range between $100 million and $5 billion a year. Another estimate sets the cost to software publishers of piracy of their products on the Internet at nearly $5 billion a year (Kaban, 1996). This report also notes that very few of these crimes are detected, and fewer still are reported to authorities. A company may not report victimization, fearing adverse publicity and a reputation for carelessness. A survey of 236 Fortune 1,000 companies disclosed that almost two-thirds had caught employees making inappropriate use of their computer systems (Zuckerman, 1996). The companies also reported major losses from computer crimes perpetrated from outside their systems, with between 15 and 18 percent of the firms indicating losses exceeding $1 million. Attempts to prevent this type of crime include new systems for encrypting data to prevent unauthorized use of computer files. Some states have also written legislation to apply sanctions to such crimes through their criminal justice systems. Enforcement remains a problem, though, because traditional police methods do not effectively detect and investigate such crimes. In 1986, the federal Computer Fraud and Abuse Act created a felony offense for computer-based fraud and theft not covered by state laws. Police agencies do not train officers to detect or solve computer crimes, however, leading to sporadic enforcement. Whistle-blowers, not police, often bring computer crimes to light. Colleagues may learn of offenses and turn in the offenders, or technically trained professionals may notice problems in systems they use.

A Brief History

There are many early precedents of crimes that occur within an occupational context. In early Roman times, the sale of goods in open markets was common. Buyers of these goods bore substantial risk, however, if the sellers misrepresented the goods. The doctrine of *caveat emptor,* or let the buyer beware, arose to characterize such transactions. Buyers were able to examine openly the goods they were about to buy, but the sellers were not required to disclose any imperfections in the fruits, vegetables, or cloth. Buyers were supposed to fend for themselves. Such a system persisted until recently and was justified in a number of ways. One such justification was economic: "The best interests of consumers are served by giving them goods and services at the lowest possible price and letting them decide whether or not to take the risk of getting unsatisfactory goods at low prices" (Rosoff, Pontell, and Tillman, 1998: 29).

As societies became more complex, however, it was not possible for buyers to protect themselves, even with full examination of the goods. While suitable for simple exchanges concerning common items, a system of caveat emptor could not exist in complicated exchanges dealing with items about which buyers might not have intimate knowledge. Few people today can really make sound judgments about the cars, computers, or appliances that they buy. We have to be informed by the judgments of others, whether by word of mouth from friends or by reviews by experts. Even then, we are often unable to make completely informed opinions about these products or services.

By the end of the 19th century, American society was experiencing a number of social problems brought about by urbanization and industrialization. The growth of cities was associated with the still-familiar problems of crime and poverty. A number of cities were also dealing with large number of immigrants, at this time mainly from Europe. The system of urban politics found in these cities was unable to deal effectively with these problems. As the country industrialized, many companies and factories experienced a need for labor that sometimes could be met by employing immigrants. These workers often toiled in terrible working conditions, frequently at very low wages and with little job security. Because they saved money on wages and working conditions, some companies were very successful during these years and grew so large that they were in danger of monopolizing their industries.

Monopolies are judged to be undesirable because this condition puts too much power into the hands of just one company. Once a company is a monopoly, it can set almost any prices it wants for its products or services. The free-market economy depends on viable competition to keep prices down and to increase the quality of products. If your company is the only one making breakfast cereal, you can charge whatever you want for it, and you don't have to develop the product. If, however, there are other companies making breakfast cereal, you need to keep your prices competitive with theirs and to continue to refine and improve your cereal so that you can continue to sell it.

One industry in which monopolization was a concern was the railroads. Controlled by rich and powerful people, the railroad industry was dominated by only a very few companies (Kolko, 1965). These companies had it pretty good. They could set their own schedules and fees and be immune from government regulation, which largely did not exist at the time. As one historian put it:

> . . . the government of the United States was behaving almost exactly as Karl Marx described a capitalist state: pretending neutrality to maintain order, but serving the interests of the rich. Not that the rich agreed among themselves; they had disputes over policies. But the purpose of the state was to settle upper-class disputes peacefully, control lower-class rebellion, and adopt policies that would further the long-range stability of the system. (Zinn, 1999:258)

Complaints from consumers could be met with indifference, and there was no control over the companies' ability to set their own policies. The railroads were important to the developing United States. There was strong need to link the East and West Coasts economically as well as politically. At the same time, it was undesirable for any company to have too much power. As mentioned earlier, the free-market economy requires competition as the primary vehicle to control prices and improve quality of goods and services. So, even though the railroads were a needed industry and provided an absolutely necessary service, the federal government was prompted to change things. Besides, those consumers were really starting to complain.

In response, the U.S. government passed special legislation to prohibit economic monopolization. Termed the Sherman Anti-Trust Act, the law was passed in 1895 to regulate precisely what had happened to those few railroad companies. The pattern of anti-monopolization seen 100 years ago continues to the present time. The federal government has intervened in a number of cases where monopolization was a possibility. In the 1990s, the U.S. Justice Department took Microsoft to court because it was said that company was monopolizing the Internet Web browser business with its product Internet Explorer. The issue was precisely the same as that faced by the United States with respect to the railroads.

The government response with the Sherman Act, however, was different from that found in other behavior that is defined as illegal. The Sherman Act is a regulatory law enforced by a special agency of the federal government, the Federal Trade Commission. It is not a state or federal law like those governing assault, rape, or theft. This is not to say that white-collar crime, that is, acts that violate regulatory laws, is not criminal. We will see shortly that it is. But because it is placed in a different body of law, and because that law is enforced differently from criminal laws, that behavior can appear noncriminal. Legislation through the 1900s also located white-collar and corporate crime in administrative and regulatory law, although the use of the criminal law in governing this conduct grew significantly.

It is disputable whether the powerful railroad companies were instrumental in placing the Sherman Act outside the criminal law. Perhaps they did, perhaps not. What is clear, however, is that in terms of content, procedures, and sanctions, regulatory laws are different in important respects from criminal laws. These differences create some problems even in defining white-collar crime.

Defining White-Collar Crime

The victims of white-collar crime are not just consumers who buy faulty products, who pay more for services than desirable, or who are bilked out of hard-earned money through swindles. Rather, white-collar crime victims are found throughout the class system and in all regions because this form of crime is so extensive. To see just how extensive, we have to consider the definition of white-collar crime.

White-collar crime is one of the most recently identified forms of criminality, perhaps because it is still ill understood. Used first by Edwin Sutherland in 1940, the term *white-collar crime* refers to a type of crime committed by persons during the course of their occupations. Sutherland was less interested in the precise meaning of the term, though, than in the severe consequences of the crimes, the documentation of which was his goal. These are, he argued, serious offenses with severe consequences to both individuals and society at large.

Initial Definitions

Sutherland defined white-collar crime as a crime committed by a person of respectability and high social standing during the course of his occupation (Sutherland, 1983: 7). What intrigued Sutherland was (1) distinguishing white-collar crime from ordinary or street crime, and (2) the fact that white-collar criminals were not "needy" criminals who committed their crimes out of necessity. At the time, a "poverty causes crime" explanation held strong sway, and Sutherland sought to demonstrate that such an explanation was not only simplistic (because not all poor people committed crimes), but wrong. His decision to concentrate his definition on the social characteristics of the offender, rather than on the offense, was to lead criminologists astray for some decades to come.

Sutherland actually began his interest in and study of "violations of law by businessmen" in the late 1920s (Sutherland, 1983: 78). His work was sporadic and was presented initially in his presidential address before the American Sociological Association. Even in his initial scholarly foray, however, Sutherland did not dwell on definitional issues, although he indicated that "white-collar crimes in business and the professions consist principally of the violation of delegated or implied trust" (Sutherland, 1940: 3). This definition was largely ignored in the empirical work on white-collar crime that followed shortly after the publication of Sutherland's book. Clinard's (1952) study of the violations of the Office of Price Administration during World War II, and Hartung's (1950) study of violations in the wholesale meat-packing industry addressed, in turn, the two levels at which white-collar crime can be said to operate—the individual and the corporate. Clinard's offenders were individuals, while Hartung's were businesses, exactly like those found in Sutherland's own research. Clinard (1952: 227) notes that white-collar crime refers to "violations of the law committed primarily by groups such as businessmen, professional men, and politicians in connection with their occupations . . . ," after which he cites virtually all of Sutherland's publications on the subject. A third work stemming from Sutherland's lead was by Cressey (1953) and is discussed later.

It appeared that the notion of white-collar crime initially resonated with many criminologists who believed that a new form of crime would extend the criminological umbrella and prove a fruitful place in which to test general theories of crime. As intriguing as Sutherland's initial message was, major work on white-collar crime, aside from the works cited, did not extend far beyond his death in 1950. Few researchers pursued the topic systematically. General criminological attention shifted to juvenile delinquency by the mid-1950s, particularly juvenile subcultures and gangs, and the topic of white-collar crime was barren throughout most of the 1950s and 1960s. Interest increased in the 1970s, but white-collar crime was hardly a major area for theory and research.

✳ **In Brief**   **Common Terms for Criminal Abuses of Power**

Type	Short Definition
Occupational crime	A crime committed in the context of the offender's job duties
Business crime	A crime committed by an organization in the course of its regular business
White-collar crime	Crimes committed during the course of job duties by middle-status or high-status offenders
Corporate Crime	Crimes usually committed by employees on behalf of the company for which they work

Sources: Sutherland, Edwin H. 1949. *White Collar Crime*. New York: Dryden; and Quinney, Richard. 1964. "The Study of White Collar Crime: Toward a Reorientation in Theory and Practice." *Journal of Criminal Law, Criminology, and Police Science* 55: 208–214.

By the 1980s, in spite of inadequate research funding, the literature on white-collar crime grew enormously. Perhaps it was related to criminologists' concern with the "greed decade" of the 1980s, the then-increasing national debt, and the widening gap of income inequality in the United States. In any case, according to traditional indicators, such as papers presented at professional meetings and publications, white-collar crime came again to be a growth area in criminology. This situation continued into the 1990s, when there was sufficient scholarly material and student and professorial interest to publish textbooks on white-collar crime (e.g., Friedrichs, 1996; Simon and Hagan, 1999) as well as collections of papers for classroom use (e.g., Ermann and Lundman, 1982; Schlegel and Weisburd, 1992; Blankenship, 1993; Geis, Meier, and Salinger, 1995). The number of courses devoted exclusively to white-collar crime also increased substantially. By the 1990s white-collar crime had become an established criminological specialty.

White-Collar Crime and Criminal Events

The study of white-collar crime is about 60 years old, and criminologists have had plenty of time to correct any misunderstandings or ambiguities that arose from Sutherland's cavalier employment of the term. Sutherland's original conception was heavily criticized, and a number of alternatives have been offered to escape the definitional confusion. These alternatives have sought to conceive of white-collar crime more in behavioral terms, and a common conception emphasizes the elements of fraud (Albanese, 1995). In this view, the meaning of white-collar crime resides in the nature of the crime, not the criminal. Other alternatives have sought to resurrect Sutherland's conception by emphasizing dimensions of the offender's occupational position, especially the elements of trust and power over others (Shapiro, 1990). This conception is more sensitive to the victims of white-collar crime than focused on the acts themselves.

A criminal events perspective alerts us to the study of neither the offender nor the crime, but to the offender, victim, and situation in which the crime takes place. As such, it is proposed here that the meaning of white-collar crime is not to be found in a narrow definition that isolates either characteristics of the offender or the nature of the crime, but in the combination of offender, victim, and context in which the crime takes place. This is called a "criminal events perspective" (see Meier, Kennedy, and Sacco, 2000).

Definitions that cater to offender characteristics, such as that offered by Sutherland, or to the nature of the crime necessarily have only part of the conceptual picture. What each misses is consideration of the other in the context of the offending. Clearly, white-collar crime is a function both of the offender and offense. A physician who commits an assault is not a white-collar criminal, nor is the auto mechanic who charges for unnecessary repairs a conventional criminal. Sutherland reminds us that white-collar criminals and ordinary criminals can commit the crimes of fraud, forgery, or bribery, but such offenses are classified as white-collar crime only when the offender is "respectable."

We need not here enter the definitional debates either to learn more about alternative definitions or to resolve the disputes. While Sutherland's conception sought to distinguish white-collar criminals from street criminals, he does provide a meaningful idea in thinking of white-collar crime as criminal violations in an occupational context.

Let us start with a simple conception of white-collar crime that views it *as a violation of law committed in the context of the offender's legitimate occupation*. This occupation, of course, must be a legitimate one so that organized and professional criminals are not considered white-collar criminals. This conception identifies two elements: law violation and occupational context, each of which is consistent with Sutherland's original formulation and the meaning found in most discussions of white-collar crime (e.g., Braithwaite, 1985). It does not specify certain laws or even bodies of law, nor does it attempt to identify the kind of occupation. Furthermore, this view is inconsistent with those definitions that find the meaning of white-collar crime only in particular kinds of illegal acts (e.g., fraud) or exclusively in the personal characteristics of the offender (e.g., high socioeconomic status). Those definitions often attempt to anticipate either the kind of acts that are involved (i.e., which laws are violated) or who can commit these crimes (i.e., who is respectable enough to be considered a white-collar criminal). The occupational context of white-collar crime is critical. More than any other factor, it provides meaning to this form of criminality; it is where criminal motivation and the opportunity for white-collar crimes come together (Coleman and Ramos, 1998). White-collar crime is crime committed during the course of a legitimate occupation. Such crimes would include, but are not limited to, misrepresentation of sales (e.g., defrauding others in sales of worthless land), violating a position of financial trust to steal other people's money (e.g., embezzlement), and participating with others in a conspiracy that would violate federal laws (e.g., price-fixing).

The proposed conception is deliberately broad, although it is able to stake out sufficient definitional ground to take in what most people mean by the term *white-collar crime*. Let's briefly examine the elements.

Occupation   The occupational dimension of white-collar crime is critically important. Occupation is a special status, one that confers power, trust, and individual identity. It is the context in which much conventional life is lived and against which many people judge themselves as successful in life. There is a large body of law governing the behavior of individuals in occupations, many of which are considered to be white-collar crimes (Brickey, 1995). The link of occupation with white-collar crime is so strong that it does little conceptual injustice to consider white-collar crime as a subset of occupational crimes generally (Quinney, 1964).

Offenders commit white-collar crimes during the course of their legitimate occupation. The adjective *legitimate* is added to the definition to reduce the chances of a particular confusion. Suppose the individual's occupation is "hit man" for a criminal syndicate. Would his or her refusal to kill someone constitute a crime under this definition? No. While the violation occurs in an occupational context, it is not a legitimate one. The occupation itself is illegal. Legitimate occupations provide a range of criminal opportunities, and these opportunities can be accessed regardless of the specific motivation of the offender. An employee who violates a position of financial trust is therefore a white-collar criminal regardless of whether the motive was to solve an unshareable financial problem (Cressey, 1953) or simply to obtain extra money for high living (Nettler, 1974).

The legitimate occupation provides a context in which the crimes occur, but it also provides the opportunity for the crimes. People entrusted with other people's money may steal or embezzle that money in the context of their jobs, but it is also the case that their jobs provide the opportunity to commit the crime. This, again, suggests that we must look at the offender, the victim, and the context in which the crime occurs if we are fully to understand white-collar crime.

Law   A broad view of white-collar crime includes both illegal and immoral acts (Simon and Hagan, 1999). Called white-collar deviance, this perspective has the advantage of being a more inclusive approach that can examine a much wider range of acts, but it suffers from a lack of consensus on what constitutes white-collar and corporate "wrongs." Simon and Hagan (1999: 154) are not guided by a normative conception, but by a universalistic one that identifies wrongs as being deviant in all places, at all times, and under all circumstances. To avoid such universalistic judgments, it is necessary to limit behavior according to an agreed-upon standard.

It is virtually impossible to imagine any definition of "crime" that does not make reference to law, although some criminologists have adopted a restricted view of what kind of law is appropriate for criminologists to use. A dispute arose after Sutherland's initial work by criminologists who argued that white-collar crime wasn't crime because it didn't violate a *criminal* law. While technically correct that many white-collar crimes are violations of administrative or regulatory law, criminologists now regard such violations as "crimes." From our earlier discussion, you might recall that many of the early laws governing white-collar crime were formulated not in criminal, but in administrative and regulatory, law. This has given the appearance, both to observers and to the white-collar criminals themselves, that these are merely technical rules, not laws backed by state sanctions. Nevertheless, violations of these administrative and regulatory rules are crimes in a very real sense. They are rules formulated by the state, and they are backed by state sanctions. Clinard (1952: 229), for example, argued that black market offenders during World War II were criminals in the sense that their behavior was socially injurious, they violated government rules, they incurred state sanctions, and they experienced social stigma. We should also point out quickly that some white-collar crimes are prosecuted and convicted in criminal courts.

The concept of professional crime represents many of the same definitional problems as that of white-collar crime. The distinction between amateur and professional theft is not the particular crime committed, since each can commit the same crimes, such as shoplifting. Rather, the meaning of professional crime is a combination of crime and characteristics of the offender: the extent, for example, to which the offender has well-developed criminal skills and attitudes, the extent to which theft becomes a way of life, the offender's associations with other criminals, and the like. Perhaps because there are few scholars working on professional crime, or because the division of labor on professional crime is scattered around different offense types, definitional disputes do not represent a hindrance to either scholars or criminal justice officials.

Fraud   Fraud is an important element in most white-collar crimes (Albanese, 1995). Fraud refers to the use of deception to acquire unlawful gain, whether in the form of money, power, or position. Fraud usually takes the form of a lie of some kind, although sometimes people lie by omission by not telling someone an important fact (e.g., the land they are about to purchase is likely to be worthless). Not all frauds are examples of white-collar crime, but most white-collar crimes are examples of fraud.

Fraudulent behavior associated with white-collar crime has an additional component: it is committed from a position of trust that is afforded the offender by occupying a legitimate occupation. For this reason, white-collar crimes are sometimes considered to be crimes where a person's power is abused.

Power or Trust   Because white-collar crimes occur within a legitimate occupational context, white-collar criminals are cloaked in an aura of respectability. In many discussions of Sutherland's conception of white-collar crime, it is assumed that respectability refers exclusively to the high social standing of the offender. But respectability is not merely a function of social status. There is another meaning of the term *respectable*, one that keys on the personal characteristics of trust and honesty—or the appearance of honesty. There are respectable auto mechanics (they perform only needed repairs) and not-so-respectable ones (who violate the law). Respectability is what makes fraud or deceit possible. Shoplifters are thieves who are posing as respectable customers. It is their appearance of respectability that creates the conditions for this crime, as opposed to robbers who strike no such pose and therefore commit the crime of theft in a different, more direct manner. In the same way, white-collar criminals are offenders who use their appearance of respectability to commit crimes.

Characteristics of White-Collar Offenders

Conventional crimes are associated with certain images—weapons, disguises, black-and-white police cars involved with high-speed chases, dead-of-night action, spur-of-the-moment emotionality, and the like. White-collar crimes have none of these. White-collar crimes have a number of characteristics, and few of them conjure up images of conventional crime. White-collar crimes are sophisticated and the result of considerable planning. They are committed by offenders who "should know better" and frequently take place over a long period of time.

We have learned a great deal about white-collar criminals over the years, and they appear to be different from many other kinds of criminals on a number of dimensions, including their self-concept and the circumstances that brought them to the crime.

Sophistication   White-collar crimes tend to be complicated events. Many of them skirt the legal boundaries in an ever-expanding system of laws and regulations. Many of them can be committed only with considerable knowledge of the occupational and legal context in which offenders find themselves. Violations of SEC regulations, for example, require not only considerable knowledge of the regulations themselves, but also the likelihood that the laws will be enforced, how they will be enforced, and by whom. For this reason, these crimes cannot be committed by just anyone; they require considerable formal training in economics, the operation of stock markets, the law, and enforcement.

Planning   Many crimes are planned in the sense that offenders anticipate their actions and the reactions of victims or the police. Virtually no offenders truly want to get caught, and such planning is conducted with the goal of minimizing the probability of detection and/or apprehension. White-collar criminals are no exception to this generalization, but the degree of planning often takes on different dimensions because of the crimes themselves. White-collar crimes are complicated events requiring not only detailed knowledge but also calculated risk taking on the part of offenders. Many offenders learn the skills associated with risk taking as part of their legitimate occupation, such as stockbrokers who frequently act on less-than-complete information. Learning the nature of financial risks and how to avoid them is good experience for those brokers who wish to risk illegal actions. White-collar crimes are not emotional acts tied to particular circumstances, as many murders are, but are significantly anticipated events.

Self-Concept   One of the most important characteristics of white-collar criminals has to do with their conceptions of themselves (Geis, Meier, and Salinger, 1995). Since these offenses take place in connection with legitimate occupations, the offenders generally regard themselves as respectable citizens, not as criminals. In regarding their actions as violations of law but not criminal acts, occupational offenders share characteristics with people convicted for such crimes as statutory rape, nonsupport, and drunken driving. Also, a typical white-collar offender maintains a relatively high social standing compared with that of a conventional criminal, so the general public seldom conceives of such a person as a participant in "real" criminality; this attitude also influences the noncriminal self-conceptions of occupational criminals (Coleman, 1997).

A white-collar offender gains strong support for a noncriminal self-concept by maintaining an otherwise respectable public image. The resulting noncriminal self-concept establishes an essential condition for occupational crime. Offenders who commit embezzlement generally occupy positions of financial trust, such as office employees, bookkeepers, or accountants. Cressey (1953: 30) identified embezzlers as trusted people who stole organizational funds as a result of three conditions: (1) They faced an unshareable financial problem. (2) They recognized the chance to solve their financial problem by violating their position of financial trust. (3) They developed suitable rationalizations for embezzling to resolve their self-conceptions as trusted people. The trust violators defined their acts as noncriminal arrangements; for example, many told themselves they would merely borrow the money or that they were not completely responsible for their acts. Similarly, researchers in the pharmaceuticals industry may deliberately falsify research findings on the side effects of drugs and still see themselves not as criminals, but as professionals working to manufacture medicines that clearly save lives and help people (Braithwaite, 1984).

Some white-collar criminals neutralize norms against their offenses and the guilt that they might otherwise feel. One antitrust offender stressed the historical continuity of his crime and the character of his offense as a common, everyday experience: "It was a way of doing business before we even got into the business. So it was like why do you brush your teeth in the morning, or something. . . . It was a part of the everyday. . . . It was a method of survival" (Benson, 1996: 68).

Criminal Careers of White-Collar Offenders

White-collar offenders do not have extensive records of juvenile misbehavior or of juvenile court appearances. Most white-collar criminals, furthermore, do not have records of conventional crime as adults either. Nor do they have histories of association with conventional criminals. The absence of a conventional criminal record should not be surprising since for some white-collar occupational positions, the absence of such a record is a necessity. Individuals will not be placed in positions of financial trust, for example, if they have a history of previous criminality or untrustworthiness. There is little evidence that white-collar offenders have had any experience with other kinds of criminality or with criminals. For this reason, most theories of criminality predict that rates of illegal violations are low among white-collar workers.

Still, it can be asserted reasonably that some white-collar offenders have extensive histories of white-collar criminality. Some white-collar violations, such as falsifying corporation records to mislead stockholders about the financial condition of a company, may take place over a long period of time. There may be many such false reports and many misleading statements. Prices can stayed "fixed" between two companies for some time, through many individual transactions. Clinard and Yeager (1980) found that rates of corporate violations were particularly high in the automobile, oil, and pharmaceutical industries, and that such high rates had persisted for some time. This means that white-collar violators persist in their behavior over time, to the point of socializing new employees into the criminal norms of the corporation. Criminal behavior can be a "way of life" for some companies and for some of the individuals in those companies.

Explaining White-Collar Crime

Some people think that white-collar crime can be explained by examining the motives of the offenders. What is the usual motive of white-collar criminals? Money. This doesn't get us very far, of course, because money, or greed, is a motive for many acts, both criminal and noncriminal. Furthermore, motives are not theories. A motive is the specific reason that prompted the offender to engage in crime. That reason, however, doesn't explain the crime because the same motive can be used to account for noncriminal behavior. People who are motivated by greed, for example, are prompted to commit their crimes because they want more than they already have. But greed is also a common motive for many noncriminal acts, such as working overtime, obtaining additional or special training for a promotion, and borrowing money from a bank. Greed, in other words, is too general a condition to provide a theoretical explanation for a specific act. Good theories of crime clearly differentiate what makes some greedy people turn to crime while other greedy people do not violate the law.

Explaining the behavior of white-collar criminals poses special problems for criminological theorists. As we have seen in earlier chapters, most theories direct our attention to the origins of law violating behavior early in life. The major theoretical traditions take as their starting point the beginnings of criminality, which are usually found in early adolescence. Learning, control, labeling, and strain theories attempt to explain both individual instances of offending and social variations in crime, such as the differential crime rates by age, sex, residence, and social class. Even theories that integrate these perspectives, such as the life-cycle theories, attempt to explain why individuals who were crime free earlier come to commit crimes later. Regardless of the specific explanation, most theories of crime take as their task the identification of the conditions that bring about crime. Since conventional offenders begin their criminality relatively early in the life cycle, this is where most theories direct their attention.

As we have seen, white-collar criminals, however, usually do not begin their criminal careers early in life. While some may have committed instances of delinquency, the great majority of white-collar criminals were relatively crime free. Most came from middle- and upper-class family backgrounds and were committed to educational careers. Aside from possibly facing family problems encountered by many of their cohort, most of these individuals experienced uneventful transitions from adolescence to early adulthood, and most entered the world of full-time work without legal incident. Had white-collar criminals not been largely law abiding, they might not have been able to secure their lawful occupations. They just don't fit the pattern envisaged by most traditional theories of criminality.

There are some specific theories of white-collar crime, and many of them emphasize a combination of opportunity and learning environments. Like most crimes, white-collar offenses require the proper physical opportunity for the crime to occur. Accountants must be in positions of trust over resources in order to steal those resources. Such persons are likely to know how to manipulate financial records and to avoid detection. Like most offenders, white-collar criminals do not wish to be detected; they assess their chances of getting caught and avoid behaviors that increase those chances.

Coleman (1997) has suggested that white-collar crime can be attributed to the coming together of three factors: (1) motivation, which is often conditioned by a sense of competition in business, the professions, and politics; (2) culturally learned neutralizations; and (3) the opportunity to commit crimes. Motivation can take the form of the desire for financial gain or status within some corporate context. This criminal motivation, according to Coleman, is generated within our society by our economic system, which gives rise to a culture of competition that fosters these motivations. The demand for profitability among people in business is at times most easily satisfied through illegal activities. The patterning of the differential opportunities to violate law accounts for the patterning of law violation rates.

This account is consistent with that of Sutherland (1949) and Clinard and Yeager (1980), who each report that corporate ethical systems vary considerably from company to company and that socialization into one or another system may determine later violations. Clinard and Yeager (1980: 132) reported that compared with nonviolating corporations, "the violating firms are on average larger, less financially successful, experience relatively poorer growth rates, and are more diversified." Furthermore, Clinard and Yeager continue, corporations in depressed industries have relatively higher rates of crime. But while these factors influenced generally the violation rates, they are insufficient to explain violation rates themselves, and it appears that the corporate ethical environment is also a key factor in explaining socialization to a "criminal" corporate culture and subsequent law violations.

People in any legitimate occupation may learn to violate the law from a large variety of sources. Employers sometimes teach their employees about theft by their own examples or by failing to take action about known violations.

> On one of many occasions, the manager [of the mall bookstore] went round with a plastic bag which she filled with about 25 romance novels [which] were worth about $2.50 each. These were taken home to her apartment but they were never returned. On another occasion, she filled a cardboard box with hardback books which came to roughly $275 worth. These were being taken home to be given to her family as Christmas gifts. They were never paid for. (Adams, 1989: 32)

Occupational criminals learn techniques as well as rationalizations, norms, and attitudes associated with their offenses. This socialization may cast violations as normal business practices and expected behavior in specific situations. As part of a larger social group, such an offender derives social acceptance, other support, and even encouragement, as occurs with many other forms of crime as well. Occupational offenders and their subcultures differ from other deviants and their subcultures in the power and social standing they hold by virtue of their other social roles (Gandossey, 1985). The occupational offender's subculture often supports both legal and illegal activities by blurring the line between accepted and deviant practices. Also, the inherent complexity of many business transactions and the importance of trust in business deals create opportunities for violations (Shapiro, 1990).

A number of characteristics may isolate people in business occupations from unfavorable definitions of illegal activity (Vaughan, 1982). First, mass-media reports decry conventional crime while often treating occupational crime, unless it involves particularly sensational circumstances, with much more lenience. Second, high-status occupations often shield offenders from severe criticism by government officials, many of whom came from the business community, accept campaign contributions from business sources, or associate socially with businesspeople in clubs and other organizations. Finally, businesspeople tend to associate chiefly with one another, both professionally and socially, a fact that may prevent objective scrutiny of the implications of white-collar crime. These arguments suggest that combined appeals to learning and opportunity may best explain such acts (Coleman, 1997).

Corporate Criminal Behavior

The notion of corporate crime requires us to think of criminal acts committed not by individuals alone but by aggregates or collections of people: corporations and other organizations. Criminologists study many other forms of aggregate crime, including organized crime and juvenile gang delinquency. The term *corporate crime* refers to the violation of laws and regulations involving corporations and their management. Often the term refers to the actions of large corporations and the executives acting on their behalf.

Large businesses make decisions through complicated organizational structures. Such a structure encompasses executives with extensive decision-making authority, perhaps including a board of directors, president, and chief executive officer (CEO), as well as a number of vice presidents. It also encompasses personnel with much less power—middle managers, supervisors, and workers. As a legal entity, a business corporation invests capital provided by stockholders, who technically own its assets. Among all of these decision makers, however, the top managers largely control the corporation.

Large business organizations often form huge economic conglomerates with assets worth billions of dollars. At the beginning of the year 2000, the lowest-rated member of the so-called Fortune 500 (the 500 largest corporations in the

United States based on revenues) had nearly $3 billion in annual revenue. The largest corporation, General Motors, posted revenues of $161 billion. As firms control such vast productive power, they create equally significant potential for social harm. This potential grows still more dangerous as companies form relationships with one another to ensure continued access to resources, such as customers or raw materials. A number of companies were combined throughout the 1990s to form larger corporations in the process of merging. The largest megadeal at the time, which took place in January 2000, occurred when Internet provider America Online bought out media giant Time-Warner for $165 billion to create a huge corporation that would combine technology and media resources. The new conglomerate was valued at $342 billion at the time of the merger. What struck many observers is that a new, Internet-based company (a dot-com) actually had the resources to buy out a long-established company.

Most corporate crimes involve economic gain. Crimes involving stocks and mergers, for example, are complicated. There is a substantial "gray" area between legal and illegal actions, and often white-collar criminals take advantage of this questionable area. Individuals and companies who engage in arbitrage, a form of investment betting, have much at stake, and it is tempting to have all the information one can get to make the right decision. Some of the best information is insider information on pending actions that will affect stock prices. Some

 Issue | **Arbitrage**

Arbitrage refers to a risky business decision. It also refers to a business practice that is often associated with illegal insider information. Suppose Company A and Company B are talking about merging. The stock for Company A currently sells for $50, while that for Company B sells for $30. The new company that results from the merger would stabilize the price of stock of Company A and increase the price of that of Company B.

An arbitrageur is someone who speculates on the likelihood that the merger will actually take place. He or she would approach the stockholders of Company B and offer them a sure profit on their stock of, say, $35 against the likelihood of the merger that could push the price even higher. The stockholders are now faced with a decision: Sell now at a sure profit, or wait and see whether the stock will go even higher. Mergers are not the guaranteed result of merger talks; things can and do go wrong, and even after considerable time, two companies might decide not to merge for a number of reasons. Some stockholders will be inclined to take the sure profit rather than risk not making any money at all.

If the merger goes through, the arbitrageur now is holding stock that will increase in value after the merger. He or she can either hold the stock until its value increases or try to sell the stock at a profit before the merger takes place. It is a speculative process, but virtually all investments in stocks are risky, and the better arbitrage specialists have usually acquired a good background in risk aversion (through insider knowledge of the companies, the merger process, and perhaps other intangibles concerning the companies).

investors are sufficiently tempted by the high stakes that they will engage in obtaining illegal insider information.

Nature and Types of Corporate Violations

Corporate crime can be big crime. The largest theft in the United States involved the collective embezzlement in the savings and loan industry (Calavita, Pontell, and Tillman, 1997). It will likely cost the United States government between $300 and $500 billion, but ultimately the victims will be the taxpayers. In 1980, Congress increased the amount for which the Federal Savings and Loan Insurance Company would insure to $100,000. Many savings and loan institutions increased their interest rates, thereby attracting wealthy investors whose money would be protected in case the savings and loan failed. A number of savings and loans did fail and had to be bailed out. Of this figure, it is estimated that perhaps as many as 80 percent of the failures were due to illegal schemes used by savings and loan managements (Beirne and Messerschmidt, 2000: 278–279).

The scope of white-collar and corporate crime continues to change. Friedrichs (1996) has noted the increase of "technocrime," or the use of computers, fax machines, cell phones, electronic surveillance, and new accounting technologies. As routine tasks done the "old way" are transferred to these new methods, there is much opportunity for crime, if only because the new technologies are complex and difficult to monitor and control. Increasingly, criminologists are suggesting that it is critical to understand the nature not only of organizations and their cultures, but also the environment in which an organization exists and the way in which work is done in the organization (Vaughan, 1999). It is possible that crimes occur in organizational contexts not so much out of willful exploitation of opportunity, but because of the risk-taking actions of organizations in an unstable environment.

There are a number of different types of corporate violations, including crimes against consumers, crimes against owners, crimes against employees, and crimes against the community as a whole.

Crimes Against Consumers   White-collar crimes against consumers can take many forms, including crimes committed by manufacturers who make unsafe products, retailers who take unfair advantage of customers in purchasing products, and repair specialists who bilk consumers in the maintenance and repair of products. These illegal acts cost consumers millions of dollars each year, and most consumers can recount instances of shoddy products that don't last or don't work as advertised. Many local television stations now have consumer segments that test claims made on infomercials and other advertisements. Toys that break quickly, cars that turn out to be "lemons," electronic items like CD players that skip, and new homes that seem to require endless repair are only a few examples. Some of these examples are instances of a manufacturer skimping on quality control during the manufacturing process, but some are instances of deliberate attempts to bolster profits at someone else's expense. Making stronger claims for the benefits of a product is sometimes explicit but often implicit. Buying a product, consumers are told, will increase their social status and sex appeal and improve their outlook on life. Such claims may be transparent to many people, but to the extent that people are fooled into thinking something that is untrue, the advertisements may be false. Such advertisements make their creators a great deal of money, but badly made products represent threats not only to the pocketbooks of consumers, but to their physical safety as well.

The manufacture and sale of unsafe products is both illegal for manufacturers and dangerous for consumers. Two of the most famous cases of this type of crime involved the manufacture of automobiles that were life threatening to their drivers and passengers. The General Motors Corvair was a small, rear engine car that was very fuel efficient. Over 1.2 million Corvairs were sold between 1959 and 1964, but the car was discontinued after a young consumer advocate, Ralph Nader, exposed the structural problems with the car in testimony before Congress and in a small but influential book called *Unsafe at Any Speed*, published in 1965. Nader pointed out that the Corvair was engineered in such a way that when the car took a sharp curve, the weight of the rear-mounted engine put enormous pressure on the rear, outside tire and wheel, thus often causing a separation of the tire from the wheel, a blow-out, and subsequent loss of control. Scores of people were injured or killed from such accidents. General Motors subsequently corrected the problem in later models, but by then the public had lost confidence in the safety of the vehicle, and sales plummeted.

The dynamics of the Corvair case were seemingly repeated with the Ford Motor Company's Pinto case (Cullen, Maakestad, and Cavender, 1987). The Pinto, also a small, fuel-efficient car, was extremely popular because it was introduced near the time of the Middle East petroleum embargo in the 1970s. Americans at the time were witnessing gas shortages, long lines at gas stations when they had fuel, and growing anxiety that their large "gas guzzlers" were part of the problem in producing the shortages. While many of the Corvair purchasers may have been influenced by the sporty look of that car, Pinto owners were more concerned with gas prices and economy. The Pinto's design was very simple, and the car was built to be light and with a small engine that consumed less gas. The design, however, also included a flaw: the fuel tank was placed very near the rear bumper and, given the construction of the tank, had a tendency to rupture in the case of even a slow rear-end collision.

This is precisely what happened with a number of Pinto cars, causing tremendous injury and death when the gasoline in the tank was ignited. Scores of persons were seriously injured, and many died as a result of these collisions. The Ford Motor Company was subsequently sued by a number of survivors of Pinto crashes or their families. Then, in a very rare event, a local prosecutor in Indiana tried Ford Motor Company in a criminal court for the wrongful deaths of three young women. The prosecutor demonstrated the design flaw and argued that Ford knew that the design was dangerous. Furthermore, he argued that Ford knew how to correct the problem but failed to do so since that would be an admission of responsibility that would make Ford liable for tremendous civil damages. The solution to the problem took the form of a rubber bladder, like a heavy balloon, that would be placed inside the fuel tank. The bladder would contain the gasoline in the event of a rupture in the gas tank, thus preventing the gas from spilling and igniting. Furthermore, the cost of including the bladder would be minimal. Ford hired the nationally known criminal defense attorney F. Lee Bailey. Ford was acquitted at the trial, although the civil suits continued in litigation for a number of years.

Clinard and Yeager (1980) found that the automobile industry was one of the most serious recidivists in terms of violations. In 1998, a judge fined Chrysler Corporation $800,000 for failing to recall 91,000 Cirrus and Stratus cars after the government had informed Chrysler the rear seat belt systems were unsafe. Chrysler had fought having the cars recalled since the repairs would be without charge to consumers. Chrysler maintained that the cars were safe but relented and finally recalled the cars after losing the court case.

There are numerous examples of serious crimes against consumers in industries other than automotives, including the pharmaceutical industry. Many people like, and need, to think that the drugs they take are safe. Prescription drugs are supposed to promote health, and physicians would not willingly prescribe anything that would make patients sick. But that is sometimes not the case. Braithwaite (1984), for example, has documented numerous instances of fraud in the safety testing of drugs. Research records on the effects of drugs on laboratory animals have been changed and false reports made about those effects. Sometimes, incomplete tests that failed to expose harmful long-term effects of particular drugs have been run. This can result in tragic consequences. In the 1950s, thalidomide was a drug to be used as a sleeping pill and tranquilizer for pregnant women. The drug, first manufactured by a German company, was marketed without systematic safety testing. Instead, the company relied only upon impressionistic testimonials from clinicians who had used the drug on a test basis. But there was, it turned out, a very serious consequence of using thalidomide: severe birth defects in the children born to women using the drug.

> About 8,000 thalidomide children are alive today in 46 countries around the world. Perhaps twice that number died at birth as a result of the drug. Some of the thalidomide children have no arms, just flippers from the shoulders; others are without legs as well—limbless trunks, just a head and a body. The physical horror of thalidomide was in some ways matched by horrible impacts on the social fabric of so many families. Mothers in particular were tragic victims. One husband told his wife: "If you bring that monster home, I leave." She did and he left her, like many other thalidomide fathers. (Braithwaite, 1984: 15)

Many consumers come into contact with faulty products through deliberately misleading or false advertising. Advertising in the United States is a multibillion-dollar industry, much of it based on the need to manipulate, not inform. In an increasingly competitive business climate, advertisers must continually think of ways to gain consumer's attention, and such efforts often skirt the borders between cold, hard facts and sensational claims, some of which are untrue. While some people fully expect advertising to mislead, many others appear to accept as fact the claims of manufacturers about their products. There are many examples of false or misleading advertising (Rosoff, et al., 1998: 41–44).

■ Beech-Nut Nutrition Corporation was cited for advertising its apple juice as being 100 percent fruit juice with no sugar. In fact, there was plenty of sugar, but no natural fruit juice.

■ Carrington Foods was ordered to pay a large fine in 1995 when its food packages proclaimed "More Crabmeat Than Ever" when it contained no crabmeat whatsoever.

■ NordicTrack agreed to settle an FTC complaint made about claims the company's ads made about how much weight the users of its exercise equipment could expect to lose.

■ Hillshire Farms agreed to pay a fine regarding ad assertions made about its "lite" meats being low in fat when they were not.

■ The Home Shopping Network reached a settlement regarding the greatly overblown retail value of the jewelry products it sells. With an exaggerated retail value, the network's "sale price" was made to seem more attractive.

One major recent case involving both misleading advertising and faulty products concerns cigarette smoking. For years, many people recognized the addictive quality of cigarettes, some assuming it was merely psychologically gratifying, others believing that there was something in the tobacco itself, such as nicotine, which produced a physiological dependency. The U.S. Surgeon General's report in the mid-1960s confirmed the serious health consequences of smoking and attributed continued smoking to nicotine. It wasn't until the mid-1990s, however, when it surfaced that the tobacco companies had been deliberately spraying additional nicotine onto the tobacco that eventually wound its way into cigarettes. Tobacco executives swore before a subcommittee of the U.S. Congress that they believed that smoking wasn't habit-forming and that its link with ill health was unproven. Eventually evidence from the company's files and the actions of whistle-blowers disclosed that the companies were indeed adding nicotine to their products to help ensure that smokers continued to smoke. By 1998, the evidence was so overwhelming that the tobacco companies were manipulating the levels of nicotine in cigarettes that some companies began to advertise their products, for example, Winston cigarettes, as "natural" with no additives other than those found directly in tobacco.

Crimes Against Owners   Crimes against owners include a multitude of offenses, some directed against the owners of companies, others against managers and other officials of the organization. These crimes are committed by people within the organization who have knowledge of organizational practices and how to get around rules and regulations governing organizational behavior.

Many of the violations in this category are against stockholders and principal owners by managers of the companies they own. Such violations would include deliberately falsifying annual reports to show a company in a financial light other than what the real condition of the company was. It would also include other crimes that misled stockholders, or potential stockholders, about the financial condition of a company (Shapiro, 1984). The Equity Funding Company was an insurance and investment company that used the premiums from the sale of insurance to invest in mutual funds to earn good interest. When sales were slow, employees began "creating" sales to impress actual and potential stockholders. Meetings were held where employees filled out insurance application forms with names drawn at random from Southern California phone directories. On paper, it looked as if Equity Funding was a thriving, growing company. Not until one of those employees blew the whistle on the operation was the truth discovered. Quickly, the stock plummeted in value, destroying the investments of many people (Soble and Dallos, 1975).

Another common type of abuse of power against owners is computer crime. Precise estimates are hard to come by, but some observers believe that computer criminals steal billions of dollars each year. Because computers are versatile and powerful tools, many businesses and other organizations have come to rely on them to perform the many tasks involved in modern business. As more people become computer literate and as more organizations come to rely on computers, abuse is not far away. Computers can be used to perform unauthorized functions, such as monetary transactions and transfers. In a large bank, for example, it might be possible to reduce all account balances by a very small amount (such as 1 percent) and place the funds into another, new account. Depending on the balances, this could result in a great deal of money.

In other computer crimes, unauthorized information could be obtained from computers, or information such as grades in the registrar's office of a university could be changed. Thieves who are able to gain access to another company's files through the computer might be able to steal company or trade secrets. Such a method is much more efficient and less risky than, for example, a burglary that might accomplish the same objective. Using the computer and modern, high-speed data transmission hardware, thieves might be on the other side of the globe so that even if their presence were detected, the thieves themselves might not be apprehended. Attempts to prevent this type of crime include new encryption systems to prevent unauthorized access to computer files and constant monitoring of who has access to which parts of a computer system's files. Some states have also written legislation to deal with such crimes in the criminal justice system, although traditional police methods are not adequate to deal with this crime. The police are not sufficiently well trained to detect or solve these kinds of crimes. Like most white-collar crime and corporate crimes, the offenses themselves are often brought to light by "whistle-blowers" (who decide to turn in the offenders) or persons with technical training who notice that things are not as they should be.

Crimes Against Employees   When an employer deliberately violates health and safety laws that protect employees because of the financial savings involved, the employer is engaging in a crime against employees. Failure to take adequate precaution against employee injury would be an example of this kind of violation. Sometimes corporations take the attitude that while there may be dangers in the workplace, it is the responsibility of the worker to take reasonable precautions against those dangers. It is sometimes difficult to discern the long-term consequences of particular working conditions. It is now known that exposure to asbestos fibers will eventually lead to serious lung diseases and death. Workers in shipyards, where asbestos is a principal raw material in the construction of ships, have been exposed to this harm for some time. Many died before it was known that workers should take precautions. If an employer knowingly permits employees to expose themselves to such dangers, such an act is clearly illegal.

A federal agency, the Office of Occupational Safety and Health Administration (OSHA), is responsible for maintaining safe working conditions. OSHA operates with inspectors who conduct surveys at work sites to ensure that employers are abiding by safety regulations in the workplace.

The problem of worker safety on the job received national attention during the Gauley Bridge disaster in 1930–1931. In this tragedy, the silica dust churned up during tunneling work for a hydroelectric plant killed nearly 500 workers. Most of the fatalities occurred well after the project had been completed, although 169 workers literally dropped dead during the construction and were buried two or three to a grave in a nearby field. In spite of such tragedies, effective occupational health and safety legislation did not develop until decades later.

Most corporate crimes victimizing workers involve injuries, exploitation, or simply job dismissal (Albanese, 1995: 66). Many labor laws are designed to reduce unfair treatment of employees. For example, the company that owns Jack-in-the-Box restaurants in California was fined for nearly 400 child labor law violations. These violations involved teenagers working more hours than permitted by law. Direct injuries can also occur in more serious violations, such as when Union Carbide Corporation was fined $1.37 million by OSHA for hundreds of violations in its West Virginia plant. Employees were required to detect the presence of deadly gas by sniffing the air after an alarm indicated a leak. They were without respirators.

Plant inattention to basic safety standards can result in deaths. A Phillips Petroleum chemical plant exploded in 1989, killing 23 workers in Pasadena, Texas. The plant had been cited for 19 safety violations prior to the explosion. In 1993, officials from Imperial Food Products were charged with 25 counts of involuntary manslaughter when its chicken processing plant in North Carolina caught fire, killing 25 people who were trapped inside and unable to escape the burning building. Another 56 workers were injured in the fire. The deaths could have been avoided had the plant officials followed accepted safety procedures.

In January 2000, it was learned that some workers in a federal uranium-processing plant in Paducah, Kentucky, participated in the 1950s in experiments that had them breathing the radioactive element. Some workers volunteered for the tests, but some workers may not have been fully informed of the dangers. While the general dangers of radiation were known at the time, many believed that uranium dust posed little or no problem to the human body. In everyday working conditions, workers faced dangers that were not explained to them. Even worker families were exposed to radiation when the contaminated clothing was taken home to be laundered.

Crimes Against the Public at Large   We are all the victims of abuses of power when we breathe polluted air or drink or use polluted water. When a company continues to pollute, in violation of Environmental Protection Agency standards and rules, it violates its position of trust and power for its own gain. For some companies, a decision not to pollute is an expensive one; anti-pollution equipment is expensive, and if that equipment is not available, the company may have to re-tool or go out of business to avoid pollution. The costs may simply outweigh the benefits.

The Hooker Chemical Company was faced with a hard decision in the early 1940s and 1950s (*Wall Street Journal*, April 29, 1980: 12). As part of its normal production process, Hooker created chemical by-products and waste material. The company decided to bury about 21,000 tons of the waste in violation of existing statutes in an area of Buffalo, New York—near the factory—known as Love Canal. Three decades later, it became clear that this was a disastrous decision. Residents of Love Canal began to exhibit an abnormally high number of cases of cancer and birth defects. The health problems were traced to the chemical waste buried by Hooker years before. That waste had eaten through the metal drums and had risen close to the surface, getting into the drinking water of the area and, in some spots, actually surfacing. By the time the problems were discovered, Hooker Chemical Company was out of business. The federal government, through the Environmental Protection Agency, was forced to evacuate the area, provide for the long-term medical care of some of the residents, and find housing in other places for those evacuated.

Abuses of power can victimize government as well. In 1977, Revco Drug Stores, a large nationwide chain, was convicted of defrauding over $500,000 from the Ohio Department of Public Welfare by double billing Medicaid (Vaughan, 1983). Revco found that although it wished to continue to serve senior citizens, the billing procedures for Medicare to receive reimbursement were terribly cumbersome. The regulations were lengthy, and the paperwork took much time. To make matters worse, Revco found that Medicare would reimburse for one drug at one time, but not the same drug for another patient at another time. It appeared that bureaucratic red tape was the culprit, and Revco decided to simply double bill on those drugs that Medicare paid for at one time but not another.

A number of well-known scandals have involved government officials. In 1974 a grand jury in Washington, D.C., returned a 13-count indictment against seven officials who held positions in the White House. In what would come to be known as the Watergate scandal, these individuals would be charged with such crimes as conspiracy, obstruction of justice, and lying under oath. All of those charged, including former Attorney General John N. Mitchell, John Ehrlichman, once assistant for domestic affairs, and Robert Haldeman, former chief of White House staff, were found to be involved in various cover-ups of a burglary at the Democratic National Committee. President Richard M. Nixon resigned his office.

The Iran–Contra scandal of the 1980s involved the indictment and conviction of a number of officials of President Ronald Reagan's administration. These persons engaged in an illegal scheme to divert profits from arm sales to Iran to support a rebel group in Nicaragua called the Contras. Nearly $4 million was transferred to the Contras in direct violation of a congressional ban on such activity. Nine former administration officials pleaded guilty or were convicted by 1992. The highest ranking official was Casper Weinberger, former Secretary of Defense, but before he was to be tried, Weinberger was pardoned by President George Bush, who in January 1993 was in his last month as president.

Costs of White-Collar and Corporate Crime

The degree of harm from white-collar and corporate crimes can be estimated, but not determined with great accuracy. The costs of these crimes tend to go beyond that for conventional crime. We can distinguish three kinds of costs from white-collar and corporate crime: financial, physical, and social.

Financial Harm   Although precise financial estimates of the economic impact of abuses of power do not exist, several estimates of such impact have been offered. In 1974, the U.S. Chamber of Commerce estimated the short-run direct cost of white-collar crime to the U.S. economy at no less than $40 billion annually, an estimate that is consistent with that quoted by Congressman John Conyers in hearings before the Subcommittee on the Judiciary in 1978. In 1976, the Joint Economic Committee of the U.S. Congress put the figure at $44 billion annually. Several observers since that time have pointed out that this estimate is very conservative and excludes a number of offenses. Senator Philip Hart, as chair of the Judiciary Subcommittee on Antitrust and Monopoly, estimated that antitrust law violations may illegally divert as much as $200 billion annually from the U.S. economy.

Congressman Peter Rodino, in hearings conducted in 1978, informed the Conyers committee that the Justice Department estimated in 1968 that the estimated loss due to violations of the Sherman Act alone was $35 billion, and a 1977 Government Accounting Office study estimated that frauds against government programs in seven federal agencies alone costs the taxpayers roughly $25 billion. Rodino placed the estimated loss from all forms of white-collar criminality at closer to $100 billion annually. Estimates of total financial loss from white-collar crimes are in the billions of dollars each year, and estimates of financial loss from specific white-collar crimes are similarly high. The American Management Association has estimated that the loss due to employee pilferage—arguably a white-collar crime, depending on the status of the employee—costs the business community $5 billion a year. A more recent FBI estimate places the total costs from white-collar crime to be between $100 billion and $200 billion annually (*USA Today*, September 1, 1987: 3A), a figure so alarming that the FBI has committed about 1,400 agents to the problem.

More recent cost estimates for these crimes range as high as $200 billion a year, as compared to estimated annual losses from street crimes of $3 or $4 billion. The cost of a single organizational crime may run into the millions or even billions of dollars. For example, Exxon ran up $2 billion in illegal gasoline overcharges from 1974 to 1981. Potential business violations include restraint of trade (price-fixing and monopoly control); fraudulent sales; illegal financial manipulations; misrepresentation in advertising; issuing fraudulent securities; income tax violations;

misuse of patents, trademarks, and copyrights; manufacturing unsafe foods and drugs; illegal rebates; unfair labor practices; unsafe working conditions; environmental pollution; and political bribery.

Perhaps the most comprehensive estimates are those presented by researchers at the National White Collar Crime Center (Helmkamp, Townsend, and Sundra, 1997). Using an exclusive definition and providing a range of estimates, these investigators estimate that white-collar and corporate crime costs between $425 billion and $1.7 trillion dollars a year. Such figures, of course, are staggering, but we are only just beginning to zero in on sound estimates for the financial costs of this form of criminality.

Most observers are quick to point out that the estimates they provide are conservative and that the actual loss is probably far greater. There is agreement, however, that the annual cost of abuses of power is far greater than that from ordinary or conventional crime. Data sources for such estimates are inconsistent and plagued by problems of reliability and validity. It seems quite safe to say that statistics on abuses of power are at a more primitive stage than were statistics on street crime prior to the initiation of the Uniform Crime Reporting system in 1930.

Physical Harm   As high as financial estimates are, by most standards, they do not include the total losses that accrue from abuses of power. For example:

> They [financial estimates] do not cover the losses due to sickness and even death that result from the environmental pollution of the air and water, the sale of unsafe food and drugs, defective autos, tires, and appliances, and of hazardous clothing and other products. They also do not cover the numerous disabilities that result from injuries to plant workers, including contamination by chemicals that could have been used with more adequate safeguards, and the potentially dangerous effects of work-related exposures that might result in malignancies, lung diseases, nutritional problems, and even addiction to legal drugs and alcohol. (Clinard, 1979: 16).

Physical harm, like financial losses, can be directed toward at least three different groups: employees of offending firms, consumers, and the community at large. Physical harm to employees includes unsafe working conditions, such as those found in many mining operations and in fiberglass plants. The effects of black lung disease and asbestos poisoning, although relatively slow to develop, can result in death.

Harm experienced by consumers includes the sale of unsafe products, such as flammable clothing for children, and impure food and drugs. Perhaps the most dramatic and significant case of physical consumer harm in recent history arose over the manufacture and sale of the Ford Pinto, discussed earlier, which had been linked with a number of driver and passenger deaths due to an unsafe fuel tank. Although the criminal trial related to this case resulted in the acquittal of the Ford Motor Company, commentators have been quick to point out that the principle of manufacturers' criminal liability for their products was more firmly established by the trial. Many other instances of severe physical harm might be cited. For several years, the Beechcraft Company allegedly used a fuel pump with a faulty design that caused a number of deaths of pilots and passengers in the Beechcraft Bonanza series aircraft; the engine would often stall when the plane was banking slightly shortly after takeoff, causing a loss of power and control.

Harm to the community at large can take many forms, such as pollution—air, water, and noise. A government report has estimated that 14,000 persons in the United States who would have died in 1978 of lung cancer and other diseases related to air pollution were spared because of improvements in air quality since the enactment of the Clean Air Act of 1970. The estimate was derived from previous studies of the impact of air pollution.

Perhaps because physical injuries are not readily quantifiable in terms of dollars and cents, these consequences of white-collar and corporate criminality are viewed as more serious by citizens than are financial or property losses. One problem is that it is often impossible to demonstrate that actions leading to physical injuries were intentional or were the result of faulty decision making or other human qualities. This evidently accounted for the court decision that found Ford Motor Company not guilty of the deaths of persons resulting from Pinto fuel tank explosions and fires. It is sometimes hard to say who is guilty, which executive or managerial decisions were the instrumental ones, and what the word *intentional* means in these cases. No one argues seriously that Ford officials wished to kill or injure Pinto drivers or passengers, but clearly their permitting the faulty fuel tank design to continue had that effect.

Social Costs or Damage to the Moral Climate   Although few dispute that the financial loss and physical harm due to abuses of power are enormous, perhaps that type of harm that has been stressed most forcefully by sociologists is the set of broader social consequences of crimes committed by persons of power. Sutherland speculated that:

> The financial cost from white collar crime, great as it is, is less important than the damage to social relations. White collar crimes violate trust and, therefore, create distrust; this lowers social morale and produces social disorganization. (Geis, 1972: 380–381).

Persons of wealth and high social standing are often held to very high standards of accountability for their conduct. The concept of noblesse oblige expresses this accountability. As one observer has put it: "It can be argued, convincingly I think, that social power and prestige carry heavier demands for social responsibility, and that the failure of corporation executives to obey the law represents an even more serious problem than equivalent failure by persons less well-situated in the social structure" (Geis, 1972: 381).

Because of the high social standing of white-collar offenders, some observers have maintained that these violations create cynicism and foster the attitude that "if others are doing it, I will, too." Tax authorities have used this interpretation of the fact that after exposure of former President Richard Nixon's tax deceits, false reporting of taxes increased substantially. More fundamentally, it is held that white-collar crime threatens the trust that is basic to community life—for example, between citizens and government officials, professionals and their clients, businesses and their customers, employers and employees, and—even more broadly—among members and nonmembers of societies. Thus, Cohen argues that "the most destructive impact of deviance on social organization is probably through its impact on trust, or confidence that others will, by and large, play by the rules" (Cohen, 1966: 4–5). Because both offenders and the offenses are "highly placed," this is a very serious consequence of abuses of power since it can affect the way in which we interact with others.

The negative impact of some abuses of power on our trust in one another assumes that (1) high-status persons serve as moral role models for the rest of the population, who, in turn, pattern their behavior after those they emulate, and (2) that the public generally views abuses of power as relatively serious, at least compared with street crime. While these assumptions may be questionable for certain abuses of power, consumers who feel they have been cheated by a merchant are certainly less likely to shop there again and, perhaps, more likely to regard other merchants suspiciously. Similarly, if one has been overcharged for car repairs or charged for car repairs that were never done, it is possible to develop an attitude that "all mechanics are crooked," even if one's sample of business relationships with car mechanics is limited.

It is possible that the public can develop negative attitudes toward social institutions as a result of crimes that involve abuses of power. While one might be suspicious of a physician who overcharges for professional services, such an attitude might easily generalize toward the entire profession of medicine. Negative experiences with retailers can lead to a negative attitude toward business in general. Illegal political contributions might create public cynicism toward government, just as bribes paid by private corporations to foreign officials can damage foreign relations. In the same way, exposure of corruption on the part of an elected official can lead to a general suspiciousness of politics and a stereotyping of all politicians.

Public opinion polls conducted during this century indicate that the public in the United States has recently been indicating more negative attitudes toward many social, economic, and political institutions. These feelings of distrust and suspiciousness may stem from direct abuses of these institutions. Such feelings may also stem from the inability of the institutions to prevent abuses of power. A national telephone survey suggests strongly that the public regards certain types of white-collar crimes as more serious than certain types of street crimes (Rebovich and Layne, 2000). For example, when asked which is more serious, 54 percent of the respondents chose embezzlement, and 27 percent picked robbery. Asked to compare armed robbery that results in an injury with deliberately selling a tainted product that results in injury, 42 percent of the respondents believed the white-collar crime was more serious, compared with 39 percent for the robbery.

Explaining Corporate Criminal Behavior

Business activities pursue many economic objectives, so analysts may reasonably expect the rates for such crimes as antitrust violations to rise during times of deteriorating business conditions and decreased profits (Simpson, 1987). But economic causes alone fail to predict the extent of a firm's illegal behavior (Clinard and Yeager, 1980: 127–132). Organizational analysis may attribute unethical business practices and law violations to a company's internal structure. One might best explain such violations as products of (1) a company culture tolerant of unethical or illegal acts, as reflected in the conduct of top executives, and (2) a structure of decision making that distorts responsibility for decisions (Shover and Bryant, 1993). Particularly in large firms, structural complexity and specialization obscure links between decision makers and the effects they cause. Decentralized decision making also complicates monitoring and control. The probability of illegal acts rises, and chances for detection and individual responsibility fall.

Every business organization builds up a distinctive cultural pattern of actions permeated with its basic position on ethical standards and obedience to the law. This culture reflects a firm's continuing emphasis on maintaining a good reputation, internal attitudes toward market expansion and power, a sense of social responsibility, and the strength of concerns for employees, consumers, and the environment. In pursuing business objectives, an organization may proceed ethically or unethically, in compliance with the laws or in violation of them.

Organizations that tolerate unethical actions tend to socialize members to accept climates of unethical behavior conducive to criminality. A former SEC enforcement chief once said, "Our largest corporations have trained some of our brightest young people to be dishonest." In a case involving a large-scale illegal price-fixing conspiracy in the large folding carton industry, one executive testified that "each was introduced to price fixing practices by his superiors as he came to the point in his career when he had price fixing responsibilities" (Clinard and Yeager, 1980: 64–65).

Widespread prevalence of unethical and illegal practices characterizes certain industries (Sutherland, 1949: 217–220). Researchers have identified such lax standards in at least four industries—autos, oil, pharmaceuticals, and defense (Braithwaite, 1984; Clinard, 1990: 21–90; Clinard and Yeager, 1980: 119–122). The auto industry has long labored under a reputation of having widespread disregard for laws designed to protect the safety of consumers, prevent consumer fraud, and safeguard the environment. The oil industry has compiled a long, industrywide history of violations that include price-fixing, illegal overcharges, illegal campaign contributions, and environmental pollution. Pharmaceuticals manufacturers have frequently produced and distributed unsafe and ineffective medications and drugs. Most of the nation's giant defense corporations have habitually charged military buyers for fraudulent cost overruns and other expenses and bribed government officials.

Along with a company's culture, the other internal determinant of organizational misbehavior comes from the role of top management, particularly the CEO. Again, the complex structural relationships within large companies sometimes create tangled combinations of delegated authority, management discretion, and the ultimate responsibility of top management. A company's top managers communicate its goals to middle managers, who accept responsibility for achieving those goals. These employees may feel intense pressure to achieve assigned goals, by legal or illegal means, since prestige, promotions, and bonuses often rest on the outcome. If investigators subsequently uncover violations, however, top managers can deny responsibility, claiming ignorance or insulation from middle managers' decisions to break the law. Clearly, top management can set the ethical tone throughout a company, though. One middle manager furnished an example:

> Ethics comes and goes in a corporation, according to who is in top management. I worked under four corporation presidents and each differed. The first was honest, the next one was a "wheeler-dealer." The third was somewhat better, and the last one was bad. According to their ethical views, varying ethical pressure was put on middle management all the way down. (Clinard, 1990: 172)

Some top executives evaluate employee performance by looking only at "bottom line" net profits. They set one ultimate test of good management: how profitably the company operates. They may not care how morally it acts. Despite predictable violations, executives do not think of themselves as criminal offenders.

Controlling White-Collar and Corporate Crime

Unlike the control of other types of criminality, the control of white-collar and corporate crime requires an understanding of the prevailing legal control philosophy. The criminal law is founded on the premise that illegal behavior should be punished and that criminal sanctions should be punitive in order to achieve certain desired ends, such as deterrence, rehabilitation, and retribution. Laws governing abuses of power are founded on a different premise. The aim of administrative sanctions is to obtain compliance from offenders, not to punish them for wrongdoing. Whereas the criminal courts may sanction to deter the offender from future violations, administrative courts may sanction, if they do, in order to get offenders to terminate their illegal behavior.

The difference is an important one. Because the object in administrative law is compliance, offenders may be given an injunction or a legal order to terminate some action. In many instances, that will be the only sanction given violators if they do terminate their illegal behavior. Thus, a paper company that is systematically polluting a river may be given an injunction to stop that action and no further penalties if the pollution stops. It is not conceivable that a criminal court would sanction a burglar with only a warning to stop that behavior; additional punishment for the burglary would also be given the burglar. The difference between the different sanction strategies behind white-collar and conventional crimes does not relate to any differences in the nature of the behavior of the crimes, but the histories of the administration of penalties. Because of the differences between the philosophies of criminal and administrative law and because of the differences in sanctioning individuals as opposed to corporations, some legal scholars have suggested that legal intervention should be used sparingly with corporations and only under certain conditions.

As with other forms of criminality, it may never be possible to completely eliminate criminal abuses of power. But there are some measures that may be helpful in reducing the incidence of these crimes. Those measures include public education, developing a stronger business ethic, legislation, and using publicity to change corporate behavior.

Public Education   By and large, most persons have little information about the extensiveness of white-collar and corporate criminality. This is also the case even when those same people have been victimized by these crimes. Many of these crimes could be reduced if individuals were knowledgeable about the crimes and could therefore reduce the chances of their own victimization. Because authorities must rely on citizen complaints, it is only through victim awareness that many of these offenses and offenders can be brought to accountability for their crimes. If citizens knew they were being victimized, they would also be more likely to cooperate in the prosecution of these cases, thereby increasing conviction rates.

Public education is not only necessary to reduce its own victimization from crimes of abuse of power, public awareness is also necessary in order to press for other changes, such as increased penalties for these offenders and more vigorous enforcement practices. As long as the public is ignorant about the range, types, and seriousness of abuses of power, there is little momentum for change. Sociologists have long noted that white-collar and corporate crimes do not seem to generate the social condemnation that other, common crimes do. This characteristic has reduced the number and kinds of reforms possible to control this kind of illegal behavior. Greater public awareness of abuses of power could lead to moving up these offenses on the public's crime agenda.

Increasing Ethical Behavior in the Workplace   Perhaps the best way to curtail abuses of power is to somehow persuade persons in positions of power that they ought not engage in that behavior. The first step in this regard is for corporations to insure that they do not engage in illegal practices and that employees know that such practices are condemned within the corporation. Many organizations have codes of ethics that help them identify unscrupulous conduct. But it appears that it is not the codes of ethics themselves that reduce abuses of power, but the demonstration of condemning illegal practices that makes the differences. It is, in other words, the deeds of top management, not the words in codes of conduct, that more powerfully determine behavior.

An additional method to increase the sense of ethics in occupations is through the development of stronger codes of conduct endorsed by influential organizations, such as the U.S. Chamber of Commerce for business and the various professional associations for professionals. While such associations typically have little to do directly with the development of ethical codes, their endorsement would serve to increase the legitimacy of existing codes and reinforce the importance of conforming to such codes. Teaching courses on ethics in professional schools and colleges would also tend to reinforce the importance of such behavior. In instances of violations of ethical codes, violators should be sanctioned in such a manner that others know of both the violation and the sanction.

One obvious way to increase more ethical behavior in organizations is to respond to instances of organizational deviance through sanctioning. Employers can use a variety of such mechanisms, including formal and informal reprimands and firing the unethical employee. The value of informal reprimands is that the individual employee can be personally shamed to change his or her behavior. Letting the employee know that such conduct will not be tolerated, yet offering to let the individual stay with the organization, may be an effective mechanism of social control (Simpson, Exum, and Smith, 2000).

Legislative Reforms   Most legislative reform takes two forms: (1) an increase in the number of abuses of power that would be incorporated into the criminal law and (2) the creation of more stringent penalties for those abuses of power that are now in criminal law. Any effort to increase the number of crimes covered by law would necessarily involve an increase in enforcement resources so that offenses could be detected and violators brought to justice. Some additional reform effort could be directed to more vigorous criminal handling of white-collar offenders. There is reason to believe that white-collar offenders are particularly deterrable, given the circumstances of their crimes, and the law may be able to make a significant impact on this behavior. Criminologists have recently turned their research attention to those official agencies responsible for sanctioning persons guilty of abuses of power. Those studies have included work on federal prosecutors, the Securities and Exchange Commission, and the Office of Surface Mining.

Legislative reforms cannot and should not be limited to reforms of increasing the criminalization of abuses of power. Laws dealing with abuses of power are terribly complicated and cumbersome. Much savings would be realized if such legislation were streamlined and more easily interpretable. The legislative answer to abuses of power is not more law, but, in a sense, less. Fewer regulations that would be easier to enforce with perhaps more biting penalties might have a greater effect than our present system of administrative and criminal laws and penalties.

It is clear that government regulation of some kind is a necessary part of any system of control over abuses of power. Many persons in the corporate world agree. Clinard found that nearly 3 out of 4 recently retired midlevel managers of large U.S. corporations believed that government regulation was necessary because industry cannot police itself. Most of the middle managers believed that top management knew about corporate violations of the law either in advance or shortly thereafter. The middle managers that Clinard interviewed also felt strongly that top management sets the ethical tone for compliance to law within the corporation.

Specific legislative reforms might include the use of special sanctions for white-collar criminals, such as community service. Some white-collar offenders as part of their judicial sentence have had to lecture to community audiences. In one well-known instance of this type of sanction, antitrust offenders were required as a condition of probation to make an oral presentation before 12 business, civic, or other groups about the defendant's case. The audiences were then sent questionnaires about the presentations, and those returned indicated that audience members reported being more informed about antitrust law and forced to reexamine their own antitrust procedures.

Increasingly, civil penalties are appropriate for corporate abuses, and people who feel they have been wronged can sue a corportation in civil court. The penalties in such actions may be substantial, as in the tobacco cases in recent years, although civil juries often have discretion to alter awards. It appears that civil juries are not necessarily influenced by an anti-corporation prejudice or sympathy for the plaintiff (Han, 2000). In fact, many civil juries may be hostile to plaintiffs in such cases, and there are only occasional instances of an anti-business prejudice. Still, civil suits do represent a viable alternative for persons who wish to attempt to settle a corporate matter in civil court.

The Use of Publicity   An issue related to the use of community service sanctions is the use of adverse publicity against offenders. Most corporations are concerned with their public image. This concern is financially motivated, and this explains why some corporations spend a good deal of money on charities and other causes where the corporation's involvement is visible. A corporation may, for example, underwrite a portion of the costs for a public television program so that it can tell viewers that it is doing so. Because of this concern over corporate image, a number of criminologists have suggested that adverse publicity may be a powerful tool in getting corporations to change their behavior. There has been some skepticism, however, about whether corporations can be stigmatized in the same way in which individuals are:

> Sociologists . . . talk about corporate recidivists, but there is very little evidence to suggest that the stigma of criminality means anything very substantial in the life of a corporation. John Doe has friends and neighbors; a corporation has none. (Packer, 1966: 361)

Fisse and Braithwaite (1983) studied the impact of negative publicity with 17 case studies of corporate crime. They concluded that adverse publicity made an impact on the corporation in each of the 17 instances, although the degree of change varied from corporation to corporation. In fact, many corporations produced changes in their operations prior to or in the absence of conviction—all as a result of adverse publicity. In general, Fisse and Braithwaite concluded: "Publicity hurts most when it challenges the integrity of a product." When products are challenged, the publicity is extracting a financial cost to the corporation. Corporate officials, however, indicated that perhaps the greatest impact of adverse publicity is the nonfinancial costs it produces—loss of corporate and individual prestige, a decrease in employee morale, distraction from the job at hand, and embarrassment about the incident that generated the adverse publicity.

Given these findings, Fisse and Braithwaite make two recommendations. First, publication of the details about an offense should be made available as a court-ordered sentence against corporate offenders. These details should be available to the mass media, not only to create the adverse publicity, but also for the purpose of creating a remedy. Second, Fisse and Braithwaite recommend that probation orders be used to require disclosure of organizational reforms and disciplinary action taken as a result of the offense.

Summary

White-collar crime is serious crime committed during the course of one's occupation. With the advent of laws regulating business, white-collar crime has come to be associated with economic crime, but virtually any organization, for example, political parties, can commit white-collar crime as well. There are different definitions of white-collar crime, but for our purposes we suggest that white-collar crime is committed by individuals, usually for self-gain, while corporate crime is committed on behalf of an organization such as a business firm. The main elements of this conception include occupation, law, fraud, and power.

White-collar crimes are characterized by sophistication and planning. White-collar criminals usually have a noncriminal self-concept. The fact that most of these crimes are not handled by uniformed police offices or criminal courts and the fact that the offenders are immersed in a web of conventional roles helps to insulate them from a criminal self-concept. One classification of different kinds of white-collar crimes distinguishes crimes based on the victim of the crime.

Selected References

Friedrichs, David O. 1996. *Trusted Criminals in Contemporary Society*. Belmont, CA: Wadsworth.

An excellent summary and review of the literature on white-collar and corporate criminality. Virtually all important topics are covered.

Rosoff, Stephen M., Henry N. Pontell, and Robert Tillman. 1998. *Profit Without Honor: White-Collar Crime and the Looting of America*. Englewood Cliffs, NJ: Prentice-Hall.

The authors cover a multitude of white-collar and corporate crimes in a readable, informative style. This is one of the best single sources on specific cases of white-collar crime.

Sutherland, Edwin H. 1983. *White-Collar Crime: The Uncut Version*. Introduction by Gilbert Geis and Colin Goff. New Haven, CT: Yale University Press.

Sutherland's original work is still worth reading. This version of the book contains the names of the corporations Sutherland was forced to omit in the first edition. The introduction contains interesting biographical information about Sutherland.

Vaughan, Diane. 1996. *The Challenger Launch Decision*. Chicago: University of Chicago Press.

The Challenger disaster has been cited as a gross example of corporate misconduct, but Vaughan argues strongly that the launch decision must be understood in the context of organizational decision making. Her analysis points to the importance of understanding the social situation in which corporate decisions are made.

CHAPTER 10

Drug Use and Addiction

Larry and Janice are a happily married couple in their 30s. They have three children and, like many parents, work hard to balance their productive work lives with family obligations. They also smoke marijuana. Larry is a lawyer who commutes each day to his workplace in a large city, while Janice works closer to their home. Larry has developed some contacts that permit him to "score" some marijuana whenever he and Janice are in the mood, which isn't all that often. They use marijuana only recreationally: on weekends, at night, and only when the kids are in bed. Marijuana is not a major part of their lives, and they don't think much about it.

Jason is a 16-year-old high school student who is obsessed with marijuana. He smokes it whenever he can. He is doing poorly in school mainly because of his persistent truancy. When he is not at school or home, he is likely to be smoking marijuana. Over the past year, Jason has found himself gravitating to a group of friends who share his interest. Jason and his friends find it exciting to do something illegal, and they like the "high" they get from the marijuana.

These two cases are very different, although the drug is the same in each case. We will see, though, that it is not so much the drug itself but the user that determines how and when drugs are used and the role that drugs have in people's everyday lives.

Drug use is a fact of daily life for most Americans. Some people may object to this shocking statement and deny that they are drug users, but their outrage only reflects their ambivalent attitudes about and selective awareness of drugs. In fact, taking drugs of various kinds has so completely pervaded the behavior patterns of people in the United States that the entire general public could never conceivably abstain from all drug use. Legal drug use is so common that virtually no one recognizes the activity as part of society's relationship with drugs. For example, many people do not consider coffee, cigarettes, or soft drinks (or their ingredients) as drugs.

Yet many nondeviant people wake up in the morning with a drug on their mind; they start every day with a dose of coffee, hoping that its caffeine will impart energetic feelings. At midmorning, they might feel the need for another drug (aspirin) as work pressures build. Some indulge in another drug, alcohol, at lunch. A midafternoon break might continue the pattern of drug use with a cigarette (nicotine) and more caffeine in a cola drink. At home, someone who abhors drug use may down a quick cocktail before supper, perhaps followed by wine during the meal. An after-dinner drink and another cigarette (how many is that for the day?) help to settle dinner. As bedtime approaches, some folks swallow pills either to fall asleep or to stay awake. (How many students pop No-Doz so that they can keep studying into the night?) At different times during the day, someone suffering from an illness may take a prescription medicine or some over-the-counter remedy. The next day, the cycle starts again.

Several features of a normal life actively promote drug use. First, most people recognize a close connection between drugs and physical well-being. For example, children learn that drugs relieve various physical discomforts. If you have a headache, take aspirin. Stomach not feeling well? Take Alka-Seltzer. Menstrual cramps? Athlete's foot? Scraped elbow? All these problems require the same kind of solution: drugs.

Also, people associate use of alcohol, an important part of the drug world, with certain social and life events. Drugs often play important roles in some people's celebration or mourning rituals. We commonly pair many events with consumption of some drug:

- Birthdays
- New Year's Eve celebrations
- Parties
- Celebrating new jobs or promotions
- "Drowning sorrows" after being fired or demoted
- Birth of a baby
- Wakes or funerals
- Sporting events
- Dates
- Religious ceremonies
- Graduations
- Weddings
- Meals
- "Sorry you're in the hospital"
- "Glad you're out of the hospital"
- Meetings with friends
- Nightcaps
- Out on the town on a Saturday (or any other) night

In addition, people think of drug taking as a way to attain desired moods or psychological well-being, perhaps a universal desire (Weil, 1996). People learn that when they fall into undesirable moods, they can alter their own feelings with drugs. A single drug may generate a new mood, such as euphoria, or alter an existing one, such as depression.

This mood-altering function of some drugs has both attracted and repelled some potential users, and it contributes to the ambivalence that many people feel about drugs. This ambivalence showed in a news magazine cover story that appeared a few years before cocaine grabbed national headlines. The cover showed a martini glass filled with white powder, and the caption read, "Cocaine: A Drug with Status—and Menace." The story described cocaine as "no more harmful than equally moderate doses of alcohol and marijuana, and infinitely less so than heroin" (quoted in Baum, 1996: 142). This combination of positive and negative feelings about drugs has promoted conceptions of drug taking as deviant behavior.

Drug Taking as Deviance

Few terms appear more commonly and with more confusing or misleading meanings than *drugs*, *drug users*, and *under the influence of drugs*. Scattered widely through media reports and public discussion, these terms actually refer to a tremendous variety of substances and behaviors. These references apply to different substances, expectations of the effects those substances will produce, and the immediate environments in which users take them. The word *drugs* covers a range of substances from aspirin and antacid pills to alcohol; hallucinogenic drugs such as marijuana; stimulants like caffeine, nicotine, and cocaine; mind-altering drugs; mood modifiers; and psychoactive narcotics like heroin that profoundly affect the central nervous system, influencing mood, behavior, and perception through action on the brain.

Many people regard drug use as one of the major problems facing society today. Yet those same people easily accept use of some drugs under certain circumstances as benign and even beneficial practices. Still other drugs gain wide public acceptance through connections to social and individual situations. Clearly, people use drugs for many different reasons. Doctors use them to treat disease, ease pain, and control emotions; drugs sedate excited patients, relax people in social situations, and relieve tension and boredom; they provide pleasure, satisfy curiosity, and, some say, open the mind to new feelings of sensitivity or spirituality; they help to create bonds of fellowship and to increase sexual performance. Whatever the reason that people use drugs, society must work to understand the nature of these substances and the situations in which people use and misuse them (and how people distinguish one from the other), as well as the social meanings of such behavior for both those who use drugs and those who do not.

The deviant character of drugs does not result only from the effects or characteristics of particular substances; the purpose for taking a drug is one criterion of deviant use. Use of a drug for medical purposes, such as an opiate that functions as a painkiller after surgery, is not considered deviant; only deviants use the same drug merely to avoid withdrawal symptoms from addiction. Similarly, the physical properties of drugs do not explain the deviant identity of use behavior. For example, among drugs that produce physical dependency, people associate some—such as heroin—with deviance, while others—such as caffeine—escape that stigma.

The concept of a *drug* refers to a substance with a chemical basis, but beyond this generality, nothing distinguishes nondeviant from deviant drug use. Consider one definition: "any substance, other than food, that by its chemical or physical nature alters the structure or function in the living organism" (Ray, 1983: 94). Such a definition includes car exhaust fumes, a bullet, perfume, a cold shower, penicillin, and ammonia. As a result, Goode (1999: 58) argues "a drug is something that has been defined by certain segments of the society as a drug." In other words, the concept is a socially created identity; attitudes toward drugs come from the same source. No combination of chemical properties distinguishes drugs from nondrugs; drugs share basically one characteristic: They have all been labeled as drugs. The term *drug* has certain socially determined and usually negative connotations.

The deviant character of using a particular drug depends on norms, which are also socially created standards. Norms may define nonmedical use of certain drugs as deviant behavior. Society takes a much different view of a physician's use of opiates to control pain in a patient and a recreational user's indulgence in opiates to get high. Because norms vary, so, too, do conceptions of deviance associated with certain drugs. In India, for example, the high castes of society display a strong, religiously associated aversion to alcohol; but not only do the same people tolerate consumption of bhang, a liquid form of marijuana mixed with milk or fruit juice, but custom and religious doctrines actually prescribe its use. Guests at weddings often expect hosts to serve bhang. Further, certain priests use marijuana and opium in ceremonies, and they may become addicted to the opium without others regarding them as particularly reprehensible or unusual deviants.

Norms can change over time, and some drugs not considered deviant at one time may open users to sanctions at another time. Opium yielded two important drugs—morphine, a potent drug developed in 1804, and heroin, about three times as powerful as morphine, developed in 1898. These drugs, as well as opium itself, became widely used in the 19th century in the United States, particularly among women, who took them in patent medicines for "female disorders." At that time, virtually anyone could easily and legally purchase opiates. In fact, pharmaceutical chemists initially produced heroin for sale over the counter as a cough remedy. Cocaine, first isolated in the late 1850s, did not become popular in the United States until the 1880s, when promoters proclaimed it a wonder drug and sold it in wine products as a stimulant (Morgan, 1981: 16).

To judge deviance, one must identify the norms that govern use of a particular drug and the situations in which they allow or prescribe its use. One must also determine who promulgates the norm, since some people, such as those who abstain completely, observe different norms than others, such as those who use drugs occasionally in social situations. Both apply different norms than addicts. Groups form different conceptions of drugs they regard as deviant. Within systems of such differences, groups can clearly "create" deviant drug use by persuading others to adopt their norms. This process of promoting one's own norms shows up in changing public attitudes about different drugs and their use.

Social Attitudes About Drugs

Most attitudes identifying drug use as deviance developed during the 20th century. Prior to this time, U.S. society widely tolerated drug use in many forms. During the 19th century, people regarded drug addiction as a personal problem, generally pitying addicts rather than condemning them. Only later did addicts experience the stigma of disreputable characters and addiction gain an association with criminal behavior. In fact, although some statutes originated in the 1800s, laws prohibiting use of drugs such as marijuana, heroin, and cocaine emerged only in the 20th century (Meier and Geis, 1997).

Changes in public opinion and subsequent changes in legal status seem to have followed public acceptance of associations between drug use and disvalued people or lifestyles. Around the turn of the century, opium smoking was prevalent among certain criminal elements. People came to associate marijuana and cocaine use with inner-city life, noting the concentrations of those drugs in urban areas, especially among immigrant groups.

Similarly, attitudes toward cigarette smoking have varied between tolerated and disvalued as people have associated it with varying groups. In the 1870s, for example, many groups and individuals strongly condemned the practice, then most common among urban immigrants of low social status and known as heavy drinkers (Troyer and Markle, 1983). Public approval of smoking increased until the middle of the 20th century, and it has declined since then. As discussed in Chapter 1, changes in public opinion have contributed to recent trends toward legal prohibitions against smoking in public places; these changes came about, not because science supports widespread agreement about the health risks of smoking, but because interested groups have lobbied legislatures and city councils for protection of individual rights not to breathe the smoke of others. Generally, drug use associated with socially marginal groups gains a deviant identity more often than drug use known as common among the well-to-do.

Negative public opinion can certainly result as information accumulates about the health hazards of drug taking, and such hazards have been known for some time. Documentation of health risks from tobacco does not explain why smoking has become deviant, however, since a social process creates standards for deviance. Consumption of a particular drug becomes deviant only when individuals and groups define it as such.

The process of creating deviant sanctions for certain drug-taking behavior involves the actions of individuals and groups who believe, for whatever reason, that no one should take the drug. Such a judgment, of course, involves a definition of *deviance*. Attitudes toward marijuana and cocaine illustrate the specific processes through which these people come to associate disreputable users with certain substances, creating standards for disreputable drugs. These cases also highlight the importance of groups in creating deviance.

The Case of Marijuana   The first major piece of legislation that stemmed from attitudes condemning drug use was the Harrison Act. Passed in 1914, this legislation, along with subsequent statutes, prohibited selling and using of opiates, cocaine, and marijuana without a doctor's prescription (Musto, 1973). As a consequence of the legislation, addicts became criminals, and drugs became recast as mysterious and evil substances. The Marijuana Tax Act, passed by Congress in 1937, was designed to stamp out use of the substance by subjecting smokers to criminal law proceedings. Brought about through the actions of special-interest groups (Becker, 1973: 138–139; Galliher and Walker, 1977), this federal law influenced states to pass similar criminalizing legislation. The Marijuana Tax Act clearly influenced not only public opinion regarding marijuana use but also the subsequent legislative climate for provisions about the drug's use, possession, and sale. Over the next three decades, marijuana gained stature as a major national problem.

Three interrelated factors had fostered the definition of marijuana as a national problem (National Commission on Marijuana and Drug Abuse, 1972: 6–8). First, the illegal behavior was highly visible to all segments of society. Some users did not attempt to hide their behavior. Second, the public came to perceive such drug use as a threat to personal health and morality and to society's well-being. Third, the perception of a significant problem grew out of broader changes in the status of youth and wider social conflicts and issues.

Marijuana use also gained importance as a symbol. In the 1960s and 1970s, a developing association identified the drug with the youth movement, defiance of established authority, adoption of new lifestyles, the emergence of street people, campus unrest, general drug use, communal living, and protest politics. Marijuana came to symbolize the cleavage between youthful protest and mature conservatism, between the status quo and change. As such, legislation targeted not only marijuana itself, but rather the style of life then associated with marijuana smoking. Similar developments affected other drugs; public attitudes toward users of other drugs, particularly heroin, seem to reflect acceptance of a stereotype termed the "dope fiend myth" that views virtually all addicts as unproductive criminals. This stereotype has inspired a widespread and highly negative reaction to drug addicts, regardless of the drugs they use and other circumstances.

In spite of the continuing negative connotations of marijuana use, two states—Arizona and California—passed referenda in 1996 permitting medical use of marijuana. Federal law remains inflexible, however, and Attorney General Janet Reno pledged to continue enforcement efforts against marijuana. State laws that conflict with this position will continue to generate controversy.

The Case of Cocaine   Cocaine has traditionally achieved popularity primarily among people of means. Drug suppliers had to devote much time and effort to harvest coca leaves, ship them through rugged South American jungles and mountains, and refine the drug to its usual crystalline form. Smuggling the drug into the United States further increased suppliers' expenses and risks. As a result, cocaine typically reached consumers as an expensive and relatively rare commodity in comparison to some cheaper and more easily available drugs. For these reasons, cocaine use generally occurred exclusively among relatively well-to-do people as an occasional practice; chronic use appeared to have remained a rare case (Grinspoon and Bakalar, 1979).

These conditions have changed over the past two decades, though. Better and easier transportation as well as a ready market in the United States boosted incentives to smugglers to ship cocaine. With better options than slow burro traffic and narrow, dangerous mountain passes, cocaine harvesters could deliver more product, and smugglers could import more of the drug to the United States. The price began to drop, and patterns of use began to change as working people, students, and others could afford the drug. Patterns of cocaine use are changing rapidly, not only in the United States but also in Europe, South America, and cities in the Far East (Cohen, 1987).

Public concern over cocaine peaked in the mid-1980s (Akers, 1992). While attention to marijuana subsided, concern over heroin and cocaine increased. In 1986, for the first time since the Gallup poll on education began (in 1968), people identified drugs as the biggest single problem confronting schools (*Lewiston* [Idaho] *Tribune*, August 24, 1986: 6A). That same year, President Ronald Reagan and Vice President George Bush supplied urine specimens to prove they were not drug users, at the same time advocating a systematic drug-testing program in schools, workplaces, and government agencies. Two well-known athletes died as a result of complications from cocaine use, further fueling public concern. Drug use among young people, always a concern, acquired special urgency when a cheap form of cocaine called *crack* flooded large, urban areas.

President Ronald Reagan typified public outrage when he proclaimed a "war on drugs." The media publicized the progress of this war with documentaries and special reports on television and special issues of news magazines. Reports and editorials about users' need for treatment accompanied publicity about the war on drugs. On September 2, 1986, a CBS broadcast, *48 Hours on Crack Street*, achieved the highest Nielsen rating of any news documentary on any network in over $5\frac{1}{2}$ years (Trebach, 1987: 13). Subsequent television shows portrayed the new threat as a pervasive problem everywhere, although some evidence contradicted this claim (Inciardi, Surratt, Chitwood, and McCoy, 1996: 2). Critics charged that the so-called war neglected alcohol, which destroys more lives than all other drugs combined, and nicotine in tobacco, which victimizes more people than cocaine and heroin (Trebach, 1987). Further, cocaine deaths actually appeared to have declined in the year prior to Reagan's declaration of war on drugs. Despite these facts, people feared cocaine and heroin more than any other drugs, and a "get-tough" attitude spread animosity toward those buying or selling them. This public outcry continues, despite evidence that the use of illicit drugs declined significantly from 1970 to the mid-1990s (Goode, 1997).

Indications of Changing Attitudes   National surveys of high school students reveal changing attitudes toward the use of various drugs and associated risks. This annual series of surveys, called Monitoring the Future and sponsored by the National Institute of Drug Abuse (NIDA), solicits responses from a sample of about 50,000 students in more than 400 schools throughout the United States. Self-reported data from surveys conducted since the early 1970s provide valuable insight into trends in use of both legal and illicit drugs. Each year's survey indicates the proportion of high school seniors who perceive some harm from the use of drugs, including marijuana, amphetamines, opiates, barbiturates, LSD, cocaine, and tranquilizers. Concern appeared to increase throughout the 1980s for most types of drugs, but these figures began to decline again in 1991 (Johnston, O'Malley, and Bachman, 1993a). In general, high school seniors perceive more potential for harm in regular use of a drug than in occasional use. While 71 percent of the seniors surveyed in 1996 perceived harm in occasional use of crack, only 56 percent agreed that people would harm themselves by using crack only once or twice (Johnston, O'Malley, and Bachman, 1996). According to the most recent survey, rates of disapproval of drugs have either maintained or declined (Johnston, O'Malley, and Bachman, 1999).

Perceptions of harm are not restricted only to use of illegal drugs; the students' concern also extends to such drugs as alcohol and nicotine. These young people, however, express perceptions of harm tempered by inexperience. For example, about 70 percent of high school seniors in 1991 agreed that regular cigarette use would harm the smoker, despite all that is now known about the health risks of tobacco use; by 1996, that figure had dropped to 68 percent. For alcohol, 9 percent of high school seniors in 1991 expected some harm from having one or two drinks, compared with only 7 percent in 1996. Similar declines were recorded for harm from "tak[ing] one or two drinks nearly every day" (33 percent in 1991 compared with 25 percent in 1996). A surprisingly high proportion of high school seniors express willingness to try even relatively dangerous or addicting drugs at least once in a while.

Public Policy and the War on Drugs

The current widespread concern over drugs has developed many signs of public hysteria. The government's continuing war on drugs has failed to acknowledge two aspects of the overall context of drug use. First, public concern over drugs varies in faddish cycles. Some drugs seem to become popular at certain times, and later they lose their allure. Attitudes change as public concern focuses for a short time on a particular drug and then moves on to another, independent of any characteristic of either drug or the result of the public attention. In the 1960s, marijuana use elicited great concern. In the 1970s, extensive national discussion centered on methaqualone or *Quaaludes* and *angel dust*. Heroin always draws attention in drug discussions, but people envisioned an epidemic of heroin use in the 1970s. Through the mid-1980s, cocaine and heroin stirred the most debate. In the 1990s, crack cocaine has replaced them in the public eye.

The war on drugs also fails to recognize links between drug-taking behavior and the general behavior patterns of people in the United States. Considering the overall makeup of society, government can never conceivably convince all segments of the general public to abstain from all drug use. Nearly everyone learns to take drugs initially by using legally available medications, remedies, and other drugs. People obtain them readily, comedian George Carlin reminds us, from legitimate places called *drugstores*. Experiences with these legal drugs introduce most people to the connection between chemical substances and desired physical benefits, such as reducing headaches, increasing bowel regularity, suppressing appetite to lose weight, clearing stuffed noses, keeping drivers or students awake at night, and so on. Two other legal drugs—alcohol and tobacco—have long histories in this country, including important links to specific social situations. The ultimate roots of the desire to take illegal drugs reside in experience and satisfaction using legal drugs.

Drug taking will undoubtedly continue throughout many segments of the population. People will use varying kinds of drugs, though. Judgments of deviance

✳ *Issue* | **Credibility and Character**

Throughout the 1990s and at the turn of the 21st century, politicians were increasingly being quizzed not only about their views of political issues, but their personal lives as well. Persons running for public office can now expect to receive questions about such personal matters as marital fidelity and prior drug use. Clearly, many people have used marijuana, and many have engaged in adulterous affairs. Is there a double standard for politicians? Are politicians who admit to having used marijuana deviant while others may only be regarded as social?

will likely target use of drugs perceived as popular among less powerful groups, including lower-class individuals, workers in socially marginal occupations, students, and people not fully assimilated into U.S. society.

Legal Drug Use

Much drug use involves legal substances with socially approved applications. The list includes alcohol, tobacco, tranquilizers for relaxation, barbiturates for sleeping, and many minor pain-killing drugs, such as aspirin. Users buy most of these drugs "over the counter" (OTC), that is, they need no physician's prescription to purchase drugs they want to take. One estimate has identified over 300,000 different drugs available for purchase OTC (Schlaadt and Shannon, 1994).

People spend much more buying caffeine-laden coffee and nicotine in cigarettes and other tobacco products. Coffee is an important commodity in international trade (the one with the second-largest volume, in fact, behind only oil), and powerful interest groups stand ready to protect trade in coffee and cigarettes. The proportion of the U.S. population who report drinking coffee has declined over the past two decades, as has the percentage of cigarette smokers, but cigarette sales have increased at the same time. Apparently, fewer smokers each consume more cigarettes now than in the past, and tobacco consumption overseas has increased. Worldwide, tobacco companies sold about $4\frac{1}{2}$ trillion cigarettes in 1981. Marlboro leads the biggest-selling brands in the United States, followed in order by Winston, Salem, and Kool. An estimated 53.6 million smokers lit up cigarettes in the United States in 1990, a decline of 11 percent from the number in 1985 (*Des Moines Register*, December 20, 1990: 6A). In spite of these declines, about 23 percent of Americans smoke.

In addition to OTC drugs, pharmacies sell many more drugs under authority from doctors' prescriptions. Many of these drugs induce sleep or help patients to relax; others stimulate patients or help them to stay awake. Regardless of their purposes, prescription drugs usually carry reputations of more dangerous addiction potential than OTC drugs. A doctor may prescribe a drug as a response to a

---

✳ *Issue* | **What Is a Drug?**

The year chain-smoking William Bennett became drug czar, tobacco killed some 395,000 Americans—more than died in both world wars. Alcohol directly killed 23,000 and another 22,400 on the highways. The Natural Resources Defense Council in March 1996 published a report, widely praised by medical authorities, estimating that as many as 5,500 Americans would develop cancer from the pesticides they ate during their preschool years. The incidence of breast cancer in American women had more than doubled since World War II, owing largely to dioxin and other pollutants.

Cocaine, on the other hand, killed 3,618 people that year, slightly fewer than died from anterior horn cell disease. Heroin and other opiates killed 2,734. And no death from marijuana has ever been recorded.

Source: Baum, Dan. 1996. *Smoke and Mirrors: The War on Drugs and the Politics of Failure.* Boston: Little, Brown, pp. 264–265.

patient's problems with stress. Clear differences separate the sexes in their choices for handling stress by taking drugs. Women are more likely than men to consume prescription drugs, such as tranquilizers (Siegal, 1987: 111), and, more generally, to call on medical services. Men, on the other hand, are more likely than women to drink alcohol to cope with stress.

In response to concerns over the effects of legal drugs, some coffee drinkers favor decaffeinated versions of the beverage; these products lower but do not eliminate caffeine. Similarly, many smokers choose cigarette brands that promise low tar and nicotine, but, again, these products reduce but do not eliminate drug content. In contrast, the late 1980s brought the introduction of a high-caffeine soft drink called *Jolt*, which quickly became popular, particularly on college campuses as an aid for late-night studying. Coca-Cola once contained a small amount of active cocaine, but the company has stripped cocaine from its ingredients since 1903 (Poundstone, 1983). But even if the company successfully removes 99 percent of the drug from its product, a pretty good effort, millions of cocaine molecules remain, just as caffeine molecules remain in decaffeinated coffee. This insignificant amount has absolutely no effect on the body, but as long as the formula for Coca-Cola includes coca leaves, it cannot completely eliminate cocaine.

Doctors write over 70 million prescriptions each year for minor tranquilizers, such as Librium, Valium, and Verstran. Legal use of amphetamines, which include many types of stimulants and pep and diet pills, has reached extensive proportions. The United States may devote the most attention to weight and beauty of any society in history, and the money spent by both men and women on diet and beauty aids (books, pills, tapes, exercise equipment, and magazines) rivals only that spent on cosmetics. Various appearance norms induce enormous stress in some individuals determined to meet certain standards for appearance, body type, and clothing style (Schur, 1983). This kind of stress may result in eating disorders, such as

TABLE 10-1

Best-Selling Drugs in the United States in 1999

Name	Purpose	Manufacturer	Sales
1. Prilosec	Anti-ulcerant	Astra Zeneca	$3.16 billion
2. Lipitor	Cholesterol reducer	Parke-Davis	$2.13 million
3. Prozac	Antidepressant	Eli Lilly	$2.04 billion
4. Epogen	For kidney failure	Amgen	$1.63 billion
5. Zocor	Cholesterol reducer	Merck	$1.53 billion

Source: IMS Health, *Washington Post.*

anorexia nervosa and bulimia, as well as consumption of amphetamines in an attempt to manipulate the body's metabolism. When one adds the cost of cholesterol reducers and other medical prescriptions, one can see how sales of these drugs amount to many billions of dollars a year (see Table 10-1).

At least through the 1960s, large numbers of intravenous amphetamine users lived in big cities (Kramer, Fishman, and Littlefield, 1967). Users experience no physical withdrawal symptoms from amphetamines, although the body can build up a tolerance that requires progressively higher dosages to achieve the desired effect. Barbiturates, a class of drugs that act as depressants in the central nervous system, sometimes supplement amphetamines in programs to achieve weight loss.

Legitimate drug companies manufacture virtually all of the barbiturates and amphetamines in use. These drugs reach illicit markets via hijackings and thefts, spurious orders from nonexistent firms, forged prescriptions, and numerous small-scale diversions from family medicine chests and legitimate prescription vials. The number of pills from such sources can add up rapidly, with over 60,000 pharmacies staffed by over 150,000 pharmacists dispensing prescriptions written by over 300,000 physicians (Ray, 1983: 59).

Advertisements stimulate amphetamine and barbiturate use. Drug manufacturers spend nearly $1 billion each year to reach physicians and convince them to write prescriptions for the advertised products. Direct mail campaigns and medical journal advertisements bring a constant barrage of literature about drugs that reflects some of the best marketing techniques that Madison Avenue can apply (Seidenberg, 1976). Photographs of attractive women (frequently wearing little clothing in ads for dermatology-related drugs), sensational situations, and slogans permeate the advertisements, all designed to guide physicians' choices of one drug over another. Drug makers target such advertising at physicians because most of them, even those who regularly prescribe drugs, are not fully expert on the many different kinds of substances and their side effects (Schlaadt and Shannon, 1994: 326–327).

Physicians are not the only decision makers targeted by drug advertisements. Drug manufacturers actively promote public awareness of their over-the-counter products. The maker of Anacin spends $27.5 million a year to persuade people of the product's superiority to aspirin, although Anacin contains only one effective active ingredient for treating pain: aspirin (Kaufman, Wolfe, and the Public Citizen Health Research Group, 1983).

In part because of such advertising, millions of Americans take nonprescription tablets, capsules, and syrups to relieve the symptoms of everyday ailments. These shoppers can choose from an estimated 300,000 different (and sometimes not-so-different) products made from only 1,000 or so different ingredients. Some of these drugs produce their effects through different means than users suspect. For example, someone who wants to relieve the symptoms of a common cold might choose a substance advertised as a "night time sniffling, sneezing, coughing, aching, stuffed head, fever, so-you-can-rest medicine." Such ads neglect to say that 50-proof Nyquil costs about $10.60 a fifth; perhaps appropriately, most users take this "medicine" with a shot glass that accompanies the package.

Types of Illegal Drugs

Drugs fall into categories depending on their general effects on the body. Depressants and stimulants each have different effects on users. Another distinction, between hallucinogens (mood-altering drugs) and narcotics (drugs that are associated with physical dependency), is also useful. As the names imply, depressants mute mental and physical activity in varying degrees, depending on dosages, while stimulants excite and sustain activity and diminish symptoms of fatigue. The depressant drugs most commonly discussed in the context of deviance are alcohol (to be discussed in the next chapter), morphine, heroin, and marijuana. Morphine and heroin, together with semisynthetic and synthetic alternatives such as methadone and ineperidine, make up the class of *opiates*. These drugs account for the greatest proportion of drug addiction in the United States.

Users most frequently take heroin and morphine, both white, powdered substances derived from opium, by injection, either subcutaneously or directly into their veins. Both produce highly toxic, or poisonous, effects, so users must prepare them quite carefully, including diluting them before use. Almost immediately after injecting either drug, the user becomes flushed and experiences a mild itching and tingling sensation. Drowsiness gradually follows, leading to a relaxed state of reverie. Soon, however, the user needs higher does of the drug to reach this state of euphoria. Thus, the addict builds up a *tolerance* for the drug as well as a physical dependence upon it. As this tolerance increases, the addict becomes comparatively immune to the drug's toxic manifestations. A morphine addict, for example, may tolerate doses as high as 78 grains in 16 hours, enough to kill 12 or more nonaddicts. Hospitals usually set the safe, therapeutic dosage of morphine at about 1 grain in the same period of time.

A heroin or morphine addict becomes physically dependent upon the drug over a varying length of time, usually quite short, after which the addiction increases slowly in intensity. Authorities generally agree that this dependence results more from regular administration than from the amount of the drug taken or the method of administration. That is, physical dependence is a consequence of regular use. While addicts receive their usual daily supply, they exhibit no readily recognized signs of addiction. Even intimate friends and family may not know of their dependence. If addicts do not receive their daily supplies, however, they show clearly characteristic symptoms, referred to as *withdrawal distress* or *abstinence syndrome*, within approximately 10 to 12 hours. The suffering begins with nervousness and restlessness and proceeds with the development of acute stomach cramps, watering eyes, and a running nose. Later, the withdrawing addict stops eating and may vomit frequently, develop diarrhea, lose weight, and suffer muscular pains in the back and legs. The "shakes" may also develop during this period. If the addict cannot obtain drugs to relieve these effects, considerable mental and physical distress results. Consequently, an addict will go to almost any lengths to obtain a supply of drugs and escape withdrawal distress. After obtaining the needed drugs, the addict will appear normal again within about 30 minutes after administration.

A related drug, *methadone* (technically known as *dolophine*), is a synthetic narcotic analgesic originally developed in Germany during World War II. This potent, long-lasting narcotic takes three forms: pill, injectable liquid, or orally administered liquid. Treatment programs for heroin addicts have used this alternative for some years. Such a methadone-maintenance program supplies one dose to an addict that may last up to a day; by contrast, heroin addicts typically take at least three daily doses. Methadone maintenance progressively increases the average dosage until the addict develops enough cross-tolerance to prevent any euphoria from other drugs, particularly the less-potent heroin. Methadone itself causes physical addiction, however, and it does not always block the euphoric effects of other opiates. It has increasingly presented problems in replacing other drugs, although chronic consumption of heroin causes worse ones (Kreek, 1979).

Cocaine, the best-known stimulant drug, appears as a white powder that users most commonly inhale or "snort" through the nose. Some prefer to "freebase" cocaine by combining it with volatile chemicals. This application method increases the potency of the drug and permits oral or intravenous injection. Traditionally considered a recreational drug that facilitates social interaction, cocaine produces a feeling of intense stimulation and psychic and physical well-being accompanied by reduced fatigue.

Crystalline cocaine once served relatively wealthy users as a very expensive drug. The high price, together with rumors of exotic properties, has contributed to cocaine's street reputation as a high-status drug. In recent years, however, cocaine has become more plentiful in the United States, and its price has declined. American use patterns emphasize infrequent consumption of small quantities (Grinspoon and Bakalar, 1979). Cocaine does not appear to produce physical dependence, but some observers claim that it causes addictive chemical changes in the brain (Rosecran and Spitz, 1987: 12–13). At present, cocaine use appears to be widespread, involving large numbers of people, although chronic use may remain limited to some subgroups in society.

Crack is a more potent derivative of cocaine, produced by mixing it with water and baking soda or ammonia. The result looks like small, ball-shaped bits about the size of large peas. In some communities, users may use crack in combination with amphetamines, and some produce a substance similar to crack by combining cocaine with amphetamines. Most often smoked in a pipe, crack makes a crackling sound when ignited. Its popularity results in part from its price. While cocaine may cost more than $100 per gram (less than a teaspoonful), crack may sell for as little as $10 per nugget. At this price, children and young users can often afford to buy it. Observers compare crack's physical addiction potential and physical harm to the user with those of heroin.

Another stimulant, *marijuana* (also spelled *marihuana*), is derived from the leaves and tender stems of the hemp (or Indian hemp) plant. Also called *bhang*, *hashish* (actually, a stronger cake form of the drug), *grass*, or *pot*, the drug is usually inhaled by smoking specially prepared cigarettes called *reefers* or *joints*. The general, technical term for this drug, *cannabis*, comes from the scientific name of the annual herb native to Asia that produces it. A marijuana smoker usually experiences euphoria, intensified feelings, and a distorted sense of time and space, all with few unpleasant aftereffects. In spite of some controversy about the effects of prolonged use, most observers see no risk of physiological addiction to marijuana, although it may, to some extent, be psychologically addicting. Research on the drug's long-term effects remains inconclusive because chronic users of marijuana also frequently use other drugs, so researchers cannot identify the effects from marijuana alone (Petersen, 1984). Marijuana use seems to affect motor-skill performance, for example, in auto driving, but some dispute claims that it causes psychotic episodes and bodily and brain damage.

Barbiturates are sedative and hypnotic drugs that exert a calming effect on the central nervous system. These synthetic drugs, when properly prescribed and taken, have no lasting adverse effects. The patient's system absorbs such a drug, and liver and kidney action render it harmless. Eventually the body passes any remaining residue. Careless use of barbiturates often leads to psychological dependence and physiological addiction, however. The direct actions of these drugs on the body may produce effects more harmful and dangerous than those of opiates. Also, overdose may well lead to death, because the drug can depress the brain's respiratory control until breathing cases (Smith, Wesson, and Seymour, 1979).

Methamphetamine (sometimes called *crank*) is a derivative of legitimate amphetamines. Made in a laboratory, this drug produces a cocaine-like high, and it resembles cocaine in other respects as well; both are white powders usually taken by snorting or injection. While a cocaine high might last for half an hour, however, an episode of crank euphoria may last all day. Crank gives users long-term energy, and some may stay awake for days at a time, always feeling full of energy. Little evidence identifies crank as a major drug in the United States, although some speculate that successful enforcement efforts to block cocaine imports may stimulate use of crank. Their argument emphasizes simple economics: Cocaine will become too expensive, helping locally made crank to compete. In any case, because methamphetamine can be homemade from readily available, and legal, substances, the consumption of this drug has increased in recent years.

Hallucinogens include marijuana, hashish, and "consciousness-expanding" drugs produced from plants such as mescaline, peyote (produced from certain mushrooms), and morning glory seeds. This category also includes LSD, a synthetic hallucinogen made largely from lysergic acid. Scientists do not clearly understand all

✳ *In Brief* **What Is "Ice?"**

Ice is a crystalline version of the stimulant methamphetamine. Users can consume it by smoking, as they might do with crack cocaine. Ice sells for about $35 to $45 per quarter gram in many places in the Midwest, while the same amount of crack cocaine sells for about $25. (Of course, prices and amounts vary by area and relationship with the seller.) What explains the popularity of the more expensive ice? While a crack cocaine high may last 2 to 6 hours, an ice high may last 16 to 18 hours.

The consumption of the drug causes insomnia for days or sometimes weeks. It disrupts the brain's ability to direct the body's movements and thoughts, and it can cause violent and psychotic reactions. Long-time users can become malnourished, and there is a much higher chance of strokes due to increases in heart rate, blood pressure, and body temperature.

of the physiological actions through which hallucinogens produce their effects; while they obviously initiate chemical effects on the brain, no one has yet explained the exact process. Observers do not regard these drugs as psychologically addicting or physically habituating substances, but the startling and sometimes pleasurable sensations they produce may lead to repeated use. Natural hallucinogens like peyote do not create effects as strong as those of LSD unless taken in prolonged dosages, but LSD is a powerful drug, indeed. A tiny amount ($1/300,000$ of an ounce) causes delusions or hallucinations, some pleasant and some terrifying. It tends to heighten sensory perceptions, often so much that they become wildly distorted. Although the experience usually lasts from 4 to 12 hours, it may continue for days.

Marijuana Use

Marijuana is the most widely used illicit drug in the United States, measured by the percentage of the population that has ever tried it. U.S. law prohibits the manufacture, sale, and possession of marijuana, although some jurisdictions have reduced violations for possession of small amounts to misdemeanor offenses. Still, many groups consider marijuana use as deviance, although others condone and even encourage it. As one marijuana smoker put it: "Even though it is not the norm of society, enough people do it to make it acceptable" (Wilson, 1989: 58).

Extent of Marijuana Use

Surveys leave little doubt about the extensive use of marijuana. Estimates of its prevalence indicate that a large proportion of the population has tried marijuana at least once. In 1972, the National Commission on Marihuana and Drug Abuse reported the results of U.S. surveys that indicated an estimated 24 million people had tried marijuana. Of this number, 8.3 million generally used it less than once a week, and about 500,000 "heavy" users reported smoking it more than once a day. The commission found that the use of marijuana had tripled in the previous $2\frac{1}{2}$ years (National Commission on Marihuana and Drug Abuse, 1972). Today, probably a third of the population of the United States has tried marijuana at least once.

The National Institute of Drug Abuse conducts annual surveys of high school students that have tracked marijuana use among this age group for nearly two decades. The surveys have documented a pattern of declining use until the early 1990s. Throughout the 1970s, reports from high school seniors suggested an increase in lifetime prevalence (i.e., the proportion of those who had ever used marijuana). Beginning in 1979, however, this proportion began a slight decline (Johnston, O'Malley, and Bachman, 1993a: 73). From 1979 to 1992, reports of use by high school seniors declined from 60 percent to 33 percent (Johnston, O'Malley, and Bachman, 1993a: 74). By 1992, the annual prevalence of marijuana use by people between the ages of 19 and 28 had fallen to 25 percent (Johnston, O'Malley, and Bachman, 1993b: 83). Further, only 5 percent of this age group had used marijuana within the previous month. In general, the rates of monthly and annual prevalence decline with age.

In the early 1990s, marijuana use ended its decline and began to rise again. In 1993, the proportions of students reporting any use of marijuana during the previous 12 months stood at 9 percent for 8th graders, 19 percent for 10th graders, and 26 percent for 12th graders (Johnston, O'Malley, and Bachman, 1996). These figures represented increases over earlier figures, although they remained well below the peak figures reached during the late 1970s. By 1995, the proportion of 8th graders who had used marijuana during the previous year had increased to 16 percent, with 29 percent for 10th graders and 35 percent for high school seniors. Undoubtedly, the overall reduction in marijuana use since the peak years has resulted in part from perceptions by high school seniors of higher risk from the practice—including physical, medical, and legal effects—as compared to perceptions during the 1970s. The increases in use from 1991 to 1996 corresponded to reductions in the percentages of high school seniors who perceived harm from the use of marijuana. The most recent survey in 1999 indicated that marijuana use has declined modestly since its peak in 1996 (Johnston, O'Malley, and Bachman, 1999). (See Table 10-2.)

Another annual study called the National Household Survey on Drug Abuse has reported parallel findings (Shalala, 1999). According to the most recent information released in 1999, marijuana continues to be the most frequently used illicit drug; about 60 percent of all illicit drug users reported using marijuana only, and another 21 percent reported marijuana use and some other illicit drug use. In 1998, 8.3 percent of youths aged 12–17 were current users of marijuana. Youth marijuana use reached a peak of 14.2 percent in 1979, declined to 3.4 percent in 1992, more than doubled from 1992 to 1995 (8.2 percent), and has fluctuated since then (7.1 percent in 1996, 9.4 percent in 1997, and 8.3 percent in 1998).

Marijuana use patterns depict a practice confined mainly to people in their 20s and younger (Goode, 1999). Statistics reveal only rare experimentation and continued use among people over 35. Marijuana use is not confined exclusively to one social class.

TABLE 10-2

Percentage of High School Seniors Who Report Using Various Drugs During the Previous Year, 1991 and 1999

Drug	1991	1996	1999
Any illicit drug	29.0%	40.0%	42.0%
Any illicit drug other than marijuana	16.0	20.0	21.0
Marijuana	24.0	36.0	38.0
LSD	5.0	9.0	8.1
Cocaine	4.0	5.0	6.2
Heroin	0.4	1.0	1.1

Source: Johnston, Lloyd D., Patrick M. O'Malley, and Jerald G. Bachman. (Dec. 1999). *Drug Trends In 1999 Are Mixed.* University of Michigan News and Information Services: Ann Arbor, MI. Available online at: www.monitoringthefuture.org; accessed March 28, 2000.

Using Marijuana and Group Support

Marijuana is essentially a social drug. People use it in groups within specific social situations. Marijuana users seem not to follow the patterns of secret drinkers or solitary opiate addicts. They come to use marijuana, either regularly or irregularly, through a learning process, bolstered by subcultural support for continued use. Someone who wishes to use marijuana must contact others with experience who show the initiate how to use the drug. In a study to determine how people become marijuana users, Becker (1973: 235–242) found three learned elements: (1) how to smoke the drug to produce certain effects; (2) how to recognize these effects and to connect the drug with them; and (3) how to interpret the sensations as pleasurable. First, uninitiated users do not ordinarily "get high," because they do not know the proper techniques of drawing and holding in the smoke. They may experience some pleasurable sensations, but new marijuana users often dismiss them as insufficient or fail to recognize their specific characteristics well enough to induce them to become regular users. Feeling dizzy and thirsty, misjudging distances, or noticing a tingling scalp may not of themselves seem like pleasurable experiences. Through associations with other marijuana users, however, new users learn to define initially annoying sensations as pleasurable and eagerly anticipated benefits of drug use.

The group association of marijuana use seems to encourage friendships and social participation. Within typically intimate groups, marijuana use may enhance functionality and interactions. Here is how one user recounts his initial experiences:

The people in my group of friends that smoke marijuana began as a result of pressure from peers combined with a general curiosity. When their friends started smoking marijuana, it became "the thing to do" and it made them wonder what it was like. (Jones, 1989: 59–60)

Marijuana can help to establish the pattern of social relations in some groups. It may contribute to such a group's long-term, continuing social relations, helping to forge some value consensus, to encourage a convergence of values through progressive group involvement over time, and to maintain the circle's cohesive nature, among other roles. A new marijuana user needs group contacts in order to secure a supply, learn the special technique of smoking to gain maximum effect, and maintain psychological support for continued participation in an illicit activity.

Continued marijuana use also requires group support. Association with others to share marijuana may also lead to use of other drugs. The use of marijuana itself does not lead to other, possibly more dangerous drugs; rather, this progression in drug use results from association with and membership in a group that condones experimentation with drugs (Goode, 1999: 229). Other drugs define their own subcultures, of course, but groups oriented toward heroin and other drugs differ from groups of marijuana users, just as the drugs themselves differ. Subcultures for marijuana differ importantly from those for heroin because marijuana use is overwhelmingly a recreational activity, but heroin plays a more dominant role in users' lives. Heroin users more often take the drug alone or with only one other person in order to share resources, such as money, "works" (drug equipment and paraphernalia), and the like. Marijuana users take the drug by themselves much less often, although some chronic, daily users engage in mostly solitary use (Haas and Hendin, 1987). These older, relatively experienced users have made marijuana an important part of their lives. As such, they use it in substantially different ways from most other users.

Marijuana and Heroin

Many people believe that marijuana use facilitates experimentation with heroin and eventual addiction; in fact, many people think that marijuana smoking leads people toward all kinds of stronger drugs. Some call this idea the *gateway theory*, because it implies that marijuana serves as a gateway to the use of other, more serious drugs. So far, research has not substantiated such a relationship between marijuana use and a tendency to use other drugs. The National Commission on Marihuana and Drug Abuse (1972: 88–89) has concluded that the overwhelming majority of marijuana users never progress to other drugs, either continuing to use marijuana or changing to alcohol: "Marijuana use per se does not dictate whether other drugs will be used; nor does it determine the rate of progression, if and when it occurs, or which drugs will be used." The user's social group seems to exert the strongest influence over decisions whether to use other drugs and which ones to choose.

Research does seem to confirm that many heroin users had earlier used marijuana, but this relationship in no way establishes a causal connection between taking marijuana and subsequent heroin use. Clearly, most people who use marijuana do not progress to heroin. National surveys indicate that about 62 million people have tried marijuana, and more than 18 million currently use the drug, compared with fewer than 6 million current users of cocaine (Trebach, 1987: 83). (The surveys found such small numbers of current heroin users that they could not yield reliable estimates.) If marijuana acted as a "gateway drug," as some claim, then many more of the 62 million who experimented with it would probably have gone on to these other drugs.

Opiate Use and Addiction

The Meaning of *Addiction*

The term *addiction* refers to physical dependence, "an adaptive state of the body that is manifested by physical disturbances when drug use stops" (Milby, 1981: 3). That seemingly clear meaning quickly clouds, however, when one considers the popular tendency to apply the term to any repeated action. This indiscriminate language robs the term of its meaning, so sociologists must distinguish between behavior that represents an addiction and behavior driven by other reasons (Levison, Gerstein, and Maloff, 1983).

Examples abound of widespread confusion over the word *addiction*. Ann Wilson Schaef (1987) regards an addiction as "any process over which we are powerless." Working from this definition, she has estimated that as many as 96 percent of all Americans either live with addicts or grew up with addicts affecting their lives.

✳ | *Issue* | **Is Everyone an Addict of Some Kind?**

Anne Wilson Schaef believes that everyone is an addict. Some people, she notes, develop chemical dependencies; others suffer from "process" addictions, defined as addictions to things or people rather than drugs. Along with alcohol and other drugs, people can become addicted, Schaef says, to things like:

- Television
- Gambling
- Work
- Caffeine
- Relationships
- Sex
- Shopping

Schaef even describes Pope John Paul II as a "sex addict," reasoning that one need not act out sexual behavior to qualify for the label but merely be obsessed with the subject or with controlling other people's sex lives.

Source: Adapted from *Des Moines Register*, October 23, 1990, p. 3T.

Furthermore, she has ascribed an addictive character to a variety of activities (making money, work, worry, religion, and sex), substances (food, nicotine), and relationships (marriage, love affairs) Thus, some might recognize addictions to spouses, to particular foods, to school or jobs, and to the color of one's bedroom walls. Some even argue for an addiction to religion:

> When, in the name of God, people hold black-and-white beliefs that cut them off from other human beings; when, in the name of God, they give up their own sense of right and wrong; when, in the name of God, they suffer financial deprivation; then, they are suffering from religious addiction. (Booth, 1991: 53)

Surely, such an expansive definition destroys the meaning of the term *addiction*—or any other term, for that matter.

This tendency to stretch the concept of addiction appears to reflect a larger trend toward conceiving of many kinds of behavior as diseases without defensible medical rationales (see Peele, 1990):

> The broadening of the meaning of addiction is part of a growing fad to medicalize many different behaviors. Drink too much? It's a disease. The child of an alcoholic? You have a disease. Overeat or gamble? Both diseases. Sex-obsessed? Definitely a disease. Workaholics, compulsive shoppers, fitness freaks, drug users, whatever the behavior: The growing trend is to call it a disease. (Sullivan, 1990: 1T)

Often, afflicted people try to treat their supposed behavioral diseases through participation in a 12-step program similar to that of Alcoholics Anonymous (see Chapter 10). These programs begin with the assertion of the addict's powerlessness over the addiction, so he or she must call for help from a larger power to overcome the addiction. The long list of such "anonymous" groups begins with the well-known Alcoholics Anonymous and extends to such organizations as Narcotics Anonymous, Sex Addicts Anonymous, Emotions Anonymous, Debtors Anonymous, and Workaholics Anonymous. A social worker has noted that "This rise in AA [style] groups has become a kind of secular religion. It's like witnessing. People come in and say 'I have a disease.' It's similar to people standing up in church and saying 'I am a sinner' (quoted in Sullivan, 1990: 1T).

For purposes of the present discussion, the chapter will adopt some basic definitions. If drug users suffer physiological consequences from the withdrawal of their drugs, those substances are physically dependence producing, or *addicting*. An *addict* is someone who experiences distress as a result of not having a drug—alcohol, heroin, or an amphetamine. Users who fail to experience distress when they do not take a drug, such as alcohol or heroin, are not addicted by this chapter's definition.

That term does not, however, capture certain differences in addictive behavior. As a result, many professionals prefer to talk about "tolerance," "dependence," and "abstinence syndrome" rather than addiction. Users of certain drugs may become physically dependent, and abstinence from the drugs may cause withdrawal distress. Such a user probably has built up a tolerance to the drug so that he or she must take larger and larger doses to produce the desired effect. This occurs in opiate addiction, which forces the addict to consume ever-greater amounts to achieve the same effect. Failure to obtain the needed dose, usually in the form of heroin, leads to withdrawal symptoms after 8 hours.

Patterns of Heroin Use

People have used forms of heroin for thousands of years; in fact, U.S. law permitted opium use until early in the 20th century. Users can experience the effects of heroin by inhaling it, smoking it, or, more commonly, injecting it into their veins. Patterns of heroin use have changed over time. In the 19th century, for example, early surveys found that about two-thirds of heroin users were women, and these sources noted a good deal of addiction in the medical profession (Morgan, 1981). Most observers described heroin use as less prevalent in the lower classes than in the middle and upper classes at the time. The average age of addicts ranged between 40 and 50, and some investigators specified addiction as a problem of middle age, since most addicts took up the habit after the age of 30 (Lindesmith and Gagnon, 1964: 164–165).

The first reports of widespread heroin use in large cities during the 20th century came to light after 1945 (Hunt and Chambers, 1976: 53). Use remained at low levels through the 1950s and then increased rapidly during the 1960s. The drug's popularity peaked in most cities during 1968 and 1969. By 1971, all large cities (those with populations over 1 million) had experienced their highest rates of use, although rates in smaller cities continued to increase during the 1970s. Information on patterns of use throughout the 1980s suggests stable levels of addiction at rates common in the late 1970s; in fact, the number of heroin addicts in the early 1980s remained close to those estimated for the late 1970s (Trebach, 1982, 1987). Although heroin addiction still has not become a growing problem, more addicts use the drug today than two or three decades ago.

The link between transmission of the disease AIDS and dirty heroin needles has generated a new fear among addicts. The extent of AIDS infection among addict populations is imprecisely understood, but a high percentage of users may have become infected. The United Nations estimates that 25 percent of those who have AIDS contracted the disease through intravenous drug use. Other heterosexuals clearly run a very high risk of contracting AIDS if they engage in sexual contact with intravenous drug users. The full influence of AIDS on patterns of heroin use has not emerged, but such figures suggest the possibility of change in user behavior in the wake of the epidemic.

While the concern over AIDS may have depressed heroin sales and use for a time, new opportunities for addiction began to arise in the early 1990s with the spread of an extremely potent form of heroin called "China White." For years, a bag of heroin (containing an amount about equivalent to a pencil eraser ground up) included less than 10 percent of the drug mixed with adulterants. In recent years, some users have bought bags containing mixtures with as much as 50 percent heroin. Such a pure concentration allows users to snort the drug like cocaine or smoke it, eliminating needles and associated risks of transmitting disease. Dangers of overdose also increase, however, since most addicts do not know the purity of a sample of the drug prior to using it. In addition, the worldwide supply of heroin seems to have increased in the 1990s, driving down the selling price and increasing the drug's availability.

Number of Heroin Users

Estimates suggest a growing number of drug addicts in the United States until the 1970s or so and relative stability since that time. For example, two experts estimated that in 1967, 108,500 addicts were dependent on heroin in the United States (Ball and Chambers, 1970: 71–73). According to an estimate from the Federal Bureau of Narcotics and Dangerous Drugs (BNDD), however, the country held 64,000 addicts in 1968 (Milby, 1981: 75). These estimates differ because they referred to different data sources. Estimates for the 1970s, however, represented drastic increases from those for the 1960s. One estimate put the number of active heroin users in the United States in 1975 at 660,000 (Hunt and Chambers, 1976: 73). More recent estimates project more than 500,000 people who use heroin regularly (Office of National Drug Control Policy, 1996: 43), defined as people who use the drug regularly, with perhaps an additional 3.5 million "chippers," or occasional users (Trebach, 1982: 3–4). So many occasional users suggests a more complex addiction process than a simple chemical result of taking heroin. The addiction process involves social influences, most of them related to membership in groups.

The United States may lead all countries in the number of addicts, but estimates for other countries give extremely unreliable results. Research scientists and physicians in 25 countries published data from an international survey in 1977 that found the largest number of opiate addicts in the general population of the United States with 620,000, followed by Iran with 400,000, Thailand with 350,000, Hong Kong with 80,000, Canada with 18,000, Singapore with 13,000, Australia with 12,500, Italy with 10,000, and the United Kingdom with 6,000 (Trebach, 1982: 6–7).

Who Uses Heroin?

Heroin users fall into a number of categories based on their frequency of use and the contexts in which they consume the substance. Some merely experiment with heroin without continuing to use it beyond an initial time or two. Other "recreational" users take heroin occasionally, perhaps only in social situations or on weekends. More frequent users generally exhibit signs of addiction. People can maintain many of these roles only through involvement in an addict subculture, which provides essential social support and access to a supply of heroin (Hanson, Beschner, Walters, and Bovelle, 1985).

Heroin use remains uncommon among people under 20 years of age. Monitoring the Future, the national survey of high school youth in the United States, found in 1996 that about 1 percent of high school seniors reported ever having tried heroin (Johnston, O'Malley, and Bachman, 1996). Even fewer students reported taking heroin in the previous year and the previous month, and still smaller percentages of students in lower grades admitted using heroin. The last survey indicated that rates of heroin use have remained stable since that time (Johnston, O'Malley, and Bachman, 1999), a result also reported by the National Household Survey on Drug Abuse (Shalala, 1999).

Addiction to opiates is heavily concentrated among young, urban, lower-class males from large cities, particularly among blacks and Hispanics of Puerto Rican descent. The high heroin addiction rates among blacks partially reflects the concentration of trafficking in black areas of many cities. The physical presence of supplies makes the drug especially available there, facilitating the development of a street addict subculture. Nearly 30 percent of all addicts are black, most of them concentrated in the northeastern part of the United States (Chambers and Harter, 1987).

Although most addicts live as street addicts, certain occupations feature relatively high rates of heroin use. As we have seen at the beginning of this chapter, the medical profession, especially physicians and nurses, includes an excessive number of addicts in relation to their proportion in the population as a whole; addicts also concentrate in the entertainment industry (Winick, 1961). Goode (1999: 403–404) estimated that as many as 12,000 physicians maintained narcotics addictions in the United States, or about 3 in every 100 physicians (based on a total of about 350,000 physicians), a figure that far exceeds the rate for the population as a whole. In fact, physicians use all drugs at much higher rates than nonphysicians (Vaillant, Bright, and MacArthur, 1970). Since doctors can prescribe drugs, they can obtain them easily and at rather low cost. Moreover, physicians know how drugs will likely affect someone who is tense or tired, an important inducement to their use of drugs. Many of these physicians do not come to the attention of authorities because they can often maintain their addictions without detection.

Many physician addicts become users as medical students. According to one study, 59 percent of a sample of physicians and 78 percent of a sample of medical students reported having used illegal psychoactive drugs sometime during their lives (McAuliffe, 1986). Medical students work under enormous pressure. They put in long hours under substantial stress, in part because they work in situations they do not control.

> There is no release from the constant bombardment of the work, no one to off-load it onto. Some want to talk but don't know how. Learning to be emotionally mature to cope with this kind of stress is not part of the medical student's education. Students come up with different ways out. Drugs provide one conscious form of release. As Frank explains, "It was never a conscious thing on my part. I never said, 'OK, enough, I'm going to get snowed.' It was gradual." (Hart, 1989: 80)

Students learn that drugs can help them by relieving stress and promoting an even disposition during hard work shifts that may last as long as 36 hours. They also learn to prefer other drugs over alcohol, because they must conceal their use while on the job. As one student pointed out: "If you drink, people can smell it. But if you're loaded they can't tell. You can explain the dilated pupils, the staggering by saying you hadn't slept or you've got the flu" (Hart, 1989: 80).

Performers in the entertainment world, such as jazz musicians, sometimes smoke marijuana, and they may use other drugs as well. This group's high rate of use seems largely connected with relatively weak disapproval by their associates as compared with attitudes common in the general population. Many well-known entertainers have died from overdoses of heroin or other problems associated directly with its use, including in recent years actor River Phoenix and rock musician Kurt Cobain. Drugs may provide very functional benefits to musicians by helping them to weather periods of unemployment, to tolerate long trips away from home, and, depending on the circumstances, to perform certain kinds of music. Studies of jazz musicians have shown that drugs form "normal" elements of some of these people's lives (Winick, 1959–1960). In fact, 53 percent of a sample of musicians reported having used heroin at least once. Many of these users attribute their drug taking to the rigors of life on "the road:"

> I was traveling on the road in 1952. We had terrible travel arrangements and traveled by special bus. We were so tired and beat that we didn't even have time to brush our teeth when we arrived in a town. We'd get up on the bandstand looking awful. The audience would say, 'Why don't they smile? They look like they can't smile.' I found I could pep myself up more quickly with heroin than with liquor. If you drank feeling that tired, you'd fall on your face. (Winick, 1959–1960: 246)

Becoming an Opiate Addict

The process of becoming an opiate addict requires more than simply using drugs derived from opium. Evidence of critical sociological and social psychological processes emerges from information pertaining to initial use and length of addiction, types of addicts and the role of addict subcultures, and various theories of opiate addiction.

People learn opiate addiction just as they learn any other behavior—primarily in association and communication with others who are addicts. As an indication of this fact, consider that such drug addiction, once common among Chinese in the United States, had almost disappeared in this group by the late 1960s (Ball and Lau, 1966). In the usual pattern of association, someone interacts with addicts for reasons other than seeking help becoming an addict. During these interactions, the individual acquires behaviors associated with addiction in much the same way as other cultural patterns are transmitted. In becoming an opiate addict, a person first learns how to use the drugs, beginning with basic awareness of them and proceeding to knowledge about how to administer them and how to recognize their effects. Beyond this information, the person must discover some motive for trying the drug—to relieve pain, to please someone, to achieve acceptance in a group, to produce euphoria, for kicks, or to achieve some other goal. This goal may relate only incidentally to the narcotic's specific physiological effects.

Most drug addicts knowingly approach their initiations into drug use, usually in their teens, by friends, acquaintances, or marital partners. Only a few of the addicts interviewed by Bennett (1986) reported feeling pressured into taking heroin. Strangers introduced fewer than 10 percent of the addicts to the drugs. Most also reported deciding to try heroin at some time before they actually took the drug for the first time. One said, "I'd heard about heroin for as long as I could remember. I wanted to try it. I knew I'd take it, but I didn't know when. I took it at the first opportunity" (Bennett, 1986: 95). Another said: "Drugs fascinated me from an early age, especially the junkie culture. Let's put it this way: I wasn't worried about becoming an addict, I wanted to become one" (Bennett, 1986: 95).

Legitimate use of drugs during illness only rarely leads to addiction. Many addicts begin their heroin use by trying the drug out of a curiosity sparked by already addicted associates. Others take their initial doses of heroin out of willingness to "try anything once." Some adolescents take drugs for the "kick" of the experience, to demonstrate disdain for social taboos imposed by others, and to heighten and intensify a moment and differentiate it from the routine of daily life (Finestone, 1957).

A chain reaction process nurtures addiction in a "sordid and tragic pyramid game" in which one addict introduces several friends into the habit, often as a means of paying for his or her own supply. Drug sellers sometimes introduce the drug at parties, offering the first "shots" free of charge (Chein, Gerard, Lee, and Rosenfeld, 1964: chapter 6). Friends, family members, or acquaintances most often introduce the behavior in this way. In particular, women's initial use very often seems to result from association with a man, especially a sexual partner, who uses heroin daily (Hser, Anglin, and McGlothlin, 1987).

The involvement of others in the initiation process does not surprise sociologists. To become an addict, one must learn techniques for injecting drugs and where to locate a supply. These associations also perpetuate rationalizations for continued use. An addict also needs social and interpersonal support from a group. Far from supporting retreatist or withdrawal behavior, the use of narcotics requires an individual to solicit the help of others. Addicts need one another.

The visibility of drugs on many inner-city streets virtually ensures that adolescents become aware of drug subcultures. Not all inner-city residents become addicts, users, or even tolerant of drugs, but addicts and their characteristic activities are visible features of some urban areas. These neighborhoods may develop climates receptive to drugs, tolerating open sales of drugs and spreading common knowledge of where to obtain them. Stephens (1991) describes street addicts as "rational actors," most of whom choose to take drugs and to adopt roles appropriate to the addict's way of life. They respond to life with severely limited chances for success in the outside world, adopting heroin use as an "expressive" lifestyle that gives dignity and a sense of belonging and success to alienated and disenfranchised individuals. Stephens outlines a role labeled a "cool cat syndrome" defined by attitudes favorable to conning or tricking others and to antisocial feelings toward the world outside the addict group. These people display little concern or guilt for their actions. Instead, they value signs of material success, such as stocks of expensive heroin and other drugs. The cool cat communicates in street language and values excitement. This role denies interest in long-term planning, condemns snitching, and promotes minimal use of violence. In addition, the street addict role reinforces feelings of persecution in a world that is populated with untrustworthy people.

While not all addicts begin opiate use within the network of a subculture, many become first-time users through acquaintances with other users and a social psychological process of learning about the availability and use of heroin (see Akers, 1992: 97–103). These interactions give an unmistakable character of group activity to heroin addiction. The processes of initial and continued use emphasize the importance of social learning in interpreting heroin addiction. Waldorf (1973) found that a sample of New York addicts reported initially using heroin, *not* as a solitary activity, but in association with others. "Persons are initiated in a group situation among friends and acquaintances. Only 17 (4 percent) of our sample of 417 males reported that they were alone the first time they used heroin; by far the majority (96 percent) reported that they used heroin the first time with one or more persons" (Waldorf, 1973: 31). Those others were almost always friends, usually of the same sex.

The Process of Addiction

Addiction often develops progressively from preliminary, experimental stages to a stable habit to termination of the addiction. Waldorf (1983) interviewed over 200 former addicts to inquire about the processes through which they became addicted and subsequently abstained. He has identified the following phases of an addiction career:

1. *Experimentation or Initiation.* Usually in the company of peers, someone tries heroin to satisfy curiosity; most users terminate opiate use after this stage, but some continue.

2. *Escalation.* A pattern of frequent use develops over a number of months, leading to daily use, physical addiction, and increased tolerance; some "chippers" continue using heroin infrequently without developing physical dependency at this stage.

3. *Maintaining or "Taking Care of Business."* Relatively stable heroin use still allows the addict to "get high"; psychologically, the addict maintains confidence about keeping up with job duties and other responsibilities, despite the unquestioned addiction.

4. *Dysfunction or "Going through Changes."* The addict may experience jail or a treatment program for the first time, and other negative effects of the habit may become evident as well; the addict may try to quit the habit, either in combination with others or alone, but fail.

5. *Recovery or "Getting Out of the Life."* The recovered addict develops a successful attitude to quit drugs, either within or outside a formal treatment program; this progression involves major life changes.

6. *Ex-Addict.* The user acquires a new social identity as an ex-addict. Successfully treated addicts adopt this role through their work in treatment programs; untreated ex-addicts very seldom adopt such a role. (Waldorf, 1983)

Different people become addicted after varying experience with the drug. Some can use heroin infrequently for long periods of time without becoming addicted. Others move to full addiction only after a year or more of heroin use, while still others progress from occasional to regular use without continuously taking drugs (Bennett, 1986). Some evidence suggests that women may become addicted more quickly than men do (Anglin, Hser, and McGlothlin, 1987), but great individual variations differentiate heroin users in their addiction careers.

After initial use and progression to full-blown addiction, addicts must learn to maintain their habits within the realities of scarce resources (money) and legal and social rejection from others. One user described a typical day:

You have to be on your toes 24 hours a day. The only time you can afford to sit back is when you're dealing and have a supply of drugs. All the rest of the time, it's a never-ending process of scoring drugs, fixing, maybe getting high—may be not—for an hour or so, and then starting all over again. You have to figure out where you are going to get your next money, who you know who can help you out, who can front you some drugs. It's an everlasting process that might go on through the night; if you don't have drugs, you can't sleep anyway. So you work through the day, through the night, and into the next morning. You've got to be on call 24 hours a day. (Coombs, 1981)

All studies evaluating the success of treatments for narcotics addiction point to high rates of relapse among withdrawing opiate addicts. Many who leave treatment programs seem to return to heroin, with figures varying from 70 to 90 percent. Yet, many addicts do eventually leave the role and refrain from opiates for long periods of time, often without the benefit of treatment. The reasons for leaving addiction remain unclear, but some reports imply a sort of life cycle of addiction through which addicts leave behind the youth-oriented drug subculture as they age. The role demands of coping and hustling become unbearable burdens, or they conflict seriously with new roles. Considerations like these may lead an addict to abandon the street addict role and seek abstinence through treatment or other means. Some simply "burn out" from playing the addict role, as people who work with addicts frequently observe.

Other addicts may consciously decide to stop using heroin. Some do so to escape the negative consequences of life as an addict, such as the need to pay for an expensive habit, the lack of meaningful goals, or rejection by nonaddicts. A 45-year-old man who had been addicted for 6 years described his decision as follows:

The one thing I did know was that I didn't want to be known as a tramp and I didn't want to feel like a nobody, a nothing. I was a very proud person, so I had to do something about those drugs. Either it was going to control me, or control my life, or I was going to stop using. (Biernacki, 1986: 51)

Such addicts might shun conventional treatment programs and stop using heroin the same way some cigarette smokers terminate their drug use: "cold turkey." The addict simply stops taking the drug and endures the physical discomfort that results until the dependence subsides. Some addicts, like some cigarette smokers, succeed in this effort, while others do not. Successful termination of a heroin habit requires high resolve, a changed or changing personal identity, and a desire to reestablish conventional social relationships. Successful termination also requires some addicts to form a conception of their bodies as "clean" and to resist a conception of self as "junkie" (Koutroulis, 1998).

Theories of Addiction

Two leading theories offer contrasting general explanations of opiate addiction. One emphasizes the attractiveness of the drug to users, and the other emphasizes the role of opiates in relieving withdrawal distress. The former theory concentrates on the important role of the "rush" or pleasure sensation that the drug provides. The latter theory concentrates more on the negative physical consequences of stopping opiate use, that is, withdrawal symptoms.

One of the leading sociologists in the area of opiate addiction, Lindesmith (1968), has explained the addict lifestyle and drug use as a consequence of the distress accompanying sudden cessation of use. Lindesmith's theory describes social psychological processes that link drugs to their effects and to the desire to eliminate withdrawal symptoms. Drug addiction results, according to this theory, when people use drugs because they fear the pain or discomfort associated with withdrawal. Addiction is *not* simply a physical process through which consumption of a drug automatically produces dependence. Users seem to escape addiction if they fail to realize the connection between the potential for distress and the opiate, Lindesmith claims, but others invariably become addicted if they come to link the distress with the opiate and use it to alleviate the uncomfortable symptoms.

In becoming addicts, users change their conceptions of themselves and of the behavior they must perform as drug addicts. They must learn from others about techniques for using drugs and sources of supply; they must also connect withdrawal symptoms with the continued use of heroin. In other words, users must become socialized to addiction. As they associate more frequently with other addicts, they become less able to free themselves from dependence on drugs. In short, users come to play the role of "addict" (Lindesmith, 1968: 194). After a while, they continue to use primarily to relieve withdrawal symptoms rather than to experience the euphoria produced by the opiates.

Lindesmith's theory encounters one problem when it dismisses completely the motivation to experience euphoria—the desire to get high—in explaining continued heroin use. McAuliffe and Gordon (1974) have found that long-term addicts do indeed experience euphoria, although they feel weaker sensations than newer addicts do. These authors discovered an important reason that relatively few long-term addicts get high when they take heroin: They usually lack enough money to buy the amount of drug needed to overcome their tolerance and produce the desired euphoria. Such addicts do appear to orient their behavior around the achievement of euphoria, however. Richard Stephens (1991) has found that street addicts use heroin, not necessarily to counter the pain of withdrawal as Lindesmith has maintained, but to experience the drug high. Research reveals a paramount role for the group or subculture in the addiction process. Street addicts gain their self-concept, their sense of personal worth, and their status from taking heroin to experience a drug high.

Lindesmith (1975) responded to such criticisms by asserting that he never claimed that the desire to avoid withdrawal symptoms supplied the only motivation for continued addiction. Physical dependence may generate its own motive for continued use, but an addict craves drugs not under the compulsion of some motive but in order to repeat a particular experience. Despite much research, sociologists cannot precisely identify the meaning or nature of an addict's craving. As one observer put it: "Researchers currently identify and measure craving on the basis of patients' statements, physical symptoms, and continuing use of illicit drugs" (Swan, 1993: 1). This method leaves the distinct possibility that craving is the result, not the cause, of addiction.

Physical dependence can determine the nature of an addict's withdrawal distress, but the interpretation of those symptoms and the subsequent continued use of heroin result from cognitive dimensions of the addiction process, not from physical ones. Lindesmith also concedes that euphoria may lure users to continued drug taking without becoming a frequent concomitant of later use. But the appetite for euphoria does not explain continued use after tolerance inhibits these sensations; only cognitive influences can overcome this limitation. In any case, Lindesmith has not phrased the theory in sufficiently precise language to permit empirical testing (Platt, 1986). While Lindesmith claims that a person must "learn" to become an addict, he does not identify this learning process sufficiently clearly to allow others to test its accuracy.

The Addict Subculture

As suggested earlier in the chapter, research has not supported the idea that people become addicts to escape or retreat from social life. Someone with a career in addiction—that is, a street addict—participates in a drug subculture largely made up of urban, slum-dwelling, male members of minority groups who adhere to a deviant, drug-related set of norms (Stephens, 1991). These users favor heroin and cocaine over other drugs, administer their drugs intravenously, frequently withdraw from drug use, and engage in many hustles to support their habits. They do not use heroin as a psychological escape or as a retreatist method, in terms of Robert Merton's anomie theory (see Chapter 5). Rather, their heroin use can grow from new membership in a close-knit society of other street addicts. Their lives revolve around the "hustle" (illegally obtaining money to pay for drugs), "copping" (buying heroin), and "getting off" (injecting heroin and experiencing its effects) (Agar, 1973: 21). Activities common in conventional society occupy little time in the lives of street addicts. They can maintain this role only through involvement in an addict subculture that provides social support and guarantees a supply of heroin (Hanson Beschner, Walters, and Bovelle, 1985).

Drug sellers must import their products illegally into the country and then distribute them through networks of suppliers and peddlers. Users participate extensively in this supplier subculture, since drug addicts must generally associate with "pushers," usually other addicts, in order to secure their supplies. The distribution system that imports heroin and makes it available to users shows a structure like those of most other distribution systems for importing and selling foreign products, except that participants worry more about reducing legal risks (McBride, 1983). For this reason, criminal syndicates have been associated with narcotics trafficking.

Just as someone must learn the role of a member of a drug-using community, one must also learn to become a drug dealer or smuggler. The individuals in this community, according to a study by Adler (1993: 3), become committed to drug trafficking in pursuit of an uninhibited lifestyle. In many respects, in fact, drug involvement provides the principal means through which these individuals express their chosen way of life. The profits from selling marijuana, amphetamines, and cocaine can fund hedonistic lifestyles with plenty of free time and cash. Participants acquire the knowledge to participate in a drug dealing subculture through either on-the-job training or guidance by someone with experience, creating a kind of sponsorship arrangement. Recruits must bring certain skills with them as they become members of a drug-distribution ring. One smuggler has described the criteria he applied in recruiting new members to his crew:

> Pilots are really at a premium. They burn out so fast that I have to replace them every 6 months to a year. But I'm also looking for people who are cool: people who will carry out their jobs according to the plan, who won't panic if the load arrives late or something goes wrong, 'cause this happens a lot . . . and I try not to get people who've been to prison before, because if they haven't they'll be more likely to take foolish risks, the kind that I don't want to have to. (Adler, 1993: 126)

Little sometimes distinguishes drug users from sellers, and in some communities, trafficking merely represents an alternative to obtaining a legitimate occupation. Drug dealing is often not a glamorous job, or even necessarily a high-paying one. A 35-year-old, unemployed male justifies his involvement in drug trafficking:

> And what am I doing now? I'm a cocaine dealer—cause I can't get a decent-ass job. So, what other choices do I have? I have to feed my family . . . do I work? I work. See, don't . . . bring me that bullshit. I been working since I was 15 years old. I had to work to take care of my mother and father and my sisters. See, so can't, can't nobody bring me that bullshit about I ain't looking for no job. (Wilson, 1996: 58)

An addict subculture organizes its norms around the supply and support of heroin use. It develops and perpetuates a series of rationalizations to justify continued use and recruitment of new members, assuring a continued supply of drugs. The group also facilitates defensive communication with its own argot for drugs, suppliers, and drug users, which initiates must learn. A complex distribution system supplies illegal drugs and supports users' habits. Information about availability, strength, and varieties of heroin for sale at specific locations is said to pass rapidly and accurately, with greater safety than that provided by telephone contacts; subculture members sift information they receive according to their consensus about its reliability.

A RAND Corporation study on the economics of drug dealing found that the typical, daily drug dealer between 18 and 40 years old nets $24,000 per year, tax-free (*Des Moines Register,* July 11, 1990: 1A). The researchers obtained data from interviews with people charged with drug crimes and on probation, as well as from other studies. Research has also discovered that many dealers need extra money to support their own drug habits. The RAND study reported that two-thirds of a sample of probationers earned an average of $7 an hour on legitimate jobs and $30 an hour moonlighting as drug dealers, working a few hours or days each week during times of peak demand.

Suppliers and most addicts live in a world defined by its own meeting places, values, and argot. None of these elements more clearly demonstrates the cultural context of addiction than the argot shared by drug users. It includes special terms for drugs, for people who supply them, and for addiction itself. It also includes special descriptive terms for users.

In summary, addicts rely on their subculture to connect with dealers, to maintain "hustles" to secure money for drugs, and to protect themselves from outside interference, particularly by the police. The addict subculture thus performs a number of important functions for addicts, not the least of which is the opportunity to associate in a mutually beneficial way with others like themselves. The drug subculture has its negative aspects as well, however, since it isolates the addict from conventional society. An addict often knows only other addicts, who tend to reinforce the addiction process rather than provide positive social support to become and remain free of drugs. Like other deviant subcultures, the drug subculture does not prepare its members to reenter the conventional world; in fact, it inhibits reentry.

Cocaine Use

When cocaine was first discovered in the 1800s, some hailed it as a new wonder drug. Among its initial, most noticeable effects, the drug suppressed fatigue, an effect that attracted the attention of Sigmund Freud. By the end of the 1880s, Freud and others had given up hope of medical applications for cocaine, but its unregulated status and pleasant effects quickly made it a staple in the patent medicine industry. Cocaine moved underground only after the Pure Food and Drug Act of 1906. At the time, users included such socially marginal groups as criminals, jazz musicians, prostitutes, and blacks. Later, cocaine would become associated with beatniks and, still later, with movie stars and professional athletes. By midcentury, cocaine was a rare and exotic commodity. It had become a rich person's drug, the "champagne of the street."

The Cocaine Highway

By the late 1960s and early 1970s, cocaine use had spread in the United States as a result of two particularly important events (Inciardi, 1986: 73–74). First, the U.S. Congress passed legislation that reduced legal limits on production of amphetamines and placed strict controls on depressants. Second, the World Bank allocated funds to build a new highway in the high jungles of Peru. Farmers had always grown coca leaves in the Peruvian Andes, but only small amounts of leaves reached locations suitable for cocaine processing. Serious obstacles limited travel from the growing slopes to population centers that could support refining operations, so exporting required long treks with pack mules through dangerous terrain. The World Bank's construction of a paved highway through the Huallaga Valley opened up transportation routes that simplified shipping of coca. At the same time, the reduced supply of amphetamines and sedatives in the United States left a ready market for the increasingly available cocaine.

Inciardi (1986: 73–74) describes a processing operation (see also Table 10-3):

> At [secret, nearby] jungle refineries, the leaves are sold for $8 to $12 a kilo. The leaves are then pulverized, soaked in alcohol mixed with benzol . . . and shaken. The alcohol-benzol mixture is then drained, sulfuric acid is added, and the solution is shaken again. Next, a precipitate is formed when sodium carbonate is added to the solution. When this is washed with kerosene and chilled, crystals of crude cocaine are left behind. These crystals are known as coca paste. The cocaine content of the leaves is relatively low—0.5% to 1.0% by weight as opposed to the paste, which has a cocaine concentration ranging up to 90% but more commonly only about 40%.

The cocaine highway leads from the jungle refineries to the Amazon River and on to the Atlantic Ocean. A number of cities serve as transportation centers. Santa Cruz, Bolivia, is a major meeting point for Colombian and American buyers of cocaine and a major point of departure for smugglers. Other important cities include Tingo Maria and Iquitos, Peru; the latter offers port facilities that can service commercial ships. The marina at Leticia, Colombia, a town of only a few thousand, hosts a suspicious number of overpowered outboard boats and other high-performance racing vessels, as well as a number of small seaplanes (Inciardi, 1986: 75).

From these beginnings, cocaine moves to markets in the United States and other places. Shipments pass through deserted airstrips and obscure combinations of air–sea routes chosen for the difficulty of patrolling them. By the time cocaine reaches its ultimate consumer, it has been diluted several times with (among other substances) baking soda, caffeine, and powdered laxatives, leaving an average purity of 12 percent that sells on the street for prices near $100. This process

TABLE 10-3
You Want To Make Some Cocaine, Eh?

What You Need	Why You Need It (The Six Cs)
1. A jungle location	*Coca leaves.* Coca grows best at altitudes above 1,000 feet in hot, humid climates with heavy rainfall. With proper care and fertilization, leaves can be harvested every 35 days or so.
2. A processing plant	*Coca paste.* Dried leaves soak in a plastic-lined pit with water and sulfuric acid. Someone wades through the mixture periodically to stir it up. After several days the liquid is removed, leaving a grayish paste.
3. Some chemicals	*Cocaine base.* Addition of water, gasoline, acid, potassium permanganate, and ammonia to coca paste forms a reddish brown liquid that is then filtered. Drops of ammonia produce a milky solid.
4. More chemicals	*Cocaine hydrochloride.* Filtered and dried cocaine base is dissolved in a solution of hydrochloric acid and acetone or ethanol. A white solid forms and settles.
5. Electrical capacity	*Cocaine.* Cocaine hydrochloride is filtered and dried under heating lights to form a white, crystalline powder. Cocaine is now ready for distribution, usually in 1-kilogram packages for $11,000 to $34,000 each.
6. Still more	*Cutting.* Before reaching the street, chemicals cocaine is diluted, or cut, with sugars, such as mannitol (a baby laxative), or local anesthetics, such as lidocaine. Usually sold in 1-gram packages for $50 to $120.

turns 500 kilograms of coca leaves worth $4,000 to the grower into 8 kilos of cocaine worth $500,000 on the street.

During the last half of 1985, a crystalline form of cocaine appeared on a large scale in the United States. Called *crack*, presumably because of the crackling sound it makes when burned, drug makers produce it by soaking cocaine hydrochloride and baking soda in water and then applying heat to create crystals. Crack had been available prior to the mid-1980s, but it became popular as an alternative to cocaine only over the past decade or so. A number of reasons have contributed to the popularity of crack (Inciardi, Lockwood, and Pottieger, 1993: 7). First, users can smoke this form of the drug rather than snorting it, which encourages rapid absorption into the body, producing a very quick high. Second, crack costs less than cocaine. A gram of cocaine for snorting might cost up to $100 depending on its purity, but the same gram could be transformed into anywhere from 5 to 30 crack rocks. Users then purchase individual rocks, some for as little as $2, depending on their sizes. Third, smugglers could easily hide and transport crack, facilitating illicit transactions.

Extent of Use

Cocaine became the illicit drug of choice for many users in the 1980s, and its increased availability has helped to introduce it to many new users. The chief of staff of the House Select Committee on Narcotics Abuse and Control lamented this traffic in 1985 (Trebach, 1987: 178): "It [cocaine] is dropping out of the skies. Literally. In Florida, people are finding packages in their driveways that have fallen out of planes." The increased availability of cocaine has supported development of new use patterns. Previously, only a small group of relatively wealthy users could afford the expensive, scarce drug. Wider availability and falling prices for some forms of the drug have encouraged the spread of cocaine among all segments of the population and throughout all areas of the country.

Many regard cocaine use as a major health problem. Estimates from the mid-1980s numbered regular users at about 10 million, with perhaps 5,000 people each day trying the drug for the first time (*Newsweek*, February, 25, 1985: 23). Other evidence implies, however, that media reports may have exaggerated the extent of the problem. Orcutt and Turner (1993) suggest that sensationalistic reporting may have contributed to unfounded concern. Graphic depictions distorted modest yearly changes to make them seem like huge jumps. Such tactics created a public perception of a serious cocaine problem in the mid-1980s in the United States.

In 1990, a survey by the National Institute of Drugs estimated that daily or near-daily users of cocaine had increased their consumption, but casual use of cocaine declined over 70 percent from 1985 to 1990 (*Des Moines Register*, December 20, 1990, p. 1A). More recent estimates suggest that cocaine use peaked in the mid-1980s and has declined steadily since that time (Office of National Drug Control Policy, 1996). The number of chronic users may have remained steady while use by occasional users has declined. The Office of National Drug Control Policy (1996: 43) put the figure of current cocaine users at about 2 million, down from an estimate of nearly 6 million in 1985.

Cocaine use by young people continues to generate major concern in the United States. The spread of crack cocaine use among inner-city children and youths has perhaps been the most troublesome issue. Crack has become widely used among young people for a number of reasons. Among them: "Crack is cocaine that is (1) so cheap that teenagers can *start* using it with the money from their allowances, (2) so widely available that, in cities, 12 year olds have no problem finding it" (Inciardi, Horowitz, and Pottieger, 1993: 178).

Monitoring the Future, the annual survey of drug use among students, found a peak in the proportion of high school seniors who had used cocaine, about 12 percent, in 1985. The number declined until 1992 and then began to increase again (Johnston, O'Malley, and Bachman, 1993a: 83; Johnston, O'Malley, and Bachman, 1996). By 1996, only 5 percent of high school seniors had ever tried any form of cocaine, with an even lower number, 2 percent, for use of crack cocaine (Johnston, O'Malley, and Bachman, 1996). These were roughly the rates found in the 1999 survey (Johnston, O'Malley, and Bachman, 1999). Thus, despite disturbing information about recent increases, current rates of cocaine use remain below those of the mid-1980s. (One should also recognize, however,

TABLE 10-4

Examples of Crack Cocaine Jargon

Term	Meaning
Dope man or bond man	Crack dealer
Cookie	A large quantity of crack, sometimes as many as 90 rocks
Bomb bag	Any bag in which drugs are carried for delivery
Deal	Sell
Sell for double	Juggle the true value
Cracks, hard white, white, flavor	Cocaine
Eight-ball	Large rock or slab of crack
Doo-wap	Two rocks
Crumbs	Small rocks or crack shavings
Kibbles and bits	Shake

Source: Chitwood, Dale D., James E. Rivers, and James A. Inciardi. 1996. *The American Pipe Dream: Crack Cocaine and the Inner City.* Fort Worth, TX: Harcourt Brace: 10–11.

that this survey may fail to capture data about crack use by young people who do not attend school.)

During the same time, a slight increase occurred in the number of eighth graders who reported using cocaine, although this statistic remained at a very low level: 1.1 percent in 1991 and 2.0 percent in 1996 (Johnston, O'Malley, and Bachman, 1996). Despite these small proportions, the trend toward rising cocaine use among this young age group may disturb some observers.

High school students' use of cocaine reflects their perceptions of the potential harm from the drug and an overall sense of disapproval of drug users. Research reveals that high school seniors have expressed declining rates of disapproval of cocaine users in recent years. In 1991, 87 percent of these students disapproved of people who had used cocaine once or twice, but by 1996, the disapproval rate had dropped to 83 percent (Johnston, O'Malley, and Bachman, 1996). The students disapproved even more strongly of crack cocaine use, but again, recent rates have shown more tolerance than those for previous years. In 1991, 92 percent of the seniors disapproved of crack use, but by 1996 only 87 percent shared that opinion.

These numbers and others confirm substantial public concern over cocaine and crack (which, like many drugs, has its own language; see Table 10-4), concern well justified by the extent of cocaine use and its negative consequences. But the antecedents to or causes of cocaine use seem to favor general rather than specific conditions. One study, for example, reported that the processes that led users to cocaine differed little from those that led others to use heroin and other illicit drugs (Newcomb and Bentler, 1990).

Methods of Use

In cocaine, drug users have found a versatile substance that they can administer in a number of ways. South American users had long smoked cocaine paste, which contained from 40 percent to 90 percent pure cocaine. Users in the United States have rediscovered this method of administering the drug. Many chronic users take cocaine through intravenous injections, and this method permits users to combine it with other drugs. A combination of cocaine and heroin, called a "speedball," intensifies the euphoric effect—and the danger—of both drugs. Speedballing killed actor John Belushi.

Users most commonly administer cocaine by inhaling it through the nostrils. They arrange a quantity of powder in a line, or several lines, and sniff it into the nose, often through a small straw (a rolled-up dollar bill will do). Cocaine administered in this way produces a relatively short sensation lasting about 20 or 30 minutes. The short duration of the drug's effects induces some users to take multiple doses. One study reported that some users might take up to 10 doses per day (Cox, Jacobs, Leblanc, and Marshman, 1983).

The user remains alert and in full mental control while under the influence of the drug, while escaping drawbacks and risks of other drugs such as hangover, physical addiction, lung cancer, infection from dirty syringes, or damaged brain cells (Inciardi, 1986: 78). The drug induces immediate sensations of euphoria regarded as pleasurable by virtually all who try it. With few apparent negative side effects, cocaine developed a reputation as a sort of ideal drug.

Widespread use of crack is a much more recent phenomenon. Most observers trace the American origins of crack to about 1985 (Chitwood, Rivers, and Inciardi, 1996: 1–3). Trebach (1987: 178) describes this development, in large part, as a response to a supply glut in the cocaine market: "In a sense, crack was a packaging and marketing strategy to deal with the economic problem of an excess of cocaine supplies." Most users smoke crack in small pipes. Many young drug users favor the drug for its relatively low cost. Its effects become apparent even more quickly than those of snorted cocaine.

Crack is not an unusual presence in the classrooms and on the streets of virtually all large U.S. cities. Some believe that it exerts stronger addicting power than regular cocaine and perhaps heroin. Yet, one observer reports: "I have searched. My assistants have searched. We have gone through many government reports. We have quizzed government statistical experts. We have yet to discover one death in which the presence of crack was a confirmed factor" (Trebach, 1987: 12).

Consequences of Use

Cocaine does not appear to produce physical dependence, and the user does not build up increasing tolerance for its effects, as a heroin user does. Most users take small amounts of the drug on infrequent occasions. Some, however, take increasing amounts and increase the frequency of their use. Some now doubt the claim that cocaine is nonaddicting, describing the assertion as a result of limited information due to the recent scarcity of the drug (Cohen, 1984; Gonzales, 1987).

Experts still vigorously debate the potential for developing physical dependence on cocaine. Stephens (1987: 35) suggests that "The current evidence favors the conclusion that physical dependency [with cocaine] may occur." In contrast, Inciardi (1986: 79) asserts that the drug is nonaddicting, blaming habitual use on "psychic dependence" driven by a motivation to avoid the feeling of depression that chronic users experience when they stop taking cocaine. This desire to use cocaine originates, not in physical conditions, Inciardi claims, but in psychogenic characteristics, "emanating from the mind" (Inciardi, 1986: 80).

Whether it results from physical or psychic dependence, chronic, heavy cocaine use does lead to a number of negative consequences, including a state of paranoid psychosis and a general emotional and physiological debilitation. Short-term effects can include heart failure, respiratory collapse, fever, and sudden death (Welti, 1987). Prolonged, chronic use can cause damage to the septum of the nose. Investigations continue to evaluate other long-term effects of the drug, which seem to include depression and heart ailments, in addition to the short-term effects (Estroff, 1987). Studies identify these especially serious consequences of cocaine use with intravenous administration and freebasing.

Perhaps the most important concern for cocaine use, as for other drugs, is not its potential for addiction in a conventional sense, but rather its effects on people's lives. This judgment might evaluate the effect of cocaine use on the user's social relationships, employment status, school performance, and general functioning in society. Clearly, as with other drugs, many users function adequately, with drugs constituting only small portions of their lives. This probability seems especially high for recreational drug users who associate cocaine with specific social situations. Other users, however, make drugs the single most important part of their lives. They spend their time seeking and taking their drugs of choice, forcing their other activities and relationships to accommodate this central interest.

Users of crack cocaine do develop physical dependence. Many crack smokers go on "missions" (3- or 4-day binges), sometimes smoking almost constantly, 3 to 50 rocks per day. During these times, they sleep and eat only rarely, and almost all activities center around the drug. Daily crack habits are not unusual, as one recovering crack user observed:

> I smoked it Thursday, Friday, Saturday, Monday, Tuesday, Wednesday, Thursday, Friday, Saturday on that cycle. I was working at that time. I would spend my whole $300 check. Everyday was a crack day for me. My day was not made without a hit. I could smoke it before breakfast, don't even have breakfast or I don't eat for 3 days. (Chitwood et al., 1996: 12)

Users find many attractive benefits in crack: It is cheap, easy to conceal, and virtually odorless while producing an intense euphoria that lasts less than 5 minutes. At the same time, crack is associated with a number of highly undesirable consequences, including digestive disorders, nausea, tooth erosion, brain abscess, stroke, cardiac irregularities, occasional convulsions, and psychological disorders.

Society's Response to Drug Use and Addiction

Efforts to control illicit drug use may target two potential objectives: control the substances themselves or the behavior of the people involved with them (e.g., users, dealers, importers). Observers often refer to these goals as attempts to alter either the supply of drugs or the demand for drugs.

Strategies directed toward controlling supplies include attempts to restrict drug imports and initiatives to decriminalize drugs. Strategies designed to reduce the demand for drugs include applying criminal sanctions to deter users and dealers, diverting drug users and addicts to treatment programs, operating addict self-help programs, and preventing drug use through measures like drug-education programs in schools.

Applying Criminal Sanctions

Two viewpoints dominate discussions on public policy toward drugs: the legalist and public health perspectives. The legalist perspective views drugs and their consumption as essentially a legal problem to be addressed by criminal sanctions and institutions. The public health or social welfare perspective views drugs as a behavioral issue to be addressed more actively by community resources than by law (Zimring and Hawkins, 1992).

 Issue | **"Cell Heads" and the Changing Nature of Cocaine Distribution**

South Florida used to be the entry site of much of the cocaine smuggled into the United States. But little cocaine now comes through Miami since the Colombian pipeline now favors the Southwest U.S. border as an entry site. There are, however, many drug smugglers in Miami, according to the *Miami Herald* (reprinted *Omaha World-Herald*, November 7, 1999: 17-A). These criminals conduct their dealings with cell phones and Internet connections. They are told to live inconspicuously, with all the trappings of middle-class normality. They are told to buy $200,000 houses, to have Saturday barbeques with neighbors, to leave the house each day by 9:00 A.M. as if they were going to work. Nevertheless, they are able to conduct their business with modern methods of communication.

Some dealers have moved their operations to the Internet. They ensure their privacy by constructing elaborate security systems so that when an unauthorized person enters their chat room, alarms go off. Only by downloading one dealer's hard drive while he was out shopping did DEA agents have a clue as to the scope of the drug trafficking enterprise. Information from that hard drive led to other computers, wiretaps, and search warrants that led to the indictments against 31 dealers. But it is a technological nightmare to try to find a way to monitor the chat rooms undetected, to trace calls coming into a local Internet service provider, and to identify individuals based on log-on nicknames.

For most of the 20th century, the United States has followed a policy favoring legal suppression of drugs (Meier and Geis, 1997). Early legislation, such as the Pure Food and Drug Act in 1906 and the Harrison Act in 1914, were direct attempts to restrict or eliminate use of undesirable drugs through legal restrictions. Supported by a number of reform groups, including the American Medical Association (Courtwright, 1982), these laws represent the principal tool for influencing drug use in this country.

Contemporary expressions of drug policy have also followed the legalist ideology. For example, the National Drug Control Strategy, adopted during the last year of the Bush administration, calls for expansion of the legal presence in the drug world and stronger federal action in the "war on drugs" (Zimring and Hawkins, 1992: chapter 1). Indeed, since the mid-1980s, the federal government has taken an increasingly active role in the response to illegal drugs.

Two important topics have dominated discussion of drug policy within the legal framework. These are the relationship between addiction and crimes committed to fund purchases, and the general question of whether to apply criminal sanctions specifically for involvement with illicit drug use.

Addict Crime   Laws affecting drug use include those that prohibit manufacturing, selling, and using certain drugs as well as those that address crimes associated with drug use. Crimes committed by addicts in order to secure money for drugs form a definite part of society's overall legal response. Observers have offered notoriously unreliable estimates of the proportions of crimes attributed to addicts as opposed to nonaddicts. Despite difficulties obtaining precise figures, the expense of a heroin habit clearly induces addicts to commit property crimes. Most addicts require between 10 and 30 milligrams of the drug per day. In addition to the direct costs of drug addiction, these crimes also impose indirect costs, such as the costs of lost property, law enforcement, drug-education programs, absenteeism, and the like.

Cocaine, another expensive drug, may also motivate some crimes. Crack cocaine costs relatively little, however, unless one needs to take a lot of it. Heavy crack users, like heroin addicts, may also commit crimes to fund drug purchases. Indeed, crack cocaine is reputed to establish a powerful hold over heavy users. Some steal, others help to distribute the drug to others, and still others trade sex for crack (Ratner, 1993).

The idea that addicts and other users often commit crimes to support their expensive habits has been called the *enslavement theory of addiction* (Inciardi, 1986: 160–169). As an addict becomes increasingly tolerant of the drug, she or he requires progressively larger doses, so the daily expenditure generally comes to exceed the legitimate funds of most addicts. Therefore, these users "hustle" to secure drugs. Simply put, the theory explains that the high cost of their habits induce most addicts to support their addictions by committing crimes. In the past, crimes committed by addicts largely involved various types of theft, such as burglary, and, for women, prostitution (Cuskey and Wathey, 1982; Rosenbaum, 1981).

Research offers little support for the enslavement theory, however. Many addicts clearly do engage in criminal activity to pay for heroin, but many of them

✳ *In Brief*   **The Relationship Between Drug Abuse and Delinquency**

A reasonable person may well wonder whether illegal drug use is a common cause of other kinds of delinquency. Such a view finds common acceptance among parents, for example, who insist that their youngsters' law-breaking behavior results from their use of drugs. But which comes first, the drug use or the delinquency?

Two recent studies do not find strong support for the view that drug use causes other kinds of delinquency. In each study, the investigators reported that drug use by youths followed observations of "conduct disorder," a psychiatric diagnosis characterized by 13 symptoms. These observations had previously identified most young drug users as delinquents in their own right based on symptoms such as stealing, truancy, fighting, arson, property destruction, cruelty to people or animals, lying, and running away from home.

Sources: See Swan, Neil. 1993. "Researchers Probe Which Comes First: Drug Abuse or Antisocial Behavior? *NIDA* Notes 8: 6–7. The studies themselves can be found in Robins, L. N., and S. McEvoy, "Conduct Problems as Predictors of Substance Abuse." Pp. 18–204 in *Straight and Devious Pathways from Childhood to Adulthood.* Edited by Robins, L. N., and M. Rutter. Cambridge: Cambridge University Press, 1990; and Young, S. E., M. S. Zoccolillo, T. J. Mikulich, and T. J. Crowley. 1992. "Relationship between Conduct Disorder and Adolescent Substance Use." Paper presented at the 54th annual meeting of the College on Problems of Drug Dependence, Keystone, Colorado, June. See also Inciardi, James A. 1986. *The War on Drugs.* Palo Alto, CA: Mayfield.

establish patterns of crime before becoming addicted. Still other addicts generate criminal behavior patterns only at certain stages of their addictions (Faupel and Klockars, 1987). With such backgrounds, addicts may view crime as a logical method for supporting heroin habits, once begun. That decision seems especially logical for female addicts, since their crimes often escape detection and arrest. A study of two groups of female addicts in Miami reported that less than 1 percent of all crimes committed by these groups resulted in arrest (Inciardi and Pottieger, 1986). One reason for this low percentage is that the women concentrated on "victimless" crimes—drug sales and vice, especially prostitution and procuring.

One way to obtain money for drugs—and other purposes, of course—is to sell drugs. In some communities, residents sell drugs simply as a financially attractive alternative to working at legitimate jobs. But if legitimate jobs are not available, people may view drug selling as a necessary activity. A 28-year-old welfare mother from a large public-housing project in Chicago explains what people in her neighborhood do when they need money:

> Shit, turn tricks, sell drugs, anything—and everything. Mind you, everyone is not a stick-up man, you know, but any and everything. Me myself I have sold marijuana, I'm not a drug pusher, but I'm just tryin' to make ends—I'm tryin' to keep bread on the table—I have two babies. (Wilson, 1996: 58)

Should the Law Prohibit Using Drugs?   Different countries establish varying applications of criminal law as a mechanism of social control over drugs. For example, in Malaysia and Singapore, a conviction for possession of 1 kilogram of hashish or 20 grams of heroin carries an automatic death penalty. Similar offenses in Denmark bring sentences of only 60 and 90 days, respectively (Rowe, 1987). Other countries impose penalties between these extremes, and these differences reflect variations in their social attitudes.

Applications of the law to suppress drug use have sparked controversy in the United States. Some advocate a strategy of *decriminalization*, eliminating or reducing legal penalties for drug use while retaining criminal sanctions for those who sell or distribute drugs (Nadelmann, 1991). Further, some believe that attempts to suppress use have actually increased the difficulties of controlling drug traffic by compelling distributors to create extensive, necessarily criminal organizations to import their supplies. In addition, addicts commit crimes to pay the high prices that illicit drug sellers charge; perhaps such crime would decline if they could obtain legally available drugs at comparatively low cost (Johnson et al., 1985). As an additional consequence of legal sanctions, drug users become "criminals" simply by using drugs. The stigma of such an arrest record additionally complicates the transition from addict to former addict.

Current skepticism about the role of law in controlling drugs results from frustration with the law's apparent inability to reduce illicit supplies (Inciardi and Saum, 1996) as well as other causes. Advocates of legalizing drugs point out that huge expenditures ($10 billion by the federal government alone in 1987) and the active employment of hundreds of thousands of agents and other personnel have not bought victory in the war against drugs. Enforcement agencies and their supporters call for increasingly large expenditures for expanded police and prosecution staffing, court trials, jails, and prisons, where large portions of the inmate populations already serve time for drug law violations. Further, by making drugs such as cocaine and heroin illicit, the government also drives up their prices, and addicts must pay these extremely high costs. Criminals benefit from this money, which addicts obtain by committing still more crimes. One study estimated that addicts in this country might commit as many as 50 million crimes a year (Ball, Rosen, Flueck, and Nurco, 1982). This figure reflects extrapolation from a listing of crimes committed by a sample of 243 addicts, who had committed an average of 178 crimes each over the course of 11 years. A medical response rather than a legal one, however, would reduce the cost of heroin and reduce both addict crime and enforcement costs.

Driven by these and other arguments, many authorities and groups have recommended decriminalizing drug use and regarding it primarily as a medical or social problem, as Great Britain and many western European countries do. Trebach (1987: 383–385), for example, proposes a program of heroin maintenance therapy that would require addicts to live productive lives or risk losing their supplies of heroin. Trebach also suggests providing affordable treatment for drug users who wish it and creating more creative legal solutions to drug problems. If successful, such an initiative would free addicts from the need to commit crimes to pay for drugs. Trebach also suggests tightening controls on currently legal drugs, such as alcohol and tobacco, and relaxing controls on currently illegal drugs. He recommends decriminalizing marijuana use and cultivation for personal use, and he would allow medical use of heroin for addicts and pain patients by prescription. He would not, however, make heroin legal for casual, recreational use.

Arguments for legalization assume that adults have the right to abuse their own bodies without interference from the law. They note that obesity causes far more physical injury than drugs do, yet no one would propose a law restricting a person's food intake. Proponents of legalization also claim that enforcement of drug laws has often led to violations of constitutionally guaranteed personal freedoms; police and courts have abused search warrants, seized property, and denied protection from self-incrimination, contrary to the protections stated in the Bill of Rights. Failing a random drug test may result in loss of employment, but it ignores indications of alcohol use, potentially a much more serious problem. Moreover, critics of current policies lament the inconsistency of allowing wide availability of two other drugs, alcohol and nicotine, with effects more devastating than perhaps all other drugs put together. Over 450,000 people die each year as a result of their use of cigarettes.

Other observers see more success in legal prohibitions than critics are willing to concede. Inciardi and McBride (1990: 1) point out a problem with comparing the death rates from use of alcohol and tobacco to those from use of other drugs: "What is summarily ignored is that the death rates for alcohol and tobacco use are high because these substances are readily available and widely used, and that the death rates from heroin and cocaine use are low because these drugs are not readily available and not widely used." Legalization might increase the supply of drugs, increasing their harmful effects, especially on vulnerable populations (Inciardi and Saum, 1996). For example, earlier discussions have mentioned high rates of drug use in inner-city areas, or ghettos, and especially among minority group members. Women head many ghetto families, and women seem more likely than men to become physically dependent on crack for reasons that are not clear. A growing threat from drug use might well further increase problems in such neighborhoods of family disruption, abuse, and child neglect.

In searching for a sound national policy, Inciardi (1986: 211) says, "It would appear that contemporary American drug-control policies, with some very needed additions and changes, would be the most appropriate approach [to control illegal drug use]." Inciardi advocates changes including assigning military units to interdict shipments of illegal drugs, continuing conventional treatment efforts with expanded funding and personnel, and promoting education programs aimed at prevention (Inciardi, 1986: 212–214). Inciardi points out that other nations have done little to control manufacturing of illegal drugs within their borders, so the U.S. government must therefore take more aggressive measures than it has previously sponsored. "Drug abuse," Inciardi (1986: 215) claims, "tends not to disappear on its own."

In a similar vein, Kaplan (1983) has advocated a legally oriented response to heroin addiction based on coercing crime-prone addicts into treatment programs (see also Inciardi, 1990). This option may promise only limited benefits, given the limited success of heroin treatment (as discussed in the next section), but Kaplan urges that even small benefits may justify the legal effort. Other observers have rejected such a policy on the grounds that legal coercion may violate civil rights, and it would not, in any case, have much impact on the street addict subculture (Stephens, 1991). Yet, research increasingly suggests that persons who enter treatment under threats of harsher penalties have lower relapse rates than other patients (Inciardi and McBride, 1990: 4). This effect may result from the relationship between treatment success and length of stay in treatment, since those who begin treatment under coercion stay there longer than those who voluntarily undertake it.

Treatment

Treatment programs offer an alternative to a legally oriented response to drug addiction. Such treatment represents a major effort in the United States, as some figures clearly confirm. Estimates for 1985 report that about 6,000 treatment facilities admitted about 305,360 clients, most of them males aged 21 to 44 who used heroin, marijuana, or cocaine (Reznikov, 1987). States reported spending more than $1.3 billion during that same time for alcohol and drug-abuse treatment and prevention services. Of this amount, nearly 80 percent paid for direct treatments, 12 percent for prevention programs, and the remaining amount for training, research, and administration. Drug-treatment programs handled most of their clients (76 percent) as outpatients, while 19 percent were admitted to residential facilities, and 5 percent were admitted to hospitals.

As its ultimate objective, treatment seeks to promote abstinence. It also pursues an intermediate goal: *harm reduction*. This effort seeks to minimize the negative consequences of ongoing drug use. If treatment cannot eliminate all drug use, as some claim, then intervention works best when it tries to minimize the harm that drugs do to individuals and communities.

A harm-reduction policy might encourage heroin addicts to use sterile needles and participate in outpatient therapy rather than or in addition to inpatient therapy. Harm reduction recognizes that not all heroin addicts—or other drug users, for that matter—will benefit from treatment, but certain steps taken while people continue to use drugs could benefit both them and their communities. One such step maintains users on drugs that do minimal damage, such as methadone.

Methadone Maintenance   Not all drug users fail to function effectively in society. Substantial numbers of addicts can participate responsibly and productively in their communities. For this reason, clinical programs in major urban areas have sought to keep thousands of opiate addicts on *methadone maintenance* as part of efforts to bring about their rehabilitation. New York City initiated methadone maintenance in the early 1960s (Stephens, 1987: 88–92). Under this program, an addict reports daily to a clinic to take a medically supervised dose of methadone, which acts as a substitute for heroin to prevent withdrawal symptoms. The methadone is usually given in a glass of orange juice after a urine test confirms that the addict has not used heroin before visiting the clinic. Methadone behaves chemically much as heroin does (including the potential for physical dependence), but it causes much longer-lasting effects because the user's body does not metabolize the drug as quickly as heroin. Therefore, one dose per day can maintain the addiction, replacing three or four or five doses of heroin. The theory explains that an addict on methadone will not return to heroin, either to relieve withdrawal distress (since the methadone prevents it) or to gain a euphoric reaction (because the methadone blocks this). The oral administration of methadone also eliminates problems associated with dirty needles—infection, AIDS, and risks of other diseases.

Some evidence supports the premise that methadone maintenance programs do reduce addict crime by providing legal sources of drugs. This benefit affects only those who voluntarily participate and remain in the program for a period of time (Newman, Bashkow, Cates, 1973). Moreover, critics worry about uncertain long-term effects of methadone; clinical research continues on this issue (Kreek, 1979). In any case, methadone clearly does not deal with the causes of addiction; in fact, supplying methadone for heroin addicts may seem like prescribing brandy to keep an alcoholic from drinking too much rum. Also, many addicts combine multiple drugs, and methadone maintenance does nothing to discourage use of nonopiates. Furthermore, such a program risks diversion of methadone from clinics into black market use in combination with, rather than instead of, heroin (Agar and Stephens, 1975; Inciardi, 1977).

Heroin Maintenance   Some observers, sensitive to the problems of methadone maintenance and similar programs, have suggested that the United States institute a heroin-maintenance policy instead. Such a program would not differ greatly from the present system in England, which emphasizes a medical response to addiction based largely on outpatient treatment by physicians and social service professionals, including prescriptions for low-cost drugs.

Addiction has been considered a disease in England since 1920, but the resulting policy experienced difficulty in the 1960s, since it excluded the most predominant type of drug user at the time: the recreational user. The most recent statement of the British drug policy attempts to acknowledge the need for treating recreational users by invoking the concept of the "problem drug taker," a term analogous to "problem drinker." The British treatment program emphasizes multidisciplinary intervention by social workers and community workers in addition to medical specialists. In fact, the current British policy reflects a departure from the strict medical or disease model that guided the country's response to drug use for so many years (Advisory Council on the Misuse of Drugs, 1982; see also Symposium on the British Drug System, 1983).

Traditionally, British physicians have prescribed minimal doses of heroin as part of their attempts to cure the addicts. Clearly, British officials and the public did not regard addiction as criminal behavior. Because legally dispensed heroin costs less than drugs purchased on the black market, the program eliminates an important motivation for addicts to commit crimes. It also prevents forced association with criminals that leads to destructive participation in a criminal subculture. Most British addicts appear to remain outside the criminal world, since they need not steal, work as prostitutes, or peddle drugs in order to obtain heroin. An apparent shift away from this medical approach may have resulted in part from increases in rates of addiction in many parts of the world, including England.

Clinical Tactics to Treat Heroin Addiction   When hospitals treat heroin addicts, the optimum treatment period is a few months. Newly admitted addicts undergo thorough medical examinations and receive treatment for any conditions other than their addictions. Treatment builds up their physical condition as it removes drugs. Currently, heroin treatment programs most commonly dispense methadone, because it subjects users to withdrawal symptoms much milder than those from heroin. After replacing an addict's dependence on heroin with dependence on methadone, the program slowly withdraws that drug. Along with methadone treatments, the addict also receives recreational and occupational therapy and vocational training. Upon release, the addict usually receives follow-up services as needed. Many different therapeutic community facilities provide this kind of inpatient care. To avoid stigma, California diverts addicts to such hospital facilities through civil commitment rather than criminal court proceedings. Either alternative forces users into treatment programs, however.

Studies of the effectiveness of drug treatment demonstrate the difficulty of correcting narcotics use, particularly heroin addiction. Its status among the most persistent forms of deviant behavior becomes clear from the high incidence of relapse among treated addicts. For example, one follow-up study found that male addicts relapsed, defined as any reuse of narcotics, at a rate of 87 percent, confirming the 80 to 90 percent relapse rates reported in most other studies (Stephens and Cottrell, 1972).

Addiction creates behavior patterns that treatment cannot break without difficulty. In fact, most treatment efforts, guided mainly by medical rationales, fail to change this behavior. Addiction results from a complex process involving sociological and psychological influences, such as drug-using associates, participation in a drug subculture, and a self-conception as an addict. For this reason, researchers continue to study how social and psychological factors affect treatment (Grabowski, Stitzer, and Henningfield, 1984).

Several studies give some reason for optimism, though. While they agree about the difficulty of terminating heroin use, they note that many users take heroin regularly without becoming addicted, and some addicts escape dependence without any treatment at all. A comparison of addicts inside and outside treatment programs reported that untreated addicts had smaller habits, more frequently stopped using heroin, lived within more cohesive families, and displayed higher levels of self-esteem; this result suggests that studies may mislead if they consider treated addicts as somehow representative of all addicts (Graevan and Graevan, 1983). Another study of a sample of 51 addicts reported that one-third had drifted between chronic use and abstinence, one-quarter had been dependent at times but had overcome these episodes without treatment, and the remainder were dependent addicts (Blackwell, 1983). Clearly, individual situations influence opiate use and addiction.

Some addicts manage to terminate their dependence on heroin themselves, not by stopping use of the drug "cold turkey," but by changing for a time to another drug, usually alcohol (Willie, 1983). In an area where progress is often difficult to define, the movement from one drug to another may strike some as no progress at all. Other observers, however, welcome cessation of heroin use under virtually any circumstances and consider it a success. Other addicts leave addiction after reaching turning points in their lives, usually profoundly moving existential crises that force them to recognize the need for self-change (Jorquez, 1983). Addicts can realize the desire to terminate their addictions only over time, however, since this change requires them to extricate themselves completely from "the life" and adjust to the world of "squares." Still, some observers have suggested that successful treatment requires nothing less than a total change of life circumstances prompted by circumstances beyond the control of the addict (see Waldorf, 1983). Other observers report that the change to nonaddiction (Bennett, 1986; Biernacki, 1986) or stopping use of cocaine (Waldorf, Reinarman, and Murphy, 1991) can result from rational decision making based on a recognized need to make a change.

Since addicts choose the street life as "rational actors," they can make better choices only after they learn that lasting change depends on taking responsibility for their own behavior. The spread of AIDS in inner-city neighborhoods has provided one incentive for heroin addicts to deal with their addictions. Research points out that people who face a genuine chance of dying because of their behavior will likely change that behavior (see Stephens, 1991). At the same time, the street addict subculture isolates addicts from conventional society. Addicts often associate regularly only with other addicts, who tend to reinforce the addiction process rather than provide positive support to stay off drugs. Like other deviant subcultures, the street addict subculture does not prepare its members to reenter conventional society; rather, it inhibits reentry.

Clinical Tactics to Treat Users of Other Drugs   By the mid-1980s, well-established programs of clinical, medical treatment for outpatients and inpatients addressed issues of marijuana and cocaine abuse. One of the best known of these is CareUnit, a program run by Comprehensive Care Corporation in California, a profit-oriented corporation that reported 1986 earnings of $192,936,000 (Gonzales, 1987: 189). CareUnit affiliates place patients in unused hospital beds in local communities, so they require no new facilities of their own. As endorsed by former first lady Nancy Reagan, CareUnit programs target their inpatient treatments for young users of any drug, frequently as a result of referrals by their parents. The rather generic treatment regimens stress drug education and family relationships. Along with these elements, the programs put patients through individual therapy along with inpatient visits, although most do not attempt to provide treatment intensive enough to result in long-term personality changes. Even a short stay can benefit the user by removing him or her from an undesirable environment.

Critics have found many faults with CareUnit treatment, including the limitation of the patient's stay to a term insufficient to effect permanent change, especially when the youngster returns to the same drug-using environment. The inpatient stay is usually fixed at 28 days, a term that coincides with the maximum covered by insurance companies, and no systematic follow-up attempts to maintain the therapeutic benefits. Critics have described CareUnit's therapy as "McTreatment," likening it to a fast-food model for promoting behavior change. This type of intervention may effectively treat the drug problems of young, inexperienced users, though.

Other private drug treatment programs work in different ways. One such program at Hazelden, Minnesota, stresses early intervention (Gonzales, 1987). The Hazelden model asserts that any treatment, even involuntary treatment, is better than none; supporters claim that such treatment cannot begin too early in a drug user's career. Intervention at Hazelden might follow a "surprise party" attended by the drug user's spouse, parents, concerned friends, neighbors, employer or teacher, and anyone else who might have some influence. A professional counselor makes the arrangements and supervises the event. Each participant makes a list of the user's recent actions that have made life miserable for him or her; each list concludes with an ultimatum: Get help or else! The presentation reinforces this point as everyone reviews a separate list of troubles and makes the same ultimatum. Inpatient treatment can begin immediately after the "party," and the patient must complete a 1-year follow-up program through Alcoholics Anonymous.

Judgments of the effectiveness of many private drug-treatment programs suffer from weak information. Without careful follow-up information on each patient or information about a comparable control group of similar drug users who did not experience the treatment programs, generalizations are seriously inhibited. These programs usually limit their services to drug users still early in their deviant careers, and one may speculate that they achieve high success rates. Some clinics acquire patients only through referrals from the criminal justice system, so patients who do not cooperate risk returning to court for criminal proceedings. A study of one such clinic reported that the clinic staff viewed control rather than rehabilitation as the fundamental objective of the program (Skoll, 1992). Such programs may provide their major benefit through continued publicizing for antidrug messages that may inhibit some use.

Similar problems cloud assessments of the effectiveness of public (tax-supported) drug-treatment programs. For one major problem, evaluators cannot know beforehand what constitutes reasonable success. How many drug users must abstain before the program is judged a success? One researcher reported an abstinence rate of over 50 percent for a program to treat cocaine use (Tennant, 1990), but it could not say whether this figure reflected a very high, very low, or simply average success rate. One study of termination from cocaine use conducted interviews with 267 heavy cocaine users, reporting that many described quitting on their own, without participation in any treatment programs (Waldorf, Reinarman, and Murphy, 1991). Furthermore, many of these users quit relatively easily without suffering the kinds of withdrawal symptoms reported by users of heroin or even cigarettes. Their reasons for quitting emphasized health problems, followed by financial and work-related problems. They ranked fear of arrest sixth in importance on their list of reasons to quit (Waldorf, Reinarman, and Murphy, 1991: 194).

Addict Self-Help Programs

Addicts may also find assistance in quitting drugs through self-help groups operated by former addicts themselves, such as Narcotics Anonymous, Synanon, and various local groups. Narcotics Anonymous (NA) was founded in 1948 by a former drug addict with a treatment model similar to that of Alcoholics Anonymous. It combats drug addiction through an informal organization. NA members recognize the difficulties that former addicts encounter as they try to refrain from drug use; they attempt to provide substantial social support that promotes this goal. Branches operate in most large cities in the United States and in Canada.

Each new member forms a close association with a more experienced member, who offers advice and responds to calls for help. Much like Alcoholics Anonymous, NA replaces norms and attitudes favoring the use of drugs with those opposed to drug use. Research offers little firm information about the effectiveness of Narcotics Anonymous, but this organization seems less successful than Alcoholics Anonymous in effecting permanent change. Weak public support, a stronger negative public attitude toward drug use as opposed to alcohol consumption, and the specific effects of drug addiction may account for this comparative ineffectiveness.

Another mutual-support organization of drug addicts got the name *Synanon* from one addict's attempt to say *seminar*. Synanon was founded in 1958 in Santa Monica, California, by a former member of Alcoholics Anonymous, Charles Dederich (Yablonsky, 1965). Typically, Synanon establishes households shared voluntarily by drug addicts who want to free themselves and one another from addiction. This group method of treating drug addiction stresses interpersonal cooperation and group support for individual addicts who need help with their problems. Addicts manage their own offices and carry out the physical operations of their establishments. Members in a Synanon group separate into three groups, defined by stages in progress toward rehabilitation. In the first stage, they live and work in the residential center; in the second, they hold jobs outside but still live in the house; in the third, members graduate to living and working outside the facility.

As an important part of the program, members meet each evening in small groups, or "synanons," of 6 to 10 members. Membership rotates so that one does not regularly interact in a single small group with the same people. No professionals participate in these sessions, which work to "trigger feelings" and precipitate "catharsis," or release of emotional energy. The discussions feature "attack therapy" or "haircuts" in which members confront and cross-examine one another; hostile attack and ridicule are expected. "An important goal of the 'haircut' method is to change the criminal-tough guy pose" (Yablonsky, 1965: 241). The group intends this method to break down defensiveness about drugs and defeat denial of addiction. The haircut also triggers feelings and emotions about addiction and the problems of coping with a drug habit.

A similar program on the East Coast is Daytop Village, a residential treatment community founded by a Catholic priest in 1963 in Staten Island, New York. This program has become an international organization with centers in the United States, Canada, Ireland, Brazil, Malaysia, Italy, Spain, Thailand, Sweden, Germany, and the Philippines. Except for Alcoholics Anonymous, this is the largest drug treatment organization in the world (Gonzales, 1987). Most centers maintain waiting lists of drug users who want to begin the relatively inexpensive treatment (about $35 a day or $13,000 a year).

The course of treatment at Daytop Village lasts about two years, and it deals with drug use in the context of family problems. Potential new members must sit in a "prospect chair" to contemplate their need for treatment. They must then stand on the chair and beg for admission to the program. After gaining admission, the new prospect is showered with hugs and encouragement and put to work. Recovering addicts besiege initiates and force them to admit their powerlessness over drugs and need for help. The program fosters group interdependence, punishing members for bad behavior and rewarding them for good behavior.

The Synanon program seems unknowingly to apply a learning or socialization theory of deviance to the treatment of drug addicts. It brings them into contact with an antidrug subculture in which they learn to play nonaddict roles. Volkman and Cressey (1963) have described the main elements of the program:

1. An individual expresses willingness to give up his or her own desires and ambitions and to accept complete assimilation into the group dedicated to "hating" drug addiction.

2. The addict discovers the effects of belonging to a group that is "antidrug," "anticrime," and "antialcohol." Members hear over and over again each day that their stay at Synanon depends upon their staying completely free from drugs, crime, and alcohol—the group's basic purpose.

3. The group maximizes the effects of a family-type cohesion, deliberately throwing members into continuous, shared activity, all designed to make each former drug addict fully realize traits common to each other member of the family-type group.

4. The program explicitly assigns each member certain status symbols in exchange for staying off drugs and even developing antidrug attitudes. The entire experience is organized into a hierarchy of graded competence unrelated to the usual prison or hospital status roles of inmate or patient.

5. The program specifically emphasizes complete dissociation from the member's former drug and criminal culture, substituting legitimate, noncriminal cultural patterns.

One cannot exactly evaluate the effectiveness of such residential organizations in bringing about changes in addiction behavior. The organization displays a secretive attitude about its clients in an effort to protect their privacy and the treatment they complete. Furthermore, a high success rate from such programs could reflect the effectiveness of the programs themselves or the fact that they generally deal only with highly motivated individuals who really want to terminate their addictions. Most patients who are admitted to programs like Synanon and Daytop Village, perhaps as many as two-thirds, do not complete the programs. In any case, recent revelations concerning the philosophy and operations of Synanon suggest that the program may not establish antidrug and anticrime subculture as effectively as once thought (Olin, 1980).

Therapeutic communities are very expensive treatment options. High success rates might justify such expenses, but the programs suffer from another problem.

Programs like Synanon and Daytop Village appeal to a very small proportion of heroin addicts. Even if they prove exceptionally effective, such programs do not offer practical replacements for other attempts to reduce addiction. These programs also run a risk of increasing dependence by addicts. Some users have a hard time completing such a program, and they may stick around its fringes for a long time.

Prevention of Drug Use

Two strategies underlie present attempts to prevent drug abuse. Some imply threats intended to scare potential users away from drugs. Others conduct education programs specially designed to alert potential users to the dangers and consequences of drug use.

Media Messages   One cannot easily evaluate the prevention strategy based on scare tactics because it lacks the formalized structure of programs of drug education. Scare tactics can contribute to a larger drug-information program, though. Media messages, most geared toward young users, stress the negative physical consequences of taking drugs, particularly cocaine and crack. One such ad shows a hand holding an egg. "Okay," the announcer says, "one more time. This is your brain." The egg is then shown frying in a pan. "This is your brain on drugs," the announcer says. By the mid-1980s, such public service announcements appeared regularly on television and in newspaper ads.

Such messages reach virtually everyone in their intended audience, but they may not achieve substantial effectiveness. Many users, like cigarette smokers, recognize the potential harm from drugs, but they may doubt their ability to control drug-taking behavior. One young cocaine user expressed this sentiment:

> I think it's a good idea that the media focuses on coke. It's good to teach young kids to stay away from it. It really is bad stuff. But as you get older, you can make your own decisions. I am old enough to make my own decisions. I don't need everyone and their brother telling me what to do. The government can try to control it, but it's impossible to stop people from using drugs. Right now it is just a passing fad to say 'no' to drugs. Next year I bet no one will remember the whole campaign. (Smart, 1989: 69–70)

Drug-Education Programs   Drug-education programs, on the other hand, define formal, structured attempts to provide objective information to potential users to help them evaluate drug use and, supporters hope, reject it. Drug-education programs generally target potential users without extensive drug experience or backgrounds. Reports note that increasing exposure to drugs also increases the likelihood of consulting peers for information about drugs, and exposure diminishes probability of accepting input from a drug-education program (Blum, Blum, and Garfield, 1976). For this reason, some studies report both retarding and enhancing effects of these programs; formal drug education inhibits drug use by some students, while it may stimulate further use by others. For example, a study in Canada found a positive relationship between drug-education programs and drug and alcohol use among adolescents (Goodstadt, Sheppard, and Chan, 1982). Clearly, these programs must cautiously choose the contents of their messages as well as the populations they seek to educate; instructors must carefully assess the prior experiences of their target groups (Dembo and Miran, 1976).

Like most problems in deviant behavior, prevention of drug use is not an easy task. The practice of taking drugs to achieve physical effects has become well ingrained in society, and only an extremely fine line separates legal drug therapies from illegal drug use in many respects and social situations. The increase in drug awareness and growing public disapproval of the use of certain drugs, such as crack, may have significant long-term effects in preventing use. However laudable the goal of teaching kids to "just say no," agreement remains elusive about the best mechanisms to ensure that refusal. Information about the negative physical consequences of using a drug seems to influence such decisions less powerfully than whether or not a young person's friends use a drug. Unfortunately, drug educators have not learned how to manipulate friendship patterns of youngsters.

Prevention of drug use will probably remain a matter of informal social control rather than formal social control. Remember, as well, that prevention efforts might do more good if they were to target the two drugs that represent the greatest problem for the most users: alcohol and tobacco.

Summary

This chapter has discussed addiction by evaluating the social and social psychological processes involved in becoming an addict and the consequences of taking drugs. Addiction develops through an essentially social process with effects independent of the physical properties of drugs and their impacts on the human organism. Merton's (1968) theory of anomie, for example, attempts to explain drug addiction in terms of cultural values (or goals) and the user's availability of illegitimate means. According to this theory, one becomes an addict by pursuing unrealistically high cultural goals, eventually turning to illegitimate means to achieve them after obstacles prevent access to legitimate means. Inability to achieve goals through legitimate means or rejection of legitimate means for other reasons leads to a "retreat" from social life and increases the probability of using drugs. A more recent version of this theory proposes to account for the increasing number of addicts in the middle and upper classes. It suggests that this new type of addict, rather than lacking intellectual and social skills required for legitimate success, may instead lack essential personal skills or encounter economic realities that frustrate achievement (Platt, 1986).

Alternatively, Lindesmith's (1968) theory of addiction emphasizes two elements. First, the addict learns how to use the drug from others; second, continued use of the drug relieves withdrawal distress that results from discontinued use. Previous chapters have commented that Lindesmith's theory has difficulty in identifying and operationalizing the role of personal cognitive activity in this process; also, it does not account for the individual's motivation to achieve euphoria, rather than simply to eliminate withdrawal distress. Still, Lindesmith's account of the development of addiction as essentially a learning process seems to make better sense of available information than other alternatives do.

Addiction begins when experienced users take the role of teachers and initiate others into the use of drugs. New users have to learn about drugs, methods for taking them, and the reactions they produce. The most reliable information comes from those who have experienced the drug at some earlier time and whose judgment the initiate trusts. Similarly, maintenance of addiction seems like a result not only of the physical properties of the drug (although users do desire sensations of euphoria and relief from withdrawal distress), but also of the support provided by the addict subculture. If a new user initially learns drug use from others, the subculture certainly perpetuates this behavior.

This interpretation finds support in research evidence from studies of changes in heroin addiction over the past decade. Trends in heroin use show that rates of addiction may now be leveling off but that substantial increases occurred in large cities through the mid-1970s. This trend eventually spread to smaller cities as well. These data also suggest a learning interpretation.

The process of learning drug use appears most strikingly in juvenile initiations into use of marijuana, crack, or some other drug. Some popular opinion blames youth drug use on "pushers" who coerce or trick youngsters into experimenting. In fact, however, initial juvenile drug use actually grows out of a complex social process determined by the availability of tutors and opportunities to take drugs. Rather than an escape from reality, much adolescent drug use represents a means of embracing the reality that many adolescents experience. This social reality incorporates drug use as an integral part of group activities. Youth drug subcultures do not arise as users band together; rather, youths begin to use drugs in the course of participating in the subculture.

Selected References

Adler, Patricia A. 1993. *Wheeling and Dealing: An Ethnography of an Upper-Level Drug Dealing and Smuggling Community*, 2nd ed. New York: Columbia University Press.

This ethnographic study depicts a drug-using and drug-dealing community in southern California. Adler presents an interesting account of the relationships among marijuana and cocaine dealers and the subculture in which they carry out their activities.

Biernacki, Patricia. 1986. *Pathways from Heroin Addiction: Recovery without Treatment.* Philadelphia: Temple University Press.

This study considers how most addicts terminate their heroin addictions, largely without outside intervention. It presents interesting case materials from addicts and readable accounts that conceive the processes of developing and escaping addiction in sociological terms.

Goode, Erich. 1999. *Drugs in American Society*, 5th ed. New York: Alfred A. Knopf.

This widely read summary describes social use of drugs in the United States. It offers little material on the physiological consequences of drugs, instead emphasizing the psychological and sociological aspects. This book presents a valuable sociological analysis of drug use.

Meier, Robert F., and Gilbert Geis. 1997. *Victimless Crime? Prostitution, Homosexuality, Drugs, and Abortion*. Los Angeles: Roxbury Press.

Controversy surrounds the question of whether the criminal law offers an effective and satisfactory tool for controlling illicit drug use. This book examines several "victimless crimes" and explores the different sides of the criminalization debate.

Office of National Drug Control Policy. 1999. *National Drug Control Strategy, 1999*. Washington, DC: Government Printing Office.

A very useful profile of all major illicit drugs, as well as a detailed statement of specific measures being taken to reduce drug use. Special attention is devoted to youthful drug use.

Stephens, Richard C. 1991. *The Street Addict Role: A Theory of Heroin Addiction*. Albany: State University of New York Press.

This systematic account of the street addict role combines sociological theory and empirical data. In spite of growing public concern over cocaine, heroin remains an almost intractable problem. This book tells why.

Waldorf, Dan, Craig Reinarman, and Sheigla Murphy. 1991. *Cocaine Changes: The Experience of Using and Quitting*. Philadelphia: Temple University Press.

This interesting and important study explores patterns of terminating use among heavy cocaine users. The authors present many first-person accounts to capture the process in the words of the users. The conception of cocaine as a drug used occasionally and intermittently conflicts with many media reports about cocaine use.

CHAPTER 11

Drunkenness and Alcoholism

Excessive use of alcohol can produce physical and psychological harm for the drinker, but it also impacts others. A husband of an alcoholic reports that he wasn't sure when alcohol became a problem for his wife but thinks that it was about 30 years after they were married (J. W. C., 1998). Sometimes she would be in bed when the husband returned from work; other times she was up and obviously intoxicated. She stopped seeing her friends, quit having coffee and lunch with acquaintances, and avoided many social occasions. She seemed unable to stop drinking. She attended AA meetings for a time and was told by many, including her physician and husband, that her drinking was out of control. Fueled by alcohol, depression, and anger, the couple fought continuously. Her depression increased, and on two occasions she took too many sleeping pills, prompting a visit to the local emergency room. On another occasion she cut her wrists and started a psychiatric program. But it didn't work. Another bout with pills and a final successful suicide attempt ended her life. "They gave me her wedding ring and necklace," the husband reported, "and then my wife of thirty-two years was gone. She was never to see the arrival of five more loving grandchildren, the marriage of our third daughter, and the graduation of our son from university. These events would have provided so much joy in her life, and she chose to miss them."

Despite expressions of substantial concern about "drug" use, alcohol is by far the most popular mood-altering drug consumed in the United States today. In fact, two drugs—alcohol and tobacco—actually cause more physical, medical, social, and psychological problems than any other drugs. People consume alcohol in the United States more than any other drug, including tobacco. Almost two-thirds of all Americans have consumed alcohol, and almost 50 percent currently drink it (Dawson, Grant, Chou, and Pickering, 1995). The widespread use and potential for harm of alcohol justify considering it as a drug separate from others, such as heroin, cocaine, and marijuana.

Not all drinking is considered deviant, and groups differ in their conceptions of deviant drinking. Society's norms determine when alcohol consumption "steps over the line" between an acceptable practice of social drinking to become deviant or a problem. Drinking itself often does not trigger sanctions for deviance; instead, the conditions under which people drink often determine society's reaction, including physical situations, the ages of the drinkers, and perhaps even the type of beverage. Klein (1991), for example, reports that wine is a relatively acceptable part of many social situations, while many people perceive more risk of harm from distilled spirits and beer. Since norms by definition apply within specific situations, drinking norms reflect conceptions of what people ought and ought not to do in particular situations. For this reason, a determination of deviance requires more information than a simple statement about the presence of alcohol; this judgment also depends on information about the conditions under which people drink.

The determination becomes still more confusing when people fail to agree about those conditions. Many U.S. residents share a fundamental ambivalence about drinking. Many regard beverages with alcohol as permissible elements of many social situations; others regard drinking as impermissible behavior virtually all the time and in every situation. Many people laugh at jokes about drunkenness and yet condemn drunken behavior in public. These contradictions suggest that many people have not yet come to grips—either morally, socially, or interpersonally—with alcohol use.

This chapter provides an overview of problem drinking and alcoholism as a form of deviance. First, it discusses the physical and behavioral consequences of alcohol consumption. It then considers the prevalence and social patterns of drinking and of alcoholism. The chapter evaluates ideas about the etiology of alcoholism in the context of its social control.

Physiological and Behavioral Aspects of Alcohol

Heavy drinking bears a well-documented relationship to a number of problems, including physical and psychological dependency, various illnesses, impaired social relationships, and poor work performance. Alcohol does not lead to a physiological habit, however, in the way that certain other drugs do. One does not become a chronic drinker after finishing the 1st, 20th, or even 100th drink. Furthermore, research has not yet convincingly demonstrated an inherited tendency toward excessive consumption of alcohol. Clearly, however, consumption of alcoholic beverages produces physical and psychological consequences.

Alcohol is a chemical substance created through processes of fermentation or distillation. Humans began relatively recently in their history to distill alcohol from grains, such as barley, corn, wheat, and others, but nearly all societies have made fermented beverages in some form, such as wine and beer, for thousands of years (Patrick, 1952: 12–39). After drinking an alcoholic beverage, the drinker's small intestine absorbs a certain amount of alcohol. The bloodstream carries this substance to the liver and then disseminates it in diluted form to every part of the body. Although the blood can never hold more than 1 percent alcohol, some evidence suggests that even relatively small amounts can affect the brain, as shown by x-rays, computerized axial tomography scans (CAT scans), and other medical research tools. In fact, the range and complexity of alcohol's effects directly or indirectly influence virtually every organ system in the body (Secretary of Health and Human Services, 2000: chapter 1).

Physiological Dimensions

The immediate effects of alcohol depend on the rate of its absorption into the body and the physical characteristics of the individual drinker. The rate of absorption, in turn, depends on the kind of beverage consumed, the proportion of alcohol it contains, how quickly one drinks it, and the amount and type of food in the stomach at the time. In addition, certain individual physiological differences, such as body weight, affect absorption. Individuals vary in their susceptibility to the physiological consequences of alcohol independent of these variables as well, suggesting that alcohol consumption can produce different physical effects in two people, even for the same amount and type of consumption (Secretary of Health and Human Services, 2000). In moderate quantities, alcohol has relatively little effect, but large quantities disturb the activities in the organs controlled by the brain and cause symptoms termed *drunkenness.*

Over long periods of time, consumption of quantities of alcohol may have a number of health-related consequences. Research has identified alcohol consumption as a contributor to numerous health hazards, such as accidents and traffic fatalities (see Hingson and Howland, 1987). Studies have also linked chronic consumption of alcoholic beverages to various gastrointestinal disorders, pancreatitis, liver disease, nutritional deficiency, impairments of central nervous system functions, disorders of the endocrine system, cardiovascular defects, myopathy, certain birth defects, and several types of cancer (Eckhardt et al., 1981). In fact, diagnosed alcoholics face a greater risk of mortality from numerous physiological disorders than nonalcoholics face (Taylor, Combs-Orme, and Taylor, 1983).

Other Health-Related Effects

According to one estimate, alcohol contributed to nearly half of all automobile fatalities in the early 1980s, but the proportion of fatal crashes involving either a drunk driver or occupant has dropped to about 30 percent (Burgess, 1998). Alcohol consumption shows a significant association with overall emergency room cases in both the United States and other countries. A study comparing emergency room causalities in the United States and Mexico reported a high rate of cases resulting from heavy drinking, drunkenness, and alcohol-related problems (Cherpitel, Stephens, and Rosovsky, 1990).

A distinct cluster of defects in newborn infants seems connected with drinking by the mother during pregnancy. Called *fetal alcohol syndrome* (FAS), these problems appear more in some groups than others. The risk of FAS among African Americans is seven times higher than that for whites, for example (Secretary of Health and Human Services, 2000: 300–321). FAS has been identified only in children born to mothers who drank heavily while pregnant and often combined alcohol with smoking and illegal drugs. The minimum criteria for diagnosing FAS include retarded growth, central nervous system abnormalities, and characteristic facial features: short eye openings and a thin upper lip with an elongated philtrum (the groove in the middle of the upper lip). Follow-up studies of children diagnosed with FAS has found a relatively high incidence of hyperactivity and short attention spans. Some adolescents with FAS achieve low IQ scores compared with the population as a whole.

Psychological Effects

Alcohol produces a number of psychological effects on emotional reactions and overt behavior. In moderate quantities, alcohol can relax tensions and worries, and it may ease the fatigue associated with anxiety. The effects of alcohol may mimic those of stimulants; in fact, this depressant reduces or alters cortical control over actions, freeing behavior from some restraint. Alcohol has a negative effect on task performance, but the strength of the effect varies, depending on the nature of the task and the drinker's experience. Inexperienced drinkers tend to overreact to the sensation of alcohol, sometimes fulfilling perceived socially expected behaviors in drinking situations. Such reactions commonly induce groups of teenagers to behave as if they were quite intoxicated under the influence of only small quantities of alcohol.

Many people may believe that alcohol releases inhibitions that restrain a drinker's behavior, perhaps leading the person to act out of control under the drug's influence. While consumption of the substance certainly impairs certain mental and motor skills, considerable evidence confirms that much so-called *drunken* comportment actually reflects behavior learned through socialization rather than the result of any automatic release of behavioral controls associated with alcohol. An anthropological survey evaluated drunken reactions in different cultural groups, showing that among some cultures, normal inhibitions remain in effect.

In many societies, people regularly consume alcohol in very large quantities without producing appreciable changes in behavior except for progressive impairment in their sensorimotor capabilities, such as coordination (MacAndrew and Edgerton, 1969: 36). For example, the culture of the Onitsha of Nigeria respects the ability to appear sober in spite of heavy drinking, while drunken behavior brings shame (Umanna, 1967). In other societies, people may display considerable physical aggression when drunk, while in yet others, drinkers may appear euphoric or happy without displaying either sexually promiscuous or aggressive tendencies. In still others, aggression may become "both rampant and unbridled, but without any changes whatsoever occurring in one's sexual comportment" (MacAndrew and Edgerton, 1969: 172).

Similar research in the United States has shown that patterns of intoxicated behavior follow norms in particular types of taverns (e.g., neighborhood bars, bars for lovers, adult sports bars) and to drinkers' desires for others to see them as philosophers, lovers, or fighters (Bogg and Ray, 1990). Research identifies no universal behavioral consequence of drinking alcoholic beverages. Drunken actions are largely learned behavior sensitive to cultural and social contexts.

Prevalence of Drinking in the United States

In the United States, alcohol drinkers favor beer, followed by wine and then distilled spirits, ranked in order by amounts consumed and total cost. Over the past century, drinking of distilled spirits has followed a downward trend as beer and wine consumption have increased, measured by tax receipts, sales from state-controlled stores, and estimates from the alcoholic beverage industry. In 1850, distilled spirits accounted for almost 90 percent of the alcohol consumed, and beer represented nearly 7 percent; by 1960, spirits accounted for only 38 percent, and beer represented 51 percent (Keller and Efron, 1961: 3). Alcohol consumption increased sharply after the repeal of Prohibition and through the 1940s, flattening out through the 1950s. After another steady rise through the 1960s and 1970s, the estimate of total alcohol consumption again declined after 1981. The estimated consumption in 1984—2.65 gallons per person—was the lowest figure since 1977, the first time since Prohibition that consumption had declined for three consecutive years (Williams, Doernberg, Stinson, and Noble, 1986).

As consumption of both beer and distilled spirits declined leading up to 1984, wine consumption increased. Consumption of spirits further extended the long decline that began in 1970, dropping to new lows of 0.94 gallons per person in 1984 (Secretary of Health and Human Services, 1987: 2) and 0.83 gallons in 1987 (Secretary of Health and Human Services, 1990: 14). Recent estimates show a continuing decline in U.S. consumption of alcohol (Secretary of Health and Human Services, 2000). A comparison with other countries reveals that U.S. per-capita consumption of alcohol trails similar figures for France and Italy, two heavy wine-drinking countries. French drinkers also consume large quantities of stronger spirits such as brandy and cognac. The republics of the former Soviet Union also seem to report extremely high per-capita consumption. One such report showed that heavy drinking had increased substantially as a result of the economic and political upheaval and was heaviest among lower-class males, a group that has suffered significantly during the transition from the former Soviet state (Carlson and Vagero, 1998).

While consumption patterns changed over the years, however, the volume of drinks consumed did not. The proportion of abstainers in the population rose slightly between 1967 and 1984 (with the increase in abstinence occurring mainly among males). Still, the proportion of people who experienced some kind of difficulty as a result of their drinking remained about the same during those 17 years (Hilton and Clark, 1987). National surveys of drinking practices report that approximately one-third of the U.S. population aged 18 and over identify themselves as abstainers, while one-third are light drinkers, and one-third are moderate to heavy drinkers. (See the summaries in Secretary of Health and Human Services, 2000: chapter 1 and Dawson, Grant, Chou, and Pickering, 1995.) In every age group, more men than women describe themselves as drinkers, and, among those who drink, more men than women are heavy drinkers. Racial and ethnic groups also differ, and whites of both sexes are the least likely to be abstainers.

Types of Drinkers

Drinkers fit certain classifications according to the extent of their deviation from cultural norms governing drinking behavior and their dependence on alcohol as part of their life organizations. Norms set standards for consumption of alcoholic beverages, indicating to drinkers which beverages suit specific occasions and times, how much they should consume, and what kind of behavior society will tolerate after consumption. This sort of classification scheme separates types of drinkers based on information about the frequency and quantity of their alcohol consumption. Researchers might gather data for this kind of judgment, such as estimates of average daily consumption (Hilton, 1988) or "volmax," the volume of monthly intake with the maximum amount consumed per occasion (Hilton and Clark, 1987). They may also look for estimates of how frequently drinkers get drunk.

Different combinations of these conditions might support various typologies of drinkers. A crude typology might begin by distinguishing a *social* or *controlled drinker* as someone who drinks for reasons of sociability, conviviality, and conventionality. Social drinkers may or may not like the taste of alcohol and the effects that it produces. Their primary characteristic is the ability to take alcohol or leave it alone. They often refrain and use alcohol only in certain social circumstances. A *heavy drinker*, on the other hand, frequently uses alcohol, perhaps occasionally consuming sufficient quantities to become intoxicated. This typology may define an *alcoholic* as someone whose frequent and repeated drinking of alcoholic beverages exceeds accepted community standards for social use to the point that it interferes with health, social, or economic functioning.

Applications of these types to specific examples would lack agreed-upon definitions of the terms *social drinker, heavy drinker,* and *alcoholic* drinker that would satisfy all observers. In practice, sociologists usually define these terms operationally, allowing for slight differences depending on who states the definitions. One source might identify a heavy drinker as someone who consumes more than 1 ounce of alcohol a day (Secretary of Health and Human Services, 2000: 3); perhaps a more meaningful criterion would identify people who take 14 or more drinks per week as heavy drinkers. Another arbitrary distinction separates social drinkers from heavy drinkers. Blue-collar workers, for example, seem to base such a judgment on an individual's work record and performance. If the individual can perform satisfactorily at work, they may not consider that person a heavy drinker, no matter how much she or he drinks (LeMasters, 1975: 161). Females often base such judgments on neglect of children rather than absence from work.

Another classification attempt applied its own categories to over 1,500 male drinkers in New Zealand (Martin and Casswell, 1987). The study formed a class of "light drinkers," 43 percent of the sample, including subjects who drank infrequently and consumed small quantities of alcohol on these occasions. "Frequent early evening drinkers," 28 percent of the sample, drank between the hours of 5 P.M. and 8 P.M. at home and before dinner. "Heavy hotel-tavern drinkers," who composed 21 percent of the entire sample, drank in public places two or three times a week. These relatively heavy drinkers began drinking after 8 P.M. "Club drinkers," who belonged to sports or business clubs where they consumed alcohol before dinner, made up 4 percent of the sample. "Solitary drinkers" made up 2 percent of the sample, and they drank virtually every day between 5 P.M. and 8 P.M. or so. Finally, "party drinkers" (2 percent of the sample) drank in the homes of others after 8 P.M. and usually only in social situations.

Even the meaning of the seemingly more clinical term *alcoholism* sparks substantial disagreement. To some, the term denotes a disease characterized by physiological dependence and uncontrolled drinking (Jellinek, 1960). Others acknowledge only a vague meaning for the term, and they doubt that it describes a uniform phenomenon (Robinson, 1972). Still others have defended the term on the grounds that it reflects some agreement on conditions, even if observers may disagree about when a particular person becomes an alcoholic (Keller, 1982).

Partly in response to this definitional confusion, researchers and treatment specialists alike have applied another term to denote someone whose drinking causes difficulties in life: *problem drinker.* Although problem drinkers are also heavy drinkers, not all heavy drinkers are problem drinkers. Some distinguish problem drinking by the consequences of alcohol consumption rather than characteristics of the drinker or the quantity and frequency of consumption. The notion of problem drinker recognizes that consumption of alcohol can result in complications in personal living. In addition to ugly hangovers, sometimes including physical collapse and intense remorse and self-disgust, heavy drinkers may experience blackouts, frequent nausea, and deteriorating interpersonal relationships with employers, friends, and family, as well as encounters with the police and other agents of social control. A simple definition describes problem drinkers as those who experience some problem as a result of their drinking, regardless of how much they consume or the circumstances surrounding that consumption.

Heavy drinkers often deny that their practices cause them problems. One study, for example, reported that heavy drinkers saw positive experiences from their consumption, such as euphoria from alcohol and facilitation of group interactions (McCarthy, Morrison, and Mills, 1983). Such beliefs may rationalize drinking behavior, or they may indicate consistency between attitudes toward drinking and drinking behavior.

Chronic alcoholics reach this condition after consuming large quantities of alcohol over long periods of time. They usually display apparent compulsions for continual drinking. Observers can interpret the term *compulsion,* or the alternative *craving,* only within specific cultural contexts, however; a desire may differ only weakly from a craving, depending on the context (Alasuutari, 1992). Along with an apparent compulsion, chronic alcoholics display solitary drinking, morning drinking, and general physical deterioration. These people live life entirely preoccupied with alcohol, sometimes devising ingenious methods to safeguard their supplies. Alcoholics deeply fear the prospect of life without a drink, so they may resort to hiding containers of liquor under pillows, under porches, and in any place likely to escape detection.

A chronic alcoholic's day may start with 8 ounces of gin or whiskey, and each day he or she may consume quantities of alcohol far in excess of the amounts customary within a comparable group. One study has showed that alcoholics usually develop their patterns of drinking over periods of 20 years or so, after becoming intoxicated for the first time at about 18 years of age. By the age of 30, many experience blackouts, or memory lapses, while intoxicated (Trice and Wahl, 1958). By 36 years of age, these people often begin drinking in the morning, and within a year later, on the average, they regularly drink alone. By 38, they begin protecting their supplies of alcohol, and by 39 they suffer their first tremors. Alcoholics in their 40s have generally experienced considerable difficulty with their drinking, perhaps changing jobs, losing relationships with family and friends, and developing health problems. Some chronic alcoholics may even die from excessive, ongoing consumption of alcohol, which leads to physical diseases, accidents, falls, fires, suicides, and poisoning (Secretary of Health and Human Services, 2000: 12–13).

Denzin (1993) argues that alcoholics cope only uneasily with their feelings about time and themselves. They drink, he argues, as a way to live in the emotional past; they want to avoid confronting the present or the future. Drinking enables them to manage how they think about themselves. In response to the question, "How long since you quit drinking?" Raymond Carver, the poet and novelist, replied:

> June 2, 1977. If you want the truth, I'm prouder of that, that I've quit drinking, than I am of anything in my life. I'm a recovered alcoholic. I'll always be an alcoholic, but I'm no longer a practicing alcoholic. (Carver, 1983: 196, quoted in Denzin, 1993: 174)

In this quote, Carver mentions his recovery and recalls the date of his last drink and the details of his drinking. He expresses pride in his recovery and a view of himself as living a new life. According to Denzin (1993: 175), "He has distanced himself from the old self of the past. He feels at home with the new, recovering self. He has made the full transition from the active alcoholic self of the past to the recovering self of the present." The point of recovery is to attain that distance, that separation between the alcoholic and the recovered self.

An alcoholic often cannot escape problems caused by alcohol consumption simply by terminating drinking. During sober periods, alcoholics recognize their physical problems and the social behavior they display to others when they are drunk. This awareness engenders self-consciousness and tension as they try to cope with the symptoms of their alcoholism while, at the same time, attempting to improve their social relationships with others, such as family and friends (Wiseman, 1981).

Drinking as a Social and Group Activity

From the beginning, drinking has been an integral part of American social life. The daily diet in New England at the time of the *Mayflower* included beer. In fact, historians have reported that a severe beer shortage in the Pilgrim community of Plymouth in 1621 inspired such pity in the ship's captain that he gave the colonists some of the ship's stores before sailing back to England. That winter, consumption of the last of the Pilgrims' beer supply motivated them to set up local operations for brewing beer and making some distilled spirits (Lender and Martin, 1982: chapter 1). Also from the earliest days, however, society frowned upon drunkenness and punished excessive drinkers. Likewise, the usual daily diet of the Spaniards who settled the Southwest and California included wine.

Many contemporary drinking patterns have come as an inheritance from previous generations. The knowledge, ideas, norms, and values concerned with consumption of alcoholic beverages, passing from generation to generation, have maintained the continuity of an alcohol-related subculture. People learn all patterns of drinking behavior, just as they learn all other behavior patterns. Drinkers display no universal way of behaving under the influence of alcohol, and they follow no universal patterns of drinking activities. People in the United States typically drink under some circumstances but not others. For example, the funeral ritual in the United States usually does not incorporate drinking, but that in Ireland does; Irish mourners commonly consume alcoholic beverages at wakes for departed loved ones.

As part of alcohol's significant role in everyday life, many people drink it in celebration of national holidays and to rejoice over victories on the football field, in war, or at the ballot box; wedding guests toast the bride and groom, and the father of a new child may celebrate by buying drinks for everyone present; job promotions, anniversaries, and other important social events often call for drinks. Businesspeople may consummate a deal over a drink, with a toast serving the same symbolic (and legal) function as a handshake. Even some religious ceremonies and mourning for dead friends are accompanied by alcoholic beverages. People encounter alcohol in a number of ordinary contexts. A study of alcohol references on prime-time television shows found that 80 percent of the network programs examined either referred to alcohol or displayed it within their scenes (Wallack, Breed, and Cruz, 1987). Characters consumed alcohol on 60 percent of the shows.

Despite widespread use of this drug in the United States, the value system implicit in American drinking patterns differs from those found in European countries. Many people there do not regard alcohol drinking as a vice or a social problem; instead, the practice has remained just one element of traditional recreation. Europeans often drink merely as one aspect of group occasions, whereas drinking all too frequently creates and dominates the occasion for a group coming together in the United States. This pattern has surely contributed to the ambivalence that many U.S. residents feel about drinking (Lender and Martin, 1982: 190–195).

Variations in Drinking Behavior

As social behavior, drinking follows a socially determined pattern. Drinking frequency varies by age, education, income, size of community, marital status, and religion (Secretary of Health and Human Services, 2000). Further, the behavior does not display random variations. These differences relate to the general positions of individuals and groups in the social structure and to the learning opportunities that those positions afford (Akers, 1992). A sociological understanding of the pattern of drinking behavior requires an understanding of norms related to alcohol and drinking.

Age   Consumption of alcoholic beverages tends to decline with increasing age, but it may begin early in life. While some reports from the United States cite consumption among children of elementary-school age, most drinkers report their first consumption experiences in adolescence. The frequency and quantity of drinking by adolescents has changed over the past decade or so. Annual national surveys show that a large proportion of high school seniors have consumed alcohol. In 1996, about 79 percent of the seniors surveyed reported having used alcohol at some time in their lives, and 62 percent of them reported having been drunk at least once (Johnston, O'Malley, and Bachman, 1996). The proportion of seniors who had used alcohol during the month preceding the survey declined from 72 percent in 1978 to 51 percent in 1996. Daily use of alcohol declined from a peak of 6.9 percent in 1979 to less than 3.0 percent in 1993.

Binge drinking among adolescents and young adults has remained stable over the past few years. The annual Monitoring the Future survey of high school students estimates that 51 percent of seniors, 40 percent of 10th graders, and 24 percent of 8th graders reported having consumed alcohol within 1 month before the survey was conducted. Those who reported having five or more drinks on one occasion were defined as binge drinkers. The proportions in each of the grades were 31 percent of the seniors, 26 percent of the 10th graders, and 15 percent of the 8th graders (Johnston, O'Malley, and Bachman, 1999).

Drinking is even more common among college students. At the same time, however, a declining proportion of students report daily drinking (Meilman, Stone, Gaylor, and Turco, 1990). The median number of drinks consumed by college students is 1.5 per week (Wechsler, Molnar, Davenport, and Baer, 1999), and surveys indicate that abstention increased 16 percent between 1990 and 1997 (Presley, Leichliter, and Meilman, 1998). In sum, research consistently finds heavy party drinking among college students, especially males, but fewer students seem to engage in this behavior every day. Not all young people who drink heavily continue drinking into adulthood. One study found that adolescents or college students classified as problem drinkers tended to drink at levels low enough to prevent problems as young adults 5 years later (Donavan, Jessor, and Jessor, 1983).

Drinking does remain common among young adults, however, and "the probability that one will drink stays relatively high up to about age 35" (Akers, 1992: 201). After that, some may continue heavy drinking, but many drinkers decrease their consumption. Both drinking and heavy drinking decline noticeably, especially among people over the age of 60.

Sex   National surveys indicate different drinking patterns for males and females. Men drink more frequently and larger amounts than women do. In one study, 25 percent of the males reported abstaining, compared with 40 percent of the females; moreover, in the heavier drinking categories, males (14 percent) outnumbered females (4 percent) (Secretary of Health and Human Services, 1993: chapter 1). Relatively heavy drinking also appears to peak at different ages for the two sexes— at age 21 to 34 for males and at age 35 to 49 for females; thereafter, drinking seems to decline for both males and females (also see Cahalan, 1982).

These drinking patterns have not escaped the attention of alcohol marketers and advertisers. Television commercials for alcohol products can promote only beer and wine, and they cannot show consumption of the beverages. These conventions represent voluntary standards, not government regulations. A Texas liquor distributor began advertising distilled liquor on cable television in 1996. It is unclear whether others will follow suit. Media advertisements tend to portray alcohol as a normal part of life or as a means for achieving success, wealth, or social approval (White, Bates, and Johnson, 1991). Advertisers concentrate their

---

✳ *Issue* | **Women and Alcohol**

Most available information on alcohol consumption and alcoholism comes from studies focused exclusively on male subjects. Recent work involving female subjects reveals differences from men in their drinking behavior and its consequences.

Fewer women than men drink, and women who do drink consume less alcohol, on the average, than men do. National surveys estimate that women make up perhaps 4.6 million (or one-third) of the total population of 15.0 million alcohol-dependent individuals in the United States. Middle-aged women experience a higher incidence of alcohol dependence than younger women, although younger women report more drinking-related problems than do older women. Researchers find a higher likelihood of heavy drinking and alcohol-related problems among women who have never married or who are divorced or separated than among married or widowed women.

Women experience more severe detrimental effects of alcohol on their livers than men do. Women develop liver disease, especially alcoholic cirrhosis and hepatitis, after shorter periods of heavy drinking compared with men, and alcoholic women die from cirrhosis in higher proportions than do alcoholic men. Studies have not yet identified clear reasons for this difference. Drinking by women has also been associated with increased risk of breast cancer, early menopause, and relatively severe menstrual distress.

One explanation for some sex-based variations may result from ineffectiveness with one sex of treatment approaches that work effectively with the other. Women are more likely than men to seek treatment, and their voluntary participation in the treatment relationship may contribute importantly to its success.

Sources: *Alcohol Alert*, October 1990. No. 10, PH 290, National Institute of Alcohol Abuse and Alcoholism. Washington, DC: Government Printing Office. See also Secretary of Health and Human Services. 1993. *Alcohol and Health: Eighth Special Report to the U.S. Congress*. Rockville, MD: National Institute of Alcohol Abuse and Alcoholism, Government Printing Office.

messages on programs and other media that interest the heaviest-drinking group—young males—especially sporting events.

Social Background   Drinking also varies with levels of education and income and between religions and regions of the country. Generally, adults with more formal education also drink more than less-educated adults. Drinking also increases directly with rising income. Religions with the highest proportions of drinkers include the Jewish faith and the Catholic and Lutheran churches; other large, Protestant denominations display the lowest proportions of drinkers among their members. Drinking is more common in large cities than in smaller towns and among unmarried people than married couples.

Regional differences in drinking also reflect interesting patterns. Reports have described low alcohol consumption in the South compared with other regions of the country and high consumption in the West. Ignoring abstainers, however, southern drinkers consume more alcohol each than drinkers in other regions. Although more residents abstain in the South than in other regions, Southerners who drink tend to consume more than drinkers in the other geographic regions do. Conversely, in the northeastern United States, a relatively high proportion of drinkers each consume less alcohol than southern drinkers to (Secretary of Health and Human Services, 1993).

Research has also identified variations in drinking behavior among various racial and ethnic groups (Welte and Barnes, 1987), to be discussed shortly. Regardless of those variations, the role of public places in which to drink has been significant.

Public Drinking Houses

People do much of their drinking in groups gathered in public drinking houses, such as bars and taverns, which serve thirsty patrons throughout most of the world. A simple definition would identify public drinking houses as places where proprietors sell alcoholic beverages for consumption on the premises. A more thorough description would mention several important characteristics: (1) They facilitate group drinking. (2) This drinking is a commercial activity, since anyone can buy a drink there (as contrasted with the bars of private clubs). (3) Public drinking houses serve alcohol rather than other drinks, which distinguishes them from coffee shops and teahouses. (4) Bartenders serve as functionaries of these institutions, and the drinking there gravitates in part around them. (5) Many customs influence these institutions, including their physical surroundings, the types of drinks they serve, their hours of operation, and the kinds of behavior considered appropriate within their facilities (Clark, 1981). Over 200,000 bars and taverns in the United States alone draw large numbers and widely varying types of patrons.

Taverns have served alcohol to paying customers for thousands of years. They date back easily to Greek and Roman times and played integral roles in early American social life. Colonial-era authorities even encouraged establishment of taverns, in part, because they believed that private drinking might well lead to excessive consumption, whereas they could regulate sales of liquor in public taverns. Therefore, the Puritan government of Massachusetts in 1656 enacted a law imposing a fine for any town that failed to maintain an "ordinary" (tavern) (Field, 1897: 11–12; see also Firebaugh, 1928). As the Industrial Revolution brought thousands of migrants, particularly single men, to the cities in search of factory work, a new type of public drinking house, the *saloon,* came to replace the wayside tavern, a facility that served drinks and tended to the needs of travelers in colonial America. Saloons opened their doors in many urban areas, commonly serving strictly male clienteles who drank at elaborate bars with "free lunches;" special family entrances allowed for separation of groups of customers.

The modern tavern made its appearance after Prohibition. Such establishments generally permitted women, and patrons enjoyed comparatively attractive surroundings, usually drinking at separate tables rather than at bars. At least five different varieties of public drinking houses developed, each one largely associated with a specific area of the city and customers of a particular socioeconomic status.

The types include the skid row bar, the downtown cocktail lounge and bar, the dining and dancing club (a category that includes singles bars), the nightclub, and the neighborhood tavern. The last type of establishment is the most numerous of all, currently comprising perhaps three-fourths of all public drinking houses.

The relationship between tavern patronage and excessive drinking has generated controversy. Some writers claim that the practice of visiting taverns does not lead to excessive drinking (Popham, 1962: 22), while others claim that it does (Clark, 1981). Neighborhood taverns, like the famed pubs of Great Britain, offer inviting places where people gather to talk, argue, play games, and relax. Some believe that social interactions in these settings encourage social control, so patrons tends to drink less there than they would drink alone at home. On the other hand, as patrons buy drinks for one another and develop increasingly enthusiastic camaraderie, some may actually drink more than they would on their own. No one should express surprise at a relationship between tavern patronage and drinking; after all, alcohol consumption, along with associated social relationships, defines the major reason that people visit taverns. As they spend more time there, they experience increasing exposure to norms and values tolerant of and even favorable to heavy drinking. Eventually, habitual drinking behavior could encourage excessive consumption in general, both within and outside taverns and bars. Research identifies the most likely patrons of taverns and those most likely to drink heavily as young, unattached, gregarious males who are not religiously oriented (Nusbaumer, Mauss, and Pearson, 1982). One survey has reported that most Canadians visit taverns occasionally, but few are regulars (Cospers, Okraku, and Neumann, 1987). These establishments tend to draw their regulars from groups of young, unmarried people who drink heavier-than-average amounts.

Tavern drinking is not, of course, the only evidence of the social and group associations of alcohol consumption. Drinking by members of the middle and upper classes may occur primarily away from public drinking houses, but it remains a similarly social activity. These drinkers often center their consumption around various group events such as cocktail parties and occasions sponsored by country clubs or private dining establishments. Such occasions usually create settings for drinking that are a good deal more private than those of public drinking houses, although they facilitate equally evident group processes.

Ethnic Differences in Excessive Drinking

While many persons consume alcohol occasionally and moderately without social effect, excessive drinking can lead to a number of personal and community problems. Pronounced differences in the prevalence of excessive drinking behavior separate various ethnic groups. Discussions of these differences will reveal relationships to some of the characteristics discussed earlier (such as religious preference). Patterns of variations like these highlight the importance of alcohol-related norms and group influence on the process of learning drinking behavior.

The Irish   Irish people have a long-established reputation for excessive drinking, and immigrants to America brought many associated behavior patterns with them.

Perhaps as a result, Irish Americans develop alcoholism at rates that probably exceed those of any other single ethnic group. Irish men drink because their culture permits and sometimes prescribes the practice, particularly consumption of strong beer and whiskey. Its norms do not confine drinking to ceremonial purposes. Traditionally, Irish society has advocated a pattern of high socioeconomic aspirations, late marriage, and strong emphasis on the virtue of abstaining from sex until after marriage. These limitations on marriage lead to sex segregation, creating "bachelor groups" that hold an important place in Irish social structure. Married men continue to participate in these bachelor groups, and they often help to socialize young men into their drinking practices.

> A boy became a man upon initiation into the bachelor group, that is, when first offered a drink in the company of older men in the local public house. Farm and marriage might be a source of male identity for a few, but hard drinking was a more democratic means of achieving manhood. (Stivers, 1976: 165)

Drinking often supplies prestige and esteem to individual men who live in male segregated social worlds. Despite high consumption, however, the hard drinker in Irish society seldom becomes a persistent drunkard because the culture does not sanction chronic drunkenness. Even today, a relatively low rate of alcoholism persists in Ireland. Still, the stereotype of heavy drinker followed Irish immigrants to the United States, where it became translated into heavy drinking presenting a serious threat of regular drunkenness and alcoholism. The problem seems isolated to Irish immigrants to the United States; studies find high rates of abstinence among Irish who have immigrated to other countries (Greenslade, Pearson, and Madden, 1995).

The Italians   Italians also maintain a tradition of high alcohol consumption, mainly drinking wine with meals. Despite this extensive use of alcohol, Italians have a very low incidence of alcoholism. Although Italian Americans similarly become alcoholics relatively infrequently, they appear to do so at higher rates than commonly occur in Italy, despite higher total consumption of alcoholic beverages there. Many Italian Americans have retained the tradition of drinking wine with meals and regard the practice as a benefit to health; such an attitude helps to prevent alcoholic excess and addiction. Italians begin drinking wine very early in life. Both men and women enjoy the beverage, and very few object to consumption of wine by young people. Few drinkers use alcohol as a way to escape (Lolli, Serianni, Golder, and Luzzatto-Fegis, 1958: 79).

These drinking patterns have carried over only partially to Italian Americans, and their absence leads to excessive drinking and alcoholism. For example, while 70 percent of Italian men and 94 percent of Italian women do all their drinking at mealtime, only 7 percent of first-generation Italian-American men and 16 percent of Italian-American women observe the same limitation.

The French   Residents of Italy and France drink about the same, relatively large quantities of alcohol each day, but the French develop alcoholism at much higher rates. In fact, France may have one of the highest alcoholism rates in the world (Sadoun, Lolli, and Silverman, 1965). This difference seems to result from a number of variations in drinking patterns. (1) Wine drinking at mealtime accounts for nearly all the alcohol intake in Italy, while a substantial amount of the French alcohol intake reflects consumption of distilled spirits and aperitifs between and after meals. In fact, French alcoholism rates are lower in the southern part of the country, where people mainly drink wine at mealtime, than in other regions. (2) Residents of the two countries adopt quite different views of exposure to alcohol in childhood. In France, rigid parental attitudes either favor or oppose wine drinking among children, while most Italians accept this kind of consumption as a natural part of a child's development. (3) The Italians set a much lower limit than the French for amounts of alcohol deemed acceptable to consume, and they tend to view drunkenness as a personal and family disgrace. (4) The French associate drinking, particularly copious amounts, with virility, while the Italians do not.

Asian Americans   In the United States, Asian Americans, about 2 percent of the overall population, use very little alcohol. Research has indicated that Asian Americans of both sexes drink less than whites, blacks, or Hispanics (Klatsky, Siegelaub, Landy, and Friedman, 1985). A comparison detected considerable differences, however, between drinking practices of several groups within the Asian population. Rates of abstention seem particularly high among California residents of Korean (67 percent), Chinese (55 percent), and Japanese (47 percent) ancestry (Kitano, Hatanaka, Yeung, and Sue, 1985; Sue, Kitano, Hatanaka, and Yeung, 1985). Among those who drink, the Chinese stand out for their low rates of heavy drinking.

Many Asians consume alcohol largely as part of social functions. These groups disapprove of public drunkenness, and they educate children to observe these patterns. In this way, drinking continues under effective social control. Physiology may also limit drinking by Asians, who often experience a "flushing response" to alcohol. This reaction produces "facial flushing, which is often accompanied by headaches, dizziness, rapid heart rate, itching, and other symptoms of discomfort" (Secretary of Health and Human Services, 1990: 35).

Native Americans   The great diversity among Native-American populations appears in the wide variations of their drinking patterns. Over 300 different tribes within the United States sometimes display differences more striking than their similarities. Some tribes mainly abstain, while others display high rates of excessive drinking. One observer has estimated that the percentage of Native Americans who drink is smaller than the same figure for the general population, but a higher percentage of these drinkers experience problems with alcohol consumption (Lemert, 1982). A tendency for heavy drinking spread across the American frontier during the 19th century with the farmers, trappers, and cowboys. The Native Americans came into contact with these groups and secured their first alcohol as a result. Unfamiliarity with alcohol, along with encouragement of white men, led to excessive drinking. When the excessive use of alcohol became common among certain tribes, federal laws addressed the problem, beginning with a general law in 1802 and a final, more specific one, in 1893. Another law enacted in

1938 made serving intoxicants to a Native American an offense. These laws were not repealed until 1953.

Because tribal groups differ substantially in their drinking behavior, one cannot make general statements about drinking among all Native Americans. In a survey of more than 280 tribes, some reported binge drinking followed by periods of sobriety, while other groups maintained almost universal abstinence (Lex, 1985). Still other groups typically practice moderate drinking. For example, the Pueblo tribe of the southwestern United States completely abstains from drinking. Some other tribes drink no more than the rest of American society. Some tribes, however, include large proportions of heavy drinkers; for example, researchers classified 42 percent of Ojibwa adults as heavy drinkers (defined as becoming drunk two to five times a week) (Longclaws, Barnes, Grieve, and Dumoff, 1980).

Native-American drinking patterns reveal excessive alcohol consumption mainly by young, unemployed males. These men usually drink wine in small groups similar to the "bottle gangs" found in skid-row areas. Both kinds of gatherings emphasize getting drunk as quickly as possible. This behavior elicits little cultural disapproval, a fact explained by some observers as evidence that some Native Americans regard excessive drinking as a form of protest against the abuses they suffer living in white society (Lemert, 1982).

Whatever the reason, alcohol-related causes contribute to large proportions of Native-American deaths. One study reported a mortality rate from chronic liver disease among Native Americans and Alaska Natives of 29.2 per 100,000 compared with 9.7 per 100,000 for the U.S. population as a whole (Secretary of Health and Human Services, 1990: 36). Further, although Native-American women drink considerably less than men, they have a higher risk of health-related problems; women account for nearly half of the liver cirrhosis deaths among Native Americans. The age-adjusted alcoholism death rate among Native Americans and Alaska Natives has decreased by 63 percent since its peak in 1973 (Indian Health Service, 1992). However, after reaching a low of 24.6 per 100,000 in 1986, it increased again to 33.9 in 1988, over 5 times the comparable rate for any other racial group in the United States.

Hispanic Americans   Hispanics are another heterogeneous group with diverse cultural, national, and racial backgrounds. Most Hispanics trace their heritage to Mexico (60 percent), and sizable Hispanic populations also recognize ancestors from Puerto Rico (12 percent), Cuba (5 percent), and other Latin American countries (23 percent). Hispanic men display relatively high rates of alcohol use and heavy drinking, while Hispanic women show high rates of abstention. The first large-scale national survey of drinking patterns among this group reported that nearly half of Hispanic women (47 percent) abstained from alcohol use, and an additional 24 percent drank less than once a month (Secretary of Health and Human Services, 1987: 39). In contrast, only 22 percent of Hispanic men abstained, and 36 percent drank heavily (at least once a week consuming five or more drinks at a sitting). Hispanic men drank more heavily in their 30s than in their 20s, and their consumption began to decline with age only after 40 years old. Drinking levels increased with increasing education and income; specifically, survey respondents in relatively high income brackets and with high levels of education abstained at lower rates and reported higher rates of heavy drinking than respondents with lower incomes and less education (Caetano, 1984). Finally, Mexican-American men had the highest rates of both abstention and heavy drinking among Hispanics of Cuban, Puerto Rican, or other Latin-American decent.

Neff (1991) compared Anglo, Mexican-American, and African-American drinkers. He has reported more frequent consumption of lower quantities by Anglos and African-Americans compared with Mexican Americans, who tend to drink higher quantities on less frequent occasions. Also, while heavy drinking seems to decrease with age for Anglo men, it tends to remain high as Hispanic and black men age (Caetano and Kaskutas, 1995).

A large proportion of Hispanics report experiencing problems associated with drinking. About 18 percent of Hispanic men and 6 percent of Hispanic women experienced at least one alcohol-related problem during a 1-year period prior to responding to a survey (Caetano, 1989), and Hispanic men encounter more alcohol-related problems than black or white men experience. Research offers little insight into the specific risk factors linked to problem drinking by Hispanics, although both drinking in general and excessive drinking seem associated with acculturation (acceptance and adaptation to the social and cultural norms of a new environment) to life in the United States. Thus, acculturated Hispanics drink more, and more often to excess, than do Hispanics in Mexico or Spain (Secretary of Health and Human Services, 1990: 34–35). Also, research finds a higher proportion of abstainers among Hispanics who have not acculturated than among those who have.

African Americans   Most available studies suggest that African Americans experience higher rates of alcoholism than do whites, but this conclusion remains tentative because relatively few researchers have studied drinking patterns among blacks (Lex, 1985; Sterne, 1967). On the basis of available evidence, black women appear to have a higher rate of alcoholism than do white women, but they also abstain at higher rates (Harper and Saifnoorian, 1991). Furthermore, rates of alcoholism are higher among black men compared with all blacks than among white men compared with all whites (Secretary of Health and Human Services, 1987: 35–37). Rates of heavy drinking for black males bear an inverse relationship to income; those in relatively low-paying jobs seem to become heavy drinkers at higher rates.

White males appear to experience their highest risk for alcohol-related problems (for example, medical, social, and occupational disruptions) between the ages of 18 and 29, but black men experience their lowest risk in this age range (Herd, 1989). As white men passed through their 30s, their rates of these kinds of problems decreased sharply, but similar rates increased for black men in their 30s. Blacks experience higher rates of problems from drinking than whites throughout the middle-aged and older ranges (Herd, 1994).

Other surveys have also compared black and white drinking patterns. In studying a national sample of 723 blacks and 743 whites, Herd's (1990) aggregate data revealed very similar drinking patterns for both groups, with comparable proportions of abstainers and frequent and heavy drinkers. But these overall similarities masked some important differences in the circumstances associated with drinking. For example, Herd found links between frequent, heavy drinking among whites and youthfulness, high incomes, and areas generally characterized by high alcohol consumption; frequent, heavy drinking among blacks exactly reversed this pattern. While the data showed lower overall rates of drinking among blacks than among whites, black males also reported higher rates of drinking-related problems than white males did. That is, as the frequency of heavy drinking increases, rates of drinking problems rise faster among black men than among white men (Herd, 1994). These findings suggest that important cultural, social, and perhaps even physiological differences create variations in the drinking behavior patterns of blacks and whites.

Some of these differences vary predictably with specific factors, such as socioeconomic status. Barr and her associates (1993), for example, found little difference in alcohol use by black and white male college graduates. At lower levels of education, however, differences in drinking increase to the point that black males who have not completed high school drink more than three times as much as comparably educated white males. Blacks and whites also differ dramatically in reported drunk driving. White men have drunk driving rates 2.5 times higher than those of black men, and white women have drunk driving rates 5.0 times higher than those of black women (Herd, 1989).

Black youths abstain from alcohol consumption at higher rates than white youths do (Lowman, Hardford, and Kaelber, 1983), suggesting that blacks come to heavy alcohol use later in life than do whites. Despite this later introduction to drinking, however, blacks enter treatment at younger ages than whites (Herd, 1985). Blacks suffer more seriously than whites from other negative consequences of heavy alcohol use. For example, blacks die from cirrhosis at twice the rate of white males and nearly four times that of white females (Secretary of Health and Human Services, 1987: 37).

Culture—not race or biology—determines patterns of alcohol consumption by blacks (Larkin, 1965). Pronounced differences distinguish lower-class black drinking patterns and rates of alcoholism from those of middle-class and upper-class blacks. Similar differences separate whites by social class. Typically, African Americans in the middle and upper classes use alcohol more moderately than lower-class blacks, and lower-class blacks drink more often than others in public rather than in private places. This tendency of lower-class blacks toward drinking in public places may also explain their comparatively high arrest rates for drinking.

Alcoholism and Problem Drinking

Alcoholism and problem drinking represent the most extreme form of drinking behavior. One determines the extent of alcoholism and problem drinking differently from the way one determines the extent of general drinking behavior. Because observers do not agree on any standard definition of "alcoholic" or "problem drinker," no one can supply completely adequate estimates of the number of such persons in the United States.

The Extent of Alcoholism and Problem Drinking

Although overall drinking patterns have changed in recent years—characterized by a decrease in consumption of distilled spirits and a slight increase in the proportion of abstainers—the proportion of heavy drinkers has remained about the same over the past two or three decades (Hilton and Clark, 1987). Although two-thirds of the U.S. population drink alcohol, actual consumption shows a very uneven distribution throughout the drinking population. The 10 percent of drinkers who drink the most (6.5 percent of the total population) account for fully half of all alcohol consumed. The other half fills the glasses of the remaining 90 percent of the drinking population, including infrequent, light, and moderate drinkers (Secretary of Health and Human Services, 1987: 4).

A definition of problem drinking that considers only alcohol consumption leads to a broad category that encompasses a fairly large proportion of the drinking population. For example, if one defines heavy consumption as 120 or more drinks per month, then 19 percent of adult male drinkers and 7 percent of adult female drinkers fit in this classification, based on responses to a national survey (Secretary of Health and Human Services, 1993: 8–9). Clearly, this group of drinkers faces the most serious risk of developing either alcoholism or serious problem drinking. More detailed analysis must allow for variations in definitions of alcoholism and the identification of most drinkers as social or infrequent users of alcohol. With these warnings in mind, the federal government's National Institute of Alcohol Abuse and Alcoholism (NIAAA) has estimated that perhaps as much as 10 percent of adult American drinkers will likely experience either alcoholism or problem drinking at some point in their lives and that at present over 7 percent of the population in the United States meet the diagnostic criteria for alcohol abuse or alcoholism (Secretary of Health and Human Services, 2000: 3).

Sometimes, analysts estimate the number of alcoholics by looking at mortality figures, particularly measures of deaths from cirrhosis of the liver. Such figures may mislead, however, since rates of cirrhosis relate only partially to alcoholism, and that condition causes less than 10 percent of all alcohol-related deaths (Secretary of Health and Human Services, 2000: chapter 1). Other analysts apply still other techniques for estimating the number of alcoholics. Some examine admissions to voluntary treatment programs, and others interview samples from the general population about their attitudes toward alcohol and any negative consequences that drinking has caused for them. Each technique invariably generates a different estimate of alcoholism.

European countries post some of the highest estimates of alcoholism in the world. France has the highest known rate, followed by Chile, Portugal, and the United States. Researchers have identified somewhat lower rates, though they remain high by world standards, in Australia, Sweden, Switzerland, the Union of South Africa, and the former Yugoslavia (Keller and Efron, 1955: 634). France also hosts the world's highest production and consumption of alcoholic beverages, and far fewer alcohol treatment and prevention programs operate there than in most other countries (Mosse, 1992). Observers describe a major problem with alcoholism and problem drinking in the former Soviet Union, although they can draw on no comparable figures, except for journalistic accounts, to those published officially in other countries (Traml, 1975). One estimate placed the number of alcoholics in former Soviet republics at 4.5 million, with three or four times more people described as alcohol "abusers" (Ivanets, Anokhina, Egorov, Valentik, and Shesterneva, 1992: 9).

Costs of Alcoholism

Industry loses large sums of money because of problems caused by excessive alcohol consumption in the form of absenteeism, inefficiency on the job, and accidents. Estimates suggest that between 3 and 4 percent of the workforce engages in deviant drinking at any one time. Such drinking creates costs not only in shoddy work performance but also in addressing the problem through alcohol treatment programs (Trice and Roman, 1972: 2).

Rough estimates of the annual economic cost of alcohol-related problems range between $85 billion and $116 billion per year (Secretary of Health and Human Services, 1993: 255). These figures reflect the costs of lost production, health-care expenditures, violent crimes involving alcohol, fire losses, research on alcohol problems, and social responses to alcohol-related problems, primarily social welfare programs for problem drinkers and their families. Lost employment and reduced productivity accounted for more than half of this amount. The health-care bill for accidents and illnesses related to alcohol abuse, including alcoholism, liver cirrhosis, cancer, and diseases of the pancreas, may exceed $15 billion.

An earlier section mentioned the very high accident rate and dangerous general mortality rates for problem drinkers. Problem drinkers experience a predictable set of health problems much more often than less active drinkers and abstainers do, including injuries from accidents, certain diseases, death in accidents or crimes, and suicide (Secretary of Health and Human Services, 2000; Eckhardt et al., 1981; Hingson and Howland, 1987). In addition, alcohol contributes to other, more subtle costs for which analysts cannot give precise dollar amounts but which nonetheless cause real damage. These costs show up in many varied ways, such as:

- Problems with infant and child care linked to drinking by one or both parents
- Marital problems and counseling for married couples struggling with alcohol problems
- Alcohol's contribution to acts of violence against children, spouses, and others
- Disruption or underachievement by school children whose parents drink excessively
- Eventual problem drinking by children of problem drinkers

This list could go on for a long time, and it omits an important cost: the simple but poignant human costs of emotional investments in relationships with people who drink too much (Straus, 1982: 146).

Alcohol-Related Crime

Arrests for public drunkenness and alcohol-related crimes constitute about 30 percent of all arrests made by the police in the United States (Federal Bureau of

Investigation, 1996: 217). These arrests apprehend members of the lower class much more often than middle-class and upper-class people, and arrests specifically for public drunkenness often detain members of certain minority groups. Those arrested for public drunkenness have a very high recidivism rate. Some have applied the term *revolving door* to the flow of public drunkenness cases through the criminal justice system. These offenders seem to perpetuate a never-ending cycle of arrest, jail, release without treatment, and rearrest (Pittman and Gordon, 1958). Thus, the high volume of arrests actually reflects somewhat less frequent arrests of a relatively constant set of offenders who go to jail over and over again.

Some arrests result from enforcement of laws against drinking by some classes of people. In addition, alcohol consumption often contributes to other crimes (Room, 1983). Drunkenness undoubtedly plays a significant role in many but not all violent crimes against the person such as homicide and aggravated assault. Wolfgang's classic study of homicide in Philadelphia determined that nearly two-thirds of the offenders had been drinking when they committed their crimes, and many other types of violent offenders perpetrate their crimes under the influence of alcohol as well (Wolfgang, 1958). Alcohol is a well-known correlate for homicide in the United States and other countries, and it may even represent an important component in a theory of homicide (Parker, 1993, 1995).

A Department of Justice study found that 54 percent of people arrested for violent crimes in 1983 had been drinking before their offenses (Bureau of Justice Statistics, 1985). Interpretation of such findings requires attention to the observation that large proportions of people drink every day and do not later commit crimes. Furthermore, some accounts describe these offenders drinking before their acts without indicating how much alcohol they consumed; statements like these do not necessarily imply that criminals act while intoxicated. Nevertheless, drinking seems to precede crimes in the highest proportion for assault (62 percent) and manslaughter (68 percent). On the other hand, people arrested for rape seldom have been drinking.

Despite an unmistakable association between alcohol and violent crime, observers cannot clearly determine what role drinking plays in the commission of these crimes. They propose no direct, causal relationship between alcohol use and violent crimes, and alcohol use does not inevitably lead to aggression. In most alcohol-related criminal behavior, the drug probably acts as a depressant, reducing the offender's awareness of the probable consequences of the act. Analysis of crimes committed under the influence of alcohol reveals no single pattern of criminality associated with drinking. Interpreting the association between drinking and crime also requires attention to another fact: Both crime and heavy alcohol consumption peak among young males (Room, 1983).

Federal statistics indicate high proportions of drinking problems among jail and prison populations. The same numbers confirm that people with drinking problems engage in criminal behavior more often than do people without such problems. One survey has reported that more than half of the 500 largest jails in the United States maintain drug and/or alcohol treatment programs for inmates; as of June 30, 1992, nearly 40,000 inmates had participated in such programs (Bureau of Justice Statistics, 1993).

Drunk Driving

Observers have developed a more detailed understanding of the exact role of alcohol in cases of drunk driving. Traffic crashes represent the largest single cause of death in the United States for people under the age of 34 (National Highway Traffic Safety Administration, 1988), and alcohol consumption plays a major role in many of these fatalities. Nevertheless, the National Highway Safety Administration estimates that the proportion of crashes in which the driver was drunk dropped almost 40 percent between 1982 and 1997 (Burgess, 1998).

More than 120 million Americans are licensed drivers, and close to 100 million of them drink. These facts suggest the probability that people drive after consuming alcohol on billions of occasions. In a sense, almost anyone can easily discern the causes of drunk driving. As Ross (1993: 4) points out:

> American society combines a near-total commitment to private automobile transportation with a positive evaluation of drinking in recreational situations. Conventional and conforming behavior in these areas implies the likelihood of people driving while impaired by alcohol. Furthermore, in certain social categories, such as younger males, the norms regarding both drinking and driving appear to be extraordinarily favorable for the creation of impaired driving. Drunk driving can thus be seen as a routine, expected aspect of American life, supported by prevailing norms and institutions.

One estimate places the drunk driver's absolute risk of a crash at about 1 in 1,000, with a much lower risk of causing injury and a still lower risk of that crash causing death (Ross, 1982: 107). Because only a small proportion of drunk driving episodes end in such tragedy, one should avoid the inaccurate premise that the average drunk driving episode always poses a life-threatening risk.

The problem of drunk driving becomes more complicated due to arbitrary and variable legal definitions of the offense. Current law in most states defines drunk driving as operating a car while one's blood alcohol level (BAL) exceeds 0.10 percent (0.08 percent in some states). The actions of groups like Mothers Against Drunk Driving (MADD) and Remove Intoxicated Drivers (RID) have drawn support from some state legislatures to lower the BAL threshold for this offense. This change in definition would, of course, increase the number of legally recognized drunk drivers on the road. Britain and Canada set a legal limit for BAL of 0.08 percent, and an emerging European standard allows 0.05 percent (Ross, 1993: 19). There is little evidence that reducing the legal limit for BAL lowers crash rates (Associated Press, June 25, 1999). Of the 10 states with the lowest alcohol-related traffic fatalities, only two (Utah and New Hampshire) have a 0.08 BAL limit. The state with the highest rate of alcohol-related deaths (New Mexico) has a 0.08 BAL limit.

Sometimes, police set up roadblocks where they screen passing drivers for alcohol consumption in a strategy to deter drunk driving. They further evaluate drivers determined to have consumed alcohol to determine their levels of impairment. Arrests may then follow based on the results of these field sobriety tests.

This procedure is common in European countries, particularly in Scandinavia and Great Britain, and in some U.S. states.

Public-interest groups have formed to advocate on the issue, playing highly visible roles in debates about drunk driving. For example, some women organized MADD and RID after losing husbands or children in traffic accidents involving intoxicated drivers. These organizations have pressed for stronger sanctions against drunk drivers and greater public awareness of the problems associated with drinking drivers. Local chapters have sprung up throughout the country, actively attempting to influence both law enforcement practices and penalties against drunk drivers. Many leaders in these local groups have experienced victimization, either personally or within their families, but others work for the cause without experiencing victimization. One analysis of the leaders of local MADD chapters has reported that victimization provides an important source of motivation. Further, these leaders often brought backgrounds in activist organizations to their participation in the group, and "MADD tends to be run by activists who have been victimized rather than victims who have become activists" (Weed, 1990: 469).

In spite of calls for harsher sanctions, many communities deal with drunk driving as a misdemeanor; such an offender who causes injury commonly faces a sentence of up to 5 years' imprisonment in many states, and one who causes a death may face charges of negligent manslaughter. In practice, sentences seldom become so harsh, although many states automatically suspend offenders' driver's licenses for periods of up to 1 year. Statistics show that nearly 90 percent of all arrested for drunk driving are males, most of them in their 20s (Federal Bureau of Investigation, 1996), a pattern confirmed by research on drunk driving (Bradstock et al., 1987). Yet, experience does not support expectations that short jail sentences and license revocations provide effective long-term deterrents to drunk driving. Furthermore, a study has found no evidence that alcohol education for convicted or accused drunk drivers effectively reduces recidivism (Jacobs, 1989).

One innovative enforcement strategy attempts to block opportunities for drunk driving. Some jurisdictions have experimented with an apparatus called an *interlock system* that prevents drunk drivers from starting their cars (Jacobs, 1989: 170–171). The factory-installed interlock system includes a tube in the car's passenger compartment into which the driver breathes for 4 minutes. Sensors evaluate her or his BAL based on this input, and acceptable results enable the car to start, indicated to the driver by a green light; a marginal BAL result triggers a yellow light; when the BAL exceeds some preset limit, the unit displays a red light. The system disables the car unless the test produces a green light. It even "codes" the driver's breath so that someone else cannot start the car. The device requires monthly checks for accuracy; without them, it prevents the car from properly operating.

Another enforcement device is a legal practice rather than a machine. States enact implied consent laws that define acceptance of a driver's license as implied consent to alcohol testing by law enforcement officers; the law then assumes drunk driving if the driver refuses to take the test. Drivers may refuse to take breath tests for many reasons, including mental confusion because of inebriation, a rational calculation of costs and benefits, and hostility toward those administering the tests (Ross, Simon, Cleary, Lewis, and Storkamp, 1995). In at least one

state with such laws, refusal to take a breath test was associated with low conviction rates for alcohol offenses.

Stiff legal penalties seem to influence drunk driving behavior, but only in the
short run. After a careful assessment of this issue, Ross has concluded that increases in the potency of legal threats, particularly enhancement of the perceived
likelihood of apprehension, does produce a statistically significant decline in
drunk driving. Such effects continue only temporarily, however, since enforcement efforts can rarely maintain the needed intensity (Ross, 1982, 1993). Severe
legal punishments and active enforcement programs may influence driver perceptions of legal risk, which, in turn, induce them to reduce drunk driving activities
(Secretary of Health and Human Services, 1990: 250–252). However, no program
has yet sustained such perceptions over time. The recent popularity of sobriety
roadblocks seems to suggest that they lead to significant reductions in alcohol-
related crashes and traffic deaths, but this method, too, provides diminishing
returns over time as drivers anticipate these police actions. For these reasons, effective deterrence would require a systematic program combining a variety of
strategies, including maintaining a high minimum drinking age (at least 21 years
of age), increased taxes on alcohol (to discourage consumption), and stringent law
enforcement (Ross, 1993).

Group and Subcultural Influences on Excessive Drinking

Group associations and subculture identification play important roles in determining who becomes an excessive drinker and who does not. Every modern society
practices its own drinking customs, and subcultures display their own behaviors as
well. Subgroups differ in the ways they use alcohol, in the extent of their drinking,
and in attitudes toward alcoholism and drunkenness. Some people believe that
frequent drinking typically leads to alcoholism; yet, some groups within the U.S.
population who drink with relatively high frequencies, such as Jews and Italian
Americans, have low rates of alcoholism (Snyder, 1978; Lolli et al., 1958). People
intuitively expect that frequent drunkenness leads to alcoholism, but the Aluets,
the Andean Indians, and those of the Pacific Northwest cast doubt upon this
conclusion. Drunkenness is common in these societies, but alcoholism is not
(Berreman, 1956; Lemert, 1954; Mangin, 1957; Washburne, 1961). Clearly, the
determinants of alcoholism go beyond frequent drinking and drunkenness.

Rates of alcoholism and problem drinking may partially reflect the integration
of drinking behavior patterns into the culture or subculture. If an entire culture or
subculture actively supports conformity to drinking standards, solidly establishing
consistent values and sanctions known to and agreed upon by all, then rates of
problem drinking tend to remain low. In contrast, high rates of alcoholism show
an association with certain cultural characteristics: conflict or ambivalence about
drinking, delays in introducing children to alcohol in social and dietary contexts,
drinking of alcohol outside of mealtime, and drinking for personal reasons rather
than as part of rituals and ceremonies or family life.

Some modern societies demonstrate marked ambivalence regarding alcohol
use. Examples include the United States, Ireland, France, and Sweden. The

resulting conflict between competing values and norms seem to contribute to high rates of alcoholism (Lender and Martin, 1982; but see Room, 1976). In other societies, permissive, positive, and consistent attitudes toward alcohol consumption seem associated with low rates of alcoholism. These conditions prevail in Spain, Italy, and Japan, as well as among Jewish groups. Attitudes that stress both positive and negative aspects of drinking may fail to regulate excessive consumption. Overall consumption of alcohol has declined in many countries throughout the world. In one survey of 25 nations, researchers found declining per-capita alcohol consumption in nearly two-thirds between 1979 and 1984 (Horgan, Sparrow, and Brazeau, 1986); alcohol consumption has continued to increase in developing countries, though (Hilton and Johnstone, 1988).

Differing opinions about appropriate conditions for drinking reflect ambivalence toward the circumstances, amounts, and motivations that define acceptable drinking. In the United States, many people display especially noticeable ambivalence about alcohol consumption. For example, 91 percent of a sample of California residents indicated that alcoholism is an illness, but 40 percent also believed that alcoholics drink because they want to do so (Caetano, 1987). Few people could completely reconcile these conflicting ideas.

The role of group and subcultural factors in producing excessive drinking and alcoholism appears in many ways: (1) gender differences, (2) choices of companions when drinking, (3) skid-row drinking, (4) occupational differences in excessive drinking, (5) religious differences, and (6) ethnic differences. In each instance, excessive drinking shows clear signs of social patterns or structures rather than characteristics of random or strictly individual behavior. Such differences point to the dominant roles of social learning and cultural values in problem drinking. A person's uses of alcohol and thoughts about it vary with group membership and feelings of identification with these groups. Groups often display low rates of problem drinking when they generally agree about drinking customs and values and when they maintain social supports for moderate drinking and negative sanctions for excessive drinking; groups without these characteristics often display high rates of alcohol-related troubles.

Gender Differences in Excessive Drinking

More men than women drink alcohol, and male alcoholics outnumber female alcoholics. Women abstain from alcohol more often than men do, and men more often become heavy drinkers. Overall, men typically consume more alcohol more frequently than do women. Men also generally weigh more than women, so they must consume more alcohol to feel its effects. This difference tends to reinforce variations in drinking behavior. Men are more likely than women to frequent public drinking houses, and they generally consume larger quantities of alcohol there (Nusbaumer et al., 1982). Clearly, the drinking habits of males differ substantially from those of females (Wanberg and Horn, 1970).

Several reasons in addition to those mentioned so far may account for related differences between men and women in rates of problem drinking. (1) A smaller proportion of women who drink implies fewer opportunities to develop problems. (2) Society attaches a stronger social stigma to women's drinking than to drinking by men. (3) Those women whose lives center more around family activities will necessarily avoid contact with norms that encourage drinking. (4) Traditional determinants of a woman's self image emphasize performance in relatively private, family-oriented roles, while men experience stresses and risks of failure in both private (family) roles and public (occupational) ones; therefore, they often suffer greater social damage than women experience (McCord and McCord, 1961: 10–11). Sex differences in excessive drinking often relate to variations in society's expectations of men and women. Men frequently engage in drinking at earlier ages, and they may participate in more social drinking than women. Observers recognize a relationship between heavy drinking by women and marital instability, but they cannot clearly determine which is the cause and which the effect.

Some research has sought to evaluate a possible convergence in the rates of excessive drinking among men and women. Changes in sex roles have led some to expect that female rates of problem drinking would approach those of men. Recent evidence does not support this idea, however. Hasin, Grant, and Harford (1990) have compared male and female liver cirrhosis rates over a 25-year period (1961 to 1985). These authors found no evidence of converging rates during this time, in spite of changes in sex roles. However, several surveys have indicated an increase in heavy drinking specifically among young women (e.g., Hilton, 1988; Wilsnack, Wilsnack, and Klassen, 1987), suggesting a need for a closer, age-specific examination of drinking behavior.

Data from two national surveys, one in 1971 and the other in 1981, also show no increases either in drinking or heavy drinking among women during the decade between the studies (Wilsnack et al., 1987: 97). The results revealed heavy drinking in certain groups of women: those aged 21 to 34, those who had never married or who had become divorced or separated, those who were unemployed, and those who cohabited with others outside of marriage. A majority of women drinkers "reported that drinking reduced their sexual inhibitions and helped them feel closer to and more open with others" (Wilsnack et al., 1987: 99). Of those women who admitted the heaviest drinking, most showed signs of role deprivation. Many had lost family roles (through divorce or abandoning their children) and occupational roles (through unemployment). Female drinking is also strongly related to drinking by their husbands or partners or by other close friends (Wilsnack et al., 1987: 105). Women's alcohol use displayed an influence by their husbands' drinking stronger than the influence on husbands by their wives' drinking.

Female alcoholics share similar background traits with male alcoholics. The family histories of women alcoholics feature more disruption than those of women as a whole, and this group reports a higher incidence of alcoholism in their families (Beckman, 1975; Bromet and Moos, 1976).

Companions and Excessive Drinking

Drinking generally takes place in small groups, so these groups provide the environments in which drinking norms develop. In fact, a person risks being identified as a deviant drinker if he or she drinks alone, not interacting with others present, even if they are drinking themselves. The development of alcoholism shows a close relationship to the types of companions with whom one associates. Most people's drinking norms appear to conform closely to those of age contemporaries and people encountered in the context of an occupation. Friends, spouses, and co-workers may strongly influence learning and transmission of norms, customs, and attitudes about drinking.

This activity clearly affects processes that determine adolescent drinking, most of it beer drinking. According to national surveys, about 25.0 percent of 10th through 12th graders reported abstaining from alcohol, 7.6 percent reported infrequent drinking, and 18.8 percent reported light drinking. Heavy drinkers constituted about 15.0 percent of that population, a proportion that has held roughly steady for the past decade or so (Johnston, O'Malley, and Bachman, 1996; Secretary of Health and Human Services, 1987: 33; Secretary of Health and Human Services, 1993). Heavy drinking appears to taper off after age 17. Further, 31.2 percent of the adolescents surveyed fit the classification of alcohol misusers (defined as self-reports of drunkenness at least six times a year or negative consequences from drinking in two of five social areas). Most adolescent drinking occurs in group contexts, so studies of youth drinking have not surprised anyone with their finding that peers play a crucial role in the adolescent drinking process. Research has found a strong reciprocal relationship between adolescent drinking and the percentage of friends (peers) who drink (see Duncan, Tildesley, Duncan, and Hops, 1995). That is, an adolescent's alcohol use affects the drinking patterns of her or his peers at the same time that it reflects influence by those peers' own drinking patterns. This relationship does not extend, however, outside groups of close friends (Downs, 1987). The number of friends who drink and perceived peer approval or disapproval of drinking act as significant predictors of adolescent alcohol use. In fact, an extensive review of the literature about peer influence on illicit drug use, and specifically on adolescent alcohol use, has led one writer to conclude that

The most consistent and reproducible finding in drug research is the strong relationship between an individual's drug behavior and the concurrent drug use of his friends, either as perceived by the adolescent or as reported by the friends. . . . Peer related factors are consistently the strongest predictors of subsequent alcohol . . . use, even when other factors are [taken into account]. (Kandel, 1980: 269)

Drinking Among the Homeless and on Skid Row

One can find only rough estimates of the size of the U.S. homeless population, but one government agency suggests a figure between 250,000 and 350,000 individuals (General Accounting Office, 1985; Secretary of Health and Human Services, 1990: 30). The traditional view of the homeless as mostly alcoholics, drug addicts, and transients must now accommodate increasing proportions of elderly people, women, unemployed people, children, members of minority groups, and sufferers with mental illness. People become homeless for many reasons, but alcohol abuse is the single most important contributor (General Accounting Office, 1985). This population often uses alcohol in combination with other drugs. In fact, as many as one-quarter of alcohol-abusing women and about one-fifth of alcohol-abusing men also use other drugs (Secretary of Health and Human Services, 1990: 31)

Group drinking plays a major role in the lives of homeless men living on skid row. Although they focus on alcohol as a major preoccupation, not all skid-row drinkers are problem drinkers. Only about one-third of skid-row residents may be problem drinkers, while another third are moderate drinkers, and the remaining third appear to drink little or not at all (Bahr, 1973: 103). Estimates into the 1990s describe heavy drinking by between 20 and 45 percent of the homeless (Mulkern and Spence, 1984; Secretary of Health and Human Services, 1987: 34–35; Secretary of Health and Human Services, 1990: 31). A study of 412 homeless and marginally homeless individuals in New York state reported a surprising proportion of abstainers (40 percent of the sample), but 13 percent reported drinking more than 20 drinks a day, consumption matched by only about 1 percent of the New York state population (Welte and Barnes, 1992).

Skid-row populations have thoroughly institutionalized the practice of drinking alcoholic beverages in pursuit of intoxication. These people drink as a symbol of social solidarity and friendship, and the culture fully accepts group drinking and collective drunkenness (Spradley, 1970: 117). The most important primary group among this population is the "bottle gang," which turns away no one who wishes to join. Homelessness forces these people to search for suitable locations to drink, and they often appear drunk in public, frequently resulting in their arrest. As its major function, such a drinking group provides social and psychological support for alcohol consumption by its members; the group also facilitates interaction among people who may not know one another. These groups exert such strong social influence that someone who wishes to deal effectively with his or her alcoholism must leave the area.

Occupation and Excessive Drinking

As mentioned earlier, the percentage of drinkers in a group tends to increase with rising occupational status, although problem drinking does not follow the same pattern. In shifting attention from low-status to high-status occupations, one observes a steady increase in the percentages of these groups who drink, but this change does not necessarily correspond to a rising proportion of problem drinkers. One study found the highest percentage of drinkers of all groups of subjects in the top occupational categories, such as lawyers and doctors, but the percentages of problem drinkers in these categories remained among the lowest of all occupational groups (Mulford, 1964; also see Biegel and Chertner, 1977).

Social patterns common in some occupations seem to encourage immoderate drinking. In particular, certain business occupations are associated with frequent and heavy drinking. Negotiators often initiate and close deals over cocktails and in other settings that involve alcohol. In fact, "work histories of sales managers, purchasing agents, and representatives of labor unions who have become alcoholics strongly suggest that their organizations tacitly approve and expect them to use alcohol to accomplish their purposes effectively" (Trice, 1966: 79). Many such people bring little or no previous histories of heavy drinking to their organizational positions; they learn drinking on the job, supported by reinforcement from the company through promotions and bonuses identified with effective social interactions.

Drinking gains an association with recreation in some occupations. In the restaurant business, for example, research has reported a higher probability of heavy drinking for people whose co-workers took end-of-work drinks at the workplace at least weekly, whose co-workers went out for drinks after work at least every week, and who worked at establishments with liberal alcohol policies (Kjaerheim, Mykletun, Aasland, Haldorsen, and Andersen, 1995).

Other occupations develop reputations for heavy drinking, but they do not necessarily include high proportions of problem drinkers. Merchant sailors demonstrate high rates of both drinking and problem drinking. They often cope with the monotony, frustration, and social isolation of life at sea by turning to alcohol for relief and entertainment. The tradition of their occupation features "bottle gangs" similar to those on skid row. Other male-dominated occupations, such as the military and the construction and building trades, also include high proportions of excessive drinkers. An important distinction separates groups with high percentages of drinkers and those with high percentages of problem drinkers, however, as illustrated in a study of two supposedly heavy-drinking occupations: naval officer and journalist. This research discovered that while naval officers and journalists did indeed drink more frequently than did people in many other occupations, they did not consume greater quantities of alcohol (Cospers and Hughes, 1983).

Still, the nature of an occupation can encourage heavy drinking, as one journalist reports:

> As in most things, you needed rules of conduct. I drank into mornings when I worked nights and at night when I worked days. When I was sent out to cover some fresh homicide, I usually went into a neighborhood bar to find people who knew the dead man or his murdered girlfriend. I talked to cops and firemen in bars and met with petty gangsters in bars. That wasn't unusual. From Brooklyn to the Bronx, the bars were the clubs of New York's many hamlets, serving as clearinghouses for news, gossip, jobs. If you were a stranger, you went to the bars to interview members of the local club. As a reporter, your duty was to always order beer and sip it very slowly. (Hamill, 1994: 227)

Within some occupations, drinking gains an identity as an acceptable and conforming practice, not a pathological one. In fact, some occupational groups not only condone drinking, but actively encourage it. This list includes longshoremen, pipeline construction workers, other building trade workers, railroad engineers, coal and mineral miners, lumberjacks, oil rig workers, tavern keepers, and alcoholic beverage workers (Sonnenstuhl and Trice, 1991: 259–261).

Religious Differences in Excessive Drinking

The drinking patterns of persons with strong religious beliefs differ markedly from those of less actively religious people and between members of different religious groups. Generally, people with certain attitudes toward religion tend not to drink, at least as adolescents. This group includes those who attend church, who perceive themselves as strongly religious, and who regard drinking as a sinful activity. In addition, adults who declare religious preferences tend to abstain more often and to drink heavily less often than those who declare no such preferences (Cahalan, 1982: 112). Fundamentalist Protestant denominations include high proportions of abstainers, while fewer Catholics, liberal Protestants, and Jews abstain. Despite quite pervasive drinking among Jews, the rates of alcoholism in this group fall far below what one might expect (Biegel and Chertner, 1977: 206; Snyder, 1978). Overall, U.S. Anglo-Saxon Protestants seem to lack the kind of clear, common standards for appropriate and inappropriate uses of alcohol that Jews share. Even when groups demonstrate substantial agreement about drinking behavior, these norms do not usually become deeply rooted in the culture, and they seldom escape some conflict between competing attitudes (Plaut, 1967: 126–127).

Observers accept certain generalizations about the effect of religious belief on drinking, although they warn about continuing differences among denominations. Among Protestants of northern European descent who drink, alcohol consumption usually lacks strong associations with other activities. Instead, these people often drink for the specific purpose of escaping or having a good time.

Orthodox Jews use alcohol in different ways than these Protestants. Almost all Orthodox Jews drink wine, as many social and religious occasions—births, deaths, confirmations, religious holidays—require it by both prescription and tradition. Jews thus become accustomed to using alcohol in moderation; they start drinking it in childhood, mostly in ritualistic contexts. Orthodox Jews have another shield against problem drinking: Their religion establishes powerful moral sentiments and anxieties that discourage intoxication through the widely held belief in sobriety as a Jewish virtue and drunkenness as a vice of Gentiles (Snyder, 1978: 182).

Society's Response to Alcohol Use and Alcoholism

Alcoholism and problem drinking violate social norms concerning moderate and otherwise appropriate use of alcohol. Drinking behavior that goes beyond accepted group practices may well draw sanctions. This is the process of social control.

Society applies rather fragmented and generally unsystematic social control efforts to govern alcohol use because the public displays ambivalence concerning appropriate standards for this behavior. This variability has inspired disagreement concerning (1) the most effective means by which to regulate manufacturing, sales, and consumption of alcoholic beverages and (2) the most accurate and useful theory of alcoholism. As a result, several models compete to explain alcoholism, and society has tried no fewer than five modes of social control—prohibition, legal regulation, education about alcohol use, offering potential substitutes, and comprehensive programs—to prevent misuse of alcohol (see Lemert, 1972).

This section begins by identifying the major strategies for regulating alcohol-related behavior. It then explores some of the major theories of alcoholism.

Strategies of Social Control

The strategy of *prohibition* develops a system of laws and coercive measures that prohibit manufacturing, distributing, or consuming alcoholic beverages. This means of control has had a noteworthy history in the United States.

The 18th amendment to the U.S. Constitution outlawed manufacturing and sales of alcoholic beverages. Passage of this measure in January 1920 represented a victory for various reform groups, which identified abstinence as a middle-class virtue worthy of protection under the law (Gusfield, 1963). The Prohibition era lasted until the passage in 1933 of the 21st amendment repealing the earlier legislation. Contrary to modern perceptions, Prohibition seems to have succeeded in reducing the nation's alcohol consumption, in spite of illegal bootlegging that brought in liquor from Canada and the operation of illicit taverns called *speakeasies* (Lender and Martin, 1982: 136–147). A return to a nationwide prohibition on alcohol seems unlikely in view of increasingly permissive attitudes toward moderate use and a continuing ambivalence about the effectiveness and desirability of legal controls (Meier and Geis, 1997).

A second strategy of social control enacts somewhat less comprehensive *legal regulation* that governs the kinds of liquor people can consume, monetary costs, methods of distribution, acceptable times and places for drinking, and selective availability of alcoholic beverages to consumers by age, sex, and various socioeconomic characteristics. Unlike a prohibition strategy, legal regulation strategy applies the law to establish standards, backed by legal sanctions, for acceptable practices in manufacturing, distributing, and consuming alcohol. Thus, laws prohibit people under a certain age from consuming alcoholic beverages. They also limit purchases of alcoholic beverages to certain places on certain days or at certain times. Some states have imposed heavy taxes on alcoholic beverages to generate revenue and to discourage excessive use by raising their prices. If one can say that U.S. society has adopted any one national strategy concerning the use of alcohol, it is legal regulation.

A third control strategy establishes a system to *educate* people about the consequences of alcohol use, with the goal of encouraging moderate drinking or even abstinence. The success of this approach depends greatly on effective presentations of factual information about the dangers of alcohol for potentially heavy or problem drinkers. This requirement raises questions about whether educational programs can reach, and then convince, people who derive most of their information and values about alcohol use from family and friends. As with formal programs to educate people about other drugs, alcohol education often establishes artificial situations divorced from the real-life situations in which people drink.

A fourth control strategy might emphasize *alcohol substitution*, perhaps promoting replacements for traditional drinks like beer with reduced alcohol content, nonalcoholic imitations of beer and wine, soft drinks, and even marijuana. Some people might justify a preference for marijuana over alcohol and cigarettes on the grounds that it causes fewer and less harmful known effects in regular use; a practical proposal for such a substitution would probably encounter powerful public resistance, however. Similarly, contemporary discoveries about possible cancer-causing chemicals in soft drinks, particularly diet soft drinks that contain saccharine as a sugar substitute, reduce the attractiveness of this substitute. Over-all, an argument for substitute products might prove intellectually persuasive but emotionally unsatisfying.

Finally, a social control strategy may work through broad programs to *prevent alcohol abuse*. Such an effort would seek to change people's attitudes toward alcohol and drinking alcoholic beverages. Americans' historically ambivalent attitudes toward drinking frown upon the practice as a violation of certain religious doc-trines and principles of abstinence; on the other hand, alcohol serves an important function at many social gatherings, and some may even resent abstinence on such occasions as wedding celebrations or New Year's Eve.

A prevention program designed to deal directly with this ambivalence would concentrate on providing public information about several subjects: the conse-quences of alcohol consumption, appropriate contexts (times, places, events) for drinking, and the symptoms of drinking problems. Civic organizations, volunteer groups, and service clubs have carried out such programs throughout the United States. Schools have instituted programs directed at youths.

Most of these programs have emphasized a public-health model of alcohol pre-vention, hoping to duplicate successful application of such a model to other health problems (Secretary of Health and Human Services, 1981: 104). The relevance of this model to alcohol problems has stirred some controversy, however, as critics have disagreed with the assumption that alcohol problems represent health or medical problems. Conventional wisdom now treats alcoholism as a disease, how-ever; to understand this medical conception of problem drinking and alcoholism, the next section considers the medical model and other models sometimes applied to explain alcoholism.

Models of Alcoholism

Many specific theories advance explanations as to why people develop drinking problems and alcoholism. Like the explanations of other forms of deviance, some theories concentrate on features or characteristics of individual deviants (psycho-logical or biological factors), while others focus on influences or causes that oper-ate outside individual deviants (sociological factors). A great deal of research in recent years has concentrated on issues of chemical dependency rather than focus-ing strictly on alcohol abuse or drug habits. Studies address the broader issue not only because alcohol is in fact a drug and both show similar patterns of use, but also because many people who report drinking problems also use other kinds of drugs.

Theories can explain alcoholism in the context of a number of models (Ward, 1990). These models present broad conceptions or perspectives of alcoholism and provide more or less detailed explanations for its occurrence. Space limitations confine this section to discussions of a few among many such models.

Psychoanalytic Model   In a psychoanalytic framework, alcoholism is a symptom of some underlying personality disorder. Many psychoanalysts blame the condi-tion on some unresolved conflict between id and ego that has its roots in early childhood experiences. According to psychoanalysis, a person develops through a series of stages; if some event arrests that development, the person can become fixated at a particular stage and make no further progress. The psychoanalytic model sometimes regards excessive drinking as evidence that the drinker became fixated at the oral stage of development. The alcoholic can then stop drinking only by resolving these mental conflicts. The model would call for a lengthy process of individual, in-depth psychotherapy to treat alcoholism.

Family Interaction Model　　The family interaction model regards alcoholism as a family problem, not an individual one (Jacob, 1987). Treatment would explore the web of relationships in the alcoholic's family, rather than looking only at individuals, such as a spouse or children. Stress often supplies one important source of family pressure on drinking behavior, and some research shows a relationship between drinking behavior and high levels of stress within families, compounded by normative systems that promote drinking to relieve that stress (see Linsky, Colby, and Straus, 1986). Research has also documented a relationship between stress and drinking at the state level as well as within families (Linsky, Colby, and Straus, 1987).

   The family interaction model emphasizes the original causes of excessive drinking less than family relationships and interpersonal forces that keep alcohol a problem in the family. By emphasizing the importance of family relationships associated with alcoholism, this model implies a need for treatment that involves the whole family. "The goal is to help each family member recognize the degree to which they contribute to the circular and degenerative alcoholic process" (Ward, 1990: 9).

Behavioral Model　　The behavioral model originates in the thinking of behaviorist psychologists, who conceive of alcoholism and treat it as a behavior (or set of behaviors) rather than as a disease. Like the family interaction model, the behavioral model looks for mechanisms that sustain drinking. In general, these theorists assert, an individual continues heavy drinking because he or she receives some reinforcement for the behavior. The process of reinforcement may include, among other things, approval from peers, euphoric sensations, and the chance to maintain certain kinds of relationships with others. In other words, excessive drinking and alcoholism are learned activities. Behaviorists believe that one can learn to abstain through manipulation of reinforcements and punishments similar to those that led one to become an alcoholic. They also believe that an alcoholic can learn to become a social or moderate drinker, and they cite some supporting evidence that former problem drinkers can and do become social drinkers under certain circumstances (Nordstrom and Berglund, 1987).

Biological Model　　One of the most active areas of research on alcoholism today searches for biological antecedents to alcoholism. Studies have not yet found a clear biological mechanism that would explain alcoholism and excessive drinking, but a number of them have suggested a possibility of some kind of biological predisposition to the condition. Much of this research has concentrated on the genetic variations at the molecular level in alcohol-metabolizing enzymes, which act to remove alcohol from the body. An extensive review of this research concludes that further investigation may well locate some inherited biological mechanism, but it also stresses the importance of other, nonbiological causes (Secretary of Health and Human Services, 1990, 1993, 2000).

Studies of biological contributors to alcoholism have identified two kinds of predisposition. *Male-limited* susceptibility occurs only in males, giving them a highly inheritable tendency toward severe early-onset drinking. *Milieu-limited* susceptibility (the more common condition) affects both sexes and produces its effect only in reaction to some kind of environmental provocation. In either instance, biological characteristics make a person more or less "vulnerable" to alcoholism (Hill, Steinhauer, and Zubin, 1987).

A model of alcoholism that emphasizes biological factors provides support for a so-called *medical model* that conceives of this behavior as a kind of disease. So far, research has not generated conclusive indications of biological antecedents, and no study to date has provided a means by which to explain the tremendous variability in rates of alcoholism and excessive drinking among various groups in society.

Medical Model   The medical profession has adopted its own model of alcoholism that likens it to any other disease, advocating treatment by medical measures (Jellinek, 1960). Medical models usually look for causes of deviance within an individual's biology or psychology.

To be sure, medical terminology and methods of analysis offer effective tools for evaluating the physiological and medical consequences of sustained drinking. The conception of alcoholism as a disease has additional effects outside the strict boundaries of medicine. The temperance movement of the 19th century found a useful political tool in a conception of alcoholism as chronic drinking by someone afflicted with a disease (Conrad and Schnieder, 1980: 73–109; Peele, 1990). These activists favored such a conception because it conveyed the desired moral condemnation of intoxication while holding forth the promise of treatment and change for the drinker; they saw these possibilities as a more palatable combination than a moralistic view of alcoholism as the result of "bad" people engaging in "sinful" behavior. The disease concept retains much popularity as a way of thinking about alcoholism, although it still generates controversy.

The adequacy of the medical model of alcoholism depends ultimately on the meaning of *disease*. If alcoholism is really a disease, it displays characteristics unlike any other presently known to medicine, mixing physical, organic indicators with psychological and sociological ones. At present, furthermore, the model does not explicitly explain whether consumption of alcohol represents a symptom of the disease or the disease itself.

The popularity of the medical model will probably continue, in view of strong public acceptance. The medical model beneficially implies a relatively humane social response: treatment in a medical facility rather than confinement in a jail. Accepting this model, however, and regarding problem drinkers as victims of illness may logically imply that they cannot accept responsibility for their behavior, however deviant (Orcutt, 1976).

This suggestion and many other aspects of alcohol consumption and alcoholism leave some people feeling ambivalent about the medical model. Most

Issue   **Is Alcoholism a Disease? Two Positions**

Yes, Alcoholism Is a Disease	*No,* Alcoholism Is Not a Disease
Alcoholism is as much a disease as other diseases that physicians treat. No one would claim that a physician should not be involved in the treatment of ulcers brought about by stress and no one should be surprised that they are involved in the treatment of alcoholism. "I would agree that a behavior, an 'activity,' even a central one, is not a disease. But I would also think that a persistent, irrational, self-destructive activity is symptomatic of a disease (p. 86).	The medical model does not fit alcoholism or problem drinking. Alcoholism is only heavy drinking and heavy drinking is not a disease. No one denies that there are medical (for example, cirrhosis of the liver) and other physical (for example, a high rate of accidents and injuries) consequences to heavy drinking; and no one denies that heavy drinking can become a central activity for some people that can dominate their lives. The question is whether such heavy drinking is a *disease*.

Sources: Keller, Mark. 1990. "Review of Fingarette, Herbert. 1988. *Heavy Drinking: The Myth of Alcoholism as a Disease.*" in *Journal of Studies on Alcohol*, 51: 86–87; Fingarette, Herbert. 1988. *Heavy Drinking: The Myth of Alcoholism as a Disease.* Berkeley: University of California Press.

acknowledge alcoholism as a disease, but they also believe that alcoholics drink because they want to do so (Caetano, 1987). Such ambivalence may lead to reluctance to adopt and fully implement the medical model of alcoholism.

Combined Perspectives   A complete explanation for persistent, heavy drinking may require a broader perspective than any of these views individually can accommodate. Some claim that a comprehensive understanding of alcohol-related problems can emerge only from an evaluation of biological, psychological, familial, social class, and sociocultural risks (Trice and Sonnenstuhl, 1990). This suggestion does not imply that alcoholism defies control or that no manipulation can affect these kinds of risks. Trice and Sonnenstuhl, for example, suggest conscious development of drinking norms in work contexts and management of constructive confrontations with problem drinkers to help reduce the impact of associated risks. But such individually focused measures also require a comprehensive understanding of public policy and broader control measures.

Public Policy and Public Drunkenness and Alcoholism

Society can seek to alleviate problems with individual drunkenness and alcoholism in a number of ways. Many have tried two techniques: community-based treatment programs and Alcoholics Anonymous. Some evidence suggests that certain alcoholics may successfully return to moderate drinking, but controversy still surrounds that claim.

Community-Based Treatment Programs

In 1971, concern over problem drinking resulted in two important effects: the creation of the National Institute of Alcohol Abuse and Alcoholism. Also, adoption of alcohol treatment programs in local communities throughout the nation has brought counseling and other services to thousands of problem drinkers. In recent years, a substantial public-health movement has encouraged development of community-based referral and treatment centers for problem drinkers, some providing outpatient counseling and some emphasizing hospitalization.

The number of alcoholism treatment services and their client lists continue to increase. American Hospital Association surveys indicate that alcoholism and drug treatment units increased 78 percent, from 465 in 1978 to 829 in 1984, while total inpatient hospital beds for drug and alcohol clients increased 62 percent over the same period (Secretary of Health and Human Services, 1987: 243). By the turn of the century, more than 700,000 people participated in both alcohol only and combined alcohol and drug treatment programs (Secretary of Health and Human Services, 2000: 427). Estimates indicate that more than 2,500 work organizations provide additional substance abuse programs for alcoholic employees (Roman, 1981).

In spite of these gains, these programs reach only a relatively small proportion of the population of alcoholics and problem drinkers. The remaining problem drinkers either participate in other programs or receive no treatment. The clients of community-based treatment programs typically come from the lower class more often than from the middle or upper classes, and these programs provide mainly outpatient services. Clients look for help in such programs when they perceive problems in their lives caused by drinking in the areas of health, social relationships, or work (Hingson, Mangione, Meyers, and Scotch, 1982).

Community-based treatment facilities spread in some communities as a result of the Uniform Alcoholism and Intoxication Treatment Act of 1971, which summarized a set of recommendations developed by the National Conference of Commissioners on Uniform State Laws. A number of states quickly adopted the recommendations, the most important of which included decriminalization of public drunkenness and establishment of publicly operated detoxification centers. These initiatives reflected the spread of the central idea, formally specified in the medical model of problem drinking, that alcoholism is a disease rather than criminal behavior. While the initiative encouraged humane handling of cases of public drunkenness, particularly by skid-row drinkers, one evaluation of its implementation found that recidivism rates in Seattle had increased fourfold over comparable levels when police had arrested and jailed people for public drunkenness (Fagin and Mauss, 1978). The increase in clients came mainly from self-referrals by problem drinkers who preferred stays in detoxification centers over terms in jail. Other evaluations of detoxification programs have echoed these findings; they find the same "revolving door" between freedom and confinement that characterized a jail-based response, but the revolving door seems to spin much faster between detoxification programs and the outside world (Rubington, 1991: 740).

Experience has never justified the optimistic belief that treatment would divert a large proportion of public drunkenness offenders from jails to hospitals. A number of factors have inhibited complete implementation of the Uniform Alcoholism and Intoxication Treatment Act, including inadequate treatment facilities and inadequate funding for alcohol programs (see Pittman, 1991: 230–232). In addition, the U.S. political climate changed in the 1980s. Society increasingly viewed people with problems as "problem people," and the public felt declining sympathy toward public intoxication. When many communities did implement detoxification programs, they developed counseling services, not medical ones, reflecting rising public resistance to tax-based funding for services by health professionals to chronically drunk homeless people.

Privately funded and operated alcohol treatment centers and programs also serve some communities. These facilities include residential "alcohol hospitals" and outpatient programs that offer services to problem drinkers and their families. Such programs may apply treatment methods like counseling, referrals to medical facilities, behavioral modification techniques, aversion-inducing drugs, and others. Attempts to evaluate the success of these programs encounter many difficulties. Their private orientation and substantial expense prevent access by the total population of problem drinkers, so these programs may serve only very highly motivated clients already determined to do something about their drinking problem. This sample hardly represents all problem drinkers (Chafetz, 1983).

Alcoholics Anonymous

Of all treatment methods for problem drinking, Alcoholics Anonymous (AA) has developed the most widely known program and, some think, one of the most successful. Two alcoholics who felt that their mutual fellowship had helped them with their drinking problems founded Alcoholics Anonymous in 1935 in Akron, Ohio. Although not known by that name at the time, the small group soon expanded, and other groups began meeting in New York City and Cleveland. In 1939, the book *Alcoholics Anonymous* was published, which was to provide the core philosophy of the organization: the 12-steps philosophy. AA implements the medical model of problem drinking, conceiving of alcoholism as a disease with symptoms that sufferers can avoid only by never drinking alcohol. AA has no formal organization as such; local chapters operate meetings of AA's members. No officers make central decisions and members pay no dues, although local chapters do support a central office in New York City that publishes a monthly newsletter called *AA Grapevine*. The newsletter, published since 1944, contains first-person accounts of staying sober as well as practical insights on how AA's members stay sober. It claims a subscription list of 117,000 persons around the world. There is a Spanish-language version. This strictly voluntary organization serves members through nearly 100,000 groups nationwide and in other countries. One report suggested total U.S. membership at approximately 630,700 in 1983, up from 476,000 in 1980 (Secretary of Health and Human Services, 1987: 259). Of this

In Brief   Alcoholics Anonymous: The 12 Steps Theory

AA articulates its philosophy in the "12 steps":

We . . .

1. Admitted that we were powerless over alcohol—that our lives had become unmanageable.
2. Came to believe that a power greater than ourselves could restore us to sanity.
3. Made a decision to turn our will and our lives over to the care of God as we understood Him.
4. Made a searching and fearless moral inventory of ourselves.
5. Admitted to God, to ourselves, and to another human being the exact nature of our wrongs.
6. Were entirely ready to have God remove all these defects of character.
7. Humbly asked Him to remove our shortcomings.
8. Made a list of all persons we had harmed, and became willing to make amends to them all.
9. Made direct amends to such people wherever possible, expect when to do so would injure them or others.
10. Continued to take personal inventory and when we were wrong promptly admitted it.
11. Sought through prayer and meditation to improve our conscious contact with God as we understood him, praying only for knowledge of his will for us and the power to carry that out.
12. Having had a spiritual awakening as the result of these steps, tried to carry this message to alcoholics, and to practice these principles in all our affairs.

figure, about 30 percent are women, and 31 percent attend meetings for problems with other drugs as well as alcohol. AA, although it does not maintain membership lists, estimates a total membership of about 2 million. Regardless of membership, it is generally agreed that 12-step programs represent the dominant approach to alcoholism treatment in the United States (Secretary of Health and Human Services, 2000: 445).

AA works to "delabel" the alcoholic and move that person back into society as a contributing, independent individual. Toward that end, the program breaks down the alcoholic's social isolation from the rest of the community. Members share life stories at meetings, and each new member interacts regularly with a sponsor—someone who has successfully coped with his or her own drinking problem long enough to develop sufficient stability to help someone else. Reciprocal obligations are particularly important in AA. The special relationship between a sponsor and a "baby" (the term for a new member) creates solidarity and identification with the group. In this way, AA resembles other self-help programs, such as those for drug addicts, mental patients, or the obese, all of which involve former deviants to help with treatment of people with conditions similar to their own. Cressey (1955) describes the process of using these people as change agents and the general outcome of such efforts as "retroflexive reformation." Essentially, the former deviant, in attempting to reform the present deviant, promotes his or her own rehabilitation, because the relationship demands learning and modeling of values and attitudes that promote rehabilitation in order to convey them to the newly treated deviant. In this respect, the AA program may achieve its greatest therapeutic success when a new member begins to bring others to AA meetings and to take some responsibility for their drinking behavior. One author has even compared the general philosophy of AA to other bodies of philosophy that stress increasing dependency on others, vulnerability, and self-acceptance (Kurtz, 1982).

Through AA, alcoholics face their problems together, with mutual support and a collective search for meaning. In fact, AA may gain its real value from this nurturing relationship—not as a treatment modality per se, but as a transitional group that promotes moral reintegration of the alcoholic into the larger society (Denzin, 1993). Control of drinking behavior is a necessary but not a sufficient condition for this reintegration, and participation in AA helps to make the difference.

Evaluating the success of this program exposes some problems. AA groups maintain no records on attendance at meetings. Also, members mention only their first names, and they carefully safeguard the confidentiality of one another's private lives. A review of the literature on outcomes for AA members suggests an overall abstinence rate between 26 percent and 50 percent after 1 year, which compares favorably with the results of other approaches (Miller and Hester, 1980).

Alcoholics Anonymous does not serve all problem drinkers. Observers have noted significant differences between problem drinkers who attend AA meetings and those who do not. One study reports that AA members tend to regard themselves, even before they attend a meeting, as likely to share their troubles with others; they infrequently report knowing others who they "believed" stopped drinking through willpower; finally, they had lost long-time drinking companions and had become exposed to positive communications about AA (Trice, 1959). Gilbert (1991) reported that a recovering alcoholic's endorsement of the AA philosophy effectively predicted days of sobriety over 1 year after treatment, but frequency of AA attendance did not. Cross, Morgan, Mooney, Martin, and Rafter (1990) identified AA participation as the only significant predictor of sobriety at 10 years after treatment. Furthermore, sponsorship of another AA member showed a strong relationship to sobriety; 91 percent of sponsors reported complete or stable remission of alcoholism. Still another study reported no predictive value for treatment outcome based on attendance at AA meetings compared with outcomes for a group of alcoholics who did not attend AA meetings (Montgomery, Miller, and Tonigan, 1995). However, those who attended AA meetings and became heavily involved in the program did achieve better outcomes than those who did not become actively involved.

Other kinds of treatment programs, including institutional programs, can also implement the AA philosophy. Research suggests that such an orientation can enhance the effects of hospital programs. Walsh and his associates (1991), for example,

 **In Brief**   **AA's 12 Steps in Practice**

What Is AA?

Two recovering alcoholics founded AA in 1935 in Ohio, and about 63,000 chapters now operate meetings for over 1 million members in almost all U.S. cities and 112 other countries. The group's headquarters in New York City publishes a newsletter. Members pay no dues, instead offering voluntary contributions to cover the costs of their self-sustaining local chapters. No national organization maintains membership rolls or conducts annual meetings. The most important organizational characteristics of AA are local autonomy and member confidentiality, which lead to the practice of sharing only first names in meetings and other interactions. The national organization promotes local decision making and authority.

How Effective Is AA?

The decentralized structure and focus on confidentiality complicate generalizations about the effectiveness of AA. Organizational records offer no help, because the group keeps none. Therefore, evaluations must focus on local groups, which do not maintain records themselves and actively protect member privacy. Further, different kinds of groups form to help different alcoholics. In any large city, for example, groups hold meetings specifically for airline pilots, attorneys, nonsmokers, senior citizens, young adults, current drinkers, and people with combined alcohol and drug problems. With concern about the applicability of generalizations, one can say that AA appears to help some alcoholics, especially those who stay with the program for some time. No other drug or alcohol treatment program provides the vigorous interpersonal support that AA does, and continued contact with the organization (by attending meetings) can surely help to reinforce antidrinking attitudes. Another contributing factor to AA's success is that no other source of support will answer a call for help at 2 A.M. without sending a sizeable bill afterward. AA members come to know they can more readily rely on—and afford—their partners in AA.

What Happens at Meetings?

AA's local chapters, or home groups, hold meetings almost anywhere and at any time. Some groups prefer breakfast meetings, others meet at lunchtime, and others at night. Most chapters meet at least every week. The meetings last up to about 90 minutes or so. No one takes attendance, and only one universal rule applies: The group admits no one under the influence of alcohol or drugs, except for first-time visitors seeking help. The meetings begin with the AA serenity prayer, a moment of silence for fellow alcoholics, a reading from the *Big Book of Alcoholics Anonymous*, and a request from the chairperson for a recovery-related topic. Members may suggest almost any topic related to drinking. Everyone can speak and, at the end, members join hands and recite the Lord's Prayer in unison. These exchanges foster a good deal of personal sharing, and members often develop common bonds despite no acquaintance prior to the meeting. Groups allow optional participation in prayers, and most meetings appear to discourage explicitly religious conversation; the program requires only that members recognize some higher authority or power than their own will.

compared inpatient treatment combined with participation in AA with AA participation alone and with patient choice of treatment. Over a 2-year period, hospital-treated patients achieved better drinking outcomes than those who participated only in AA.

As a basic element of the AA program, members must recognize the existence of some higher power, and they must admit to themselves that they are powerless to cope with their own alcoholism. These positions have recently faced challenges from a number of alcoholic groups that closely resemble AA but deny the need for a spiritual basis for recovery, as has the idea that an alcoholic must become a submissive and powerless person (Marchant, 1990). Others are Secular Organization for Sobriety (SOS), International Association for Secular Recovery Organizations (IASRO), and Rational Recovery (RR). These groups try to protect members from AA's doctrine of submissiveness, which they regard as damaging to the self-esteem of the recovering alcoholic. One alternative to AA's 12 steps calls for a member to "assume responsibility for one's own life, though at times choosing to seek the help of others." Among Rational Recovery's 11 tenets is one that expressly denies the value of submission: "The idea that I need something greater than myself upon which to rely is only another dependency idea, and dependency is my original problem."

A recent edition of AA's main publication, called the *Big Book*, seeks to meet the needs of problem drinkers in the 1990s. Written to eliminate sexist language and to broaden the conception of spirituality beyond formal religions, the new publication introduces drinkers to the 12-step philosophy ("J," 1996). The new updating of this book, written by an alcoholic who wishes to remain anonymous, is called *A Simple Program*.

The Continuing Controversy: Can "Recovering" Alcoholics Ever Return to Drinking?

Alcoholics Anonymous promotes serious disagreement with its claim "once an alcoholic, always an alcoholic." Even after remaining sober for very long periods of time, AA members still describe themselves as *recovering alcoholics*. This claim has stimulated extensive interest and debate about whether alcoholics can ever return to drinking without encountering problems.

According to Alcoholics Anonymous, an alcoholic may never return to drinking again without becoming a problem drinker. Indeed, conventional wisdom among treatment specialists holds that subjects must refrain *completely* from drinking in order to avoid alcoholic behavior. One study challenged this belief by recruiting 40 alcoholics for voluntary hospitalization and subsequent participation in a controlled drinking experiment (Sobell and Sobell, 1973; see also Sobell and Sobell, 1993). The subjects participated in a behavior therapy program designed to promote responsible drinking without getting drunk. The study converted a room in the hospital to simulate a bar. Then 20 alcoholics participated in 17 treatment sessions involving aversive stimuli (such as drugs that induced nausea when combined with alcohol or electrical shocks), while a control group of the remaining 20 alcoholics received group therapy and other services. The investigators reported significantly better treatment outcomes after release from the hospital for subjects who went through the behavior therapy program than for those who did not. The findings seemed to confirm the possibility that some alcoholics could learn controlled drinking, contradicting the medical model of alcoholism as a disease.

A follow-up evaluation contacted the same subjects several years later, however, and found no evidence that alcoholics could return safely to controlled drinking, even after participating in the behavior modification program (Pendery, Maltzman, and West, 1982). The follow-up evaluation found subsequent excessive drinking by eight of the subjects, while six were abstaining completely, four had died alcohol-related deaths, one seemed to have practiced continuing controlled drinking, and one could not be found. These results offered no support for the idea that alcoholics can learn to drink in a controlled manner.

An ability to drink moderately would suggest no biological origin for alcoholism, as some claim, even in its later stages. Rather, such data would suggest that environmental factors powerfully influence heavy drinking patterns. Some research dealing with other addictive behavior, such as smoking cigarettes and using heroin, supports the idea that some addicts may manage to refrain from the addicting substance even without treatment (Biernacki, 1986; Waldorf, 1983). Other studies have found evidence for a possibility of continued use of addicting substances without returning to addictions (Glascow, Klesges, and Vasey, 1983).

One recovering alcoholic has recounted his journey in this way:

> I didn't join Alcoholics Anonymous. I didn't seek out other help. I just stopped. My goal was provisional and modest: 1 month without drinking. For the first few weeks, this wasn't easy. I had to break the habits of a lifetime. But I did some mechanical things. I created a mantra for myself, saying over and over again: *I will live my life from now on, I will not perform it.* I began to type pages of private notes, reminding myself that writers were rememberers and I had already forgotten material for 20 novels. I urged myself to live in a state of complete consciousness, even when that meant pain or boredom. (Hamill, 1994: 261)

The Sobells (1995) have recently claimed an established justification for managed-drinking techniques supported by increasing recognition that not all problem drinkers develop physical dependency on alcohol. Treatment programs that recognize such a possibility will produce better results than they would achieve if they were to require absolute abstinence of those patients. Others, however, warn that unclear evidence reinforces the need for caution in recommending controlled-drinking programs to problem drinkers (see "Reactions to the Sobells," 1995).

Summary

Alcohol is the most commonly used mood-altering drug in the United States. People consume it as an integral part of many social situations; indeed, norms for many occasions actually prescribe its use. Drinking may become deviant, depending on the norms of the groups to which an individual belongs. People become socialized into the drinking norms of their groups, and this socialization process explains differences in drinking behavior among different groups, such as that between males and females and the practices of various ethnic groups.

Observers have not agreed upon any one definition of *alcoholism*, and many treatment specialists now substitute the term *problem drinker*. Alcoholics are heavy drinkers who consume significant quantities of alcohol frequently over long periods of time. Problem drinkers are those who experience some difficulty as a result of their drinking—in their jobs, family relationships, or other areas of their lives. Most definitions of the term *alcoholic* combine amounts and frequency of drinking, but the notion of *problem drinker* emphasizes the consequences of drinking, regardless of the amount or frequency. No one can say exactly when someone becomes an alcoholic, and individual differences may determine variations in when different people reach this state.

A number of subcultural and group influences affect excessive drinking. Companions exert particularly important effects for adolescent and homeless drinkers, although for different reasons. The percentage of drinkers in a population increases with occupational prestige, and a high percentage of adults who declare a religious preference also abstain from alcohol consumption. Religious affiliations also differ among themselves, with many abstainers among fundamentalist Protestant groups and fewer abstainers in the Catholic, liberal Protestant, and Jewish faiths. Rates of drinking do not always show strong relationships to rates of alcoholism. Rates of alcoholism are high among the Irish and French but low for Italians (despite very high rates of drinking) and Asian Americans. Larger proportions of men than women drink, and more men are heavy drinkers.

Social control of drinking in the United States reflects a fundamental ambivalence about alcohol consumption. People value drinking in some situations but not in others, and many regard drinking as permissible in moderate quantities but not in excess.

Various models of the origins of alcoholism have not resolved this apparent ambivalence. One major perspective, the medical model, views alcoholism as a disease, a view bolstered by recent biological research; in contrast, the behaviorist perspective views alcoholism as learned behavior. Each of these models has implications for efforts to control problem drinking. Regardless of the relative merits of these views, the public has now accepted a disease model of alcoholism and the resulting attitude that treatment offers a more appropriate response than punishment to problem drinking. At the same time, the law prohibits many forms of drinking, and alcohol has been implicated as a cause in a number of crimes.

Society has developed a number of specific treatment measures to control alcoholism. Community-based and inpatient treatment programs rely on a combination of counseling and detoxification. Alcoholics Anonymous groups offer informal, voluntary programs based on mutual self-help by members. Many difficulties complicate efforts to determine the precise success rates of various control programs, but even successful programs deal with only small percentages of all problem drinkers. Subsequent research dealing with the causes of alcoholism, including interactions between biological and sociological causes, may point to more effective means of dealing with excessive drinking and alcoholism.

Selected References

Clark, Walter B., and Michael E. Hilton, eds. 1991. *Alcohol in America: Drinking Practices and Problems*. Albany: State University of New York Press.

The papers in this collection review the nature, extent, and distribution of drinking behavior. They offer excellent material on changes in alcohol consumption patterns over time, as well drinking problems, the social context in which much drinking takes place, and drinking patterns among major ethnic groups.

Fingarette, Herbert. 1988. *Heavy Drinking: The Myth of Alcoholism as a Disease*. Berkeley: University of California Press.

Many sources support a disease conception of alcoholism, but Fingarette makes a strong case against regarding alcoholism as a disease; he defines it simply as behavior: heavy drinking. When heavy drinking becomes a central activity for someone, others regard him or her as an alcoholic. At the same time, this author argues, medical problems likely emerge as a result of heavy drinking, but these characteristics do not amount to a "disease" of alcoholism.

Jacobs, James B. 1989. *Drunk Driving: An American Dilemma*. Chicago: University of Chicago Press. Also Ross, H. Laurence. 1993. *Confronting Drunk Driving: Social Policy for Saving Lives*. New Haven, CT: Yale University Press.

These two books represent excellent accounts of the behavioral and criminal issues connected with drunk driving. Jacobs displays sensitivity to a variety of social, political, and legal issues surrounding this behavior, and his account brings them all together. Ross has done a great deal of original research on deterring drunk driving and presents an excellent analysis of policy issues.

MacAndrew, Craig, and Robert B. Edgerton. 1969. *Drunken Comportment: A Social Explanation*. Chicago: Aldine.

This classic book reports on two anthropologists' study of drunken behavior. It documents the cultural rather than physiological forces that determine drunken comportment. Evidence from a number of cultures supports a culturally diverse understanding of the differences in drinking behavior, providing fascinating reading.

Secretary of Health and Human Services. 2000. *Alcohol and Health: Tenth Special Report to the U.S. Congress*. Rockville, MD: National Institute of Alcohol Abuse and Alcoholism, Government Printing Office.

This most recent edition reports on a valuable, ongoing research project by the National Institute of Alcohol Abuse and Alcoholism. It updates much information on alcohol use and its causes, associated behaviors, and consequences. The report also documents patterns in drinking behavior among minority groups in the United States. The previous reports also contain valuable discussions of the biological basis of alcoholism. Each version of this reports contains so much information about patterns of alcohol use that all are absolutely indispensable reference sources for any serious student of alcohol problems.

CHAPTER 12

Sexual Deviance

Sex work has increasingly gone cyber. One such worker, Veronica, paid an Internet site $100 to list her ad. The site, called L.A. Exotics, provides a menu of options to those who visit, including massage outcall (they come to you), massage incall (you go to them), blonde, escorts, entertainers/strippers, fetish, and personals, among others. Veronica provided a photograph and a short description and included a phone number (Shuger, 2000). She received more than 250 phone calls and three new men who hired her more than once. Veronica is expensive; she charges an introductory fee of $500, $500 per hour, and $2,500 for the evening. Most of her customers are married, over 35, and high-end corporate executives. She lives in Los Angeles, but some of her customers are from other areas and visit the city on business or vacations. She says she only works two nights a week.

Unlike Veronica, some sex workers were not prostitutes pre-Internet. These women find the safety and income of the Internet irresistible. One such worker made an investment of $100 a month for listing an ad on an escort Web site and $1,000 a year for a webmaster to maintain her own personal Web site (Shuger, 2000). She estimates that she gets about 700 cybervisitors per day. She started out charging $4,500 a day with a two-day minimum, but the response was so good that she gave herself a raise to $5,800 a day, still with a two-day minimum.

While some forms of sexual expression have been facilitated by new technology, what constitutes sexual deviance depends on the nature of sexual norms that regulate this activity.

The sex act is a natural part of human life and a necessary one to perpetuate the species. It also can bring the most pleasurable sensations of all human experience. Sexual behavior, like other forms of human activity, is governed by norms that regulate socially acceptable practices and general orientations. Observers can judge the deviance or conformity of a sexual act or condition only with reference to these norms.

This chapter discusses the nature of sexuality and distinguishes between preferences or orientations and behavior. People become sexual beings by acquiring "sexual scripts" that depict sex roles learned in a process of sexual socialization. That socialization, along with responses from others, determines the sexual stimulation that someone feels in response to a person or another element of experience. Despite the biological origin of the human sexual drive, sexuality emerges from a social process that one can explain largely in social terms (Plummer, 1982).

A sociological understanding of sexual deviance requires an awareness of the contents of sexual norms and the sanctions that society applies for violating those norms. This chapter discusses sexual norms that regulate heterosexual activity as well as major types of deviance from those norms.

Sexual Norms

Sexual behavior is normative behavior. Society characterizes a particular sexual act as deviant according to its norms surrounding that act. Traditional religious beliefs have powerfully influenced those norms. Indeed, religion may well affect judgments about no other activity more strongly than it affects the understanding of appropriate and inappropriate sexual behavior. Some religious people regard morality largely as an evaluation of acceptable sexual behavior. As religious beliefs change, so, too, do group norms concerning the deviant identities of various sexual activities. Premarital sexual intercourse, for example, no longer draws as much moral disapprobation as it did in earlier decades. Acceptance of masturbation has also increased, and some even regard the practice as a necessary part of normal sexual development (Janus and Janus, 1993: 106).

Sexual norms differ from those regulating other activities only in content. In other respects, they are the same. People learn norms governing sexual behavior through interactions with others by symbolic communication, direct interaction, and example. They specify what people *ought* to do in given situations and then elicit conformity through a complex system of social rewards and punishments.

Sexual behavior encompasses a variety of acts made up of combinations of participants, situations, statuses, and physical surroundings. Sexual intercourse by an unmarried couple might violate some people's norms. Sexual intercourse by a married couple would conform to most expectations of appropriate sexual behavior, unless the couple performed the acts in public. Some people might object to certain sexual acts, such as sodomy or sadomasochism, even by married people in private.

As these few examples show, sexual norms vary according to such factors as the relationships between the participants (although some sex acts, such as masturbation, involve only one person), the physical settings, the social situations, and the precise behaviors. Science has complicated this evaluation further by eliminating the need for sexual intercourse in procreation (since a laboratory can now perform in vitro fertilization and impregnate a woman by implanting a growing fetus). This development has challenged some sexual norms, such as those concerning sexuality in the marriage relationship that arose mainly to guarantee this once-necessary process of societal regeneration. Further, sexual relationships offer more than physical gratification to participants; they also help to fulfill desires for intimate physical contact and communication, which form part of sexual activity both outside and inside the marriage relationship. Like other relatively infrequent and temporary conditions, then, some sexual gratification comes from areas outside those prescribed by sexual norms.

Deviating From Sexual Norms

Deviations from sexual norms encompass many different types of behavior, some of them prohibited by law and some of them likely to draw other kinds of negative reactions. They share a risk of violating the norms of certain groups or legal codes or both. Many such offenses do little harm to others; in fact, people identified as victims may have willingly participated.

One convenient method for evaluating the deviance of a sexual act focuses on social reactions to the act. Laws and explicit group positions may reflect the contents of sexual norms, revealed by public expressions such as stigmas promoted through the mass media, informal sanctioning efforts by individuals, and the activities of organizations set up to promote or discourage sexuality. Actually, the term *sexual deviant* may mislead, since sex often forms a minor part of a person's total life activities for both heterosexuals and homosexuals. In fact, evaluation of the time people spend in such activities confirms that sex plays a brief role in a person's life.

Norms vary in different societies, but most judgments of sexual deviation consider a few important characteristics (DeLamater, 1981):

1. The degree of consent, such as norms that prohibit forcible rape
2. The identities of the participants, such as norms that restrict legitimate sexual partners to human beings and exclude animals
3. Relationships between participants, such as norms that restrict legitimate sex partners to people in certain age ranges and with acceptably distant kinship bonds
4. Certain kinds of acts and conduct
5. The settings in which sex acts occur

Of course, legal codes may define certain standards for sexual deviance, but these formal limits do not necessarily imply agreement by various groups in society to consider the sexual acts as instances of deviance. As with other areas, individuals and groups may disagree with the sexual guidelines established in law (McWilliams, 1993).

Many prohibitions govern sexual behavior, but they provide far from uniform guidelines in large, modern, industrial societies, such as the United States. Some norms directly prohibit certain sexual activity, while some prohibit conditions related to sexual activity. Some groups' norms prohibit many kinds of sexual and sexually related acts: forced sexual relations (forcible rape), sexual relations with members of one's own family (incest), sexual intercourse with a person under a certain age (statutory rape), sexual molestation of a child, adultery, sexual relations in a group setting with multiple partners, sex between unmarried adults, co-marital sexual relations between two or more married couples (swinging), abortion to terminate an unwanted pregnancy, deliberate exposure of one's sex organs (exhibitionism), watching others who are undressed or in the act of sexual intercourse (voyeurism), sexual relations between persons of the same gender (homosexuality or lesbianism), and sexual intercourse with an animal (bestiality). Even this long list omits many standards. For example, most people also accept normative prohibitions on public displays of the naked human body, presumably because this exposure involves display of the genital organs or women's breasts. Norms often prohibit or discourage sales of materials deemed indecent and obscene (pornography).

A simple definition might define sexual deviance as an act contrary to the sexual norms of the group in which it occurs. A number of limitations cloud this evaluation, however. Some may draw the line differently between acts only slightly at variance with applicable norms, while many may agree about other acts shockingly at variance with acceptable practices. The degree of variance imputed to an act varies with the sexual norms in different groups. For example, some groups advocate acceptance of what they call "intergenerational sex," or sex with children. The motto of the Rene Guyon Society is "Sex before eight and then it's too late." The Pedophiliacs Information Society advocates intergenerational sex at even earlier ages (Janus and Janus, 1993: 129). Such positions, of course, stimulate strong negative social opinions. A sexually oriented act that meets with acceptance in one group or subcultural context may represent a serious breach of law in another, as the following three situations illustrate (Gebhard, Gagnon, Pomeroy, and Christenson, 1965):

1. A truck driver seats himself in a booth at a roadside cafe. He gives the waitress his order, and, as she turns to depart, pats her on the buttocks. Other drivers witness this act without becoming offended, nor does the waitress object, either because she has become accustomed to such behavior or because she interprets it as a slightly flattering pleasantry.

2. The same behavior occurs in a middle-class restaurant. The waitress feels that she has suffered an indignity, and many diners deplore the gesture as an offensive display of bad manners. The restaurant manager reprimands the offender and asks him to leave.

3. A man bestows the same pat upon an attractive but unknown woman on a city street. She summons a nearby police officer as some indignant witnesses gather to voice their versions of the offense. Ultimately, the man faces charges of a sexually motivated offense.

The social circumstances surrounding an action determine whether people consider it as part of everyday, acceptable behavior, an offensive display worthy of mild disapproval, or a very discourteous act requiring bystander and police intervention. In the other words, the normative structure of the act determines society's reaction to it.

Social Change and Sexual Behavior

Sex has become a dominant aspect of life in many societies. Few other topics occupy so much of the leisure time (in fact, so much of the entire waking life and, perhaps, dreaming life as well) of such large portions of society. "Entire industries spend much of their time trying to organize presentations around sexual themes or try to hook products onto a potential sexual moment or success. That there has been a radical shift in the quantity and quality of sexual presentations in the society cannot be denied" (Gagnon and Simon, 1970b: 1).

People's willingness to approve of any given sexual act—and even the probability that they will define it as sexual behavior—has changed over time. In a national survey conducted in 1970, the Kinsey Institute found extremely conservative attitudes toward many different forms of sexual expression (Klassen, Williams, and Levitt, 1989: chapter 2). Most respondents disapproved of homosexuality, prostitution, extramarital sex, and most forms of premarital sex. Almost half of the sample even criticized masturbation.

During the past few decades, however, some indicators suggest a shift in U.S. society as well as in most European societies, toward greater openness about and tolerance toward sexual variations. Increasing tolerance seems to promote greater freedom in the mass media (especially the motion picture industry), in public discussions of sex, and in presentations of sexually explicit themes. An increasing number of plays, novels, and motion pictures have featured homosexual situations and characters. Softening sanctions apply to many forms of sex, such as premarital sexual relations and homosexual behavior, and accepted mass media portrayals now feature the naked body, often showing genitalia and pubic hair. Some groups

 In Brief   **Kinsey's Determination of Sexual Orientation**

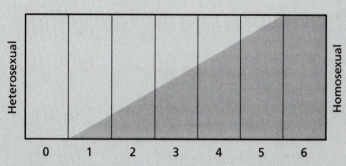

Key:

0 = Exclusively heterosexual with no homosexual
1 = Predominately heterosexual, only incidentally homosexual
2 = Predominately heterosexual, but more than incidentally homosexual
3 = Equally heterosexual and homosexual
4 = Predominately homosexual, but more than incidentally heterosexual
5 = Predominately homosexual, but incidentally heterosexual
6 = Exclusively homosexual

Alfred Kinsey defined this range of homosexual identity in his study to reflect both behavior and orientation or preference. He also considered subjects' self-definitions or self-conceptions (how they defined themselves).

Kinsey's analysis implies a range of identities for all males: 50 percent fit in Category 1, exclusively heterosexual throughout their lifetimes, while 4 percent are exclusively homosexual throughout their lifetimes. The others fall somewhere between, justifying the term *bisexual.* One may expect considerable change in the proportions over time, however, and within subgroups.

Source: Kinsey, Alfred C., Wardell B. Pomeroy, and Clyde E. Martin. 1948. *Sexual Behavior in the Human Male.* Philadelphia: W. B. Saunders: 638.

have enthusiastically welcomed such openness, while others have bitterly condemned the trend as morally unacceptable laxness. Still others regard these changes with ambivalence, perhaps feeling uneasy about the general tendency toward progressively looser standards.

Attitudes toward specific sexual practices have changed as well. People associate one set of changes, sometimes called the *sexual revolution*, with increased permissiveness concerning a number of sexual acts, including premarital sexual intercourse, cohabitation, spouse swapping, sexually explicit telephone conversations, open marriages, and sexually oriented nudity. While talk of a revolution undoubtedly overstates the social effects of increasing tolerance, attitudes about some forms of sexuality, such as sex before marriage, have certainly transformed some people's lives. After reviewing data from a long-term national survey, Smith (1990) has reported a drop in the percentage of respondents who agree that premarital sex is always wrong from 35 percent in 1972 to 25 percent in 1990; also, the percentage who express the belief that such practices are not wrong at all rose from 26 percent in 1972 to 42 percent in 1985. However, the same survey has revealed little if any shift away from disapproval of extramarital and homosexual relations.

Many factors account for changes in contemporary attitudes toward sex. Despite reports of a revival of religious feeling in recent years, patterns of church attendance show a steady decline, recently reaching a plateau (Harris, 1987: 67–71). Church attendance in the United States fell from 49 percent in 1958 to 40 percent in 1980. More refined estimates combine self-reports with observational methods to estimate an even lower figure for church attendance; one source indicates that about 20 percent of Protestants and 28 percent of Roman Catholics attend church services at least once a week (Hadaway, Marlet, and Chaves, 1993). Still, church attendance alone may not accurately indicate a general decline in religious orientation in the United States. Most people claim a belief in God and subscribe to a normative system that clearly regulates many forms of sexuality. Many people may hesitate, however, to impose their own norms on others who take different views.

Changes in sexual norms reveal themselves in actual behavior, as do other aspects of sexuality. For example, women now routinely wear pants, and men wear earrings. Not all women wear traditionally male clothing, and not all men wear earrings, but broader standards of acceptable gender-related expression—called *gender bending*—suggest a move toward individual choice. These choices are found in styles of dress, hair, and body adornment, such as tattoos. Still, some segments of the population continue to condemn all but the most traditional expressions of sexual identity. Exceptionally strict or puritanical views of sex often encourage censorship of sexual thoughts and expressions as part of a wider-ranging return to traditional practices and values.

These views contrast sharply with a widespread desire for individualized expression in personal lifestyles. Changing sexual norms play a role in a number of other developments, such as demands for equal treatment by women (the women's movement); shifting gender roles and related standards for public displays, leading to tolerance of flamboyant hairstyles and dress, and even use of cosmetics, by heterosexual men; growing recognition of the importance of women's orgasm during sex; introduction and widespread use of contraceptive devices to protect against unwanted pregnancies and sexually transmitted diseases; and greater tolerance, especially by young people, of variations in behavior once regarded as deviant. The erosion of rigid gender roles has also contributed to an increasingly relaxed sexual atmosphere.

The gradual liberalization of sex norms reflects changes in U.S. society's definition and understanding of sex (D'Emilio and Freedman, 1988). The shift in sexual attitudes reflects many broadly social forces, including the effects of urbanization, which loosened small-town social control and created new opportunities for sexual experimentation. The development of capitalism has also contributed to these changes. As the U.S. and world economies have come increasingly to depend upon production and sale of consumer goods, they have encouraged an ethic that favored consumption of those goods; this idea, in turn, has contributed to an acceptance of self-indulgence, immediate pleasure, and personal satisfaction.

Sexual norms have not changed in uniform patterns throughout society; all groups do not show evidence of the same changes. For example, some conservative political and religious groups resist the trend toward tolerance and actively espouse traditional sexual norms, which they regard as moral edicts rather than social guidelines (Janus and Janus, 1993). Further, sexual norms do not change

 Issue │ **Dancing on the Margins of Sex**

The occupation of nude dancing provides an example of behavior on the margins of sex. While nudism is not necessarily related to sexuality (see Chapter 1), there is a close connection between nudism and sexuality in "gentlemen's clubs" that provide female strippers as entertainment. These strippers are paid minimum wage or nothing at all and must make their money from customers' tips. As a result, strippers must "play the customer" and establish a relationship with customers to elicit good tips. This relationship, which can be termed "counterfeit intimacy," is necessarily temporary and is designed to give the illusion that sexual intimacy is possible.

The object of the interaction is different for both the stripper and the customer. The stripper must put forward the idea that she is sexually available and maintain this posture as long as possible. Customers, on the other hand, may be genuinely interested in sex with the stripper or with at least being aroused by her, but customers are also interested in paying as little as possible during the interaction.

Many of the customers of nude dancing clubs appear to be engaging in vicarious sexual arousal that is similar to that produced by pornographic pictures and movies. As with other forms of pornography, stripping is unconnected to the development of close relationships, authentic affection, and personal intimacy. Sex can occur between stripper and customer, but its cash basis likens this behavior to prostitution, not relationship building. The monetary basis for the relationship is inescapable because the stripper will not perform without the prospects of tips, and the customer expects to have to pay for her visual or physical services.

Source: Forsyth, Craig J., and Tina H. Deshotels. 1997. "The Occupational Milieu of the Nude Dancer." *Deviant Behavior*, 18: 125–142.

quickly, and developments like these always meet resistance. Major social changes require time and an attitude of tolerance, two major preconditions for all normative changes. Thus, while one can relatively easily document changes in the contents of sexual norms, one can trace the progress of the change—how it originated, who advocated it, the relative political power of interested groups, and how it eventually defined a permanent new norm (if it did)—only with difficulty.

Remember these warnings about intergroup variations in attitudes while reading the sections that follow, which discuss several forms of sexual behavior now regarded as examples of deviance. The necessarily selective discussion largely emphasizes areas for which sociological research supplies a relatively solid basis for conclusions regarding theory and social policy.

Selected Forms of Sexual Deviance

Clearly, the meaning of *sexual deviance* varies in different situations. Society may or may not regard sexual intercourse between a male and a female as deviant behavior depending on such factors as the ages of the partners, their marital status, and the time and place of the act. Heterosexual deviance includes acts like premarital and extramarital sex and prostitution. The strength of one observer's perception of sexual deviance in a given act depends on the strength of the norm governing that behavior. Because such evaluations differ, people often dispute about how deviant they regard some acts, such as premarital sex and extramarital sex.

Extramarital Sex (Adultery)

The marriage relationship establishes one of the most important contexts for sexual norms. Those norms sometimes permit certain sexual activities between married couples that they sanction outside marriage. *Extramarital sex* refers to sexual behavior by a married person with someone who is not his or her spouse.

In 18 B.C., the Emperor Augustus turned his attention to social problems at Rome. Extravagance and adultery were widespread. Among the upper classes, marriage was increasingly infrequent, and many couples that did marry failed to produce offspring. Augustus, who hoped thereby to elevate both the morals and the numbers of the upper classes in Rome and to increase the population of native Italians in Italy, enacted laws to encourage marriage and having children *(lex Julia de maritandis ordinibus)*, including provisions establishing adultery as a crime. The law against adultery made the offense a crime punishable by exile and confiscation of property. Fathers were permitted to kill daughters and their partners in adultery. Husbands could kill the partners under certain circumstances and were required to divorce adulterous wives. Augustus himself was obliged to invoke the law against his own daughter, Julia, and relegated her to the island of Pandateria.

Norms against adultery have changed since that time. Currently, no society's norms give its members complete freedom to engage in extramarital sex, although some tolerate such liaisons more easily than others do. In the Kinsey Institute survey in 1970 (Klassen, Williams, and Levitt, 1989: chapter 2), respondents expressed the strongest disapproval for any form of heterosexuality in reactions to questions about extramarital sex; 87 percent of them considered adultery as an offense. Respondents' evaluations did indicate softer attitudes toward participants who were in love. They disapproved less strenuously of extramarital sex between people in love than of purely sexual relationships outside marriage.

People often describe extramarital sex as an *adulterous relationship* or *having an affair.* An affair may or may not include sex, however; it can involve only a romantic or platonic relationship. One report described a careful distinction by married people who were having affairs between dating relationships and sexual ones (Frost, 1989). These participants associated the term *cheating* with sex but not with dating relationships. However, the participants also would not inform their spouses of the relationships, despite the lack of sexual involvement, suggesting that they anticipated negative reactions.

In the 1950s, Kinsey and his associates (1948, 1953) reported that about one-half of all married men and one-quarter of all married women engaged in extramarital sexual relations at some time. Hunt (1974) found that the figure for males had not changed much after 25 years. (In fact, it had fallen slightly from levels reported by Kinsey to 41 percent.) Wives, however, had increased their involvement in extramarital affairs. In 1983, *Playboy* magazine reported in a series of issues throughout that year on the results of a survey of 100,000 readers. Although these subjects did not form a representative sample of the American public as a whole, the *Playboy* survey reported that 36 percent of the married men and 34 percent of the married women admitted to extramarital affairs. In fact, among the young age categories (people 29 years of age and younger), more females than males reported involvement in extramarital affairs. A different survey that year reported that 30 percent of the men and 22 percent of the women in a sample of couples married 10 or more years reported having participated in at least one affair (Blumstein and Schwartz, 1983). Interpretation remains difficult, however, because none of these estimates, including Kinsey's, reflects answers from random samples of the population.

The Janus report, on the other hand, gave results of a national survey of 2,765 adults. Of these respondents, 65 percent of the married men and 74 percent of the married women declared that they had never had an extramarital affair (Janus and Janus, 1993: 196). Conversely, 35 percent of the men and 26 percent of the women admitted having experienced sex with someone other than their spouses at least once while married.

The most extensive survey of sexual behavior, the National Health and Social Life Survey (NHSLS), obtained information from 3,400 randomly selected respondents in 1992. Of this group, only about 16 percent of the men indicated that they had sex with two or more extramarital partners (Laumann, Gagnon, Michael, and Michaels, 1994: 208). In all age groups, the survey found only limited evidence of extramarital sex. Over 90 percent of the women and over 75 percent of the men reported fidelity over the entire course of their marriages (Laumann et al., 1994: 214).

The reasons for engaging in extramarital affairs vary by gender. The *Playboy* survey found, in order of frequency, that females often felt that affairs offered reassurance of their desirability, better sex, a change of routine, sexual variety, and sex without commitment. Males, on the other hand, participated in affairs for sexual variety, reassurance of their desirability, a change of routine, better sex, and sex without commitment. One of the Janus respondents, a self-employed business-man in his 60s, articulated yet another reason: "What kept me [married to my wife] were other women. I really believe that more marriages have been kept to-gether by extramarital affairs than by all the marriage counselors in the country" (Janus and Janus, 1993: 197).

Participants in extramarital sex differ from faithful married people in a number of respects (Buunk and van Driel, 1989: 102–105). Males in general express more permissive attitudes than females do about extramarital sex. Older people tolerate sex outside marriage less often than younger people do, and members of the upper-middle class with higher educational achievement report more permissive attitudes than do members of other classes, although people from all social classes participate in extramarital sex.

Studies reveal a link between dissatisfaction in one's marriage and relatively permissive views of extramarital sex. They also show a strong relationship between actual involvement in extramarital sex and approval of this behavior by friends and acquaintances. The frequency of affairs also bears a relationship to the proportion of one's reference group who have actually had such affairs. Physical opportunity, such as temporary separation between a married couple, also plays a role.

Sex and the Computer

New forms of sexual activity have accompanied changes in technology and lifestyles. Heightened concern over sexually transmitted diseases and a general in-terest in sexual variety can lead some people to experiment with new forms of sex. Phone sex, for example, is a more recent variation of techniques to provide remote sexual stimulation, either within or outside of marriage. The NHSLS survey found that only 1 percent of the men and almost none of the women in the sample had ever called a telephone sex number (Laumann et al., 1994: 135). Those who do complete a relatively simple process: By calling such a service (and paying a fee), one gains access to another person to discuss one's own sexual fantasies and/or to hear sexy talk from a stranger. The auditory incitement, coupled with an active imagination, can produce sexual satisfaction, either by itself or in con-nection with masturbation. Some services ask callers to supply credit-card num-bers for payment, while others accept calls over 900-type numbers that direct charges to callers' telephone bills.

Phone sex owes its popularity to the advantages of this form of sexual stimula-tion: strictly impersonal encounters under the control of the callers. The parties exchange no commitments, they feel no need to "perform" or play unwanted roles, they need not engage in foreplay, and they maintain complete control of the nature and length of the encounter. Furthermore, this form of sex allows participants to escape much of the guilt and worry that personal encounters may involve; since the parties experience no physical contact, they take no risk of contracting sexually transmitted diseases, including AIDS. This potential for sex without touching may seem attractive to individuals who might feel intimidated at the prospect of visiting a sex club or having an extramarital affair.

But it is the computer that has changed substantially the nature of impersonal sex. In the personal computer, the pornography industry found a new medium through which to distribute its product. As personal computer systems equipped with communications hardware and software became widely available, people gained capabilities for quick, private communications. They also gained the ability to communicate with other computers called *servers*, which functioned as reservoirs of information. Some servers contained sexually oriented messages exchanged between computer users or digital pictures of pornography. Because the Internet has continued to function as an open link to all kinds of information, many people have gained unregulated access to these messages and pictures. Frequently relying on voluntary compliance, the server operators have asked people under the age of 18 to avoid accessing the messages or pictures. Few believe in the effectiveness of such voluntary compliance.

Net access has become widespread not only in the United States, but in other countries as well. A U.S. computer user may easily access a server in another country. The user may not even recognize this international link, given the speed of modern computer and communications equipment.

Observers have coined the term *cybersex* to denote the use of computers to flirt, exchange romantic messages, and even acquire sexual satisfaction—all online. "I have heard from many of my clients," reports one social worker, "that their husbands are spending too much time on the Internet and they're worried about them having affairs" (Charkalis, 1996).

People do develop relationships over the Internet, despite well-publicized dangers in the practice. Participants in cybersex usually meet in "chat rooms" on the Internet. In the course of these conversations, they adopt aliases, called *nicks* (short for *nicknames*) to allow others to identify them while preserving their anonymity. Most of the exchanges involve text-based messages typed at computer keyboards. Some cybersex participants post messages on bulletin boards instead, but several software programs allow truly interactive, real-time conversations. People who might fear personal contact often prefer the safety of on-screen conversations from their homes with others thousands of miles away. New video technology also allows people to see, hear, and talk with others over the Internet.

Interactive exchanges through chat rooms give participants the impression that they are learning about other people, and sometimes this assumption correctly states their relationship. People can share ideas, jokes, feelings, and the like over the computer, but they can learn only so much about others with whom they interact. Without physical presence, people do not see one another's body language, gestures, surroundings, and other nonverbal communication cues. As a result, carefully managed keyboard communication can convey desired—and possibly misleading—impressions. People can really be whoever they want to be on the

Internet, assuming whatever name, physical appearance, social status, and other characteristics they choose.

People who meet initially on the Internet sometimes arrange for physical meetings. Many parents worry that their children will communicate with people who pretend to inaccurate identities, especially since most such communications mask the age and sex characteristics of participants, allowing dishonest people to lure innocent chat room participants into dangerous situations. Prodigy, an online computer service through which people connect to the Internet, conducted a poll of 6,000 members; 47 percent reported participating in cybersex, and 39 percent said they had pretended to be someone or something else (Charkalis, 1996).

Sex Work and Prostitution

There are a number of different kinds of sex work, including exotic dancing, erotic massage, escort services, telephone sex operations, prostitution, and any other type of work that deals with impersonal sexual activity. The term *sex work* highlights the commercialized nature of some forms of sex. In the discussion that follows, the terms *sex worker* and *prostitute* will be used interchangeably.

Prostitution seems to appear in virtually all societies, but at the same time, most groups disapprove of the practice. The extent of prostitution and people's reactions to it have fluctuated over many years, but the essential facts surrounding exchanges of sex for money have remained the same. A basic definition of *prostitution* identifies it as promiscuous and mercenary sexual behavior with emotional indifference between the partners. A more precise definition becomes difficult because conceptions of this activity vary (Aday, 1990: 104). Some regard prostitution as a dehumanizing indulgence and a gross violation of the laws of nature, while others regard it merely as an alternative occupation that may actually perform a useful service to society. Prostitutes do exchange sex for money, but varying definitions of *sex* might lead people to include other activities in the same category, such as acting in a sexually explicit movie, visiting a sex therapist with one's spouse, and paying for a sex massage.

Prostitution is a publicly created category that is defined by social control efforts more than by the behavior itself. That is, public concern and moral panic concerning prostitution is more a function of arrests and media portrayals than it is the acts of selling sex themselves (Brock, 1998). Prostitution is generally a crime, but in the United States and most other countries, laws prohibit not the sex acts themselves but solicitation to perform sex acts in exchange for money or other things of value. Although some prostitutes may choose customers selectively on the basis of age, sex, race, or physical attractiveness, most engage in sex with anyone who can afford to pay their prices. Most of these exchanges result in only temporary and impersonal relationships, but some prostitutes develop stable interactions with certain clients. Prices vary for different sexual acts, possibly including both oral and anal sex as well as sadistic, masochistic, and exhibitionist acts in addition to traditional forms of heterosexual relations. Clients may also pay prostitutes to engage in role playing (pretending to be exotic characters) without intercourse or other sexual activity, and prostitutes may provide other services in addition to sex, such as simply listening to the troubles of clients.

Sex work is linked to several important values in American society. The general culture stimulates images of the importance of sexual values in life, but many men have trouble satisfying the resulting values or desires. Advertising messages often incorporate sexual references to attract attention, and sexual references and examples abound in the mainstream mass media, let alone in explicitly pornographic material. As sexually stimulating input floods people from a number of everyday sources, it clearly promotes reactions expressed in needs for overt sexual behavior. Individuals cannot always act out their sexual feelings as they wish, however, and prostitution may serve as an outlet for expressing these desires.

Prostitution also shares some characteristics with the crime of forcible rape (Meier and Geis, 1997). Both reflect illicit sex-related behavior, both frequently involve brutality along with the sex itself, and both can occur in an atmosphere of humiliation of female participants. Neither behavior views women participants as people with needs of their own for sexual satisfaction or as people likely to achieve such satisfaction. Neither places demands on the male partners for sophisticated sexual performance. Perhaps because of these connections, at least one study has found that prostitutes become rape victims at high rates (Miller and Schwartz, 1995). A number of myths surround rapes of prostitutes, including the idea that forced sex with a prostitute does not constitute rape, that such an act causes no harm when the victim is a prostitute, and that prostitutes in some sense deserve to be raped.

In a sense, no great mystery obscures the motivations of most clients. Prostitutes provide easy and certain sex without later interpersonal responsibilities (Kinsey et al., 1948: 607–608). Such an exchange may offer relief from an unsatisfactory marital relationship without the entanglements of an extramarital affair. Moreover, prostitutes may provide services that clients cannot obtain elsewhere. Some clients patronize prostitutes in pursuit of variety in their sexual contacts, and others wish some special service other than regular intercourse (James, 1977: 402–412). Some clients' physical characteristics, such as disfigurements, limit their access to other sexual partners. Also, the simple idea of hiring a prostitute induces sexual excitement in some men by fulfilling a fantasy.

Not all clients, however, patronize prostitutes strictly for motivations of physical sexual attraction. Holzman and Pines (1982) interviewed 30 clients to determine the meaning they found in this kind of sexual activity. Many of them reported more concern about the prostitute's personality than her physical appearance. They looked for "personal warmth and friendliness. Although they wanted to pay for sex, it seemed that they did not want to deal with someone whose demeanor constantly reminded them of this fact" (Holzman and Pines, 1982: 112). Among other things, information from clients characterizes prostitution as a desired service that inspires widespread demand, sometimes constrained by limited supply.

Nature and Extent

The basic idea of prostitution involves exchanges of sex for money. Only the client's imagination and financial resources limit the available range of sexual acts.

Clients, of course, want as much as they can get for the smallest possible price. Prostitutes effectively operate businesses; they share the attitude of most business-people that "time is money." If they spend a long time with each client, they can offer less time to additional clients. To maintain the largest possible incomes, therefore, they try to maintain the shortest possible contacts with clients, unless those clients offer to pay for the extra time.

One study asked a sample of 72 southern California prostitutes about their activities and incomes (Bellis, 1990). Altogether, the sample served about 560 customers a day, ranging from 2 to 30 each, with an average of 8 clients every day. Fees ranged from $20 to $100, the most common being $30. The most popular act was a combination of intercourse and fellatio (called *half and half*), the choice of 75 percent of the customers.

Some prostitutes earn all or most of their incomes from prostitution, while others work only part-time. Part-time prostitutes generate additional income from a range of sources, such as welfare, part-time jobs, and other illicit activities. Prostitutes can engage in this activity selectively, perhaps participating only for short periods of time. They may work for a time and then quit, returning to prostitution only when economic necessity compels them. During the interim, they may work at legitimate jobs.

No one can determine for certain exactly how many people obtain at least part of their income from selling sexual services, but some have made estimates. A 1971 study of 2,000 prostitutes estimated that this activity involves between 100,000 and 500,000 women in the United States who earn over $1 billion each year (Winick and Kinsie, 1971). A book on deviance has confirmed the 500,000 figure (Little, 1983: 35). While such figures represent only guesses, they may give the most reliable available estimates. Arrest statistics provide another source of information. In 1998, U.S. police agencies arrested 94,000 people for prostitution and commercialized vice (Federal Bureau of Investigation, 1999: 210). Most of these arrests took female prostitutes into custody, but a number of them apprehended males. Of males arrested, some faced charges for offering prostitution services, but the formal definition of the crime—solicitation—applies to clients of prostitutes as well, and most males arrested fit in this category. Most prostitution arrests targeted people in their early 20s, and about 60 percent apprehended white offenders.

While arrest data provide some statistical indicators and suggest some relationships, they prove notoriously poor indicators of the number of prostitutes actually operating. Police arrest some prostitutes only once, while they frequently haul others to jail several times. Moreover, they arrest some prostitutes under charges other than solicitation, such as vagrancy, disorderly conduct, loitering, or other sex offenses. Prostitution arrests also vary substantially, depending on periodic decisions by police to crack down on this activity for short periods of time. These sweeps produce high arrest rates, perhaps creating the appearance of an increase in prostitution when arrest figures really indicate nothing more than increasing police attention.

Janus and Janus (1993: 347) have estimated the prevalence of prostitution and concluded that more than 4.2 million women in the United States aged 18 to 64 may have engaged at some time in their lives in selling sexual services for money. Adding "baby pros" (explained in a later section) and others not included in the basic estimate could boost the figure as high as 5.0 million.

Although prostitution activity remains extensive, it appears to have declined steadily over the past four or five decades, except for periodic increases in wartime. More than 50 ago, Kinsey and his associates (1948: 597) estimated that prostitution accounted for less than 10 percent of total nonmarital sexual activity by males, and the figure is undoubtedly much lower today. The Kinsey study further reported that prostitutes participate in not more than 1 percent of extramarital sexual intercourse. In another report, Kinsey and his associates (1953: 300) estimated a drop by 1050 in the frequency of visits to prostitutes by American males to about one-half of comparable levels prior to World War I. Much of this decrease in patronage of prostitutes seems to have resulted from increases in sexual freedom for women, allowing greater sexual access for males. Still, a large number of men have participated in commercial sex. Janus and Janus (1993: 348) have estimated that as many as 20 percent of all adult men have had some experience with prostitutes.

One can evaluate the extent of prostitution in another way, by estimating the number of clients that prostitutes serve. The NHSLS survey, mentioned earlier, found that only 16 percent of men reported that they had ever paid for sex and that this percentage declined in successive age groups (Michael, Gagnon, Laumann, and Kolata, 1994: 63). That is, older men reported having paid for sex more often than younger men did.

International Dimensions of Sex Work

International variations of sex work expand on the forms of prostitution commonly known in the United States (Davis, 1993). Illicit trafficking of girls and women from country to country produces less visible criminal activity than trade in other illicit commodities, such as drugs, but it nevertheless pervades many regions. Organizations or individuals may coerce women, persuade them with offers of money, or just kidnap them against their will and then send them to other countries. These traffickers resell the women to private parties as sexual slaves or to others who run prostitution services for their own profit.

Women co-opted in this way usually recognize later opportunities to escape from those who control them, but often they do not escape. Many need what little money the organizers might allow them, and others follow their socialization to accept the decisions of men, even if those decisions openly contradict the women's own self-interest. Government inattention has allowed the problem to continue. As one observer has explained:

> Usually, many factors coalesce to create conditions of female sexual slavery. Often but not always, the conditions of poverty combine with female role socialization to create vulnerability that makes young girls and women susceptible to procurers. Social attitudes that tolerate the abuse and enslavement of women are reinforced by governmental neglect, toleration, or even sanction. At levels of

government and international authority where action could be taken against the slave trade, one finds at best suppression of evidence and at worst complicity in it. (Barry, 1984: 67)

Some women, particularly those who are widowed early or from poor countries such as India and Southeast Asian nations, may see prostitution as an opportunity to make money for their families. Sometimes, relatives of the girls and women encourage this attitude, and some sell female relations into sexual slavery for initial payments well as a percentage of later profits. The number of children engaged in prostitution is hard to estimate, but observers have reported the following figures: 800,000 underage prostitutes in Thailand, 400,000 in India, 250,000 in Brazil, and 60,000 in the Philippines. One source estimates that more than 90,000 underage prostitutes provide sexual services in the United States (Serrill, 1993). Child prostitution commonly occurs in many Asian countries, and it may be increasing, in part because of the mistaken belief of clients that children are less likely than adult women to have AIDS. Yet, one survey found that more than 50 percent of the child prostitutes in Thailand were HIV positive (Serrill, 1993).

In Thailand, many child prostitutes work unwillingly in a large and booming sex industry (Hodgson, 1995). The country has gained a reputation for extensive child prostitution, and many visiting businesspeople, tourists, and other foreigners ask about the practice. "It is an accepted activity for even respected businessmen, just as frequenting brothels is accepted pastime" (Kunstel and Albrights, 1987: 9). Girls under 15 years of age, and some 11 years old and younger, work as prostitutes in Thailand. Poor girls from the economically deprived northern part of the country experience pressure to go south to engage in prostitution with the expectation that they will send their earnings back to their families.

Some houses of prostitution do not permit the girls to leave the premises, requiring them to remain always "on call," waiting for a customer. When a customer arrives, a host may ring a bell, calling the girls to leave whatever they are doing and assemble in a main room. There, they arrange themselves before the customer in a specially lit area to permit easy visibility. The customer calls out one of the numbers pinned to the girls' clothing to select a particular prostitute.

A sample of prostitutes in India reported that many never-married women chose this way of life because it suited their particular goals and interests. Women who had married became prostitutes because their in-laws had refused to support them after the deaths of their husbands (Chattopadhyay, Bandyopadhyay, and Duttagupta, 1994). Prostitution in the Sudan, on the other hand, seems related to modernization processes. Forces of economic development, natural catastrophe, and civil war have uprooted individuals from rural society and freed them from traditional restraints on prostitution (Spaulding and Beswick, 1995). Prostitution also increased in Turkey after the opening of its border with the former Soviet Union, reflecting a similar assault by circumstances on traditional moral codes (Beller-Hann, 1995).

The 1990s witnessed an increase in attention to international sex work. Some of this attention was from law enforcement agencies, but much was from human rights and feminist groups concerned about the coercive aspects of sex work and extending basic human rights to sex workers (Kempadoo and Doezema, 1998).

In Brief **Organizations and Groups Concerned With International Sex Work**

Many organizations have arisen in the 1990s that deal with sex work. Many of these are advocacy groups for the rights of sex workers, not only in countries with emerging economies but in developed countries as well. Some of these groups operate within specific countries, while others are more international in scope. Below is a sample of such organizations:

- The Coalition Against Trafficking in Women
- End Child Prostitution in Asian Tourism
- The Global Alliance Against Traffic in Women
- Indonesia Resources and Information Project
- International Women's Development Agency
- Joint Committee on the National Crime Authority (Canberra)
- Prostitutes' Collective of Victoria (Melbourne)
- Prostitutes' Rights Organization for Sex Workers
- Queer and Esoteric Workers' Union
- Sex Workers Outreach Project
- Workers in Sex Employment in the ACT (Canberra)
- Network of Sex Work Projects
- Sex Worker Education and Advocacy Taskforce
- Sex Workers! Encourage, Empower, Trust, and Love Yourselves!

Source: Kempadoo, Kamala, and Jo Doezema, eds. 1998. *Global Sex Workers: Rights, Resistance, and Redefinition*. New York: Routledge.

Such groups recognize that sex work reflects both voluntary and involuntary dimensions, with some workers not willingly choosing to participate in sex work and others engaging in sex work as an occupational choice. Even the term *sex work* emphasizes that prostitution is a job, much like any other, and that people who participate in it often freely choose to do so.

Types of Prostitutes

One can classify prostitutes according to their methods of operation, the degree of privacy their work allows them, and their incomes. These criteria define three major kinds of prostitutes as well as several collateral types who work in particular settings.

A *streetwalker*, the most common type of prostitute, solicits directly for clients in a relatively public place, such as a street corner or bus station. This type of prostitution goes on mainly in large, urban areas, often compiling long arrest records for participants. Streetwalkers often provide the cheapest available prostitution services. They generally operate within established territories and develop contacts among members of the local communities, such as desk clerks at cheap hotels or motels.

A *bar girl* also solicits clients in a public place, but one that offers more protection from public view than a street corner allows. Some of these prostitutes solicit clients alone. Others work with pimps or, more usually, bar employees such as bartenders. A bar girl's success usually depends more on physical attractiveness than that of a streetwalker, because she must build up business from the relatively small number of potential clients who visit the tavern.

A *call girl* enjoys the highest status of any type of prostitute. She works out of an apartment or hotel room and takes clients strictly through referrals from known and trusted sources. While she may serve customers in her home, usually she meets clients in hotel rooms or their own residences. Earning the highest incomes of any type of prostitute, call girls also enjoy the greatest immunity from arrest and the stigma of prostitution. A particularly successful one may build up a regular, wealthy clientele, perhaps by providing interpersonal services other than sex such as "counseling" or just listening. Call girls represent an increasingly common type of prostitute. Their methods of operating ensure relative privacy from the police and little attention from conventional people. This type of prostitution also suits part-time work better than other types do.

In addition to streetwalkers, bar girls, and call girls, other types of prostitutes work through massage parlors, photographic studios, or commercial escort agencies. Not all women who work in such places are prostitutes, but these businesses offer covers for conducting prostitution along with legitimate services. Some prostitutes travel with a specific group of clients; sometimes called *road whores,* such women often cater to working-class migrant labor camps or visitors to urban conventions (James, 1977). Some reports cite cases of prostitution associated with truck stops, serving long-distance truckers and other travelers who recognize the activity (Aday, 1990: 113–115; Diana, 1985). Prostitutes who work through escort services can choose when they take clients and for how long, thus providing greater flexibility than other prostitutes experience.

One report has also identified three kinds of prostitutes who work in business offices (Forsyth and Fournet, 1987). None fit the image of streetwalkers, but they nevertheless use sex to further their economic ends. *Party girls* have sex with clients for money. *Mistresses* form sexual relationships with their bosses, motivated principally by the desire to ensure job security. A *career climber* extends this idea by forming continuing sexual relationships with a series of bosses in an effort to promote her own career mobility and advancement.

Organized houses of prostitution once flourished in so-called *red-light districts* of many cities, but they have become uncommon today (Heyl, 1979). Nevada is the only U.S. state with laws that allow prostitution as a local county option. There are presently 35 licensed brothels in 10 counties of Nevada (Housebeck and Brents, 2000). Customers of a typical establishment travel outside town to a rural trailer complex called a *ranch.* The women who work there may pay up to 50 to 60 percent of their income to the brothel owner. One of the best-known of these brothels, the Mustang Ranch, filed for bankruptcy in 1990. Thousands of souvenir seekers attended an auction that sold off the brothel's assets (pictures, furniture, etc.). Most of the prostitutes come from outside the area of the brothel, and there is considerable movement by prostitutes between different brothels, depending on the volume of work. This type of legalized prostitution, however, probably never will completely replace illegal prostitution within Las Vegas and Reno, if only because some customers will probably always want prostitutes to visit their hotel rooms.

Many other cities throughout the world also allow legal prostitution, among them Bombay, India. The best-known examples in the Western world are Hamburg, Germany, and Amsterdam, the Netherlands. In Amsterdam's red-light district, prostitutes work from small apartments with street-level windows through which they can see and be seen by potential customers. They charge largely standardized fees, and the facilities assure their personal safety. The women require their clients to wear condoms, and the women submit to regular health inspections. The prostitutes look upon their work as a business, forming a union and expecting certain recognized employee benefits.

Some U.S. observers have advocated a similar system to avoid the problems of unregulated prostitution. These problems include potential for corruption of young females, the spread of AIDS, and a substantial risk of physical abuse of prostitutes by their customers. Although some criticize such legalized prostitution on the grounds that it degrades women, supporters maintain that a woman should enjoy the right to use her sexual capabilities commercially or noncommercially, as she herself wishes.

Prostitution and Deviant Street Networks

Prostitution does not take place within a social vacuum. Instead, discernible sets of relationships called *deviant street networks* provide the settings for much prostitution (Cohen, 1980). These networks reflect links among people engaged in a number of illicit activities, or *hustles*, prostitution among them. Other network activities might include petty theft, forgery, credit-card fraud, embezzlement, drug trafficking, burglary, and robbery. The prostitutes work their own hustles, including selling sex, within these networks. They engage in whatever activity seems likely at the time to yield the most income. Generally, however, most networks specialize in only one or two of these crimes (Miller, 1986: 36). Members of the street network share information about criminal opportunities, and they generate support, self-esteem, and courage for participants.

Miller (1986) has studied a series of deviant street networks in Milwaukee, and Cohen (1980) has studied similar networks in New York City. Most networks operated under the control of black males who typically already had compiled extensive police records. These men did not act formally as pimps for the women. Each man formed a group somewhat like a pseudo-family with one, two, or three women who worked at prostitution. These men functioned more like "deviant managers" than pimps (Cohen, 1980: 55–59), working closely with the women to provide on-the-spot protection and supervision. Separate networks operated relatively autonomously from one another.

These networks recruit women as prostitutes from a number of sources. Black women who join such groups largely come from family structures and backgrounds that resemble the deviant street networks in some respects. Many leave households composed of kin, near-kin, and unrelated persons who live together mainly for economic reasons (Angel and Tienda, 1982; Stack, 1974). Females head most of these family groups, which lack traditional sources of family authority. Some black females actually begin working various hustles before they enter the deviant street networks of prostitution (Valentine, 1978). At 16 or 17 years of age and against their parents' wishes, these girls become acquainted with activities like shoplifting. Later, they might welcome invitations to participate in prostitution and other illegal activities.

> Often the initial recruitment as well as recruitment thereafter is described by women as rather low-key and offhand. Just as someone might approach a group chatting in a kitchen to ask for help in moving a newly acquired couch into an apartment in exchange for a beer or simply as a gesture of friendship, so might someone be asked to help lug copper tubing pilfered from an abandoned building in exchange for a share of the profits or to help sell some stolen merchandise or some marijuana in exchange for a portion of what was being sold. (Miller, 1986: 79)

The deviant street networks recruit other girls from the city's population of runaways to become prostitutes. The group offers these girls shelter and money in exchange for their labor. They become part of the extended "families." Along with prostitution, the women within such a network may also engage in other forms of crime, such as theft and drug dealing. In this manner, prostitution goes on not by itself, but as part of a larger web of illegal activities and participants in the deviant street networks.

These networks have changed somewhat with the increase in drug trafficking in some urban areas. Links between prostitution and drug sales in some communities have brought prostitutes into more frequent contact with people in drug subcultures than with those in criminal subcultures (Miller, 1995). Individual prostitutes initiate sex-for-crack exchanges more often than do those who participate in larger webs of illegal activities.

Becoming a Prostitute

Hollywood has provided a caricature of prostitution in the form of the "happy hooker myth" (Davis, 2000: 139). In this myth, a prostitute is a sexy, "pretty" woman who freely enters prostitution until the right man comes along. In the meantime, she enjoys high income, freedom, attention, and high earnings. She is portrayed as being empowered by her position and having an edge over men by reversing traditional gender roles. She is freed from having to report to a boss and from working a standard 40-hour-a-week job filled with the numbing repetition of routine. But the myth does not approach the reality of the lives of the great majority of prostitutes.

Youth and some physical attractiveness do play important roles in the success of a prostitute, so most are between the ages of 17 and 24 or so. They reach their peak earning ages probably at around 22. Some older women continue to work as prostitutes, but most of them remain in the activity for special reasons, such as raising funds for drug addiction, alcoholism, or some other expensive habit. Most prostitutes appear to come primarily from the lower socioeconomic classes, many from inner-city areas. Still, no evidence indicates that they enter this profession simply to escape poverty, although most share a desire to improve their economic positions. At one time, a disproportionate number of prostitutes came from disadvantaged foreign-born groups; today, a disproportionate percentage comes from racial minority groups. Prostitutes clearly pursue economic motivations and engage in this activity mostly for the money.

Some 50 or 60 years ago, many accepted the image of prostitutes as victims of "white slavers" who had induced sexually inexperienced women to enter the profession against their will. Studies indicate, however, that the process of becoming a prostitute differs in many ways from this old stereotype. A study of 30 prostitutes, for example, found several important precursors to this role: early sexual activity, a history of school delinquencies, and, frequently, prior commitments to training schools for girls (Davis, 1981). The mean age of first intercourse for this sample was about 13 years old, although this fact in itself does not determine a life of prostitution. Many women who become prostitutes experience geographic mobility and disruption prior to taking up this deviant identity, and a number of prostitutes report past family difficulties such as divorce. While the transition to prostitution may bring substantial trauma for some women, others may find much more stable lives than the ones they leave behind (Gagnon and Simon, 1970a: chapter 7).

Those sex workers who connect with clients through the Internet have a number of advantages that make this kind of work attractive to them. There is no violent control of women, and the sex workers don't have to fork over any profits to a middle-man (pimp). They operate through the security of traceable e-mail and high-end hotels, and the prices they charge virtually guarantee a certain kind of "respectable" client (Shuger, 2000). The combination of autonomy, security, control, wealth, and lifestyle represents a major inducement to engage in this kind of work.

Child Prostitutes   Child prostitutes, or "baby pros" (Bracey, 1979), participate in a little understood and studied area of prostitution. Inciardi's (1984) study of child prostitutes aged 8 to 12 has indicated that the girls often receive their introductions to this activity through their parents or other family members. None of the girls worked full-time as prostitutes, and all were attending elementary school at the time; none were runaways. Their backgrounds frequently included casual nudity, sexual promiscuity, pornography, and prostitution by family members. One of the baby pros explained:

> My sister would take me to work with her [at a massage parlor] sometimes when she couldn't get a baby sitter. I can't remember the first time I saw a dude get on top of her, but it didn't seem to bother her. She said it was fun and felt good too. (Inciardi, 1984: 75)

In general, Inciardi had described an almost cavalier attitude by these girls about participating in prostitution. They generally continued their involvement because they earned a lot of money. Also, many seemed to fear rejection by their families if they stopped, especially those whose parents were involved in producing pornography. "Now I'm used to it," one girl reported, "and the spending money is real nice" (Inciardi, 1984: 76).

A similar pattern seems to carry children into prostitution throughout the world: These children run away from home to escape physical abuse, neglect, or other undesirable family conditions and then become prey for pimps or other criminals, who divert the children for their own illegal purposes.

Adolescent Prostitutes   Adolescent prostitutes generally begin their careers at about the age of 14 (Weisberg, 1985: 94). Most come from dysfunctional homes marked by family separation, divorce, conflict, weak parental affection, and substantial sexual abuse. The backgrounds of many adolescent prostitutes reveal past trouble in a number of social settings: in school, at home, in the community. They seem to drift into this activity through a pattern of contributing behaviors. Living on the margins of society, some girls may associate with others with similarly marginal or stigmatized traits. While a number of these girls report unconventional early sexual experiences, little evidence suggests that these experiences determined their later deviance. One prostitute described the significance of her initial sexual experience with her father:

> You don't have words to say how it felt. I really didn't feel anything. But your own father doing that to you—I felt like dirt for years. It didn't hurt but it didn't feel good either. At that age [11 years old] I wasn't sure exactly what was going on. . . . was no big deal. I didn't exactly like it but I didn't exactly not like it either. After we did it for a while . . . it still was no big deal. (Diana, 1985: 65)

Most adolescent male prostitutes enter into prostitution in a similarly unplanned, almost accidental way. Some describe the process in almost fatalistic terms. One youth recalled his first experience:

> Then this man came up and he offered me $25 if I would do his little thing with him. And I said "sure." That was the first one I ever did. I thought, "that was great for 20 minutes of my time, to get that and then go party on something." (Weisberg, 1985: 52)

Two distinct subcultures create varying environments for male prostitution. Within a peer-delinquent subculture, prostitution forms one of many illegal activities as part of a larger routine or lifestyle of hustling. Within the gay subculture, however, male prostitution represents more of a participant's identity than simply performing sexual acts to generate income. Almost all clients of male prostitutes from either subculture are other males. Some men and adolescents work as male prostitutes only occasionally, while others participate actively in inner-city street life and prostitution (Weisberg, 1985: 40). In one study of male prostitutes, only 8 of 35 street hustlers described themselves as exclusively homosexual; the others, like their female counterparts, participated for the money (McNamara, 1994).

This is the case in other countries as well. A study of male prostitution in Costa Rica reported that most of the young men were neither homosexual nor bisexual (Shifter, 1998). Most were not street kids, but most were products of one-parent households with a history of abuse and alcoholism.

Prostitution by females shows a strong association with running away from home and needing to support oneself at an early age. A 16-year-old female explained her situation this way:

> I left home in the first place because I was being hassled all the time about comin' home stoned, about my dates, the way I looked, missing school, and all the rest. I can't blame them for bein' upset, and in some ways it was easier bein' home. . . . Now it don't matter what time I come in, or who I bring in with me, but I gotta be having a lot more johns [customers]. Everything is so darn expensive. Forget about the drugs, just the cost of my clothing keeps me on the stroll [locations for soliciting customers] and doing car tricks [providing sex to motorists] much later every night. (Inciardi, Lockwood, and Pottieger, 1993: 121)

"Street kids" engage in prostitution at extremely high rates. A study of street kids in Miami reported that 5 percent of the boys and 87 percent of the girls engaged in prostitution at least some of the time (Inciardi, Lockwood, and Pottieger, 1993). Female adolescent prostitutes report negative experiences in male parental relationships (Weisberg, 1985: 89). Like older female prostitutes, many adolescent female prostitutes express generally negative attitudes toward men.

Male and female adolescent prostitutes serve remarkably similar customers: 30- to 50-year-old men, usually white, from a variety of social and occupational backgrounds (Weisberg, 1985: 161). Male prostitutes generally perform their acts in cars or clients' residences rather than hotel rooms to avoid raising the suspicions of staff members. Females prefer places near the streets where they meet customers, like hotel rooms or a nearby apartment.

Prostitution as a Career   Few prostitutes begin their involvement in this practice in childhood. An association with people who live on the fringes of prostitution seems an important contributor to participation in this activity. In order to fully participate in "the life," a potential prostitute must learn the trade from others who know how it works. In the United States, such contacts largely involve relationships with women who themselves practice prostitution. While pimps identify some candidates and influence them to become prostitutes, most do not enter the profession in this way. Most prostitutes who form relationships with pimps do so after entering the profession. Nor does the stereotype of novice prostitutes as sexually inexperienced innocents hold true. Most girls who eventually become prostitutes began to associate with other prostitutes or pimps only after they became sexually active and had previously explored the idea of selling sexual services (Gray, 1973).

The career of a call girl includes at least three developmental stages: entrance into the career, apprenticeship, and development of contacts. The mere desire to become a call girl does not allow one to attain this status; instead, one needs training and a systematic arrangement for contacts. One call girl explains, "You cannot just say get an apartment and get a phone and everything and say, 'Well, I'm gonna start business,' because you gotta get clients from somewhere. There has to be a contact" (Bryan, 1965: 289). One study has concluded "the selection of prostitution as an occupation from alternatives must be sought in the individual prostitute's interaction with others over a considerable time span" (Jackman, O'Toole, and Geis, 1963: 160). A study of 33 Los Angeles call girls found that half had begun moving toward the career through contact with a call girl, some over long periods of time, others for shorter times (Bryan, 1965). Some reported solicitation by pimps with offers of love and managerial experience. When a working call girl agrees to aid a novice, she assumes responsibility for her training; women brought into prostitution by pimps may accept training either by him or by another call girl of his choice.

After making contact and deciding to become a prostitute, the apprentice begins her apprenticeship. This period hardly involves a formal process of instruction, but some women report spending an average of 2 or 3 months working in others' apartments in training situations. They perform as prostitutes during this time, in the process learning the value structure of the profession and important guidelines for handling problem situations. By acquiring a set of norms and values, the apprentice develops a feeling of solidarity with the group and at the same time becomes alienated from "square" society. Among other norms, this training transmits the belief that prostitution simply represents more honest sexual behavior than that of most people; moreover, the learning promotes an image of men as corrupt and exploitative, a value of "fair" dealings with other prostitutes, and fidelity to the pimp.

Contacts   Since a successful call girl needs a clientele, training provides equally important input about acquiring contacts. An apprentice can buy books or "lists" with clients' names from other call girls or pimps, but some include unreliable information. Most frequently, a trainee develops her own contacts during the apprenticeship period. For an initial fee of 40 to 50 percent, the trainer call girl often agrees to refer customers to the apprentice and oversee her activities. This fee compensates the trainer, as does the convenience of dispatching another woman to meet unexpected or conflicting demands or simply to take care of her own contacts. Over time, however, the new call girl must develop her own list of clients.

After compiling a list, she must also keep clients. The prospects for forming sound, professional relationships depend on the way in which the prostitute manages her interactions with clients. Some accepted rules govern interpersonal contacts with customers, including what the prostitute should say on the phone during a solicitation; most repeat standard "lines" such as needing money to pay rent, buy a car, or pay doctor bills. Additional rules guide social interactions while collecting fees, acceptability of specific customer preferences and types of customers to avoid, how to converse with customers, cautious use of alcohol, and knowledge of physical problems associated with prostitution. Apprenticeship seems to provide little instruction in sexual techniques. In spite of the importance of rules that define successful practices, one study found considerable variation in their adoption (Bryan, 1965).

Exchanging Sex for Drugs   Changes in the availability and use of drugs have brought related changes in prostitution practices in some areas. In some neighborhoods, use by prostitutes of cocaine, crack cocaine, and heroin has promoted exchanges of sex for drugs rather than for money. Evidence for this trend comes not only from impressionistic reports of growing numbers of such trades, but also from wider reports of new sex-selling patterns emerging over time. Drug-motivated competition has apparently decreased prices for prostitution services as addicts undercut other prostitutes to ensure they make incomes sufficient to support their habits (see generally Ratner, 1993). One Chicago prostitute, not a crack user herself, complained:

> I ain't hardly workin' at all anymore. There ain't nothing out there anymore. . . . Those rock stars [females who habitually smoke crack], whatever they call themselves, with them you can't make money anymore. Any decent 'hoe gonna ask for twenty dollars, but these girls, they'll give head for five dollars. Half-and-half, ten dollars! [sarcastic laugh] Can you believe it? . . . Now these johns comin' up and that's all they want to pay. (quoted in Ouellet, Wiebel, Jimenez, and Johnson, 1993: 69)

Prostitutes who exchange sex for drugs often think of themselves as freer than those who work for pimps (Miller, 1995). Independence of relationships with others brings its own restraints, however, especially since crack-addicted prostitutes must maintain ties to their local drug scenes. As a result, they often can manage only illusory perceptions of freedom, since they often exchange one form of dependency (on pimps) with another (on drug dealers and fellow users).

Self-Concept

Prostitution requires a new conception of oneself. Societal reactions, arrests, and associations with other prostitutes promote change in the prostitute's self-concept. Research has found a relationship between the self-image of the urban prostitute and the extent of her social isolation; especially isolated women tend to view their behavior in a more acceptable light (Jackman et al., 1963: 150–162). Some call girls, however, particularly those who cater to wealthy clients and work in other occupations such as secretary or modeling jobs, may look down upon "prostitutes" and associate that word strictly with streetwalkers.

Like other deviants, prostitutes recognize the reactions of others to their work, but they often justify their practices with three arguments:

1. Prostitutes are no worse than other people and often less hypocritical.
2. Prostitutes achieve certain dominant social values such as financial success and supporting others who depend upon them.
3. Prostitutes perform a necessary social function.

One study gathered input about the philosophies of 52 call girls with an average age of 22 and length of experience of 27 months. This research found that virtually all respondents believed in the importance of prostitution as an outlet for the varied and extensive sexual needs of men and the necessity to protect other social institutions (Bryan, 1965: 287–297). These reasons resemble those given by sociologists who explain the existence of prostitution by the functions that these deviants provide (Davis, 1937).

Prostitutes support their favorable self-concepts particularly by regarding clients as legitimate targets for exploitation. This sense of exploitation is enhanced by the feeling by prostitutes that they are taking financial advantage of men. Secure in the knowledge that they have what their customers want, prostitutes work in a "sellers market" of services. Most prostitutes develop anti-male attitudes as part of their training, and job experiences reinforce these attitudes. A prostitute regards her exploitation of men as no more immoral than the actions of her customers and the rest of the world. Another view regards most interpersonal relations between the sexes as acts of prostitution in some form. Wives and other women use sex to achieve their materialistic objectives, the argument goes, and at least prostitutes honestly admit to this deception. Statements like these do not imply, however, that prostitutes form uniformly positive images about their work. Most adolescent prostitutes, for example, express negative attitudes about their practices (Weisberg, 1985:163).

An interview with a prostitute in Spokane, Washington, has illustrated these points (Murphey, 1987). She reports working alone, without participating in a deviant street network and without a pimp. She did not become sexually active until 18 and has a high school diploma. At 23 years old, she has worked the streets for 2 years. She has no illusions about her job. She works strictly for the money and nothing else. (She admitted to an income between $4,000 and $5,000 per month—from which, of course, she pays no taxes.) She partially justified her entry into prostitution by expressing the belief that prostitution differs little from a lot of behavior by women who frequent singles bars. "I'm just not giving it away," she said.

Like most of her co-workers, she gained the information and understanding that permit her to operate through interactions with knowledgeable others. The basics of her own method of operating include many restrictions: no contact without a condom, no customer may tie her up, no oral contact of any kind, and careful judgment about clients before joining them in their cars or rooms. Like virtually all prostitutes, she expresses a jaundiced view of men: "When I look at these guys, they're like such fools to me, you know? Anybody who would have to go out and pay for it must really be out of it, or really be hurting or something." One of the baby pros that Inciardi (1984: 77) interviewed echoed this woman's feelings: "You have to be awfully fucked up to want to be pissed on or screwed by a kid." The female adolescent prostitutes interviewed by Weisberg (1985: 89) also reported negative attitudes toward men.

Despite these general attitudes, some prostitutes develop personal relationships with clients. In fact, many report thinking of some clients as a kind of friend.

Not infrequently, personal friendships with customers are reported: "Some of them are nice clients who become very good friends of mine." On the other hand, while friendships are formed with "squares," personal disputations with colleagues are frequent. Speaking of her colleagues, one call girl says that most "could cut your throat." Respondents frequently mentioned that they had been robbed, conned, or otherwise exploited by their call girl friends. (Bryan, 1966: 445)

This difference between the shared ideology and actual beliefs of specific prostitutes may trace back to the lack of cohesiveness among them and perhaps to stronger personal feelings of society's stigma than the ideology implies. Nevertheless, the activities of learning the ideology, along with the other norms and values associated with prostitution, constitute an important part of becoming a prostitute.

Some prostitutes engage in this form of deviance without developing deviant self-images, without progressing in their acquisition of deviant norms and values, and without extensive identification with prostitution as a career. The transition to career prostitution occurs as the person comes to acquire the self-conception, ideology, social role, and language of prostitution.

After beginning in prostitution, these women tend to develop attitudes and behavior patterns that contribute to their social role. In particular, they come to share an argot, or special language, with others in their line of work, with terms for special acts and services, patterns of bartering with their customers, and many rationalizations for their activities. While many prostitutes eventually leave the occupation for marriage or other employment, a few manage to achieve and maintain high standards of living through their activities. For others, however, age, venereal disease, alcoholism, or drug addiction pull them toward derelict lives punctuated by more or less regular arrests and jail terms.

Many prostitutes feel a strong economic motivation to continue prostitution. Other inducements uncovered by one study included loneliness, entrapment under the control of pimps, and drug addiction (Davis, 1981: 312). Women who participate in deviant street networks engage in prostitution as part of their "family" activities and simply another hustle. Undoubtedly, many women continue in prostitution because they see no other, more legitimate, and equally high-paying alternatives. These women may feel trapped by economic necessity and even lament the absence of more attractive alternatives. These possibilities raise questions about how willingly prostitutes participate in this occupation (Hobson, 1987).

Prostitution and AIDS

Contemporary concern about AIDS has serious implications for prostitution. Most AIDS cases in the United States still afflict homosexuals, and many of the rest show heavy associations with intravenous drug use. Still, AIDS is increasingly frequently transmitted by heterosexual contact. Most prostitutes who contract AIDS do so by using drugs, and many believe that they can avoid AIDS through regular use of condoms. The Spokane prostitute mentioned earlier, for example, denied that she faced any unusually high risk for AIDS because she regularly visited a physician. Clearly, such visits would not prevent the disease, although they might lead to early detection and perhaps effective treatment. These benefits would, of course, offer little consolation to either her or her clients.

Municipal areas' rates of AIDS among prostitutes vary from an estimated 4 percent in Los Angeles to over 80 percent in some eastern U.S. cities (Bellis, 1990: 26). Prostitutes generally face their highest risk not from sex with strangers, but from the relationship between prostitution and intravenous drug use. In fact, of all women with AIDS, most have contracted the disease through intravenous drug use. Bellis's study of 72 prostitutes has found substantial concern and knowledge about AIDS, but the prostitutes failed to protect themselves or their customers by, for example, abstaining from heroin, injecting themselves only with unused needles, or requiring condom use. "Yeah, I'm concerned," reported one, "until I stick the needle in. When I'm hurting, dope's the only thing on my mind" (Bellis, 1990: 30). The prostitutes reported acting under compulsion by the need for money to supply their drug habits. As another put it: "I'm afraid of AIDS but I've got a drug problem. Drugs drive me. If it weren't for heroin, I wouldn't be out here doing this. Dope pushes everything else out of my mind" (Bellis, 1990: 30).

Studies have identified the greatest danger of AIDS for prostitutes, other than their own drug use, in sexual relations with high-risk, nonpaying partners with whom they have formed romantic attachments, weakening their resolve to use condoms (Campbell, 1991). As an Australian prostitute noted: "I can't use a condom with my boyfriend. What'll he think—that I'm gonna charge him next?" (Waddell, 1996: 81).

In a sophisticated study of 350 licensed Nevada prostitutes, required by law to use condoms since 1988, Campbell (1991) found that not one prostitute had yet tested positive in mandatory monthly screenings, although tests had identified the infection in 13 applicants for brothel positions. Campbell contrasts the Nevada rate with that in a number of other cities, including Colorado Springs (3.8 percent HIV positive), Los Angeles (3.7 percent), San Francisco (9.9 percent), New Jersey (57 percent statewide), and Miami (26.6 percent). One study of AIDS rates among male prostitutes in the Netherlands reported that a minority of those who had practiced anal intercourse in the previous year had not consistently used condoms (De Graaf, Vanwesenbeeck, Van Zessen, Straver, and Visser, 1994). Many of the prostitutes tried to avoid even the possibility of infection by performing only oral or manual sex acts.

Prostitution and Social Control

Attitudes toward prostitution have varied over time, and today they vary between countries. These attitudes depend in part upon women's roles in a society and whether prostitutes provide services in addition to sexual ones. In ancient Greece, for example, prostitutes commanded generally high respect. Similarly, Indian devadasis, or dancing girls, plied their trade in connection with that country's temples for centuries; besides singing and dancing, they practiced prostitution. Japanese geishas, traditionally trained in the arts and music as well as in conversation and social entertaining, offer another example of women who could often engage in prostitution yet still maintain high status in society.

Through the Middle Ages, societies did not regard prostitution as criminal activity but rather as a necessary evil. Both civil and religious officials attempted to regulate prostitution, but the strength of social condemnation did not approach that now common among some segments of U.S. society.

Anglo-American criminal law has revealed strong disapproval of prostitution, particularly soliciting activity. Such strident attitudes derived largely from the Protestant Reformation. Even today, many Catholic countries, such as those in Latin America, display comparatively tolerant views of prostitution. Where laws prohibit prostitution, they represent efforts to control certain private, moral behavior through punitive social control. Undoubtedly, even vigorous enforcement discovers only a small proportion of acts of prostitution.

Critics oppose prostitution on many grounds:

1. It involves a great deal of promiscuity, particularly with strangers.
2. The prostitute damages other social relationships by commercializing sexual relations with emotional indifference outside of marriage, in which one partner participates solely for pleasure and the other solely for money.
3. The women who engage in prostitution experience unwholesome social effects.
4. It threatens public health by facilitating the spread of venereal diseases.
5. Prostitutes need police protection in order to operate, and their arrangements with officers reduce the general quality of law enforcement.
6. Sexual acts with prostitutes generally allow no possibility for marriage and procreation, making them inferior to ordinary premarital sex relations.

Law enforcement of prostitution cases often creates a sordid business. Because such cases generally lack complainants, the police must employ aggressive tactics to detect these crimes. They may resort to entrapment and other questionable enforcement practices in their zeal to enforce these laws. The demeanor a prostitute adopts toward a police officer substantially affects her vulnerability to arrest (Skolnick, 1975: 112). Many such arrests result from responses to solicitations by police officers or to other lures. Sometimes, police rely on informers to locate the rooms where they can find illegal activity in progress. In order to escape prosecution, an arrested prostitute may offer to serve as an informant in apprehension of her pimp or a narcotics peddler. Although the males who frequent prostitutes also technically violate the law, at present many jurisdictions still focus primarily on the prostitutes. They rarely also arrest customers. Some communities have attempted to draw attention to prostitution clients by printing offender's pictures in newspaper ads, but such innovations are rare.

Laws against prostitution generally discriminate against women. Many detractors of these standards argue that a woman should be able to engage in intercourse for money if she so desires. A national organization of U.S. prostitutes called *COYOTE*—for *Call Off Your Old Tired Ethics*—was founded in 1973 in San Francisco. Along with a similar group in France, this organization has attempted to change public attitudes toward prostitution by calling for decriminalization of the practice. Through its newsletter, *COYOTE Howls*, and other outlets, the group called for states to repeal laws against prostitution, maintaining that this activity could actually benefit a community by providing an outlet for male sexual activity.

COYOTE identified an important goal of dissociating prostitution from its historical link with sin and crime, while substituting an image of prostitution as a kind of work and a behavior worthy of civil rights protection (Jenness, 1993). By attempting to repeal existing laws and engage community leaders in debate over questionable enforcement practices, COYOTE members have attempted to redefine prostitution, transforming it from a social problem facing the community to merely an occupational choice that suits some women. Supporters underscore these claims by asserting that (1) not all prostitution is forced, (2) prostitution represents merely a service occupation in the community, and (3) denial of the choice to engage in prostitution violates a woman's civil rights to work as she pleases. These arguments have not led, however, to stampedes to repeal of laws against prostitution.

A group of prostitutes formed a subsequent organization in 1985 called *WHISPER* (for *Women Hurt in Systems of Prostitution Engaged in Revolt*). This organization has pursued a purpose slightly different from that of COYOTE. WHISPER depicts the downside of prostitution, with graphic accounts of women restrained by their customers in chains and ropes, burned with cigarettes by their pimps, and generally degraded sexually for money they need in order to live. WHISPER deemphasizes the notion of entirely voluntary participation in prostitution; the group maintains that women need money to live, and prostitution offers them one way to get it (Hobson, 1987: 221–222). WHISPER decries the victimization of women who engage in prostitution.

There are a number of international organizations for sex workers (Kempadoo and Doezema, 1998). Most groups are concerned with the working conditions and human rights of sex workers; others are more directly concerned with the exploitation of sex workers. A male prostitute in Costa Rica echoes the hopes of many who wish to be seen as regular workers rather than prostitutes:

> I believe one becomes a degenerate because only degenerates accept prostitution. If people respected us for the work we do, instead of denigrating it, we would be associated with other people and would even have the opportunity to form a union. (Shifter, 1998: 23)

Transvestitism

A wide range of sexual norms specify many combinations of appropriate objects, times, physical situations, and motivations for sexual relationships. This complex network of standards may sanction many behaviors associated with heterosexuality as examples of deviance. One such behavior is transvestitism, or cross-dressing.

A transvestite is someone (either a man or a woman) who dresses in clothing generally considered appropriate for the other gender for reasons that include sexual satisfaction (not for money, entertainment, etc). The element of sexuality distinguishes transvestitism from appearing in costume onstage, for example, and observers believe that most transvestites are men. Many, but not all, transvestites are also transsexuals (people who identify with the opposite sex or who have undergone sex change operations).

The deviant character of transvestitism mainly results not from wearing clothing of the other sex, but from the primary importance of this practice in supplying sexual pleasure and satisfaction to participants. If this view correctly interprets the behavior, many regard transvestitism as deviant behavior because it promotes essentially antisocial activities.

Other heterosexual activities, such as masturbation, draw unfavorable attention for the same reason. Masturbation usually provides only a short-term outlet, however, rather than becoming a primary means of satisfaction for people who can choose other practices. Under these conditions, some regard masturbation as a natural behavior. When, however, it becomes a person's only or primary source of sexual expression, it, too, will likely acquire an identity as deviance.

Most women who cross-dress today do so for fashion rather than to gain public acceptance as men. As a result, women who gain sexual satisfaction from cross-dressing may do so without stigma. Women's fashion standards permit them to wear pants and even men's suits. Many female transvestites take advantage of this flexibility to indulge their identification with the opposite sex. People have practiced transvestitism for centuries; it is not a new sexual form. Dekker (1989) describes a number of cases of women who cross-dressed for a variety of reasons—including patriotic, economic, role-related, and sometimes sexual ones. Lesbians practiced transvestitism during the early 18th century because the norms of the day regarded sex almost exclusively as intercourse with a penis; therefore, some lesbians cross-dressed and used artificial penises in relations with their lovers. Beginning in 1800, Parisian women needed police permission to wear pants at any time other than carnival. Few women, however, requested these permits (Matlock, 1993: 42). One argument explains women's cross-dressing in the 19th century as a way of challenging male superiority in society, as an act of political rebellion. But changes in women's dress did not effectively advance women's rights; in fact, instances of transvestitism sparked only ridicule, harassment, and, at times, legal action (Steele, 1988: 162).

 Issue | **The Ambiguity of Sex**

A transsexual inmate who filed a complaint with the Canadian Human Rights Commission in 1999 to be transferred to a women's prison and undergo a sex change operation was granted her request. The inmate, Synthia Kavanagh, called the decision by Corrections Canada "a great victory for all transgendered people." Kavanagh, 37, was born Ricky Chaperon but has dressed and lived as a woman since she was 13 and legally changed her name to Synthia when she was 19 years old. It was not clear who would pay for the procedure. Kavanagh's complaints against the corrections system began in 1993 because correctional officers discontinued her supply of female hormones, refused to permit her access to sex change surgery, and insisted on keeping her in a male facility.

Source: Wattie, Curtis. 1999. "Murderer to Have Sex Change, Move to Prison for Women." *National Post*, Vol. 21, November 19, pp. 1–2.

In modern times, transvestitism goes on no more visibly than it did 300 years ago. Most male cross-dressers today maintain secrecy and privacy, although many male transvestites are married men. The reactions of their wives upon learning of these preferences provide instructive examples that reveal the deviant nature of this behavior. In one account, June, an English woman married for 21 years to George, a cross-dresser, describes their relations:

> First of all when he told me, I let him have the go-ahead and he could dress. The children were a lot younger then, so when they'd gone to bed, he could come down dressed; but then something inside you rebels and is repulsed and says: "This isn't right." You married a man and you've got this man dressed as a woman and enjoying the role. I just cracked up and kept crying. And it was more and more tablets [antidepressant pills]. (Woodhouse, 1989: 103)

At a distance from the behavior, the deviant nature of transvestitism may show clearly, but those close to the transvestite may not recognize its character, especially in the early stages. Common activities include some elements of cross-dressing, for example, among small children while condoned by parents. Even in adulthood, women today often wear clothing traditionally associated with men. Because of wide variations in fashion, some circumstances actually encourage cross-dressing, such as when trends boost the popularity of a "unisex" look among both some men and some women and when adolescent girls try to achieve a "tomboy" look.

Stage shows with female impersonators represent a seeming legitimate outlet for cross-dressing, but this apparently respectable behavior risks a strong potential of sanctions for deviance. Some people may express feelings of ambivalence in hostility toward the impersonator. Anxieties may surface when experiences undermine gender certainty, raising issues like threats to masculinity. Star impersonators historically have felt constantly compelled to prove their masculinity to avoid allegations of effeminacy (Ullman, 1995). One study found that female impersonators viewed themselves not as women, but as imitators of women, and their reasons for engaging in this work included recreation, involvement in gay activities, attention, and/or to break into the entertainment industry (Tewksbury, 1994).

Many wives of transvestites seem initially to regard the behavior as rather unimportant, but they report a growing sense over time that their husbands are engaged in deviant behavior. Polly, another transvestite's wife, has explained:

> At first, I couldn't see anything odd in it at all, because I dressed as I wanted to—jeans or dress, I couldn't see what the big hang-up was. Obviously, I realize it now, but at the time it seemed very cut-and-dried. So he puts on a dress occasionally, what's that to me? It's when you learn you get a change of personality at times as well that it starts to worry you. (Woodhouse, 1989: 106)

An atmosphere of public ambivalence obscures clear perceptions of transvestitism. On the one hand, the behavior, in and of itself, causes no harm. Some may regard the practice as an attempt merely to extend the frontiers of masculinity and to engage in some gender-bending. Additionally, some find humor in transvestitism, as in media presentations like television programs by Britain's Monty Python troupe and the comedy group "Boys in the Hall." On the other hand, transvestites sooner or later experience conflicts with a variety of sexual norms, as do their partners. The extent of the cross-dressing behavior, especially its importance to the participants' sexual satisfaction, proves particularly troublesome for some persons. In the Janus survey cited earlier in the chapter, only 6 percent of the men and 4 percent of the women considered cross-dressing an acceptable practice (Janus and Janus, 1993: 121). Further, only 2 percent of the men and 1 percent of the women regarded it as a very normal one.

Male transvestites want to appear feminine, but they seek an image of femininity divorced from the everyday lives of most mothers, wives, and other women; they enter a fantasy world that even women cannot understand. Wives of transvestites, like those quoted above, eventually have difficulty accepting their husbands' behavior partly for this reason. A 42-year-old married father has reported:

> I belong to a transvestite support group . . . a group for men who cross-dress. Some of the group are homosexuals, but most are not. A true transvestite— and I am one, so I know—is not homosexual. . . . They are a bunch of nice guys . . . really. Most of them are like me. Most of them have told their families about their dressing inclinations, but those that are married are a mixed lot; some wives know and some don't, they just suspect. . . . I have been asked many times why I cross dress, and it's hard to explain, other than it makes me feel good. There is something deep down that it gratifies. (Janus and Janus, 1993: 121)

Pornography

Current debates exhibit a good deal of confusion over the activities and materials that constitute pornography, the problems it poses for society, and the role of law in resolving these disputes (Jackson, 1995). Provisions in law might seem to offer a logical starting point for the debate, but legislators and courts have encountered extreme difficulty in trying to decide exact, legal standards for obscenity that separate pornographic materials from others. In the 1973 *Miller* v. *California* case, the Supreme Court ruled that states can class material as obscene and ban it if an average person, applying contemporary community standards, would find that it (1) appeals to "prurient interests in sex," (2) describes sexual conduct "in a patently offensive way," and (3) "taken as a whole, lacks serious literary, artistic, political, or scientific value."

This test allows sanctions for obscenity only for materials that exhibit all three conditions. The Supreme Court decision had the important effect of basing the final decision about what constitutes obscenity on community norms. It clearly established local courts, applying local community standards rather than national standards, as the final judges. Obviously, Las Vegas would apply different standards than Topeka, Kansas. Questions remain, however, about the best manner of gauging this community feeling. In a 1978 decision, the Supreme Court ruled that juries sitting in judgment over a publication or film must "determine the collective view of the community as best it can be done." Furthermore, the Supreme Court also declared that the evaluation could not consider children's sensitivity for a publication or film directed only at adults.

Just as legal conceptions of pornography vary, so, too, do social conceptions. While many groups, particularly those with conservative political views, have almost automatically taken strong stands against pornography, many others with different political views have also condemned it. Many feminists, for example, vilify pornography for its perceived contribution to the sexual objectification of women through demeaning portrayals of them. In this sense, critics of pornography deplore its perceived effect of promoting female submission and male domination (Leuchtag, 1995). Still other feminists expressed concerned that pornography actually has the same effect as sex discrimination, classing it as a form of verbal violence against women (MacKinnon, 1993).

Other feminists disagree, complaining that the would-be censors also view sex as inherently degrading to women. They question any assumption that elimination of pornography would reduce sexism and violence against women (Strossen, 1995). Furthermore, depictions of women as sexually active people may actually contribute to further liberation by freeing them from limiting, conventional stereotypes. In any case, research has not established a clear link between pornography and destructive actions, and the suspicion that exposure to pornography may lead to some harm does not necessarily imply a need for censorship (Ferguson, 1995).

Nevertheless, critics continue to advocate relationships between pornography and some forms of sexual deviance, and they also describe other problems that it poses for society. The word *pornography* derives from the Greek work *porn*, which referred originally to prostitutes and their trade (Barry, 1984: 205). Clearly, pornography no longer relates strictly to descriptions of the behavior of prostitutes and their clients; the concept has broadened to include virtually any sexually explicit material. Some pornography displays a violent quality, including certain recurring themes such as sexual slavery. Pornographic depictions appear in motion pictures, written works such as books and magazine articles, and newer media. For example, Dial-A-Porn is a service that allows people to call a special telephone number and receive sexual messages.

Evaluating the popularity of sexually explicit materials requires an uncertain judgment. One estimate (Malamuth and Donnerstein, 1984: xv) numbers the adult male readership of the two most popular sexually oriented magazines *(Playboy* and *Penthouse)* higher than the combined readership of the two most popular news magazines *(Time* and *Newsweek)*. Even if these particular publications do not fit a strict definition of pornography, such figures, if true, reflect strong interest in sexually oriented materials.

Social and political concern over the potential for harmful effects of *obscenity* (a legal term) and *pornography* (a popular term) has built over time. During the late 1960s and early 1970s, interest in the harm that might result from viewing pornographic movies and magazines led to the formation of a national commission to study the nature and effects of pornography (Commission on Obscenity and

Pornography, 1970). In addition, Supreme Court decisions, discussed later in the chapter, set guidelines for regulating pornography within constitutional guarantees of free expression. A more recent surge of interest during the 1980s resulted in formation of another national commission to examine changes in pornography since the time of the previous commission and to make another set of recommendations regarding its uses and regulation (Attorney General's Commission on Pornography, 1986).

The 1970 commission attributed many problems of pornography to a less-than-open prevailing atmosphere regarding sexual matters. That body recommended more systematic public discussion of the subject as well as a program of sex education in the schools. This tone is conveyed in an important statement:

> The Commission believes that much of the "problem" regarding materials which depict explicit sexual activity stems from the inability or reluctance of people in our society to be open and direct in dealing with sexual matters. This most often manifests itself in the inhibition of talking openly and directly about sex. Professionals use highly technical language when they discuss sex; others of us escape by using euphemisms—or by not talking about sex at all. Direct and open conversation about sex between parent and child is too rare in our society. Failure to talk openly and directly about sex has several consequences. It overemphasizes sex, gives it a magical, nonnatural quality, making it more attractive and fascinating. It diverts the expression of sexual interest out of more legitimate channels. Such failure makes teaching children and adolescents to become channels for transmitting sexual information and forces people to use clandestine and unreliable sources. (Commission on Obscenity and Pornography, 1970: 53)

The more recent pornography commission did not concur with this conclusion. The 1986 report took a more serious view of pornography, with numerous statements about potentially harmful effects of exposure to pornography and language promoting greater, not less, regulation (Attorney General's Commission on Pornography, 1986). The differences of opinion between these two national commissions can be best understood within the larger context of pornography and social reactions to it.

Social Regulation of Pornography

One can find examples of explicit sexual references intended to entertain or arouse audiences in many forms created by diverse societies throughout history: Greek and Roman mosaics, poetry, and drama; Indian writings such as the *Kamasutra;* medieval ballads and poems such as those by Chaucer; farcical French plays of the 14th and 15th centuries; and Elizabethan poetry and art, along with many present-day examples. Only fairly recently, however, have governments begun trying to regulate such references and themes by law (Attorney General's Commission on Pornography, 1986: 235). Medieval religious institutions—such as the Catholic church—established the first formal regulations, applicable only to descriptions of sex that accompanied attacks on religion or religious authorities.

Even common law courts in England were reluctant to directly address the issue of pornography.

Contemporary legal concern with pornography dates from the early 1800s in England. As changing technology allowed increasingly economical printing, thereby increasing the availability of printed materials to the masses, sexually explicit materials that once achieved only limited circulation began to reach wider audiences. This growing exposure boosted demand, which in turn prompted an increase in the supply (Attorney General's Commission on Pornography, 1986: 241). This change, occurring right before the Victorian era, accompanied a social trend of increasing willingness to condemn perceived violations of sexual morality. The development of citizens' groups, such as the Organization for the Reformation of Manners and its successor, the Society for the Suppression of Vice, paralleled this emerging social concern.

In the United States, the same concerns motivated organizations like the New York Society for the Suppression of Vice, which pressed for legislation tightening restrictions against pornography. Such groups succeeded in securing legislation against sexually explicit materials, forcing the production of and market for pornography to become almost entirely clandestine through the first part of the 20th century. Subsequent legal skirmishes raised issues of First Amendment protections of free speech, and the most recent laws governing pornography have reflected the outcomes of these disputes. They also reflect deep ambivalence regarding pornography and questions about how or whether society should regulate it.

U.S. Supreme Court decisions on pornography have usually reflected close votes among the justices, suggesting that jurists themselves have not resolved the disputes about what is and what is not pornography that still divide segments of the public. Among U.S. states, Massachusetts in 1711 enacted a statute prohibiting distribution of pornography, and Vermont followed suit in 1821. The first federal statute prohibiting importation of pictorial pornography was enacted in 1842. The first federal statute prohibiting obscenity in 1865 forbade distribution of such materials through the mails. Continued legislation coincided with a wider social decline in direct influence of religion over community life, a spread of free universal education, and increases in literacy.

Two Supreme Court decisions, both from 1973, embody modern U.S. legal opinion on regulation of pornography. Important standards emerged from *Paris Adult Theaters* v. *Slaton* (413 U.S. 49, 1973) and *Miller* v. *California* (413 U.S. 15, 1973). The court decided that public displays of pornography represented appropriate candidates for regulation, while displays in one's own home did not. It also defined local community standards, as explained earlier, as the appropriate criteria for determining obscenity. These decisions returned the questions of what is and what is not pornography and what to do about it back to local jurisdictions. As a result, some local officials pursued prosecutions, while others declined to do so, for the same material. Controversies continue over such matters as which sets of principles constitute local community standards and even the basic *definition* of pornography.

The nature of such controversy is depicted in the 1997 film *The People v. Larry Flynt*. The movie depicted the personal and legal difficulties of Flynt, publisher of

---

❋ *Issue* | **The New Policing: Cybercops**

The Internet is a huge, largely unregulated means of communication. Businesses and private individuals have found using the "Net" to be a boon for shopping, doing research, and entertainment. It is also a significant means of distributing pornography.

America Online (AOL) is the largest Internet service provider in the United States, with more than 20 million users. Despite a zero-tolerance policy, AOL has had difficulty regulating child pornography. AOL employs scores of online patrol cops to monitor the activities of users. They eject disruptive members from chat rooms, enforce an anti-vulgarity code, and generally look for sites that might offend sexually.

But there is only so much one can do. By the end of 1997, federal and local law enforcement agencies had identified more than 1,500 suspected pedophiles in 32 states primarily through AOL chat rooms. One mother sued AOL for $8 million with the assertion that AOL is a "home shopping network for pedophiles and child pornographers" after her son's social studies teacher began serving a 22-year federal prison sentence for distributing and receiving child pornography through AOL.

Source: Malkin, Michelle. "Cyberspace Cops Can't Keep Up." *Omaha World Herald,* October 23, 1999, p. 13.

---

the magazine *Hustler.* Some hailed the movie as a brilliant promoter of free speech, while others, including Flynt's daughter, Tonya, railed against the idea that Flynt was any kind of hero. The mission statement of the Tonya Flynt Foundation indicates

> The Tonya Flynt Foundation is a not-for-profit corporation formed for and dedicated to eradicating child sexual abuse, rape and all forms of violent crime related to pornography. We intend to expose the pornography myth that pornography is a victimless crime. We also intend to raise community awareness of the effects of pornography on our society in order to protect the women and children who are so commonly abused by pornography. (www.tonyaflynt.com)

Tonya Flynt believes there is a direct relationship between the consumption of pornography and violent crimes, a view we will examine shortly.

Pornography in Everyday Life

People in the United States, like those in many other Western nations, experience many sexually explicit materials in everyday life. Mass media outlets, such as films, novels, television shows, periodicals, and newspapers, graphically portray sexual images and behaviors today that never appeared in those media in earlier times. The motion picture industry instituted a rating system in 1968 to self-regulate access to sensitive images in movies and to alert potential viewers to their sexual and violent content. In an effort to keep pace with changing social sentiment, the motion picture industry changed its rating system in 1990 to eliminate the X-rating, which limited access to patrons 18 years old or older, and substituted NC-17, indicating materials not suitable for children under 17. The motive behind the change in the rating system was to attempt to remove the stigma of pornography from films whose contents earned an X rating. In so doing, however, the ratings do provide potential viewers with some information about the contents of the film so that they may make more informed viewing choices. Network television has also practiced self-regulation by assigning in-house censors to evaluate program content and ensure compliance with federal regulations. In 1997, the television industry launched a program-rating system much like that for motion pictures. The ratings reflect the producers' judgments of the appropriate ages for the programs rather than specific types of content. Cable television systems, which are not subject to the same Federal Communications Commission regulations as broadcast or network television, permit more latitude in presentations of sexually explicit materials.

The spread of cable television, along with the wide availability of videocassette recorders (VCRs), has broadened the market for pornography. By renting or buying X-rated and NC-17-rated videocassettes or laserdiscs, people can view pornography in any dwelling equipped with a suitable player. Many video retail outlets maintain "adult" sections full of sexually explicit movies for rent. Just as printing technology expanded the market for printed pornography a couple of hundred years ago, home video systems and cable television technology have expanded the availability of video pornography during the past decade.

As mentioned earlier, a more recent innovation has brought pornography to myriad computer systems. Communications networks can transmit digital images in the same way that they handle data, from one computer to another. Appropriate software then translates this data into on-screen images. It is not difficult to find pornography on the Internet, although many such sites are now using age-check systems to try to ensure that viewers are all adults. Some sites will display various images or stories that are pornographic, and there is little or no attempt to check the age of the viewer. The data for such images fill large files, sometimes creating storage problems. CD-ROMs have helped to solve this problem by offering large storage capacity. Companies now sell pornographic CD-ROMs much as they sold magazines only a few years ago.

Some evidence indicates narrowing social conceptions of pornography and increasing tolerance for materials universally considered pornographic only a few decades ago. One national survey, for example, found increasing acceptance by many adults of access to and displays of materials depicting genitalia and many kinds of sexual activity (Winick and Evans, 1994). Whether this finding reflects an emerging national consensus for acceptance remains to be seen. The NHSLS survey found that men viewed pornography more often than women did. Those results revealed that 23 percent of the men, but only 11 percent of the women, reported experiencing X-rated movies or videos, and 16 percent of the men, compared with only 4 percent of the women, indicated that they had viewed sexually explicit books or magazines (Laumann et al., 1994: 135).

Production of pornography employs an extensive industry. Organizations that make films, videotapes, and magazines are heavily concentrated in certain areas, with 80 percent of them located in and around Los Angeles (Attorney General's Commission on Pornography, 1986: 285). Professional companies may employ 50 or more employees for production, advertising, and sales (Abbott, 2000: 18). Each company may release 20 or more titles a month and makes use of the most glamorous and popular talent in the industry. Although laws no longer prohibit this activity, a substantial portion of the pornography industry still operates "underground." Many of these companies are amateur and operate on low budgets; a single person frequently performs the functions of writer, producer, and director. The industry overlaps little with the mainstream film industry, and performers in X-rated movies rarely become well known for mainstream movie roles. Teams of writers, rather than single individuals, often generate text for sexually explicit novels, pooling different sections of the old book and often reusing old material by altering it slightly to fit circumstances and characters in the new stories.

Actors and actresses are drawn to this type of work for a variety of reasons. Like prostitution, some actresses earn quite a bit of money, but most make meager wages. Actresses are paid by the "scene," and fees vary with the popularity of the actress and the nature of the scene. An average scene will earn an actress $500, which amounts to a high hourly wage but fails to take into account the long periods of time between working opportunities (Abbott, 2000: 20). Nevertheless, actresses can make $5,000 a month in the beginning and more later if they become popular. There are also many expenses that must be borne by actresses, including the costs of cosmetic plastic surgery (e.g., liposuction and breast augmentation are the most common) and HIV testing, which must be performed every month in order to work. Other reasons for working in the pornography industry include freedom, flexible hours, and fun.

Research has found connections between the pornography industry and organizers of other forms of vice and between the pornography industry and organized criminal syndicates. Many retail pornography stores visited in one observational study in Philadelphia also provided prostitution, illegal gambling, and illicit drug sales (Potter, 1989). In nearly 40 percent of the stores, customers could gamble or find referrals to gambling operations, and in 70 percent of the stores, customers could obtain drugs or information on where to get them. Many of the establishments offered prostitution services on-site. Their owners often carried on other related businesses, including publishing a sex tabloid that carried explicit personal advertisements, managing a massage parlor, and running an escort service.

The Effects of Pornography

Observers have divided the effects of pornography into two classes: direct and indirect effects. Direct effects might include arousal of the pornography's audience and changes in their behavior that result from exposure to it. Studies have examined the relationship between exposure to pornography and sex crimes, for example, looking specifically for direct effects. Indirect effects would include subtle, long-term changes caused by exposure to pornography, such as redefinitions of sexual objects or sexual accessibility. Some observers argue for a long-term consequence that strengthens identification of women as objects for sex or violence and weakens their identities as people. Another indirect effect might divorce the context for sexual relations from partners' emotions and feelings. Some see a substantial long-term danger that pornography tends to reduce sex to a purely physical act rather than a component in a richer human relationship.

Harmful Effects   Evaluations of harm caused by pornography generate controversy because few observers agree about what constitutes harm and how to measure it. Further, evaluators cannot always say that some harmful effect results specifically from exposure to pornography. As a result of these problems, social science research has not yet provided definitive answers to the many questions about the harmful effects of exposure to pornography (Fisher and Grenier, 1994). Laboratory studies have tried to assess the link between exposure to pornography and subsequent acts of aggression; they have produced inconclusive but primarily negative results. Despite claims of some studies, the body of research supports the general conclusion that exposure to nonviolent pornography does not seem to lead to instances of aggression (Donnerstein, Linz, and Penrod, 1987: 38–60; Smith and Hand, 1987). Nevertheless, other studies report an association between exposure to "hard-core" pornography and sexual aggression. Boeringer (1994) has found that exposure to violent pornography among a sample of college men shows a link to sexual coercion and aggression. Still other research suggests no causal role for pornography in the development of pedophiles (Howitt, 1995).

The 1986 national pornography commission highlighted sexually violent material, and, according to the commission, this focus explains differences between its conclusions and those of the 1970 commission. The 1970 pornography commission concluded that exposure to pornography does not promote sexually aggressive behavior, either interpersonal violence or sexual crimes (Commission on Obscenity and Pornography, 1970: 32). This commission did not conclude that such exposure produces no effects, however; indeed, most consumers of pornography reported feeling sexually aroused by their experience.

Materials that depict themes of sexual violence may express sadomasochistic themes, such as use of whips, chains, and torture devices. Some of these materials also follow a recurrent story line of a man making some sort of sexual advance to a woman, suffering rejection, and then raping the woman or in some other way forcing himself violently on her. Most of these materials, including those in magazine and motion-picture formats, depict women characters as eventually becoming sexually aroused and ecstatic about the sexual activity. Exposure to such material, the 1986 commission has suggested, may lead directly to variations of this sort of behavior. Further, it also might promote attitudes that perpetuate the "rape myth" (Attorney General's Commission on Pornography, 1986: 329). This rape myth holds that women say "no" but really mean "yes"; therefore, men can feel justified in acting upon the woman's refusal of sex as an indication of willingness. After all, the myth continues, even if the woman really does not want sex at the beginning, once the forced contact begins, she will change her mind and enjoy it.

The content and imagery of pornographic films confirms stereotypes of male superiority and female passivity. These films also reinforce an attitude that male sexual desires and satisfaction eclipse the importance of female needs. One study, for example, examined the content of pornographic movies made from 1979 to 1988. Over time, it noted some shift in the context of the sex—progressively fewer movies involved prostitutes and fewer took place in the workplace, while comparatively large numbers depicted sex in other settings (Brosius, Weaver, and Staab, 1993). The movies continued to highlight the sexual desires and prowess of men while consistently portraying women as sexually willing and available under virtually any set of circumstances.

The 1986 commission assessed materials that portrayed highly explicit sexual acts along with violent content. "It is with respect to material of this variety," the 1986 commission concluded, "that the scientific findings and ultimate conclusions of the 1970 Commission are least reliable for today, precisely because material of this variety was largely absent from the Commission's inquiries" (Attorney General's Commission on Pornography, 1986: 324). With respect to sexually violent material, the Attorney General's Commission on Pornography (1986: 324) concluded that: "In both clinical and experimental settings, exposure to sexually violent materials has indicated an increase in the likelihood of aggression," especially aggression toward women. Other research has supported this contention (Donnerstein et al., 1987).

The commission qualified its conclusions, however, in the following manner:

We are not saying that everyone exposed to [sexually violent] material . . . has his attitude about sexual violence changed. We are saying only that evidence supports the conclusion that substantial exposure to degrading material increases the likelihood for an individual and the incidence over a large population that these attitudinal changes will occur. And we are not saying that everyone with these attitudes will commit an act of sexual violence or sexual coercion. We are saying that such attitudes will increase the likelihood for an individual and the incidence for a population that acts of sexual violence, sexual coercion, or unwanted sexual aggression will occur. (Attorney General's Commission on Pornography, 1986: 333)

The 1986 commission mentioned some additional less obvious but no less important effects. For one, most pornography depicts women in a "degrading" manner. It shows sexual partners, usually women, solely as tools for the sexual satisfaction of men. Other materials depict women in decidedly subordinate roles or engaged in sexual practices that many would consider humiliating. As a consequence of exposure to such materials, viewers, particularly young ones, may define both women and sexual behavior in general in a callous manner that objectifies women and removes the emotional content of sex (Weaver, 1992).

Positive Effects, or Potentially Beneficial Functions of Pornography   The continuing presence of pornography in modern society suggests to some sociologists that it serves an important social and personal function. One sociologist has compared the function of pornography with that of prostitution: In a society that negatively labels impersonal, nonmarital sex, people can achieve gratification mainly in two ways, by hiring prostitutes for relations with real sex objects and by using pornography, which can lead to "masturbating, imagined intercourse with a fantasy object" (Polsky, 1967: 195). If such an interpretation correctly explains these phenomena, then one might suppose that the frequency of pornography use would decrease as recreational sex occurs more frequently and carries a lighter stigma. No evidence currently confirms this possibility.

Furthermore, numerous studies have suggested that since pornography arouses both males and females in its audience, no negative consequences follow from that condition (see the reviews in Commission on Obscenity and Pornography, 1970). Exposure to pornography may even encourage healthy behavior and prevent crime. Eysenck (1972), for example, reports that sex criminals first view pictures of intercourse at ages several years older than noncriminals. Other studies suggest that sex criminals come disproportionately frequently from sexually restrictive families; as a result, they often receive less information and exposure to sexual subjects than other people get (Goldstein, Kant, and Hartman, 1974). Rapists, in particular, seem to have come unusually frequently from sexually repressive environments. Other research has found that the availability of pornography, including violent pornography, is not necessarily related to aggressive criminality, such as forcible rape (Abramson and Hayashi, 1984). Other research has, however, discovered a relationship between sales of pornographic magazines in various states and rates of reported crimes against women in these same states (Baron and Straus, 1984). Data like these, as well as results from other studies, have led one observer to a potentially surprising conclusion:

> Contrary to what common sense might suggest, there is a negative correlation between exposure to erotica and development of a preference for a deviant form of sexuality. The evidence even indicates that exposure to erotica is salutary, probably providing one of the few sources in society for education in sexual matters. (Muekeking, 1977: 483)

One can probably safely say that media portrayals of sex and violence do affect some people in ways other than sexual arousal. Social science does not currently provide a basis for accurately predicting such effects because individual and cultural differences complicate this judgment. Also, not all effects of such exposure prove negative, such as provoking instances of sexual aggression. Research has yielded conflicting findings about both short-term and long-term effects of pornography (see Malamuth and Donnerstein, 1984), and it will probably take some time yet to account for all of the many factors that influence the relationship between pornography and subsequent behavior. The more subtle effects of pornography—including the imagery of women, sex, and physical relationships without emotional context—may represent its most important consequences because they persist for long periods of time (Itzin, 1992; Weaver, 1992).

Clearly, sociologists need much more information than they now have about the positive and negative effects of exposure to pornography. This exposure may produce a range of effects, depending on personal differences and specific situations. Absolutist definitions of problems and solutions from either censors or zealous libertarians should give way to compromises on production, sales, and distribution of pornographic materials (see also Downs, 1989).

Summary

Sexual norms represent the guidelines for determining sexual deviance. Many sexual norms define complicated combinations of appropriate objects, times, places, and circumstances for sexual behavior. Sexual norms can change over time, as indicated by the increasing tolerance for premarital sexual relations in the United States. Society feels less tolerance for other forms of heterosexuality, however, such as adultery. Sex in the 1990s has gone high-tech, with a reliance on modern forms of communication such as the telephone and computer. Transvestitism represents another example of sexual behavior that violates subtle norms about gender-appropriate clothing and behavior. Because sexual norms usually apply to particular groups, and because such norms change over time, different groups disagree on the deviant character of some acts.

Society has long (and often unsuccessfully) sought to regulate one form of deviance—prostitution—by applying both sexual norms and formal prohibitions such as laws. Clients patronize prostitutes because they desire sex without subsequent responsibilities and entanglements. These desires evidently transcend international boundaries, since prostitution occurs in most countries. Different types of prostitutes form a stratified system with variations in income and privacy. Much prostitution takes place in a larger urban context of "hustling," in which women practice prostitution as only one illicit means by which to earn their livings. Prostitutes who are members of deviant street networks often participate in other forms of illegal and legal work to supplement their incomes.

Many women, and some men, engage in prostitution part-time, but the transition to life as a prostitute involves elements of learning and opportunity. Many prostitutes have sexual experiences prior to participating in this activity, but they must also learn a particular set of attitudes and values conducive to success in their line of work. These attitudes support using one's body for the pleasure of others in exchange for money, and they encourage certain feelings toward clients and the law. Other attitudes include an understanding that prostitutes do only what other women do in singles bars and on dates, but they deal more honestly with exchanges of sex and things of value. Maintenance of a nondeviant self-concept relies heavily on such attitudes. Prostitutes also must learn how to develop contacts with clients as well as how to avoid problems with disease and the law.

Media displays commonly include sexually oriented materials and messages. Many uncertainties cloud the definition of obscene material, however. Religious organizations established the first regulation of pornography, but more recent control efforts have emphasized legal regulations. Pornography may be more prevalent today than it was even a decade ago, and concern over an increasing market for pornography has prompted national inquiry into the issue. Widespread use of videocassette recorders has increased opportunities to bring pornography into homes and other private places. Interest in pornography reflects a concern for potentially harmful effects from exposure, leading to calls for tighter regulation as a way to eliminate or reduce these harmful effects. Actually, however, some observers have perceived beneficial effects as well as harmful ones, but two national pornography commissions, as well as much social and behavioral science research, have disagreed on the nature of these effects.

Selected References

Donnerstein, Edward, Daniel Linz, and Steven Penrod. 1987. *The Question of Pornography: Research Findings and Policy Implications*. New York: Free Press.

This compendium of research findings includes reports related to the nature and impact of pornography from a behavioral science perspective. Donnerstein's research has provoked wide discussion and debate.

Itzin, Catherine, ed. 1992. *Pornography: Women, Violence, and Civil Liberties*. New York: Oxford University Press.

This collection of papers, most within a British context, discusses the nature and regulation of pornography. The results reflect the unexpected political alliance between far right (very conservative) and far left (radical feminist) positions.

Klassen, Albert D., Colin J. Williams, and Eugene E. Levitt. 1989. *Sex and Morality in the U.S.* Middletown, CT: Wesleyan University Press.

This report from the Kinsey Institute of Indiana University details sexual behavior based on data from a national sample. The account helps to document the extent of some behavior as well as attitudes that contribute to those actions.

Laumann, Edward O., John H. Gagnon, Robert T. Michael, and Stuart Michaels. 1994. *The Social Organization of Sexuality: Sexual Practices in the United States*. Chicago: University of Chicago Press. Michael, Robert F., John H. Gagnon, Edward O. Laumann, and Gina Kolata. 1994. *Sex in America: A Definitive Survey*. Boston: Little, Brown.

The first book reports the results from the most extensive survey on sexual attitudes and behavior to date. Unlike magazine polls, this study polled a representative sample of adults in the United States. The second book is a popular version of the first that, surprisingly, contains some information not in the more academic version.

McWilliams, Peter. 1993. *Ain't Nobody's Business If You Do*. Los Angeles: Prelude Press.

This book offers a highly readable discussion of the role of the law in areas of behavior by consenting partners, more of them sexual, that do no harm to others.

Miller, Eleanor M. 1986. *Street Woman*. Philadelphia: Temple University Press.

This description of prostitution challenges conventional wisdom on the subject. It depicts prostitutes as street hustlers who engage in other forms of crime and deviance as well. Most work without pimps but live life heavily submerged in a deviant subculture.

Weitzer, Ronald, ed. 2000. *Sex For Sale: Prostitution, Pornography, and the Sex Industry*. New York: Routledge.

A collection of papers concerning various aspects of the sex industry. Topics range from reasons for participating in pornography, the nature of street prostitution, and the impact of AIDS on sex work to issues of social and legal control over sex work.

CHAPTER 13

Suicide

In January 2000 some leading members of the Christian Democratic Party in Germany were facing a financial scandal. Christian Democrats were being asked to explain several million dollars in undeclared, and thus illegal, payments to the party. Former German Chancellor Helmut Kohl would resign from the pressure, but Wolfgang Huellen had another response. On January 20, 2000, he retired to his apartment in Berlin and hanged himself as an investigation was getting under way into the web of illicit payments to the party. Huellen indicated in a suicide note that he was aware of crimes of embezzlement associated with the payments.

Police Sergeant Solomon Bell was off duty on January 26, 2000, and engaging in his favorite "hobby": gambling. Things weren't going well. At one casino in Detroit, Sergeant Bell lost between $15,000 and $20,000. Thinking his luck might be different at another casino, Bell went to the high-stakes blackjack tables and tried one last hand. He lost $4,000. He stood up from the table, shouted "Nooooo!", drew his gun, and shot himself in the head before anyone could intervene.

While observers can easily identify the immediate conditions that precipitated Huellen's and Bell's deaths, no one should assume that such personal traits and circumstances uniquely motivate their or any other suicides. Despite the appearance of an individualistic act, people decide to commit suicide within a larger social context. As with other forms of deviance, suicide varies between groups, situations, and time periods. Suicide is a process, not just a single act. As with other forms of deviance, the norms that define social expectations involving suicidal death must underlie any understanding of its deviant character and social meanings.

Suicidal Behavior

Statistics record at least 500,000 known suicides each year throughout the world (United Nations, 1996: 13). The true figure may approach 1.2 million, however, because the act goes underreported in all countries. In the United States, suicide ranks eighth on the list of causes of death organized by frequency (Sanborn, 1990). Still, interpretations determine whether specific observers class such acts as part of a serious problem. Most people evaluate specific suicides in different ways, depending on the circumstances surrounding those acts. Almost everyone regards the suicide of a teenager as a tragedy, while they may understand and even condone suicide by a terminally ill person.

Suicide is the deliberate destruction of one's own life. Always an intentional act, it can cause death either through the individual's own deliberate acts or from his or her choice not to avoid a threat to life. In his classic study of suicide,

Durkheim (1951: 44) has even included acts of public altruism performed by religious martyrs, defining *suicide* as "all cases of death resulting directly or indirectly from a positive or negative act of the victim himself, which he knows will produce [suicide]."

This apparently simple statement masks some ambiguity in the terms *suicide* and *suicidal*, however, that results from the wide range of situations to which they can refer (see Farberow, 1977: 503–505). In some situations, people may actually take their own lives to fulfill perceived social obligations rather than as voluntary choices, as in the traditional practice among Japanese nobility and samurai warriors of hara-kiri. In some instances, people direct others to kill them—as when the Roman emperor Nero ordered an attendant to kill him so that he would not die by his own hand. More recently, media reports have highlighted cases of terminally ill patients asking their doctors or families to terminate their lives. The term *euthanasia* refers to these suicides motivated by desire to avoid suffering that results from disease or injury.

Death caused indirectly by actions without immediate lethal consequences may also amount to suicide. In fact, some observers have argued for a conception of suicide that includes both relatively quick acts and self-destruction that takes place over long periods of time. For example, some observers have discussed such behavior and conditions as alcoholism, hyperobesity, use of certain kinds and quantities of drugs, and cigarette smoking as forms of slow, relatively indirect suicide (see Farberow, 1980). Although such activities do not derive directly from suicidal intentions, people consciously associate them with relatively short lifespans, so some might undertake them as indirect suicidal behavior.

A conception of indirect suicides hardly seems implausible in light of intensive research. One source has estimated that more than a quarter of all suicides end personal histories of heavy alcohol use or follow intensive drinking bouts (Secretary of Health and Human Services, 1993: chapter 10). Some suicides relieve problems associated with drinking, physical deterioration, and increasing medical problems, while others seem like impulsive rather than premeditated acts. A number of studies have reported the relationship between alcohol consumption and suicide (Stack and Wasserman, 1995), although this link seems stronger in some countries than others (Norstrom, 1995).

This chapter does not concern itself with so-called *indirect suicides*, primarily because these people do not directly pursue death as their main motivation. Rather, the chapter confines its discussion to deliberate, immediate acts that lead directly to termination of one's own life.

Suicide as Deviant Behavior

Societies with western European backgrounds, including the United States and Canada, generally condemn suicide so strongly that their members might assume that this attitude prevails everywhere. It does not. Today's norms continue a history of wide variations in attitudes toward self-destruction.

Historical Background

Prevailing attitudes in Islamic countries strongly condemn suicide. The Koran expressly condemns the practice, and suicides remain generally rare (although not unknown) in Islamic countries (Headley, 1983). The people of the Orient, however, have not disapproved of suicide under all circumstances. In fact, *suttee*, or suicide by a widow throwing herself on her husband's funeral pyre, continued as a common occurrence in India until well into the 19th century, even after the law prohibited it in 1829 (Roa, 1983: 212). Priests taught that such a voluntary death would award a woman with a passport into heaven, atone for the sins of her husband, and give social distinction to surviving relatives and children.

Chinese society formerly accepted suicide, especially as an effective tool for revenge against an enemy, because it exposed the enemy to embarrassment and enabled the dead person to haunt this enemy from the spirit world. Voluntary death has held an honorable place in Buddhist countries, but devout Buddhists acknowledge neither birth nor death; they must meet any fate with stoical indifference.

For many centuries, people in Japan regarded suicide favorably in some situations, and the suicide rate there remains very high by world standards today. Members of all classes, but particularly nobles and military figures, traditionally learned that every individual must surrender to the demands of duty and honor. Hara-kiri developed over 1,000 years ago, originally as a ceremonial form of suicide to avoid capture after military defeat. Later, it gained the status of an appropriate response to condemnation by superiors, in contrast with the fate of ordinary people in comparable circumstances, who were hanged in public squares (Tatai, 1983: 18). So the nobility condemned to die had a choice of hara-kiri or public execution; hara-kiri was often the preferable alternative. Japan still experiences suicide pacts by lovers who wish to terminate their existence in this world and reunite in another, and some suicides continue to serve motives of revenge and protests against the actions of enemies.

The attitudes toward suicide of contemporary western European peoples originated mainly in the principles of the Jewish and Christian religions. The Talmudic law of the Jewish religion takes a strong position against suicide (Hankoff, 1979). The Christian condemnation of suicide reflects basic concepts such as the sacred status of human life, the individual's subordination to God, and the conception of death as an entrance to a new life under conditions determined by one's behavior in the old. In particular, the concept of a life after death strengthened the moral position of the church against suicide.

Although early Christians sanctioned suicide connected with martyrdom or to protect virginity, this attitude shifted to disapproval of self-destruction for any reason. People in Christian countries came to regard it not only as a sin, but also as a crime against the state. Authorities might confiscate the property left behind by a suicide and subject the corpse to various mutilations.

In the Middle Ages, leaders of the Christian church strengthened their denunciation of suicide. In particular, Augustine stated in *The City of God* that no argument could ever justify suicide, since that act precludes any possibility of repentance. He also described it as a form of murder and therefore prohibited by the sixth commandment as well as noting that no one could do anything worthy of death. Similarly, Thomas Aquinas opposed suicide as an unnatural act and an offense against the community. Above all, he claimed that a suicide usurped God's power to grant life and death. Throughout the Middle Ages and well into modern times, few had the temerity to take their lives in the face of such strong religious opposition, condemnatory public opinion, and severe legal penalties for survivors. Infrequent, sporadic outbreaks of mass suicide have diverged from this pattern on certain occasions, such as epidemics, religious fanaticism intended to gain martyrdom, or social crises (Dublin, 1963).

Religious condemnation of suicide has sparked some challenges, particularly among philosophers in the Age of Enlightenment, who stressed the importance of individual choice in all matters concerning life and death. In his essay "Suicide," David Hume argued that people have the right to dispose of their own lives without sanctions as a sinful act. Other writers, such as Montesquieu, Voltaire, and Rousseau in France, challenged laws on suicide as denials of individual choice about life and death. Other philosophers disagreed, however. From Germany, Kant described suicide as contrary to reason and therefore an offense.

England punished suicide as a felony for centuries, offenders forfeiting their property to the Crown. These provisions were abolished only in 1870. In his famous *Commentaries*, Blackstone (1765–1769: 188) had given these reasons for forfeiture: "The suicide is guilty of a double offense; one spiritual, in evading the prerogative of the Almighty and rushing into his immediate presence uncalled for; the other temporal, against the King, who hath an interest in the preservation of all his subjects."

In early America, a Massachusetts law forbade interring a suicide in the common burying place of Christians. Instead, the law mandated burial in some common highway, with a cartload of stones laid upon the grave to stand as a brand of infamy and as a warning to others. This law was repealed in 1823, but it and others like it helped to shape attitudes toward suicide in the United States.

Public Attitudes Toward Suicide

Public attitudes today generally condemn suicide, but the strength of that condemnation varies depending on the circumstances surrounding the death. Some observers also have found evidence of increasing tolerance of suicide over time (Marra and Orru, 1991). This slowly evolving change in attitudes has not yet challenged most people's image of suicide as a wrongful act or, at least, an unfortunate one. These attitudes may not regard successful suicides as worthy of religious or moral castigation, but they do reflect a general feeling of undesirability for suicide. Some may regard suicide as the result of a psychological disturbance or a more serious and persistent mental disorder, but many see some suicides as rational acts tied to particular circumstances (Ingram and Ellis, 1992).

Still, public attitudes maintain some negativity against successful suicide, and studies show that social sanctions extend to those who attempt self-destruction as well. Such an act will most likely elicit criticism if it appears to be a less-than-serious attempt to die, perhaps a gesture to attract attention (Ansel and McGee,

1971). Legally, prosecutors could bring charges against people who attempted suicide in New Jersey and in North and South Dakota until 1950. Many such attempts, serious or not, may endanger the lives of other people or rescuers. Unintended harm to others may result from an attempt to fill a room or garage with carbon monoxide, to drown oneself, or to discharge a firearm at oneself. No European country, including the Soviet Union, legally prohibits suicide, although England had such a law from 1854 until its repeal in 1961. That country prosecuted few offenders under the law, however. Still, some feared that repealing it might encourage suicide pacts, so the law gained criminal penalties for aiding, abetting, counseling, or procuring the suicide of another.

Social acceptance or condemnation of suicide varies, depending on many characteristics, including religious background and education. Generally, increasingly strong religious beliefs correspond to less-accepting attitudes toward suicide, for any specific religious affiliation (Johnson, Fitch, Alston, and McIntosh, 1980). Many people with strong religious beliefs assert that only God can choose to terminate a life; they acknowledge no human authority for such a decision. Further, relatively young, well-educated males seem to accept suicide under special circumstances and euthanasia than other people do. Groups may vary in the severity of their condemnation of suicide, but sometimes, no clear relationship links those attitudes to suicidal behavior. Markides (1981) found more fatalistic views about life and death among Mexican Americans than among Anglos, but these attitudes showed no clear relationship to suicide rates in either group. Clearly, however, such views contribute to general explanations of differences in suicide rates.

Attempted Suicide

Suicide attempts, both in the United States and elsewhere in the world, may number as many as 20 times the total number of successful suicides (United Nations, 1996: 13). Most suicide attempts occur in settings that encourage or at least allow intervention by others. The strong possibility, and perhaps the probability, that others will prevent these acts suggests that most represent not serious attempts to die, but calls for attention and intervention.

One study compared 5,906 attempted suicides in Los Angeles with 768 people who succeeded in committing suicide. It developed a profile of the typical (modal) suicide attempter: a native-born, white female in her 20s or 30s, either a married housewife or single (not divorced or separated), who attempted suicide by taking an overdose of barbiturates, purportedly to escape marital difficulties or depression (Schneidman and Farberow, 1961). In contrast, the typical successful suicide was a married, native-born, white male in his 40s or older who worked in a skilled or unskilled job before ending his life by gunshot, hanging, or carbon monoxide poisoning to escape problems with ill health, depression, or marital difficulties. Most people who attempt suicide are adolescents and young adults.

Women attempt suicide more frequently than do men. A study of suicide attempts in England found that females outnumbered males in this category by as much as $2\frac{1}{2}$ times (Hawton and Catalan, 1982: 8). U.S. studies have reported similar results, noting that females initiate 90 percent of all adolescent suicide attempts (Stephans, 1987: 108). This fact invites at least two interpretations: Women apply less successful measures when they want to commit suicide, or, more likely, women more frequently use threats of suicide to accomplish their goals. The English study found significant differences by age category in the ratio of female to male attempters, with young females much more likely to attempt suicide than younger males, compared with older males and older females (Hawton and Catalan, 1982).

The reasons for attempting suicide appear to vary by race. Young black females seem to carry out such acts more often than comparable whites in response to the loss or threatened loss of love relationships; depending on the nature of such a relationship, however, this kind of event can provoke a major crisis for any adolescent (Bush, 1978). Comparisons with successful suicides have also found that attempters tend to see themselves as too weak to cope with life's difficulties (Leenaars et al., 1992) and dissatisfied with their social integration, while those who successfully commit suicide reported to others prior to their death that they see themselves as immature and/or antisocial individuals.

The kinds of persons who only contemplate suicide may resemble those who actually attempt it. One study of Australian adolescents found similarities between both groups, such as high levels of depression, general anxiety, sleep disorders, and irritability (Kosky, Silburn, and Zubrick, 1990). The study found associations for attempters with chronic family discord and substance abuse. The odds of suicide attempts by boys increased substantially if they had experienced loss.

Extent of Suicide

Many people commit suicide each year, but far larger numbers engage in other forms of deviant behavior, such as property crimes, mental disorders, illicit drug use, and problem drinking. Unfortunately, observers encounter very serious problems when they try to evaluate statistics on suicide. Official statistics—those maintained by local, state, and federal government agencies—give no insight into the decision-making process that classifies deaths as suicides, rather than attributing them to other causes. Such judgments implement no uniform set of standards, seriously complicating analysis. One study sampled 191 coroners in 11 states to determine the processes by which they certified cases as suicides (Nelson, Farberow, and MacKinnon, 1978). This research revealed extensive variation among the resources, philosophies, procedures, statutes, and backgrounds that the coroners applied to this judgment. Thus, one coroner might count a death as a suicide, while another might recognize and record it another way. No one has yet elaborated a precise relationship between statistics for officially recorded suicides and actual suicides. Like crimes, officials do not learn about all suicides, and they record some known cases in misleading ways. More than half of a sample of 200 medical examiners agreed that reported numbers might reflect less than half of the actual number of suicides (Jobes, Berman, and Josselsen, 1986). Researchers find even greater problems with statistics in other countries, such as those in Asia (Headley, 1983).

These statistical problems limit the precision of estimates that state the number of suicides. In 1987, one official estimated that 25,000 suicides occurred annually in the United States (McGinnis, 1987: 21), a figure close to that in 1981, when 26,010 suicides were recorded in the United States (Hacker, 1983: 70). Another source placed the number of U.S. suicides in 1990 at about 30,800 (Bureau of the Census, 1992: 84–85; Bureau of the Census, 1993: 91), and estimates continued to project similar figures at the turn of the century (Bureau of the Census, 1999). Undoubtedly, these figures underestimate the total of all suicides, perhaps by as much as one-fourth to one-third, because people hide the truth to blunt the stigma attached to such deaths. Relatives and others may deliberately conceal the true circumstances of a death, and doctors and officials may fill out death certificates in ways that protect the feelings of survivors. Some suggest standardizing the methods by which medical examiners report deaths in order to improve statistics and the analysis they support (Jobes, Berman, and Josselsen, 1987), but such an initiative would obviously only partially solve the problem.

Absolute numbers give a misleading impression because populations have increased, so most sources report suicide figures as rates, that is, the number of suicides for a given population size. The suicide rate in the United States has held fairly stable over the past few decades at between 12 and 13 suicides per 100,000 population (Bureau of the Census, 1993: 91). At the turn of the century, the rate had declined slightly to 11.8 (Bureau of the Census, 1999).

The suicide rate does fluctuate somewhat, however, with particular responsiveness to changes in the economy (generally accelerating during periods of depression and falling during periods of prosperity, but see Yang, 1992). The suicide rate reached its highest level in the United States in 1932, the depth of the Great Depression, at 17.4 per 100,000; the lowest rate, 9.8, accompanied widespread prosperity in 1957. National suicide rates also decline during wartime, a trend noted by Durkheim more than 80 years ago. From 1938 to 1944, during the period of World War II, rates declined 20 to 50 percent in all warring nations (Sainsbury, 1963: 166; but see Marshall, 1981). In the United States, the rate declined by about one-third, from 15.3 in 1938 to 11.2 in 1945, during the years of World War II. Several conditions may account for this wartime decline. The feeling of public unity that prevails during most wars works against the social isolation that typically contributes to suicide. Wars that do not generate widespread national support, such as the war in Vietnam, show no associations with declines in the suicide rate. Wars may also bring abundant economic opportunities, another social deterrent to suicide.

Some investigators argue that the relative stability of the overall suicide rate in recent years masks some interesting and important changes. Seiden and Freitas (1980) argue that the steady national rates hide decreases in suicide among older age categories offset by increases in suicide among younger people (see also McGinnis, 1987). Suicide varies strongly with age, increasing as ages advance, but suicide rates among young people have increased significantly during the past two decades.

Suicide rates differ substantially among countries, as well as within and between social categories, such as age, sex, and race. Observers have identified other variations as well, but some remain unexplained by any particular theoretical perspective. For example, a study of over 18,000 suicides from 1973 to 1979 has reported that, contrary to popular belief, suicides decline around major national holidays (Phillips and Wills, 1987). The rate declined before, during, and after Memorial Day, Thanksgiving, and Christmas, and it declined before and after New Year's, the Fourth of July, and Labor Day, with normal rates on those days. The variations reported here reflect consistent patterns of suicide, so some sociological theory should explain them.

Variations in Suicide by Country

For many years, Japan annually experienced the highest suicide rate in the world, but in recent years the highest annual rate has been in Hungary, with a suicide rate more than three times higher than that of the United States (see Table 13-1). The suicide rates in Finland, Belgium, Denmark, and Austria average about twice as high as that of the United States. The Canadian suicide rate (not included in the table) also runs above that of the United States (Leenaars and Lester, 1992).

TABLE 13-1
Estimated Suicide Rates in Selected Countries

Country	Suicide and Self-Inflicted Injury Rate per 100,000 Population
Hungary	30.9
Finland	26.1
Denmark	20.4
Austria	20.2
France	19.3
Czech Republic	18.8
Japan	15.1
Sweden	14.2
United States	11.8
Netherlands	10.0
Portugal	7.5
Italy	7.2
Spain	7.1

SOURCE: Bureau of the Census. 1999. *Statistical Abstract of the United States, 1999.* Washington, DC: Government Printing Office, p. 837. Data collected by the United Nations World Health Organization.

The previously mentioned unreliability of suicide statistics complicates precise international comparisons.

Methods of certifying deaths vary between countries, and some nations have established reputations as better, more careful recordkeepers than others. Reasons for suicide seem to vary from country to country less than recognized rates do. A Danish writer has attributed the high suicide rate in Denmark to causes independent of individuals, such as political and economic factors (Paerregaard, 1980). Another source has implicated similar causes in the relatively high French suicide rate (Farber, 1979), emphasizing low social integration reflected in high rates of alcoholism, a large elderly population, high immigration and low emigration, and high urbanization. In general, predominantly Catholic countries report low suicide rates, with some exceptions. (Austria, a Catholic country, has a high rate.) Research has not identified a strong relationship between religious preference and suicide, as a later section will explain. At this point, however, certain national variations merit attention.

Sweden and Norway   Residents of Sweden commit suicide at relatively high rates compared with those of its neighbor, Norway. One explanation for differences among Scandinavian countries cites differences in child-rearing patterns. Some have claimed that a Norwegian child's upbringing stresses open expressions of emotions and aggressive feelings, preventing him or her from carrying pent-up hostility into later life (Hendin, 1964). More likely, the greater strength of the primary group in Norway may contribute to strong relationships, which help to prevent suicide (Farber, 1968).

Australia   Although the country's overall annual suicide rate has remained within a relatively constant range of between 12 and 14 per 100,000 population, suicide is the second most common cause of death among young Australian males (United Nations, 1996: 4). In 1991, suicides exceeded motor vehicle accident deaths in this population for the first time in 50 years. Evidence also suggests that suicide, once uncommon among Aborigines, is occurring with increasing frequency among these people.

China   Once an accepted practice in China, as explained earlier, suicide is now discouraged by strong attitudes there. Estimates indicate a varying suicide rate between 8 and 12 per 100,000 population in urban areas and from 20 to 30 per 100,000 in rural areas. The country's highest suicide rates prevail among the young and elderly. Suicide prevention efforts and discussions began in the 1970s, and increasing research targets this problem.

Estonia   From the beginning of this century, the suicide rate in Estonia has fluctuated in conspicuous coordination with social-political pressure and disturbance. Low annual rates held during the social democratic movement of 1905 and during most of the period of Soviet occupation. Rates dropped between the two world wars of the 20th century and increased again between 1947 and 1953 as political pressure on Estonians increased under Soviet rule. The rate dropped to 14 per

100,000 population in 1955, but it rose again during the Khrushchev era to 32, falling back to 25 in 1989. Since that time, the rate has been increasing.

Finland The Finnish suicide rate ranks among the highest in the world, as it has for some time. The relatively steady rate throughout the 1990s masks increases among young people, especially young men, like those experienced in many countries. A strong antisuicide movement in Finland now addresses the problem in an attempt to change the trend.

Hungary The suicide rate in Hungary has led the world for the past 100 years. The highest ever recorded, 46 per 100,000 population, occurred in 1985, followed by a slight decline in recent years. Suicide deaths have run three times as high as road-accident deaths. The highest rates occur among the elderly, who develop the weakest social support networks.

United States The U.S. suicide rate has stabilized between 11 and 13 per 100,000 population over the past several years. Elderly people kill themselves at the highest rate, but the rate for young people raises concern, as it does in other countries, because it shows faster increases than those for other groups display. (So far, however, it has not yet overtaken the rates for some other groups.) Research suggests substantial social differentials in suicide among different subgroups and in varying social circumstances.

Social Differentials in Suicide Rates

Years ago, a leading study found an association between customs and traditions accepting or even condoning suicide and large numbers of individuals taking their own lives; where the state, church, and/or community severely condemn that act, however, it seldom occurs (Dublin and Bunzel, 1933: 15). Such a generalization about society's reaction to suicide shows no clear relationship, however, to variations by sex, race, marital status, and so forth within particular countries. For example, nothing suggests that unusually severe disapproval accounts for the low observed suicide rates among blacks and young people. Moreover, no evidence attributes all of the increases or decreases in specific countries' rates to corresponding changes in their norms governing suicide (Gibbs, 1971: 302).

Instead, differentials in suicide rates show extremely strong and independent variations. Some have concluded, therefore, that no social status or condition, including widespread acceptance or criticism, acts to generate a constant rate in all populations. According to one example, "an occupation with a high suicide rate in one community may have a low rate in another; and rates for countries or religious groups change substantially over time" (Labovitz, 1968: 72).

Sex Differentials Suicide occurs more commonly among men than among women in almost all countries. In fact, men's rates generally average three to four times higher than women's, although women *attempt* suicide more often than men do (Canetto and Lester, 1995). Of approximately 26,000 U.S. suicides in 1981, men committed 73 percent and females the remaining 27 percent (Hacker, 1983: 73). In Finland, almost four times as many men as women commit suicide; in Norway, South Africa, and France, the ratio approximates the U.S. differential: three to one. In Hungary and Austria, countries with very high overall rates, the difference runs only slightly higher than two to one. Among older people, the gap between men's and women's rates widens even further, while it shrinks among adolescents. Women in Asia commit suicide much more frequently than do men in Western Europe and America; thus, the difference in the ratio substantially declines. In Japan, for example, male suicides exceed those of females by only one and a half to one.

Differences also separate the sexes in the means by which they attempt or accomplish suicide. Males tend to use more dangerous or immediately lethal means, such as firearms, while women more often favor chemicals or knives. These differences undoubtedly reflect gender socialization experiences of each sex.

Observers have advanced a number of hypotheses to explain these gender-related differences in suicide rates. Wilson (1981), for example, argues that males experience failure according to obvious and clearly defined standards, but the comparatively diffuse female sex role lacks established standards for success and failure; therefore, feeling of failure less frequently lead women to commit suicide because they may still doubt what constitutes failure. As female roles become better defined, this explanation implies the probability of a corresponding change in the suicide rate for females. Davis (1981) has confirmed this expectation somewhat after looking at changes in suicide rates by sex over time; this research has concluded that increased labor force participation by women has contributed to an increase in the female suicide rate.

Age Generally, suicide rates increase with age in developed countries, as mentioned earlier, and this relationship holds for men more strongly than for women. Men's rates continue increasing with age in each successive range, while women's rates peak at 45 to 54 years of age. Evidence points to increases in rates of suicide among young age groups in recent decades. In particular, suicide rates for children between 10 and 14 years of age showed an increase of 75 percent between 1979 and 1988, according to a 1993 study by the U.S. Centers for Disease Control and Prevention in Atlanta (*Des Moines Register*, June 19, 1993, p. 6A). Days after the CDC issued the report, a 6-year-old stepped in front of a moving train to become the youngest U.S. suicide victim on record. Increasing availability of firearms seems to have influenced the probability of young people committing suicide; in 1989, nearly 59 percent of child suicides involved firearms, compared with 53 percent in 1983, a change that may be attributable to the increased availability of firearms.

The suicide rate for those aged 15 to 19 increased 140 percent between 1960 and 1975 (Hawton, 1986), although evidence seems to suggest stabilization at a rate comparable to that for adults (Males and Smith, 1991). Still, suicide has become the second-leading cause of death, behind accidents, among people in this age group (Spirito, Brown, Overholser, Spitz, and Bond, 1991).

The suicide rate for those aged 20 to 24 also increased 130 percent (Hawton, 1986) between 1960 to 1975. The rate for those aged 25 to 34 nearly doubles that for the 15-to-24 age group, although it shows a slower increase (Sanborn, 1990).

✳ *In Brief*   **Suicide, Age, and Sex**

There is a relationship between suicide and age and sex. Generally, the higher the age, the higher the suicide rate. But there is also a relationship between suicide and sex. Although females are more likely to attempt suicide, males are more likely to successfully complete a suicide in every age category, including younger persons. The chart below shows these relationships.

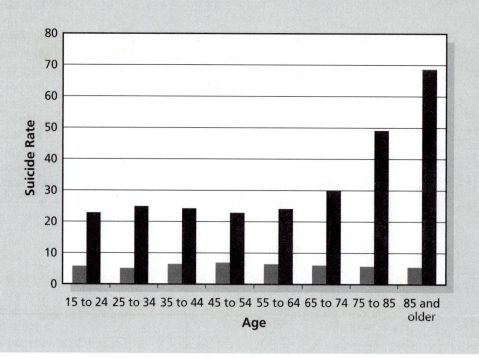

Source: Bureau of the Census. 1999. *Statistical Abstract of the United States, 1999*. Washington, DC: Government Printing Office, p. 106.

A Chicago-based study concluded that people become progressively more socially and physically isolated as they age, with a corresponding increase in the wish to die (Maris, 1969: 15). This result may help to explain the suicide rate for people over 65 in the United States, which runs almost twice as high as the rate for people between 25 and 34.

A study of mental patients aged 45 to 60 linked suicide specifically to feelings of isolation, knowledge of other suicides, pride, belief in an afterlife, and history of depression. It described suicides as

more socially isolated; more knew others who had committed suicide; more felt no pride in aging and predicted poor treatment from relatives when they became even older; more approved suicide in some circumstances and did not believe in an afterlife; more had been depressed severely and/or frequently and had a family history of depression. (Robins, West, and Murphey, 1977: 20)

Different combinations of age and race relate to differences in suicide rates. In the United States and several other countries, the male suicide rate continues to increase into old age (Headley, 1983; Stafford and Gibbs, 1988), but the rate for U.S. white females increases until it peaks at about age 50. The suicide rate for nonwhite females, however, peaks between 25 and 34, and that for nonwhite males peaks between 25 and 29. The large proportion of whites in the U.S. population obscures these variations, producing an overall suicide rate that generally increases with age. For similar reasons and a changing age distribution, overall suicide rates may increase in the future along with the general aging of the population (McIntosh, 1992).

Many conditions may help to explain why suicidal behavior becomes increasingly common among progressively older groups, particular among those who must endure unpleasant life circumstances (Lester and Tallmer, 1994). Declining social contacts and mounting health problems can severely affect quality of life for many elderly people. Administrators of long-term care facilities (either nursing homes or "retirement homes") report substantial suicidal behavior among their residents, especially white males (Osgood and Brant, 1990). Refusal of food, drink, and medications defined the most common suicidal behaviors in this group. Members cited varying reasons for these behaviors, including depression, loneliness, feelings of rejection by family members, and loss. Among the most elderly population (aged 85 and older), the suicide rate for males exceeds that of females by 12 times; loss of their wives may powerfully influence these suicides (Bould, Sanborn, and Reif, 1989: 67–68).

Patterns of suicide differ slightly between developed and less-developed countries. One study of suicide in 49 countries reported that rates for countries with emerging economies show either single peaks in early adulthood or double peaks in early adulthood and again in older age groups (Girard, 1993). Furthermore, the suicide rates of young women approached or exceeded those of young men in many less-developed countries. No one can yet say whether these differences from developed countries result from variations in general economic conditions or from some other factor.

Race   Previous research on the distribution of suicide by race has concluded that whites kill themselves at substantially higher rates than occur among African Americans. One 1969 study found suicide rates for white subjects twice as high as those for nonwhites (Maris, 1969). A 1970 study of suicide among African Americans assessed the validity of an often-repeated claim that suicide rates among young African American women had risen sharply during that decade. The results characterized men in their 20s as the most suicide-prone group within the African-American population, and no substantial increase in suicide had occurred among young African-American women (Davis, 1979). Both African-American males and females in the young age categories through middle age commit suicide at higher rates than comparable whites do, although the overall African-American rate remains below that of whites (Kirk and Zucker, 1979).

Varying explanations have sought to establish reasons why the suicide rate for African Americans peaks in younger age categories than that for whites. Hendin (1969) has attributed African-American suicide to self-hatred as a result of experience as a minority group in this country. Kirk and Zucker (1979) have found lower racial consciousness and weaker feelings of group cohesion among those who attempted suicide than among those who did not, a finding consistent with Hendin's interpretation.

Marital Status   Married people commit suicide at lower rates than single, divorced, or widowed people do. In particular, married people post lower suicide rates than do single people in all age groups. About three times as many widowers as married men and five times as many divorced men take their own lives. Marriage shows an association with low suicide rates among women as well (Gibbs, 1982). The conditions of marital dissolution and the duration of separation from marriage also appear to act as important variables in this relationship. Specifically, final divorce and long-term separation relate more strongly to suicide than do short-term separations for any reason (Jacobson and Portuges, 1978).

As with age, racial differences show up in the relationship between marital status and suicide. Davis and Short (1978) have reported an inverse relation between marital status and suicide among blacks like that for whites, but the relatively weak relationship accounts for little variation in suicide rates.

Extensive research has documented the relationship between marital dissolution and suicide in the United States (e.g., Brealt, 1986), but only limited research has targeted similar relationships in other nations. One exception, a study by Stack (1990), has examined the effects of divorce on suicide in Denmark. Despite a much lower rate of divorce there than in the United States, the study duplicated the results of research in the United States: A higher divorce rate corresponded to a higher suicide rate over time. Specifically, Stack has associated a 1.00 percent increase in the divorce rate with a 0.32 percent increase in the suicide rate. In fact, divorce trends predicted both adult and youth suicide rates.

Religion   Suicide rates vary greatly among the main religious groups in Western civilizations. In general, both in Europe and in the United States, suicide rates for Catholics fall below those for Protestants, although this differential has diminished. In the past, the Jewish rate has typically remained lower than the Catholic rate, except during periods when persecutions have made life particularly difficult or hopeless for Jews, leading to waves of suicide. In recent years, the Jewish suicide rate has risen considerably, perhaps reflecting changes in religious influence and greater participation in general society. Furthermore, both Protestant and Catholic rates have increased during the past century (Maris, 1981), and one study has reported a positive relationship between suicide and membership in the Catholic church (Yang, 1992).

Some observers have linked religious differences in suicide rates to variations in those groups' integration into larger society (Stack, 1982). Protestant religious groups tend to promote more individualistic values than the Catholic church does. Further, the Catholic position on suicide gives specific guidance about the destructive effect of suicide on an individual's afterlife. One should avoid placing too much emphasis on religious affiliation alone, though. The rate of suicide in northern Italy almost doubles that of southern Italy, despite poorer economic conditions, less education, and stronger adherence to Catholic doctrine in the southern regions (Ferracuti, 1957: 74). An individual's particular religious affiliation may affect the tendency toward self-destruction less powerfully than active versus casual religious participation. In general, as people involve themselves more actively in their religions, they become less likely to regard suicide as an acceptable act (Hoelter, 1979).

While strong religious feeling appears to discourage suicide, one study found contradictory indications in suicide notes and victims' diaries. Religion may help unhappy people to *justify* their decisions to commit suicide (Jacobs, 1970). These materials often mentioned meeting loved ones in the afterlife, with language about "happy reunions" and release from worldly problems to reach "the final rest." Suicidal people may look to religious teachings for confirmation that they bear no blame or sin and that self-destruction represents their only option for dealing with their lives. The following two notes, reported by Jacobs (1970), give typical examples:

> I am sad and lonely. Oh God, how lonely. I am starving, Oh God, I am ready for the last, last chance. I have taken two already, they were not right. Life was the first chance, marriage the second, and now I am ready for death, the last chance. It cannot be worse than it is here.

> My dearest darling Rose: By the time you read this, I will have crossed the divide to wait for you. Don't hurry. Wait until sickness overtakes you, but don't wait until you become senile. I and your other loved ones will have prepared a happy welcome for you.

Occupation and Social Status   In his classic study of suicide, Durkheim (1951: 257) found links between occupational status and suicide, which occurs unusually frequently among people in the upper ranks of various occupations as well as in positions of high status. Most research since that time, however, has strongly associated suicide with membership in the lower social classes (Maris, 1969). One of the most comprehensive studies of differential mortality in the United States found inverse relationships between suicide and education (a common indicator of social class), at least for white males; the research gathered incomplete data for other groups (Kitagawa and Hauser, 1973). In fact, the least-educated group had twice the suicide rate of the most-educated group. Studies in other countries have confirmed this relationship (e.g., Li, 1972).

High suicide rates tend to characterize occupations with substantial uncertainty and economic insecurity compounded by lack of social cohesion among individuals. Workers who enjoy secure employment and support from other workers generally commit suicide at low rates. This finding suggests the possibility of an important subcultural component in suicide that operates much like other forms of deviance.

Some occupations have long-established reputations for high suicide rates. Policing is a dangerous occupation filled with alternating periods of boredom and anxiety. Anecdotes and articles in the popular press give the impression of a high suicide rate among police officers (Violanti, 1995), but careful analysis finds little support for such a claim. Lester (1992a) examined Interpol data for 16 countries and compiled a suicide rate for police officers not consistently higher than that for men in general in each of the countries.

A study of a sample of 166 cases from Chicago reported that occupational status, by itself, does not effectively predict suicide (Maris, 1981: chapter 6). Citing data from an unpublished study, Maris (1981: 146) has indicated that members of the clergy and dentists, with roughly comparable occupational prestige, commit suicide at quite different rates (10.6 and 45.6 per 100,000, respectively). At the low end of the occupational scale, machinery operators kill themselves at a rate of 15.7 per 100,000 compared with 41.7 for mine operatives and laborers. The study's nonrandom sampling techniques prevent clear generalizations, though.

Types of Suicide

Suicidal behavior bears a relationship to a society's organization, occurring especially frequently in urban societies. In fact, some reports have suggested no experience of self-destruction among some preliterate societies. One observer has reported that Australian aborigines invariably responded to questions about suicide by laughing and treating the possibility as a joke (Westermark, 1908: 220). Another researcher has reported a similar response from natives of the Caroline Islands. A survey of some 20 sources dealing with the Bushmen and Hottentots of southern Africa has revealed no references to suicide among these people (Faris, 1948: 148). Suicide occurs among some folk societies, however, and at much higher rates in some than in others. Observers have reported suicides among the natives of Borneo, the Eskimos, and many African tribes. Some have also described it as fairly common among the Dakota, Creek, Cherokee, Mohave, Ojibwa, and Kwakiutl tribes as well as Fiji Islanders, the Chuckchee, and Dobu Islanders.

Durkheim (1951) displayed a particular interest in self-destructive acts among preliterate societies. As a result of his studies, he classified suicides according to three types: *altruistic, egoistic,* and *anomic* suicides. Durkheim also mentioned another type, *fatalistic* suicide, but he made little of it, and those who have commented on his theory have generally ignored it. Suicides of this nature reflect reactions to excessive regulation perceived to block future improvements in one's life. A slave's suicide would provide a good example of fatalistic self-destruction.

After laying out his classification scheme, Durkheim then examined the specific motives underlying each type of suicide. In spite of his interest in personal motivation, however, his explanation of suicide evokes social concepts. Later theorists have refined Durkheim's theoretical argument to ground it entirely in structural relationships within societies rather than individualistic arguments (for example, Bearman, 1991).

In general, Durkheim regarded suicide as a result of the degree of social interaction and regulation in a society, the prevalence of group unity, and the strength of ties binding people together. He treated it not as an individual phenomenon, but as a consequence of social organization and structure.

Altruistic Suicide

On the whole, according to Durkheim, members of preliterate societies committed suicide for reasons considerably different from those in modern society. Members of folk societies who killed themselves often did so for *altruistic* reasons. By taking their own lives, they hoped to benefit others, an expression of the emphasis in such societies on group welfare over individual prosperity or even survival. When someone's actions or continued survival caused harm to the group, that person felt bound to commit suicide so that the group would have one less mouth to feed or to protect it from the gods. Altruistic suicides include those arising from physical infirmities, connected with religious rites or warfare, or intended to expiate violations of certain norms, mores, or taboos. Under such conditions, suicide does not constitute a deviation; in fact, an individual might commit a transgression by refraining from the act.

Old or infirm members of certain preliterate societies facing limited food supplies may pose real burdens to their tribes. Among the Eskimos and the Chuckchee, for example, old people who could no longer hunt or work killed themselves so that they would not consume food needed by other adults in the community who would contribute to its production. Some suicide occurs in warfare when combatants kill themselves to avoid capture and slavery or because they feel disgraced by defeat. Probably the most common form of altruistic suicide, however, represents an attempt to achieve expiation for violation of society's mores, often in the form of taboos. Violators who fail to commit suicide in atonement risk the group's imposition of other sanctions, such as perpetual public disgrace. Trobriand Islanders who violated certain taboos, for example, generally committed suicide by climbing palm trees, from which they gave formal speeches before jumping to their deaths (Malinowski, 1926: 97).

In modern societies, some elderly people or patients with incurable illness sometimes end their lives to avoid burdening others, but others generally do not approve this type of altruistic suicide. One group in the United States, the Hemlock Society, promotes access to humane methods of suicide, but only under certain medical circumstances. These conditions advocate limiting euthanasia to mature adults who have made considered decisions after receiving medical advice (Humphry, 1993). Even carefully evaluated choices of suicide still invoke considerable ambivalence, with some people strongly condemning them and others approving.

Modern society offers examples of individuals giving their lives in time of war in order to accomplish some goal involving group values; sometimes this behavior draws social approval as a heroic act. During World War II, many Japanese soldiers engaged in behavior judged as suicidal by outside observers; faced with certain death, large numbers of Japanese troops charged or held out to the last soldier. Late in the war, legendary Japanese kamikaze pilots disregarded their own lives to fulfill their missions. They consciously guided planes loaded with explosives into Allied ships, giving their own lives in attempts to destroy their targets. (For a general discussion of suicide in Japan, see Tatai, 1983 and Pinquet, 1993.)

Egoistic Suicide

People commit *egoistic* suicides for individualistic reasons. Some describe these suicides, the most common in modern societies, as consequences of lacking identification with others or group orientations. In societies commonly characterized by such attitudes, individualistic motives of many kinds regularly lead to suicide: financial problems, health difficulties, marital or relationship struggles, and occupational setbacks. Egoistic suicides are products not of tightly integrated societies like those that produce altruistic suicides, but of societies without close or group-oriented interpersonal relationships.

Anomic Suicide

Anomic motives lead to suicide when individuals feel "lost" or normless in situations governed by confused or disrupted social values. Such suicides also occur when individuals experience downward social mobility or so much success that they feel they have achieved everything, so that life no longer holds any meaning.

In one example of an anomic suicide, a wealthy, middle-aged businessman with no apparent financial, health, or marital problems took his own life. He had devoted all his efforts to building up his company to achieve a lifelong goal: a merger with a larger company. As the fruits of this long-sought merger, the man retained the presidency of his own concern and became a vice president in the larger company. After the conclusion of the agreement, however, he immediately went into a depression and eventually committed suicide. The coroner's report described the act as the reaction of a man who had focused exclusively on building his business and making the desired deal; the realization that he had sacrificed his position as the single, direct owner of his business left him feeling as though he had lost his objectives in life.

A study of white male suicides in New Orleans has provided additional examples of anomic motivations for self-destruction. Breed (1963) has associated such suicides with substantial work-related problems such as demotion and downward social mobility, reduced income, unemployment, and other job and business difficulties. In the study of Chicago suicides discussed earlier, Maris (1969) found that many anomic "suicidal careers" included common elements such as employment difficulties and feelings of hopelessness.

Some anomic suicides result from severe disruptions of society's equilibrium. Such an upheaval creates a social void without any social order adequate to satisfy an individual's social desires, leaving him or her uncertain about which way to turn. Such anomic suicides in modern society commonly follow severe and sudden economic crises. For example, many suicides followed soon after the U.S. stock market crash in 1929. Similarly, sudden, abrupt changes in one person's standard or style of living may produce a sense of normlessness. These experiences contribute to high rates of suicide, particularly for people who face prospects of downward social mobility or breakups of long-term relationships such as marriages.

Adolescent Suicide

While people generally react to any suicide with sorrow, most regard adolescent suicide as an especially tragic event. Adolescents "just starting out" in life seem poised to explore limitless new opportunities, but they often cannot evaluate their immediate difficulties in the perspective of lifelong events. An adolescent may feel burdened by a major problem after an event, such as a broken romantic relationship, that older people recognize as a normal experience of growing up (see Berman, 1991).

Such events have inspired particular concern in recent years as evidence mounts of dramatic increases in youth suicide rates over the past two decades. The five leading causes of death for adolescents and young adults (people aged 15 to 24) are, in descending order, accidents, homicide, suicide, cancer, and heart disease (U.S. Department of Health and Human Services, 1985). While death rates due to accidents, cancer, and heart disease have declined since 1950, the rate of death due to homicide doubled, and the death rate from suicide almost tripled from 1950 to 1982. The youthful suicide rate has declined again in recent years, but the change has not alleviated concern over the trend for the past three decades (McGinnis, 1987). One source reports a suicide rate in the age group 15 to 24 of about 9.0 per 100,000, and it gives a substantially lower rate for the range 15 to 19: 0.4 per 100,000 (Iga, 1981). For comparison, recall that the national suicide rate approximates 12.0 per 100,000.

Statistics about adolescent suicides, however, probably underestimate their true incidence even more dramatically than those for adult suicides. Children under the age of 10 almost never commit suicide; people younger than 15 only occasionally do so. These assertions do not imply that children do not occasionally "wish they were dead" as they grow up. These feelings do not lead to suicides partly because children have only incompletely formed self-identities, status attachments, and social roles, which sometimes induce suicide in adults when certain situations endanger them. Moreover, childhood crises usually persist only temporarily and seldom promote the sort of long-range "brooding" that accompanies comparable adult crises.

High adolescent death rates appear to persist in countries with high overall suicide rates. In Japan, for example, Iga (1981) has reported that 40 percent of male and 60 percent of female junior high school students indicated a wish to die, and an additional 24 percent and 23 percent, respectively, had occasionally entertained thoughts of death. Some have attributed these figures, along with a recorded adolescent suicide rate twice that of the United States, to elements of the Japanese educational system. Young people must take a one-time college-entrance examination that substantially determines their future educational opportunities and, therefore, their life chances. This exam creates enormous stress on young people in a society where suicide is a highly institutionalized adjustment mechanism.

Research has traced another class of suicides by young Japanese victims to school-yard bullying, which generates considerable despair because parents and school authorities tolerate the practice (Sakamaki, 1996).

The significance of increasing adolescent suicide rates has not yet yielded to precise interpretation. While young people today feel great pressures as they grow up, earlier generations appear to have known similar anxiety. Generational worries have risen and diminished as people have responded to such issues as nuclear holocaust, war (such as Vietnam), illicit drug use, and violence; no one has established the real relationship of such concerns to youthful suicide rates. Undoubtedly, some blame parents for weak moral upbringing, while others blame society for the cumulative effect of its ills (see Table 13-2).

Nevertheless, culture influences suicide, and its elements bear predictive relationships to self-destructive behavior. For example, some conditions certainly increase the risk of suicide, and teenagers experience these risk factors as combinations of social pressures. Chances of suicide rise for adolescents with histories of previous suicide attempts, depression, illicit drug use, various kinds of deviant behavior, and suicidal behavior in the family (Gould, Shaffer, and Davies, 1990). These direct relationships highlight an important need for a theory to make sense of them.

TABLE 13-2
Causes of Death by Suicide, United States, 1993–1995

Rank	Cause	Deaths
1	Firearm	56,208
2	Poisoning	15,691
3	Suffocation	14,589
4	Fall	2,037
5	Cut/pierce	1,506
6	Drowning	1,148
7	Other specified	939
8	Fire/burn	531
9	Other specified, NEC	373
10	MV traffic	327
11	Unspecified	156
12	Natural/environ	21
13	Transport, other	2
		93,528

SOURCE: Centers for Disease Control, Web site accessed August 17, 1999.

Sociological Theories of Suicide

Social scientists have made several attempts to formulate general theoretical explanations of suicide in contemporary society. Separate perspectives have emphasized social integration, degrees of social constraint, status integration, status frustration, community migration, and socialization. Each one attempts to explain both patterns of suicide within society (its distribution by sex, race, age, marital status, and so forth) as well as individual occurrences.

Social and Religious Integration

In his classic 1895 study, Durkheim (1951) stated that an explanation of the suicide rate in any population should focus not on the attributes of individuals but on variations in social cohesion or *social integration*. Simply put, this theory asserts that greater social integration produces a lower suicide rate, a view still widely accepted today.

Durkheim believed that suicidal behavior displays an inverse relationship to the stability of social relations among people and the extent to which social institutions, whether religion, family, or others, integrate them together. He cited many examples as evidence for his thesis, including the lower suicide rate of Catholics compared with that of Protestants and the lower rate of married people compared with single, divorced, or widowed people. He attributed all of these differences to more complete social integration in the less suicide-prone groups.

While Durkheim demonstrated his thesis in many ways, others have criticized him for failure to establish a set of rigorous criteria for measuring social integration (Pope, 1976). Maris (1969) has modified Durkheim's thesis by proposing an inverse relationship between suicide and the degree of social constraint on an individual's actions. Low external constraints allow individuals to act without regulation either by other people or by shared ideas. For example, this version of the theory attributes the high probability of suicide by men, particularly those in advanced age categories, as a result of social isolation or role failure that weakens their perceptions of outside limits on their actions.

Some dispute the integrative effects of religion, predicted by Durkheim's theory to discourage suicide, at least in the United States (see Pescosolido, 1990; Stack, 1982). Some research does suggest, however, that religion influences suicide in ways largely predictable by Durkheim's views (Brealt, 1988). A number of major changes since Durkheim's day, may, however, have altered the effect of religion on suicide. Protestant groups observe more diverse practices than they did when Durkheim wrote, as the process of modernization has brought a number of changes in the teachings of today's denominations. For one result, the integrative effect of membership in particular religious denominations varies from region to region in the United States. Religious networks may define unique patterns of interactions in each area; the relationship between religion and suicide probably varies by region as well. Numerous, similar changes in the Roman Catholic church might also explain increased suicide rates among its members (Yang, 1992).

Finally, empirical evidence supports only mixed conclusions regarding the Durkheim hypothesis. While Durkheim himself provided substantial supporting evidence, at least one study of preliterate tribes found differing integration without accompanying variations in suicide (Lester, 1992b).

Status Integration

Another perspective forms an argument related to Durkheim's theory of social integration. Based on an interpretation of suicide differentials in the United States, this theory holds that a particular pattern of status occupancy or *status integration* in a society powerfully influences suicide (Gibbs and Martin, 1958, 1964). Basically, this theory asserts that a society's incidence of suicide varies inversely with a measure of status integration in the population. Conflicting attributions of status, such as a young person working in a high-status occupation, cause difficulty in forming and maintaining social relationships; this problem inhibits social integration and contributes to high suicide rates for such people.

In contrast, suicide remains relatively rare in populations that hold status positions closely associated with one another. Stable status relationships protect members from role conflict and help them to conform to the demands and expectations of others. Such situations encourage stable and durable social relationships with others.

Other observers have found this theory extremely difficult to test adequately because society includes so many possible combinations of status characteristics. Moreover, most statements of the theory attribute equal importance to all potential status roles. An adequate empirical test would have to cross-classify suicidal behavior against all conceivable combinations of age, sex, race, occupation, marital status, education, and so forth; such an evaluation would require extensive data about every member of a huge sample.

Among tests conducted so far, the theory has received limited empirical support in a study of white women (Gibbs, 1982). In a more extensive test, however, Stafford and Gibbs (1985) detected little support for the theory when they applied a more sophisticated measure of status integration to 1970 census data for employment, household, marital, and residential characteristics. Another investigation has questioned the assumption that all status roles exert equally important influences, and it has described changes in the importance of some status roles over time (Stafford and Gibbs, 1988). Occupational status in particular has gained importance in recent decades, and marital status plays a more important explanatory role for males than for females.

Status Frustration

Several sociologists have tried to link suicide and homicide within a framework of varying adjustments to aggression generated by *status frustrations*. According to Henry and Short's (1954) theory, the best-known example, suicides direct this aggression against themselves, whereas perpetrators of homicides direct it at others. Specifically, according to this perspective, individuals become so frustrated through "the loss of status position relative to others in the same status reference system" that they feel like killing someone (Henry and Short, 1954: 26). In this view, some frustration, often over failure to maintain a constant or rising position in a status hierarchy relative to others, arouses aggression, and suicide expresses this aggression through an attack on oneself.

The theorists offer several kinds of evidence to support this relationship. They explain that married people commit suicide at lower rates than nonmarried (divorced, widowed, or single) people because marriage establishes a strong relational system that constrains people to conform to the demands and expectations of others regardless of the amount of stress and resulting frustration they experience. Similar ideas about involvement with other people also explain low rates of suicide in rural areas, high rates in inner-city areas, and the general tendency for suicide rates to increase for older age groups in which close relations with others diminish.

A major portion of Henry and Short's theory rests on the assumption of a higher suicide rate for high-status people than for low-status people. As pointed out earlier, however, some indicators suggest the reverse relationship (also see Levi, 1982). Henry and Short's theory also implies that women should commit murder more frequently than men do, but they admit that strong evidence contradicts this expectation. Still, homicide and suicide remain closely connected in some theoretical ways, and few theories address these relationships.

Community Migration

All of the sociological theories considered in this section so far emphasize the individual's integration with society in some way. Stack (1982) has suggested evaluating integration as a local phenomenon rather than in a global, societywide concept. Rather than looking at integration with society as a whole, he would examine individuals' participation in their surrounding communities, using migration rates as an index of these ties. He has reported that areas with high rates of interstate immigration also post correspondingly high suicide rates, even when controlling for other differences between them (Stack, 1980). High rates of migration probably indicate that people in an area lack established ties with their neighborhoods and the other people who live there; this figure thus implies poor integration of residents into their immediate surroundings.

Stack cites additional evidence to suggest a particularly close relationship between the suicide rate for females and their ties to local communities. One might expect this result, since males move to established jobs more often than females, who often accompany their husbands in such transfers. Work relationships should, in some sense, help to offset the loss of personal relationships; females, on the other hand, frequently must move *away from* familiar situations rather than *into* new ones.

Socialization for Suicide

Some may initially feel uncomfortable thinking about suicide as a learned behavior. After all, a successful suicide precludes repeated practice, while learning often implies repeated, progressively improving performances. Yet, people can learn suicidal behavior just as they do any other behavior.

Learning establishes suicide as an alternative to apparently hopeless situations and conditions. Learning perpetuates norms and values that accept suicide, either for the good of the group (such as in altruistic suicides) or for personal reasons (as in egoistic and anomic suicides). For example, Akers (1985: chapter 24) has argued that while experience of suicide cannot reinforce later suicidal acts, reinforcement can promote behavior sometimes considered suicidal by others. In this way, a person can come to commit suicide by completing a process of social learning.

> There are at least two learning paths to suicide: (1) learning to behave suicidally, but not fatally, and ultimately reaching the point of suicide, and (2) learning about and developing a readiness of suicide and completing it without prior practice in specifically suicidal behavior. (Akers, 1985: 299)

Through socialization in suicidal behavior, an individual may acquire an attitude of acceptance toward suicide through tolerance of others' decisions to commit suicide in certain situations. Formation of such attitudes may accompany occasional self-injurious behavior, such as banging one's head, cutting or scraping the skin, or self-punching, sometimes producing serious injuries. Some exhibit these behaviors to manipulate others and achieve some end, including gaining attention, reducing anxiety, or controlling someone else. People can learn that suicide attempts help them to achieve their goals. They also learn that they cannot employ such threats too often without escalating the stakes.

Individuals can acquire learning that encourages suicide in subtle ways. People who work in suicide prevention frequently interpret such actions as a form of communication, a "cry for help," especially those communicated by children and adolescents (Hawton, 1986). Responses to suicide attempts may reinforce acceptance of such acts as methods of communication; if a response fails to provide the desired help, the suicide may ask again in the same way.

Some claim that highly publicized cases of suicide by public figures, such as Marilyn Monroe, contribute to socialization in suicide, potentially triggering chain reactions of successful imitations. Based on data from the United States and Great Britain between 1947 and 1968, Phillips (1974) has asserted a direct relationship between the extent of newspaper publicity about suicide and the suicide rates in communities where the news stories appeared. The largest increases followed the death of Marilyn Monroe; in the following month, suicides increased by 12 percent in the United States and by 10 percent in England and Wales. The suggestion of the possibility seems to raise consciousness of the option of suicide, giving it certain legitimacy. In subsequent investigations, Phillips (1979, 1980) has reported a relationship between publicity about murder-suicides and increases in airplane accidents, and between publicity about suicides and motor vehicle fatalities, particularly accidents killing only the drivers of the vehicles.

Every child learns to gain attention or to solicit help for some problem by simulating illness or injury. Socialization establishes techniques like these for gaining sympathy early in life through experiences with parents and friends, and techniques that prove most successful—that is, those that receive the greatest social attention and rewards—become the most established behaviors for the individual. Suicide may culminate a progressive sequence of certain techniques, such as feigning illness, deliberately injuring oneself, or intentionally inducing illness by taking drugs. This obviously complicated process involves learning techniques and rationales for using them over a long period of time. This possibility does not to deny the real problems and crises that provoke people to make serious attempts at suicide; it merely recognizes that certain difficulties and crises do not always force people to turn to suicide.

A socialization perspective recognizes the relationship of suicide to the nature and pattern of one's interactions with others. For example, it indicates why people who are well networked in their communities, schools, workplaces, or churches should commit suicide less often than those who lack these characteristics. These relationships increase the likelihood that others will help potential suicides to solve personal problems and point out other alternatives. At least one study suggests just such a role of positive social support networks in maintaining low suicide rates (Stack and Wasserman, 1992).

In the final analysis, the validity of any theory of suicide rests on the validity of the statistics on which it bases empirical tests. The chapter has already described well-known problems with those statistics, however. Recall the faulty assumption that coroners everywhere function in the same way, recording similar deaths in similar ways regardless of jurisdiction. In addition, perhaps the most significant error in applications of official statistics duplicates the error involved in formulating the theories themselves: Theories and blanket evaluations of data both assume that suicidal actions carry the same meanings throughout Western societies. Officials and theorists imagine, perhaps incorrectly, that they share the same definitions. All observers assume some uniformity, when actually "an *official* categorization of the cause of death is as much the end result of an *argument* as such a categorization by any other member of society" (Douglas, 1967: 229). In other words, relatives of the deceased person may persuade a coroner to list the cause of death as something other than suicide to avoid the stigma of such a death.

Consequently, two researchers have looked not at official statistics, but at suicide notes to find significant insights about why people commit suicide (Douglas, 1967; Jacobs, 1967). One observer has estimated that 1 in every 16 suicides leaves a note (Schneidman, 1976). Analysis of these notes can look for statements that confirm the expectations of various perspectives on suicide, what the writers seem to have experienced and how they viewed these experiences, the social constraints that seem to have restrained the individual from suicide, and how successfully or unsuccessfully the writers seem to have overcome these constraints. Folse and Peck (1996), for example, report that suicide notes frequently mention perceived failures, either by the suicides themselves or by others. While such feelings of failure do not directly cause suicide, they do contribute to it. Schneidman, however, has challenged the methodology of studying suicide notes. He observes:

That special state of mind necessary to perform a suicidal act is one which is essentially incompatible with an insightful recitation of what was going on in

---

❋ *Issue* | **A Suicide Gene?**

A report in the February 7, 2000, issue of the *Journal of Medical Genetics* suggested that there might be a biological component to suicide. Blood from three groups of patients were tested: a group who were depressed and had suicidal thoughts, a group who were depressed but did not display suicide ideation, and a group who was neither depressed nor suicidal. The results suggested that the group who was depressed and had suicidal thoughts was more likely than the other groups to have a mutation in the gene encoding for the serotonin 5-HT2A receptor, a protein that transmits brain signals.

The results raise enormous moral questions. What would happen if companies forced their employees to take the test and then discriminated against those found to carry the mutation? Would sufferers be denied life insurance or barred from flying planes or driving school buses?

---

one's mind that led to the act itself. . . . To commit suicide, one cannot write a meaningful note; conversely, if one could write a meaningful note, he would not have to commit suicide. (Schneidman, 1976: 266)

For another problem, many notes challenge efforts to interpret them. An investigation of this problem asked a group of psychology students to distinguish genuine suicide notes from simulated ones and to tell how they made the distinction; they frequently made mistakes in this process (Leenaars and Lester, 1991). The genuine notes often referred to previous traumatic events and expressed a common general tone of hopelessness. But the simulated notes did not differ so much that the raters could tell the difference. In another study, four experienced raters evaluated 20 real and 20 false suicide notes (Black and Lester, 1995). They, too, failed to distinguish the genuine notes from the fake ones.

To date, the controversy regarding the validity and meaning of suicide notes has prevented any systematic use of them to develop or test theories of suicide. Still, such notes undoubtedly contain important clues concerning causes of individual suicides.

The Suicide Process

People in modern society recognize death as an alternative to meeting or solving life's problems. While relatively few people actually commit suicide, many others undoubtedly contemplate doing so. But even prolonged frustrations and crises by no means always result in suicide, and no one has clearly determined just why some people do kill themselves. People face innumerable unpleasant events and crises in different ways; some drink, some seek religion, some make light of their situations, and others evade the issues or even consciously try to avoid them. In the most general sense, a person commits suicide generally because he or she cannot find a satisfactory alternative solution.

People define different events and conditions as problems, but some certainly experience more difficulties than others. Negative life changes like terminated interpersonal relationships, job disruptions, family conflicts, and economic stress can also generate not only immediate troubles but cumulative feelings of unhappiness over time. The individuals involved must seek solutions to these and other problems, and people who contemplate suicide may simply represent those who lack other alternatives (see Cole, Protinsky, and Cross, 1992).

The suicide process involves an unsuccessful search for possible alternatives to deal with problems culminating in the final decision that death represents the only possible solution. Ringel (1977) has identified three principal components of a predictable suicide syndrome: (1) constricting or narrowing alternatives, leaving problems with an all-consuming image and no way out except suicide; (2) a certain aggressive response directed toward oneself, perhaps leading to self-blame for an unfortunate accident or some other trauma in one's life; and (3) indulgence in suicidal fantasies that construct and mentally play out suicidal acts.

Social isolation appears as a consistent feature in many suicidal situations (Trout, 1980). Staff members of suicide prevention centers work to identify this condition and some others that Ringel identifies and then act constructively to improve them. Jacobs (1970: 233) has summarized the process in more detail. First, potential suicides usually compile long histories of problems. Second, they frequently encounter immediate escalations of problems that add new difficulties to still-unresolved ones. Third, failure to solve both new and old problems further increases isolation from meaningful social relationships. Fourth, and finally, the individual experiences a profound sense of hopelessness in the face of his or her problems, promoting an acute sense of failure and termination of remaining meaningful social relationships. This process is illustrated in the experience of an adolescent girl who recounted her suicide attempt:

> As a teenager, I basically had no friends, no interests at all. I stayed home. I felt very insecure around people, like I wasn't worthy to be around them. I'd skip classes; I'd be in the john crying. It finally got to the point where I begged my parents to let me quit. My grades were suffering terribly. So my father signed the papers and after that, it's all I heard, "You flukey, jukey bird," from my father because I quit school. Well, I loved my father, but he drank and beat my mother and would bust up the house. She left with us kids several times. Basically, I stayed in my room and I reached the point where I didn't want to be alive. (Stephens, 1987: 113)

Additional conditions sometimes make important contributions in this last stage of the suicide process. Every individual approaches such a crisis with a different personal history and set of experiences. The potential suicide may come to regard the unmet need (for example, desire for a relationship or promotion) or a loss of status as the final blow that destroys all future hope. The person learns attitudes toward self and others that contribute to depression and withdrawal behavior. Some become fixed on specific goals until they become obsessions. Someone whose fiancee breaks off the engagement, for example, may value nothing else—parents, career, friends, hobbies, or other interests. Many suicidal people lack objective views of their surroundings and situations; they can see their problems only from their own points of view. The potential suicide's interpretation of the situation's difficulty may also affect the decision to die. Circumstances such as economic losses or other difficulties may seriously disturb one person but only mildly affect another.

While many suicides terminate patterns of increasingly withdrawn behavior, others show the opposite kind of behavior pattern. A study of female suicide attempters, for example, has documented a number of suicide attempts from girls who exhibited patterns of early rebelliousness and defiance at home and in the community (Stephens, 1987). Many of these girls became involved in drugs and sexual promiscuity, along with minor delinquencies of other kinds. Like girls who exhibited the withdrawn pattern, however, they experienced unhappiness in adolescence and arrived at suicide as the result of changes in their self-concepts and attitudes.

People sometimes develop definite plans to commit suicide but never carry them to completion. Some may repeatedly plan to kill themselves but delay the final act after removal of the original, precipitating cause, discovery of an alternative solution, or reinforcement of some attitude opposed to self-destruction. Other people often contribute to the potential suicide's definition of the situation and the progression of the suicide process:

> An individual comes to feel that his future is devoid of hope; he, or someone else, brings the alternative of suicide into his field. He attempts to communicate his conviction of hopelessness to others, in an effort to gain their assurance that some hope still exists for him. The character of the response at this point is crucial in determining whether or not suicide will take place. For actual suicide to occur, a necessary (although not sufficient) aspect of the field is a response characterized by helplessness and hopelessness. (Kobler and Stotland, 1964: 252)

Social Meanings of Suicide

Clearly, suicide results from a social process, and one that reflects substantial social meanings and antecedents. Like other types of behavior, people can learn suicidal behavior, and they may associate certain social rewards or advantages with the self-destructive act. In their notes, suicides often use the communication opportunity to substantive advantage by pointing out to others sentiments like "you were wrong about me," or "see, I really do love you." The suicide helps them to portray this presentation of self in the most dramatic manner possible.

Studies of attempted suicide have often noted helpful reactions from others that result. People may obtain these desired reactions most readily only under threat of suicide and gain extremely helpful benefits when others do react (Stengel, 1964: 37). One reason for this reaction, of course, is the recognition that voluntary self-destruction is deviant conduct. This widely held and strongly maintained norm ensures a reaction of confronting the problems that apparently drove the individual to attempt or commit suicide.

The Purpose of Suicide

Suicidal actions seldom represent irrational behavior. Rather, they express meaning, usually about some fundamental flaw in a situation that makes suicide appear to that person as a rational solution, or the only one, to those problems. In fact, the real *meaning* of a suicide may differ from that recognized by friends, family, or the coroner. Outsiders may regard the suicide as a senseless and irrational act of a distraught, lost, or depressed person. As mentioned earlier in another context, some observers have looked for clues to the purposes of suicides in notes they have left behind (Leenaars, 1988). The contents of these notes may directly disclose the mental states or motives of the writers.

Studies of suicide notes and diaries, as well as interviews with those who have attempted suicide, indicate a powerful role of suicidal death as a way to transform or affirm one's essential, substantial self in many ways (Douglas, 1967: 284–319). Suicides generally construct a number of *patterns of social meanings* for themselves and in relation to others, according to Douglas:

- *A means of transforming the "soul" from this world to another.* Some people feel motivated to end their own lives to fulfill promises of life after death. They speak of life as a property of the physical body, not as an attribute of the individual. Suicide notes suggest that these people expect to return to God or move on to a new world after death.

- *A means of transforming the substantial "self" in this world or another world.* With an act of suicide, some people try to show others that their true identities differ from others' opinions of them. Through suicide, these people intend to show themselves as committed, loving, trustworthy, and sincere individuals; surely, only someone with such qualities would go to the extreme of suicide to prove them, or so such people sometimes reason. These suicides believe that they are giving up their lives to ask forgiveness of some wrong or to prove what kind of people they were all along.

- *A means of obtaining "revenge" by blaming others for one's death.* These suicides clearly assign blame and make definite connections between their own suicides and the others' actions. For example, a 22-year-old clerk killed himself because his bride of 4 months wanted a divorce so that she could marry his brother. The letters he left showed plainly the desire for revenge by bringing unpleasant publicity upon his brother and his wife, as well as a desire to attract attention to himself and his miserable condition. In the letters, he described his shattered romance and told reporters to see a friend to whom he had forwarded correspondence. The first sentence in a special message to his wife read: "I used to love you; but I die hating you and my brother, too." This was written in a firm hand, but as his suicide diary progressed, it expressed thoughts in more erratic handwriting. Some time after turning on the gas, the young man wrote: "Took my panacea for all human ills. It won't be long now. I'll bet Florence and Ed are having uneasy dreams now." An hour later, he continued: "Still the same, hope I pass out by 2 A.M. Gee, I love you so much, Florence. I feel very tired and a bit dizzy. My brain is very clear. I can see that my hand is shaking—it is hard to die when one is young. Now I wish oblivion would hurry." The note ended with those words (Dublin and Bunzel, 1933: 294).

■ *Escape patterns to avoid the responsibilities of continued life.* These suicides express great restlessness, although the nature of this dissatisfaction often remains unspecified. The person cites feelings of disgust with life or uselessness. A married woman of 24, for example, left this suicide note: "I've proved to be a miserable wife, mother and homemaker—not even a decent companion. Johnny and Jane deserve much more than I can ever offer. I can't take it any longer. . . . This is a terrible thing for me to do, but perhaps in the end it will be all for the best" (Schneidman and Farberow, 1957: 43–44). Similarly, a divorced man of 50 left this suicide note:

To the Police—

This is a very simple case of suicide. I owe nothing to anyone, including the world; and I ask nothing from anyone. I'm fifty years old, have lived violently but never committed a crime.

I've just had enough. Since no one depends upon me, I don't see why I shouldn't do as I please. I've done my duty to my country in both world wars, and also I've served well in industry. My papers are in the brown leather wallet in my gray bag.

If you would be so good as to send these papers to my brother, his address is: John Smith, 100 Main Street.

I enclose five dollars to cover cost of mailing. Perhaps some of you who belong to the American Legion will honor my request.

I haven't a thing against anybody. But I've been in three major wars and another little insurrection, and I'm pretty tired.

This note is in the same large envelope with several other letters—all stamped. Will you please mail them for me? There are no secrets in them. However, if you open them, please seal them up again and send them on.

They are to people I love and who love me. Thanks, George Smith (Schneidman and Farberow, 1957: 44).

■ *Self-destruction after killing another person.* A Philadelphia study has determined that about 4 percent of those who committed homicide then took their own lives (Wolfgang, 1958: 274). Other studies in the United States have shown an incidence of such suicides between 2 and 9 percent of the total. In England and Wales, they make up a much greater proportion; each year suicide follows about one-third of all murders there, and 1 in every 100 suicides follows this pattern (West, 1965). When discovery of other crimes, such as embezzlement, brings major personal disgrace, suicide occasionally follows. Frequently, perpetrators perceive such homicides as acts of love, as indicated in a note from a divorced woman who shot her two young sons and then took a drug overdose.

Dear Mommy and Daddy,

I'm sorry to do this to you'all but I can't take this life any more. I'm taking my boys with me. Please put one on my right and one on my left side. I love my boys and hope God forgives me and lets me be with them. I know in my heart that my boys will be with God. God please forgive me for I have sinned.

I love you, Mother and Daddy,

Eileen

[P. S.] Please dress the boys in blue. They look good in it. Please put me between them. I love them and want them to be in heaven, God's heaven. Please put with Monty Jay his night, night blanket, one that Mom made. Please put with Jeff his little tiger that he got on his first Christmas on my bed. . . . (Daly and Wilson, 1988: 79)

Rescuers found the despondent mother who wrote this note in a drug-induced coma and revived her, only for her to face trial and a life sentence in prison.

Such notes suggest a major residual problem in a suicide: its effect, especially guilt feelings for harm to significant others—family, friends, employers, and so forth. Survivors of suicides frequently apply a number of explanations to reduce or manage guilt feelings they experience as a result of these acts. Some survivors deal with suicide by describing the suicides as results of outside causes or impersonal forces, as inevitable outcomes, and as beneficial events, despite appearances. Such a rationalization also recasts the meaning of a suicide into terms that survivors can both understand and learn to tolerate. Elderly people's suicide notes differ, however, from those of younger suicides. As one observer noted: "Older suicidal adults wish to die. Their notes, compared to individuals in young and middle adulthood, are less likely to show evidence of redirected aggression, complications, and unconscious implications. There are fewer contradictions and distortions and less confusion" (Leenaars et al., 1992: 72).

Notes alone give weak support for substantiating elaborate interpretations of suicide processes. Only about 12 to 15 percent of all suicides leave notes (Leenaars, 1988: 35), and any generalization from those who do risks substantial distortion. Analysts cannot determine whether those who leave notes accurately represent all suicides.

Suicide, Mental Disorders, and Hopelessness

People who commit suicide generally do not express any identity as mentally deranged or suffer from temporary insanity. The association between suicide and mental illness developed from the notion that no sane person would take her or his own life. Yet, suicide, as pointed out earlier, may seem entirely understandable in light of a certain set of circumstances. In fact, because suicide is, by definition, an intentional act, many mental disorders even control or inhibit it. According to one author, "Mental disorders or developmental deficiencies that reduce the capacity for planning and deliberation, and that prevent the psychological organization of sequential actions, greatly reduce the potential for suicide" (Litman, 1987: 90).

Actually, most suicides follow rational planning and careful execution with no more evidence of mental disorder than would appear in so-called *normal people*. The goals that motivate suicides, no matter how exaggerated, generally target real benefits; these people usually suffer real personal losses rather than psychotic hallucinations or delusions with little or no basis in reality. Analysts sometimes struggle to determine the circumstances surrounding a suicide, especially when reports from family or friends about a suicidal person's state of mind reflect those people's misunderstandings.

Among mental disorders, severe depression seems to bear the most common association with suicide. Further, the percentage of suicide victims with psychotic disturbances, although not large, is significant enough to justify some concern. Certainly, when cases of attempted suicide feature some severe mental disorder, the same disturbance might lead to a repetition of the attempt without discovery and treatment.

Maris (1981: chapter 8) has reported on a study of all officially recorded suicides in Chicago from 1966 to 1968. This research found that the percentage of people who attempted and completed suicide and who displayed psychological problems exceeded the comparable percentage for a control group of people who died from natural causes. In fact, about 40 percent of suicide victims had been hospitalized at some time in their lives for mental disorders, compared with about 50 percent of those who attempted suicide but failed and 3 percent of those who died of natural causes. Furthermore, while the study found a relationship between depression and suicide, a specific sense of hopelessness proved a more important indicator than general depression. An English study has also identified certain mental states that significantly influenced adolescent suicide attempters:

> The main feelings that appear to precede attempts by adolescents are anger, feeling lonely or unwanted, and worries about the future. A sense of hopelessness is a major factor distinguishing depressed adolescents who make attempts from similar adolescents who do not. (Hawton, 1986: 99)

For these reasons, one could not accurately say that all suicidal people suffer from mental disorders. Instead, the relationship seems to depend on other conditions like social stress and ability to cope with that stress.

Many observers currently focus on the precise roles of depression and hopelessness in creating suicidal thoughts (called *suicide ideation*) or behavior. A number of investigators have identified the development of suicide ideation as an essential step in the suicide process. For this reason, they promote it as an important object of intervention or prevention efforts (Rich, Kirkpatrick-Smith, Bonner, and Jans, 1992). Some reports indicate that hopelessness defines the main component of suicide ideation, while others target depression as the more important indicator

(Rudd, 1990). In either case, negative life stress may precede both depression and hopelessness, resulting from such events as significant loss (the death of someone close) or failure (Rudd, 1990).

Some research has suggested gender differences in the social sources of suicide ideation. One study of this activity in adolescents has speculated, for example, that females, following their socialization to express emotion openly, should relate suicide ideation more strongly to affective disorders, such as depression, than should males, who act on the basis of different socialization experiences (Yagla Mack, Hoyt, and Miller, 1994).

Such experiences extend outside the ranks of urban people. One study of suicide ideation among rural adolescents has reported the strongest predictive value for family characteristics (Meneese and Yutrzenka, 1990). Specifically, "disorganized" families were associated with such ideation. These families featured little or no structure, indicated by general family rules, responsibilities of various family members, financial planning, and the like. Stressful life events can affect people of any age, although the nature of those stresses can vary (home, school, employer, spouse) depending on individual ages and other circumstances.

Preventing Suicide

Efforts to prevent suicide rely heavily on identification of social forces that ultimately produce suicide. If suicide results directly from the conditions of modern life and stresses that individuals feel as they live in complex, industrialized societies, then a remedy would require a broad-based response. Until prevention efforts overcome the difficulties of such extensive social change, most must continue to work toward individual intervention, often without essential guidance from a coherent and valid sociological theory. Without such an accepted theory, policy implications for reducing suicide necessarily vary depending on the theories to which particular programs subscribe. A perspective based on control theory (see Chapter 6) may emphasize social and/or status integration, implying different measures than an orientation based on learning theory (also discussed in Chapter 6) that emphasizes socialization to suicidal situations and social isolation of potential suicides.

Such confusion about theoretical principles and associated practical responses has not blunted the substantial and continuing interest in prevention of suicide. For example, a number of social scientists have attempted to develop predictive instruments that would assess individuals' risk of suicide. No instrument developed so far enables intervention workers to predict suicidal behavior with any accuracy, although research has identified a number of risk factors (see Spirito et al., 1991).

Suicide Prevention Centers

The lack of agreement concerning a theory of suicide also has not deterred organization of many community agencies in the United States and various other countries. These programs work to prevent suicides by offering counseling and other assistance. Much of this work was pioneered after World War II.

In Vienna, Austria, Caritas, a Catholic organization, carries on most suicide prevention work, in part through a preventive clinic for attempted suicides at Vienna hospitals. A special suicide prevention telephone service provides facilities through which people contemplating self-destruction can converse with others about their problems. If desired, a social worker can visit the caller's home. In Great Britain, too, a network of volunteer organizations works to prevent suicide. The Good Samaritans, established in 1953, combine elements of religion and psychotherapy in counseling suicidal people.

In the United States, suicide prevention centers operate in most large communities, usually in association with community mental health clinics. The first such center in the United States was founded in 1958 in Los Angeles (Farberow, 1977: 543). Numerous organizations have set up suicide prevention programs in most states under a variety of names, such as Suicide Prevention Service, Call-for-Help Clinics, Crisis Clinics, Crisis Call Centers, Rescue, Inc., Dial-a-Friend, and Suicides Anonymous. A Center for Studies in Suicide Prevention was established in the National Institute of Mental Health in 1967.

Staff members at suicide prevention centers carry out much of their initial contact with clients by telephone. Telephone directories in most large cities prominently display these numbers, as do public-service advertisements in the mass media. Publicity encourages distraught people to call the numbers and talk about their problems, and perhaps to visit the centers if possible. Suicide prevention centers practice crisis intervention; that is, they offer services geared toward clients' immediate needs rather than long-term ones, which may require intensive counseling and advice. Centers offer these short-term services 24 hours a day, usually in the form of telephone counseling directed at immediate situations.

Staff members answer phone calls and try to establish rapport and maintain contact with callers. They then attempt to assess the potential danger in the situation—that is, the caller's imminent resolve to commit suicide. Someone who mentions a potential future act of suicide constitutes a less immediate threat than someone who has developed a plan and arranged the means to carry it out at the moment of calling. Prevention center staff members also evaluate the resources available to clients, such as friends, to help ease the situation. They then try to set up a treatment plan of some kind, in the process urging the suicidal person to commit himself or herself to future activities that will diminish the current probability of suicide.

Callers to suicide prevention centers fit no convenient stereotype. They bring varied problems, each of which may require a different solution. As one might expect, callers commit suicide at a rate substantially higher than that for the population as a whole; estimates range as high as 1,000 per 100,000, compared with about 12 or 13 actual suicides per 100,000 for the general population (Litman, 1972). A number of an agency's contacts come from chronic callers. One study of 67 such chronic callers identified 51 percent as drug or alcohol dependent. "The prototype of the chronic caller to the Suicide Prevention Center is a divorced female in her late 30s who is alcohol or drug dependent and intermittently suicidal" (Sawyer and Jameton, 1979: 102).

One study found that most callers to the Los Angeles Suicide Prevention Center indicated suffering from depression, and about two-thirds of them reported contemplating suicide, while the others called in the act of suicide (Wold, 1970). Two-thirds of the callers were women, and about half had histories of suicide attempts. More than half of them called the center themselves, while the others involved family members, friends, or professionals in the calls. The study reported assigning high ratings of suicidal potential to about 20 percent of the callers, moderate ratings to 40 percent, and low ratings to the remaining 40 percent. While the center prevented many callers from carrying out their intended acts, a comparison found a number of differences between a sample of all callers and a group who actually committed suicide after contacting the center. The group who killed themselves reversed the proportion of males to females in the sample of all callers: Two-thirds of the completed suicides were males and one-third female. Also, depression seemed more marked in those who actually committed suicide.

Effectiveness of Suicide Prevention Centers

Precise evaluations of the effectiveness of suicide prevention centers encounter serious analytical problems, since many callers who do not commit suicide remain unidentified after the calls. Several studies have focused on evaluating the impact of the English suicide prevention organization mentioned earlier, the Good Samaritans, but they have reported mixed results (Hawton, 1986: 133–134). Some studies have associated such agencies with declines in the suicide rates of the areas they serve, while other studies have reported no change. A study from the United States compared suicide rates in North Carolina communities served by suicide prevention centers with rates in communities without such centers. The research concluded that the centers produced minimal effects on suicide rates after comparing rates before the centers opened to those after they had begun operating (Bridge, Potkin, Zung, and Soldo, 1977). Despite results like these, some feel that a center provides a valuable service if it prevents even one person from committing suicide.

Miller, Coombs, Leeper, and Barton (1984) compared suicide rates between counties in the United States served by suicide prevention centers and those that lacked such centers. The results indicated little overall difference in those rates. However, the authors reported significant differences in counties' suicide rates of young girls (females up to 24 years of age). Over time, the cumulative suicide rate among this population declined by 55 percent in counties that had suicide prevention centers, while it increased (by an estimated 85 percent) in counties that lacked such agencies. This finding gains importance based on results that identify young girls as by far the most frequent callers to suicide prevention centers. Another study has reported more encouraging results for the general population. It found a negative association between the presence of a suicide prevention center in a state in 1970 and changes in the suicide rate from 1970 to 1980 (Lester, 1993). This result suggests that such centers did indeed produce some preventive effect.

Some facilities work with people involved in suicide ideation or who have previously attempted suicide. One study to evaluate such a program focused on the effectiveness of its method of controlled confrontations to explore inner experiences and life difficulties related to suicidal behavior (Orbach and Bar-Joseph, 1993). The program emphasized instruction in coping strategies as a way to immunize clients against self-destructive behavior. These methods suggested the program's principle that many of those who consider suicide need to enhance their skills for dealing with some of life's problems. The study's comparison of experimental and control groups yielded encouraging results, although the authors did not suggest that this type of program would work with all potential suicides.

Other research has reinforced claims for the effectiveness of programs to improve interpersonal and coping skills. One study compared 67 successful adolescent suicides with a group of 67 adolescents in the community (Brent et al., 1993). In contrast to the adolescents in the community, the suicides more often had experienced interpersonal conflict with parents and with male or female friends, disruption of romantic relationships, and/or legal or disciplinary problems. Although no one can say whether intervention could have prevented the suicides, better skills for handling these particular stressful life events may have helped them to identify and pursue other alternatives.

These centers effectively formalize society's suicide prevention efforts, providing an example of how formal social controls can back up informal controls. Most people who seriously consider committing suicide decide not to do so, often because strong moral or religious convictions oppose self-destruction. Some actually attempt suicide for reasons other than killing themselves, for example, to get attention for some problem that might otherwise pass unnoticed or to get revenge on another. Still others seriously intend to complete their suicidal acts, but they turn to prevention centers for access to resources that they lack on their own. Suicide prevention centers offer alternatives for people who do not know where else to turn for help with serious problems.

Physician-Assisted Suicide

Assisting a suicide is considered a serious, and often criminal, act. Yet, some believe that there are circumstances in which this is not only desirable, but also necessary. The topic of whether and to what extent physicians should be involved in assisting patients in committing suicide has raised many questions in the community as well as among physicians. Some believe that such a practice violates the Hippocratic oath and its prohibition on hastening death, while others believe that physicians must ultimately take personal responsibility for their actions (Nuland, 2000).

Oregon is the only state in the United States that now permits physician-assisted suicide. First approved in 1994 and reaffirmed by voters in 1997, the "Death with Dignity" law permits a physician to prescribe a lethal dose of barbiturates to patients who have less than 6 months to live. In 1999, 27 terminally ill Oregonians ended their lives this way. This figure was up from 16 the year before (Sullivan, Hedberg, and Fleming, 2000).

Much of the controversy surrounding physician-assisted suicide stems from the actions of an activist physician who self-admittedly assisted in a number of suicides. In Michigan in the 1990s, Dr. Jack Kevorkian instituted a completely different and extremely controversial response to suicidal desires by terminally ill patients. This medical doctor openly helped a number of hopelessly ill persons to end their lives in a dignified manner. Dr. Kevorkian intended these actions to establish a legal precedent allowing doctors to assist certain patients in voluntarily ending their suffering and choosing dignified deaths. Authorities arrested him several times under a newly passed Michigan law when his actions in these suicides gained media attention. As of 1994, the ultimate legal outcome remained in doubt

---

❋ *Issue* | **The Morality of Physician-Assisted Suicide**

Dr. Jack Kevorkian, a Michigan physician, has indicated that he has assisted a number of people in committing suicide. In one publication, Kevorkian described the process:

> I started the intravenous dripper, which released a salt solution through a needle into her vein, and I kept her arm tied down so she wouldn't jerk it. This was difficult as her veins were fragile. And then once she decided she was ready to go, she just hit the switch and the device cut off the saline drip and through the same needle released a solution of thiopental that put her to sleep in ten to fifteen seconds. A minute later, through the same needle flowed a lethal solution of potassium chloride. (quoted in Denzin, 1992: 7)

The topic of physician-assisted suicide has generated substantial public attention. Public opinion was generally in support of assisted suicide for terminally ill patients, but there are a variety of ethical concerns.

Assisted Suicide Is Wrong	Assisted Suicide Is Right
To some, Dr. Kevorkian was a killer who was turning medicine on its head by reversing the code of life. "Now medicine kills," (Denzin, 1992: 7), some said. Even if assisted suicide is not murder because the offender does not benefit, it is a strange way to use medicine because it does not defeat the traditional enemy of medicine: death. Assisted suicide is still suicide and should not be condoned.	To others, Dr. Kevorkian represented a moral answer to the question of what medicine could do for people who were terminally ill and beyond the reach of healing. In these cases, physician-assisted suicide helps them maintain some dignity and control over the remainder of their lives (Weir, 1992). In this view, medicine should not blindly strive to maintain human life without attention to the quality of that life.

This has been a very controversial issue. Dr. Kevorkian has been arrested and released a number of times for violating a Michigan statute (created because of his work) that prohibited assisted suicides. In 1999, he was convicted of second-degree murder and is currently serving a prison sentence in Michigan.

for medically assisted suicide in certain types of cases. At the same time, the Netherlands has made legal changes that allow doctors to assist in the voluntary deaths of terminally ill patients.

Summary

Most Western societies strongly condemn suicide, but this attitude has not prevailed universally and for all time. Some societies at some times have treated suicide as a permissible—and even honorable—option under certain circumstances. Strong negative attitudes in the United States show associations to religious beliefs; young, well-educated males generally accept suicide more willingly than others do under certain circumstances.

Estimates of suicide rates often differ because of differences in the way official statistics record deaths and differences in the way coroners determine causes of death. As a result, no one can determine with absolute accuracy how many suicides occur. The U.S. suicide rate has, as far as observers can determine, remained stable over the past several decades at about 12 per 100,000 population. Many more people, perhaps 20 times as many, attempt suicide than complete it. Attempters are more likely to be young women.

Men successfully commit suicide more often than women do, and whites do so more often than blacks do. Generally, suicide rates increase for progressively older age groups, although suicide rates among blacks peak in the 20s and 30s and decrease in the older age categories. Adolescent suicides remain relatively rare events (70 percent of all suicides are older, white males), but the rate of these suicides has increased over the past three decades. Suicide among young people generates substantial attention and public reaction. Married people kill themselves at lower rates than unmarried people do, and people with strongly religious attitudes (regardless of denomination) have lower suicide rates than those without such powerful affiliations. Generally, suicide rates are high among members of the lower classes, but high rates characterize some middle- and high-income occupational categories.

Observers have identified different types of suicide. Some people commit altruistic suicide to provide perceived benefits to others. An anomic suicide may occur when someone loses his or her purpose for living or feels cut off from relevant social groups' norms. Such suicides may also feel considerable frustration when their group values become confused or break down. Egoistic suicides occur for a variety of personal reasons, including failing health, financial losses, or difficulties in social or interpersonal relationships. In the United States, egoistic suicides make up the largest category.

Sociological theories of suicide have traditionally emphasized the individual's integration with the larger society or social group. Theories of social integration and status integration pursue such lines of reasoning. Status frustration and community migration theories, which develop from social disorganization and socialization perspectives, evoke social psychological arguments. The usual suicide process begins with increasing social isolation and continues with identification of suicide as a possible solution to perceived problems. Social interactions determine the meanings that people attach to suicide, and those meanings suggest that suicide offers an important means by which to affirm or transform a personal identity. People who commit suicide are not necessarily mentally disordered.

Most communities perceive prevention of suicide as an important social responsibility, and the spread of suicide prevention centers reflects that concern. These agencies now serve most large cities and many smaller ones. They establish most contacts with suicidal clients by telephone, although many follow up calls with personal contact whenever possible. Researchers dispute the effectiveness of suicide prevention centers, and resolution of the question remains difficult because centers never learn the identities of many callers, preventing follow-up investigations. If no suicide results after a call, some factor other than the agency contact might have accounted for that fact. Some studies, however, have detected some reduction among adolescent suicides as a result of intervention by suicide prevention centers.

Selected References

Akers, Ronald L. 1985. *Deviant Behavior: A Social Learning Approach*, 3rd ed. Belmont, CA: Wadsworth.

This book applies a social learning perspective to explain a number of forms of deviance, including suicide. It shows how suicidal behavioral can result from social learning, just as other kinds of behavior do.

Canetto, Silvia Sara, and David Lester, eds. 1995. *Women and Suicidal Behavior.* New York: Springer.

This volume is a collection of papers on suicidal behavior by women. It focuses on the reasons and conditions under which women engage in this behavior, covering both research and clinical approaches.

Hawton, Keith. 1986. *Suicide and Attempted Suicide among Children and Adolescents.* Beverly Hills, CA: Sage.

This study of English suicide attempters by one of the best-known clinicians in the field also contains information about suicide and attempted suicide in other nations. It offers essential information for those interested in youthful suicides.

Headley, Lee A., ed. 1983. *Suicide in Asia and the Near East.* Berkeley: University of California Press.

This interesting collection of papers documents suicidal behavior in a number of countries in Asia and the Near East. It offers valuable discussions about the statistical picture as well as the social meaning of suicide in each country.

Stack, Steven. 1982. "Suicide: A Decade Review of the Sociological Literature." *Deviant Behavior* 4: 41–66.

This paper systematically reviews the empirical and theoretical literature on suicide, identifying and evaluating different explanatory perspectives.

PART IV

Studies in Stigma

THE 1977 film "The Elephant Man" was based on a true story. John Merrick was born with a disease that horribly disfigured his head. The precise disease is not known, but it could have been Proteus syndrome, which causes an overgrowth of bone and other tissue. He had to sleep in a chair; otherwise, the weight of his head would have killed him. His head was large, misshapen, and horrible to look at; as a result, Merrick often ventured into public wearing a hood over his head. Nevertheless, he experienced a lifetime of stigma. At the end of the film, beaten down by stares, comments, and social isolation, Merrick decides he would rather be a "normal" person than continue his existence as an outcast. With Samuel Barber's poignant "Adagio for Strings" playing on the movie's soundtrack, Merrick lies down to sleep in his bed, on his back, knowing full well the consequences of so doing.

In Part I, we talked of the nature, meaning, and context of deviance. In Part II, we discussed some major theories of deviance, and in Part III, we identified some major forms of deviance. In this part, the focus is on the stigma that some acts and conditions can elicit. In this part we discuss physical disabilities and eating disorders, homophobia and homosexuality, and mental disorders. In each instance, we concentrate on the reactions of others and the consequences of these reactions, as well as social dimensions of these behaviors and conditions. ■

CHAPTER 14

Physical Disabilities and Eating Disorders

The discussion so far has emphasized deviant *behavior*, nonconforming actions that people take. This chapter considers a set of *conditions*, nonconforming characteristics that people display. This topic bears an important relationship to the book's definition of *deviance:* behavior or conditions that violate society's norms. Recall from Chapter 1 that deviance constitutes only those deviations from norms that elicit, or will likely elicit if detected, disapproving reactions or negative sanctions. This chapter deals with two such conditions: physical disabilities and eating disorders.

Deviance and Physical Disabilities

In late 1999, Richard and Dawn Kelso left their 10-year-old son at a Delaware hospital with his toys, medical supplies, and a note saying they could no longer take care of him. Steven Kelso has cerebral palsy, breathes through a tube, and uses a wheelchair. Until his arrest shortly after abandoning Steven, Richard was the chief executive officer of a successful $500-million-a-year chemical company. Dawn Kelso served on a state advisory council for people with disabilities.

Part of the stress for Richard and Dawn Kelso comes from the 24-hour-a-day care that such children require. Even having assistance is sometimes not enough. Dreams of having healthy children are destroyed, and many such parents go through many of the stages associated with a terminal illness: denial, anger, sadness, bargaining, and, finally, acceptance.

Adhering to societal norms of appearance can have its dangers as well. In a national report, one of the top liposuctionists in the country estimated that perhaps as many as 10 percent of all cosmetic surgeries require revision (Kalb, 1999: 58). The cost of "botched" physical surgery, performed because patients want to change some part of their bodies, includes not only high fees, but in some cases deformities and disabilities as a result of the surgery.

So it is with people who have disabilities. They are often regarded as not only physically, but emotionally different. They are often denied access to normal social interaction with others. Many conversations center on the nature of the disability and what the person cannot do, rather than on what the person can do. The disabled are sometimes denied employment and housing based on their disability. The disabled are regarded as "outsiders."

Physical disabilities can be seen in the larger context of deviance because many people treat others with physical disabilities as deviants. While most people do not regard physical disabilities as evidence of deviant conditions, people who live with these disabilities do experience many of the same social reactions as deviants, including stigma and social rejection. A wide range of physical disabilities and impairments elicit such severe stigmatizing and discriminatory reactions that many societies, and this chapter, class them as examples of deviance.

People with disabilities do not uniformly elicit negative reactions from others. We shall see that some people are openly accepting, and others are even overly nice to a disabled person. In that regard, we are here speaking more of a "sociology of difference" than a "sociology of deviance." These different reactions reflect a basic ambivalence regarding physical disabilities. While some people with physical disabilities may experience stigma in the form of behavioral discrimination and avoidance from the able-bodied, others clearly form stable, accepting relationships with people who do not have physical disabilities. And the disabled can experience both kinds of reactions at different times.

Goffman (1963) has listed several classic examples of conditions that have often elicited social stigma, including the blind, the deaf, the mute, the epileptic, the crippled, and the deformed. One may expect such reactions in a society where physical appearance matters, and people frequently respond with wariness or even hostility to others whose appearance does not conform to their expectations. To Goffman's list, one might add the mentally retarded, the obese, those afflicted with cerebral palsy, and those with severe stuttering in their speech patterns.

People with physical disabilities often encounter isolation, segregation, and discrimination in their interactions with others (Nagler, 1993a). For example, people with physical handicaps have confronted denial of access to public transportation and educational facilities. Such discrimination causes extremely important consequences, since transportation and education powerfully affect life chances.

Disabled men and women also must overcome obstacles to form personal relationships on many occasions. Others may avoid contact with them because their appearances violate normative guidelines for "acceptable" appearance. Additional limits on interpersonal interactions sometimes result because others associate problems in understanding with physical problems.

Some may perceive further confirmation of disabled people's deviant status if they cannot fully care for themselves, suggesting incomplete assumption of adult roles. People often interpret such personal limitations as evidence of more encompassing dependency roles. In addition, those with physical disabilities may experience wage discrimination compared with able-bodied workers who perform the same kind of work (Baldwin and Johnson, 1995). For these and other reasons, people with disabilities experience social reactions and processes similar to those targeted for people who voluntarily commit deviant acts.

Quite a large number of people live life with some form of disability. Some observers have estimated that mental, physical, or sensory impairments cause disabilities for more than 500 million people in the world. According to the National Organization of Disability, more than 40 million Americans live with disabilities that interfere with their completion of at least one major life task (Blaska, 1993: 25). A smaller, but still large number of Americans must contend with severe physical disabilities. According to one estimate, for example, 3 out of every 100 children born in the United States and Great Britain are diagnosed as mentally retarded at some time in their lives, and between 20 and 25 percent of these children show signs of such severe retardation that they are diagnosed at birth or in early infancy (Edgerton, 1979: 1). Another estimate places the number of people in the United States who can neither speak nor hear at 2 million, and another 11 million people have severe hearing impairments (Higgins, 1980: 33). Another source cites about 1.5 million epileptics and more than 3 million who suffer from crippling impairments (Scott, 1969: 42). This group of deviants also includes more than 1 million blind or partially sightless people, of whom probably 50,000 live with total blindness (Scott, 1980: 8). Clearly, such rough estimates of those with disabilities requires agreement on what is and what is not a disability, a topic of continuing debate (see Francis and Silvers, 2000).

Definitions and Distinctions

The chapter must begin with important distinctions among the terms *impairment*, *disability*, and *handicap*. *Impairment* refers to the loss of some ability, usually caused by some physical reason. Sometimes, physical conditions present at birth inhibit functions in the optic nerve, the portion of the brain that controls talking, a limb, or the inner ear; each constitutes an example of an impairment. *Disability* refers to a loss of function that accompanies an impairment. While *impairment* refers to a physical condition, *disability* describes the effect of that loss on the affected person's activities—seeing the surrounding environment, speaking, using limbs, maintaining social relations, and so forth. The term *handicap* is often used to refer to the limitation on normal activities of self-care and mobility that results from some impairment (Thomas, 1982: 3–8). Thus, while physical conditions determine impairments, disabilities and handicaps represent social and behavioral consequences of those impairments, including one's inability to meet, among other things, social responsibilities and obligations.

Social context establishes the meanings of these terms. Clearly, everyone is disabled from some activities. Most people cannot dunk a basketball, open a locked safe without the combination, or perform brain surgery, but no one thinks of people who cannot do these things as disabled. Some people cope with obvious disabilities (e.g., an inability to walk that makes a wheelchair necessary), but other disabilities remain unseen, such as severe arthritis. Persons with visible disabilities encounter different social expectations from those with nonapparent disabilities, despite similar levels of impairment (Stone, 1995). Clearly, *disability* implies a socially constructed category, one that may describe some people some of the time.

Physical impairment differs from illness. Society exempts the sick from many social role responsibilities, such as normal occupational and familial duties, and no one holds them responsible for their illnesses. Others do, however, expect sick people to want to get well and to seek competent help to promote that end. In these respects, important attributes separate sickness from the relative permanence of physical impairment. "The label of sickness, although it may imply severity, also implies a temporary condition which can, through some kind of intervention (usually medical), be made to disappear. It is only the label of disability that carries the connotation of permanency and irreversibility regardless of the degree of severity of the condition" (Safilios-Rothschild, 1970: 71).

Individuals do, of course, experience degrees of physical disability and impairment. In all probability, extreme cases likely draw some kind of stigmas. For example, the extent of sight loss varies greatly between cases of blindness, as do the seriousness of crippling or facial disfigurements and the severity of mental retardation. While the legal definition of blindness establishes a 20/200 vision rating, Scott (1969: 42) has severely criticized this determination as an arbitrary gesture insensitive to most important determinants of a person's functional vision; he asserts that such a standard "lumps together people who are totally blind and people who have a substantial amount of vision." For this reason, the label *blind* applies to a very diversified, heterogeneous, and shifting group of people. Similarly, the category *deaf* applies to people with hearing loss ranging from slight impairment to total deafness. Obesity, too, varies in degree and definition; one study has defined the condition as 30 to 40 percent above "normal" weight (Cahnman, 1968). Others set this criterion at 20 percent. People can be overweight to varying degrees, but many regard a visibly fat person as *obese*, and this label likely implies discrimination against the person (Millman, 1980). Likewise, one stutterer may speak only with a severe and constant impairment, while another may stutter only under certain conditions. Mental retardation encompasses an extremely varied range of cognitive impairment; the American Association on Mental Deficiency lists classifications based on a "normal" IQ (intelligence quotient) of 90 to 100, while borderline impairment applies between 70 and 90, and IQs below 70 indicate various degrees of retardation. About 85 percent of all retardates fall in the category of "mildly retarded" with IQs between 55 and 70 (Edgerton, 1979: 4–5).

Disabilities and the Idea of Deviance

In a sense, people with disabilities constitute a minority group. As pointed out earlier, they experience isolation, segregation, and many kinds of discrimination as a result of their disabilities. Unlike homosexuals and other identified groups, however, they have not yet established a common sociological identity (but see Deegan, 1985). As a result, they have not successfully pressed for political power to remedy common concerns and problems. There are some local examples of disabled people who have organized and made political demands. For example, an organization of disabled people in Denver successfully agitated for remodeling at the state capitol to give them access.

The central difference between a physical disability and another form of deviance is its identity as a condition rather than a behavior; a disabled person exerts no control over this condition. Thus, society regards criminal offenders and drug addicts as choosing these forms of deviance, but society imputes deviant status to people with visible physical handicaps, physiologically obese people, and the men-

TABLE 14-1

Talking About Disabilities

Observers have offered some suggestions on appropriate language for discussions of disabilities. Inappropriate choices perpetuate the stigma that society applies to people with disabilities.

Don't Use	Suggested Phrase	Why?
Handicapped person	Person with a disability	A disability does not automatically imply a handicap. Disabilities often mean only that people perform familiar activities in ways that differ somewhat from those of people without disabilities, but with at least equal participation and results.
The retarded	People with retardation	Avoid categorizing people. Many differences distinguish people with similar disabilities.
Crippled boy	Boy with a physical disability	The word *crippled* has a negative or judgmental connotation, so discussions of the topic should avoid it. Additionally, the preferred phrase puts the person first, rather than the disability.

SOURCE: Blaska, Joan. 1993. "The Power of Language: Speak and Write Using 'Person First.'" Pp. 25–32 in *Perspectives on Disabilities*, 2nd ed. Edited by Mark Nagler. Palo Alto, CA: Health Markets Research.

tally retarded as a result of causes outside their immediate control. Such a status reflects ascribed characteristics, not achieved ones. Physical disabilities imply more than simple biological facts; they are social constructs as well. We even talk differently about disabilities than we do other conditions (see Table 14-1).

Society considers both visible disabilities, like those caused by physical disfigurements, and less evident ones, like mental retardation, as deviance because they depart from normative conceptions of "normal" conditions, and affected people experience sanctioning processes that lead to social stigmas. Thus, a handicap brings "an imputation of difference from others, more particularly, imputation of an *undesirable* difference. By definition, then, a person said to be handicapped is so defined because he deviates from what he himself or others believe to be normal or appropriate" (Friedson, 1965: 72). Many disabled people violate the norm of physical well-being and wholeness. Goffman (1963: 126–130) discusses the importance of *identity norms*, which depict images of socially ideal people, or shared beliefs about appropriate conditions or appearances for people. In contemporary U.S. society, these ideals embody a number of characteristics: "young, married, white, urban, northern, heterosexual Protestant father of college education, fully employed, of good complexion, weight, and height, and a recent record in sports" (Goffman, 1963: 128). People with physical disabilities acquire identities as deviants not necessarily because of anything they have done, but because others impute to them undesirable differences from these kinds of images. In this way, society defines conditions, as well as behavior, as examples of deviance.

The social reactions of others to disabled people seem to depend on certain characteristics of the person, disability, and social situation. But to the extent that disabilities entail departures from social expectations, disabled people always risk experiencing social stigma (Stafford and Scott, 1986). They readily recognize such reactions by others, as appears clearly in a report from a hemophiliac boy, who must not participate in some physical activities and who sometimes must use crutches because of soreness in his legs:

> I'll be on my crutches sometimes and I'll be walking down the street, and people will get off away from me and walk around me instead of walking beside me because some people think that they might catch something. Nobody ever sat down and told them that they shouldn't be afraid of people with handicaps, that we are people just like them, except that we got a problem. (quoted in Roth, 1981: 93)

Disability as Deviant Status

As a socially defined category, disability establishes a master status that tends to override all other status characteristics of an individual. Someone with a physical disability often experiences "a personally discreditable departure from a group's expectations" (Becker, 1973: 33). Groups in general expect a person to have two legs and two arms, functional ears and eyes, and other features that enable them to carry on daily activities in a commonly recognized fashion. Physical disabilities violate these normative expectations of society. In Sagarin's (1975) words, they give a "disvalued" character to people.

Disability may even reflect personal discredit on people. Others often define their public identities primarily according to their handicaps. They lose status as individuals and become only "cripples" or "deaf-mutes." This status as discredited and stigmatized members of society causes difficulty for some disabled persons in securing jobs that they could perform, as well as engaging in normal social interactions with others. Some may even have to pay for companionship by dealing with prostitutes and hostesses in bars.

Some observers have argued that deviance and disability, like sickness, differ primarily in the extent of responsibility imputed to individuals for their conditions. Deviants can choose to enter deviant roles; the disabled do not choose this status. A distinction separates deviance, as willful behavior subject to an individual's control, from physical disability, which clearly does not fit this model (Haber and Smith, 1971). Some may further distinguish disabled people from one another according to their perceived responsibility for their own conditions; for example, someone may have suffered injuries in a car accident that he or she caused, or an incurably ill person may have procrastinated in seeking medical care and aggravated a previously treatable condition. Obese people encounter rather universal judgment, often unfair, that they create their own overweight conditions, and others offer them little of the sympathy that blind people or paraplegics receive. The obese, moreover, tend to internalize that viewpoint, often feeling guilty themselves about their conditions.

As in other forms of deviance, people who commit deviant acts often take on deviant identities, sometimes leading others to regard them primarily as deviants. Both occasional and professional criminals are deviants, but society frequently applies a well-defined criminal status to professional offenders different from the status determined for occasional offenders. The deviant identities of physically disabled people differ from those of criminals and other kinds of deviants because society ascribes this status to them although they have not committed deviant acts. They acquire deviant status for other reasons, many of them associated with the concept of the sick role.

Disability and the Sick Role

Attitudes toward physically disabled people relate to conceptions of the sick role. According to Parsons (1951: 428–479), the sick role emerges from two interrelated sets of exemptions. Most people exempt individuals defined as ill from certain obligations and responsibilities. No one blames them for their illnesses, nor does anyone expect their conditions to improve due to their motivation alone. Others view sick people as suffering from temporary impairment in their capacity to function. For these reasons, illness relieves people of normal family, occupational, and other duties.

In exchange for these exemptions, however, others also impose certain expectations on people occupying the sick role. For example, they must define that role for themselves as an undesirable condition and do everything within their power to facilitate their own recovery. Someone suffering from a physical illness should seek medical help, usually from a physician, and cooperate with resulting recommendations. Following a doctor's orders means precisely that: The advice of a physician on medical matters reflects not only the recommendation of someone with presumed knowledge about these conditions but also society's expectation that ill persons will try to move to more conventional roles.

So stated, four elements make up the sick role: (1) no attribution of responsibility of individuals for their own conditions, (2) exemptions from normal role obligations, (3) recognition of the undesirable character of illness despite the benefits of these role exemptions, and (4) an obligation to seek help to heal the sickness. Thus, the sick role normally protects people from blame for their conditions and excuses them from many obligations on the condition that they attempt to move to other roles as quickly as possible. Society gives some rewards to a sick person, but only on condition that he or she decline to enjoy them. In the final analysis, individuals and groups expect sickness to remain only a temporary condition. (See Table 14-2.)

Physicians often legitimate people's assumption of the sick role. Some aspire without success to the sick role when doctors refuse to certify them as sick. Those who successfully obtain this legitimation gain society's permission to assume the sick role; those denied entry by doctors must assume normal roles or face the social stigma of illegitimate usurpation of its benefits (Wolinsky and Wolinsky, 1981). Sometimes legitimation comes from the presence of someone with evident authority to validate the role. One person in a wheelchair has observed, "Salespeople are—if I'm alone—salespeople treat me very well. If there's someone else with me

TABLE 14-2
Society's Do's and Don'ts for the Sick Role

Do	Don't
Follow doctor's orders.	Ignore the advice of health professionals.
Try to move away from the sick role (try to not perform this role any longer than necessary).	Enjoy too much the privileges of the sick role (e.g., release from occupational, educational, and family obligations).
Tell others that you do not wish to occupy the role any longer than necessary.	Tell others you wish you could remain ill for a longer time.
State a desire to perform the role only if you have a legitimate claim on the privileges of the role.	Fake a claim just to enjoy its benefits.
Conceive of the role as only a temporary exemption from obligations.	Expect to enjoy the role's benefits on a long-term basis.
Assume the role because you have no choice.	Make yourself sick just to gain exemptions from obligations.

often they'll assume that other person is an attendant, and that they must deal with that person, which I don't like very much" (Cahill and Eggleston, 1995: 685).

Common ambivalence about physical disabilities reflects many people's conflicting view of disabilities as both unavoidable and at the same time undesirable conditions. While they define illness as a departure from the normal and desirable state, they do not regard it as a reprehensible fault in the same way as they judge either sin or crime; they seek to neither blame nor punish the sick person. "So long as he does not abandon himself to illness or eagerly embrace it, but works actively on his own and with medical professionals to improve his condition, he is considered to be responding appropriately, even admirably, to an unfortunate occurrence. Under these conditions, illness is accepted as *legitimate* deviance" (Fox, 1977: 15). But because most people with disabilities will never be able-bodied, the benevolence most feel toward those who are ill does not quite apply to the disabled. The relationship between the disabled and the able-bodied is strained because the disabled can never leave the sick role. This results in feelings of ambivalence.

Societal Reaction and Ambivalence Toward Disability

Appearance norms govern conceptions of "ideal" appearance—the sizes, shapes, and functions of human bodies and features. Conversely, standards for deviant physical conditions vary dramatically from culture to culture. Some societies favor slimness, while others judge body size with more latitude; some tolerate infirmity easily, while others expect vigor and a full range of functions, imputing deviance to people who do not conform.

Of course, judgments of deviance reach beyond conscious violations of behavioral norms to encompass many kinds of conditions that people expect. Few regard others with physical or mental disabilities as voluntary deviants, but they nevertheless apply some sanctions for any deviation from implicit conceptions of "normal" physical characteristics or functions. People who, even involuntarily, violate such conceptions qualify as deviant for the purposes of this discussion.

Involuntary membership in the class of deviants does differentiate physically disabled people from others. Deviance does not lose its undesirable character, even for conditions outside the deviants' control, however, so even these involuntary violations of social norms draw negative sanctions; the severity of those sanctions varies according to the degree to which the violations represent voluntary choices.

This section reviews various forms of physical disability with different images as more or less voluntary conditions. Totally blind people without reasonable hope of recovering their sight, the crippled, and the mentally retarded most accurately represent the case of involuntary deviants. Obese people and stutterers move toward the voluntary category because they may conceivably move from the deviant role to the other, more conventional one, even though the cause of the disabling condition may remain outside their control. This inability to move from the disabled role, then, still essentially constitutes deviance in society's judgment.

While cultural variables affect this judgment, almost all cultures appear to have rejected some kinds of physical disabilities throughout human history (Myerson, 1971). Not all impairments justify sanctions for disability; images of disability vary from culture to culture and from age to age. A facial disfigurement stigmatized in one country may seem like a sign of beauty or even supernatural powers in another. American adolescents with physical blemishes may face rejection or teasing from their peers, but other societies purposely apply scarring marks or tattoos at this age to enhance youthful beauty. The infirmity of old age, regarded negatively in contemporary U.S. society, commands respect and admiration in other societies, which often regard advancing age as a sign of wisdom and power. Similarly, some cultures value plumpness and even obesity in women as criteria for feminine beauty, while American society associates thinness, often in the extreme, with female beauty.

Such socially defined reactions as well as behavioral or biological characteristics determine the extent of any deviance-related sanctions applied to physical disability. This section reviews the effect of cultural conceptions of disability on reactions to the blind, the mentally retarded, the visibly physically handicapped, the obese, and people with eating disorders. In each of these categories, cultural stereotypes clearly exert a powerful influence on societal reactions.

Of course, disabled people experience many positive reactions as well as negative ones. One study of public reactions to people in wheelchairs found that many people genuinely wished to help. A woman in a wheelchair described the following situation at a restaurant:

> I had to go to the bathroom and I was with three guys. And I had to go bad. It wasn't accessible. Usually I can do it if there's something I can hold on to, but

not with this one. So, two women helped me I didn't even know. They go, "Well, honey, we'll help you." They were drinking. They were about fifty. I was more embarrassed than anything. (Cahill and Eggelston, 1995: 691)

Blindness

As mentioned earlier, no one can accurately and comprehensively estimate the number of blind people in the United States because too many divergent definitions specify varying degrees of blindness. An effort to complete such a count would also encounter problems locating and identifying blind individuals. For these reasons, estimates of the number of blind people in the United States range from somewhat less than 500,000 to more than 1.5 million. These figures should include people with sufficient impairment in their eyesight to be classified as legally blind, although the higher figure adds others to the group of people so classified (Koestler, 1976: 46).

In fact, observers often specify three categories of blindness (Koestler, 1976: 45–46):

1. Totally blind, determined by total absence of any light or image perception
2. Legally blind, determined by central visual acuity of 20/200 with corrective lenses in the better eye along with restriction of the central visual field that allows the individual to see objects only within a 20-degree arc
3. Functionally blind, determined by inability to read ordinary newspaper print, even with perfectly fitted glasses

To resolve these rather technical distinctions, most observers apply the label *blind* to people who are totally or nearly totally blind (that is, they lack all or almost all light perception). The label *visually impaired* refers to those with less disabling but still serious visual deficiencies.

People have long recognized the blind as one of the most conspicuous groups of disabled people in society. Since the eyes communicate much human expression, some feel extremely disturbed when they confront blind people. A blind person's gaze does not transmit the same psychological or emotional cues as that of a sighted person communicates. Facial expressions provide less information to others. Various behavioral mannerisms and other visible clues increase the social conspicuousness of the blind, including odd postures, rocking of the head or tilting it at odd angles, and touching objects in a groping manner, as well as distinctive paraphernalia such as thick glasses, white canes, and Seeing Eye dogs.

Historically, societies have relegated blind members to inferior roles as outcasts and beggars. Sighted people have often sought to gain religious merit by giving alms to blind beggars. In England under Queen Elizabeth I, poor laws and charities grouped the blind together with paupers, orphaned children, and mentally disordered people. Only in the 18th century did Western societies distinguish different types of dependents and develop specialized welfare services targeted for their needs.

During the 19th century, urbanization and industrialization brought a different type of stigmatizing reaction to the blind:

> The humanitarianism and organized philanthropy of the second half of the century in England and the United States introduced the conception of character defect as a middle-class explanation of pauperism, which had the concomitant result of associating physical defect with personality weakness and lack of self-resolution. . . . At the same time, the humanitarian movement was responsible for the special regard for the blind which gave them a more secure relief status than other dependent groups. (Lemert, 1951: 114)

The humanitarian movement also promoted a new attempt to restore the self-confidence and self-reliance of blind people by creating special schools for them. These schools promoted ideas that blind people could achieve educational and employment goals as productive members of society. One blind man reported gaining an important capability for mobility by attending a blind school:

> [Being blind] broadened my lifestyle because all of a sudden, I went to the blind schools, I could be free from my family. I didn't have to have my parents say yes or no every time I wanted to go into the city, I could just go. I had a little money in my pocket that I would hustle up by washing dishes in the school, so I would go and do my trip on weekends. I learned my own mobility and my own social skills. (quoted in Roth, 1981: 184)

Today, public stereotypes of the blind include expectations of "helplessness, docility, dependency, melancholia, and serious-mindedness" (Scott, 1969: 21). One writer has reviewed literature, historical records, mythology, and folklore and found a number of recurring themes concerning blindness and blind people, characterizing them as deserving of pity and sympathy, miserable, helpless, useless, compensated for their lack of sight, suffering punishment for some past sin, maladjusted, and mysterious (Nonbeck, 1973: 25). Such ideas form the core of social stereotypes about blind people, often promoting negative images that reinforce the social stigma on blindness. When blind people encounter sighted people, this stigma influences the interaction. Each party in such an interaction may prefer an avoidance reaction to the discomfort of the encounter.

> The effects of these reactions on a blind man are profound. Even though he thinks of himself as a normal person, he recognizes that most others do not really accept him, nor are they willing or ready to deal with him on an equal footing. . . . The stigma of blindness makes problematic the integrity of the blind man as an acceptable human being. Because those who see impute inferiority, the blind man cannot ignore this and is forced to defend himself. (Scott, 1969: 25)

Mental Retardation

People with mental retardation are limited as to intelligence and ability to perform cognitive tasks. Three components define this disability: (1) Mentally retarded people do not learn as quickly or as much as nonretarded people do; (2) mentally retarded people do not retain as much information as others do; and (3) mentally retarded people have weak powers of mental abstraction, which limit their use of the information they do retain (Evans, 1983: 7). Beyond these simple generalizations, even experts disagree about the full implications of such cognitive impairments.

The exact number of mentally retarded people in the United States remains unknown; nor can observers determine reliably and exactly which individuals they should so classify. Agreement suffers from wide variations in the meaning of intelligence and in current methods for measuring it. An extensive literature has debated the nature and meaning of IQ, ultimately suggesting that IQ tests produce only relative and fallible measures. Lacking reliable measurement devices and a widely accepted determination that states which IQs constitute retardation, observers can make only educated guesses about the number of individuals who might fit the criteria for mentally retarded people.

Of all human assets, people probably set the highest value on the cherished ability to think; it helps them to plan and arrange their lives, to manage all their affairs. Those who lack these attributes, leaving them without the mental capacity commonly recognized as "normal," experience a stigma among the most devastating of all in society, at least in modern Western cultures. No other stigma implies such general unfitness as that associated with mental retardation, since many people regard a mentally retarded person as lacking in even *basic* competence. Other disabled people may have to refrain from some activities, but they retain some competencies, if only for limited activities. On the other hand, a retarded person, by definition, lacks competence even to handle any of his or her affairs. The social situation of mentally retarded people degenerates further when others recognize this disability as a permanent condition. "As everyone 'knows,' including the ex-patients, mental retardation is irremediable. There is no cure, no hope, no future. If you are once a mental retardate, you remain one always" (Edgerton, 1967: 207).

No physically identifiable traits distinguish most mentally retarded people from others. Their disabilities may become evident only in certain situations—when they face difficult social interactions, when they try to converse with others of normal intelligence, and in varying circumstances that bring them face to face with others. Upon discovering their cognitive incompetence, others may react in several ways. They may "talk down" to the retardates or speak in slow and deliberate tones of voice. This recognition reduces the interaction to the lowest plane, limiting exchanges to the simplest possible vocabulary and avoiding complexities, such as humor. Others may assume that retarded people lack any knowledge of commonplace objects or events and try to explain even the simplest ideas during the course of these conversations. Moreover, since normal people frequently fear embarrassing retarded people, they exercise a great deal of tact. Such careful treatment slows down any interaction to the point of virtual cessation; it changes dramatically from the kinds of interactions enjoyed by people of normal intelligence (Edgerton, 1967: 215–217).

The pervasive stigma of mental retardation prompts others to interpret almost every action of a retarded person as a result of the disability. As a result, when a person of low intellectual capacity gets into trouble, almost automatic reactions attribute the problem to the retardation. On the other hand, if a mentally retarded person does something that would earn a reward for a person of normal intelligence, others may attribute the event to chance or random behavior. In fact, others sometimes decline to attribute even common human traits to mentally retarded people. For example, a sociologist studying mental retardation has reported a simple truth that opened the door to explaining other social aspects of retardation:

> One of the older female [mentally retarded] students returned from a visit to a beauty shop in a state that could be conservatively described as euphoric. I was surprised. I did not expect them to want to be attractive—an all too common social misconception that militates against many retarded people being given a chance to be attractive. This attitude explains why some dentists counsel parents against orthodontia for their retarded children, why retarded people in many institutions are shorn. (Evans, 1983: 120)

Physical Handicaps

Some people live with visible physical handicaps from birth, perhaps because of some deformity during development or accidents or illnesses that may force amputation of a limb, cause serious burns, or produce partial or total paralysis. Depending upon the severity of such impairment, others likely will react toward the person as a deviant, particularly in a society that actively emphasizes physical health, attractiveness, and basic competence. Inability to perform physical tasks for oneself frequently reduces a person's social status to that of a child. Within this context, society recognizes a physically handicapped person in "someone who perceives himself/herself and is perceived by others as unable to meet the demands or expectations of a particular situation because of some physical impairment—i.e., an anatomical and/or a physiological abnormality" (Levitin, 1975: 549).

Societies have commonly separated pronounced cripples from "normal" people, and this tendency has resulted in varying degrees of isolation, persecution, and ridicule. The blind, the crippled, and lepers traditionally have made their livings through begging. Preliterate societies commonly exposed newly born, deformed infants to the elements, allowing them to perish. Ancient Sparta, a society that stressed physical perfection, actively eliminated deformed children from its ranks. Many groups have regarded physical deformities as punishments from God for unknown sins. Hebraic law, for example, treated physical abnormalities as signs of physical degradation, and it specifically prohibited cripples from approaching temples. The Old Testament contains many specific passages:

> For whatsoever man *he be* that hath a blemish, he shall not [serve as a priest]: a blind man, or a lame, or he that hath a flat nose, or anything superfluous, or a man that is broken-footed, or broken-handed, or crookbackt, or a dwarf: or that hath a blemish in his eye, or be scurvy, or scabbed, or hath his bones broken; No man that hath a blemish of the seed of Aaron the priest shall come nigh to offer the offerings of the Lord made by fire: he hath a blemish; he shall not come nigh

to offer the bread of his God. He shall eat the bread of his God, *both* of the most holy, and of the holy: Only he shall not go in unto the vail, nor come nigh unto the altar, because he hath a blemish; that he profane not my sanctuaries: for I the Lord do sanctify them. (Leviticus, 21: 16–23)

Cripples often suffered ridicule during the Middle Ages, frequently serving as court jesters. During the 16th and 17th centuries, many believed that evil spirits afflicted cripples, and many were burned as witches. Cripples were often regarded as evildoers, or at least as poor wretches, as seen in several of Shakespeare's plays and in subsequent literature such as Victor Hugo's *The Hunchback of Notre Dame.*

Contemporary attitudes still tend to regard people with visible physical handicaps apart from other human beings; many people today look on them with pity or avoid them altogether. Crippled people face problems with occupational roles, social relationships, and general social participation. Severe physical disabilities strain marriages and often cause job changes and social isolation. Seldom, for example, do retail stores hire people with physical disabilities as sales personnel, and barriers often keep them out of similar positions of public contact in which their external appearance may turn away customers. Even when someone with a severe physical handicap finds a suitable job, such as working with computers, she or he often lacks pay high enough to offset the costs of life with a disability. As one handicapped person has put it:

As a paraplegic, you have to pay for attendants and you have to pay for medical supplies—all the extra needs that the nondisabled person doesn't have. I have to hire people to help me up in the morning and help me into bed at night, and to do the cooking, cleaning, whatever. . . . You also have to buy certain medical supplies, which are outrageous in price. (quoted in Roth, 1981: 45)

Social reactions to people with physical handicaps often combine pity and fear, and such attitudes proceed from sources that are not difficult to discern. Children's socialization emphasizes the desirability of independence, good health, and an attractive appearance. As a consequence of such an emphasis, people regard with suspicion others who do not fit shared conceptions of physical success.

The more effectively people in general are socialized to respect individual achievement, so heavily emphasized by the modern rejection of nepotism in favor of more equal opportunities for individual competition for success in life, the more likely it is that physical or mental disabilities which limit personal independence and restrict achievements will be seen as a badge of imperfection and inadequacy affecting all spheres of life. (Topliss, 1982: 109)

Many of these reactions were observed in January 2000 when actor Christopher Reeve "walked" across the television screen during a Super Bowl commercial. The computer-generated virtual walk sparked hope and concern from many sources. Some said the message was uplifting: There is hope for all conditions, and one must never give up. Others had a different reaction: "It's nice to give hope to victims of spinal-cord injuries," said Dr. Robert D'Ambrosia, president of the American Academy of Orthopedic Surgeons. "But something like this happening anytime soon is far-fetched" (Cuthbert, 2000).

Obesity

People who are excessively fat, tall, or short, particularly dwarfs and midgets, often face stigma and ridicule. Current appearance norms value thinness for women and slim, muscular builds for males (Schur, 1984). Generally speaking, in the Western world, whenever obese people "have existed and whenever a literature has reflected aspects of the lives and values of the period, a record has been left of the low regard usually held for the obese by the thinner and clearly more virtuous observer" (Mayer, 1968: 84).

With their highly visible traits, others react strongly, frequently perceiving obese people as deviants; such reactions often impose great social stigma because other members of a group often feel "contaminated" by association with them. This negative response may produce a formidable barrier to full social participation and acceptance (Maddox, Back, and Liederman, 1968). Such attitudes have become common only recently, however, since excessive body weight has provoked serious concern and deviant stigmatization in recent times, partly because of increased health concerns and increasingly strong norms regarding what constitutes an attractive body in Western society (see Sobal and Maurer, 1999). For example, Goffman's (1963) classic work on social stigma did not include obesity among the physical causes he discussed. But contemporary concern over various eating disorders, such as anorexia nervosa and bulimia, has highlighted the importance of cultural stereotypes of beauty and ideal body shapes in the causation of these disorders. "All my friends were beautiful," reported one anorexic. "They were skinny and perfect so I felt I had to be just like them" (Corkrean, 1994: 1).

Beginning in the 1940s, Metropolitan Life Insurance Company began to publish tables that listed the "ideal" weight for a given height. The tables were not adjusted for age, so gaining weight after age 20 was not taken into account (Maine, 2000: 32–33). The tables were obtained from information from Northern European immigrants who resided in the eastern United States. They have been updated since that time, but the implication is that increased weight is associated with greater risk of death-causing conditions. Even using more modern methods, like the body mass index, there is no agreed-upon definition of obesity and little evidence that excessive weight or thinness (except at the extremes) are by themselves related to increased death risk (Maine, 2000: 36).

Whatever their source, today's culture incorporates attitudes of rejection toward obese persons, and children begin forming them at an early age. One study showed various drawings of children with deformities of various types to subjects 10 and 11 years old. When asked to indicate their preferences among the pictured children as friends, most subjects selected the obese child as their last choice (Richardson, Goodman, Hastorf, and Dornbush, 1961). The drawings also depicted a normal child, a child with crutches and a brace on one leg, one confined to a wheelchair, one with an amputated forearm, and one with a facial disfigurement. The study also found significant sex differences in these ratings, however: Boys showed more wariness than girls of the amputated child than the obese child, while girls always ranked the obese child last.

Obese people experience intensified feelings of rejection when their sizes complicate even everyday activities, such as buying clothing. Unable to shop where most people do, obese people must choose from limited selections and styles. Sometimes, their clothing reinforces the difference in their appearance from the social norm.

The strength of the stigma attached to obesity may depend on the blame or responsibility that others assign to these people for their appearance (DeJong, 1980). Contemporary theories of obesity sometimes ascribe the condition to psychosomatic causes, overeating to reduce anxiety (Kaplan and Kaplan, 1957), but more recent research evidence does not strongly support this theory (Ruderman, 1983).

Attitudes toward obesity rest upon moral foundations. Some chastise obese people as gluttons unwilling to control their behavior regardless of the consequences. Many people feel that sanctions simply give these people what they deserve, assuming that they could have prevented their problems through basic self-restraint. One observer has remarked that "The obese teenager is thus doubly and trebly disadvantaged: (1) because he is discriminated against; (2) because he is made to understand that he deserves it; and (3) because he comes to accept his treatment [by others] as just" (Cahnman, 1968: 294). As a result, obese people may withdraw from social interactions to escape negative sanctions, in the process adding to their difficulties rather than reducing them.

While many obese people manage to achieve substantial, deliberate weight loss, they do so only by struggling with physiological, psychological, and social causes. Certainly, basic principles of diet, exercise, and self-monitoring (careful awareness of food intake and its relationship to one's weight) play critical roles in successful weight loss efforts (Colvin and Olson, 1983). Still, some obese people seem unable to develop the necessary habits to reduce their weight and to maintain that weight loss. They face particularly difficult problems when social pressures encourage them to try to lose weight to maintain their past physiques. A study of members of weight loss therapy groups found that dieters who lost weight often felt disappointed at the negative effects of their new identities on friends and family (English, 1993). These reactions, in turn, exerted pressure on the dieters to return to "normalcy," which for them meant obesity.

Eating Disorders

In August 1996, Miss Universe, Alicia Machado, became an object of controversy. After she won the title in May 1996, approval quickly changed to criticism. In August, Machado, at age 19 and 5 feet, 7 inches tall, reportedly weighed 135 pounds, about 18 pounds more than when she had won the Miss Universe title. No matter that this weight is about ideal given her height; pageant officials contemplated removing her as Miss Universe. She announced that she hadn't eaten in the 3 weeks prior to the contest to achieve the winning figure. While contest officials denied planning to strip Machado of the title, some had expressed concern that she looked "chubby" in a swimsuit and would not present a good image of Miss Universe.

Machado engaged in a practice that is common among those in certain occupations, such as the entertainment field and athletics, to achieve a certain body type or weight. But eating disorders are often means to no end. That is, some people engage in them on a continual basis because they never feel thin enough. Not only are there health implications from such practices, there are social and interpersonal implications as well.

Eating disorders are related to physical disabilities because they frequently impair social functioning. Public revelations about eating disorders also subject people to stigma from others. Research may offer some justification for classing these problems as results of some sort of mental disability, but accumulated evidence remains far from clear on this point. Regardless of the origins of eating disorders in mental or personality problems, physiological problems, or attempts to adhere to cultural values of thinness, afflicted people frequently experience social stigma, and many consider them deviants. As a result, these people must manage their deviance, often by concealing it from others.

The term *eating disorders* encompasses a variety of behaviors associated with patterns of consuming food. Currently, anorexia nervosa and bulimia probably generate the most discussion of all such disorders. Anorexia drives people to purposeful starvation, eating small amounts on only infrequent occasions. Bulimia sets up an episodic pattern of binge eating marked by rapid consumption of food in a short time. The bulimic often follows this behavior either with forced vomiting or laxative abuse in an attempt to prevent feared weight gain that would otherwise result from the binge-eating behavior.

Contemporary interest in these disorders has followed reports suggesting relatively high incidence among certain populations, particularly among college-age women. However, current attention does not imply that these conditions only recently developed. Diagnoses of anorexia nervosa date to the 19th century, when London physician Sir William Withey Gull and Parisian neuropsychiatrist Charles Lasegue both described anorexia for the first time in 1873 (Vandereycken and Lowenkopf, 1990). The first American references to the condition emerged in 1893.

Anorexia and bulimia prevail most commonly in the United States among young, white, affluent women, although definitional inconsistencies have produced widely varying estimates of the extent of the disorders from study to study (Connors and Johnson, 1987). The basic indication for bulimia involves binge eating over a period of time, but observers have not reached agreement about how much or what kind of food constitutes a binge; one person's binge may seem like a satisfying meal to another (see Fairburn and Wilson, 1993). Furthermore, researchers have not agreed about the period of time that defines binge eating, and estimates vary from a few minutes to several hours. Most estimates on bulimia derive from self-reports, and variations in phrasing of questions about eating behavior often prevent any reliable comparability. One investigator might ask whether a research subject has *ever* engaged in binge eating, while another might

This ad, which appeared in a Rochester, New York, newspaper in 1957, illustrates that era's ideals regarding body shape—ones far different from those of contemporary American society.

458

ask about binge eating in the past year, and still a third might ask about *current* binge eating. Each phrasing will lead to a different estimate of the incidence of bulimia.

Because of such differences in definition and measurement techniques, estimates of the prevalence of binge eating vary between a low of 24 percent in a sample of adults (Zincand, Cadoret, and Widman, 1984) to 90 percent in a sample of college women (Hawkins and Clement, 1980). Estimates of binge eating among males range from 8 percent (Hamli, Falk, and Schwartz, 1981) to 64 percent (Hawkins and Clement, 1980). Most bulimics and anorexics report serious concern with dieting in their teen years, and most developed stilted self-images of their own bodies that kept them from identifying ideal body weights and shapes for themselves; their eating behavior always reflected a simple desire to look thinner than they looked at any particular time.

Not all bulimics have thin physiques. In fact, many look normal, and some seem overweight. Overweight bulimics have reported less frequent cycles of binge eating and fasting than others display, but they will more likely abuse laxatives and report histories of self-injurious behavior and suicide attempts (Mitchell, Pyle, Eckert, Hatsukami, and Soll, 1990).

People develop eating disorders, such as bulimia, over time. A subtle socialization process promotes learning about social criteria for some "ideal" body size and shape from others. Childhood experiences may foster a concern over appropriate gender role behavior regarding dieting and a sense of guilt over food consumption (Morgan, Affleck, and Solloway, 1990). Cultural stereotypes spread by the mass media find reinforcement in playground interactions and everyday conversations. While many young females experience some physical victimization, research has not found a direct relationship between bulimia and such experiences (Bailey and Gibbons, 1989). Some women learn as girls to value thinness and to overestimate their true weight in relation to common measures. The constant pressure to become ever thinner, coupled with a low rate of success for dieting, drives some women to persistent weight-loss efforts. This desire for slimness often goes beyond physical attractiveness goals and seems to encompass attempts to gain some control over one's life (Taub and McLorg, 1990).

People with eating disorders generally agree with the cultural values that define slim bodies as attractive and excessive weight as a physically and morally unhealthy condition (DeJong, 1980). Appearance goals clearly motivate eating habits as significantly as do concerns for health (Hays and Ross, 1987), and most people with eating disorders report deep concern about physical appearance. In this sense, these disorders ultimately result from socially determined motivations (Fallon, Katzman, and Wooley, 1993). Over time, however, the eating disorder seems to dominate behavior. "When I first started losing weight," an anorexic has said, "I thought I'd be so pretty all the guys would like me. Ironically, once I got into it, I didn't care about guys. I didn't care about anything but losing weight" (Corkrean, 1994: 5).

Most anorexics and bulimics value conformity, and they report high educational and occupational aspirations (Humphries, Wrobel, and Wiegert, 1982). People who exhibit tendencies toward anorexia and bulimia also report that their families emphasized eating patterns and exercise as they grew up. Many report fathers' preoccupation with exercise, mothers' engrossment with nutritious food preparation, and development of "friendly rivalries" with others to achieve slim body sizes (McLorg and Taub, 1987). These people perceived encouragement and rewards for slimness, even at an early age, an emphasis repeatedly reinforced in national advertising, fashion news, and the desire for popularity among peers.

Under almost constant bombardment by messages promoting the value of thinness, one study reports, many people with eating disorders experience difficulty differentiating

> between socially approved modes of weight loss—eating less and exercising more—and the extremes of these behaviors. In fact, many of their activities— cheerleading, modeling, gymnastics, aerobics—reinforced their pursuit of thinness. Like other anorexics, Chris felt she was being "ultra-healthy," with "total control" over her body. (McLorg and Taub, 1987: 186)

In this sense, participants vividly experience the ambivalence of social attitudes surrounding eating disorders. Many initially see themselves as simply conforming to cultural norms that support thinness. They view their eating habits not as deviant behavior, but simply as energetic efforts to conform to prevalent appearance norms. Studies of bulimics report that they consistently overestimate their own body sizes and shapes, that they deeply fear obesity, and that they describe themselves as fat when, in fact, they are at or below average body weights (Powers, Schulman, Gleghorn, and Prange, 1987). One study among college students reported that important predictors of bulimia included feelings of guilt over food and a sense of weak control (Morgan et al., 1990).

Female athletes may be especially susceptible to eating disorders. Christy Henrich, a top national competitor in gymnastics, barely missed making the U.S. Olympic team in 1988. By 1990, eating disorders had left her so weak that she withdrew from competition. By 1993, she weighed only 60 pounds, and she died in July 1994 from multiple organ failure brought about by her disorder. Henrich began a strict dieting regime after a judge at an international competition told her that she needed to watch her weight. Her coach described the incident as an offhand comment, but Henrich interpreted the warning to mean that she was too fat to compete as an Olympic gymnast. Her daily food intake after that time often amounted to a single apple, later declining further to an apple slice a day. Ultimately, no one—her family, fiancé, or psychiatrist—could help her (*Des Moines Register*, July 28, 1994: 1–2A). Many other well-known female athletes have publicly acknowledged their eating disorders, including Olympic gymnasts Cathy Rigby, Nadia Comaneci, and Kathy Johnson; diver Megan Neyer; and tennis player Zina Garrison.

Many people with eating disorders can admit their problems only after they recognize the disruptions that their behavior causes. In general, bulimics freely admit their condition more often than anorexics do. Both conditions involve both behaviors and attitudes, sometimes in combinations that seriously interfere with people's lives. "Individuals who binge eat and purge frequently, feel out of control around food, have a high drive for thinness, tremendous fear of fat, severe body dissatisfaction, and chronically diet evince a combination of attitudes and behaviors which significantly interferes with their life adjustment" (Connors and Johnson, 1987: 178).

Explaining Eating Disorders

Because eating disorders seem to affect individuals, some may look only for individualistic causes for these conditions. Psychologists have identified a number of psychological conditions associated with eating disorders, including low self-esteem and feelings of helplessness (Shapiro, Blinder, Hagman, and Pituck, 1993). Sociologists, however, often analyze these behaviors from another point of view, noting that eating disorders follow socially influenced patterns rather than emerging at random. They affect women more often than men and younger women more often than older women.

The medical literature began to document cases of anorexia nervosa only a century ago, and during the decade before the First World War, references to anorexia "in aid of modish thinness and romantic acceptance begin to proliferate" (Shorter, 1987: 82). Society valued thinness during the 1920s, but such norms did not create either social hysteria or the consequences for women that seem evident today. "The positive associations with plumpness were too recent [at the time] and too well entrenched to be easily eradicated" (Seid, 1989: 97).

A number of observers associate eating disorders with features of late-20th-century American culture, particularly social standards of female beauty that dominate conversations, media images, and clothing fashions (Bordo, 1993). For example, the pictures in most magazines targeted at young women appear to stress a thin physique as a basic element of female beauty. Slender women dominate movies, television, and popular fiction. Girls play with dolls, especially the extremely popular Barbie dolls that continually reinforce the link between attractiveness and thinness. Introduced in 1959 by Mattel Toy Company, Barbies have been a part of the playtime of most young girls. Most girls have a Barbie by their third birthday, and many will have an average of seven dolls during their childhood (Lord, 1994). Barbie is exceptionally tall but with a child's size three foot. Her permanent measurements are 39-18-33.

> Contoured with no body fat or belly, a human Barbie would not menstruate. Her indented rib cage could only be achieved through plastic surgery and the removal of ribs. In fact, Barbie has lost weight since she was created in 1959. Barbie's accessories . . . reinforce her messages about women's bodies. Slumber Party Barbie (1965) came with a bathroom scale permanently set at 110 pounds and a book, "How to Lose Weight," with the direction inside, "Don't eat." (Maine, 2000: 210–211)

This too-thin image finds further support in the physical characteristics of successful models, actresses, and public figures, most of whom actively maintain and flaunt thin bodies. One writer has described thinness as a cult: The pursuit of thinness requires intense, everyday devotion not unlike that expected of participants in a religious cult (Hesse-Biber, 1996).

In the 1950s, one of the sexiest women in the world stood 5 feet, 5 inches tall and at her skinniest measured 37-23-36. Her dress size was 12. For comparison, television stars Teri Hatcher of the series *Lois and Clark* and Courtney Cox Arquette of *Friends* both wear size 2! These days, Marilyn Monroe might find herself shopping from "plus" or "queen" size clothing racks.

Even Miss America has become thinner. Using data collected from 1922 to 1999, two researchers (Rubinstein and Caballero, 2000) reported that the winners of the Miss America pageant were not only thinner, but some were dangerously thinner. In the 1920s, contestants had body mass indexes with the range that is now considered medically normal (between 20 and 25). The decline in body mass indexes occurred at a time when contestants had grown slightly taller; there was an increase of 2 percent in height but a decline of 12 percent in body mass index. Some of the winners were in the range of malnutrition as defined by the World Health Organization (body mass index of less than 18.5). Some winners had an index as low as 16.9.

Observations of such thin role models may well depress almost anyone, except, perhaps, Paris designers and very young women. In fact, feelings of depression often follow recognition of one's distance from the physical ideal represented by these small sizes. Further, clothing promotion provides only one example. Girls and young women daily view a steady parade of thin bodies coupled with such desirable qualities as success, love, and acceptance. Kiernan (1996) asks, "Is it any wonder that 8 out of 10 young women, by the time they turn 18, say they dislike their bodies?"

The overall social emphasis on thinness pressures young women to achieve a standard that only a few can attain with any comfort. Under such pressure, some women understandably view themselves as too heavy or too big, even if they closely approach average sizes for healthy adults. They may feel the need to undergo rigorous and continuous dieting and exercise. But even the most diligent dieting and exercise program may fail to produce the physical results that some women want. Under such social pressure, some women develop eating disorders as part of their efforts to achieve their thinness goal.

In previous decades, women did not endure such close scrutiny of their bodies. By the 1960s, however, the ideal of the hourglass figure, which approximated the builds of most women, gave way to a thinner, more sticklike shape with fewer accents on the waist. By the end of that decade, fashionable miniskirts shifted visual emphasis to the legs. Today, society expects women to maintain thin shapes on their own through exercise and diet rather than through artificial devices, such as the girdles that once restrained and smoothed unwanted bulges. They must work hard to "get into shape," less for health reasons than to approximate appearance standards. Fitness has become a $43-billion-a-year business in the United States, channeling wealth to health clubs, exercise videos, home exercise equipment, workout clothing, and accessories (Hesse-Biber, 1996: 47).

Ancillary businesses, such as elective plastic surgery, have also grown. Nearly 3 million plastic surgery procedures were performed in 1998 (Maine, 2000: 128),

❋ *Issue*   **Makeovers and Second Chances**

The desire to improve oneself, including one's physical appearance, is a common one. That desire propels some women toward eating disorders, some toward radical diets and untested drugs, and others toward plastic surgery. Self-improvement is a commendable goal, but sometimes it can take strange turns.

Paula Jones, Monica Lewinsky, and Linda Tripp have something in common besides their connection with Bill Clinton: They all experienced "makeovers" after the scandal in 1999. Plastic surgeons reduced Paula Jones's nose, Monica Lewinsky shed 30 pounds in her new career as a Jenny Craig spokesperson, and Linda Tripp received virtually head-to-toe plastic surgery and liposuction. "How weird," one columnist wrote (Dowd, 2000: 23), "a scandal that investigated whether the president was treating women as sex objects has ended with most of the women involved getting makeovers to look sexier."

Source: Maureen Dowd. 2000. "Scandal Advances to Cosmetic Surgery." *Omaha World-Herald,* January 5: 23.

and people spend more than $5 billion a year on plastic surgery, an increasingly common and accepted form of "self-improvement." According to a survey by the American Society of Plastic and Reconstructive Surgeons, women make up about 94 percent of their patients (Hesse-Biber, 1996: 53), although other estimates place the percentage of procedures by women at 90 percent (Kalb, 1999). Rhinoplasties (nose jobs), breast augmentation and reduction, liposuction, eyelid surgery, and face lifts are some popular procedures today.

Plastic surgeons must endure lengthy training to do their precise work. Not surprisingly, then, the cost of plastic surgery is high. Liposuction procedures averaged about $3,500 in 1998, breast augmentations $3,000, breast reductions $5,500, tummy tucks $4,000, nose jobs $3,500, and face lifts $5,000 (Kalb, 1999: 53; Maine, 2000: 132). In 1998, 25 percent of cosmetic surgery patients were repeat patients. And the cost of these procedures must be borne entirely by the patient; insurance companies do not pay for cosmetic plastic surgery. The American Society of Plastic and Reconstructive Surgeons, which represents the country's 5,000 board-certified plastic surgeons, offers a finance program for its members that will provide loans to patients and collect the interest for physicians.

The desire to remake nature springs from the same sources as eating disorders do, and neither ends at the borders of the United States. In a number of countries, thinness has come to symbolize an ideal associated with Western culture. Images worldwide associate slender bodies with self-discipline, control, attractiveness, assertiveness, and independence (Pate, Pumareiga, Hester, and Garner, 1996). In the Dominican Republic, people closely associate female beauty with the virtues of a good wife and mother; thinness is not a necessary criterion for beauty. In Argentina, on the other hand, a woman's weight has become increasingly bound up with standards of beauty, as one Argentinean woman has observed:

The only people who see being fat as a positive thing in Argentina are the very poor or the very rural people who still consider it a sign of wealth or health. But as soon as people move to the bigger cities and are exposed to the magazines and the media, dieting and figures become incredibly important. (Thompson, 1994: 29)

Standards of beauty vary in different cultures. Some associate beauty with wearing certain items of apparel or adornment (Furio, 1996). Moslem women, for example, hide their bodies and hair under veils when in public. Indian women pierced their noses with rings from which they hang jewelry long before Generation X discovered the practice. Ndebele women in southern Africa wear extremely large and brightly painted beads to enhance their beauty. Other cultures emphasize the importance of body shape, however. Fat equates to beauty among the Tuaregs of Saharan Africa. Some doting mothers force-feed daughters, since others will think them unfit for marriage if they don't have 12 rolls of fat around their bellies. Force-feeding also goes on in Mauritania, where others judge a husband's machismo according to his wife's size.

Western standards favor bigness only for breasts, and then only if they ride high on a woman's body. Many associate large breasts that droop with maternal obligations, not a source of sexual attractiveness in America, although other cultures disagree. Senegalese women pull on their breasts with ropes to achieve maximum droop, while in Papua New Guinea, mothers begin pulling and rolling their daughters' breasts as soon as they appear, knowing that no man there would marry a woman with high breasts (Furio, 1996).

The Western cultural emphasis on thinness applies almost without exception to young women, but only in a limited way to older women and men. This distinction suggests an important reason that eating disorders affect young women in the United States much more often than other groups. At the same time, not all young women develop such disorders. Some individuals either achieve socially approved body size goals, adjust to their failure to do so, or develop problems other than eating disorders, such as depression, as a result of this frustration. It is also true, of course, that some may actually develop realistic body size goals, accepting the thin image only in part. But if eating disorders actually reflect the effects of these cultural features of American society, then solving related behavior problems in some women will not resolve the social difficulties of eating disorders.

Normative standards of beauty and acceptable body size and shape are culturally determined. In the United States the idea that the body is somehow flawed and must be corrected pressures many women to diet, experience eating disorders, and choose elective plastic surgery. Adhering to current norms of appearance produces considerable anxiety and at times social isolation.

Disability as a Socialization Process

Recall that *impairment* refers to a physical loss, while the concepts of disability and handicap have distinctly social contexts. To some extent, patterns of physical disability follow patterns common to other forms of deviance. Some data, for example, show a disproportionate likelihood of physical disability among older, male, lower-class, and working-class populations (Nagi, 1969). As one observer has stated, "It is not facetious to think of the ghetto and the factory as the major settings in America in which disability is manufactured" (Krause, 1976: 206). Elements of the social context in which disability appears—the influence of norms and values on conceptions of disability, the importance of social psychological adjustments to life with disabilities, individual acquisition of the sick role, and attention to social considerations in the rehabilitation process—define conditions as important as the nature of the physical problems.

Socialization and Disability

The values of achievement, independence, and activity are deeply ingrained in American society and its norms. "The good person is judged to have health, youth, beauty, and independence and to be productive" (Albrecht, 1976: 13). Such stereotypes seem to orient many social attitudes toward people with disabilities. The influence of such stereotypes also suggests a lack of psychological or social preparation in people who become disabled. Because they subscribe to norms that emphasize vigor and independence, disabilities confront people with great difficulty in accepting their own violations of those norms.

Most physical disabilities result from some physical trauma or disease. After an accident or a similar event, victims often realize with shock that they have become disabled people. They ask many questions, and society provides few answers. After medical professionals determine that an accident victim will live, paramount questions concern the future quality of and activities in that life. Disabling injuries leave people in very ambiguous situations because they often lack satisfying answers to such crucial questions as: "Why did this happen to me?" "Who will take care of me?" "Can I work?" "What will my life be like?" They will discover answers to these questions over time, but they feel pressing needs to resolve such questions immediately after the onset of disability.

Some must also begin to assume the sick role. Many find this process a difficult one because the demands of this role contradict their own values. In fact, the process by which people become socialized to their disabilities begins by denying them. Victims typically refuse to adopt the sick role, responding as though the disability would last only a short time, eliminating the need to make permanent plans that take into account the disability.

Over time, they learn the sick role, however, as they learn any other role. People with disabilities must assume the sick role for the long term, however, since it represents a more or less permanent role acquisition. No amount of cooperation with medical authorities, not even the strongest motivation to move to more conventional roles, will enable them to regain lost functioning. They must resign themselves to permanent deviant status and to their own lack of choice about this status.

Disabled people adapt to their disabilities by going through developmental stages similar to those suggested by Kübler-Ross (1969) for adjustments to death and dying: denial and isolation, anger, bargaining, depression, and acceptance. These stages stretch over different periods of time for individual patients, and they move from one stage to the next in a complicated process. During this time, medical professionals, as well as family and friends, act as important socializers. Newly disabled people must make a number of role adjustments similar to those of patients with physical illnesses. Many find ways to improve daily life only by creating supportive environments, learning coping strategies, and educating family members and employers (Mechanic, 1995).

Depression and acceptance often continue to alternate for some time after the onset of disability. Physically disabled people, as a group, report more problems with depression than do nondisabled people at all age groups and for both sexes (Turner and Beiser, 1990). This tendency toward depression, including major depression, seems related to chronic stress produced by disabilities. The stress comes not only from the loss of specific capabilities, but also from worries concerning issues of long-term physical care, such as finding and retaining good medical and physical assistance, managing insurance coverage for costs, and evaluating one's prospects for a "normal" life. But not all disabled people suffer from depression, and their tendencies toward this problem vary, depending on the nature of their disabilities. Severe obesity, for example, relates to depression, less as a direct result of overweight than as an indirect result of dieting and physical health problems associated with obesity (Ross, 1994). In other words, overweight people suffer depression due to the distress of trying to conform to social norms for body size and attractiveness, not specifically from a particular body size.

Some regard rehabilitation as a socialization process as well (Albrecht, 1976). During this time, disabled people learn new skills and adjust to the handicaps that they will experience. Considerable variability prevents standardization of the rehabilitation process. Various types of disabilities affect different functions with different degrees of social visibility, so they impose different rehabilitative needs. Serious facial burns, for example, create highly visible impairments likely to elicit dramatic responses from others; yet they do not cause extreme disabilities in physical functions. Rehabilitation for such an impairment would likely focus less on adjusting a disabled person's expectations for physical achievement toward physical realities than in matching expectations for social achievement with the social reality. Moreover, the disabled person must deal not only with the reactions of others, but with self-reactions as well.

Self-Reactions of the Disabled

An individual may react in basically three ways to the social stigma that results from a disability (Safilios-Rothschild, 1970): Deny it, accept it, or seek indirect benefits from the situation. People who have always placed a singularly high value on physical appearance may attempt to deny that any impairment troubles them; deaf people may pretend to hear, for example. Others may attempt to "mask" the disability, for example, by wearing prostheses.

Some may view their impairments as acceptable although not ideal conditions. They accept their conditions without becoming depressed. Many respond to stigma this way, but not immediately after suffering impairment; instead, they reach this accommodation to their disabilities after practicing denial for a certain period of time. Persons with different disabilities reach this stage over different periods of time. Some may remain at the denial stage for long periods of time before moving to acceptance. Family and friends provide important support in this transition.

Some disabled people adapt to the changes in their lives all too eagerly and actively seek to benefit from their conditions. They may maximize physical limitations and restrictions that they could overcome, while minimizing their remaining capabilities. Obese people and those with physical disabilities, for example, may win attention by emphasizing their predicaments. People sometimes gain financial benefits by incurring disabilities. One study analyzed changes in federal legislation concerning disabled people; it found a strong association between economic conditions and self-reports of disability with accompanying claims for compensation: Self-reports of disability peaked during periods of high unemployment and generally depressed economic activity (Howards, Brehm, and Nagi, 1980).

Stereotyping simplifies interactions for nondisabled people when they meet others with disabilities. In fact, assessment of a disabled person tends to stop when someone perceives the disability. From then on, that condition tends to shape all communication. For example, interactions between nondisabled and disabled people often must span wide spatial and social separations. The parties tend to maintain more physical space than in comparable interactions between nondisabled people, and communications tend to remain within less personal limits (Safilios-Rothschild, 1976). This distance can powerfully affect disabled people, since it seems to force them into defensive postures. Many disabled people feel compelled to defend themselves against imputations of moral, psychological, and social inferiority in such interactions.

Disability as a "Career"

Disabled persons constantly encounter reminders of their difference from norms about appearance and physical functioning. In essence, they become identified, certified, and derogated just as any other recognized deviant does. These characteristics of disability create conditions for its sociological analysis as a deviant career. Career disability, or disability that amounts to secondary deviance, results from role adaptation rather than formation of a new role. Once society legitimates or validates the disability (usually through medical intervention), role expectations may change to correspond with judgments about the seriousness of impairment. Society's reaction contributes crucial input to the process of forming the stable pattern of disabled behavior that defines career deviance. In turn, this social process helps to motivate the disabled person to create a new, deviant self-concept.

The stigmatization process sometimes works in subtle ways. People's comments may not disturb someone with a disability, but the way they speak and respond often communicates disapproval; someone may convey a stigma in gestures, facial expressions, and behavior. Excessive kindness and concern may constitute a labeling reaction, despite good intentions (Hyman, 1971). On the other hand, disabled people may have to call attention to their own deviant status if inability to do unavoidable tasks forces them to ask others for assistance (Myerson, 1971). Often, perceived help contributes to the transformation of physical disability into a deviant career.

Professionals and Agencies

Interactions with professional assistance workers, such as doctors, counselors, physical therapists, and social workers, provide extremely significant input to help shape disabled people's self-concepts and their movement toward career disability. Doctors, for example, must explain any impairment and the extent of the disability that it will cause. Similarly, social workers may work with retarded persons in the community, and counselors may aid paraplegics with employment and personal problems.

One study of mental retardation points out significant problems when psychologists diagnose children as mentally retarded on the basis of IQ tests that tend to give high scores to those from white, middle-class backgrounds (Mercer, 1973). If performance standards define normal scores according to the average scores of white, middle-class children, then many members of minority groups who take such tests automatically fall into categories below normal, increasing their chances of acquiring social identities as retarded people. Mercer (1973) claims that the test process itself contributes more than other structural conditions to the disproportionate labeling of minority children as mentally retarded. While such an intelligence test may offer one diagnostic tool with which to identify retardation, those who administer it should not confuse *detection* of the condition with the condition itself, since most operational definitions of retardation select some relatively arbitrary IQ level as a cutoff. Too often, the label becomes the reality.

Interactions with professional groups determine critical aspects of a disabled person's future role status and self-conception, and sheer chance in selection of specific agents may exert an important influence on that person's career. For example, Scott (1969: 119) has observed that blind people often gained rewards when they conformed to the expectations communicated by staff members at agencies devoted to assisting them. Blind people receive praise for their insight when they describe their problems in the same terms that their rehabilitators prefer, and they receive criticism for blocking or resisting when they do not. Gradually, the behavior of blind people comes to correspond with the beliefs of the support workers, particularly in the isolated, sheltered environments that many agencies create for their clients. Clients who live and work within such environments perform quite well there, but this experience leads to maladjustment for life in the larger community.

Many rehabilitation agencies and other organizations make substantial contributions to society's efforts to prevent, treat, and control disabilities and to individual clients. This agency support and contribution often depends, however, on the prevalence and seriousness of the disabilities they serve; thus, caseloads full of accredited disabled people strengthen agency requests for funding to pay for additional staff and other resources. Some claim that this relationship actually creates many handicaps because treatment personnel define them as such (Friedson, 1965: 74). Frequently, people with handicaps recognize these weaknesses in their interactions with treatment professionals. Firsthand accounts from people with physical handicaps, for example, often contain repeated, disparaging references about medical personnel (e.g., Roth, 1981). Some have reported failure by their initial encounters with doctors and therapists after incurring their impairments to prepare them for their handicaps, and many reactions describe medical personnel as insensitive to emotional conditions associated with physical disabilities.

Agency involvement may not always serve the best interests of people with disabilities. Edgerton (1979: 29) has pointed out that children with Down's syndrome develop less competence when raised in institutions than when raised by their parents. Obviously, many differences distinguish individual patients, but these and similar results argue against assuming that rehabilitation professionals always operate in the best interests of their patients. Albrecht (1992) has argued, in fact, that rehabilitation has become an enormous business controlled more by economic criteria than health-care priorities. As health-care bureaucracies have grown and financial interests have become more complex, banks, government agencies, and insurance companies often influence the content and duration of rehabilitation programs as strongly as health-care professionals do. The current system lacks accountability and competition.

Subcultures and Groups

The formation and growth of subcultures provide important support for the maintenance of patterns of career deviance. Subcultures institutionalize customs, recruit new members, and support the social needs of current members; as a result, they facilitate management of a deviant identity. Some groups establish formal organizations, such as foundations to aid the blind, the deaf, and others with specific physical disabilities. These groups may hold regular meetings and conventions, and they generally serve to some extent as interest groups to promote the priorities of their constituent members. Disabled people may form such groups themselves out of a need for self-defense against exclusion from participation in conventional society.

Disability subcultures serve specific functions, such as providing social and recreational outlets for members, educating the public about the nature of particular disabilities, pressing public officials for favorable policies toward disabled persons, and helping members to find marriage partners and jobs. In general, these groups emphasize efforts to change disabled people themselves rather than to change the society that has labeled and discriminated against them. Even small, informal groups may arise spontaneously among patients who share a common waiting room in a doctor's office, where they share information and offer mutual support.

Participation in a subculture may enable an individual to better manage a particular disability. Those with some disabilities, however, must do without such benefits. For example, some obese people may belong to organizations with others like themselves, but they often do not closely associate with these people, so they seldom form specific subcultures. Moreover, some organizations devote themselves to ridding obese members of their impairments rather than offering social support. Further, even successful organizations experience changes in membership over time that limit group cohesiveness (Warren, 1974).

Disabled people establish many "communities" within the larger society. Membership depends, however, on more than simply having a disability. People must satisfy three necessary conditions to become members of deaf communities: (1) identification with the deaf world, (2) shared experiences associated with hearing impairment, and (3) participation in the community's activities (Higgins, 1980: 38–77). A deaf person must have accepted this impairment and developed a self-concept that includes deafness. Subculture membership also depends on a desire to associate with other deaf people and share common aspects of their lives. Many deaf people remain outside such communities. Higgins has pointed out that many such communities join members selected by race, age, and preferred method of communication (signing or speaking). Even within the communities of the disabled, members form subcommunities to better share problems, deflect stigma from the outside, and enjoy opportunities for mutual social support.

The nature of a community of disabled people depends, in part, on its membership and their problems. Aged, deaf people, for example, commonly encounter specialized problems that associations with others help them to solve. In this instance, the community serves as the context for exploring and solving problems concerning aging, social isolation, and approaching death (Becker, 1980). A subculture for disabled people functions much like one for other deviants by establishing a locus of social activity.

The Role of Stigma in Disability Careers

No invariant career path leads to a standard form of deviance for people with physical disabilities, just as no standard pattern of behavior leads other deviants to acquire their nonconforming identities. Still, common pejorative labels characterize and group physically impaired people. Other children often mock youngsters with disabilities; obese adolescents often experience only limited social lives; people commonly misunderstand anorexia; myths and stereotypes afflict the blind, along with their sightlessness; others may ridicule or shun people with mental retardation, eating disorders, and stuttering problems; virtually all disabled people experience occupational discrimination. These and many other consequences result from the conditions of disabled people.

Once they develop stable identities incorporating their disabilities, they must accommodate changes in their interactions with others. Family and friends closely scrutinize the eating habits of someone with a history of anorexia or bulimia, and these wary reactions may under some circumstances actually encourage further instances of the behavior. Some anorexics and bulimics have described certain expectations of others that they should act like people with anorexia or bulimia. For example, some anorexics have complained that friends never offer them food or drink, assuming continued disinterest in nourishment.

While being hospitalized, Denise felt she had to prove to others she was not still vomiting, by keeping her bathroom door open. Other bulimics, who lived in dormitories, were hesitant to use the restroom for normal purposes lest several friends be huddling at the door, listening for vomiting. In general, individuals interacted with [anorexics and bulimics] largely on the basis of their eating disorder; in doing so, they reinforced anorexic and bulimic behaviors. (McLorg and Taub, 1987: 185)

Groups and organizations of disabled persons work effectively to move some members to conventional roles. Some, for example, use stigma in a positive way to encourage normal role acquisition. A study of several groups devoted to aiding obese people in weight reduction have reported that stigmatizing or labeling inappropriate behavior encourages "normalization." Specifically, "groups who used ex's [ex-obese persons, in this case] as change agents, all used strategies of identity stigmatization in order to facilitate normalization of members' behavior" (Laslet and Warren, 1975: 79). This therapeutic use of stigma moved members away from deviant roles rather than pushing them further along deviant career paths. Another study has found that people who had visible physical handicaps actually participated actively in the labeling process that established their identities as deviants, and they could, through appropriate behavior and verbalization, negotiate the deviant label (Levitin, 1975). In many instances, these people actively resisted the negative labels of others.

Perhaps more clearly than other forms of deviance, physical disability creates a deviant identity through causes other than formal or informal social labels. Society's stigma alone does not create the deviance associated with physical handicaps; this identity results in part from an objective reality, such as blindness, obesity, extreme thinness, or a crippling condition or disease. Social support provides an important element of successful adjustment after disability. Research among cardiac patients, for example, has generated abundant evidence that informal support and integration with family members enhance a patient's chances of recovery from, or successful adaptation to, the physical condition (Garrity, 1973). Conversely, patients who lack such support encounter increasingly difficult processes of adaptation.

Most people's ambivalence toward physical disabilities themselves thus tempers their conceptions of individuals who exhibit such conditions. People with disabilities alternatively experience love and hate, pity and scorn, fear and welcome, attraction and repulsion. While members of society do distinguish these involuntary deviants from others whose deviance results from voluntary behavior, disabled people still experience stigmatizing reactions, and they must deal with those responses along with their impairments.

Managing Disability

Certain problems of living inevitably confront people with disabilities as a result of their impairments. These problems vary with specific impairments and their severity, and they create many kinds of challenges, including mobility, securing and retaining suitable employment, dealing with the medical demands of the impairments, and managing the physical activities of everyday life. Another set of problems, however, comes from society's stigma. The reactions of significant others like family and friends may engender so much stress that disabled people feel the need for specific management techniques. Blind people may struggle harder to deal with the reactions of sighted people than to accommodate themselves to their blindness. The social effects of stigma on top of the physical limitations of disability produce what some have called a *second affliction* in addition to the original condition (Wang, 1993). This "double disease" has important implications for the disabled person's chances of reassuming conventional roles and interacting in conventional ways with others. This adjustment may become particularly challenging for a disabled person who is also a minority member or a woman. Such people may experience additional discrimination (Habib, 1995).

While their deviant identities stem from conditions rather than behavior, people with disabilities employ the same general processes of stigma management that serve other deviants, such as criminal offenders, homosexuals, and survivors of unsuccessful suicides. These management techniques work to minimize the stigma that might otherwise result from deviant conditions. In fact, disabled people adopt a variety of coping techniques (Safilios-Rothschild, 1970). Some react with hypersensitivity to imputations related to their conditions; others deny any assignment of deviant status, seek to "normalize" their conditions, and withdraw from the nondisabled world as much as possible; some even identify with the dominant group's reaction and perhaps even hate themselves, perhaps displaying prejudice against others with the same disability; still others become militant, attempt to make up for deficiencies by striving in other areas, or retreat into mental disorders or alcoholism. This range of responses resembles that available to anyone who feels a social stigma for any reason.

The question of disability management raises two problems. First, when does disability become a problem for an individual? More importantly, how does the disabled person manage society's stigma? Goffman (1963) has dealt with the first issue by differentiating *discredited* people from *discreditable* ones. While the former term refers to individuals with apparent or readily recognized disabilities, the latter designates those with conditions that others neither know about nor immediately recognize. Increasingly visible disabilities create progressively more difficult management problems for affected individuals. Therefore, techniques that help these people to hide their disabilities from social view offer important advantages for them. Various management techniques suit different forms of disability and rejection. These methods include passing, normalizing, coping, and dissociation.

Passing

A disabled person can attempt to pass as a normal member of society and thus completely avoid playing the deviant role. Passing involves disguising the disability so that others will not notice it and hence refrain from stigmatizing the disabled person. Such people have developed many ways to pass in normal society, but the success of this technique depends upon the visibility of the impairment. A study of mentally retarded subjects, for example, has concluded that society applies such a strong stigma to this disability that affected people feel a powerful need to manage its effect in some way (Evans, 1983). Many mentally retarded people have difficulty accepting this diagnosis, and some never do accept identities as retarded people. As a result, many attempt to pass by feigning normal mental abilities, although in many situations they cannot succeed. Some retarded people carry pens and pencils although they cannot write, and some wear wristwatches even though they cannot tell time. Their most common technique of passing, however, involves communicating unrealistic aspirations (Evans, 1983: 126–127). Some tell stories of the cars they plan to drive or houses where they plan to live, while others brag about occupational positions to which they aspire.

People with hearing difficulties may try to pass by denying their impairments. People sometimes try to deceive others about their vision impairments by wearing contact lenses or pretending to see things in the presence of others that they cannot really see. Obese people may try to wear inconspicuous clothing that partially

 Issue | **Living Large**

Figures from the National Center on Health Statistics show that more than 1 in 3 Americans is overweight, up from 1 in 4 a decade ago (Klein, 1996). As one response to this trend, society might resolve to fight harder in the "battle of the bulge." According to estimates, Americans spend more than $40 billion each year on diet-related products, not including pharmaceuticals (Hainer, 1996). Those dollars pay for special diets, exercise equipment, health-club memberships, and sports lessons and equipment, adding up to a considerable sum.

In another increasingly popular alternative, however, some people simply surrender to the trend and declare a victory. "Reclaiming the word fat has given me incredible power," one woman says. "If more fat people would use that word with pride, it could change the world" (Hainer, 1996: 1D). Sally Smith is executive director of the national Association to Advance Fat Acceptance, based in Sacramento, California. Smith (5 feet, 5 inches tall and 330 pounds) is fat and proud of it.

Marilyn Wann is the creator of *FAT!SO?*, a magazine for large people. "Some people think men who are attracted to extremely fat women have a fetish," according to Wann. "Excuse me? You think 6 feet tall, 100 pound supermodels aren't extreme? Talk about a fetish!"

Some evidence indicates increasing accommodation of overweight people. *Glamour* magazine began in 1996 to publish a "Fashion That Fits" column, giving fashion advice for large women. Designers like Dana Buchman, Givenchy, and Emanual have added plus sizes to their lines. *Mode*, a beauty magazine for large women, began publication in 1997.

Source: Hainer, Cathy. 1996. "The Renaissance of Fat Pride." *USA Today*, November 21, pp. 1D–2D; and Klein, Richard. 1996. *Eat Fat*. New York: Pantheon.

hides their body shapes; they may also try to remain in dark surroundings. Because anorexics lack accurate perceptions of their own body weights and sizes, they often do not feel conspicuous until others point out their adverse behaviors.

Bulimics may pass more easily than people with other physical disabilities can, because they generally maintain normal physical appearances; their disorder becomes evident only in the execution of the associated behavior (e.g., purging). Bulimics sometimes try hard to hide their bulimia by bingeing and purging in private. In a study of an eating disorder self-help group, McLorg and Taub (1996: 398) gave two examples of this strategy, already cited, in which a hospitalized patient felt compelled to leave her bathroom door open and other bulimics, who lived in dormitories, feared using the restroom at all to avoid the appearance of vomiting in secret.

Normalizing

Rather than trying to pass, disabled people can try to manage society's stigma by normalizing their deviance. This process essentially supplies socially acceptable explanations for their disabilities. For instance, physically handicapped people may assert in conversations with normal people that they really do live normal lives, or they may avoid taboo words like *cripple*. This strategy seeks to minimize the debilitating effects of a disability and generally to disavow deviant status. An anorexic may make an excuse for missing a meal by deliberately scheduling other commitments at that time; some simply explain that they are dieting.

The normalizing strategy often encounters certain difficulties. If disabled people succeed in disavowing deviance and persuading others to react to them as normal members of society, they must then sustain this normal role in the face of myriad small amendments, qualifications, and concessions that a disabled person cannot avoid (Davis, 1961). In many instances, others willingly participate in this management technique as much as they can. Special physical arrangements often help physically disabled students to gain admission to universities, for example, but once admitted, the schools expect them to perform on the same intellectual level as nondisabled students. The effort of a normalization strategy to move away from the sick role seems to inspire social groups to aid the process.

Coping

Someone with a physical disability may repeatedly encounter stigma from others, in the process developing ways of coping with this situation (Eisenberg, Griggins, and Duval, 1982). Wright (1960: 212–217) has outlined three general categories of situations in which a disabled person perceives an intrusion by normal people, each instance creating a need for coping behavior. First, the disabled person may treat others, usually strangers, as intruders if their reactions seem focused only on the disability; this situation may lead to a desire to retaliate in some manner, usually with biting sarcasm. Second, the disabled person may resent others' tendencies to incorporate the disability into a social situation; this attitude may lead to an "ostrich reaction" marked by completely running away or pretending the disability does not exist, or it may provoke an attempt to redirect the interaction to another subject. In the third type of situation, the disabled person desires to exclude the condition from an interaction while preserving the relationship; coping techniques include good-natured levity and embarking upon a superficial conversation. In each of these situations, the disabled person tries to avoid acknowledging or talking about the deeper and more personal meanings of the disabling condition.

Some people with physical disorders cope with their conditions by acknowledging them and applying a positive evaluation. Some people with eating disorders, for example, may prefer the stigma of bulimia or anorexia over the stigma of obesity.

Dissociation

Davis (1972: 107) has identified another technique of deviance management termed *dissociation*. Dissociation represents a retreat from social confrontation and a passive acceptance of the deviant role. Such a rejection of conventional roles and activities increases the likelihood that a person will avoid interactions with nondeviants. Handicapped children, for example, learn quickly that interactions with other children may bring pain, so they may avoid as much of this contact as possible. Obese people may reduce their social activities because they fear the attention and scorn of others. Stutterers often avoid situations likely to focus attention on their speech impediments.

A dissociation strategy involves avoiding situations likely to expose the disabled person to stigma. All too often, this excessive caution leads to social isolation that only adds to other, unavoidable difficulties. By avoiding social situations of all kinds, the disabled person forsakes positive social experiences while deterring negative ones.

Certain other general techniques also help disabled people to manage their roles as deviants. The list includes secrecy, manipulating the physical setting, rationalizing, changing to nondeviance, and participating in deviant subcultures (Elliott, Ziegler, Altman, and Scott, 1982). Bulimics, for example, might dismiss their purging as a temporary weight-loss technique. Some of these devices do not work for people with certain physical disabilities. Those with severe physical impairments, such as blind people, may have no reasonable hope of changing their conditions and hence escaping deviant identities. Some disabilities defy any attempt to keep them secret (confinement to a wheelchair), although some disabled people do attempt to conceal certain disabilities and to "pass" as nondisabled people. Those who practice dissociation try to replace interactions with nondisabled people by participating in a subculture populated by others with difficulties similar to their own. Whatever management techniques they use, most disabled people feel the need to deflect some of the stigma directed at them by others.

The Americans With Disabilities Act

Many authors have clearly documented social sanctions, and even overt discrimination, against people with physical disabilities:

Although people with disabilities have witnessed significant positive changes in quality of life and integration, they still find themselves victimized by long-standing and traditional social, psychological, physical, fiscal, architectural, and political barriers—all of which are inhibitors to the acceptance and participation by the disabled in mainstream society. (Nagler, 1993a: 33)

One critical area of concern is education. The 1975 Individuals with Disabilities Education Act was intended to stop discrimination against children with disabilities. Before passage of the act, many children had been excluded from public schools, institutionalized, or placed in programs with little or no learning component. But compliance with the law has proven troublesome, as the National Council on Disability reported in January 2000. Surveys of compliance showed that most states failed to guarantee that children with disabilities were not segregated from regular classrooms, that most failed to follow rules requiring schools to help students find jobs or continue their education, and most failed to ensure that local school authorities abided by nondiscrimination laws. It has been only when parents pressed the issues in federal courts that school systems have accommodated them. But such actions are typically undertaken by individuals, not groups, demanding political justice.

While some, such as blind and hearing-impaired people, have managed to command public attention and support for their unique disabilities, no political movement has galvanized a broad effort for inclusion that cuts across all forms of disabilities. Even the passage of important federal legislation protecting the rights of disabled people—the Americans with Disabilities Act (ADA) of 1990 (see Table 14-3)—did not necessarily reflect growth in their political power. One observer has claimed, however, that George Bush advanced his first presidential campaign by garnering the majority of the political support of the disabled population through a promise to pass the ADA (Nagler, 1993b: 481).

When it became effective on July 26, 1990, the ADA established federal laws covering many of the conditions discussed in this chapter. The act sets up civil-rights protections intended to safeguard Americans with certain mental or physical disabilities against discrimination in employment, public accommodations, transportation, and telecommunications. These provisions reinforce the Rehabilitation Act of 1973, which prohibited programs and activities that received federal funds from discrimination on the basis of physical handicap.

The act's provisions apply to three kinds of people: (1) anyone with a physical or mental impairment that substantially limits one or more major life activities, (2) any person with a record of such an impairment, and (3) any person perceived as having a physical or mental impairment. Major life activities include caring for oneself, performing manual tasks, walking, seeing, speaking, breathing, learning, and working.

ADA requires that others, especially employers, make "reasonable accommodations" to a disabled person's condition, such as accessible designs of existing physical facilities, job restructuring, modification of training devices, or relocating activities. Case-by-case judgments will have to determine exact criteria for *reasonable*

TABLE 14-3
Americans With Disabilities Act of 1990

Some Covered Medical Conditions	Excluded Conditions
Speech and hearing impairment	Compulsive gambling
Epilepsy	Homosexuality
Muscular dystrophy	Kleptomania
Multiple sclerosis	Pyromania
Back problems	Transvestitism
Mental retardation	Current illegal drug use[a]
Diabetes	
Arthritis	
Emotional illness	
Sensitivity to smoke	
Heart disease	
HIV disease	
Drug addiction	
Alcoholism	

[a]Nor does the act protect someone who completes a supervised rehabilitation program and no longer uses drugs.

accommodations and the provisions through which employers can meet the needs of someone covered under ADA.

One of the most important elements of ADA is the remedies it offers for someone who brings suit under its provisions. People who file charges alleging discrimination under the ADA can ask for remedies including monetary damages (such as back pay), but they can receive no punitive damages. Individuals can also request equitable individual relief such as physical accommodations and attorney's fees. An additional provision sets up a $50,000 penalty for the first violation by an employer and $100,000 for subsequent violations.

The ADA has helped increase awareness of the problems faced by disabled people among the able bodied. It may have also increased the amount of acceptance that disabled people experience. Disabled people are increasingly visible in advertisements and the media. As pointed out earlier, the Barbie doll may have done much to reinforce for young girls a connection between thinness and beauty. In 1997, the Mattel Toy Company introduced a version of the Barbie doll in a wheelchair. Public attitudes are changing with respect to physical disabilities, and the ADA may have played some role in that process.

✳ *Issue* | **Blind Sue AOL**

The National Federation of the Blind filed a lawsuit against America Online in November 1999, claiming that its system is incompatible with the software needed by visually impaired persons. Most aids that translate computer graphics and text into Braille or sounds do not work with AOL's software, thereby constituting a violation of the Americans with Disabilities Act. Several other Internet service providers use compatible software that contains the special screen-access scanners the blind use to "read" graphics on line. The graphics must be tagged with words that describe the pictures. AOL has indicated that it is presently working on such an adjustment to its system software.

The ADA has been both challenged and broadened since its inception. In 1999, two commercial airline pilots challenged their employers through the courts to permit them to retain their job under ADA because, while they wore glasses contrary to airline policy, their eyesight was corrected to normal vision. In spite of such challenges, by July 2000 it was clear that the first decade of the ADA was successful. In a special ceremony marking the 10-year anniversary of the act, President Bill Clinton had much praise for the accomplishments and progress of the disabled as a result of the ADA.

Summary

Someone with a disability may well experience social stigma and other reactions normally directed at deviants, but others react this way to a condition rather than to behavior. Disabilities illustrate how society defines deviance as either a condition or a behavior. Millions of people in the United States have physical impairments of various kinds, although not all of them draw sanctions as equally disvalued forms of deviance.

Society reacts as it does because people with physical disabilities violate norms of appearance, wholeness, or health, as well as certain expectations of the sick role. These involuntary deviants often feel disvalued and bear substantial social stigma for their conditions. The passage of the Americans with Disabilities Act in 1990 expanded legal protections for them, but it has not eliminated subtle, informal sanctioning and stigmatization processes.

The sick role varies society's expectations in a way that temporarily exempts people from certain responsibilities and obligations on the condition that they resume conventional roles as soon as possible. Physically disabled people occupy the sick role more or less permanently, however, depending on their impairments. Others recognize physical differences between those who have disabilities and those who do not. This distinction appears to create substantial social ambivalence concerning physical disabilities.

People with eating disorders face many of the same problems as those with severe physical disabilities. Faced with powerful images and expectations of thinness, particularly directed toward women, some react to these forces by engaging in unhealthful practices such as anorexia and bulimia nervosa. Most people with eating disorders begin simply dieting, but after a time the dieting is not directed toward the goal of thinness. The eating disorder itself in a sense becomes the goal.

Disabled people and those with eating disorders must manage the stigma they experience in some manner. Historically, some of them, such as blind people and those with visible physical disabilities, have suffered discrimination and forced placement in undesirable social roles, such as that of beggar. Other, more subtle stigmatizing activities include social isolation and ridicule, such as that experienced by obese people and those with anorexia. To protect themselves, disabled people sometimes establish or join subcultures, including both formal organizations and informal groups, that provide social support and opportunities for interactions with other, similarly stigmatized people. For example, a number of communities create supportive networks for deaf people, although not all (or even most) people with hearing impairments belong to them. The nature of the stigma facing physically disabled people differs with their impairments. Others react to mental retardation with a strong stigma, since they perceive an important deficiency in the mentally retarded person: a lack of social or mental competence.

While disabilities result from physical impairments, affected people require socialization into the disabled role. This learning begins with acceptance of a long-term sick role and adjustment of other social roles to accommodate the disability. People may pass through developmental stages in learning disabled roles much like those described for people adjusting to death. Such professionals as physicians and social workers make important contributions to this socialization process because they validate disability and confer legitimacy on someone who must accept the sick role. In the process, however, they sometimes demand that the disabled person conform to certain social stereotypes.

A disabled person must adopt certain management techniques by which to reduce or deflect society's stigma. He or she may self-impose some stigma after carrying over attitudes formed in normal roles prior to becoming impaired. One may try to pass as a nondisabled person, cope with stigma by redefining social situations, and dissociate or retreat from social situations that might expose one to stigma. Other techniques may work better for people with different disabilities. Many other kinds of deviants practice some of the same stigma management techniques as well.

The Americans with Disabilities Act of 1990 represents the most significant federal legislation concerning disabled people. It mandates reasonable accommodation for them in living and working situations, and it provides legal remedies for those who suffer discrimination that violates provisions of the act.

Selected References

Albrecht, Gary L. 1976. "Socialization and the Disability Process." Pp. 3–38 in *The Sociology of Physical Disability and Rehabilitation.* Edited by Gary L. Albrecht. Pittsburgh: University of Pittsburgh Press.

Although somewhat dated, this paper remains one of the best sociological discussions of the disability process. The book also contains chapters on different aspects of the sociology of physical disabilities. All of these discussions view physical disabilities as sociological, not individualistic, phenomena.

Albrecht, Gary L. 1992. *The Disability Business: Rehabilitation in America*. Newbury Park, CA: Sage.

Rehabilitation has now become a commodity bought and sold in a marketplace dominated by financial interests from banks, insurance companies, and government agencies. These groups have as much influence over service delivery to disabled people as health-care professionals have. This important topic brings disturbing implications for people with disabilities.

Edgerton, Robert B. 1979. *Mental Retardation*. Cambridge, MA: Harvard University Press.

This book provides an excellent examination of social, psychological, and legal issues surrounding mental retardation. The author also examines the social policies that have developed regarding mental retardation.

Francis, Leslie Pickering, and Anita Silvers, eds. 2000. *Americans with Disabilities: Exploring Implications of the Law for Individuals and Institutions*. New York: Routledge.

A large and comprehensive collection of viewpoints on which conditions ought properly to be considered as disabilities. For example, should people who are HIV positive qualify as being disabled? The moral and policy implications of disability definitions are explored.

Hesse-Biber, Sharlene. 1996. *Am I Thin Enough Yet? The Cult of Thinness and the Commercialization of Identity*. New York: Oxford University Press.

This resource offers an excellent chronology of the "cult" of thinness in American society and its implications, especially for women.

CHAPTER 15

Homosexuality and Homophobia

Matthew Sheppard was 22 years old when he died. An openly gay student at the University of Wyoming, Sheppard had been burned, battered, and lashed to a fence in 1998 near Laramie to die. A passing bicyclist stopped because it looked as if someone had tied a scarecrow to the fence. Sheppard had hung there 18 hours in near-freezing temperatures. The incident produced vigils, demonstrations, and calls to President Bill Clinton to pass federal hate crimes legislation. Sheppard died because he was gay.

Few people refrain from acts of sexual gratification during their entire lives. Most experience sexual activities with members of the opposite sex, some partner with members of their own sex, and a smaller proportion interact sexually with members of either sex (bisexuals). Regardless of the types of sexual activities an individual may favor, most remain within their customary sexual preferences. But most people are not persecuted for their sexual preference, let alone killed.

The word *homosexuality* describes a general preference for sex with other members of one's own gender, whether male with male (gay) or female with female (lesbian). Homosexual relationships, like heterosexual ones, generally extend beyond sexual activities to include companionship and affection for each other. Some relationships, of course, retain purely sexual characteristics.

Our focus in this chapter is on homosexual behavior and the reactions to this behavior. In order to identify the nature and intensity of these reactions, it is necessary to trace the development of the social understanding and definition of homosexuality.

At times, in the discussion that follows, we follow popular usage that considers homosexual and heterosexual to be nouns indicating types of persons. Actually, these terms are more accurately adjectives describing the sexual preference of an individual or group. The term *homosexuality* is a generic term used to describe these sexual preferences, whether they are among men or women. Here, we follow the strict meaning of the term *homosexuality*, where the prefix *homo* means *same*, not male. Thus, we intend the term to refer to both male homosexuality and lesbianism. When used in a specific context, however, the term *homosexuality* refers to the sexual preference of men for other men. The term *lesbianism* refers specifically to the sexual preference among women for sex relations with other women.

The Development of the Notion of Homosexuality

The term *homosexuality* became popular in 1869 when K. M. Benkert defined it as failure to achieve "normal erection" during contact with a member of the opposite sex (Money, 1988: 9). Although Benkert meant the term to apply to both males and females, the decision to define it based on sexual functioning rather than a general sexual preference had significant effects. The criterion for genital homosexuality rejects alternative criteria such as falling in love with members of one's own gender (homophilic) or simply feeling attracted to them (homogenic). The professional literature abandoned both of the latter terms in favor of Benkert's term: *homosexual*.

Yet, this language emphasizes the sexuality within such relationships, perhaps conveying the impression that they rest primarily or exclusively on sexual activities. This chapter also uses the term *homosexual* in consideration of its widespread acceptance, although it also makes a number of distinctions about homosexual behaviors and activities that go far beyond narrow, physical conceptions based only on sexual acts.

Homosexuals exhibit this identity in a number of ways, including attitudes that express sexual or erotic preference, acquiring homosexual self-concepts, or participating in actual sexual relations with members of one's own gender, whether male or female. This last expression may represent the most commonly recognized indicator of homosexuality. Between males, physically sexual practices can take a number of forms: sodomy (anal contact), fellatio (oral–genital contact), and mutual masturbation. Homosexual relations between women can consist of oral stimulation of the clitoris (cunnilingus), mutual masturbation, and vaginal intercourse using vibrators or other implements as artificial penises. Members of all social classes engage in homosexual behavior, as do both males and females, people with varying educational achievement, participants in a wide range of occupations and professions, those with varied interests and avocations, and either married or single people.

While most people define homosexuality strictly in terms of the genders of sexual partners, an alternate conception based on sexual or erotic preference supports explanations for certain otherwise uncertain sexual behavior patterns (Langevin, 1985: 2–3). In one of the best-known studies of sexuality in the United States, Kinsey and his associates (1948) have reported that 37 percent of men experience homosexual contact at some time in their lives, but only 4 percent remain exclusively homosexual, expressing a consistent erotic preference for men throughout their adult lives. The remainder generally engaged in such practices out of a desire for variations in sexual activities or while living only with males in situations such as all-male schools or prisons. Further, some men have experienced homosexual relations with male prostitutes, but their involvement in this behavior did not reflect their primary sexual preference (Luckenbill, 1986).

Merely participating in homosexual relations does not itself constitute a homosexual identity, at least in the sociological sense of that term. Sociologists place more emphasis on the notion of a *homosexual identity*, that is, a person's self-concept as a homosexual. Such a self-concept affects a person's identification as a homosexual more significantly than does the type of sexual behavior in which the person might engage. Homosexuals may prefer activities associated with this identity and consider themselves as members of the category even if they occasionally engage in heterosexual relations. A sociologist identifies a true homosexual as any adult who regards himself or herself as a homosexual and who willingly acknowledges the label before another person.

Why Do Some People Regard Homosexuality as Deviant?

Social groups define certain acts as deviant by creating norms that regulate such behavior. Most societies designate what some term *sexually appropriate* and *sexually inappropriate* roles according to a person's age, social status, and other criteria. Some societies consider homosexual roles and behavior as inappropriate, while others condone or even encourage such behavior. Ample historical evidence confirms that cultural attitudes toward homosexual behavior have differed from one period to another. People certainly practiced homosexuality in Greek and Roman times and, in some circumstances, society may even have approved (Dover, 1978). Ford and Beach studied 76 folk societies and found that among 49 of them, or 64 percent, "homosexual activities of one sort or another are considered normal and socially acceptable for certain members of the community" (Ford and Beach, 1951: 130). As for other conduct, however, behavior tolerated in one society may represent intolerable deviance in another.

Society creates a deviant category of homosexuality by initiating and reinforcing sexual norms that pertain to orientations and behavior involving same-gender sex. In this sense, homosexuality, like other activities discussed in this book, is not inherently deviant, but it becomes so as a result of a purposive social process that establishes such a definition (Greenberg, 1988; see also Murray, 2000). Since, as later sections will show, society develops complicated sets of norms to regulate sexual behavior, the determination of homosexuality as deviance within a given society sometimes requires qualified statements. No known society has generally accepted persons who have played, or who have wished to play, homosexual roles exclusively or for indefinite periods; some condone homosexuality as a phase through which certain people pass, but this norm implies an expectation that everyone eventually accepts the heterosexual standard.

Critics have raised a number of objections to homosexual behavior, most of them derived from religious prohibitions (Murray, 2000). Such activity cannot lead to reproduction or to a "normal" family situation. In a sense, it distorts the general distribution of complementary sex roles among members of society. One study found negative reactions not to homosexual acts as such, but to sex role stereotypes associated with such acts (McDonald, 1976). As with many issues, perceptions are often more important than reality, and the perceptions that some people have of homosexuality are strongly negative. Another basis for negative social attitudes toward homosexuality comes from the conception of "normal" sexual functioning and development. However, discussions of sexuality later in the chapter will reveal the lack of substance in beliefs about inherently "normal" sexual behavior.

Schur (1984) has argued that gender itself forms a basis for deviance in U.S. society. The concept of sexism, or the belief in the superiority of one gender over another, contributes to this threat of gender-based sanctions. Specifically, women experience a systematic disvaluation process involving such behavior as sexual harassment and discrimination in the workplace. Historically, some have considered women to be inferior in many ways to men. In daily interactions, women have encountered reactions based on their category membership as females, rather than their individual identities as unique human beings. Female attributes and features have gained a disvalued image, as when people criticize men for weakness if they act like women. Increasing awareness of sex role socialization and efforts by the women's movement to change some of this socialization show clear examples of how processes of deviance, stigma, and disvaluation emerge in both everyday life and the larger political situation.

Social Dimensions of Homophobia

Regardless of one's personal view of homosexuality as deviance, people with homosexual orientations frequently become targets for social stigma and rejection due to differences from the heterosexual orientation of the dominant society. Such stigmatization affects even young people whose sexual identities may not yet have fully formed. This process appeared clearly in one study of 329 adolescents who expressed concern about their sexual orientations and contacted a gay support group in New York City (Hetrick and Martin, 1987). Most of the adolescents described their most pressing worries as isolation, family violence, educational issues and school relationships, emotional stresses, shelter problems, and potential sexual abuse. Fears like these would pose substantial problems for adults, but they create especially worrisome situations for relatively vulnerable people like adolescents. Many of these kinds of problems persist for homosexuals, although most members of this group eventually manage to resolve many problems with stigma and social rejection.

Homophobia is both an attitude (prejudice) and a motivation for behavior (discrimination). Negative attitudes about homosexuality are likely to be expressed in public opinion polls and private conversations, while discriminatory behavior against homosexuals can be expressed in a variety of ways, including job and housing discrimination, the denial of medical benefits to partners of gay people, and, most extreme, the physical assault of homosexuals because of their sexual orientation. Some observers regard homophobia as a psychological problem, a kind of emotional disorder with distinct symptoms (Kantor, 1998). In this view, homophobia requires the kind of clinical understanding and therapeutic care required of any other psychological abnormality. Here, we regard homophobia not just as an individual characteristic but a social manifestation of the processes that create and maintain deviant categories.

Homophobic Attitudes

The term *homophobia* refers to a fear and dislike of lesbians and gay men. Some people and groups treat homosexuality with more tolerance than others display. Studies have associated female gender, acquaintance with homosexuals, and parental acceptance with more tolerance of homosexuality (Glassner and Owen,

1976). Within families, mothers tolerate their sons' homosexuality more easily than fathers do. Groups associated with relatively strong rejection of homosexuality include members of the working class and lower class, religious fundamentalists, and people without college education (Hammersmith, 1987). Evidently, homosexual stereotypes relatively frequently appear misleading or incorrect to people who have experienced past nonsexual contact with homosexuals and to those whose backgrounds include exposure to diverse social roles compared with people who have not had these experiences. Society's increasing tolerance of homosexuality, comparatively free circulation of information about the behavior, and the militancy and openness of gay organizations define important trends currently at work to alter the strongly negative stereotype of the homosexual.

Homophobic behavior includes avoidance of contact with gays and lesbians and with anything associated with homosexuality. Such behavior might also include overt discrimination, such as refusing to hire an otherwise qualified homosexual. Homophobia takes many forms, expressed in both attitudes and behaviors. Many homophobic people have little difficulty identifying and expressing their attitudes. These feelings generally emphasize clear awareness and strong feelings. Some people may, however, express homophobic attitudes only reluctantly, but they may reveal their feelings by displaying some kind of homophobic behavior.

Some have attributed the origins of homophobia to religious doctrine linked to homosexuality and to theories of psychological maladjustment. Observers find many examples of the association between strong Christian beliefs and intolerance toward homosexuality, but some dispute the origins of such an association. Fone (2000), for example, argues that the biblical story of Sodom is the basis for subsequent prohibitions against homosexuality and that with its religious basis, homophobia is the last acceptable prejudice. Others reject that view (Soards, 1995). Greenberg (1988) argues that the rejection of homosexuality strengthened the Christian community at a time of struggle within that group. Increasing references to homosexuality as an example of moral perversity galvanized belief and defined an identity for believers. Theologian Karl Barth (cited in Soards, 1995: 43) sought to stiffen the resolve of Christians in their faith by describing homosexuality as "physical, psychological, and social sickness, the phenomenon of perversion, decadence, and decay, which can emerge when man refuses to admit the validity of the divine command." A similar transformation took place in the Latter-Day Saints (Mormon) church, which tolerated homoeroticism until the mid-1950s, when it began to express strong condemnation (Quinn, 1996).

Empirical efforts to study homophobia have examined such antecedents as religious background, strength of religious identification, and political conservatism. One researcher adds another factor—homosociality—defined as the social preference, but not necessarily an erotic attraction, for one's own gender (Britton, 1990). This study has determined that general religious and social conservatism relate strongly to homophobia and that "conservatism about the proper roles of men and women seem to be the source of this relationship" (p. 436). Another study by Ficarrotto (1990) confirmed these findings when it determined that sexual conservatism and social prejudice (discriminatory attitudes related to race and sex) served as independent and equal predictors of antihomosexual sentiment. Furthermore,

Britton finds that people who favor gender segregation in social institutions (for example, Boy Scouts for boys only, Girl Scouts for girls only, and all-male social organizations such as lodges and clubs) tend to express the strongest homophobic attitudes, but only against gay men. Homophobia may serve as an important boundary that helps to maintain the distinction between appropriate social and sexual interaction in these settings. Therefore, even religious and social conservatives—people likely to oppose homosexuality in general—may thus exhibit greater tolerance for homosexuality among females than among males.

Homophobic Behavior

It is one thing to have homophobic attitudes; it is quite another to express them in some concrete behavior. Few homosexuals regard themselves as criminals or deviants, as being "sick" or immoral. The negative feeling of others as expressed in stigmatizing efforts, however, is not without effect. Many homosexuals often feel it necessary to conceal their homosexuality from others. Homosexuals have reported that they sometimes feel guilty for their behavior, and they fear negative social sanctions from persons, such as family, friends, and employers, with whom they wish to continue to associate. Often, the homosexual outwardly may be gregarious and popular but inwardly feel rejected and alone (Harry, 1982). Often these feelings are motivated by the desire to avoid homophobic attitudes and behavior, including in some instances arrest and conviction for a criminal act.

Discrimination against homosexuals takes place in many arenas, including the military, where there has long been a prohibition against accepting gays. Senior military officials point to the inescapability of close living quarters and the potentially negative reactions on the part of heterosexual soldiers. In 1993, President Bill Clinton advocated a new policy in the military, the so-called don't ask, don't tell policy. This policy was adopted after Congress rejected a proposal to end sep arate treatment of gays and lesbians in the military. The new policy admitted that homosexuality was prohibited, and it asked commanders not to ask whether soldiers were gay and asked gay soldiers not to tell their commanders that they were gay. It was an awkward situation at best. By the end of 1999, President Clinton admitted in a press conference that the policy was a failure. No one seemed to like it, neither those who wanted to retain the ban nor gay rights advocates who wanted it lifted entirely. Even later statements from the Pentagon that the policy was going to be "fine tuned" during the year 2000 failed to satisfy everyone. Those guidelines included a policy that all troops would undergo periodic antiharassment training and would assign investigations of homosexual activity to more senior leaders.

Gay people have also been subject to severe legal discrimination because of their sexual orientation. On September 30, 1999, the Associated Press reported that President Yoweri Museveni order the arrests of homosexuals in Uganda (*Omaha World Herald*, September 30, 1999, p. 6). Homosexuality is illegal in Uganda and carries a maximum penalty of life imprisonment. Museveni was quoted in the government-owned newspaper *New Vision* as saying "I have told the Criminal Investigation Department to look for homosexuals, lock them up and charge them."

The action may have been prompted by two recent and well-publicized gay marriages in Uganda. The Uganda episode may seem like an extreme example, but negative reactions to homosexuality vary along a continuum from tolerance to repression.

Law and Public Attitudes

Homosexuality and the Law

Laws forbidding homosexual behavior began in the ancient Jewish sex codes, later formalized by the Christian church into the ecclesiastical laws that governed medieval Europe. In turn, these provisions later formed the basis for English common law (see Katz, 1976, 1983). An edict of the Emperor Justinian condemned homosexual offenders to die by the sword in 538, and this prohibition in Justinian Code constituted the foundation for legal punishments of homosexuality in Europe for 1,300 years. For hundreds of years ecclesiastical courts mandated punishments for homosexual acts, often including torture or death. By 1533, however, royal courts had assumed jurisdiction over such offenses in England; the English statute enacted then provided no more tolerant standard: On conviction, the accused was put to death without benefit of clergy. This punishment remained in effect until the 19th century, when reforms reduced it to life imprisonment.

In ancient times, societies did not recognize the concept of homosexuality as current ones do (Meier and Geis, 1997: Chapter 4). The earliest Hebrews and the ancient Greeks had no word for homosexuality. Greek and Roman cultures permitted sexual relations irrespective of gender, although they may have remarked on exclusive same-gender sexuality as a rare and even rather unusual occurrence (McWilliams, 1993: 605).

Some observers have described the negative position of the Christian church on homosexuality as a long-standing and consistent attitude (Soards, 1995). Other writers claim that homosexuality found tolerance in the Christian tradition until the mid-13th century, when the church moved toward a more negative view (Boswell, 1980). These positions do not imply that early societies encouraged homosexuality. Soards (1995: 38–40) has cited regulations in Spain about 700 A.D. calling for castration of homosexuals, an edict reinforced later by declarations of the king of Spain at the Council of Toledo. By the 12th century, homosexuals faced orders to show through confession and penance that they had become worthy of redemption after their "shameful sin of sodomy."

As late as the mid-18th century, homosexuals were burned at the stake in Paris. Liberalization of legal attitudes brought more tolerance after the French Revolution, and the later Napoleonic Code omitted provisions about homosexual acts from the legal structure, an arrangement that persists today in many European countries. While many modern societies seek to protect young people from experiencing or encountering homosexual acts and to protect "public decency," most European countries do not consider homosexual acts committed in private by consenting adults to be criminal violations (Geis, 1972). Under the British law of

1956, sodomy with an underage partner still justified a sentence of life imprisonment, but the law applied a lighter sentence for similar contact among adults. After long debate, in 1965 that country dropped legal penalties for private homosexual acts between adults over 21 years old. Penalties remained for acts with younger partners and for those who procured for homosexual acts.

Many states now apply criminal sanctions only to public homosexual acts (as they do for public heterosexual acts); they exempt all homosexual acts performed in private from legal penalties. In the United States and in other countries, no one ever commits a crime simply by *being* a homosexual; criminal prohibitions that persist apply specifically to homosexual *acts*, such as sodomy, fellatio, and mutual masturbation. Soliciting for such acts is also generally a crime. Practically, the law punishes only homosexual acts between males; although sex between women remains by no means uncommon, authorities seldom punish it.

In 1986, the U.S. Supreme Court ruled on the constitutional standard for state laws against sodomy. In August 1982, Michael Hardwick was cited for carrying an open bottle of beer in public. He failed to appear in court to face the charge, and the police obtained a warrant for his arrest. A police officer went to Hardwick's apartment to serve the warrant and entered at the invitation of a guest. The officer subsequently saw Hardwick—who was in his own bed in his own bedroom at the time—engaged with another male in an act of sodomy. The Supreme Court upheld Hardwick's conviction for violating the state's law. A subsequent Georgia case *(Power v. Georgia)* decided in November 1998 found that laws against consensual sexual activity among adults did not violate the state's constitution.

The changes in European laws mentioned earlier influenced U.S. laws concerning homosexual acts. At one time, all states had laws against sodomy. While these laws applied to both heterosexuals and gay people, they were primarily used against gays. As a result of the Hardwick case, states were allowed to criminalize sodomy. As of October 1999, sodomy laws were repealed or struck down by courts in 32 states. The remaining states have laws with varying penalty levels from a small fine to years of imprisonment. Some states' laws do specify relatively stiff penalties for certain homosexual behavior (up to 10 years); in actuality, however, authorities rarely if ever enforce these criminal felony laws. When they arrest offenders under these laws, they generally target solicitation, a misdemeanor that draws less than a 1-year sentence.

Changing Public Attitudes

Continuing intolerance of homosexuality stems, in part, from historical roots. Greenberg (1988), for example, has extensively documented attitudes toward homosexuality from medieval times. Repression of homosexuals spread in the 13th century as an unanticipated consequence to organizational reforms in the church and of class conflict in society. Campaigns for celibacy encouraged condemnation of sodomists along with witches.

The pitch of public disapproval of homosexuality has declined markedly from previous decades, although it and certain other forms of sexual behavior continue to draw fairly vigorous criticism (Stephan and McMullin, 1982). For example, the percentage of people who favored legalizing homosexual relations between consenting adults increased only slightly from 1977 to 1989, from 43 percent to 47 percent (cited by Posner, 1992: 202). More than half of the population does not favor relaxation of such sanctions. In fact, 54 percent of those polled in 1986 agreed that government should outlaw homosexual relations between consenting adults. The research found even stronger negative opinion among older age groups, but evidence also confirms that the majority of college students maintain negative views of homosexuality (Endleman, 1990: 52). Also, 51 percent of respondents to the 1986 poll expressed the belief that the U.S. Constitution does not protect even private homosexual acts (Gallup Poll, 1986).

From 1972 until at least 1991, public opinion surveys showed that "over 70 percent of Americans believed that homosexuality was always morally wrong" (Michael, Gagnon, Laumann, and Kolata, 1994: 172). A Kinsey Institute survey in 1970 asked over 3,000 adult respondents selected from a nationwide population

 Issue | **Two Reactions to Homosexuality**

Antihomosexual feelings are motivated by a number of sources, including the attitude that homosexuality is a sin and it is inappropriate even if it is not a sin. The two reactions below, both from the same source, illustrate each of these rationales:

From a person reacting to a series of newspaper articles on homosexuality: "In the beginning, God . . . created Adam and Eve—not Adam and Steve."

From a nationally known advice giver:
Dear Ann Landers:

My partner and I made a formal commitment five years ago. Although our union is not recognized legally, the wedding ceremony was deeply spiritual. All our friends and family members shared in our happiness. We are gay.

"Denny" and I are invited everywhere as a couple. Everyone thinks of us that way. The problem we have been struggling with came to a head a few weeks ago at the wedding of Denny's brother.

Denny is a great dancer. He never sits out a number. The women love to dance with him, which means I am left alone a lot. At his brother's wedding, I insisted that Denny dance with me and he finally did. No one reacted, at least we didn't notice any stares.

We enjoy dancing together but we don't want to make others uncomfortable. Do you think it's OK? We need an outside opinion.

Jerry in D.C.

Ann says: In Eastern European countries, men traditionally dance together. Nobody thinks it's strange. If you and Denny want to dance together in the company of family and friends who are aware of and accept your relationship, I see nothing wrong with it. I assume, of course, that you mean conventional dancing—no Lambada, no cheek to cheek and no slow dancing with erotic overtones.

Source: *Des Moines Register*, October 26, 1990, p. 3T.

about a variety of sexual acts; it found the strongest disapproval for homosexuality among partners who had no special affection for each other (Klassen, Williams, and Levitt, 1989. 18). Fully 88 percent of the respondents categorized such acts as "always wrong" or "almost always wrong." Fewer respondents applied the same categories to homosexual acts between partners who were in love, but the percentage remained high at 79 percent. A 1988 national survey asked young men about their attitudes toward homosexuality, determining that 89 percent found sex between two men "disgusting" and only 12 percent thought they could be friends with a gay male (Marsiglio, 1993). These views showed associations with perceptions of a traditional male role, religious fundamentalist beliefs, and low formal educational achievement by parents.

Attitudes toward homosexuality within the medical profession also have changed, specifically those in the field of psychiatry. Before 1973, many psychiatrists viewed homosexuality as a disturbance in an individual's mental health, and even a category of mental illness. In 1973, however, the American Psychiatric Association reversed itself and declared that homosexuality by itself does not necessarily constitute a psychiatric disorder. This decision appears to have resulted from efforts by activist groups to change the organization's perception (Spector, 1977). In fact, almost 70 percent of the approximately 2,500 psychiatrists who responded to a survey opposed the action (Greenberg, 1988: 430).

In the 1970s, several cities initiated a trend by adopting antidiscrimination laws to protect homosexual rights. Such laws usually established violations for discrimination against homosexuals in employment, housing, and so on. These ordinances have encountered a mixture of resentment and acceptance, raising political issues associated with gay rights in some communities. In Florida, demonstrations in several major cities accompanied such legislation. In San Francisco, an estimated 200,000 people marched in protest, partly precipitated by the fatal stabbing at about the same time of a homosexual boy by four youths shouting "faggot, faggot." Police estimated that this march and the Gay Pride Week celebration that followed eclipsed in size even the protests against the Vietnam war in the 1960s. In New York City, about 20,000 people marched, and Seattle's mayor proclaimed a citywide Gay Pride Week, describing the event as consistent with the city's standard for treatment of all people as equals. Similar demonstrations occurred in other cities as well.

The debate over the morality or immorality of homosexual behavior continued through the 1980s among several Christian groups. Fundamentalist Christians displayed and actively promoted intolerant attitudes that other, more liberal Christians declined to endorse. As the AIDS scare spread in the mid-1980s, many fundamentalist groups cited the disease as affirmation from God of their condemnation of homosexuality. The Reverend Jerry Falwell helped to lead these efforts through his group, the Moral Majority, in the early 1980s. This group expanded the traditional role of religious organizations by lobbying for antihomosexual laws. It and groups sympathetic to its message constituted a significant backlash against the prevailing trend toward increased toleration of homosexuality. The movement sponsored public forums for debates on such issues as hiring and retaining homosexual teachers and other public servants. Some of these initiatives

TABLE 15-1
Public Opinion and Being Gay

How do you feel about homosexual relationships?

	1978	1998
Acceptable for others, but not self	35%	52%

Are homosexual relationships between consenting adults morally wrong or not a moral issue?

	1978	1998
Yes, morally wrong	53%	48%
Not a moral issue	38%	45%

SOURCE: Adapted from a Time/CNN poll, October 1998.

appeared on political ballots through state referendums, while others simply stirred public discussion.

Well into the 1990s, the fear of AIDS continued to polarize opinion about homosexuality in a number of communities. Few doubt that the AIDS epidemic increased fear about homosexuality, since the disease showed close ties to the sexual practices of gay men. Some found that the threat of AIDS provided a convenient justification to advocate their moral judgments against homosexuality. Others saw reactions to the AIDS epidemic as examples of bigotry by those who condemned homosexuals on this ground alone. These debates have continued despite strong links between AIDS and illegal drug use, which helped to transmit the disease to new populations, making it an increasingly heterosexual affliction.

Noteworthy resistance notwithstanding, research has found a modest but measurable increase in tolerance of homosexuality among most groups and in the general population (see Table 15-1). On the whole, fewer people are objecting to homosexuality on moral grounds, and an increasing number views homosexuality as not a moral issue at all. Still, strong condemnation continues more among conservative Christians than others. The issue retains its divisive effect, both through broad topics, such as the morality of homosexuality in general, and narrower ones, such as the appropriateness of allowing known homosexuals to become members of the clergy. Further, increasing tolerance has not brought a substantial majority of Americans to agree that homosexuality is just another lifestyle. When 70 percent of a sample of Americans agree that homosexuality is immoral behavior (Shapiro, 1994), one might reasonably question how fully the respondents agree in their definition of *morality*.

The Attribution of Homosexuality

A common myth asserts that one can readily identify adult male homosexuals based on their physically effeminate traits and lesbians based on their masculine appearances. Actually, most homosexuals display no distinguishable physical differences from heterosexuals. When they become socially visible—and many do—people make the distinction because gays and lesbians perform homosexual roles.

While some homosexuals report that they can identify other homosexuals, their success at this effort often reflects enhanced consciousness of sexual identity; heterosexuals, on the other hand, often prefer not to find or notice them. One study evaluated questionnaire responses from 1,900 gay men and 1,000 lesbians, revealing that only 6 percent of the lesbians felt that others could recognize their sexual identities; 68 percent felt that others could not tell, and 27 percent were unsure (Jay and Young, 1979). As one lesbian put it: "Don't be silly. I can't tell other lesbians are lesbians. How can *most people*, which would indicate *straight people*, tell I'm a lesbian?" (Jay and Young, 1979: 188). A more recent study confirmed this point after evaluating videotaped interviews with both homosexual and heterosexual men and women (Berger, Hank, Ravzi, and Simkins, 1987). The researchers showed the taped interviews to 143 male and female raters and asked them to identify the homosexual interviewees. Less than 20 percent of the raters could successfully do so.

Many exceptions complicate any effort to generalize about a characteristic homosexual appearance. As an indication of the great variations in the social and physical characteristics of homosexuals, one may review the long list of important historical figures who were homosexuals—philosophers, military leaders, artists, musicians, and writers.

> Socrates and Plato made no bones about their homosexuality; Catullus wrote a love poem to a young man whose "honeysweet lips" he wanted to kiss; Virgil and Horace wrote erotic poems about men; Michelangelo's great love sonnets were addressed to a young man, and so were Shakespeare's. There seems to be evidence that Alexander the Great was homosexual, and Julius Caesar certainly was—the Roman senator Curio called Caeser "every woman's man and every man's woman." So were Charles XII of Sweden and Frederick the Great. Several English monarchs have been homosexual. . . . About some individuals of widely differing kinds, from William of Orange to Lawrence of Arabia, there is running controversy which may never reach a definite conclusion. About others—Marlowe, Tchaikovsky, Whitman, Kitchener, Rimbaud, Verlaine, Proust, Gide, Wilde, and many more—there is no reasonable doubt. (Magee, 1966: 46)

Other well-known homosexuals have included Andy Warhol, Florence Nightingale, Susan B. Anthony, Emily Dickinson, Leonardo da Vinci, Tennessee Williams, and Rock Hudson (Russell, 1996).

Less well-known individuals often gain reputations as homosexuals as a result of how others define them and whether or not they publicly exhibit a sexual preference for others of their own genders. Regardless of overt evidence of homosexuality, others may impute this identity after retrospective interpretation of a person's behavior, a process that reviews and reevaluates past interactions with the individual. People make such a judgment by reinterpreting past interactions in view of other evidence of homosexuality; they search their memories for subtle cues and indicators that might justify an attribution of homosexuality. Indirect evidence includes rumor, general information about the person's behavior, characteristics and inclinations of his or her associates, expressions of sexual preferences, and experiences, including unverified ones, reported by acquaintances. Direct observation may confirm the judgment by detecting behavior that "everyone knows" implies homosexuality, such as effeminate appearance and manners for a male and masculine appearance and manners for a female. Such observations may or may not lead to true conclusions about an individual's sexual orientation.

Prevalence of Male Homosexuality and Homophobia

Any estimate of the number of homosexuals in the population depends in large part on criteria for defining homosexuality and how one tries to count people who fit this definition. Estimates of the frequency of homosexuality have varied for these two reasons.

Prevalence of Homosexuality

To date, observers have worked from inadequate data when estimating the incidence, prevalence, and increases or decreases of homosexuality. The principal problems with data come from differences in homosexual practices and variations in people's commitments to homosexual identities. These conditions influence who is counted as a homosexual and limit that applicability of data. Any count may vary substantially, depending on the definition that one adopts.

For example, a study released in 1989 reported that 1.4 percent of men have participated "fairly often" as adults in homosexual relations (Fumento, 1990: 207–208). A researcher would find a much higher incidence of homosexuality if the count were to include everyone who has ever had sexual contact with a member of the same gender; a much lower figure results if the count includes only those who have publicly identified themselves as homosexuals, because many people who share a strong homosexual orientation still do not publicly declare their preference. Observers who have grouped people with homosexual identities report that "according to which definition of 'homosexual' one uses, homosexuals represent the first, second, or third most common minority in the United States today" (Paul and Weinrich, 1982: 26).

Other researchers have made various attempts to estimate the size of the male homosexual population, particularly in the United States and in Great Britain. Both early and contemporary estimates generally define about 4 to 5 percent of the male population as homosexuals. As reported above, Kinsey has reported homosexual experiences or behavior resulting in orgasm by 37 percent of the white, male population at some time between adolescence and old age. The same source indicates that only 4 or 5 percent considered themselves as exclusively homosexual throughout their adult lifetimes. Another earlier study set a conservative estimate of the extent of homosexuality in the United States at roughly 4 to 6 percent of the total male population over 16 years old (Lindner 1963: 61).

A 1993 study conducted at the National Cancer Institute has suggested that homosexuals comprise about 2 percent of the population (Henry, 1993). This study counted exclusively or predominantly gay males with preferences known to family members. In a national sample of the population, Harry (1990: 94) asked the following question in a telephone interview:

> I have only one question. You may consider it somewhat personal to answer but most people have been willing to answer it once we remind them that this is a totally confidential survey. We reached you on the phone simply by chance and don't know your identity. Here's the question: Would you say that you are sexually attracted to members of the opposite sex or members of your own sex?

In this survey, 3.7 percent of the respondents identified themselves as homosexuals or bisexuals.

The National Life and Social Life Survey studied a random sample of 3,432 Americans in 1992. It has reported similarly low numbers, but the estimates varied depending on the wording of survey questions. About 2 percent of the men in the survey reported that they had experienced sex with a man in the past year, and 5 percent indicated that they had participated in homosexual sex at least once since they had turned 18 years of age (Michael et al., 1994: 175). The survey found lower estimates of homosexuality, however, when questions asked whether respondents considered themselves heterosexual, homosexual, bisexual, or something else. Looking just at behavior, Laumann and his colleagues (Laumann, Gagnon, Michael, and Michaels, 1994: 294) have reported that 2.7 percent of their sample of males and 1.3 percent of the females reported having sexual relations with someone of the same gender in the previous year (see also Michael et al., 1994: 176).

Researchers encounter even greater difficulties in estimating homosexuality in other societies, although some have ventured guesses. Whitam and Mathy (1985) assert that homosexuals have formed relatively small elements of all cultures at all historical moments (approximately 4 to 5 percent of the total male population). Other estimates contradict this blanket statement, noting near-universal participation in homosexuality among certain cultural groups in New Guinea, while other cultures offer few or no examples (Herdt, 1981). Homosexual practices do appear to vary substantially from culture to culture.

Variations of Male Homosexuality

Like heterosexuality, homosexuality encompasses many social and behavioral variations. Its social structure and practices among certain individuals both vary.

Some evidence suggests relatively common homosexual behavior within certain occupational groups, but no conclusive indicators allow firm judgments. Homosexuals participate in almost every occupation and reach all educational levels. Some occupations may attract homosexuals by freeing them from the need to conceal their behavior among colleagues who accept them. Members of certain occupational groups seem to accept others' definition of themselves as "effeminate." Schofield concludes that

[F]or whatever reason, it is in fact now probable that there is a higher proportion of revealed homosexuality in certain job categories—such as interior decoration, ballet and chorus dancing, hairdressing, and fashion design—than in others. The adjective *revealed* is important, because the true proportions for those occupations in which greater concealment is necessary is not known. (Schofield, 1965: 209)

Short-Term Relationships   Many male homosexuals adopt wide-ranging sexual practices, generally confining their relations with other homosexuals to brief and relatively transitory sexual encounters. Homosexual males often move between varied relationships before settling down; in contrast, lesbians appear to form relatively long-lasting relationships (Troiden, 1989). This tendency toward impermanent relationships among male homosexuals often overshadows even limited affectional-sexual ties within the predominant pattern of "cruising" and one-night stands. Some indications suggest a decline in cruising, however, due to the frighteningly high probability of acquiring AIDS through this promiscuous behavior.

Such impersonal sexual relations attract the participation of certain male juveniles, seldom homosexuals themselves, who offer sexual services in exchange for monetary payments (Reiss, 1961). Specifically homosexual prostitutes also offer transitory sexual relations. These "hustlers" serve other homosexuals, providing services, particularly for physically unattractive and aging customers, that they could not obtain without great effort (Luckenbill, 1986). The adult homosexual prostitute plays an active role in homosexual life. He learns his behavioral role—such as gestures, vocabulary, clothing, and even makeup—from others with experience, just as heterosexual prostitutes become a part of heterosexual life (Rechy, 1963: 36).

Some homosexuals meet others in bars, parks, clubs, cafes, baths, hotels, beaches, movie theaters, toilet facilities, and other public places (Weinberg and Williams, 1975). Many of these encounters lead to highly impersonal sexual relations sometimes carried out in a "tearoom" or "T-room," a homosexual term for a public toilet (Humphreys, 1975). These tearoom experiences offer a venue readily accessible to the male population in locations near public gathering places—department stores, bus stations, libraries, hotels, YMCAs, and, in particular, isolated sections of public parks. A third person may act both as voyeur and lookout, and the participants frequently exchange little conversation.

Studies of homosexuals have found, contrary to a widespread belief, that a substantial proportion of them have lived within heterosexual marriages at one time or another. In fact, one study has given a figure as high as one-fourth for this group (Dank, 1972; Lewin and Lyons, 1982). Admitted homosexuals marry for many reasons: in response to social pressures, in reactions against homosexual life, or as commitments to home-centered lives (Ross, 1971). Some homosexual men seem to view marriage to women as demonstrations of their normality to themselves and others. Most have not developed self-concepts as homosexuals when they become married, usually in their 20s, even though they may have engaged in homosexual acts; they typically develop this identity only later.

The sociological definition of *homosexual* does not fit many married men who engage in impersonal sexual relations with other men; they participate in the behavior without expressing homosexual identities. Humphreys's (1975) study of tearoom sex identified the largest group of participants (38 percent) as currently or previously married men, largely truck drivers, machine operators, or clerical workers. Most of them wanted not homosexual experiences, but rather quick orgasms through means more satisfying than masturbation, less involved than love affairs, and less expensive than prostitutes. Humphreys described another group as "ambisexuals," mostly relatively well-educated members of the middle and upper classes, many of them married or otherwise participants in heterosexual behavior. These people liked the "kicks" of such unusual sex experiences. The gay group of openly confessed homosexuals constituted only 14 percent, and the last group—"closet queens"—made up an even smaller proportion of the tearoom trade. Closet queens are homosexuals, unmarried or married, who keep their homosexuality secret.

Long-Term Commitments Discussions of impersonal sex do not imply that homosexuals cannot or do not want to form more or less permanent bonds with other homosexuals. A study of 190 gay men has found that they do establish long-lasting love relationships with other men (Silverstein, 1981). A more extensive study assigned 485 male homosexuals to one of five different types, depending on a number of variables (Bell and Weinberg, 1978: 132–134). The 67 homosexuals assigned to the *close-coupled* type lived with homosexual partners in relationships that resembled marriage. They tended not to interact sexually with other partners nor to engage in cruising. They also reported few sexual problems and few regrets over their homosexual identities. The 120 individuals assigned to the *open-coupled* type also lived with homosexual partners, but they also engaged in cruising and maintained active sex lives outside those relationships. The 102 individuals assigned to the *functional* type had remained single (not joined with partners in couples), and they actively pursued sex with a number of partners. These subjects also experienced few sexual problems and few regrets about their homosexual identities. The 86 subjects in the *dysfunctional* type reported frequent homosexual relations, but they experienced many sexual problems and regrets over their homosexuality. The 110 persons in the final, *asexual* category reported few homosexual contacts, and they experienced many problems and regrets associated with their sexual orientation.

Male homosexuals who live together in more or less permanent unions may develop quite stable relationships that integrate their sexual identities into long-standing affectional, personal, and social patterns. One study evaluated the relationships of 156 male couples, one-third of whom had lived together for more than 10 years (McWhirter and Mattison, 1984). The researchers found the largest age differences between partners in couples who had stayed together the longest periods of time. All of the couples who had stayed together longer than 30 years (8 couples) reported age differences between 5 and 16 years. Just as in heterosexual relationships, the reasons that initially bring partners together differ from those that keep them together in the 5th, 10th, or 20th year. While physical attraction, sexuality, and compatibility provided important initial links, later strengths included companionship, economic benefits, and lack of possessiveness.

Many homosexuals want legal protection for these benefits, particularly if one partner dies, so they agitate for legal changes to include them within civil marriage provisions, and some want religious recognition of their unions.

One study of 92 male homosexual couples reported on relationships lasting from 1 to 35 years (Berger, 1990). Few of the couples had staged any commitment ceremonies, although many wanted such public recognition. Most of the couples' close friends were also gay, and about two-thirds of their families supported the relationships. The couples experienced their most persistent conflicts over money and relations with family members, issues that plague heterosexual couples as well. The similarity between gay and heterosexual families also appeared in a study of 24 gay and 29 heterosexual fathers who answered survey questions on parenting (Bigner and Jacobsen, 1992). Both groups gave very similar responses about parenting styles and attitudes.

Sex Role Socialization and Becoming a Homosexual

A complete understanding of homosexuality requires information about the meaning of *gender* and the processes by which individuals come to identify with one gender over another. These processes lead to feelings of sexual preference and identity through subtle and complex effects that sociologists have only recently begun to understand. A full explanation undoubtedly requires information about biology and psychology, but this section seeks mainly to identify a sociological perspective on the development of sexual identities.

A few individuals claim that homosexuality merely reflects purposeful behavior: "Although a person may be a homosexual . . . he can make a deliberate choice to come out of it," one asserts (Bryant, 1977: 69). Such views do not coincide with the weight of scientific evidence, so this section will not review them. As one observer has said: "Gay men are not free to invent new objects of desire any more than heterosexual men are—their choice is structured by the existing gender order" (Connell, 1992: 747). Homosexuals do not appear to choose their identities anymore than heterosexuals do. People can indeed choose whether or not to commit homosexual or heterosexual acts, but this determination differs from choosing one's basic sexual preference.

Sexual Development

Sexual development occurs through a complicated process about which understanding continues to unfold. Physical traits and activities affect sexual development, such as the structure of primary and secondary sex characteristics. Psychological and social psychological influences also produce their effects, such as the development of a sexual self-concept and identity.

Biological Perspectives on Sexual Development   Research has emerged, particularly during the 1990s, to suggest a relationship between homosexual orientation and certain biological structures. Some of this work has compared brain structures between homosexuals and heterosexuals. While this work has so far reached only preliminary stages, some studies have concluded that physical brain anatomy differs between homosexual and heterosexual people (LaVay, 1991). One study has reported, for example, that the homosexual brain features an enlarged suprachiasmatic nucleus compared with that of the heterosexual brain (Swaab and Hofman, 1990). Another has reported that male homosexuals have a smaller interstitial nucleus of the anterior hypothalamus than that in heterosexual men (LaVay, 1991). No one can currently define the significance of such findings for sexual behavior. Until science clearly implicates such structures in sexual preference, people should wait to express complete faith in such a conclusion.

Other studies have examined additional physical differences between homosexuals and heterosexuals. Allen and Gorski (1992) have found that another brain structure, the midsagittal plane of the anterior commissure, is 34 percent larger in homosexual men than in heterosexual men. Further, this structure is 18 percent larger in homosexual men than in heterosexual women. Brain anatomy principles relate this structure to bilateral communication between the right and left hemispheres rather than to sexual function, however, so the importance of these physical differences remains open to interpretation. Certainly, the research results support one interpretation that gays as a group differ from nongays, but no one asserts that this single physical structure causes homosexuality. A second interpretation, however, suggests a dissociation between the biological structures related to sexual functioning and those that affect sexual preference, supporting the idea of a distinct difference between sexual preference and sexual behavior. Still other researchers have explored chemical and hormonal makeups, looking for evidence of homosexuality. Roper (1996), for example, believes that the action of testosterone on the brain may determine homosexual orientation.

In another line of biological research, studies have searched chromosomes for influences on sexual preference. Hamer and Copeland (1994) conducted a study at the National Cancer Institute, placing ads that solicited responses from 114 gay men. The investigators found intriguing evidence that the families of 76 gay men included much higher proportions of homosexual male relatives than occur in the general population. Because most of the homosexual relatives came from the maternal sides of these men's families, the study concentrated on the X chromosome, which comes from the mother. The researchers examined DNA from 40 pairs of homosexual brothers. The laws of inheritance indicate that two brothers each have a 50 percent chance of inheriting the same single copy of their mother's X chromosome (the chromosome with the theorized DNA "marker" for homosexuality). Thus, the researchers reasoned that 20 of the 40 pairs of brothers would have the chromosome. Instead, they found that 33 pairs of brothers shared five different patches of identical genetic material, suggesting that they had all inherited the same X chromosome from their mothers. The investigators concluded that at least one gene inherited by a son from his mother may help to determine the son's predisposition toward heterosexuality or homosexuality. Presumably, a common version of the gene increases the likelihood that the son will become a heterosexual, while an uncommon version increases the likelihood that he will become a homosexual.

✱ *Issue* | **Genes and Homosexuality: The Reaction**

Reactions to the finding of a possible genetic basis for homosexuality were predictably mixed, with some professing enthusiasm and others skepticism.

Cautious Acceptance	Lingering Skepticism
J. Michael Bailey, an assistant professor of psychology at Northwestern whose own work suggests a genetic link to homosexuality, was quoted as saying that the work by Hamer and his associates was a "terrific study" and, "if replicated, [the study will] be a genuine breakthrough in sexual-orientation research." (Wheeler, 1993: A6)	Evan Balaban, an assistant professor of biology at Harvard, thought Hamer overstated the conclusions of the study. "They've shown little more than a group of highly selected men who happen to be homosexual share among them a certain region of the X chromosome at a higher rate than would be expected due to chance." This does not show that a gene determines sexual orientation or that it necessarily influences it. (Wheeler, 1993: A6)

While many gay leaders greeted the study with enthusiasm, Eric Juengst of the National Center for Human Genome Research summarized a more moderate view: "This is a two-edged sword. It can be used to benefit gays by allowing them to make the case that the trait for which they being discriminated against is no worse than skin color. On the other hand, it could get interpreted to mean that different is pathological" (Henry, 1993: 39).

Assuming the research is replicated, the role of the gene is unclear. For example, one observer remarked:

If past experience with the genetics of human behavior is any guide, inheritance of the mysterious gene wouldn't, in and of itself, cause a man to be homosexual. Instead, he simply would be more likely to be tipped into homosexuality by some environmental factor than a man who integrated the gene for a heterosexual orientation. (Bishop, 1993: B1)

Others pointed out it is not possible to know what kind of sample Hamer employed because they were not selected randomly. As a result, the research subjects may represent 1 percent or 99 percent of the adult male homosexual population, or some figure in between. Generalization of the findings is impossible.

The scientists explicitly indicated that the gene had only "significant influence" on sexual orientation rather than absolutely determining it. They foresaw no possibility of testing preadolescent boys to determine their eventual sexual orientations. Although the results do suggest some kind of X-chromosome linkage, the researchers failed to find evidence of any direct inheritance (Hamer and

Copeland, 1994: 104). Because the research targeted an area of the X chromosome large enough to accommodate several hundred genes, any attempt to isolate a so-called *homosexual gene* would require much more testing of a large number of families with homosexual males. Further research on the genetic basis of lesbianism is also required, since this research concentrated solely on males.

Another concern results from the fact that 7 of the 40 gay brothers did not share the characteristic bits of DNA, and the study did not test the DNA of the heterosexual males to determine whether they also carried the fragment of DNA. If some homosexual brothers lacked the genetic material, then it clearly did not act as a biological determinant of sexual orientation. For this reason, the researchers remained cautious about suggestions that they had isolated a specific gene that "caused" homosexuality.

Only limited research so far supports biological claims for an inherited tendency toward homosexual behavior. Significantly, no study has provided evidence that lesbianism results from some biological preference gene. Logically, the idea remains implausible that a gene could program behavior so specifically that it could predetermine a same-gender sexual *preference*. The term *homosexuality* indicates a behavioral role composed of attitudes, norms, and practices that allows no possibility of biological transmission, at least by any mechanism yet known. A review of historical and comparative cultural data on sexual behavior does not support a theory of biological predetermination of sexual preference, although the sex drive has biological roots.

Sociological Perspectives on Sexual Development Although it flows from biological antecedents, sexuality reflects learned behavior. Sexuality is a social construction "that has been learned in interaction with others" (Plummer, 1975: 30). It is dictated not by body chemistry, but by social situations and expectations. Even distinctions of male and female refer to socially constructed categories, as does the conduct that arises from these roles. One learns that some people or objects should stimulate arousal, but others should not. One also learns at what age one should gain the capability of arousal and sexual intercourse.

In fact, one can learn to regard virtually anything as a sexual stimulus if it becomes paired with a sexual response. For this reason, some observers have described the sex drive as "neither [inherently] powerful nor weak; it can be almost anything we make it" (Goode and Troiden, 1974: 15). A person acquires the social meaning of sexuality, then, in the same manner as one acquires other social acts, as part of the overall socialization process. He or she learns sexuality not all at once, but over a period of time and according to principles of learning and social interaction as discussed in other chapters. Some people learn to become homosexuals through the same general learning processes by which others learn to become heterosexual. This learning differs only in content.

The typical conception of sexuality, however, varies substantially from this sociological portrait. Many people have become accustomed to thinking about sexuality as a totally innate characteristic that depends exclusively upon certain vague, biological determinants. Actually, while unmistakable biological limits set boundaries for sexual development, a better concept would probably describe a largely socially determined continuum from very masculine on one end to very feminine on the other. In fact, Kinsey and his associates (Kinsey, Pomeroy, and Martin, 1948) operationalized their classic conception of sexuality on a 7-point scale, with completely heterosexual at one end and completely homosexual at the other. In between lie different orientations distinguished from one another by individuals' varying socialization experiences.

Much of that socialization takes place within the context of sex roles. Sex roles (sometimes called *gender roles*) are collections of norms that define socially accepted male and female behavior. Clear distinctions in the elements of sex roles provide useful separation and elaboration of people's individual and group identities. Some criticism of homosexuals and transvestites may grow out of perceived threats to this distinction and therefore to important components of individual identities. In cultures with relatively high tolerance of homosexuality (e.g., ancient Greece), people recognized no strong collective boundary that needed maintenance. The Greeks felt no need to maintain any collective identity by enforcing a strict moral code. The Judeo-Christian ethic, in contrast, maintains strong boundaries between male and female roles to reinforce the historically strong emphasis on a separate Jewish identity (Davies, 1982). This concern, articulated in the Old Testament, promoted the survival of the Jews under serious threats to their continued existence.

People begin to learn sex roles at birth through experience of the behavioral expectations of parents and others. Society's ongoing instruction sometimes begins as early as the choice of the color of the baby's blanket, meant to convey certain expectations about his or her eventual sex role. The baby may miss the significance of a blue or pink blanket, but others recognize the symbol and react to the baby on the basis of the blanket color more than anything else. A blue blanket leads people to detect (expect) masculine attributes, while a pink one induces them to detect and expect feminine attributes.

Boys and girls receive guidance in sex roles from the decorations in their nurseries; "boy things" may include footballs or other sporting equipment, while "girl things" may emphasize dolls. Over time, boys recognize and fulfill expectations to act aggressively, while girls learn to provide passive responses to the sexual "scripts" presented to them. Others' responses reinforce behavior that conforms to their expectations, based in important ways on the child's sex, while those responses punish unexpected behavior. Evidently, if gender brings any inherited, biological tendencies, learning can modify them extensively over time. Thus, individuals come to identify themselves as males or females as a result of recognizing and performing the sex roles assigned to them.

Like role socialization, learning of sexual behavior itself begins early in life (Akers, 1985: 184–185). People can learn to associate sexual satisfaction with virtually any object or person, but sexual behavior is always embedded in a web of normative constraints that set limits on acceptable objects and people. Rewards and punishments experienced from early childhood onward help an individual to define acceptable sexuality. Most learn to adopt heterosexual roles and to derive sexual satisfaction from objects and persons that society considers "conventional," that is, within the norms of the applicable group.

But the sexual socialization process sometimes works imperfectly, and some individuals come to derive sexual satisfaction from objects and people outside their groups' normative structures. This occurs for at least two reasons. First, those who provide socializing feedback bring ambiguous feelings to erotic acts. Many parents and others feel uncomfortable providing sex education that includes information about gender-appropriate objects for satisfaction. Most socializers become embarrassed when they encounter the topic of sex. Second, needed instruction in sexuality covers much ground, from appropriate partners to appropriate times, objects, places, and ages. In fact, sexual norms define some of society's most complicated standards because people must learn so many different combinations of contingencies.

No one should express surprise, therefore, that the socialization process sometimes fails to adequately prepare an individual for socially appropriate sexual growth and maturation. Some individuals remain open to disvalued sexual alternatives, such as receiving sexual gratification from a prostitute or engaging in unusual sexual practices like sadism or masochism. For similar reasons, some people, not surprisingly, feel sexual attraction to members of the same sex. Even in the face of complex sexual norms and an ambiguous socialization process, socialization prepares the large majority of people to become heterosexuals.

On another note, the possibility exists that learning about sexual preferences may interact with certain biological factors that research has not yet identified. This possibility adds further complication to the effort to understand the process of acquiring a particular sexual preference.

Becoming a Homosexual

The general theoretical perspective presented here views the development of sexual preference in the larger context of sexual socialization (see also Plummer, 1981). Individuals develop their own sexual preferences, or orientations, by learning to favor certain objects or practices, or, alternatively, by not learning to favor other alternatives.

This discussion requires a reminder of the distinction between *homosexual behavior*, which refers to sexual practices with partners of one's own sex, and *homosexual preference*, which refers to subjective feelings of stronger sexual attraction to a person of the same sex than to a person of the opposite sex. These terms refer to different things. A person may engage in homosexual activities but still feel primarily attracted to others of the opposite sex. On the other hand, some married males may feel stronger attractions to other men than to their wives, gaining most of their sexual stimulation from men. The particular combination of homosexual attraction and active homosexual behavior may result from participation in a homosexual subculture; the extent of a person's participation in that subculture may determine the strength of acceptance of the role of homosexual. For this reason, research does not identify *the* model of homosexuality; rather, varying involvement with homosexuality changes with different levels of behavior and attraction.

Many children engage in experimental sex play involving homosexual activities, particularly when obstacles discourage or prevent experimentation with members of the opposite sex. One study of a group of homosexuals in Great Britain found that a male's initial homosexual experience usually involved a fellow schoolboy of the same age (Westwood, 1960). These experiences did not, however, necessarily lead to later homosexuality or associated patterns of behavior because little emotional feeling may accompany such sex behavior among boys.

Early homosexual experiences may become significant if they involve adult partners or repeated acts carried out with the same boy over a year or so. Over two-thirds of such experiences involve other boys as partners. Only one-fifth of boys introduced to homosexuality experienced initial contacts with adults, and a further 11 percent had no homosexual experiences of any sort until they had become adults, all of them with adult partners. Contrary to a popular view, seduction by adults does not lead important proportions of boys toward homosexuality. Another British study focused on six groups of 50 homosexuals each. By the time they had reached adulthood, nearly all subjects who later became homosexuals had had at least one exposure to sex (Schofield, 1965). Three-fourths of the men in three groups reported their first exposure occurring before the age of 16, and 16 percent reported that this contact had involved adult partners. Most homosexuals struggle for long periods of time against their inclinations toward homosexual activity before recognizing the orientation as a permanent part of their behavior and developing self-conceptions as homosexuals. A number of studies of adult gay men confirm such findings (see the summary in Savin-Williams, 1999: 16). The mean age for first same-sex attractions was between 10 and 13, and the mean age of first homosexual sex was between 13 and 15. The mean age across studies where the youth labeled himself as gay was between 15 and 21.

Earlier discussion emphasized the identity of sex roles as learned behavior. One learns actions associated with masculinity and femininity as part of this sex-role socialization; they do not reflect any biological inheritance. An explanation of the emergence of homosexuality and heterosexuality thus invokes three concepts: (1) sex-role adoption, (2) sex-role preference, and (3) sex-role identification. *Sex-role adoption* refers to active choices to adopt behavior patterns characteristic of one sex or the other, rather than simply the desire to adopt such behaviors. Sociologists sometimes refer to this process as *sex-role* or *gender-role nonconformity*. *Sex-role preference* describes the desire to adopt behavior patterns associated with one sex or the other, or the perception of this behavior as preferable to alternatives. Finally, *sex-role identification*, a crucial process in developing homosexuality, indicates the actual incorporation of a given sex role and the unthinking reactions characteristic of it. In other words, the person internalizes the sex role and develops a self-concept consistent with associated expectations. Some people may identify with the opposite sex and adopt many associated behavior characteristics.

People who eventually become homosexuals seem to acquire a sex-role identification, also called *sex-role assimilation*, toward members of their own gender in childhood. Some research has suggested a link between the effectiveness of traditional sex-role learning in children with the sexes of their siblings, the presence or absence of a father in the home, and their birth-order positions in the family. One study, for example, reported that most homosexuals in the sample came from backgrounds of physically or emotionally absent fathers (Saghir and Robins,

1973). Similarly, a comparison study found that 84 percent of a group of homosexuals reported that their fathers seemed indifferent to them, while only 18 percent of a group of heterosexuals expressed the same sentiment. Only 13 percent of the homosexuals but two-thirds of the heterosexuals identified with their fathers; while 18 percent of the homosexuals perceived satisfactory relationships with their fathers, 82 percent of the heterosexuals reported satisfactory relationships (Saghir and Robins, 1973: 144–145). One homosexual, when asked about the causes of his homosexuality, implicated the absence of a father figure:

> Well, for one, because I was never raised around a man, and I never had my father there, you know. My brothers were there off and on, very more off than on. And like, when I was away, I was with my grandmother and my auntie. When I came out here to L.A., I was with my mother and grandmother. I never really had a male image to enforce in me this and that, you know, so I guess that might have had a strong influence on the future. (Green, 1987: 355)

Other research has failed to document the supposition that homosexuality results from identification with the parent of the opposite sex (Bell, Weinberg, and Hammersmith, 1981). Most research, however, does point to the importance of childhood sex-role development, especially any behavior that deviates from sex-role expectations. Researchers note this sex-role nonconformity ("sissy" behavior in boys and "tomboy" behavior in girls) in large numbers of subjects who have developed homosexual preferences (Bell et al., 1981; Green, 1987). In one study of homosexual couples, 75 percent reported that other children had called them "sissy" as boys (McWhirter and Mattison, 1984: 130). But severe methodological problems limit the reliability in these studies; one such design—that by Green cited above—experienced a sample loss of one-third before it reached completion (Paul, 1990).

Still, early childhood experiences do not by themselves determine an individual's eventual sexual orientation. Sex-role learning continues throughout adolescence and into early adulthood. Adolescence is a particularly important time, because young people change from homosocial contacts (primarily with others of their own sex) to heterosocial ones (contacts with members of the opposite sex). By the end of adolescence, people have become fully aware of the contexts for sexual behavior—which others make sexually desirable partners, when and with whom they can appropriately have sexual relations, and so on. By this period in their lives, most people have developed sexual identities. Such an identity becomes a deep-seated trait, reflected in feelings of sexual preference and orientation regardless of specific behavior (Harry, 1984). In this sense, adult homosexuality is "just a continuation of the earlier homosexual feelings and behaviors from which it can be so successfully predicted" (Bell et al., 1981: 186). As some individuals learn to identify with male roles, some also learn to identify with female ones. Again, both pass through the same development process; the content of what they learn determines the difference.

Many gay men are aware of earlier experiences where they felt different from other boys because of their identification with female roles and behavior. One youth recalled his childhood in the following way:

I knew that a boy wasn't supposed to kiss other boys, although I did. I knew it was wrong, so this must be an indication that I knew. I also knew that I wasn't supposed to cross my legs at the knees, but I wouldn't like quickly uncross my legs whenever that was the case. So this is certainly of a young age that I noticed this. I think I knew that it was sort of a female thing, sort of an odd thing, and I knew that boys weren't supposed to do that. (Savin-Williams, 1999: 28)

Observers can encounter problems completely identifying the web of specific influences that affect the determination of sexual orientation. Clearly, however, the definitions of the situation offered by others form part of that web. People define certain objects and situations as male and others as female, and these definitions temper their expectations for behavior in those situations. Berger (1986: 179) notes, "Sexual orientation is a complex phenomenon. Becoming homosexual is the result of both personal and social variables and is determined in part by how one's behavior is labeled by others." Many adult homosexuals describe the early formation of their sexual identities as confusing and lonely processes, with an important role for the reactions—either real or anticipated—of others. Mike, a 19-year-old British male, conveys the isolation he felt through this period in his life:

I went through such hell. I thought I was going to have a breakdown. Gradually you attach the label *gay* to yourself because if you don't you really crack up. I did it gradually after years of torment, but still hated myself for it. Accepting that it could be real was the hardest part of my life. I felt lonely, couldn't turn to anyone through fear of what would happen to me. I didn't know any gays so how could I know that we are just ordinary people? I felt I would only be alone as I wasn't straight but also I wasn't the kind of gay my mates used to laugh and joke about. (quoted in Plummer, 1989: 207)

Developing a Homosexual Identity

People acquire homosexual identities, often through long, interactive processes that depend heavily on the actions and reactions of others. Some of these reactions are positive, while many are negative. Such a process shows up in the distinction between career or secondary homosexuals, those who actively perform a homosexual role, and those who engage in homosexual behavior without developing related self-concepts. Recall from Chapter 6 an important difference between primary and secondary deviance: the extent of the link between deviant behavior and a person's deviant self-concept. Primary deviants commit deviant acts without acquiring deviant self-concepts, while secondary deviants recognize and define themselves as deviants.

Primary homosexual behavior often results from involvement in a particular situation. It may occur, for example, in single-sex communities like prisons, isolated military posts, naval ships, and boarding schools. Some male prostitutes may commit homosexual acts only for money without developing self-concepts as homosexuals (Luckenbill, 1986). These people and others who commit homosexual acts are not homosexuals in the full sociological sense; they do not establish identities as career deviants for the central reason that they never develop homosexual self-concepts.

The acquisition of a gay identity is a subtle and private process that occurs at different times for different individuals. In one study, youths identified their sexuality as early as third or fourth grade and as late as early to mid-20s (Savin-Williams, 1999: 123). The largest number reported the development of a gay identity as during high school or college years. Some gay youth report that making the transition to a gay identity was facilitated by increased knowledge of what it is like to be gay:

> I was reading sort of all this educational stuff about homosexuality and it portrayed it in a positive way and that is, that they have their own culture and their own heroes and models. So at this point then I was able to say to myself that I am gay myself. (Savin-Williams, 1999: 132)

Secondary homosexuals tend to seek sexual gratification predominantly and continually with members of the same sex. These acts represent expressions of homosexual self-concepts and roles. In fact, Goffman (1963: 143–144) limits the term *homosexual* to "individuals who participate in a special community of understanding wherein members of one's own sex are defined as the most desirable sexual objects and sociability is energetically organized around the pursuit and entertainment of these objects." Association with other homosexuals provides an important feature of career homosexuality, particularly in homosexual bars, which tend to serve as vital meeting places for homosexuals in many cities. One almost inevitably develops a homosexual self-concept after association with other homosexuals, in both sexual and nonsexual contexts, for a period of time. The gay bar facilitates maintenance of a homosexual self-concept by limiting and defining contacts with nongays and by reinforcing priorities of homosexual life in a situation controlled by homosexuals (Reitzes and Diver, 1982). Gay bars also provide important support for homosexual liaisons, both short-term encounters and long-term relationships. One survey of 92 homosexual couples determined that gay bars provided the most common meeting place for these partners (Berger, 1990). Such places offer social support as well.

Homosexual masculinity traits form essential parts of the homosexual identity (Connell, 1992). Gay men construct this image of masculinity through interaction with others in a process of gradually increasing sexual involvement with males and declining contacts with members of other groups. This involvement can also lead to participation in a gay community. Masculinity represents not a single characteristic, but a range of different conceptions of elements that men value and associated with maleness. The amalgamation changes over time and from group to group. Behavior considered manly in one group (e.g., urban gang members) may strike another (e.g., physicians) as barbaric or inappropriate excess.

Several crucial conditions influence development of a homosexual identity. The expectations of others and the extent of one's identity with available role models both contribute to or impede this process, as do the reactions of others—their attribution or imputation of homosexuality to specific individuals. Generally, a homosexual identity initially grows from the realization of a homosexual preference; subsequent development results from continued participation in single-sex activities and environments. Official definitions of a person as a homosexual by medical doctors, psychiatrists, or even the police may promote the development of a homosexual identity.

A person can recognize a particular sexual orientation at almost any time, but it often occurs in early adolescence. The precipitating conditions, that is, events that immediately proceed the realization, also vary from person to person. A 40-year-old male reported that he achieved insight about his sexual identity at age 14 from reading a chapter in a book on sexual development:

> When I read the chapter, I knew immediately that's who I was. I'll never forget it, as it was one of the most traumatic evenings I've ever had in my life. *I just knew.* I had to go through this entirely alone. There was simply no one to talk to. Oh, I sort of considered briefly discussing it with the family doctor. I just felt very alone. I wondered if there were anyone else like me. After that night, I continued to participate in school activities and to date and all that. But it was all a facade and I knew it. (Lynch, 1987: 40)

Many homosexuals report that experiences had given them some inkling of a difference between themselves and other people early in life. One writer recounts staying in from recess in elementary school because he didn't like to play soccer:

> [A] girl sitting next to me looked at me with a mixture of curiosity and disgust. "Why aren't you out with the boys playing football?" she asked. "Because I hate it," I replied. "Are you sure you're not a girl under there?" she asked, with the suspicion of a sneer. "Yeah, of course," I replied, stung and somewhat shaken. (Sullivan, 1995: 3–4)

When someone says, either out loud or privately, "I am a homosexual," it expresses quite a different reality from engaging in a homosexual act. One can engage in a homosexual act and still think of oneself as a homosexual, heterosexual, or bisexual, just as one who engages in a heterosexual act may maintain a self-image as a heterosexual, homosexual, or bisexual. The recognition of a homosexual identity results in an extremely important element of self-awareness, and the completion of a change to this homosexual identity often culminates in publicly acknowledging it, that is, "coming out."

The Coming Out Process

Coming out involves a public declaration of a deviant, homosexual identity (Dank, 1971) and action to convey that identity to heterosexuals (Plummer, 1975). Coming out is a process, not a single announcement. It involves several elements: recognition of one's sexual preferences, experiences with others in sex-role socialization, a process of realizing that these elements form part of a sexual identity, and resulting behavioral commitments to a homosexual lifestyle (see Coleman, 1981–1982;

Dank, 1971). The coming out process begins and reaches completion over many years in a tenuous sequence of steps, since not all homosexuals progress from one fixed point to another. Descriptions of the coming out process emphasize the continuing importance of sex-role socialization and the expectations of others.

Stages in Coming Out   The coming out process moves through four stages: (1) sensitization, (2) identity confusion, (3) identity assumption, and (4) commitment (Troiden, 1989). In the *sensitization stage,* the individual begins to become aware of differences from others of the same sex. By high school, the individual has developed a distinct sense of contrasts with other people. The individual feels marginal but cannot understand these feelings at such a young age.

The *identity confusion stage* represents a separation of behavior from the person's sexual feelings and recognition of an individual sexual orientation. A boy may experience sexual attraction for other males but either fail to act upon those feelings or try to deny them. By middle or late adolescence, a perception of self as "probably" homosexual begins to emerge. Researchers explain that homosexual males begin to suspect they might be gay at an average age of 17 (Troiden, 1989: 53). The social stigma on homosexuality discourages an open discussion of these changes, however, and ignorance and a lack of awareness of others encourage further confusion.

The *identity assumption stage* brings important events like defining oneself as homosexual and acknowledging a shared identity with other homosexuals. During this stage of coming out, the individual may have some contact with homosexual subcultures (e.g., gay bars). Initial contacts with other homosexuals provide important input for resolving some internal conflicts. Coming out involves a clear self-definition as homosexual, initial involvement in a homosexual subculture, and redefinition of homosexuality as a positive and viable lifestyle. It also involves disclosure to another person:

> I came out to my aunt, mother's sister. I had always had a close relationship with her. I saw her more as a friend. I felt more comfortable to tell her than any other family member because she was so liberal and accepted everybody. . . . She was very positive about the thing too. (Savin-Williams, 1999: 151)

In the final stage, *commitment,* the individual takes on homosexuality as a way of life. Sexual activity may become combined with emotional life, such as forming a stable homosexual relationship with a single partner (Warren, 1972). Taking a homosexual lover confirms a gay identity, as does disclosing one's homosexual identity to family members and other heterosexuals.

Coming Out and Social Support   The process of coming out requires that an individual make some difficult decisions. One source of concern centers on the reaction of others to these decisions. In one publication, two authors have described possible variations in parents' reactions:

- Your parents will accept you as you are. This is not common, and with the best intentions in the world they may take a long time to come to terms with your situation.

- Your parents will try to understand but the news will make them feel guilty, as if your gayness is their fault. They probably will think your life is headed for ruin if you persist in your homosexuality, and therefore will pressure you to change.

- Your parents will not react. They will refuse to believe you, and the subject will never be brought up again.

- Your parents will reject you. Melodramatic as this may seem, gay people do get thrown out of their family home, disowned, and told never to come back. (Muchmore and Hanson, 1989: 73–74)

Individuals do, however, experience many variations in the coming out process. One study reported that 18 percent of a sample of 199 gay men labeled themselves as *homosexual* without participating in overt sex with other males; 22 percent established their homosexual identities while participating in long-term relationships with other males; and 23 percent developed their identities only after ending involvement in such a relationship (McDonald, 1982). Homosexuals generally come out in the social context of contact with other homosexuals. This fact creates an important role for homosexual subcultures.

Homosexual Subcultures

Like other subcultures, homosexual subcultures represent collections of norms and values. Such a subculture creates conditions that permit or condone homosexuality. Members come to learn these norms as part of the coming out process, and exposure continues in social situations that involve other homosexuals. Most homosexual men participate in a gay community to some extent. Even heavy involvement in this subculture does not mean, however, that homosexuals have contacts only with other homosexuals. In fact, they also have contacts with the "straight" world, particularly with family members and employers.

No one homosexual subculture or gay community defines all possibilities for such involvement, just as no single lifestyle fits all homosexuals (Bell and Weinberg, 1978). Instead, individuals encounter variations on a common theme—social networks that protect and facilitate homosexual relations through the formation of common bonds with similar others around the homosexual role. A local homosexual community may link overt members and secret ones ("closet queens") through bonds of sex and friendship. These groups, which often cut across social class and occupational lines, serve to relieve anxiety and to promote social acceptance.

The actual number of such gay communities varies by region, depending on the outside community's tolerance of homosexuality. Homosexual communities gather individuals in many cities, large and small, and in rural areas as well (Miller, 1989). In fact, locations throughout the world feature such communities (Miller, 1992). In New York City and San Francisco, well-developed gay communities define established social networks; similar subcultures in other cities remain less visible to outsiders. Some of these communities feature well-organized interactions, while others do not; some are interracial, such as the Black and White Men

Together (BWMT) organization. Regardless of place, gay communities express concerns about such issues as discrimination, legal sanctions, AIDS, mutual support among members, and maintaining homosexual relationships.

The development of a homosexual community seems to depend in large part on society's intolerance for homosexuality and the resulting desire to weaken the stigma imposed by the outside society. In this sense, these communities provide very functional support for the participants; they provide "training grounds" for establishing norms and values, milieus in which people may live every day, social support, and information for members. Homosexual subcultures may have become strong enough in some U.S. cities to redefine them as true cultures, or locally dominant systems of norms and values (Humphreys and Miller, 1980). Homosexuals in the United States rely more on subcultures than do those in, for example, the Netherlands or Denmark; homosexuals there felt less need to organize their own communities because those societies imposed less severe repression and negative attitudes toward homosexuality (Weinberg and Williams, 1974: 382–384). Homosexual communities, as such, seem to develop when individuals feel the need for supportive and learning environments.

Miller (1992: 360) speculates that certain necessary characteristics define essential conditions for development of homosexual communities:

1. Some personal freedom and social tolerance
2. Economic independence and social mobility
3. Relatively high status for women
4. Declining power of the family and religion in defining and determining every aspect of an individual's life

In Argentina, for example, the high cost of living prevents many young people from living openly homosexual lives. Many must remain in their parents' homes, leaving them little personal freedom.

Subcultural activities have led to increasing numbers of formal homosexual organizations and gay clubs throughout the world. Members frequent familiar gathering places in many parts of the world, and some even visit spots promoted specifically to homosexual tourists (Whitam and Mathy, 1985).

The homosexual rights, or "homophile," movement in the United States began on the West Coast after World War II. The first major organization within this trend was the Mattachine Foundation, established in 1950 in Los Angeles as a secret club to promote discussion and education about homosexuality. The club later moved its headquarters to San Francisco and changed its name to One, Inc. A national organization of homophile societies approached reality in 1966 with the establishment of the North American Conference of Homophile Organizations. Among other functions, this group organized meetings for local clubs. Some homophile organizations have sponsored militant social involvement, such as the New York Gay Liberation Front and the Gay Activist Alliance, also in New York (Humphreys, 1972). Thousands of grass-roots homosexual groups continue to operate in the United States.

Unlike organizations such as Alcoholics Anonymous and Synanon, homosexual organizations espouse no desire to change the behavior of their members. Such organizations wish instead to ease some of the social and legal stigmas surrounding homosexuality; in effect, they wish to reinforce and legitimize homosexuality. These organizations engage in a number of educational and political activities. They furnish information, distribute literature, and publish periodicals on topics that interest homosexuals. They also vigorously reject any idea that homosexual behavior represents a sickness or pathology, and most argue against regarding homosexuality as deviant in any sense.

The homosexual rights movement became a visible part of society's political environment on June 28, 1969, when patrons of Stonewall, a gay bar in Greenwich Village in New York City, refused to cooperate with police who were carrying out a routine raid. The patrons, composed at the time of the raid mainly of flamboyant drag queens and prostitutes, escalated their protests against the police into nearly 5 days of rioting that eventually drew participation from hundreds of sympathetic supporters. The rioting appeared to accomplish little; no laws changed, "gay bashing" continued, and homosexuals retained their image as socially and sexually marginal people. This episode of resistance achieved significance, however, by influencing the imaginations of homosexuals throughout the country and elsewhere. Many began eagerly to resist the rejection and shame heaped upon them by conventional society (Bawer, 1996: 4–15). *Stonewall* became synonymous with resistance to that oppression.

The first post-Stonewall generation of homosexuals worked hard to promote development of a community that would command respect, but some determined that the effort demanded extremist and aggressive tactics. The movement sought to establish a public community of homosexuals, requiring activities to entice potential members to come out of the closet. Groups such as Queer Nation and other activists promoted gay pride marches, celebrations of Stonewall, and organized events meant to shock, annoy, retaliate, and educate straight society— all at the same time. Through this activity, a portion of the gay community "developed a radical direct-action movement among men and women who are no longer interested in dwelling only within the safe ghettos of gaydom" (Browning, 1994: 25).

The homosexual rights movement and political activism have emphasized two elements since Stonewall (Meier and Geis, 1997). First, highly militant tactics emphasize fighting back and taking on the straight community. This approach denies the deviant identity of homosexuality and affirms the importance of homosexual relationships, families, and values. A second, more accommodating set of tactics seems targeted at long-term results. They emphasized applications of law to achieve specific political gains, such as antidiscrimination legislation and legal provisions allowing homosexual marriages.

There are various models of groups who wish to change their status in society. While the civil rights movements serves as a model for acquiring gay rights, perhaps religious movements are a closer analogy to the gay movement because in each case there must be an identity change (Richards, 1999). Throughout the 1990s gay identity has changed from a political to a lifestyle category in the mind of many (Valocchi, 1999). Increasingly, the hetero/homo binary has become somewhat blurred insofar as each group is treated alike by Madison Avenue. At the beginning of the 21st century, major corporations found gay people to be of particular interest in advertising, although the nature of the ads is frequently designed to speak both to homosexuals and heterosexuals. Such images tend to reinforce a lifestyle interpretation of homosexuality.

Female Homosexuality (Lesbianism)

Lesbians are female homosexuals. "The essence of lesbianism is preference for women, not rejection of men. If a woman chooses other women for her sexual partners because her deepest feelings and needs can only be satisfied with women, she is a lesbian" (Cruikshank, 1992: 141). The term *lesbianism* comes from the name of the Greek island of Lesbos, where the Greek poetess Sappho (600 B.C.) led a group of women in a network of homosexual relationships and behavior.

Researchers have paid less attention to female homosexuality, or lesbianism, than to male homosexuality. The difference may reflect the balance of scientific interest or variations in relationships that make lesbianism harder to study than male homosexuality. In many ways, the activities of female homosexuals resemble those of males, but certain differences distinguish the two groups. Two similarities deserve special notice at this point. First, lesbians, like male homosexuals, do not choose their sexual orientation. Individual lesbians may make some choices about how they express their sexual orientation, but they do not themselves determine whether they have that orientation (see Card, 1992). Second, lesbians, like male homosexuals, have been and are subject to homophobic attitudes and behavior.

The Nature of Lesbianism

Like male homosexuals, lesbians encounter public stigma and social rejection. Laws generally do not specifically prohibit sexual acts between women, and few jurisdictions ever try to apply other laws that might bring some sanctions for related behavior. In fact, Kinsey and his associates (Kinsey, Pomeroy, Martin, and Gebhard, 1953: 484) did not find a single recorded case of a female convicted of homosexual activity in the United States from 1696 to 1952. In that study's large sample of women who reported homosexual experiences, only four had encountered any difficulties with the police. Still, many lesbians feel continuing fear of disclosure, both on the job and among their nonhomosexual friends.

Tension between the "straight" and lesbian worlds presents a particular problem for lesbians who work. Many female homosexuals appear to display stronger commitments to their jobs than most women show because they do not depend on male partners for financial support. Many lesbians attempt to manage their employment settings by adopting heterosexual manners, behaviors, and expressions. In this respect, their social interactions may resemble those of male homosexuals.

The attribution of homosexuality has a different effect on lesbians than it has on gay men. Lesbians exhibit fewer identifiable characteristics that male homosexuals show, so they usually become targets for adverse public opinion only when they actively proclaim their sexual identities, for example, adopting unique styles of dress or associations. Women friends and even acquaintances commonly display signs of close personal relationships and a certain amount of physical demonstration, such as hugging and kissing. This demonstrative behavior obscures the distinctions that define deviant lesbian relations, making that judgment more difficult to establish than a comparable evaluation of male homosexual relations. Females more often exhibit bisexual and otherwise inconsistent sexual behavior than males (Blumstein and Schwartz, 1974). Only one-third of the lesbians Rust (1992) interviewed stated they felt attracted 100 percent to women and never to men. The remainder provided different figures, ranging from 50 to 95 percent. Lesbians do, however, display a range of behaviors and identities similar to that of male homosexuals. Some lesbians are married women, but their primary sexual orientations incline them toward other women; others remain unmarried and participate in bisexual relationships; still others do not marry and form strongly lesbian identities.

Extent of Lesbianism in Society

The difficulties of estimating the number of male homosexuals expand dramatically when one attempts to estimate the number of lesbians. A number of reasons contribute to this difference, including less openness and cohesion in lesbian subcultures compared with those of gay men. These characteristics may be changing as the women's movement continues to encourage organization and focus in many local lesbian groups. In any case, female homosexuals maintain outward appearances of heterosexuality more easily and more frequently than gay men do, seriously complicating efforts to identify them.

More than 40 years ago, the Kinsey study (1953: 512) estimated that, at the time of their marriages, one-fifth of single women and 5 percent of all women had experienced homosexual relations leading to orgasm, a figure lower than the comparable one for men. One can find other estimates, but they may not reflect scientifically accurate procedures. Cruikshank (1992: 163), for example, says, "Because lesbians remain a hidden population, their exact numbers cannot be known." She goes on, however, to speculate without corroboration that, "They probably make up at least 15 percent of the female population."

A study based on a representative national sample of Americans in 1992 found lower estimates of lesbianism than these. Less than 2 percent of the women who responded to the study said they had experienced sex with another woman in the previous year, and about 4 percent said they had participated in sex with another woman after age 18 (Michael et al., 1994: 174). When asked whether they considered themselves heterosexual, homosexual, bisexual, or something else, less than 2 percent of the women identified themselves as lesbians or bisexuals (Michael et al., 1994: 176).

Most studies of sexual orientation do not even attempt to estimate the number of lesbians. Perhaps the only agreement about the extent of lesbianism has come from the popular speculation that it occurs more often than people have commonly thought, perhaps making it more prevalent than male homosexuality. In a number of respects, lesbianism represents one of the least studied forms of sexual behavior.

Becoming a Lesbian

Certain values and norms may prove conducive to female sexual experimentation in important ways, although they might not lead to full female homosexuality.

Women typically value themselves to varying degrees as heterosexually desirable because mass media outlets extensively portray physical appearance as a critical aspect of female identity, and men develop their expectations based on media images of female desirability. Sexuality, therefore, becomes a part of a woman's self-evaluation, and women recognize sexuality, both their own and that of others, in strongly emotional terms. Traditionally, society has regarded and treated women as sexual objects for males, and this attitude may also shift to other females.

Women's norms, in contrast with those of men, also permit them to touch one another physically and to form and maintain strong emotional relationships. Although such norms usually regard this kind of physical intimacy as an exclusively social practice, they make it both more accepted and more common among women than among men in U.S. society. In some situations, these standards may allow shared discussions of sex and sexual fantasies that lead to behavioral experimentation.

Women drift casually into homosexuality more often than men do, generally starting with vague romantic attachments to other women. Lesbians' initial physical contacts with other women generally occur before the age of 20, a large percentage of them before 15. However, little evidence supports the idea that others seduce females in any sense into lesbianism. As with males, recognition of a female-centered sexual orientation precedes lesbian physical relations, and most lesbians report having experienced heterosexual relations before lesbian ones (Bell et al., 1981). A study of one group of lesbians indicated variations in initial sexual contacts, but most involved only manual stimulation. Almost a third of the women studied said that their first contact included oral sex unrelated to the achievement of orgasm (Hedblom, 1972: 56). While most of the lesbians explained that they maintained clear-cut boundaries between homosexual and heterosexual worlds, nearly two-thirds of them reported experiencing sexual relations with men, a third of them within the previous year.

Most lesbians first recognize their homosexual feelings in late adolescence or early adulthood, and overt homosexual behavior frequently develops in a late stage of an intense emotional relationship with another woman. By middle or late adolescence, a self-perception as "probably" homosexual begins to emerge among women who later will form lesbian identities; lesbians begin to suspect that they "might" be homosexual at an average age of 18 (Troiden, 1989: 53).

The general processes of sexual development and sexual socialization discussed earlier occur in both males and females. Females in most cultures learn early in life about appropriate female sex roles and related expectations (Reiss, 1986). Lesbians tend to practice sex-role behavior patterns that resemble closely those of heterosexual females. Generally consistent cultural expectations allow closely comparable sexual learning for both homosexual and heterosexual females.

This statement does not imply, however, that both always share the same early experiences. One important study found that a significant number of homosexual females had displayed "tomboy" attitudes as young girls, behaving somewhat as boys did (Saghir and Robins, 1973: 192–194). In fact, girls who later became lesbians exhibited boylike behavior more commonly than boys who later became homosexual men exhibited girllike behavior. Other studies have documented similar findings of a higher probability of sex-role nonconformity among girls who eventually become lesbians than among boys who eventually become homosexual men. One major study found boyish behavior and interests in the backgrounds of over three-quarters of the lesbian research subjects (Bell et al., 1981: 188).

Like many homosexual males, early lesbian experiences may grow out of experimentation and curiosity—the major reasons for such activities given by females who identify themselves as heterosexuals but who have experienced lesbian relations. One study has reported that women often had their first experiences as a result of "male orchestration," that is, group sex situations in which the male members of "spontaneous threesomes" encouraged the two females to engage in sexual intimacies (Blumstein and Schwartz, 1974: 282). Other than such sexual experimentation, women do not become lesbians because others seduce them into that orientation; rather, the development of a lesbian sexual orientation or preference generally precedes lesbian behavior. No evidence indicates that females become lesbians as a result of negative experiences with men in excess of those reported by heterosexual women (Brannock and Chapman, 1990).

Male and female homosexuals differ substantially in promiscuity. Lesbians tend to view themselves as less promiscuous than male homosexuals (Hedblom, 1972: 55), and their behavior confirms this self-perception. Lesbians participate less frequently than gay men do in cruising for casual sexual partners, even in bars; they more often maintain committed relationships with others or even form long-term, emotionally based bonds that approximate those of marriage.

> Being a female homosexual is like being a female generally, both sexually and socially. There is a tendency to greater conformity, stability of relationships, and an absence of indiscriminate sexual involvements. There is also a general emphasis on relationships, romantic involvements, and faithfulness in relationships. (Saghir and Robins, 1980: 290)

This difference is woven into the basic female sex role, which links sexual gratification to emotional or romantic involvement. In comparison, the average male homosexual tends to experience less stable relationships, while the lesbian acts with more reserve and pursues selective involvements that exclude interest in multiple sex partners or varieties of sexual practice. Both male and female homosexuals can and do develop long-term relationships with others, but such relationships seem to emerge more frequently among lesbians. Moreover, the incidence and causes of relationship problems among lesbian couples parallel those in heterosexual marriages; the participants in both kinds of relationships may need counseling over similar issues, including unequal power, duties, or other complaints (Boston Lesbian Psychologies Collective, 1987).

Only limited studies have examined the role of occupation in the development of lesbianism, although rates of lesbian activity seem high within some occupations. A study of striptease dancers found that one-fourth of them engaged in lesbian relationships (McCaghy and Skipper, 1969). Moreover, the study found common bisexuality in this group, estimated in a range from 50 to 75 percent. The study attributed this high percentage to several potential causes, including the dancers' limited opportunities for stable sexual relationships with males. Also, many develop negative attitudes toward men based on their behavior while viewing sex shows, leading many dancers to prefer to associate primarily with women.

Lesbian Self-Concept

Homosexual women often achieve lesbian self-concepts within relational contexts (Gilligan, 1982). Their initial attractions toward other women reflect not sexual feelings, but emotional ones of friendship or closeness on the basis of mutual interests. A woman who recognizes her feelings of attraction to other women may "try on" the label of lesbian to see how it fits (Browning, 1987). During the course of that process, lesbians often establish emotional links with "special" women (Troiden, 1988, 1989). These relationships tend to last longer than comparable ones among male homosexuals.

Because women tend to emphasize emotional rather than physical aspects of their mutual attractions, this self-labeling process generally occurs in the context of friendships with other adult women. The close personal relationship that establishes the context for a lesbian encounter provides a crucial condition for the development of a lesbian identity. "The majority pattern appears to be one in which self-identification as a lesbian develops before or during genital contact itself, and as a late stage of a close, affectionate relationship" (Cronin, 1974: 273).

Some have long thought that lesbianism combines elements of both masculinity and femininity (Greenberg, 1988: 373–383). Such stereotypes form the basis for many people's conceptions of lesbians. They imagine a lesbian as a masculine looking, acting, and/or thinking woman. In any pair of lesbians, social expectations often cast one as feminine and the other as masculine. In reality, some lesbians may fit such stereotypes, but others do not. Only a minority of lesbians actually commit themselves to the "butch" role with its masculine traits, although others may experiment with it, particularly during the "identity crisis" period that follows their entry into a homosexual subculture after coming out. Those who do adopt male-oriented roles within lesbian relationships may see themselves differently during sexual interactions. One study, for example, found that lesbians perceived themselves as more feminine during sexual interactions than during their overall sex-role activities (Rosenzweig and Lebow, 1992).

Nearly all lesbians interviewed by Simon and Gagnon (1967: 265) wanted to become emotionally and sexually attached to other women who would respond to them as women. Still, when they abandon the world of men, lesbians must take on many social responsibilities carried out in heterosexual life by men. A homosexual female must overcome difficult obstacles to develop an acceptable self-concept and identity, although the emergence of the women's movement and the homosexual rights movement have provided helpful support.

This statement does not portray lesbians as generally unhappy persons overwhelmed with personal problems. One study of 127 lesbians, in fact, reported that most of them seemed happy and satisfied with their lesbian role (Peplau, Padesky, and Hamilton, 1982). The study noted a link between this satisfaction and characteristics of the relationships the subjects experienced with other women, such as equality of involvement and equality of power in the relationships, important benefits stressed by the women's movement in general.

Lesbian Subcultures

Lesbians do not immerse themselves in the homosexual world as much or participate as actively in its subculture as do male homosexuals. Homosexual subcultures act as functional networks organized to give support and a context for social relationships. Observers find fewer examples of lesbian subcultures, for one reason, because the lesbian role proves less socially alienating than that of the male homosexual. For another reason, lesbians can mask their sexual deviance behind typically asexual responses to other women more effectively than can men, whose gender roles call for more sexually active and aggressive behavior. Because lesbians form relatively permanent relationships more often than gay men do, these women manage to keep their social and sexual lives more private than the public social whirl of gay bars and other homosexual meeting places. Such stable relationships limit turnover of partners for lesbians and reduce their need to "make the gay scene" to search for partners. Still, despite long-lasting relationships, lesbians experience many problems (Blumstein and Schwartz, 1983).

This explanation has cited lesbians' less compelling need for subcultural support than male homosexuals experience, but it does not suggest that lesbians experience essentially no significant deviance-related problems. Lesbians must sometimes deal with issues of homophobia and heterosexism, as well as general sexism in society that all women must confront (Dooley, 1986). The stigma against lesbianism compounds instances of discrimination known to all women in their search for occupational success, housing, and other social benefits. This combination can result in substantial social rejection. Male homosexuals also tend to live in more comfortable economic situations than lesbians do; in fact, some lesbian couples encounter pronounced economic hardship.

The homosexual rights or gay liberation movement has actively encouraged gay males to proclaim and defend their civil and human rights. The resulting public uproar has largely overshadowed a similar effort among lesbians. Adam (1987: 92) observes:

> From the beginning of gay liberation, lesbians often found themselves vastly outnumbered by men who were, not surprisingly, preoccupied with their own issues and ignorant of the concerns of women. Many women became increasingly frustrated as gay liberation men set up task groups to counter police entrapment, work for sodomy law reform, or organize dances that turned out to be 90 percent male.

In an effort to address some of the problems of their daily lives and their lack of political power, certain lesbians have sought and received benefits from membership in a general homosexual community (Simon and Gagnon, 1967). Particularly in large cities, some lesbians tend to congregate in certain bars, usually those also patronized by male homosexuals, and these places facilitate lesbian sexual relationships. Research suggests that lesbians begin to participate in gay community activities around the ages of 21 to 23 years old (Troiden, 1989: 59). This community can also provide contacts for females without ongoing relationships as well as social support. In such a milieu, a lesbian can express herself fully and openly with others who share feelings and experiences much like her own. Like the male homosexual subculture, lesbians' interactions call on a special language and ideology that encourage common attitudes and rationalizations that help members to resist society's stigma.

One important requirement for the development of a lesbian subculture is economic independence for its members. This condition benefits all women, but it proves especially important for those lesbians who need access to resources that allow them to function in society; women who depend on men for money seldom display openly gay behavior and identities (Miller, 1992).

As noted earlier, homosexual subcultures arise to meet the unique personal and social needs of their members. Because lesbians feel less powerful stigmas and rely less on organized subcultures for social support, lesbian subcultures form in smaller numbers and remain less organized than those for gay males. Few female groups have organized outside their own regions, and only one—the Daughters of Bilitis—has any claim to national representation. A number of local lesbian organizations have, however, begun to promote educational goals and provide counseling (Simpson, 1976). Often such groups maintain associations with those organized for male homosexuals.

Homosexual women established early affiliations with the broader women's movement that began in the 1970s, further reducing their need for a distinct, well-developed subculture. The women's movement advances the political interests of lesbians along with those of heterosexual women, greatly reducing the need for separate organizations to perform this function. The women's movement has provided important support, including new ideas about female sexuality. Among other attitudes, it has suggested female-female associations as welcome alternatives to unsatisfying heterosexual relationships. While the ideology of the women's movement does not overtly promote sexual experimentation, membership in related organizations and commitment to the shared ideology may lead to such encounters. In fact, a woman may feel the need to have at least one sexual experience with another woman in order to widen her perspective on sexual and political liberation. One woman reported: "I wanted to go to bed with a radical lesbian; I just had to know what it was like" (Blumstein and Schwartz, 1974: 287). Some relatively militant advocates of women's rights have even suggested that heterosexual relations are politically incompatible with the ideology of the movement:

> [T]he purpose of feminist analysis is to provide women with an awareness of their servitude as a class so that they can unite and rise up against it. The problem now for strictly heterosexually conditioned women is how to obtain the sexual gratification they think they need from the sex who remains their institutional oppressor. It is the lesbian who unites the personal and political in the struggle to become freed of the oppressive institution. (Johnston, 1973: 275–276)

By the 1990s, observers could identify two kinds of lesbian feminists: "those who have found a place for themselves somewhat apart from but somewhat connected to mainstream America and the more radical women known as separatists" (Cruikshank, 1992: 159). Separatists dissociate themselves from men, including gay men, in every way possible; they also prefer to avoid associations with heterosexual women. They regard the aim of lesbianism not as an expression of sexual identity or a source of personal satisfaction or happiness, but as a particular, overt political stance. Some may regard the objective as the overthrow of patriarchy, not simply an expression of an alternative sexual orientation or lifestyle.

As the 21st century begins, gay and lesbian organizations continue to make political gains. Antidiscrimination initiatives now commonly appear on statewide election ballots, and some groups are pressing for legal acknowledgment of gay and lesbian marriages. This activity may result in increased tolerance for male and female homosexuality or a backlash from political and moral conservatives. Possible backlash can take the form of proposed legal changes that prohibit same-sex marriage or deny rights and benefits to gays because of their sexual orientation. In June 1997, the Southern Baptist Convention, the largest single Protestant denomination in the United States, voted to boycott the Disney company. The action was prompted by Disney's earlier decision to provide health care benefits to "partners" of employees, thus recognizing both homosexual and heterosexual relationships for receiving company benefits.

AIDS and the Homosexual Community

Homosexual communities have experienced great change in recent years. Subcultures organized to resist stigma from nonhomosexuals and laws with discriminatory effects must now confront a new threat with a more internal origin: a disease known as *acquired immune deficiency syndrome* (AIDS). Obviously, a diagnosis of AIDS does not in itself indicate deviance, and this section's discussion focuses on exploring an important force for change in gay communities rather than comprehensively documenting the illness.

Since the beginning of the AIDS epidemic, an estimated 50 million individuals worldwide have been infected with HIV, of whom more than 33 million are still alive and more than 16 million have died (World Health Organization, 1999).

The Disease and Its Transmission

People with fully developed AIDS suffer from strings of unusual, life-threatening infections and rare forms of cancer (American College Health Association, 1987). The virus that causes AIDS also produces a set of milder but often debilitating illnesses called *AIDS-related complex* (ARC), characterized by enlarged lymph nodes, chronic fatigue, fever, weight loss, night sweats, and abnormal blood counts. The disease progresses at different rates in different people. Symptoms may show up years after initial exposure to the virus, and some people may not develop any symptoms after such an exposure.

Many ARC patients improve without treatment, and others develop AIDS itself. A single virus causes both AIDS and related conditions: human immunodeficiency virus (HIV). This virus can live and reproduce itself only inside living cells. Infected people can spread it through its presence in certain body fluids, notably blood, semen, and vaginal secretions. The only tests available for AIDS evaluate blood for the presence of an antibody to HIV rather than testing directly for the disease.

Over half of AIDS patients are men exposed to HIV through unprotected sex with other men. An additional 25 percent are intravenous drug users exposed by sharing needles with infected people, and 7 percent fit both categories. A small percentage of AIDS patients became infected through heterosexual contact, most of them males. The bulk of newly infected patients have contracted the disease from sex with drug users. Estimates suggest that less than 1 percent of all U.S. adults are infected with the AIDS virus. (See Figure 15-1.)

AIDS shows a relationship with race and class, largely because of the connection between race and certain kinds of drug use. As one observer put it:

> In New York City, more than half of the adults with AIDS are African-Americans and Hispanics, largely as a consequence of the racial composition of intravenous drug users. Nine out of ten children who died in 1987 of AIDS in New York were minority children. The Centers for Disease Control has reported that a black child is fifteen times more likely to be born with AIDS than a white child. (Price, 1989: 65)

AIDS is a problem throughout the world (see Table 15-2). This fact led to the formation in January 1996 of the Joint United Nations Program on AIDS

FIGURE 15-1
Aids Risk by Category

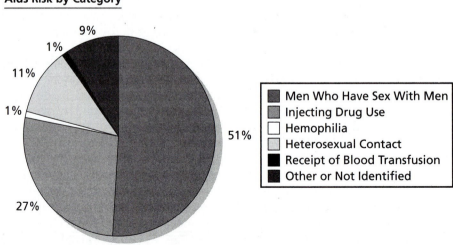

■	Men Who Have Sex With Men
■	Injecting Drug Use
□	Hemophilia
■	Heterosexual Contact
■	Receipt of Blood Transfusion
■	Other or Not Identified

Source: Centers for Disease Control, 1999. *HIV/AIDS Surveillance Report,* Vol. 11, No. 1.

TABLE 15-2
December 1996 World AIDS Estimates

Persons living with HIV/AIDS	22.6 million
New HIV infections in 1996	3.1 million
Deaths due to HIV/AIDS in 1996	1.5 million
Cumulative number of HIV infections	29.4 million
Cumulative number of AIDS infections	8.4 million
Cumulative number of deaths due to HIV/AIDS	6.4 million

SOURCE: Joint United Nations Program on HIV/AIDS, December 1996.

(UNAIDS), which coordinates several U.N. health and policy efforts on AIDS. Several countries in Africa have experienced substantial numbers of AIDS cases. For example, 62 percent of the world's AIDS cases come from sub-Saharan Africa. Another 23 percent come from southern and southeastern Asia, while nearly 4 percent lived in North America at the beginning of 1997. In contrast with conditions in the United States, AIDS spreads in Africa mainly through heterosexual intercourse, facilitated by long-neglected epidemics of venereal diseases that promote viral transmission.

Sub-Saharan Africa is the epicenter of the epidemic. Here, 55 percent of infected adults are women, which means more than 6 HIV-positive women for every 5 HIV-positive men. UNAIDS and WHO estimate that 12.2 million African women and 10.1 million African men aged 15–49 were living with HIV at the end of 1999 (World Health Organization, 1999). Life expectancy at birth in southern Africa, which climbed from 44 in the early 1950s to 59 in the early 1990s, is expected to drop back to 45 sometime between 2005 and 2010.

The world's steepest HIV curve in 1999 was recorded in the newly independent states of the former Soviet Union, where the proportion of the population living with HIV doubled between 1997 and 1999. In the larger region comprising these nations and the remainder of Central and Eastern Europe, the number of HIV-infected rose by more than a third in 1999 alone, to reach an estimated 360,000.

Medical science offers no cure for AIDS, but appropriate precautions can often prevent infection. In fact, people cannot easily contract AIDS. HIV is a very fragile virus spread only through just the right conditions. Infected people cannot transmit it by casual contact. Rather, AIDS is transmitted by intimate, usually sexual, contact and by exposure to contaminated blood. Some sexual activities involve more danger of infection than others. Anal intercourse exposes the partners to especially high risk, but observers have found a low risk of AIDS transmission via vaginal intercourse (Brody, 1997). Oral sex, on the other hand, is the kind of sexual contact perhaps least likely to transmit AIDS.

The first reported cases in the United States in 1981 afflicted homosexual men. Researchers first identified the human immunodeficiency virus in 1983. Estimates

✳ *Issue* | **Variations in AIDS Occurrence**

The United Nations Joint Program on AIDS reports that in North America, although there has been an overall slowing in the increase in AIDS incidence, there has been substantive variation in the populations affected. For example, in the United States, the increase in AIDS incidence in the 1990s has been greatest for women compared with men, African Americans and Hispanics compared with whites, and persons infected through heterosexual contact compared with those infected through other modes of transmission. As a result of these trends, AIDS incidence in 1995 was 6.5 times greater for African Americans and 4 times greater for Hispanics than for whites, 20 percent of persons diagnosed with AIDS were women, and 15 percent were infected heterosexually.

The HIV infection rates are also high among certain groups, such as incarcerated persons. In 1994, 2.3 percent of nearly 1 million prisoners in the United States were known to be infected with HIV, the rate of AIDS among prisoners was 7 times the rate of the nonincarcerated population, and AIDS was the second leading cause of death among prisoners.

Among Canadian prisoners, HIV prevalence is higher in women, between 2 and 10 percent versus 1 to 4 percent for men; for both sexes, transmission is primarily related to use of injected drugs.

in 1993 placed the number of people in the United States infected with the AIDS virus at about 1.5 million, all of them capable of spreading the disease through sexual contact or by sharing needles. Of this number, perhaps 20 or 30 percent will develop full-blown AIDS within 5 years. Out of 161,073 AIDS cases reported in the United States since June 1981, 100,777 people had died as of the end of December 1990 (*Des Moines Register*, January, 25, 1991, p. 7A). The Centers for Disease Control estimate that as many as 215,000 people will die from the disease in the next 3 years.

Patients with AIDS may require little medical intervention or extensive treatment, depending on the stage of the disease. But all AIDS patients, regardless of their stages in the disease, feel emotional and psychological needs, and these needs may extend to their friends and families as well. Health-care professionals are becoming increasingly sensitive to these psychological needs, but society still has much to learn about the psychological and social impacts of AIDS (Baum and Temoshok, 1990).

Impact of AIDS

Many people remained unaware of AIDS until October 2, 1985. On that day, almost every household in the Western world became familiar with the disease through news of the death of movie star Rock Hudson. AIDS was not unknown before that time; indeed, Hudson became one of some 12,000 people who had died or were dying from the disease. Rather, until Hudson's death, AIDS had seemed little more than a localized affliction largely confined to a few marginal groups in society. The death of a famous movie star raised the status of the disease and made it something that might affect anyone.

Soon after, other celebrities received AIDS diagnoses, although many felt too embarrassed to admit it (Shilts, 1987: 585–586). Initial reports indicated that Broadway choreographer Michael Bennett suffered from heart problems. A representative of designer Perry Ellis described him as dying from sleeping sickness. Lawyer Roy Cohn, who gained fame during the McCarthy hearings, insisted he had liver cancer, while conservative fund raiser Terry Dolan said he was dying of diabetes. As entertainer Liberace lay on his deathbed, a representative said he was suffering ill effects of watermelon poisoning. Only later, when the disease became more prevalent and others became willing to discuss it, would victims admit to infection with AIDS.

Since those early days, a rapidly growing number of people, including celebrities, have admitted having AIDS. The surprise over Rock Hudson's death became shock when one of the best-known professional athletes of his day, Ervin "Magic" Johnson, admitted having tested HIV positive. Johnson's disclosure revealed the disease as a potential affliction of young people; wide audiences could identify with him and through him with all AIDS patients. The impact of this announcement multiplied through publicity that the infection might prevent Johnson from becoming a member of the U.S. Olympic basketball team in 1992 (although he did subsequently play). Zimet and his colleagues (1993) surveyed more than 100 junior high school students about their awareness and knowledge of AIDS following Johnson's announcement and the attendant publicity. They found that the news had spread awareness of AIDS among these students and that nearly 60 percent wanted to learn more about the disease. Further, 21 percent reported increased anxiety about AIDS compared with only 12 percent who felt less anxiety.

Since homosexuals make up about two-thirds of U.S. AIDS patients, one might predict an enormous impact of the disease on gay life. Observations of homosexual activities in many cities, reports from homosexuals, and input from other sources support this expectation.

Impact on Sexual Behavior   One study has reported some change in the sexual behavior of gays caused by fear of AIDS, especially reductions in casual sexual contact (Quadland and Shattis, 1987). The biggest change took the form of a reduction in sexual relations among homosexuals, previously unknown to one another, who met primarily in gay bars and other places where gays congregated for the purpose.

The study offered no evidence, however, of a reduction in monogamous sex or sexual relationships in general. Participants in couple relationships reported no changes in their sexual behavior as a result of AIDS. These results suggest the impacts of the extensive programs of sex education undertaken in some communities whose members face high risk of AIDS. Similar results have been reported in studies of homosexual behavior in other communities (Coates, Stall, and Hoff, 1990).

At the same time, some research also suggests that gay men tend to underestimate the risk of sex without condoms. One study reported that 83 percent of gay males who reported engaging in at least one high-risk behavior in a typical month continued to evaluate their behavior as relatively safe (Bauman and Siegel, 1990). The researchers attributed this high percentage to a misperception that sex with a limited number of known partners would greatly reduce the risk of AIDS. While facts confirm this general perception, limited numbers and known partners still leave intact the absolute risk of unprotected anal intercourse. In another study of 92 gay couples, Berger (1990) found that virtually all described their relationships as "monogamous," but only about half practiced safe sex.

Another study in 1991 determined that 31 percent of a sample of gay and bisexual men from 16 small cities and towns reported having engaged in unprotected anal intercourse within the previous 2 months. Further, research by the San Francisco Department of Public Health that same year found that 43 percent of homosexual men aged 17 to 19 admitted to anal intercourse without condoms (Signorile, 1993: 24). Among African-American men, the number may rise as high as 53 percent.

Perhaps even more disturbing is the deliberate practice of unprotected sex ("barebacking") in an attempt to contract AIDS. Gauthier and Forsyth (1999) provide information on "bug chasing," the practice of unprotected sex between HIV-positive and HIV-negative men, with the latter knowingly seeking infection from the former. Making contact in chat rooms on the Internet, these men offer complex motivations for their behavior. One participant, for example, indicates that

> For years now, poz [HIV-positive] men have had to bear a double burden, the burden of the virus itself and the burden of preventing the spread of the disease. Correctly, I believe, many of us have come to the conclusion that within ourselves there is little need to practice this self imposed sexual martyrdom. . . . For us, it's too late. So why shouldn't we party? (Gauthier and Forsyth, 1999: 90)

Some barebackers believe that latex ruins intimacy, while others display a fatalistic attitude and believe that infection will come to them regardless of their sexual practices. "I think about AIDS," one barebacker reported, "and the possibility of contracting HIV but that fear is not nearly as strong as the fear of not being with the man I love" (Ames, Atchinson, and Rose, 1995: 70). For some, AIDS has become destigmatized in gay communities and is a symbol of belonging, a special status. Some "bug chasers" may be motivated by the special status of being HIV-positive, which may have become a special badge of membership in the gay community.

Political Impact   The other major impact of the AIDS epidemic appeared in political life. From the early days of 1980, when the first gay men began to fall ill, 5 years passed before major health-care institutions recognized and began dealing with the outbreak as a serious crisis. Medical services, public-health institutions, federal and private research establishments, mass media organizations, and the gay community's leadership all recognized the problem only after a substantial delay (Shilts, 1987). AIDS research and medical facilities for victims waited until late 1985 for funding, and the surgeon general of the United States initiated action during the following year.

In the early days of the outbreak, AIDS maintained a reputation as a local problem largely confined to deviant groups and without much effect on the larger society. Only when heterosexuals began to develop AIDS did official action begin. In 1986, Dr. C. Everett Koop, the U.S. surgeon general at the time, decided to address the problem directly after interviewing many people. Koop recommended avoiding mandatory testing (which he thought would scare away those most at risk) and called for a federally funded research program together with emphasis on preventive tactics such as condom use and limiting sexual contacts within monogamous relationships. Koop's program finally initiated a major public-health movement directed at the disease.

Public opinion polls indicated widespread public awareness of the AIDS problem. A Gallup poll in November 1987, 2 years after Rock Hudson's death, found that 68 percent of Americans identified AIDS as the country's most serious health problem (*New York Times*, November 29, 1987: 6). The next largest segment, 14 percent, indicated cancer as the most serious health problem, and 7 percent cited heart disease. Not only homosexuals, but most people, have taken precautions to avoid AIDS. According to the Gallup poll, more than half of American adults (about 55 percent) described taking specific precautions, such as using condoms, carefully choosing sexual partners, or avoiding blood transfusions if possible. Sexually active young people may not show evidence of such concern, however.

Although no one can identify all consequences of the AIDS epidemic, one seems clear: The spread of AIDS has produced a backlash of public opinion about homosexuality. Public opinion polls indicate highly stigmatizing attitudes toward both AIDS patients and gay men (St. Lawrence, Husfeldt, Kelly, Hood, and Smith, 1990). Some members of the public blame homosexuals for the disease, and this attitude may also prevail among some health-care workers, including

---

❊ *Issue* | AIDS Overwhelms South Africa

South Africa is so overwhelmed by the AIDS epidemic that some public hospitals are turning people away, limiting treatment and forcing physicians to choose whom to save, according to an Associated Press report (*Omaha World-Herald*, January 9, 2000: 16-A). One out of every 10 people in South Africa has AIDS, and the medical system is simply not up to the challenge of providing able and complete treatment for all.

At some hospitals, babies who are HIV positive are admitted for treatment only once. The head of the pediatrics department there is quoted as saying, "When a baby gets bad, like with pneumonia, we won't admit it for the second time but will tell the mother to take it home and let it die."

"We can't afford to spend money on people who are going to die anyway," another staff member said. Limited resources mean there are few alternatives for some patients. Funds are in short supply; the government of South Africa spends less than $85 per person on medical and health care. The country has appealed for outside assistance.

doctors and nurses. One study categorized images of homosexuality in 59 articles from medical periodicals (Schwanberg, 1990). The largest proportion (61 percent) represented negative connotations, a substantial change from a previously neutral position.

For another clear impact, AIDS undoubtedly represents a personal tragedy for thousands of people, homosexual and heterosexual, both direct victims and their friends and family members. Kelly (1990) describes a "stigma fallout" that spreads public disapproval among intimates and close friends and associates of AIDS victims who are not themselves infected with the disease. These secondary victims, some of them gay, may tend to abandon an AIDS sufferer who shows unmistakable signs of the illness. This tendency limits access of many AIDS sufferers to supportive social networks during the time they really need the help. Regardless of the other policy consequences, AIDS clearly has taken a prominent place in the national policy agenda.

Current Controversies Over Homosexuality

Conflicts over the reasons for and implications of homosexuality have spilled over into the courts and other forums within society. Two recent issues involve whether states should legally recognize marriages between people of the same sex and whether the law should regard homosexuality as a legally protected status.

Same-Sex Marriages

In 1990, a lesbian couple in Hawaii decided to get married. Like other marrying couples, they were in love, but they wanted legal recognition of their union in order to extend coverage by one woman's medical insurance to the other. Only married people could receive such coverage. When the state denied the couple's application for a marriage license, they filed suit in court. The following year, the Hawaiian legislature passed a law defining marriage as a union only between a man and a woman, and the state supreme court delayed rehearing the case until a special commission issued a report. In December 1995, that commission recommended legalizing same-sex marriages because members could identify no good reason not to recognize them.

No case concerning gay marriages has progressed so far in any other state or federal court system. Although the case originated in Hawaii, the issue reached national significance because the full faith and credit clause of the U.S. Constitution virtually requires other states to recognize legal Hawaiian marriages. In other words, a couple legally recognized as married in Hawaii can claim married status in the other 49 states as well, unless those states pass laws specifically withholding recognition and the courts allow such statutes to stand. The same principle affects divorces; a divorce granted in one state is recognized in all other states.

Some states responded to the Hawaiian court ruling by introducing legislation withholding recognition of same-sex marriages. Utah, despite its long history of nontraditional marriages, passed a law in 1995 that denies recognition of marriages that do not conform to Utah law. In 1996, South Dakota defined marriage in that state as a union between a man and a woman. A number of states considered similar legislation, but the outcomes remained far from certain. Supporters have justified legislation of this sort with the rationale that the state has an interest in encouraging heterosexual marriages as the best guarantees of satisfactory environments for raising children. Related legislative activity will likely continue throughout the 1990s.

National opinion surveys reveal considerable opposition to same-sex marriages. A poll conducted by NBC News in 1995 reported that 60 percent disapproved of legal recognition for same-sex marriages, while just 26 percent approved. The rest of the respondents were undecided. A Gallup poll released on April 4, 1996, obtained similar results. Overall, two-thirds of respondents indicated that the law should not allow same-sex marriages, compared with 27 percent who favored legal recognition. Even among those who regarded homosexuality as an "acceptable lifestyle," 43 percent opposed legal provisions for homosexual marriages.

These polls reflect differences between civil and religious marriages. Civil marriage is a legally protected relationship available to all within a political jurisdiction who meet constitutional standards and procedures. Religious marriage represents church recognition of a union subject to the rules and policies of individual churches and denominations as well as those of the state. The boundaries of acceptable civil marriage have changed considerably throughout U.S. history. At one time, laws did not permit African-American and Asian-American couples to marry at all; until 1967, laws in some states allowed criminal prosecution of those who married outside their own race; in earlier times, women implicitly became the property of their husbands when they married. Civil marriage provisions have changed all of these policies. Still, restrictions remain; the Roman Catholic church does not recognize a second marriage after divorce, although civil law does provide for such a marriage. Also, many gay couples have held marriage ceremonies in churches and temples in many states. The Hawaii case does not determine whether individual churches will recognize same-sex marriages, but whether they will gain civil acceptance.

Most objections to same-sex marriages reflect negative social attitudes toward homosexuality. Further, some people worry that official recognition of these marriages might constitute society's endorsement of homosexuality and perhaps also sodomy. Such a concern, of course, assumes that a state marriage license represents a gesture more important than merely filling out a form and paying a fee. States issue licenses to ride bicycles on public streets, chauffeur auto passengers, or run restaurants without directly endorsing those activities; the licensing requirements merely attempt to establish order and control over them. While some may debate whether legal acceptance of same-sex marriage would make society more civilized, as argued by Eskridge (1996), states certainly require more effort to obtain driver's licenses than to obtain marriage licenses. Candidates wait in longer lines, and they must pass written and driving tests. States establish no such concern or barriers to protect their interests in marriage.

Many people find a persuasive argument for same-sex marriage in the equal protection clause of the Fourteenth Amendment. This constitutional provision guarantees "equal protection under the law" to all citizens. Some interpret this guarantee as allowing governments to prohibit marriage among consenting adults only if they can show that the restriction serves a "compelling state interest." Such a prohibition raises the issue of sex discrimination, a violation of the guarantee of equal protection, since it limits an individual's choice of marriage partners. A woman cannot marry a woman, and a man cannot marry a man. According to Appiah (1996: 54):

> Some will object that this is preposterous: the current law treats men and women equally in requiring both to marry someone of the other gender. But, by that line of reasoning . . . we could defend anti-miscegenation laws: for all those require both whites and blacks to marry within their "races."

Clearly, U.S. society has long rejected this line of reasoning, but the legal system has not yet fully embraced its implications for homosexuality.

Homosexuality as a Protected Status

The law gives access to powerful resources that can confer important political advantages on selected groups. No one should feel surprised, then, that some have applied it to restrict and regulate homosexual conduct. Of course, homosexuals can try to apply legal resources, just as other groups do, by advocating antidiscrimination legislation. Such legislation assumes unmistakable importance to those affected by it.

Civil rights do not confer absolute protections; instead, these provisions require interpretation within moral and political contexts (Dworkin, 1977). Therefore, one group's protection of its right may look like deprivation of a right for another. On November 3, 1992, Colorado voters found a referendum on statewide ballots that would, if passed, add this amendment to the state's constitution:

> Neither the State of Colorado, through any of its branches or departments, nor any of its agencies, political subdivisions, municipalities or school districts, shall enact, adopt or enforce any statute, regulation, ordinance or policy whereby homosexual, lesbian or bisexual orientation, conduct, practices or relationships shall constitute or otherwise be the basis of or entitle any person or class of persons to have or claim any minority status quota preferences, protected status or claim of discrimination.

The amendment passed by a vote of 53 percent to 47 percent, effectively barring all units of state and local government from providing any protection against discrimination to homosexuals, lesbians, and bisexuals. Passage of the amendment led to substantial legal maneuvering. Ruling on a suit asking for a decision declaring the amendment unconstitutional, a state court invalidated the amendment. This decision prevented the amendment from invalidating ordinances or policies prohibiting discrimination based on sexual orientation in Denver, Boulder, and Aspen.

The debate over the amendment centered on the question of whether homosexuals should enjoy legally protected status. Supporters argued that the amendment merely denied "special rights" to homosexuals and put them in the same legal and social position as everyone else. Members of the homosexual community and other critics of the amendment argued that it represented unjust and unconstitutional discrimination since it blocked homosexuals, but no other group, from legal protections for their housing, insurance, health benefits, welfare, private education, and employment.

The case reached the Colorado Supreme Court (see *Evans v. Romer*, 854 P.2d 1270, Colo. 1993), where the justices ruled that the amendment violated the U.S. Constitution because

> [T]he Equal Protection Clause of the United States Constitution protects the fundamental right to participate equally in the political process. . . . Any legislation or state constitutional amendment which infringes on this right by "fencing out" an independently identifiable class of persons must be subject to strict judicial scrutiny.

The state appealed this decision to the U.S. Supreme Court, which ruled in a 6-to-3 vote on May 21, 1996, that the Colorado amendment did indeed violate the national Constitution. This decision, of course, did not settle the larger social issue of whether homosexuals constitute a group in need of special protection, the same one that underlies the same-sex marriage case in Hawaii. The court in the Hawaiian case argued that the lesbian couple should be permitted to marry, among other reasons, because homosexuals constituted an oppressed group to whom the state could not deny equal protection of the law, in this case the marriage law (*The Wall Street Journal*, June 19, 1996: A5).

A related but different issue arose in the state of Vermont in 2000. The issue had to do with whether health and insurance benefits that are available for married couples should be provided for the partners of gay people. Actually, before this time a number of both private and public organizations had decided to provide benefits to gay partners, but the issue had been legally challenged in Vermont. The state's Supreme Court ruled that benefits extended to married heterosexual couples could not be denied to gay couples under state law. Continued legal action and social debate are virtually guaranteed.

Summary

Homosexuality encompasses both specific kinds of sexual behavior and a more general sexual orientation. Homosexual behavior involves sex relations between members of the same gender. A homosexual orientation is an attitude of preference for sexual gratification through contact with members of one's own gender. Many more people participate in homosexual behavior than ever develop homosexual orientations. Normative codes and laws have prohibited homosexuality from ancient times to the modern day, but public opinion in recent years has moved toward an increasingly accepting attitude. Homosexual organizations in many U.S. states and cities have successfully pressed for antidiscrimination legislation, although a conservative backlash has threatened stiffer sanctions in some areas. While some deny that homosexuality is deviant, it continues to draw significant public stigmatization, and most people in the United States probably consider homosexuality as an example of deviance.

A person develops a homosexual orientation within a biological context, but its full meaning becomes evident only through analysis of the sexual socialization process in which individuals acquire and identify with sex roles. Recent research has attempted to identify a biological component of sexual development. Studies working to locate particular genetic structures associated with homosexuality have offered suggestive but not conclusive evidence. Sociologists recognize sexual socialization as a complex process that begins with initial learning of very complicated norms. Sexual norms identify appropriate objects, times, places, and situations for sexual activity. The myriad possible combinations of these contingencies create substantial opportunities for errors in socialization, especially when the embarrassment linked with the topic inhibits candid guidance from many socializers to young people experiencing sexual development.

People appear to form their sexual preferences by adolescence, although a few homosexuals report participating in related behavior only later in life. They acquire homosexual identities even later, after a process of increasing homosexual behavior, perhaps leading to participation in a homosexual subculture or community. The coming out process entails a series of stages that represent increasing commitment to homosexuality. Sociologically, a homosexual is someone who has adopted a homosexual identity.

Lesbians acquire their sexual orientation in the same general manner as male homosexuals do—through socialization in lesbian norms and values. More lesbians than gay males develop their sexual orientation in the context of friendship relationships with others of their own gender. In other words, lesbian sex takes place more often than sex between male partners in the context of ongoing social relationships. Lesbian relationships commonly last over considerable durations, as do some unions between male homosexuals. However, research suggests that lesbians participate in casual sex with strangers less frequently than male homosexuals do, and they also participate less actively in generally less fully developed lesbian subcultures compared with gay males. Lesbianism generally defines a more private role. Lesbians, therefore, have less to fear from any social stigma or legal sanctions because their behavior and orientation remain less visible to others. For these reasons, lesbians feel less need than gay men experience for the supporting atmosphere of homosexual subcultures.

The AIDS epidemic will continue to generate changes in gay communities and the lives of homosexuals. Some evidence indicates that the threat of AIDS has changed the behavior of many homosexuals. It has reduced the frequency of casual sex, but the behavior of homosexuals committed to monogamous relationships shows few changes. The federal government, including public health officials and researchers, began to address the disease only in 1986. Revelations about its effect on celebrities and heterosexuals other than intravenous drug users elevated AIDS to the status of a national problem.

Selected References

Bell, Alan P., and Martin S. Weinberg. 1978. *Homosexualities: A Study of Diversity among Men and Women*. New York: Simon and Schuster.

This study reports tremendous diversity among homosexuals, their backgrounds, the extent of their commitments to homosexual subcultures, and involvement in a homosexual lifestyle.

Browning, Frank. 1994. *The Culture of Desire*. New York: Simon and Schuster.

This journalistic account chronicles the emergence of the gay movement over the past two decades. Browning records the major changes in the drive toward social equality and records an important diversity of opinion among gays on the best way to attain political and social goals.

Greenberg, David F. 1988. *The Construction of Homosexuality*. Chicago: University of Chicago Press.

This book focuses on the central theme of how homosexuality gained its identity as deviance. No serious student of homosexuality can ignore this major reference on the subject, with its wide scope and impressive detail.

Kelly, Robert J. 1990. "AIDS and the Societal Reaction." Pp. 47–61 in *Perspectives on Deviance: Dominance, Degradation, and Denigration*. Edited by Robert J. Kelly and Donal E. J. MacNamara. Cincinnati, OH: Anderson.

This sociological account explores the association between the stigma of AIDS and that of homosexuality.

Michael, Robert T., John H. Gagnon, Edward O. Laumann, and Gina Kolata. 1994. *Sex in America: A Definitive Survey*. Boston: Little, Brown.

This book reports on a poll of a national, representative sample of U.S. adults concerning their sexual attitudes and practices. While it presents a popular summary of the survey results rather than a scholarly discussion, the book gives much information regarding gay and lesbian sexuality, including the impact of AIDS.

Shilts, Randy. 1987. *And the Band Played On: Politics, People and the AIDS Epidemic*. New York: St. Martin's Press.

This well-written and compelling journalistic account details the emergence of the AIDS problem in the United States. Shilts documents inaction by the federal government and general public in the early stages of the epidemic. Along with its role as a social history, this book also represents a contribution to the sociological literature on social movements.

Troiden, Richard R. 1989. "The Formation of Homosexual Identities." *Journal of Homosexuality* 17: 43–73.

This author describes the process of coming out from a sociological perspective. His system of stages and sequence may not occur invariably for every individual, but they effectively capture the process.

CHAPTER 16

Mental Disorders

In December 1999, the U.S. Surgeon General's Office released an extensive report on mental health and mental illness. Among the conclusions of that report were the following:

- About 1 in 5 Americans experiences a mental disorder in the course of a year.
- Approximately 1 in 5 children and adolescents experience the signs and symptoms of a disorder during the course of a year.
- Fifteen percent of the adult population use some form of mental health service during the year.
- In 1996, the direct treatment of mental disorders, substance abuse, and Alzheimer's disease cost the nation $99 billion; direct costs for mental disorders alone totaled $69 billion. In 1990, indirect costs for mental disorders alone totaled $79 billion.

The report also indicated that the stigmatization of people with mental disorders has persisted throughout history to the present time. Persons with mental disorders must deal not only with their disorder, but also with the negative reactions of persons around them.

The Stigma of Mental Disorders

Society will likely impose strong sanctions on individuals known or suspected to suffer from mental disorders, depending on the specific behaviors that these people exhibit. Many people will certainly react differently to someone with a mental disorder than to someone with a physical illness. Most people feel generally sympathetic toward patients with physical illnesses because they understand and recognize the symptoms those patients display or because their own past experiences help them to identify with such afflictions. Mental disorders, on the other hand, may involve intangible and potentially frightening feelings and ideas beyond the ability of others to comprehend.

Attitudes toward the mentally ill range from avoidance to ridicule and revulsion. Stigma is also manifested by bias, distrust, stereotyping, fear, and embarrassment. People with mental disorders may experience difficulty in gaining access to housing and jobs, resulting in low self-esteem, isolation, and hopelessness. Public attitudes about mental illnesses have changed over time, but these changes have not resulted in a lessening of stigma (U.S. Surgeon General, 1999: 7). Surveys show that increasingly, people are able to distinguish mental disorders from ordinary unhappiness and worry. People now are also more likely to attribute mental disorders to a mix of physical and social abnormalities. In spite of this greater understanding, however, social stigma against people with mental disorders is in some respects stronger now than it was 50 years ago.

People with mental disorders have been stigmatized for a number of reasons. First, the prevailing attitude tends to hold people with mental disorders as somehow responsible for their own conditions while viewing sufferers with physical illness as victims of circumstance. Second, the system of care for those with mental disorders has been separate from that for the care of those with physical illnesses. Third, there is a prevailing view that people with mental disorders are violent and otherwise disruptive or unpredictable. In fact, the public increasingly seems to believe that people with mental disorders, especially serious disorders, are more violent than in the past. Selective media reporting that continues to link violence with mental disorders reinforces this fear of violence.

A person with a mental disorder may display behavior inappropriate to the situation and even commit bizarre acts. A mental disturbance may prevent someone from meeting normal expectations in particular situations or everyday life. These social expectations, or norms, form the basis for any assessment of deviance and the subsequent administration of social sanctions. Even mild mental disorders may draw society's sanctions, perhaps in the form of labels such as *mentally ill, insane,* or *crazy,* while more extreme cases may even lead to commitment to a mental hospital. While most observers regard a mental disorder as a serious form of deviance because it affects others in addition to the person with the disorder, extremely difficult problems of social definition hamper efforts to further understand the nature of mental disorders and their causes and cures. As sociologists and social anthropologists have paid increasing attention to this topic, they have come to implicate social and cultural causes to explain the origins of mental disorders and their distributions from group to group and place to place.

A full understanding of the broad sociological perspectives on this form of deviance requires some understanding of the various types of mental disorders. Some derive from organic or physical origins, while others show stronger relationships to social situations, such as reactions to stress.

Psychiatric View of Mental Disorders

Psychiatrists have not adopted any single approach to all mental disorders. They continue to dispute, for example, whether mental disorders result from physical causes, which would make them exclusively medical problems, or whether such disorders are the result of environmental causes. Some psychiatrists regard conditions that society calls *mental disorders* as basic "problems in living" or instances of socially aberrant behavior (Szasz, 1974). This view does not deny that such people may act oddly, but it refuses to treat behavior called *insanity* as a medical problem (Szasz, 1987). Other psychiatrists class mental disorders as medical problems as real as any other physical afflictions, so they advocate a medical model or psychiatric approach to their treatment (Roth and Kroll, 1986). Although a great deal of confusion persists throughout the psychiatric profession about the identity of mental disorders as disease, medical training predisposes most psychiatrists to think in these terms and to look for medical diagnoses and treatments (Goffman, 1959: 320-386).

For these reasons, psychiatrists do not always agree among themselves about diagnoses. Many people classed as normal sometimes exhibit behavior that some observers might consider evidence of disorders. Probably everyone can remember some experience with behavior like hallucinations, phobias, persecution complexes, and emotional extremes of elation and depression. Almost everyone at one time or another has experienced irrational fears, daydreams, flights of ideas, and lapses of memory. By themselves, such experiences do not reflect mental disorders; in combination with other experiences or when they interfere with basic social processes, however, these behaviors may contribute to psychiatric diagnoses of mental disorders. Unfortunately, psychiatric diagnoses have achieved a poor record of reliability, since different psychiatrists may evaluate the same or similar symptoms and reach different diagnoses.

The Psychiatric Diagnostic Manual

In an effort to limit the range of shifting definitions and establish reliable criteria for accurate diagnoses, the American Psychiatric Association (1994) has developed the *Diagnostic and Statistical Manual* (known as *DSM-IV* to denote its fourth edition). This manual represents an ongoing attempt to standardize psychiatric diagnoses of specific mental disorders. First published in 1952, subsequent editions have incorporated many changes in psychiatric language over the years. The current edition has dropped the term *neurosis* as a reference to a minor psychiatric ailment that does not interfere with daily functioning; it has also eliminated homosexuality as a diagnostic category of mental disorder. Of course, the words *neurosis* and *homosexuality* continue in common use, but they no longer constitute precise diagnostic categories.

Over the years, the DSM has replaced the expression *mental illness* with *mental disorder*. Rather than providing specific criteria to support precise diagnoses, DSM-IV identifies broad correlates that identify features frequently associated with the disorders it details. In this manner, the manual leads practitioners to conclude not that a person is a schizophrenic, for example, but that he or she displays some typical characteristics of schizophrenia.

Like its predecessors, DSM-IV avoids the flavor of a dispassionate, scientific document. In late 1985, for example, a group of psychiatrists working on a revision that would become DSM-III-R proposed a new diagnosis of *paraphilic rapism* for behavior associated with rape. The conception of rape as a disease deeply upset feminist leaders, who worried that it might justify an insanity defense for rapists, and their prompt protests caused the head of the psychiatric committee to say: "We probably will withdraw the diagnosis of rapism" (Goleman, 1985). Few physicians could even conceive of withdrawing a type of medical diagnosis to avoid the threat of a lawsuit, and this episode does not mark the first time that political considerations have affected the content of DSM. In 1973, lobbying by various groups achieved withdrawal of materials identifying homosexuality as a disorder. Clearly, the characteristics and definitions that designate certain mental disorders have changed in response to political and other influences.

Even after such changes, many psychiatrists remain dissatisfied with other features of the DSM. DSM-IV, for example, recognizes two kinds of depression, major depression and dysthymic depression, each of which covers a variety of syndromes and symptoms. A close examination reveals severe difficulties in identifying the symptoms and treatments for depression. According to one psychiatrist:

> Today's depression classification is as confusing as it used to be 30 years ago. All things considered, the present situation is worse. Then, psychiatrists were at least aware that diagnostic chaos reigned and many of them had no high opinion of diagnosis anyway. Now, the chaos is codified [in the DSM] and thus much more hidden. (Van Praag, 1990: 149; see also Zimmerman, 1988)

The DSM is not a scientific document. It presents a descriptive rather than theoretically validated approach to diagnoses and treatment. As a result, DSM does not provide reliable knowledge about the nature, causes, and cures of most mental disorders.

> In most cases the causes of the conditions are unknown (or disputed) and the treatments uncertain. Diagnoses are grouped empirically in terms of their manifest similarities, but even those diagnoses (like schizophrenia) that are based on much research and experience probably include many different specific entities that lead to similar types of bizarre behavior. (Mechanic, 1999: 15)

Psychiatry traditionally classifies mental disorders according to two types—those derived from organic causes and functional, or nonorganic, disorders. Organic disorders usually trace their causes to specific organisms, brain injuries, or other physiological factors, perhaps including inherited conditions. Clear biological or physiological antecedents justify definition of these disorders as medical conditions. On the other hand, observers identify no organic or physical basis for *functional disorders*, and analysis has not broken them down into such specific categories.

Organic Mental Disorders

Organic disorders trace their origins to identifiable organic or physiological problems. They include senile (old age) psychoses, paresis, and alcoholic psychoses. Some of these cases show associations with arteriosclerosis (hardening of the arteries), which causes poor blood circulation that affects brain functions. These disorders are characterized by loss of memory, particularly for recent events, inability to concentrate, and certain delusional thoughts. Alzheimer's disease is a kind of senile psychosis.

Paresis is a condition caused by syphilis. It develops at least 10 years after the initial syphilitic infection, often leading to progressive brain degeneration in untreated patients. The mental symptoms of paresis often include complete alteration in personality characteristics—for example, a typically neat, well-dressed person may appear slovenly in public. Eventually, memory about time and place becomes defective, and some cases feature depression. Chronic alcoholism in association with vitamin and nutritional deficiencies may sometimes produce such severe physical and psychological deterioration that it causes alcoholic psychoses. Such psychoses resulting from alcoholism, while classified as organic, do not result from definite physical causes. Relatively few alcoholics develop psychoses.

Functional Mental Disorders

The category of mental disorders that most actively interests sociologists are the functional disorders. The name comes from the process through which, according to many psychiatrists, these disorders "function" to adjust individuals to their particular difficulties. No conclusive evidence about most functional mental disorders reliably indicates that they arise from organic causes, although some have implicated conditions such as heredity, physiological disorders, and other organic deficiencies. Unlike organic mental disorders, the lack of a standard test by which to make a diagnosis has severely hampered research on functional mental disorders. The DSM provides particularly appropriate help in classifying and treating functional disorders. Because it encompasses such a large number of disorders, however, this section can outline only a few of the more common ones.

Minor Disorders   The unreliability of diagnoses for functional disorders shows up most clearly in various minor psychiatric conditions, historically called *neuroses*. *Compulsive behavior*, for example, indicates repeated actions over which people think they have little or no control, such as stepping on or avoiding cracks in the sidewalk, excessive hand washing or bathing, counting telephone poles or other objects, dressing in a set manner, and insisting on certain meticulous arrangements, such as leaving all drawers carefully closed or shoes and other objects lined up in a particular order. Neuroses may also appear as *obsessions*, or persistent ideas that often represent emotional fears of objects, acts, or situations. Diagnoses of *phobias* often result from severe anxiety generally associated with certain situations, such as fear of confinement (claustrophobia), fear of open places (agoraphobia), and fear of high places (acrophobia).

Manic-Depressive Behavior   Manic-depressive behavior, sometimes called *bipolar disorder*, consists of attitude swings ranging from extreme elation in the manic stage to extreme depression. Although manic-depressives do not necessarily pass cyclically through stages of mania and depression, they shift very abruptly and noticeably from one mood to the next. In the manic stage, the person talks quickly, moving in a rapid but understandable progression from one topic to another. In the depressed stage, the person engages in extensive brooding, but little serious mental deterioration results. He or she minimizes activity, so frenetic during the manic phase, and talk turns to feelings of dejection, sadness, and self-depreciation. The manic-depressive maintains contact with reality as well as undiminished memory and place–time orientation.

Paranoid Behavior   Paranoia no longer represents a widely used diagnostic category. Mental health professionals now class most paranoid disorders as forms of schizophrenic behavior. Paranoid behavior shows up as extreme suspicion of people and conditions, with ideas of personal persecution reinforced by an intellectual defense that often appears to have some basis in reality. One person's paranoid delusions usually focus on a few areas, and they may even center on a single person. Most paranoids continue to function, and their personalities do not deteriorate over time. They do not hallucinate.

Schizophrenia   Mental-health professionals diagnose schizophrenia more frequently than any other serious functional disorder, and for this reason, it warrants more than a brief mention. Schizophrenia generates no single behavior; it may show up in many ways. Some common symptoms include hallucinations, delusions, diminished ambition, and social withdrawal. These behaviors often cause difficulties for schizophrenics as they try to continue working, studying, or even maintaining close interpersonal relationships.

Descriptions of schizophrenia cite a detachment of the emotional self from the intellectual self, leading to an interpretation of the term as implying a "split personality." Perhaps the most obvious symptoms of schizophrenia are withdrawal from contact with others and inability to play expected roles. The schizophrenic may build up thoughts of an imaginary world, sometimes reinforced by false perceptions and hallucinations of many kinds. The schizophrenic may contend

---

✳ *Case Study*   **A Short Interview With a Schizophrenic**

Freddie is a long-stay mental patient in a hospital in England. He is talking with his psychiatrist, who is attempting to determine his mental state. Following is a short excerpt from a longer interview.

Psychiatrist: Tell me, Freddie, how do you feel in your nerves today?
Freddie: I feel upset.
Psychiatrist: You feel upset, Freddie? Why is that? Tell me.
Freddie: Control of me. I feel like a thick sound from a plate. Dolphin been drove into the cooking room. There must be some kind of connection. . . .
Psychiatrist: A sound from a plate? What sort of sound is it Freddie?
Freddie: Sound. Teacup. Rattle of a teacup. Hold on my body.
Psychiatrist: Can other people read your thoughts, Freddie?
Freddie: They do. They do, doctor. Like when I'm smoking.
Psychiatrist: Do you ever feel there's any external force controlling you, Freddie?
Freddie: I would speak of Germany, doctor, as a force toward me.
Psychiatrist: Do you ever go to any therapies, Freddie?
Freddie: I seem to arrive at Villa 11, doctor. In a physical sort of way, you know.

Source: Lindsay Prior. 1993. *The Social Organization of Mental Illness.* London: Sage.

with a daily barrage of uncontrolled ideas, imagined voices, and urges that seriously disrupt daily living. Schizophrenics may become exceedingly careless in their personal appearance, manners, and speech; sometimes their behavior involves pronounced silliness and situationally inappropriate actions.

No clear understanding or agreement on the precise boundaries of schizophrenia supports a diagnosis of the disorder. For example, DSM-IV specifies that a patient must show some evidence of an active phase of psychotic symptoms, such as hallucinations, delusions, thought disorders, and the like, for a period of 1 week in order to fit the classification of schizophrenic. Research has examined the effect of setting this time period (Flaum, 1992), and subsequent editions of the DSM will likely alter the requirement (Raine, 1993: 300). In fact, the regular stream of changes every few years in the definition of schizophrenic disorders reflects the difficulty in understanding them.

Yet, schizophrenia remains the most common diagnosis for serious mental disorders. About one-fourth of all first admissions to mental hospitals cite it as a cause, and schizophrenics occupy over 30 percent of all mental hospital beds. Treatment costs exceed $7 billion a year, and indirect costs (counseling, loss of productivity, premature mortality, etc.) add another $14 billion, making the total cost of schizophrenia roughly equivalent to that of all cancers combined (Keith, Regier, and Rae, 1991: 34).

Estimates indicate that schizophrenia affects about 1 percent of the general population (Bassett, 1991: 189). This disorder, sometimes called *dementia praecox*, develops primarily between the ages of 15 and 30, especially in male patients, and few people develop it after age 50 (Flor-Henry, 1990). While physical disorders that affect mental functions, such as Alzheimer's disease, usually accompany aging, schizophrenia occurs more commonly in young adults. Its characteristics and extensive reach recommend it as a subject for sociological research probably more than any other disorder.

No single theory of schizophrenia has yet achieved dominance among all observers. Some psychiatrists claim that all schizophrenics suffer from a single pathophysiological process expressed with individual variations, while others describe it as a syndrome that incorporates a number of different diseases, each with its own pathophysiological process (see Kirkpatrick and Buchanan, 1990). Still others claim that research on biological contributors to schizophrenia has "very clearly demonstrated a genetic predisposition for [schizophrenia], and this basic finding is not seriously disputed today" (Raine, 1993: 51; see also Torrey, Bowler, Taylor, and Gottesman, 1994). A simple agreement that inherited characteristics contribute to schizophrenia still does not identify the specific mechanism or enumerate other factors associated with the disorder. Furthermore, the disorder shows clear evidence of social components related to the patient's ability to perform and change social roles (Carter and Flesher, 1995).

Until practitioners and theorists agree on the definition of *schizophrenia*, all will struggle to interpret accurately a number of dimensions of this disorder. Of course, even if indisputable evidence eventually implicates genetic characteristics in the etiology of schizophrenia, it will not rule out important roles for environmental, psychological, and sociological influences.

Problems of Definition

Any assessment of the social deviance associated with mental disorders encounters basic obstacles caused by uncertainties about the meanings of both *mental health* and, consequently, *mental disorder.* While most definitions cite deviations from "normal" behavior, they simply raise more questions about standards for normality. Certainly, many people judge some individuals' behavior as strange or inappropriate, and they may perceive the rationales behind these actions as bizarre; still, actions that seem normal in some situations may represent abnormal deviance in others. For example, dentists may wash their hands 50 times or more during the day as an appropriate and necessary protection for their patients; the same behavior would seem extremely odd and even compulsive for an office worker or business executive.

Variations in standards for behavior defy any effort to establish a single definition for mental illness or disorder. Different groups of observers have adopted a number of broad methods for solving this problem, including (1) a statistical definition, (2) a clinical definition, and (3) a sociological definition based on residual norms.

Statistical Definition of Mental Disorder

Some might measure mental health according to statistical statements of the most frequently encountered behaviors. Perhaps a numerical average could evaluate the incidences of various behaviors, with the highest scores highlighting the practices in the middle of the range, which then represent normality. Just as statistical definitions of deviance fail because they ignore norms, however, statistical definitions of mental illness neglect an important component of disordered behavior. They fail to capture the process that produces judgments of such actions as strange, inappropriate, or wrong behavior, given the circumstances. Just like statistical conceptions of deviance, statistical conceptions of mental illness can indicate only what most people do, not what they ought to do.

Clinical Definition of Mental Disorder

A clinical definition of mental health relies on the individual judgment of a practitioner. Actions that a clinician regards as healthy become the standards for acceptable behavior; other behavior represents mental illness if a mental health professional, such as a psychiatrist, says that it does.

Even with the aid of standardizing diagnostic devices like DSM-IV, the clinical method encounters problems identifying mental disorders. Rosenhan (1973), for example, has reported on a study that placed normal people in psychiatric wards under diagnoses of schizophrenia after falsely telling emergency room doctors that they were hearing voices or other sounds. Once on the ward, these people behaved normally, but the psychiatric staff continued to perceive evidence of mental disorder in those actions. Clinical judgments sometimes show influence of previous conceptions of a disorder, even when subsequent behavior contradicts those conceptions. Clinical or physical medicine often must resolve difficult, complex problems in defining "normal" or healthy behavior; these limitations cannot compare with the vast complexities that accompany any clinical definition of *mental health*, compounded by problems related to normative definitions and value judgments.

The U.S. Surgeon General's (1999) report resorts to rather general language to define mental health and mental illness: "*Mental health* is a state of successful performance of mental function, resulting in productive activities, fulfilling relationships with other people, and the ability to adapt to change and cope with adversity" (U.S. Surgeon General, 1999: 4), and "*Mental illness* is the term that refers collectively to all diagnosable mental disorders. Mental disorders are health conditions that are characterized by alterations in thinking, mood, or behavior (or some combination thereof) associated with distress and/or impaired functioning" (U.S. Surgeon General, 1999: 5).

Value Judgments   A clinical definition subjects conceptions of mental health to practitioners' value judgments. Leading psychiatric writers have advanced many standards for mental health, including striving for happiness, effectiveness, sensitive social relations, freedom from symptoms, functioning unhampered by conflict, and the capacity to love another person, to mention a few. Menninger's (1946: 1) definition, still widely quoted, states:

> Let us define mental health as the adjustment of human beings to the world and to each other with a maximum of effectiveness and happiness. Not just efficiency, or just contentment, or the grace of obeying the rules of the game cheerfully. It is all of these together. It is the ability to maintain an even temper, an alert intelligence, socially considerate behavior, and a happy disposition. This, I think, is a healthy mind.

Such criteria might bar almost anyone from the ranks of normal people. Such conceptions regard a state of emotional health as par, to borrow a term from golf, for attaining mental health. They establish ideals, rather than practical guidelines, and these images of perfection often contradict one another.

Evidence of these contradictions comes from the widely varying estimates based on clinical definitions of the prevalence of mental disorder in the general population. A review of 25 studies has reported percentages ranging from 1 percent to over 60 percent (Dohrenwend and Dohrenwend, 1969). Moreover, studies after 1950 identified disorder at a median rate seven times higher than the same rate for studies before 1950; such an enormous difference probably does not reflect a real change in underlying trends but rather a revision of the criteria for including specific cases in estimates of mental disorder.

For example, the Midtown Manhattan survey, one of the best-known national assessments of the need for psychiatric services, asked questions about psychological disorders, feelings of "nervousness" and "restlessness," and difficulties in interpersonal relationships (Srole, Langner, Michael, Opler, and Rennie, 1962). The study then abstracted this information and submitted it to a team of psychiatrists, who rated the amount of each respondent's "impairment" in psychiatric terms. The Midtown Manhattan survey found mental disorder in an estimated four-fifths of the population. Clearly, only the broadest imaginable conception of mental disorder would class such a large number of persons as mentally ill. Other, more reasonable assessments have placed the figure at between 16 and 25 percent of adults under 60 years of age (Dohrenwend et al., 1980).

Normative Definitions   Redlich (1957) has advocated another method for making the clinical classification of behavior as "normal" or "abnormal" based on three important sociological criteria. (1) The *motivation* of the behavior affects this distinction, perhaps separating normal hand washing before cooking from a neurotic washing compulsion. (2) The *situation* in which the behavior occurs also influences the judgment, for example, distinguishing appearing dressed only in swimming trunks on an Alaskan street in winter versus the same dress on a sunny California beach. (3) *Who makes the judgment*—experts, such as qualified psychiatrists, or the general public—also affects the clinical classification of abnormal behavior. In the absence of universal clinical criteria, many propositions regarding standards for normal behavior clearly lack both reliability and validity, subjecting them to challenge by the public.

Behavior that contradicts one society's norms of ideal mental health may seem like a perfectly normal practice to members of another society. People of some cultures believe that transgressions in personal life lead to disease. Members of other cultures may act on fears that seem "irrational" to outside observers, such as concerns that humans will become transformed into cannibals. An outsider might easily evaluate these fears as symptoms of neuroses or other mental disorders, since they seem to arise from sheer fantasy. One must distinguish, however, between individually generated fears and culturally induced ones. This problem frequently influences clinical diagnoses of mental disorders, even within complex societies, when people's roles in numerous, varied subcultures lead them to display behavior considered normal in their own groups but classed as clinically abnormal by psychiatrists. A French sociologist has commented on the artificial line between mental normality and mental disorder:

> The dividing line between the two realms varies . . . from group to group within the same society. Thus it is never entirely possible to escape from relativity. The function of the psychiatrist is to search for the "causes," to report on the "whys" of the illness, but society decides who the patients will be. There is a subtle play of influences between doctor and the public. The doctor, through the mass media or other agencies, tends to enlarge the field of mental illness, to make the public more aware of disturbances that are minor and have been until then attributed to "oddness" or "eccentricity." On the other hand, he accepts the lay definition of mental illness, and his work is limited to refining or making more explicit this definition by introducing categories of "insanity." . . . But these categories never extend beyond the boundaries of insanity as defined by public opinion. (Bastide, 1972: 60)

Residual Norms and Societal Reaction

A sociological definition would specify mental disorder with reference to *residual norms* and societal reactions to them. Groups establish norms that designate certain behaviors as examples of deviance, including crimes, sexual deviations, drunkenness, bad manners, and other, more specific acts. Each norm, in other words, applies to a specific behavior. A similar view of mental disorder could regard it as any residual violation of norms, that is, a violation not covered by other specific behavioral expectations (Scheff, 1999). This definition would lay out various forms of mental disorder such as withdrawal from association with others, reacting to hallucinations, muttering, posturing, depression, excited behavior, acting on compulsions and obsessions, and auditory disturbances. Like other normative violations, this definition identifies mental disorders by observing violations of society's behavioral expectations. Unlike other instances of deviance, however, the normative violations of mental disorder often prove difficult to specify ahead of time. Sociologists describe the related norms as *residual*, or left over, for this reason.

Judgments of such residual deviance must view actions as normative behavior, not only in and of themselves, but also according to expectations for the social contexts in which they occur. Someone might gain an identity as a residual deviant by talking to spirits, for example, but not if the behavior were to occur within a religious or spiritual context that regularly involves such activities. Similarly, residual deviance does not include someone who sees visions of nonexistent events while under the influence of hallucinogenic drugs (although norms for drug use may lead to a judgment of a specific kind of deviance). These normative violations qualify as mental disorders when they occur outside socially acceptable contexts.

An operational definition of *mental abnormality* would look for answers to questions like "normal under what conditions and for whom?" These considerations seem to contribute helpful guidance for any adequate definition of *mental disorder*. They highlight the difficulty of drawing a sharp, operational line between mental health and mental illness and the need to refer to norms in making this distinction. They clearly cast the problem as one of setting social limits on eccentricity. An English writer has noted the apparent lack of any clear-cut criterion for behavior that constitutes a psychiatric case; judgments about whether a person needs treatment always amount to "a function of his behavior *and* the attitudes of his fellows in society" (Carstairs, 1959: 156). People may or may not receive diagnosis and treatment for mental impairments, depending how others evaluate their behavior. Some psychiatrists have even gone so far as to deny the identification of mental disorder as an illness; rather, they assert that it merely represents defective strategies for handling life situations that cause difficulties for certain individuals (Szasz, 1974; Torrey, 1974).

Such a view is consistent with the history of mental health treatment. As causes and treatments for some conditions became known, the condition was transferred to another medical specialty (Grob, 1994). For example, dominion over hormone-related disorder was moved to endocrinology and made more "medical" in etiology and treatment. As a result of the tendency to remove known medical conditions from the list of mental disorders, the field of mental health and illness is left to deal with conditions that remain a mystery (U.S. Surgeon General, 1999: 9).

As a person's behavior conforms more closely to the expectations of others, they tend to form more favorable evaluations of him or her. On the other hand, behavior outside the expected range often provokes negative evaluations. Collective action, then, by a family, neighborhood, or community to hospitalize an individual for treatment of a mental disorder always emerges from interaction between the behavior itself and the groups tolerance and standards for such behavior. Medical and psychological perspectives on mental disorders differ from a sociological approach, which seeks to understand mental disorders by referring to the social roles and normative expectations that they offend.

Social Stratification and Mental Disorder

Attention to social interactions offers important insight for any attempt to understand mental disorders. The urgency of such considerations becomes apparent in a single glance at the distribution of mental disorders, either altogether or by specific types, throughout the population. Like other forms of deviance, incidences of mental disorders display clear social patterns. Evidence indicates variations in diagnosed mental disorders by social class, sex, and occupation. The Epidemiological Catchment Area Program surveys provide some of the most useful information about the distribution of mental disorders. These surveys gather information from noninstitutionalized adults about how closely they match diagnostic criteria from the DSM. Much of the information from these surveys, covering five areas of the United States, appears in Robins and Regier (1991). The following sections summarize some important findings.

Social Class

Diagnoses of severe psychiatric disorders show a disproportionate concentration in the lowest social classes. A large study in New York found "well" ratings for one-third of respondents in relatively high socioeconomic status (SES) groups but for less than 5 percent of those from the lowest strata. In the highest SES group, the study rated only 12.5 percent as "impaired," while it assigned nearly one-half of the lowest SES group to that category (Srole et al., 1962: 138).

Another study evaluated nearly every patient of any psychiatrist or psychiatric clinic, as well as residents of psychiatric institutes, in New Haven, Connecticut (Hollingshead and Redlich, 1958). This comparison of 1,891 patients to a 5 percent random sample of the normal population, or 11,522 people, revealed rather decided class differences. When the researchers divided both groups into five classes, Class I at the top and Class V at the bottom, they found that lower socioeconomic class corresponded to increasing prevalence of diagnosed disorders. Class I contained 3.1 percent of the population and only 1.0 percent of the mental patients, whereas Class V, with 17.8 percent of the population, had almost twice that percentage of the mental patients. Even after additional divisions according to sex, age, race, religion, and marital status, social class remained the most important variable associated with diagnosed mental disorders. Diagnoses of minor psychiatric ailments, however, showed a concentration at the upper socioeconomic levels.

The relationship to social class does seem stronger for some disorders than for others, though (Eaton and Muntaner, 1999). The Catchment studies have revealed a strong relationship between schizophrenia and social class. People at the bottom of the class system develop schizophrenia at a rate five times higher than that of people at the top. The studies show only a slight relationship, however, between affective disorders and SES. Among women of North African decent, depression shows a positive relationship to education level (Dohrenwend et al., 1992), and bipolar disorder occurs relatively frequently among young people with few years of education. Despite minor variations like these, affective disorders emerge in all social classes. Still, taking the rates of all diagnosed mental disorders together, rates rise in progressively lower SES categories, whether one measures SES by education level, income, or employment status.

The fairly strong relationship between social class and mental disorder leaves room for interpretation. Most analysis focuses on a single, general question: Does membership in the lower class cause mental disorder, or do mentally disordered people slide into the lower class? This question raises complicated issues about potential explanations for a strong negative relationship between social class and mental disorders. These results might mean higher rates of mental disorders among lower-class people than among members of other classes, or they might suggest more frequent diagnosis and hospitalization of mentally disordered members of the lower class compared with equally disordered people from the middle and upper classes. Some writers suggest that the concentration of schizophrenia in the lower class has resulted from genetic selection accompanied by either downward social mobility or failure to move along with peers to higher strata, both caused by the debilitating consequences of mental disorder (Mechanic, 1972). In any case, the relationship between social class and mental disorders has generated more than one interpretation.

Another explanation cites broader social forces. A social selection process may propel healthy and able individuals upward through the class system, while it carries mentally disordered people downward from higher socioeconomic statuses. One study has indicated that social selection (downward mobility of unhealthy individuals) offers the best explanation of the high incidence of schizophrenia in the lower class, while high rates of depression among women and antisocial personalities among men more often result from social causation (as reactions to adversity and stress associated with low SES) (Dohrenwend et al., 1992). In another study, Link, Lennon, and Dohrenwend (1993) interviewed more than 500 psychiatric patients and institutional residents from a community in New York. Their results support the idea that the relationship between mental disorders and social class results more from social causation than from social selection. The weight of the evidence suggests that social class causes disorders more often than disorders determine membership in social classes (see also Cockerham, 1995).

Sex Differences

Along with variations by social class, the characteristics and frequency of mental disorders also differ by sex. Observers have found no consistent sex differences in rates of functional psychoses, in general, but females do develop manic-depressive disorders at higher rates than do men. A thorough review by Zigler and Glick (1986: 240–250), however, reports higher rates of other disorders among men than among women. Depressive symptoms appear unusually commonly among females and male schizophrenics, and females are also diagnosed as paranoid more often than males are. Explanations for the high rates of depression among women have cited stronger status pressures on women than on men, particularly their tendency to find marriage less satisfying than males do (Gove, 1972).

The results from the Catchment surveys support these findings. They reveal no statistically significant gender-related difference in incidence of schizophrenia, but females may develop the disorder at rates slightly higher than those of men. Overall, more men (36 percent) than women (30 percent) experienced any disorder over their lifetimes, perhaps because they develop different kinds of disorders. So, while there are no differences in their overall *rates* of disorders, men and women do differ in terms of the *type* of disorder experiences. Specifically, men display antisocial and substance abuse disorders more often than women do, but women more frequently develop affective disorders (Rosenfield, 1999).

Females may seek help for psychological disorders more willingly than men do. Additional evidence for Western societies other than the United States seems to indicate more psychiatric disorders among women than among men, but one study has attributed between 10 and 30 percent of this excess to women's greater willingness to seek help (Kessler, Brown, and Broman, 1981). Studies in the United States have consistently confirmed that females contact medical personnel, even ordinary physicians, about mental health problems more often than men do (Leaf and Bruce, 1987). This difference likely reflects men's stronger negative attitudes about the label of mentally disordered that might result if they were to seek professional help.

Age

Schizophrenia bears a definite relationship to age. The age group from 18 to 29 shows the highest concentration of schizophrenia. The pattern of high rates among young age groups holds for both sexes and every ethnic group. The rates of affective disorders, such as depression and bipolar disorder, also decrease with increasing age. This finding may, however, reflect inaccurate memories by older survey respondents. Overall, young persons experience mental disorders of all sorts at higher rates than do older people.

Race and Ethnicity

The Catchment surveys indicate a lifetime rate of schizophrenia for African Americans twice as high as that for whites, but this result may reflect the strong relationship between race and SES. On the other hand, Hispanics develop the disorder at lower rates than whites do. African Americans may experience a slightly lower rate of affective disorders, but they show about the same rates as whites in the youngest and oldest age categories. For all disorders, African Americans have higher rates of both lifetime histories and active disorders than whites have, but differences in demographic characteristics may explain this variation; African Americans average younger and poorer than whites, and they receive less education.

Marital Status

The rates of various mental disorders display relationships to marital status. For example, never married, divorced, and separated people develop schizophrenia at rates two to three times higher than married people do. The rate of bipolar disorder is also higher among never married and divorced people compared with married people. A similar but weaker relationship also holds for major depression. Taking all disorders together, divorced and widowed people develop problems at comparatively high rates, as do those who have never married. This difference may result because strong marital relationships shelter people from some sources of stress and provide emotional support during difficult times. This possibility suggests that social stress may exert an important influence on the distribution of mental disorders.

Social Stress in Mental Disorders

The previous section indicated potential links between the distribution of mental disorders in society and other conditions, including the amount of stress that people experience. Occasional emotional stress serves a useful function, and individuals normally encounter some problems adjusting to stressful situations. Intense and persistent stress sometimes results, however, from social situations that cause anger, fear, frustration, worry, and so forth, potentially threatening physical and emotional health. Medical researchers, for example, have shown much interest in the relationship between excessive social stress and physical conditions like hypertension and digestive disorders.

Social stress may exhibit similar links to mental disorders. In fact, social stress seems directly related to behaviors frequently defined as elements of mental disorders. Stressful situations in life like marriage, divorce, and illness or death of a close relative or friend may sometimes precede the onset of mental disorders, as may more minor but still stressful events such as marital disputes, coping with troublesome children, or even minor conflicts with other people. (See Table 16-1.)

Observers have incorporated these situations and others like them into a single scale weighted according to the strength of the stress that each one will likely generate. One review of studies based on these scales has found a clear indication that certain stressful "life events tend to occur to an extent greater than chance expectation before a variety of psychiatric disorders" (Paykel, 1974: 147). Further, certain types of disorders seem associated with specific proportions of such stressful life events. People who attempt suicide report the highest number of these events, depressives the next highest, and then schizophrenics (Paykel, 1974: 148).

Stress in Modern Life

Ample evidence confirms the expectation that social stress contributes substantially to the mental distress that individuals experience. Everyone hears regularly about how much stress has increased compared with earlier times. Some people nostalgically wish for the "good old days," fondly remembering a stress-free, pastoral existence that they imagine characterized rural areas years ago. Such a time may never really have existed, and each age certainly has its own sources of stress, yet few people would deny that stress plays an active role in modern life.

TABLE 16-1
Relative Weights of Stressful Life Events

Below is an example of some life events and their relative degree of stress. There are other scales with different weights, but the idea is the same: to gauge the extent to which the accumulation of eventful things in one's life adds up to a certain level of stress. Note that even "positive" events produce stress.

Event	Life Change Score
Married	500
Widowed	771
Divorced	593
Separated	516
Pregnancy	284
Birth of child	337
Illness or injury	416
Death of loved one or other important person	469
Started school or job	191
Graduated from school or training	191
Retired from work	361
Changed residence	140
Took a vacation	74

SOURCE: Adapted from Robert E. Markush and Rachel V. Favero. 1974. "Epidemiologic Assessment of Stressful Life Events, Depressed Mood, and Psychophysiological Symptoms—A Preliminary Report." In Barbara Snell Dohrenwend and Bruce P. Dohrenwend, eds., *Stressful Life Events: Their Nature and Effects.* New York: Wiley, p. 174.

People regularly encounter almost too many sources of stress to list. News stories routinely report on links between favorite foods and cancer in laboratory animals. These and similar news accounts fill audiences with fear of crime, diseases, accidents, and personal misfortune; media outlets sometimes seem to define *news* as recent, unhappy events. Divorce continues at high rates, and blended families sometimes don't blend. In many families, both parents—if two parents remain in the home—must work to maintain reasonable standards of living. As a result, many parents struggle to meet both the economic demands of living in the late 20th century and their special responsibilities associated with raising children. Parents often complain that the demands of parenting have increased, forcing them to compete with the lure of television and movies that often portray American life in dangerous terms and tempt children to commit various deviant acts. Additional stresses inherent in modern life often strain people's resources to cope. Loss of a loved one to death, dissolution of a marriage, an injury suffered in an automobile accident—these and other stresses burden many people and their families.

Even daily living brings significant stress. Men and women hurrying to work must rush breakfast and navigate through traffic jams. Simple mistakes such as mismatched clothing or a "bad hair day" pose challenges. Some work lacks any interest, leading to terrible frustration with invariable routine. Scheduling conflicts like dental or medical appointments sometimes require tense negotiations at work. Stress rises as one grabs a hurried lunch and returns to work in a rush to meet some deadline. After fighting the rush-hour traffic to get home, evenings may leave only a few moments to catch up with one's spouse's and children's lives, although household chores never seem to do themselves. By itself, each event does not cause a particularly important jump in stress, but their cumulative effects may produce enough stress to contribute to mental disorders.

Adults are not the only ones who experience stress. Children may put up with bullies at the playground and high expectations from their parents for grades or athletic performance. They often compare themselves with other, perhaps more fortunate children. They may feel a desperate need for the right kind of jeans, bicycle, video game, or doll if their peers have those things. Children may also feel frustrated by inability to control their own activities as parents schedule them for dance lessons, Little League baseball, soccer, and church youth activities in addition to their regular schooling.

Adolescents experience acute stress from many sources. After-school activities, including sports and academic clubs, take their time. Other interests, such as dance or piano lessons, soccer, part-time jobs, and even spending time with friends, may become burdensome chores. The formation of adolescent identities sometimes seems like a painfully slow and fragile process hampered by arguments with parents and teachers, conflicting demands of friends and school work, and the whirl of romantic relationships. Adolescents sometimes fear the future, worried about how their lives will come out. Many adolescents may not envy their parents' lot, but parents at least know what has happened to them and much about what to expect; they need not live with the anxiety about the future—or at least not the same kind of anxiety—as their adolescent children must face.

Senior citizens occasionally feel disappointed in their "golden years." Retirement income often remains fixed while most prices can rise. Medical matters gain significance, partly because such problems arise increasingly frequently, and partly because the cost of medical care genuinely concerns most senior citizens. Eventually, seniors watch as their friends die off and their social circles shrink. With fewer friends to share experiences, many stay at home, not making the effort to engage the world beyond their front doors. After all, some seniors must expend great energy and time just to get around; those who don't look forward to interesting encounters will likely decline to put out the effort. The passage of time leaves many seniors feeling as if they don't belong. They have trouble making new friends with diminished opportunities to socialize and little in common with younger strangers. As young adults, they met other adults through their children's activities at school or on sports teams, but those links no longer remain. Some seniors also find that their children develop their own lives, far away from them, so they visit only infrequently and with difficulty.

At no age do people live stress-free lives.

Stress and Anxiety

Social interactions frequently create conflict situations marked by stress, particularly if the conflicts threaten self-images, roles, or values. These stress factors tend to reproduce a certain amount of anxiety (Blazer, Hughes, and George, 1987). Anxiety resembles fear in many ways; like fear, it is an emotional reaction produced by unmanageable stimulation. Fear reactions, however, may end in avoidance or even flight from a real danger, while anxiety may seem to continue without hope of reaching completion. Fear produces overt stimulation, whereas anxiety acts in covert ways. It leaves people in undefined emotional states with which they would like to cope but cannot.

Stress plays a visible role in neurotic compulsive behavior, in which practices such as excessive orderliness and obsessional ideas help people to relieve anxiety. A wide range of compulsive acts, words, and thoughts may effectively relieve stress and anxiety including preoccupation with certain obsessions, tapping, counting, or saying set words. In another example, hypochondria, a constant preoccupation with health, represents a more general hunt for solutions; in this way, the preoccupation diverts and releases anxiety. Observers have noted that an effort to resist such a compulsion brings only mounting anxiety, while indulging the compulsion provides at least a temporary respite (Cameron, 1947: 277). Society's reaction to certain noticeable forms of neurotic behavior may tend to further increase stress and anxiety and thus compound the problem. The following excerpt illustrates such a case:

> A 33 year old married woman discovered some beetles while cleaning out an old cupboard in her house. She immediately had to wash her hands and to repeat the washing three times. Each time she cleaned and dusted the house she began to wash her hands three times, and thereafter in increasing multiples of three. She was soon washing her hands hundreds of times a day and thereafter felt compelled to bathe herself between six and nine times daily. All the time, she recognized that these compulsions were morbid but felt helpless against them. In the next stage of the disorder, she developed the belief that every object that might have come into contact with hair had become contaminated. She began to dispose of her own and her husband's personal possessions and thereafter to sell articles of furniture ridiculously cheaply. At the time of her admission [to a psychiatric hospital], her entire suite of furniture . . . had been sold and the patient came in covered by an unused bedsheet, the only uncontaminated object in the home that could be used to cover her naked body. (Roth and Kroll, 1986: 9)

Stress in Social Situations

Many functional mental disorders appear to arise out of continuous series of events that unfold over long periods of time. Acute stress situations often precipitate major events, bringing such a process to a climax. Immediate stress situations exert particularly important effects in the manic-depressive disorders. One hypothesis connects mental disorders, particularly those that do not interfere with the ability to function in society, with irreconcilable internal conflicts caused by intense striving for material goods and the competitive emphasis in present-day industrial urban society. Horney (1937) characterized life in modern Western societies as a highly individualistic process marked by great competitive striving for achievement and social status. According to her, these forces lead to conflicts between competitive, materialistic desires and efforts to fulfill them and between competitive striving and the desire for the affection of others. Such conflicts produce stress and neuroses.

The effect of psychological stress on mental disorders appears to vary with social class, a finding that helps to explain the relationships detailed earlier between class status and certain mental disorders. Observers note that members of the lower class experience more unpleasant events than others, and they also experience the greatest difficulty in dealing with these problems (Myers, Lindenthal, and Pepper, 1974). The Midtown Manhattan study of mental disorder found an association between mental disorder and the number of stress-inducing factors, independent of their natures. That study also determined that low-status groups encounter the most stress of any in society (Srole et al., 1962, 1977).

A review of results from eight epidemiological surveys further underscores the importance of the relationship between stress and social class. Kessler reports a link between various indicators of social class—income, education, and occupational status—and stress, although this relationship operates differentially for different populations (Kessler, 1982). Income provides the strongest predictive power for stress among men, while education predicts stress most effectively among women, both those in the labor force and homemakers. Clearly, the relationship between stress and social class operates through more than merely economic effects, confirmed by findings about the importance of noneconomic variables for stress among women.

Events and relationships within families also generate stress, as do individuals' activities. A common source of stress like economic difficulties can have consequences for both individuals and the group. Low family income or unpredictable work and income streams can generate substantial economic pressures on a family. In particular, economic hardship can generate marital stress that, in turn, can lead to disharmony between husbands and wives and disruption of parental relations with children (Conger and Elder, 1994). An individual's psychological distress can become modified or expanded within a family context. Mounting economic problems sometimes alter relationships by changing individual behavior, either directly or by changing family relationships.

Individual ability to manage and control stress also exerts an important influence. In one study, people who could not effectively solve their job or relationship problems displayed relatively extensive psychological symptoms, while the symptoms of successful problem solvers did not differ from those of individuals who had fully resolved their situations (Thoits, 1994). This result suggests that detrimental psychological effects may result not from the impact of stress itself, but from the individual's ability cope with it (see also Pearlin, 1999).

One important source of stress is role conflict. One study focused on a group of schizophrenic married women, concluding that they had repeatedly experienced severe marital difficulties over the years (Rogler and Hollingshead, 1965). In fact, evaluation of the husbands or wives who developed schizophrenia could find "evidence that they were exposed to greater hardships, more economic deprivation, more physical illness, or personal dilemmas from birth until they entered their present marriage than do the mentally healthy men and women" (Rogler and Hollingshead, 1965: 404).

This study intensively evaluated schizophrenics and nonschizophrenics in a representative sample of lower-class husbands and wives between the ages of 20 and 39 living in the slums and housing projects *(caserios)* of San Juan, Puerto Rico. It carefully eliminated childhood experiences, social isolation, and occupational history as explanations of the disorders. Rather, the study traced the mental disturbances to the stress created by conflicts and problems associated with lower-class life and neighborhood situations. The schizophrenics had experienced many more problems, and more severe ones, than the nonschizophrenics had known. Culture and low socioeconomic status clearly forced some people to deal with tension. Typical problems in the Spanish culture of Puerto Rico included courtship, women's adjustment to sexual and other roles in marriage, disparities between achieved and desired standards of living, conflicts with neighbors, and limits on privacy in the housing projects. Various additional problems of role fulfillment and performance compounded the difficulties. These stress problems continued to mount, imposing contradictory claims and leading to conflict, mutual withdrawal, and alienation among neighbors, deteriorating until trapped individuals reached a breaking point and began to show signs of schizophrenia.

Stress as a Precondition for Mental Disorders

Although considerable evidence links stress with much mental disorder, particularly minor disorders, questions remain about how different people perceive and respond to stress-inducing situations. Stress, in itself, often even when it becomes severe, does not inevitably produce mental disorder. Studies have confirmed this principle by evaluating the stress associated with modern living, wartime civilian bombings, soldiers under combat, prisoners in Nazi concentration camps, and people with several physical illnesses or injuries. Some people cope very well with stressful situations, while others do not. As Mechanic (1989: 73) has observed:

> In its simplest form, stress conceptions suggest that all people have a breaking point and that mental illness and psychiatric disability are the products of the cumulation of misfortune that overwhelm their constitutional makeup, their personal resources, and their coping abilities. Stated in this way, the perspective is not very useful for it cannot successfully predict who will break down but in retrospect can explain everything.

Clearly, many people withstand considerable stress without developing significant mental difficulties, creating profound questions about the exact relationship between them. Other questions concern whether all life changes, favorable or positive ones as well as unfavorable or negative ones, cause stress in the same way.

Some theorists attempt to explain how life changes may encourage mental disorders by triggering a process in vulnerable individuals. One study, for example, found a significant causal role for life events in the occurrence of depression, but it noted that only certain types of events, those involving long-term threats, exert important influence (Brown and Harris, 1978). A related view holds that events in general, not only adverse ones, contribute to mental disorders, particularly schizophrenia. Some people may be especially prone to the effects of stressful life events (Dohrenwend and Dohrenwend, 1980).

Research has not clearly detailed how people perceive life events and how they may handle those developments within given social situations. Some may call on more social support than others can, and some can simply cope more effectively than others can with life events, particularly adverse ones. This difference highlights the importance of coping strategies as methods for intervening between life events and mental disorders.

Coping and Social Adaptation

Successful adaptation to stress involves a number of measures. First, an individual needs so-called *coping capabilities* and skills to deal with social and environmental demands. Coping requires both anticipation of and successful reaction to environmental demands; another important capability focuses on influencing demands when possible (Mechanic, 1989: chapter 7). Second, adaptation to stress demands motivation to act and react to situations. Someone may escape anxiety by withdrawing or lowering aspirations, but he or she may pay a psychological cost for such behavior, especially when social expectations and roles further constrain it.

Successful coping also depends on both physical and social resources (Turner, 1999). Economic resources can relieve stress from many sources. Social resources, in the form of social support by family, friends, neighbors, specialized support groups, and others, can alleviate stress from other sources by providing an atmosphere conducive to solving interpersonal problems.

Any adequate explanation of mental disorders must consider the interaction of stress and individual coping skills (Wheaton, 1983). Chronic stressors may produce more important effects than acute ones create, suggesting an important effect from the period of time over which one experiences stressors. The continuing influence of some stressors further reinforces the value of long-term individual coping skills and more or less constant social support to ward off the disabling effects of stress.

Social science clearly needs to learn more about the nature of stress and the best methods for dealing with this problem. Accumulating evidence relates stress more closely to conflicts in social roles than to other conditions such as gender or marital status (Thoits, 1987). Thus, someone's identity as a woman or part of a married couple may bring certain stressors, and these may indeed contribute to mental distress, but the most serious conflicts emerge from combinations of roles that certain people—such as married women—must perform.

The marital relationship, in particular, may act as an important buffer against many disorders. The Catchment Area Program results point strongly to the importance of marital status as a correlate of many community-related mental disorders. While marriage itself sometimes involves stressful elements, many marriages obviously help to shield the partners from stressors and to resolve unavoidable stressors.

Social Roles and Mental Disorders

Social science has elaborated four general theoretical perspectives within which to examine mental disorder. First, one can regard it as the product of some biological or genetic deficit. The limitations discussed early in the chapter justify omitting studies related to that perspective from this section's discussion. Recognize, however, that sociologists do tend to ignore biological forces when they interpret the findings of their own research, and future developments may force recognition of a powerful role for biological causes of mental disorders.

Psychology suggests a second perspective on mental disorder as the outcome of certain personality types determined through conditioning or learning experiences. Such types may display behavior inappropriate to certain situations, and the impact of various kinds of interpersonal relations and cultural patterns may trigger social recognition of the consequences as mental disorders. From this perspective, such disorders ultimately result from mental conflicts, superego defects, or traumatic events in early childhood.

For a third, more sociological possibility, one might view mental disorders largely in behavioristic terms. This perspective regards what others might call *mental illness* as a kind of behavior defined as deviant and unacceptable by the significant others who surround the person now identified as disordered. This way of thinking ties mental disorder to the values and social preferences operating in a given cultural system. Therefore, society maintains an extremely changeable interpretation of mental disorder, with substantial variation following changes in cultural values, normative expectations, and social preferences.

A fourth, primarily sociological conception of mental disorder may look for problems with social roles, especially an inability to shift between or adapt to roles, an active effort to play the role of a mentally ill person, and a sequence of self-reactions. One needs an understanding of the nature and performance of social roles in order to assess deviance. Inadequate role performance violates normative expectations, thus increasing the probability that society will impose a negative sanction. Like other forms of deviance, mental disorders elicit negative sanctions from a number of specific societal sources, including family, friends, employers, and relatives, as well as from such outside sources as the police and mental health professionals. In addition to reactions from others, self-reactions will also likely influence the disordered person's social role and behavior.

Inability to Shift Roles

Many people who develop mental disorders appear to lose the ability to shift easily or at all from one social role to another. Everyone normally plays many roles, even in a single day, depending on the situations she or he encounters and the expectations that others express. One view regards schizophrenics as individuals who experience problems playing the roles expected of them in normal social relations, leading to social isolation. Under stress, they fail to change their role performance as social situations demand.

Cameron's (1947: 466–467) early theory, for example, describes paranoid behavior as a product of inappropriate role playing and role taking. Such deviants evaluate events in inflexible ways; they cannot shift roles or see alternative explanations for the behavior of others. Gradually, they build up private worlds defining themselves as the central social objects. This process leads to development of an imaginary "pseudocommunity" that rationalizes their unique interpretations of persecution in the ordinary behavior of others toward them. Unable to interpret accurately the roles of others, they therefore lose the social competence to interpret motives and intentions.

Lemert (1972: 242–264) has challenged this interpretation of a "pseudocommunity," however. After studying a number of cases of paranoia, this research concludes that the paranoid reacts to a real community and not to a false one or a symbolic fabrication; Lemert notes that the disordered person actually does experience, for example, unfair treatment by others. Further, in addition to the inability to shift social roles, an understanding of the delusion and associated behavior must recognize a process of exclusion that disrupts the paranoid's social communication with others.

Performing the Mentally Disordered Role

Scheff (1999) has advanced a sociological theory of mental disorder that describes such conditions as results of playing the role of a mentally disordered person. Scheff's argument recalls the earlier discussion of residual norms. Society establishes many rather clear expectations (norms) to govern behavior in many social situations, and it clearly recognizes a wide range of violations of these norms. Criminal sentences punish violations of the law, any use of illicit drugs amounts to drug abuse, and people deplore the waste caused by suicide.

But other norms establish only vague expectations less specifically understood in social groups. Such provisions determine conditions like "the appropriate length of time for staring into space, about the proper way to imagine or fantasize" (Aday, 1990: 134). These expectations constitute residual rules or norms of everyday living. Actions and conditions that characterize mental disorder such as withdrawal, depression, compulsions, obsessions, and hallucinations violate these common norms, described by Scheff as *residual rule breaking* to distinguish them from the violations of better established sets of norms like criminal law or social etiquette. Residual violations may arise from diverse sources, such as organic difficulties, psychological problems, external stress, or willful acts of defiance against some person or situation.

People very commonly transgress in these casual ways, and no one notices many of their offenses. The average person, for example, may see an illusion or hear an occasional odd sound or voice and simply forget the experience. Most people most of the time, Scheff (1999) argues, fail to recognize their own residual deviance or ignore or rationalize it, depending on the circumstances. On the other hand, when others explicitly identify and label such residual behavior, they begin to organize it into a comprehensive role of a mentally disordered person. The culture defines this role, so people who play it must learn it from the culture. Many societies, including the United States, develop shared conceptions of the criteria for insanity, creating what Scheff (1974) has termed the *social institution of insanity*. Popular conceptions of mental disorders (behavior labeled *crazy*) spread through everyday conversations and mass media content, including advertising. All adults probably know how to "act crazy."

Just as an actor may become typecast, someone may come routinely to play the mentally disturbed role in response to expectations imposed by and role taking of others. "Treatment" by psychiatrists, psychologists, counselors, and other mental health professionals may strengthen the attachment of the *mentally disordered* label to a person, enhancing the stability of the associated behavior. People thus confirmed as mentally disturbed may have difficulty in turning to other, more socially acceptable roles. They may fully adopt the deviant role as the only one available for them.

This explanation of mental disorder brings many advantages: (1) It takes into account the normative effects of mental disorder, describing it as deviance with reference to residual norms. (2) It focuses attention on how people become aware of these norms and the imagery associated with "crazy" behavior. (3) It suggests a compelling view of mental disorders as extensions of familiar role-playing activity, varying only the content of the role to emphasize behavior expected of a mentally disturbed person. The perspective does not deny that people display disorders; it merely identifies a social context for them.

In addition to the central idea of residual deviance, the theory set forth by Scheff (1999) attached great importance to input from others that confirms the *mentally disordered* label. Research has provided mixed support for this view. One study showed that people generally tend to measure their rejection of others as mentally disordered according to the source from which those deviants seek help. Respondents applied varying rejection scores for identical descriptive cases of mental disorder treated in different ways; they gave the lowest score, indicating the weakest rejection, to someone who sought no help, followed in order by someone who received help from a member of the clergy, someone who was treated by a physician, outpatient treatment by a psychiatrist, and finally a stay at a mental hospital. Another study indicated a positive relationship between other people's views of mental patients and the length of hospitalization (Greenley, 1972). Clearly, the process of institutionalization brings its own stigma, regardless of the stigma associated with the disorder itself (Goffman, 1961).

Such supporting evidence has not protected the labeling perspective on mental disorder from criticism. Some claim that Scheff has exaggerated the importance of labeling. The same criticism targets others who argue that society automatically applies a negative stereotype to individuals identified as mentally disordered, in the process encouraging secondary deviation characterized by pronounced mental disorder.

In another criticism of these ideas, Gove (1970) asserts that real, serious disturbances drive the vast majority of psychiatric patients toward treatment. No arbitrary label classes them as mentally ill, according to this argument, since most families deny mental illness in loved ones until they can no longer avoid recognizing it. In fact, Gove found that the nature of psychiatric symptoms affects others' responses more strongly than the attitudes of patients' families do. Moreover, Gove has indicated that the evidence on stigma, while it remains far from conclusive, suggests that most ex-patients encounter few serious problems with social stigma, and any stigma they do experience relates more directly to their current psychiatric status or general social ineffectiveness than to their histories as patients in mental hospitals.

Nevertheless, no one should dismiss the role of labeling in mental disorders. A review of the literature shows that social factors influence the tendency to label specific people as mentally ill beyond the labels warranted by the nature and severity of their psychopathology (Link and Cullen, 1990). These social influences even guide the labeling practices of professionals, such as psychiatrists, and the type and amount of treatment a person receives. Labels such as *mentally ill* and *mental patient* carry such powerful implications, according to one study, that most former mental patients continue to suffer negative effects of this labeling, despite strategies to hide their status (Link, Mirotznik, and Cullen, 1991). In fact, efforts to hide their status as former mental patients appeared to produce more harm than good, especially among those former patients who tried to hide this fact. Culture powerfully reinforces the stigma of mental disorder, and individual coping actions may not always ameliorate its effects.

Scheff (1999) describes mental disorders as deviance from norms that society cannot identify in advance; this determination of deviance after the fact gives meaning to the term *residual norm*. However, Scheff inconsistently maintains that people learn residual norms via interpersonal communications and input from the mass media and that society establishes clear stereotypes of disordered behavior. A comparative study has found no support for the labeling hypothesis that popular stereotypes of mental disorders primarily determine the symptoms that mentally disordered people exhibit (Townsend, 1975). Further, this conception fails to explain some types of disorders that people have developed in the absence of labeling of any kind, sometimes becoming sufficiently serious to require hospitalization (Roth and Kroll, 1986: 15–16).

Labeling theorists argue that application of the deviant label by others, particularly psychiatrists, subjects individuals to an important and frequently irreversible socialization process that leads them to acquire deviant identities as mentally disordered people. The empirical literature gives only sketchy information about whether this does indeed occur. According to an accurate summary of such findings by Gove (1982: 295), "a careful review of the evidence demonstrates that the labeling theory of mental illness is substantially invalid, especially as a general theory of mental illness."

Self-Reactions and Social Roles

Everyone experiences a self-reaction to his or her own appearance, status, and conduct. All come to conceive of themselves not only as physical objects, but as social objects as well (Shibutani, 1986). This universal capacity of self-conception plays an important role in mental disorder. Mentally disordered persons may develop distorted self-conceptions that reflect difficulties in interpersonal relationships and continuing anxiety. Some may lose confidence and become preoccupied with their own thoughts, while others, without logical reason, may adopt egocentric self-images as either great successes or great failures. In response to ongoing difficulties in interpersonal relations, mentally disordered people may learn to use self-reactions in fantasies, dreaming of themselves as different from their true selves in order to overcome conflicts. Such self-centered reactions obstruct their capacity to communicate with and relate to others, leading to further magnification of their own concerns about symptoms and conflicts and further diminishing their ability to act with emotional feeling.

Along with social effects, a person's actions can result in self-approval or self-reproach. People may sometimes praise themselves for what they have done or said, or they may feel disturbed by their own actions and rebuke themselves, producing frustration and conflict. Adults with depressive psychosis may perpetuate this self-punishment to unhealthy extremes in the process of internalizing their difficulties with their social situations, perhaps creating a "tragic melodrama, where the depressed self-accused lashes himself so mercilessly in talk and fantasy that death seems the one promise of penance and relief" (Cameron, 1947: 101). In such mental disorders, the self-image may become so detached from the individual that it becomes not a social object but a physical object suitable for mutilation and punishment. In certain forms of neurotic behavior, this dissociation may proceed so far that the person may even forget his or her own identity. Some hysteria and amnesia patients come to identify with roles they have played previously in life or with another self; they attempt to escape their conflicts by changing themselves, sometimes by adding complementary selves that do not acknowledge one another.

Disturbances in language and in meaningful communication, often symptoms in schizophrenia, indicate the connection of these processes with interpersonal relations. The schizophrenic invents a world of fantasy that lifts her or him out of conflict, at least in some personal estimation. Language eventually expresses such disorders in thought processes. Language becomes a private tool and ceases functioning as a social one; the schizophrenic does not recognize whether another person understands such language. The patient may invent words and link them together in a fashion that makes speech incoherent to others. For example, consider one schizophrenic patient's response to the question "Why are you in the hospital?"

> I'm a cut donator, donated by double sacrifice. I get two days for every one. That's known as double sacrifice; in other words, standard cut donator. You know, we considered it. He couldn't have anything for the cut, or for these patients. All of them are double sacrifice because it's unlawful for it to be donated any more. [Question: Well, what do you do here?] I do what is known as the double criminal treatment. Something that he badly wanted, he gets that, and seven days criminal protection. That's all he gets, and the rest I do for my friend. [Question: Who is the other person who gets all this?] He's a criminal. He gets so much. He gets twenty years' criminal treatment, would make forty years; and

he gets seven days' criminal protection and that makes fourteen days. That's all he gets. (Cameron, 1947: 466–467)

Culture influences the nature of self-reactions. It acts through religion, the strength of its emphasis on material success, and the perception of individual control over events that affect oneself. One study has reported a tendency for mental patients of Asian origin to perceive the label *mentally ill* in rather magical terms, reacting to the power of that label and the process that transfers that power to the individual (Rotenberg, 1975). The study found that patients from Western cultural backgrounds, on the other hand, tend to react to the label as an indication of their own separation from their groups or their self-concepts; the Western patients had developed self-concepts as "sick" people. Cultural differences, however, do not explain all differences in self-reactions. Serious mental disorders, such as psychotic schizophrenia, appear to display similar symptoms in a number of widely differing cultures, suggesting that cultural molding may remain within some limits, at least for this malady (Murphey, 1982).

Mental disorders may fall along a continuum of behavior, influenced by personal resources, symptoms, and social expectations (Gove and Hughes, 1989). In this sense, a mental disorder may resemble a career path, with various potential routes depending upon many contingencies, sometimes acting in combination. At an early stage, acute distress may develop, followed by a reaction that involves some sort of mental disorganization. As the disorganization progresses, the person may accept psychotic episodes as a real part of the world. At this stage, the person has become quite removed from reality and suffers severe isolation. Gove and Hughes maintain that the boundaries of such a theory begin with certain biological conditions, and medical science still knows little about many of them. Regardless of the precise role of biological causes, according to these authors, people express their mental disorders in real symptoms that observers can interpret within a broader sociological framework.

Social Control of Mental Disorders

Some people recognize their own mental disorders and voluntarily seek help; society forces treatment involuntarily upon others. In either case, the assistance comes in outpatient treatment administered by an individual psychiatrist, psychologist, or social worker; similar treatment at a community psychiatric clinic; or residential care in a local hospital or a mental hospital. Most involuntary treatment takes place within mental hospital settings. Once widespread, inpatient treatment in large mental hospitals has decreased sharply in recent decades with the development of local, voluntary facilities for outpatient, community treatment of mental disorders.

It is not the case that all people who need assistance receive help. Because of the significant stigma associated with mental disorders, many people are unwilling to disclose to others that they suffer symptoms associated with mental disorders. To do so, they risk social disapproval, prejudice, stereotyping, and distrust. They are also inhibited in seeking help by the restricted coverage provided by many insurance companies for the treatment of mental disorders (U.S. Surgeon General, 1999: 454).

Mental Hospitals

Although mental hospitals originated only fairly recently, they have lost popularity as treatment alternatives over the past several years. For centuries, society traditionally dealt with pronounced mental disorder by punishing or isolating affected people, or both, often subjecting mentally disordered people to severe and even cruel treatment. Beginning in the 18th century, these people faced confinement in institutions known as *lunatic* or *insane asylums*, later more politely called *mental hospitals*. Rothman (1971) has explained that psychiatrists at the time attributed insanity to chaotic social forces, implying that order, discipline, and social stability would produce mental health. Therefore, the asylum worked to bring discipline to those with mental disorders. Psychiatrists lobbied for public financing to build asylums, and by 1860 almost 85 percent of the then 33 U.S. states had done so. By the end of the 19th century, the asylums functioned as custodial facilities and not as treatment centers.

Nevertheless, mental hospital admissions escalated during the first half of the 20th century. Most had become large and overcrowded, and primarily custodial goals dominated their operating routines. Pilgrim State Hospital in New York reached a population of about 12,000 in 1955, the year that the total county and state mental hospital population in the United States peaked at over 558,000 patients (Conrad and Schneider, 1980: 62). Such a large number of patients combined with the small number of trained professionals available to administer treatment dictated many practices that had nothing whatsoever to do with psychotherapy or other treatment. In fact, the facilities set up routines of daily living closer to those of prisons than to practices appropriate for medical facilities. St. Elizabeth's hospital, near Washington, D.C., housed a patient population of over 7,000, monitored by a staff of several thousand within a physical plant that extended to several hundred acres.

By 1956, however, mental hospital populations had begun to fall, and reductions in inpatient counts had quickened their pace by 1965. By 1981, state mental hospitals housed 125,000 inpatients (Kiesler and Sibulkin, 1987: 46–47), a 78 percent decrease from 1955. By the beginning of the 1990s, about 115,000 patients lived in such hospitals (Mechanic, 1989: 161), and the figure had declined further by the beginning of the 21st century to less than 80,000. Observers have cited several reasons for this decline in mental hospital populations, including economic and fiscal constraints (Scull, 1977). Mental hospitals had become too expensive for most counties and states to continue funding them at previous levels. The large physical plant of such an institution, combined with expensive daily upkeep, constituted an enormous drain on limited resources.

Perhaps the most significant reason for the decline of mental hospitals as primary treatment alternatives was the dramatic increase in use of psychotropic drugs, which exert their principal effects on people's minds, thoughts, or behavior (see Figure 16-1). In the 1950s, these drugs eliminated the need to hospitalize a

FIGURE 16-1

Where Are People With Mental Disorders Treated?

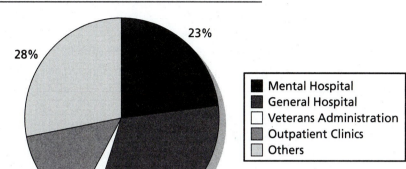

SOURCE: Bureau of the Census. 1999. *Statistical Abstract of the United States, 1999.* Washington, DC: Government Printing Office, p. 144.

large proportion of patients (Conrad and Schneider, 1980). One new drug, chlorpromazine, was developed in France in 1952 and eventually reached the United States under the trade name of Thorazine. Psychiatrists greeted this drug with enthusiasm, as they have nearly 1,000 others developed since, after discovering that it contributed to quieter wards, decreased delusions, and more smoothly functioning institutions. Moreover, drugs allowed many patients to remain at home rather than enduring commitment to institutions. Psychiatrists believed that these drugs would greatly facilitate psychotherapy, although critics charged that pharmaceutical treatments would merely mask psychiatric symptoms without fundamentally changing patients' conditions. Despite the important changes that resulted from widespread use of chlorpromazine, significant reductions in state mental hospital populations did not occur until the next decade.

Mental hospitals appeared to perform two main functions for their remaining patients and for the society that built and maintained them: treatment that would enable mentally disordered people to return to normal society and protection for both patients and society. Within a state mental hospital's typically forbidding exterior, large numbers of patients follow daily schedules intended to facilitate certain institutional routines. Custodial atmospheres still pervade most public mental hospitals; indeed, they have often served as little more than dumping grounds for aged victims of senile disorders, chronically ill patients who have exhausted other options, and mentally disordered members of the lower class. According to one author, people live in mental hospitals "not . . . because they are mad, but because they have been rejected by society and have no suitable place in it" (Perucci, 1974: 30). The use of psychotropic drugs, however, has changed some aspects of life in these facilities. They can leave fewer wards locked and more of their facilities open to the outside.

Social Structure of a Mental Hospital

Goffman (1961) has referred to mental hospitals as "total institutions," a class of organizations—also including monasteries, prisons, schools, and hospitals—that change people's lives and identities. Like prisons and most other total institutions, mental hospitals represent unique communities with special social structures for assigning status and making decisions. Generally, patients populate the lowest-status group within a public mental hospital, below attendants and clerical staff. Life there often subjects patients to degrading experiences such as "confessional" exchanges in group therapy, where they must publicly reveal intimate information about themselves, and total and continual surveillance of their activities by staff. Such experiences compound the stark humiliation of social failure implied by hospitalization itself. A patient lives with status "so low that moral norms, appropriate outside, may be relaxed in the hospital. For example, extramarital encounters may be permitted without censure. Patients are so low on the totem pole that they have little in the way of reputation to lose" (Grusky and Pollner, 1981: 282). While such conditions surely do not prevail in all mental hospitals, a grim reality defines patient status in far too many institutions.

At the top of a mental hospital's prestige hierarchy are superintendents and professional staff members, including psychiatrists, psychologists, occupational therapists, and social workers. This social system of status and power relationships often suffers from serious breakdowns in formal and informal communications among staff members and between them and patients (Stanton and Schwartz, 1954: 193–243). Social distance between nursing staff and patients commonly inhibits treatment efforts, and ineffective communication among staff members frequently interferes with patient recovery.

Patients in large mental hospitals generally do not experience relationships with treatment professionals like those between medical patients and their physicians. Mental hospitals appear to develop more superficial and impersonal patient-doctor relationships. As one patient has said, "The doctor just comes through one door and goes out the other. He spends no time with the patients" (Weinberg and Dunham, 1960: 41). Patients have felt that doctors' authority to shorten or prolong their terms of commitment constitutes great control over their future lives. Patients therefore try to cultivate the friendship of doctors and even learn to feign symptoms of recovery. This process resembles somewhat the interactions of prison inmates with parole board members.

Although the situation is improving as hospitals close and reduced numbers of patients remain confined, many public mental hospitals still function with inadequate staffing and high turnover among attendants. Frequently, low-echelon staff, especially attendants, lack proper professional motivation, and they receive poor training. These deficiencies cause significant problems because patients have their most extensive contacts with attendants, who play largely custodial roles.

The continual renewal of the *mental patient* label may encourage hospital inmates to accept the "sick" role. Some sociologists have worried that the processes of role assignment and role playing increase the likelihood that mentally disordered people will develop stable identities as mental patients, initiating what Goffman (1961: 125–169) and others term their *careers* as mental patients. Observers continue to dispute the contribution of staff reactions to acquisition of the role of a mental patient in contrast to the effects of other contributors such as psychiatric symptoms (see Gove, 1982). In any case, some patients clearly come to occupy this role, and it interferes with their performance of other roles upon their release. Still other patients make more effective transitions to outside communities. This process of shifting roles, as pointed out earlier, exerts important influence on any understanding of mental disorders generally, and those persons who experience the greatest difficulty in shifting from the role of mental patient to nondeviant roles may also experience the greatest impairment.

A well-known study arranged for the placement of eight "normal" people in 12 different hospitals in five states (Rosenhan, 1973). None of the hospitals recognized these patients as sane people after admission, although they exhibited perfectly normal behavior. The mental hospital itself appears to impose a special kind of environment that changes the meaning of behavior. The routine activities and behavior that it demands constitute symptoms of mental disorders. Staff members and others may simply assume that only mentally disturbed persons come to such places. Even normal behavior counts as craziness in an environment capable only of interpreting disordered behavior.

The Deinstitutionalization Movement

As mentioned earlier, mental hospitalizations have declined significantly from the mid-1950s to the present time. As analysis has increasingly revealed limited value of treatment in large mental hospitals for many patients, new methods have emerged.

Community Mental Health Programs Observers have often called the reduction of hospital populations the *deinstitutionalization movement*, reflecting the trend away from mental health treatment within large public facilities and toward community mental health centers. The replacement of hospitalization with community services received its greatest impetus in the 1960s (Joint Commission on Mental Illness and Health, 1961; Kennedy, 1967). A commission studied the questions and recommended two principles for future treatment efforts: directing treatment primarily to help people with mental disorders sustain themselves in the community and returning most patients in mental hospitals to the community as quickly as possible to avoid the isolating effects of hospitalization. Local communities should retain the main responsibility for people with mental disorders, according to these ideas. Congress subsequently initiated a program that provided federal aid to establish community mental health centers and fund other treatment improvements. Plans called for these centers, located in patients' own communities, to emphasize prevention as well as treatment.

The deinstitutionalization movement intended to offer outpatients, including those who have previously experienced hospitalization and those who have not, a variety of services through these local clinics. Psychiatrists, psychologists, social workers, and nonprofessional workers would staff these facilities, running programs based on both individual counseling and group therapy. Some large centers would offer specialized programs such as occupational therapy. Where necessary, staff members would prescribe antidepressant drugs. Any patient who required hospitalization would usually stay in a local general hospital, which would provide various therapies for short periods of time. Only if more extensive hospitalization were required would staff members arrange transfers to county or state mental hospitals.

The Reagan administration did not maintain this commitment to community mental health services. The Omnibus Budget Reconciliation Act of 1981 (PL 97-35) cut federal funds for mental health services by 25 percent. The law provided for continued mental health services but not with the same impact (Kiesler and Sibulkin, 1987: 18). States and local communities had to make up much of this cost reduction to maintain their current mental health services, and some jurisdictions could not afford these costs.

Along with the development of community mental health facilities, increasing numbers of organizations have participated in the deinstitutionalization movement, most of them representing the interests of patients. For example, the Network against Psychiatric Assault, Mental Patients' Liberation Front, and Mental Health Consumer Concerns all stress protection of legal and social rights for mental patients. Others, such as Diabasis House, Recovery, Inc., and the American Schizophrenia Association, deal with or advocate specific therapies or approaches to treatment (Grusky and Pollner, 1981: 356).

Problems With Deinstitutionalization   The development of community mental health programs has not proceeded without difficulties. Alternative care facilities may act in the same custodial and repressive ways as the large mental hospitals they have replaced. No evidence indicates that community mental health centers rehabilitate patients any more effectively than mental hospitals do. While such facilities *intend* to increase rehabilitative services to people who need them, they have frequently produced different *consequences*. For example, some have reintroduced mental patients into community-based institutions, such as nursing homes, without any obvious therapeutic benefit to those patients.

Throughout the 1960s and 1970s, the criterion of danger to others became the main benchmark for cases that required involuntary hospitalization (Stone, 1982). This standard creates important difficulties because psychiatrists (or anyone else) can never reliably predict how much danger a person poses. In most large cities, large numbers of homeless and dysfunctional people now struggle to provide for themselves outside mental hospitals, which represent inappropriate treatment alternatives for them judged according to the single accepted criterion (Herman, 1987). The increase in the urban homeless population in recent years also suggests that community health centers have failed to meet the problems typical of these mentally disordered people. Many of them turn to panhandling or crime to subsist. The persistence in large numbers of mentally disordered people among the urban homeless and the inability of community mental health centers to meet their needs may represent the most significant failure of this movement (Torrey, 1988). Estimates suggest that chronic mental disorders, such as schizophrenia, may afflict from 10 percent to 50 percent of homeless people (General Accounting Office, 1988).

Concern over patients' civil rights helped to propel the deinstitutionalization movement, but efforts to protect these rights also left some of the most dangerous disordered people in the community (Isaac and Armat, 1990). In effect, mental patients sometimes gained rights at the expense of those of the community — families, physicians, and others — to decide who should remain in confinement for their own good and for the good of others. As mental patients left hospitals or never went there in the first place, they did not necessarily reach the attention of treatment specialists in the community. Sometimes, in fact, life outside the hospitals took these people to even worse places, where they became victims of a variety of street predators — hustlers, con artists, runaways — no more concerned for their welfare than the caretakers in the hospital (Johnson, 1990).

In addition to a concern over civil rights, the deinstitutionalization movement represented a response to certain questionable assumptions: faith in the therapeutic benefits of Thorazine and other psychotropic drugs, the belief that reduced hospital admissions would also reduce taxpayer costs, and the expectation that many patients would live better lives in the community than in mental hospitals. Questions have arisen about all of these beliefs. In fact, some critics charge that the deinstitutionalization movement retains direct responsibility for a large proportion of the homelessness problem (Isaac and Armat, 1990). Further, some claim, community mental health advocates focus narrowly on the needs of relatively able, only moderately troubled patients rather than the more intractable problems of the urban homeless mentally ill (Torrey, 1988). Furthermore, increases in availability and use of illicit drugs among the urban poor may have swollen the ranks and exacerbated the problems of this homeless population (Grob, 1994).

For reasons like these, mental health professionals increasingly favor mental hospitals as not only appropriate but necessary to provide adequate care for some kinds of patients, namely those with severe disorders. Such institutions may sometimes have negative effects on people, but they also may have positive effects. Since drugs comprise a large part of the treatment of chronic mental disorders, mental hospitals represent the only alternative for providing adequate care to some patients. Mechanic (1989: 164) claims, "many patients find their hospitalization experience a relief." Many enter the hospital from disruptive and stressful environments. Many patients cannot take care of themselves, and some pose risks of harm to others. Furthermore, many patients report helpful effects of hospitalization and deeply appreciate the care they receive there.

Regardless of future solutions, some people with mental disorders still require some kind of confinement while others do not. Inadequate care of those who have remained outside hospitals has resulted in the same kind of difficulties as the inhumane treatment of hospital patients, leaving a system in need of reform and, perhaps, complete redesign.

Reducing Stigma

The public regards people who have mental disorders with considerable suspicion. The dominant stereotype of those with mental disorders is that they are unpredictable and potentially quite dangerous. The best available evidence suggests that even people with serious mental disorders are not especially dangerous, but the myth continues. Unfortunately, this leads some segments of the population to react negatively when confronted with situations involving people with mental disorders. These reactions may result in job and housing discrimination or in avoidance that increases the social isolation of those with mental disorders. Stigma also leads many people with mental disorders to avoid treatment for their disorders. Nearly two-thirds of all persons with mental disorders do not seek treatment (Kessler et al., 1996), and part of the reason is that the social stigma surrounding mental illness puts up a barrier to seeking help.

The Surgeon General's report (1999) concedes that there is no simple solution to the stigma faced by people with mental disorders. Given that at least some of the negative feelings such people face stems from generalized fear, one measure would be to increase public knowledge of mental disorders. There is some reason to believe that better information about mental disorders, particularly severe disorders and their relationship with violence, could reduce social stigma (Penn and Martin, 1998). But it must also be admitted that the public has become better informed about issues of mental health and illness over the past several decades, yet the degree of social stigma appears to have increased (U.S. Surgeon General, 1999). Public information alone will not reduce stigma.

Perhaps the most effective way to reduce the stigma that surrounds mental disorders is to find the causes of such disorders and apply appropriate treatment (U.S. Surgeon General, 1999). Some disorders in the past have been redefined as "medical" problems and hence appropriate for medical rather than behavioral intervention. Some conditions, notably neurosyphilis and pellagra, produced delirium among their patients. Until it was clear that such disorders have a biological basis, patients were regarded as mentally disordered. Once the cause of the disorders was clarified and appropriate medical treatment was brought to bear on them, the stigma surrounding them was much reduced.

In spite of extensive and suggestive research, the biological basis of most disorders has not been established. Such work is continuing, and it is possible that the biological antecedents to some disorders may be clarified in the future. But it is also likely that we might find that many disorders are the result of a combination of social, psychological, environmental, and physical factors that will further complicate effective treatment.

Summary

Ambiguities cloud judgments about the behaviors and conditions that constitute mental disorder, so standards for mental illness reflect society's tolerance of eccentricity. Organic causes produce a limited number of disorders, justifying medical or chemical intervention. Chemical or genetic causes may also contribute to the functional disorder of schizophrenia. Other functional disorders, however, show no known relationships to physical or biological causes. As a result, those suffering from mental disorders experience considerable stigma in the form of social isolation, distrust, prejudice, and discrimination.

Some sociologists consider such mental disorders as examples of residual normative violations, that is, offenses against previously unelaborated norms. Mental disorder amounts to behavior inappropriate to applicable social situations or to the expectations of a group. Some also view mental disorder as role behavior. People with mental disorders have difficulty fulfilling role expectations and changing roles. Through contact with treatment, they may also adopt the role of a mentally disturbed person. In fact, some sociologists believe that once a person has been labeled as *mentally disordered* or a *mental patient*, he or she may experience difficulty in resuming normal roles.

The reactions of others to a person's behavior reveal how seriously it violates norms and the tolerance with which society will treat these violations. People who act in unusual ways seem strange because they violate expectations of "normal" behavior. In this sense, people judge what madness is on the basis of their conceptions of appropriate behavior. Clearly, however, any standards for mental disorders include actions that endanger oneself or others, as well as inability to take basic care of oneself.

Culture exerts important influence on mental disorders. Certain acts may lead to judgments of mental disorder in one society but not in another. Society structures criteria for mental disorders. Society attributes the most severe disorders to the lowest socioeconomic classes, especially various forms of schizophrenia. Questions remain, however, about whether this fact reflects differences in actual disorders or differences in class-based standards that guide diagnoses. Females develop certain disorders, such as depression, at higher rates than males do, although this difference results in part from the higher likelihood that females will seek help for problems with emotional or psychiatric connotations.

Many psychiatric disorders seem related to aspects of social stress. People who experience especially significant life events or changes seem increasingly likely to develop some kinds of mental disorders than do others who experience fewer changes. High stress associated with life in the lower class helps to explain observed class differences in disorders. Current research is concentrating on the sources of stress, its relation to particular social roles or combinations of roles, and coping mechanisms for reducing stress.

Society's effort to control mental disorders has shifted its focus from institutional treatments to noninstitutional, community-based contexts. During the 1950s, the introduction of mind-altering drugs and a growing awareness of the economics of mental hospitalization led to declining confinement of inpatients. Mental hospitals still treat many patients, but outpatient care and service has become a more common alternative. Many have questioned the role of mental hospitals as institutions of healing. Critics have charged that they function more like prisons than hospitals, but some disorders still require some form of institutionalization.

The community mental health movement began in the 1960s and continues to the present day, although funding cutbacks have jeopardized the range and intensity of available services. Community mental health centers attempt to apply local resources and facilities to treat mental disorders. Increases in urban homeless populations may have resulted from a lack of other options by many who might have lived in mental hospitals in earlier decades. This trend indicates deficiencies in many of the assumptions that underlie deinstitutionalization and the community mental health movement.

Better information and greater sensitivity to the nature of mental disorders may reduce the degree of social stigma toward mental disorders. However, the greatest hope may be that causes are eventually discovered that place disorders into a medical rather than mental category. Society is less likely to stigmatize those whose conditions are primarily medical and biological.

Selected References

Gove, Walter R., ed. 1982. *Deviance and Mental Illness.* Beverly Hills, CA: Sage.

This collection gathers papers assessing various aspects of the labeling perspective of mental disorders, including its theoretical and empirical status. The volume adopts the view that most mental disorders reflect real behavioral or psychiatric problems rather than simply reactions of others to supposed problems.

Grob, Gerald N. 1994. *The Mad among Us: A History of the Care of America's Mentally Ill.* New York: Free Press.

This interesting and sensitive history chronicles the mental health movement in the United States. Grob traces the role of psychiatry in defining mental disorders and the care of those with disorders up to modern times.

Horwitz, Allan V., and Teresa L. Scheid, eds. 1999. *Handbook for the Study of Mental Health: Social Contexts, Theories, and Systems.* New York: Cambridge University Press.

An extremely valuable collection of papers covering many important dimensions of mental health, as well as causes, management, and public policies on mental disorders. Each paper is written by an acknowledged expert on that topic.

Mechanic, David. 1999. *Mental Health and Social Policy: The Emergence of Managed Care,* 4th ed. Boston: Allyn & Bacon.

This book has become, in some respects, conventional wisdom concerning sociological approaches to mental disorder—including theoretical, empirical, and policy implications. The most recent edition of this book also explores the consequences of managed care for those with mental disorders.

Scheff, Thomas J. 1999. *Being Mentally Ill: A Sociological Theory,* 3rd ed. Hawthorne, NY: Aldine de Gruyter.

This book articulates one of the most influential sociological views on the causes of mental disorders based on a labeling perspective. Scheff argues that mental disorders reflect learned role behavior and that the reactions of others critically influence assumption of that role.

REFERENCES

CHAPTER 1 The Nature and Meaning of Deviance

Adler, Patricia A., and Peter Adler, eds. 2000. *Constructions of Deviance: Social Power, Context, and Interaction*, 3rd ed. Belmont, CA: Wadsworth.

Alwin, Duane F. 1986. "Religion and Parental Child-Rearing Orientations: Evidence of a Catholic–Protestant Convergence." *American Journal of Sociology* 92: 412–440.

Becker, George. 1977. *The Genius as Deviant*. Beverly Hills, CA: Sage.

Becker, Howard S. 1973. *Outsiders: Studies in the Sociology of Deviance*, enlarged ed. New York: Free Press.

Ben-Yahuda, Nachman. 1990. *The Politics and Morality of Deviance: Moral Panics, Drug Abuse, Deviant Science, and Reverse Discrimination*. Albany: State University of New York Press.

Best, Joel. 1990. *Threatened Children*. Chicago: University of Chicago Press.

Best, Joel, and David F. Luckenbill. 1994. *Organizing Deviance*, 2nd ed. Englewood Cliffs, NJ: Prentice Hall.

Birenbaum, Arnold, and Edward Sagarin. 1976. *Norms and Human Behavior*. New York: Holt, Rinehart and Winston.

Blake, Judith, and Kingsley Davis. 1964. "Norms, Values, and Sanctions." Pp. 456–484 in *Handbook of Modern Sociology*, Robert E. L. Faris, ed. Chicago: Rand McNally.

Boccella, Kathy. 1996. "Parents Pushing Panic Button." *Des Moines Register*, November 3, pp. 1–2.

Brothers, Joyce. 1974. "The Liberated Child." *Los Angeles Times*, June 25, Section C, p. 1.

Bryant, Clifton D., ed. 1990. *Deviant Behavior: Readings in the Sociology of Deviant Behavior*. New York: Hemisphere.

Chambliss, William J. 1976. "Functional and Conflict Theories of Crime: The Heritage of Emile Durkheim and Karl Marx." Pp. 1–28 in *Whose Law? Whose Order? A Conflict Approach to Criminology*, William J. Chambliss and Milton Mankoff, eds. New York: Wiley.

Cohen, Albert K. 1955. *Delinquent Boys: The Culture of the Gang*. New York: Free Press.

Cohen, Albert K. 1966. *Deviance and Control*. Englewood Cliffs, NJ: Prentice-Hall.

Cohen, Albert K. 1974. *The Elasticity of Evil: Changes in the Social Definition of Deviance*. Oxford: Oxford University Penal Research Unit; Basil Blackwell.

Cohen, Stanley, ed. 1971. *Images of Deviance*. Baltimore: Penguin Books.

Crouch, Ben M., and Kelly Damphousse, 1991. "Law Enforcement and the Satanism–Crime Connection: A Survey of 'Cult Cops.'" Pp. 191–204 in *The Satanism Scare*, James T. Richardson, Joel Best, and David G. Bromley, eds. New York: Aldine De Gruyter.

Cox, Jenny. 1989. "Naturist Nudism." Pp. 122–124 in *Degrees of Deviance: Student Accounts of Their Deviant Behavior*, Stuart Henry, ed. Brookfield, VT: Avebury.

Davis, Fred. 1961. "Deviance Disavowal: The Management of Strained Interaction by the Visibly Handicapped." *Social Problems* 9:120–132.

Dinitz, Simon, Russell Dynes, and Alfred Clarke, eds. 1975. *Deviance: Studies in Definition, Management and Treatment*, 2nd ed. New York: Oxford University Press.

Dodge, David L. 1985. "The Over-Negativized Conceptualization of Deviance: A Programmatic Exploration." *Deviant Behavior* 6: 17–37.

Durkheim, Emile. 1982. *The Rules of Sociological Method*. Edited with an Introduction by Steven Lukes. New York: Free Press; originally published 1895.

Edgerton, Robert. 1976. *Deviance: A Cross-Cultural Perspective*. Menlo Park, CA: Cummings.

Erikson, Kai T. 1965. *Wayward Puritans*. New York: Wiley.

Erlanger, Howard S. 1974. "Social Class and Corporal Punishment in Childrearing." *American Sociological Review* 39: 68–85.

Faris, Ellsworth. 1937. *The Nature of Human Nature*. New York: McGraw-Hill.

Forsyth, Craig J., and Marion D. Oliver. 1990. "The Theoretical Framing of a Social Problem: Some Conceptual Notes on Satanic Cults." *Deviant Behavior* 11: 281–292.

Geis, Gilbert, and Ivan Bunn. 1990. "And a Child Shall Mislead Them: Notes on Witchcraft and Child Abuse Accusations." Pp. 31–45 in *Perspectives on Deviance: Dominance, Degradation, and Denigration*, Robert J. Kelly and Donal E. J. MacNamara, eds. Cincinnati: Anderson.

Gibbs, Jack P. 1965. "Norms: The Problem of Definition and Classification." *American Journal of Sociology* 70: 586–594.

Gibbs, Jack P. 1981. *Norms, Deviance and Social Control*. New York: Elsevier.

Gibbs, Jack P. 1989. *Control: Sociology's Central Notion*. Urbana: University of Illinois Press.

Goode, Erich. 1993. "Positive Deviance: A Viable Concept?" *Deviant Behavior* 12: 289–309.

Goodin, Robert. 1989. *No Smoking: The Ethical Issues*. Chicago: University of Chicago Press.

Gouldner, Alvin W. 1968. "The Sociologist as Partisan: Sociology and the Welfare State." *The American Sociologist* 3: 103–116.

Greenberg, David F. 1988. *The Construction of Homosexuality*. Chicago: University of Chicago Press.

Gusfield, Joseph R. 1963. *Symbolic Crusade*. Urbana, IL: University of Illinois Press.

Hall, Jerome. 1952. *Theft, Law and Society*, 2nd ed. Indianapolis: Bobbs Merrill.

Hawkins, Richard, and Gary Tiedeman. 1975. *The Creation of Deviance: Interpersonal and Organizational Determinants*. Columbus, OH: Charles E. Merrill.

Henslin, James M. 1972. "Studying Deviance in Four Settings: Research Experiences with Cabbies, Suicides, Drug Users, and Abortionees." Pp. 35–70 in *Research on Deviance*, Jack D. Douglas, ed. New York: Random House.

Herman, Nancy J. 1987. "'Mixed Nutters' and 'Looney Tuners': The Emergence, Development, Nature, and Functions of Two Informal, Deviant Subcultures of Chronic, Ex-Psychiatric Patients." *Deviant Behavior* 8: 235–258.

Hicks, Robert D. 1990. "Police Pursuit of Satanic Crime." *Skeptical Inquirer* 14: 276–286.

Higgins, Paul C., and Richard R. Butler. 1982. *Understanding Deviance*. New York: McGraw-Hill.

Hill, David. 1990. "Nothing but the Bare Essentials in this Nudists' Camp." *Des Moines Register*, October 18, p. 9E.

Hopper, Columbus B., and Johnny Moore. 1990. "Women in Outlaw Motorcycle Gangs." *Journal of Contemporary Ethnography* 18.

Johnson, Robert. 1996. *Hard Time: Understanding and Reforming the Prison*. Belmont, CA: Wadsworth.

Jones, Angela Lewellyn. 1998. "Random Acts of Kindness: A Teaching Tool for Positive Deviance." *Teaching Sociology* 26: 179–189.

Lemelle, Anthony J., Jr. 1995. *Black Male Deviance*. New York: Praeger.

Lemert, Edwin M. 1951. *Social Pathology*. New York: McGraw-Hill.

Lemert, Edwin M. 1972. *Human Deviance, Social Problems, and Social Control*, 2nd ed. Englewood Cliffs, NJ: Prentice-Hall.

Lemert, Edwin M. 1982. "Issues in the Study of Deviance." Pp. 233–257 in *The Sociology of Deviance*, M. Michael Rosenberg, Robert A. Stebbins, and Allan Turowetz, eds. New York: St. Martin's Press.

Lewis, Oscar. 1961. *The Children of Sanchez*. New York: Vintage.

Liazos, Alexander. 1972. "The Poverty of the Sociology of Deviance: Nuts, Sluts, and Preverts." *Social Problems* 20: 102–120.

Lofland, John. 1969. *Deviance and Identity*. Englewood Cliffs, NJ: Prentice-Hall.

Manis, Jerome G. 1976. *Analyzing Social Problems*. New York: Holt, Rinehart and Winston.

Mansnerus, Laura. 1988. "Smoking Becomes 'Deviant Behavior'." *New York Times*, April 24, Section 4: 1, 6.

Meier, Robert F. 1981. "Norms and the Study of Deviance: A Proposed Research Strategy." *Deviant Behavior* 3: 1–25.

Meier, Robert F. 1989. "Deviance and Differentiation." Pp. 199–212 in *Theoretical Integration in the Study of Deviance and Crime: Problems and Prospects*, Steven F. Messner, Marvin D. Krohn, and Allen E. Liska, eds. Albany: State University of New York Press.

Meier, Robert F., and Gilbert Geis. 1997. *Victimless Crime? Prostitution, Drugs, Homosexuality, Abortion*. Los Angeles: Roxbury.

Moynihan, Daniel Patrick. 1993. "Defining Deviance Down." *The American Scholar* 61: 17–30.

Murphy, Elizabeth. 1999. "'Breast Is Best': Infant Feeding Decisions and Maternal Deviance." *Sociology of Health and Illness* 21: 187–208.

Orcutt, James D., and J. Blake Turner. 1993. "Shocking Numbers and Graphic Accounts: Quantified Images of Drug Problems in the Print Media." *Social Problems* 40: 190–206.

Parrillo, Vincent N. 1996. *Diversity in America*. Los Angeles: Pine Forge.

Pfuhl, Erdwin H., and Stuart Henry. 1993. *The Deviance Process*, 3rd ed. New York: Aldine de Gruyter.

Quinn, James F. 1987. "Sex Roles and Hedonism among Members of 'Outlaw' Motorcycle Clubs." *Deviant Behavior* 8: 47–63.

Quinney, Richard. 1981. *Class, State and Crime*, 2nd ed. New York: Longman.

Ranulf, Svend. 1964. *Moral Indignation and Middle Class Psychology: A Sociological Study*. New York: Schoken Books; originally published in Denmark, 1938.

Rushing, William A., ed. 1975. *Deviant Behavior and Social Process*, 2nd ed. Chicago: Rand McNally.

Sagarin, Edward. 1975. *Deviants and Deviance: An Introduction to the Study of Disvalued People and Behavior*. New York: Holt, Rinehart and Winston.

Sagarin, Edward. 1985. "Positive Deviance: An Oxymoron." *Deviant Behavior* 6: 169–181.

Schwendinger, Herman, and Julia Schwendinger. 1977. "Social Class and the Definition of Crime." *Crime and Social Justice* 7: 4–13.

Schur, Edwin M. 1984. *Labeling Women Deviant: Gender, Stigma, and Social Control*. New York: Random House.

Scott, Robert A., and Jack D. Douglas, eds. 1972. *Theoretical Perspectives on Deviance*. New York: Basic Books.

Shupe, Anson. 1995. *In The Name of All That's Holy*. Westport, CT: Praeger.

Shupe, Anson. 1998. "Future Study of Clergy Malfeasance." Pp. 230–237 in *Wolves Within the Fold: Religious Leadership and Abuses of Power*. New Brunswick, NJ: Rutgers University Press.

Shupe, Anson, William A. Stacey, and Susan E. Darnell, eds. 2000. *Bad Pastors: Clergy Misconduct in Modern America*. New York: New York University Press.

Spector, Malcolm, and John I. Kitsuse. 1979. *Constructing Social Problems*. Menlo Park, CA: Cummings.

Stafford, Mark C., and Richard R. Scott. 1986. "Stigma, Deviance, and Social Control: Some Conceptual Issues." Pp. 77–91 in *The Dilemma of Difference: A Multidisciplinary View of Stigma*, Stephen C. Ainlay, Gaylene Becker, and Lerita M. Coleman, eds. New York: Plenum.

Story, Marilyn D. 1993. "Personal and Professional Perspectives on Social Nudism: Should You Be Personally Involved in Your Research?" *The Journal of Sex Research* 30: 111–114.

Thompson, Hunter. 1966. *Hell's Angels*. New York: Ballantine.

Trebach, Arnold S. 1987. *The Great Drug War*. New York: Macmillan.

Troyer, Ronald J., and Gerald E. Markle. 1983. *Cigarettes: The Battle Over Smoking*. New Brunswick, NJ: Rutgers University Press.

Victor, Jeffery S. 1990. "The Spread of Satanic-Cult Rumors." *Skeptical Inquirer* 14: 287–291.

Victor, Jeffery S. 1993. *Satanic Panic: The Creation of a Contemporary Legend*. Chicago: Open Court.

Watson, J. Mark. 1982. "Outlaw Motorcyclists: An Outgrowth of Lower Class Cultural Concerns." *Deviant Behavior* 4: 31–48.

Wilson, William Julius. 1987. *The Truly Disadvantaged: The Inner-City, the Underclass, and Public Policy*. Chicago: University of Chicago Press.

Winnick, Charles. 1990. "A Paradigm to Clarify the Life Cycle of Changing Attitudes Toward Deviant Behavior." Pp. 1–14 in *Perspectives on Deviance: Dominance, Degradation, and Denigration*, Robert J. Kelly and Donal E. J. MacNamara, eds. Cincinnati: Anderson.

Yinger, J. Milton. 1982. *Countercultures: The Promise and Peril of a World Turned Upside Down*. New York: Free Press.

Zeitlin, Marian, Hossein Ghassemi, and Mohammed Mansour. 1990. *Positive Deviance in Child Nutrition*. New York: United Nations Publications.

CHAPTER 2   Deviant Events and Social Control

Abadinsky, Howard. 1988. *Law and Justice*. Chicago: Nelson-Hall.

Arthur, Linda B. 1998. "Deviance, Agency, and the Social Control of Women's Bodies in a Mennonite Community." *NWSA Journal* 10: 75–99.

Bayley, David H. 1983. "Accountability and Control of the Police: Some Lessons for Britain." Pp. 145–160 in *The Future of Policing*, T. Bennet, ed. Cambridge: Cambridge University Press.

Becker, Howard S. 1973. *Outsiders: Studies in the Sociology of Deviance*, enlarged ed. New York: Free Press.

Berk, Richard, Harold Brackman, and Selma Lesser. 1977. *A Measure of Justice: An Empirical Study of Changes in the California Penal Code, 1955–1971*. New York: Academic Press.

Bierne, Pierre, and Richard Quinney, eds. 1982. *Marxism and Law*. New York: John Wiley & Sons.

Blumstein, Alfred, Jacqueline Cohen, Jeffrey A. Roth, and Christy A. Visher, eds. 1986. *Criminal Careers and "Career Criminals."* Washington, DC: National Academy Press.

Braithwaite, John. 1989. *Crime, Shame, and Reintegration*. Cambridge: Cambridge University Press.

Brison, Karen J. 1992. *Just Talk: Gossip, Meetings, and Power in a Papua New Guinea Village*. Berkeley: University of California Press.

Brooke, James. 1996. "Idaho County Finds Ways to Chastise Pregnant Teens: They Go to Court." *New York Times*, October 28, p. A26.

Bryant, Clifton D., ed. 1990. *Deviant Behavior: Readings in the Sociology of Deviant Behavior.* New York: Hemisphere.

Chambliss, William J. 1964. "A Sociological Analysis of the Law of Vagrancy." *Social Problems* 11: 67–77.

Chambliss, William J., and Robert Seidman. 1982. *Law, Order, and Power,* 2nd ed. Reading, MA: Addison-Wesley.

Coleman, James William. 1989. *The Criminal Elite: The Sociology of White Collar Crime,* 2nd ed. New York: St. Martin's Press.

Forsyth, Craig J. 1992. "Parade Strippers: Being Naked in Public." *Deviant Behavior* 13: 391–403.

Gelles, Richard. 1985. "Family Violence." *Annual Review of Sociology* 11: 347–367.

Gibbs, Jack P. 1989. *Control: Sociology's Central Notion*. Urbana: University of Illinois Press.

Gottfredson, Michael, and Travis Hirschi. 1990. *A General Theory of Crime*. Palo Alto, CA: Stanford University Press.

Green, Edward C. 1977. "Social Control in Tribal Afro-America." *Anthropological Quarterly* 50: 34–77.

Hagan, John. 1980. "The Legislation of Crime and Delinquency: A Review of Theory, Method, and Research." *Law and Society Review* 14: 603–628.

Hall, Jerome. 1952. *Theft, Law and Society,* 2nd ed. Indianapolis: Bobbs Merrill.

Horwitz, Allan V. 1990. *The Logic of Social Control*. New York: Plenum.

Karmen, Andrew A. 1981. "Auto Theft and Corporate Irresponsibility." *Contemporary Crises* 5: 63–81.

Kelling, George, and Catherine M. Coles. 1996. *Fixing Broken Windows: Restoring Order and Reducing Crime in Our Communities*. New York: Free Press.

Kimbrough, David L. 1995. *Taking Up Serpents: Snake Handlers in Eastern Kentucky*. Chapel Hill: University of North Carolina Press.

Lemert, Edwin M. 1951. *Social Pathology*. New York: McGraw-Hill.

Lemert, Edwin M. 1972. *Human Deviance, Social Problems, and Social Control,* 2nd ed. Englewood Cliffs, NJ: Prentice-Hall.

Lesieur, Henry R. 1977. *The Chase: Career of the Compulsive Gambler.* New York: Doubleday Anchor.

Lofland, John. 1969. *Deviance and Identity.* Englewood Cliffs, NJ: Prentice-Hall

Lowman, John, Robert J. Menzies, and T. S. Plays. 1987. "Introduction: Transcarceration and the Modern State of Penalty." Pp. 1–15 in *Transcarceration: Essays in the Sociology of Social Control,* John Lowman, Robert J. Menzies, and T. S. Plays, eds. Aldershot, U.K.: Gower.

Meier, Robert F. 1982. "Perspectives on the Concept of Social Control." *Annual Review of Sociology* 8: 35–65.

Meier, Robert F., and Gilbert Geis. 1997. *Victimless Crime? Prostitution, Drugs, Homosexuality, Abortion.* Los Angeles: Roxbury.

Miethe, Terance D., and Robert F. Meier. 1994. *Crime and Its Social Context.* Albany: State University of New York Press.

Oliver, William. 1994. *The Violent Social World of Black Men.* New York: Lexington.

Packer, Herbert A. 1968. *The Limits of the Criminal Sanction.* Palo Alto, CA: Stanford University Press.

Reiss, Albert J., Jr., and Jeffrey A. Roth, eds. 1993. *Understanding and Preventing Violence.* Washington, DC: National Academy Press.

Rosecrance, John. 1990. "You Can't Tell the Players without a Scorecard: A Typology of Horse Players." Pp. 348–369 in *Deviant Behavior: Readings in the Sociology of Deviant Behavior,* Clifton D. Bryant, ed. New York: Hemisphere.

Rule, James B. 1988. *Theories of Civil Violence.* Berkeley: University of California Press.

Sacco, Vincent F., and Leslie W. Kennedy. 1996. *The Criminal Event.* Belmont, CA: Wadsworth.

Santee, Richard T., and Jay Jackson. 1977. "Cultural Values as a Source of Normative Sanctions." *Pacific Sociological Review* 20: 439–454.

Scott, John Finley. 1971. *The Internalization of Norms.* Englewood Cliffs, NJ: Prentice-Hall.

Shapiro, Susan P. 1984. *Wayward Capitalists: Target of the Securities and Exchange Commission.* New Haven, CT: Yale University Press.

Shibutani, Tamotsu. 1986. *Social Processes.* Berkeley: University of California Press.

Skipper, John K., and Charles M. McCaghy. 1970. "Stripteasers: The Anatomy and Career Contingencies of a Deviant Occupation." *Social Problems* 17: 391–404.

Stark, Rodney. 1987. "Deviant Places: A Theory of the Ecology of Crime." *Criminology* 25: 983–909.

Tewksbury, Richard. 1996. "Patrons of Porn: Research Notes on the Clientele of Adult Bookstores." Pp. 278–286 in *Social Deviance: Readings in Theory and Research,* 2nd ed., Henry N. Pontell, ed. Upper Saddle River, NJ: Prentice-Hall.

Thomas, Charles W., and Donna M. Bishop. 1987. *Criminal Law: Understanding Basic Principles.* Newbury Park, CA: Sage Publications.

Thompson, William E., and Jackie L. Harred. 1996. "Topless Dancers: Managing Stigma in a Deviant Occupation." Pp. 268–278 in *Social Deviance: Readings in Theory and Research,* 2nd ed., Henry N. Pontell, ed. Upper Saddle River, NJ: Prentice-Hall.

Troiden, Richard R. 1989. "The Formation of Homosexual Identities." *Journal of Homosexuality* 17: 43–73.

Weisheit, Ralph A., David N. Falcone, and L. Edward Wells. 1996. *Crime and Policing in Rural and Small-Town America.* Prospect Heights, IL: Waveland Press.

Willcock, H. D., and J. Stokes. 1968. *Deterrents and Incentives to Crime among Youths Aged 15–21 Years.* London: Home Office, Government Social Survey.

Williams, Kirk, and Richard Hawkins. 1986. "Perceptual Research on General Deterrence: A Critical Review." *Law and Society Review* 20: 545–572.

CHAPTER 3   The Social Context of Deviance

Adler, Patricia A. 1993. *Wheeling and Dealing: An Ethnography of an Upper-Level Drug Dealing and Smuggling Community,* 2nd ed. New York: Columbia University Press.

Adler, Stephen J., and Wade Lambert. 1993. "Just About Everyone Violates Some Laws, Even Model Citizens." *The Wall Street Journal,* March 12, pp. A1, A4.

Akers, Ronald L. 1985. *Deviant Behavior: A Social Learning Approach,* 3rd ed. Belmont, CA: Wadsworth.

Anderson, Elijah. 1999. *Code of the Streets: Decency, Violence, and the Moral Life of the Inner City.* New York: Norton.

Attorney General's Commission on Pornography. 1986. *Final Report.* Washington, DC: Department of Justice, Government Printing Office.

Baker, Robert, and Sandra Ball, eds. 1969. *Violence and the Media.* Washington, DC: Government Printing Office.

Bandura, Albert. 1973. *Aggression: A Social Learning Approach.* Englewood Cliffs, NJ: Prentice-Hall.

Baron, Larry, and Murray A. Straus. 1989. *Four Theories of Rape in American Society: A State-Level Analysis.* New Haven, CT: Yale University Press.

Barr, Kellie, Michael P. Farrell, Grace M. Barnes, and John Welte. 1993. "Race, Class, and Gender Differences in Substance Abuse: Evidence of Middle-Class/Underclass Polarization among Black Males." *Social Problems* 40: 314–327.

Becker, Howard S. 1973. *Outsiders: Studies in the Sociology of Deviance,* enlarged ed. New York: Free Press.

Benda, Brent B. 1987. "Crime, Drug Abuse, Mental Illness, and Homelessness." *Deviant Behavior* 8: 361–375.

Benson, Michael L. 1996. "Deny the Guilty Mind: Accounting for Involvement in White-Collar Crime." Pp. 66–73 in *In Their Own Words: Criminals on Crime,* Paul Cromwell, ed. Los Angeles: Roxbury.

Boccella, Kathy. 1996. "Parents Pushing Panic Button." *Des Moines Register,* November 3, pp. 1–2.

Brantingham, Paul, and Patricia Brantingham. 1984. *Patterns in Crime.* New York: Macmillan.

Bursik, Robert J., Jr., and Harold Grasmick. 1993. *Neighborhoods and Crime: The Dimensions of Effective Community Control.* New York: Lexington.

Calhoun, Craig. 1993. "Nationalism and Ethnicity." *Annual Review of Sociology* 19: 211–239.

Chambliss, William J. 1999. *Power, Politics and Crime.* Boulder, CO: Westview Press.

Clinard, Marshall B. 1976. "The Problem of Crime and Its Control in Developing Countries." Pp. 47–71 in *Crime in Papua New Guinea,* David Biles, ed. Canberra: Australian Institute of Criminology.

Clinard, Marshall B. 1990. *Corporate Corruption.* New York: Greenwood/Praeger.

Clinard, Marshall B., and Peter C. Yeager. 1980. *Corporate Crime.* New York: Free Press.

Cogan, John F. 1982. "The Decline in Black Teenage Employment: 1950–70." *American Economic Review* 72: 621–638.

DeFleur, Melvin L., and Sandra Ball-Rokeach. 1982. *Theories of Mass Communication*, 4th ed. New York: Longman.

Dohrenwend, Bruce P., and Barbara Snell Dohrenwend. 1969. *Social Status and Psychological Disorder*. New York: Wiley-Interscience.

Dornbusch, Sanford M. 1989. "The Sociology of Adolescence." *Annual Review of Sociology* 15: 233–259.

Duneier, Mitchell. 1999. *Sidewalk*. New York: Farrar, Straus, and Giroux.

Elliott, Delbert S., and David Huizinga. 1983. "Social Class and Delinquent Behavior in a National Youth Panel." *Criminology* 21: 149–177.

Feldman, M. Philip. 1977. *Criminal Behavior: A Psychological Analysis*. New York: John Wiley & Sons.

Felson, Marcus. 1998. *Crime in Everyday Life*, 2nd ed. Beverly Hills, CA: Pine Forge Press.

Feshbach, Seymour, and Robert D. Singer. 1971. *Television and Aggression*. San Francisco: Jossey Bass.

Finkelhor, David, Gerald Hotaling, and Andrea Sedlak. 1990. *Missing, Abducted, Runaway, and Throwaway Children in America—Numbers and Characteristics, National Incidence Statistics*. Washington, DC: Bureau of Justice Statistics, U.S. Department of Justice.

Fischer, Claude S. 1975. "The Effect of Urban Life on Traditional Values." *Social Forces* 53: 420-432.

Fischer, Claude S. 1984. *The Urban Experience*, 2nd ed. New York: Harcourt Brace Jovanovich.

Friday, Paul C., and Jerald Hage. 1976. "Youth Crime in Post-Industrial Societies: An Integrated Perspective," *Criminology* 14: 347–367.

Gibbs, Jack P. 1989. *Control: Sociology's Central Notion*. Urbana: University of Illinois Press.

Gillis, A. R. 1996. "Urbanization, Sociohistorical Context, and Crime." Pp. 47–74 in *Criminological Controversies: A Methodological Primer*, John Hagan, A. R. Gillis, and David Brownfield, eds. Boulder, CO: Westview Press.

Glassner, Barry, and Julia Loughlin. 1987. *Drugs in Adolescent Worlds: Burnouts to Straights*. New York: St. Martin's Press.

Gomez Buendia, Hernando. 1990. *Urban Crime: Global Trends and Policies*. New York: United Nations Publications.

Goode, Erich. 1999. *Drugs in American Society*, 5th ed. Boston: McGraw-Hill.

Gottfredson, Michael R., and Travis Hirschi. 1990. *A General Theory of Crime*. Palo Alto, CA: Stanford University Press.

Grinnell, Richard M., Jr., and Cheryl A. Chambers. 1979. "Broken Homes and Middle-Class Delinquency: A Comparison." *Criminology* 17: 395–400.

Hacker, Andrew. 1992. *Two Nations: Black and White, Separate, Hostile, Unequal*. New York: Charles Scribners' Sons.

Hacker, Andrew. 1996. "Modest Proposals." Review of William J. Wilson, "When Work Disappears." *New York Review of Books* 43 (November 23): 8–12.

Harris, Louis. 1987. *Inside America*. New York: Vintage.

Hawley, Amos. 1971. *Urban Society: An Ecological Approach*. New York: Ronald Press.

Herd, Denise. 1994. "Predicting Drinking Problems among Black and White Men: Results from a National Survey." *Journal of Studies on Alcohol* 55: 61–71.

Herman, Nancy J. 1987. "'Mixed Nutters' and 'Looney Tuners': The Emergence, Development, Nature, and Functions of Two Informal, Deviant Subcultures of Chronic, Ex-Psychiatric Patients." *Deviant Behavior* 8: 235–258.

Hirschi, Travis. 1969. *Causes of Delinquency.* Berkeley: University of California Press.

Hirschi, Travis, and Michael Gottfredson. 1980. "Introduction: The Sutherland Tradition in Criminology." Pp. 7–19 in *Understanding Crime: Current Theory and Research,* Travis Hirschi and Michael Gottfredson, eds. Beverly Hills, CA: Sage.

Hodges, William F., Helen K. Buchsbaum, and Carol W. Tierney. 1983. "Parent-child Relationships and Adjustment in Preschool Children in Divorced and Intact Families." *Journal of Divorce* 7: 43–58.

Hogan, Dennis P., and Evelyn M. Kitagawa. 1985. "The Impact of Social Status, Family Structure, and Neighborhood on the Fertility of Black Adolescents." *American Journal of Sociology* 90: 825–855.

Hunter, Albert. 1974. *Symbolic Communities: The Persistence and Change of Chicago's Local Communities.* Chicago: University of Chicago Press.

Johnson, Elmer H., ed. 1983. *International Handbook of Contemporary Developments in Criminology,* Volumes I and II. Westport, CT: Greenwood Press.

Kaplan, Robert M., and Robert D. Singer. 1976. "Television Violence and Viewer Aggression: A Reexamination of the Evidence." *Journal of Social Issues* 32: 35–70.

Kobrin, Solomon. 1951. "The Conflict of Values in Delinquency Areas." *American Sociological Review* 16: 653–661.

Liebert, Robert M., John M. Neale, and Emily S. Davidson. 1973. *The Early Window: Effects of Television on Children and Youth.* New York: Pergamon Press.

Loeber, Rolf, and Madga Stouthamer-Loeber. 1986. "Family Factors as Correlates and Predictors of Juvenile Conduct Problems and Delinquency." Pp. 29–149 in *Crime and Justice: An Annual Review of Research,* Volume 7, Michael Tonry and Norval Morris, eds. Chicago: University of Chicago Press.

Lowery, Shearon A., and Melvin L. DeFleur. 1988. *Milestones in Mass Communication Research: Media Effects,* 2nd ed. New York: Longman.

Lowry, Carol R., and Shirley A. Settle. 1985. "Effects of Divorce on Children: Differential Impact of Custody and Visitation Patterns." *Family Relations* 34: 455–463.

Maccoby, Eleanor E., and Robert H. Mnookin. 1992. *Dividing the Child: Social and Legal Dilemmas of Custody.* Cambridge, MA: Harvard University Press.

Mann, Coramae Richey. 1993. *Unequal Justice: A Question of Color.* Bloomington: Indiana University Press.

Mayhew, Bruce H., and Roger L. Levinger. 1976. "Size and Density of Interaction in Human Aggregates." *American Journal of Sociology* 82: 86–110.

McAuliffe, William E. 1986. "Psychoactive Drug Use by Young and Future Physicians." *New England Journal of Medicine* 315: 805–810.

Meier, Robert F. 1982. "Perspectives on the Concept of Social Control." *Annual Review of Sociology* 8: 35–55.

Miller, Eleanor M. 1986. *Street Woman.* Philadelphia: Temple University Press.

Miller, Neil. 1992. *Out in the World: Gay and Lesbian Life from Buenos Aires to Bangkok*. New York: Random House.

Moore, Joan. 1987. "L.A. Gangs: Threat of a Nightmare." *Los Angeles Times*, December 9, Part II, p. 7.

Murray, Charles. 1984. *Losing Ground: American Social Policy, 1950–1980*. New York: Basic Books.

National Television Violence Study. 1999. *National Television Violence Study*, Vol. 3. Thousand Oaks, CA: Sage.

Peek, Charles W., and George D. Lowe. 1977. "Wirth, Whiskey, and WASPs: Some Consequences of Community Size for Alcohol Use." *Sociological Quarterly* 18: 209–222.

Pins, Kenneth. 1990. "Census Report Shows Breakup of Black Family." *Des Moines Register*, July 12, p. 1A.

Potter, W. James. 1999. *On Media Violence*. Thousand Oaks, CA: Sage.

Putnam, Robert D. 2000. *Bowling Alone: The Collapse and Revival of American Community*. New York: Simon and Schuster.

Rankin, Joseph H. 1983. "The Family Context of Delinquency." *Social Problems* 30: 466–479.

Reiss, Albert J., Jr. 1986. "Why Are Communities Important in Understanding Crime?" Pp. 1–33 in *Communities and Crime*, Albert J. Reiss, Jr. and Michael Tonry, eds. Chicago: University of Chicago Press.

Reiss, Albert J., Jr. and Jeffrey A. Roth, eds. 1993. *Understanding and Presenting Violence*. Washington, DC: National Research Council.

Reiss, Albert J., Jr., and Michael Tonry, eds. 1986. *Communities and Crime*. Chicago: University of Chicago Press.

Reiss, Ira L., and Gary R. Lee. 1988. *Family Systems in America*, 4th ed. New York: Holt, Rinehart and Winston.

Rosecrance, John. 1988. *Gambling without Guilt: The Legitimation of an American Pastime*. Pacific Grove, CA: Brooks/Cole.

Rowe, David C. 1994. *The Limits of Family Influence: Genes, Experience, and Behavior*. New York: Guilford.

Sampson, Robert J. 1986. "Crime in Cities: The Effects of Formal and Informal Social Control." Pp. 271–311 in *Communities and Crime*, Albert J. Reiss, Jr. and Michael Tonry, eds. Chicago: University of Chicago Press.

Schur, Edwin M. 1973. *Radical Non-Intervention*. Englewood Cliffs, NJ: Prentice-Hall.

Sennett, Richard. 1977. *The Fall of Public Man: On the Social Psychology of Capitalism*. New York: Alfred A. Knopf.

Shelley, Louise I. 1981. *Crime and Modernization: The Impact of Industrialization and Urbanization on Crime*. Carbondale: University of Southern Illinois Press.

Short, James F., Jr. 1990. *Delinquency and Society*. Englewood Cliffs, NJ: Prentice-Hall.

Shover, Neal, and Kevin M. Bryant. 1993. "Theoretical Explanations of Corporate Crime." Pp. 141–176 in *Understanding Corporate Crime*, Michael Blankenship, ed. New York: Garland.

Sparks, Richard. 1992. *Television and the Drama of Crime: Moral Tales and the Place of Crime in Public Life*. Buckingham, U.K.: Open University Press.

Spates, James L., and John J. Maciones. 1987. *The Sociology of Cities*, 2nd ed. Belmont, CA: Wadsworth.

Stack, Steven. 1982. "Suicide: A Decade Review of the Sociological Literature." *Deviant Behavior* 4: 41–66.

Stephens, Richard C. 1991. *The Street Addict Role: A Theory of Heroin Addiction*. Albany: State University of New York Press.

Stephens, Richard C. 1987. *Mind-Altering Drugs: Use, Abuse, and Treatment*. Newbury Park, CA: Sage.

Sutherland, Edwin H., Donald R. Cressey, and David F. Luckenbill. 1992. *Principles of Criminology*, 11th ed. Chicago: General Hall.

Tittle, Charles R. 1989. "Urbanness and Unconventional Behavior: A Partial Test of Claude Fischer's Subcultural Theory." *Criminology* 27: 273–306.

Tittle, Charles R., and Robert F. Meier. 1990. "Specifying the SES/Delinquency Relationship." *Criminology* 28: 271–299.

Tittle, Charles R., and Raymond Paternoster. 1988. "Geographic Mobility and Criminal Behavior." *Criminology* 25: 301–343.

Tittle, Charles R., Wayne J. Villemez, and Douglas Smith. 1978. "The Myth of Social Class and Criminality: An Empirical Assessment." *American Sociological Review* 43: 643–656.

Vaillant, George E., Jane R. Bright, and Charles MacArthur. 1970. "Physicians' Use of Mood-Altering Drugs." *New England Journal of Medicine* 282: 365–370.

Weisheit, Ralph A., David N. Falcone, and L. Edward Wells. 1996. *Crime and Policing in Rural and Small-Town America*. Prospect Heights, IL: Waveland Press.

Wilson, Thomas C. 1991. "Urbanism, Migration, and Tolerance: A Reassessment." *American Sociological Review* 56: 117–123.

Wilson, William Julius. 1987. *The Truly Disadvantaged: The Inner City, the Underclass and Public Policy*. Chicago: University of Chicago Press.

Wilson, William Julius. 1991. "Studying Inner-City Social Dislocations: The Challenge of Public Agenda Research." *American Sociological Review* 56: 1–14.

Wilson, William Julius. 1996. *When Work Disappears: The World of the New Urban Poor*. New York: Alfred A. Knopf.

Winick, Charles. 1961. "Physician Narcotic Addicts." *Social Problems* 9: 174–186.

Wirth, Louis. 1938. "Urbanism as a Way of Life." *American Journal of Sociology* 44: 1–24.

CHAPTER 4 Becoming Deviant

Adler, Patricia A., and Peter Adler, eds. 2000. *Constructions of Deviance: Social Power, Context, and Interaction*, 3rd ed. Belmont, CA: Wadsworth.

Agar, Michael. 1973. *Ripping and Running*. New York: Academic Press.

Akers, Ronald L. 2000. *Criminological Theories: Introduction, Evaluation, and Application*, 3rd ed. Los Angeles: Roxbury.

Andrews, D. A., and James Bonta. 1994. *The Psychology of Criminal Conduct*. Cincinnati, OH: Anderson.

Becker, Howard S. 1973. *Outsiders: Studies in the Sociology of Deviance*, enlarged ed. New York: Free Press.

Bennett, Trevor. 1986. "A Decision-Making Approach to Opioid Addiction." Pp. 83–102 in *The Reasoning Criminal: Rational Choice Perspectives on Offending*, Derek B. Cornish and Ronald V. Clarke, eds. New York: Springer-Verlag.

Bouffard, Jeffrey, M. Lyn Exum, and Raymond Paternoster. 2000. "Whither the Beast? The Role of Emotions in a Rational Choice Theory of Crime." Pp. 159–178 in *Of Crime and Criminality*, Sally S. Simpson, ed. Thousand Oaks, CA: Pine Forge.

Brennan, Patricia A., Sarnoff A. Mednick, and Jan Volavka. 1995. "Biomedical Factors in Crime." Pp. 65–90 in *Crime*, James Q. Wilson and Joan Petersilia, eds. San Francisco: ICS Press.

Byrd, Richard E. 1966. *Alone*. New York: Putnam.

Clarke, Ronald L., and Marcus Felson, eds. 1993. *Routine Activities and Rational Choice*. New Brunswick, NJ: Transaction.

Cohen, Albert K. 1965. "The Sociology of the Deviant Act: Anomie and Beyond," *American Sociological Review* 30: 5–14.

Cohen, Albert K. 1966. *Deviance and Control*. Englewood Cliffs, NJ: Prentice-Hall.

Cornish, Derek B., and Ronald V. Clarke, eds. 1986. *The Reasoning Criminal: Rational Choice Perspectives on Offending*. New York: Springer-Verlag.

Cromwell, Paul, ed. 1996. *In Their Own Words: Criminals on Crime*. Los Angeles: Roxbury.

Douglas, Jack D., ed. 1970. *Observations on Deviance*. New York: Random House.

Douglas, Jack D. 1972. *Research on Deviance*. New York: Random House.

Elliott, Gregory C., Herbert L. Ziegler, Barbara M. Altman, and Deborah R. Scott. 1982. "Understanding Stigma: Dimensions of Deviance and Coping." *Deviant Behavior* 3: 275–300.

Ellis, Lee. 1989. *Theories of Rape*. New York: Hemisphere.

Eysenck, Hans. 1977. *Crime and Personality*. London: Routledge and Kegan Paul.

Fishbein, Diane H. 1990. "Biological Perspectives in Criminology." *Criminology* 28: 27–72.

Gay, Peter. 1988. *Freud: A Life for Our Time*. New York: Norton.

Gecas, Viktor. 1982. "The Self-Concept." *Annual Review of Sociology* 8: 1–33.

Gibbons, Don C. 1994. *Talking About Crime and Criminals: Problems and Issues in Theory Development in Criminology*. Englewood Cliffs, NJ: Prentice Hall.

Gottfredson, Michael, and Travis Hirschi. 1993. "A Control Theory Interpretation of Psychological Research on Aggression." Pp. 47–68 in *Aggression and Violence*, Richard B. Felson and James T. Tedeschi, eds. Washington, DC: American Psychological Association.

Hakeem, Michael. 1984. "The Assumption that Crime Is a Product of Individual Characteristics: A Prime Example from Psychiatry." Pp. 197–221 in *Theoretical Methods in Criminology*, Robert F. Meier, ed. Beverly Hills, CA: Sage.

Hanson, Bill, George Beschner, James M. Walters, and Elliott Bovelle. 1985. *Life With Heroin*. Lexington, MA: Lexington Books.

Harding, Christopher, and Richard W. Ireland. 1989. *Punishment: Rhetoric, Rule, and Practice*. London: Routledge.

Heiss, Jerold. 1981. "Social Roles." Pp. 94–129 in *Social Psychology: Sociological Perspectives*, Morris Rosenberg and Ralph H. Turner, eds. New York: Basic Books.

Herman, Nancy J. 1987. "'Mixed Nutters' and 'Looney Tuners': The Emergence, Development, Nature, and Functions of Two Informal, Deviant Subcultures of Chronic, Ex-Psychiatric Patients." *Deviant Behavior* 8: 235–258.

Hogan, Dennis P., and Nan Marie Astone. 1986. "The Transition to Adulthood." *Annual Review of Sociology* 12:109–130.

Inciardi, James A. 1984. "Professional Theft." Pp. 221–243 in *Major Forms of Crime*, Robert F. Meier, ed. Beverly Hills, CA: Sage.

Kitsuse, John I. 1980. "Coming Out All Over: Deviants and the Politics of Social Problems." *Social Problems* 28: 1–13.

Knoblich, Guenther, and Roy King. 1992. "Biological Correlates of Criminal Behavior." Pp. 1–21 in *Facts, Frameworks, and Forecasts: Advances in Criminological Theory*, Vol. 3, Joan McCord, ed. New Brunswick, NJ: Transaction.

Levison, Peter K., Dean R. Gerstein, and Deborah R. Maloff. 1983. *Commonalities in Substance Abuse and Habitual Behavior*. Lexington, MA: Lexington Books.

Lilly, J. Robert, Francis T. Cullen, and Richard A. Ball. 1989. *Criminological Theory: Context and Consequences*. Newbury Park, CA: Sage.

MacFarquhar, Larissa. 1994. "The Department Department." *Lingua Franca* 4: 20–23, 56.

Matza, David. 1969. *Becoming Deviant*. Englewood Cliffs, NJ: Prentice-Hall.

McCord, Joan. 1992. "Understanding Motivations: Considering Altruism and Aggression." Pp. 115–135 in *Facts, Frameworks, and Forecasts: Advances in Criminological Theory*, Vol. 3, Joan McCord, ed. New Brunswick, NJ: Transaction.

Merton, Robert K. 1972. "Insiders and Outsiders: A Chapter in the Sociology of Knowledge." *American Journal of Sociology* 78: 9–47.

Nack, Adina. 2000. "Damaged Goods: Women Managing the Stigma of STDs." *Deviant Behavior* 21: 95–121.

Pallone, Nathaniel J., and James J. Hennessy. 1992. *Criminal Behavior: A Process Psychology Analysis*. New Brunswick, NJ: Transaction.

Parsons, Talcott. 1951. *The Social System*. New York: Free Press.

Pearson, Geoffrey. 1993. "Talking a Good Fight: Authenticity and Distance in the Ehnographer's Craft." Pp. vii–xx in *Interpreting the Field: Accounts of Ethnography*, Dick Hobbs and Tim May, eds. Oxford: Clarendon Press.

Prus, Robert, and Styllianoss Irini. 1980. *Hookers, Rounders, and Desk Clerks*. Toronto: Gage.

Rado, Sandor. 1963. "Fighting Narcotic Bondage and Other Forms of Narcotic Disorders." *Comprehensive Psychiatry* 4: 160–167.

Raine, Adrian. 1993. *The Psychopathology of Crime: Criminal Behavior as a Clinical Disorder*. New York: Academic Press.

Ratner, Mitchell S., ed. 1993. *Crack Pipe as Pimp: An Ethnographic Investigation of Sex-for-Crack Exchanges*. New York: Lexington.

Reid, Scott A., Jonathon S. Epstein, and D. E. Benson. 1996. "Does Exotic Dancing Pay Well But Cost Dearly." Pp. 284–288 in *Readings in Deviant Behavior*, Alex Thio and Thomas Calhoun, eds. New York: HarperCollins.

Reitzes, Donald C., and Juliette K. Diver. 1982. "Gay Bars as Deviant Community Organizations: The Management of Interactions with Outsiders." *Deviant Behavior* 4: 1–18.

Rich, Grant Jewel, and Kathleen Guidroz. 2000. "Smart Girls Who Like Sex: Telephone Sex Workers." Pp. 35–48 in *Sex for Sale: Prostitution, Pornography and the Sex Industry*, Ronald Weitzer, ed. New York: Routledge.

Rowe, David C. 1994. *The Limits of Family Influence: Genes, Experience, and Behavior*. New York: Guilford.

Sagarin, Edward. 1975. *Deviants and Deviance: An Introduction to the Study of Disvalued People and Behavior.* New York: Holt, Rinehart and Winston.

Scheff, Thomas A. 1983. "Toward Integration in the Social Psychology of Emotions." *Annual Review of Sociology* 9: 333–354.

Secretary of Health and Human Services. 1993. *Alcohol and Health: Eighth Special Report to the U.S. Congress.* Rockville, MD: National Institute of Alcohol Abuse and Alcoholism, Government Printing Office.

Stephens, Richard C. 1987. *Mind-Altering Drugs: Use, Abuse, and Treatment.* Beverly Hills, CA: Sage.

Sutherland, Edwin H. 1937. *The Professional Thief.* Chicago: University of Chicago Press.

Sutherland, Edwin H. 1950. "The Sexual Psychopath Laws." *Journal of Criminal Law, Criminology, and Police Science* 40: 540–549.

Sutherland, Edwin H., and Donald R. Cressey. 1978. *Criminology,* 10th ed. Philadelphia: Lippincott.

Szasz, Thomas. 1987. *Insanity: The Idea and Its Consequences.* New York: Wiley.

Thompson, William E., and Jackie L. Harred. 1996. "Topless Dancers: Managing Stigma in a Deviant Occupation." Pp. 268–278 in *Social Deviance: Readings in Theory and Research,* 2nd ed., Henry N. Pontell, ed. Upper Saddle River, NJ: Prentice-Hall.

Troiden, Richard R. 1989. "The Formation of Homosexual Identities." *Journal of Homosexuality* 17: 43–73.

Tunnell, Kenneth D. 1992. *Choosing Crime: The Criminal Calculus of Property Offenders.* Chicago: Nelson Hall.

Turner, Ralph H. 1972. "Deviance Avowal as Neutralization of Commitment." *Social Problems* 19: 308–322.

Waldorf, Dan, Craig Reinarman, and Sheigla Murphy. 1991. *Cocaine Changes: The Experience of Using and Quitting.* Philadelphia: Temple University Press.

Walters, Glenn D. 1992. "A Meta-Analysis of the Gene-Crime Relationship." *Criminology* 30: 595–613.

Wilson, James Q., and Richard J. Herrnstein. 1985. *Crime and Human Nature.* New York: Simon and Schuster.

Yamaguchi, Kazuo, and Denise B. Kandel. 1985. "On the Resolution of Role Incompatibility: A Life Event History Analysis of Family Roles and Marijuana Use." *American Journal of Sociology* 90: 1284–1325.

Zinberg, Norman. 1984. *Drug, Set, and Setting: The Basis for Controlled Intoxicant Use.* New Haven, CT: Yale University Press.

CHAPTER 5   General Theories of Deviance

Aday, David P., Jr. 1990. *Social Control at the Margins: Toward a General Understanding of Deviance.* Belmont, CA: Wadsworth.

Agnew, Robert. 1992. "Foundation for a General Strain Theory of Crime and Delinquency." *Criminology* 30: 47–87.

Agnew, Robert. 1995. "The Contribution of Social Psychological Strain Theory to the Explanation of Crime and Delinquency." Pp. 113–137 in *The Legacy of Anomie Theory,* Freda Adler and William S. Laufer, eds. New Brunswick, NJ: Transactions Publishers.

Agnew, Robert. 2000. "Strain Theory and School Crime." Pp. 105–120 in *Of Crime and Criminality: The Use of Theory in Everyday Life*, Sally S. Simpson, ed. Thousand Oaks, CA: Pine Forge.

Akers, Ronald L. 2000. *Criminological Theories: Introduction, Evaluation, and Application*, 3rd ed. Los Angeles: Roxbury.

Attorney General's Commission on Pornography. 1986. *Final Report*. Washington, DC: Department of Justice, Government Printing Office.

Beck, Allen J., and Darrell K. Gilliard. 1995. *Prisoners in 1994*. Washington, DC: Bureau of Justice Statistics.

Becker, Howard S. 1973. *Outsiders: Studies in the Sociology of Deviance*, enlarged ed. New York: Free Press.

Beckett, Katherine, and Theodore Sasson. 2000. "The War on Crime as Hegemonic Strategy: A Neo-Marxian Theory of the New Punitiveness in U.S. Criminal Justice Policy." Pp. 61–84 in *On Crime and Criminality*, Sally S. Simpson, ed. Thousand Oaks, CA: Pine Forge.

Beirne, Piers, and James Messerschmidt. 2000. *Criminology*, 3rd ed. Boulder, CO: Westview.

Beirne, Piers, and Richard Quinney, eds. 1982. *Marxism and Law*. New York: John Wiley & Sons.

Block, Alan A., and William J. Chambliss. 1981. *Organizing Crime*. New York: Elsevier.

Bulmer, Martin. 1984. *The Chicago School of Sociology: Institutionalization, Diversity, and the Rise of Sociological Research*. Chicago: University of Chicago Press.

Bursik, Robert J., Jr. 2000. "The Systemic Theory of Neighborhood Crime Rates." Pp. 87–104 in *Of Crime and Criminality: The Use of Theory in Everyday Life*, Sally S. Simpson, ed. Thousand Oaks, CA: Pine Forge.

Bursik, Robert J., Jr., and Harold Grasmick. 1993. *Neighborhoods and Crime: The Dimensions of Effective Community Control*. New York: Lexington.

Cain, Maureen, and Alan Hunt, eds. 1979. *Marx and Engels on Law*. New York: Academic Press.

Chambliss, William J. 1976. "Functional and Conflict Theories of Crime: The Heritage of Emile Durkheim and Karl Marx." Pp. 1–28 in *Whose Law? Whose Order? A Conflict Approach to Criminology*, William J. Chambliss and Milton Mankoff, eds. New York: John Wiley & Sons.

Chambliss, William J. 1999. *Power, Politics and Crime*. Boulder, CO: Westview Press.

Clinard, Marshall B., ed. 1964. *Anomie and Deviant Behavior*. New York: Free Press.

Clinard, Marshall B., and Peter Yeager. 1980. *Corporate Crime*. New York: Free Press.

Cloward, Richard, and Lloyd Ohlin. 1960. *Delinquency and Opportunity*. New York: Free Press.

Colvin, Mark, and John Pauly. 1983. "A Critique of Criminology: Toward an Integrated Structural-Marxist Theory of Delinquency Production." *American Journal of Sociology* 89: 513–551.

Cooley, Charles H. 1918. *Social Process*. New York: Charles Scribner's Sons.

Davies, Christie. 1982. "Sexual Taboos and Social Boundaries." *American Journal of Sociology* 87: 1032–1063.

Davis, Nanette J. 1980. *Sociological Constructions of Deviance: Perspectives and Issues*, 2nd ed. Dubuque, IA: William C. Brown.

Davis, Nanette J., and Bo Anderson. 1983. *Social Control: The Production of Deviance in the Modern State*. New York: Irvington.

Foucault, Michel. 1965. *Discipline and Punish*. New York: Pantheon.

Gibbons, Don C., and Joseph W. Jones. 1975. *The Study of Deviance: Perspectives and Problems*. Englewood Cliffs, NJ: Prentice-Hall.

Gouldner, Alvin W. 1980. *The Two Marxisms: Contradictions and Anomalies in the Development of Theory*. New York: Seabury Press.

Greenberg, David F. 1988. *The Construction of Homosexuality*. Chicago: University of Chicago Press.

Hamilton, V. Lee, and Steve Rytina. 1980. "Social Consensus on Norms of Justice: Should the Punishment Fit the Crime?" *American Journal of Sociology* 85: 1117–1144.

Hanson, Bill, George Beschner, James M. Walters, and Elliott Bovelle. 1985. *Life with Heroin*. Lexington, MA: Lexington Books.

Hester, Stephen, and Peter Eglin. 1992. *A Sociology of Crime*. London: Routledge.

Higgins, Paul C., and Richard R. Butler. 1982. *Understanding Deviance*. New York: McGraw-Hill.

Horton, John. 1981. "The Rise of the Right: A Global View." *Social Justice* 15: 7–17.

Inverarity, James M., Pat Lauderdale, and Barry C. Feld. 1983. *Law and Society: Sociological Perspectives on Criminal Law*. Boston: Little, Brown.

Kornhauser, Ruth Rosner. 1978. *The Social Sources of Delinquency*. Chicago: University of Chicago Press.

Krisberg, Barry. 1975. *Crime and Privilege*. Englewood Cliffs, NJ: Prentice-Hall.

Lea, John, and Jock Young. 1986. "A Realistic Approach to Law and Order." Pp. 358–364 in *The Political Economy of Crime: Readings for a Critical Criminology*, Brian McLean, ed. Englewood Cliffs, NJ: Prentice-Hall.

Lemert, Edwin M. 1951. *Social Pathology*. New York: McGraw-Hill.

Liska, Allen E. 1987. *Perspectives on Deviance*, 2nd ed. Englewood Cliffs, NJ: Prentice-Hall.

Liska, Allen E. 1992. "Introduction." Pp. 1–33 in *Social Threat and Social Control*. Albany: State University of New York Press.

Lowman, John. 1992. "Rediscovering Criminology." Pp. 141–160 in *Rethinking Criminology: The Realist Debate*, Jock Young and Roger Matthews, eds. Newbury Park, CA: Sage.

Matthews, Roger. 1987. "Taking Realist Criminology Seriously." *Contemporary Crises* 11: 371–401.

Matthews, Roger, and Jock Young. 1992. "Reflections on Realism." Pp. 1–23 in *Rethinking Criminology: The Realist Debate*, Jock Young and Roger Matthews, eds. Newbury Park,CA: Sage.

Matza, David. 1969. *Becoming Deviant*. Englewood Cliffs, NJ: Prentice-Hall.

McWilliams, Peter. 1993. *Ain't Nobody's Business If You Do*. Los Angeles: Prelude Press.

Meier, Robert F. 1976. "The New Criminology: Continuity in Criminological Theory." *Journal of Criminal Law and Criminology* 67: 461–467.

Meier, Robert F. 1982. "Perspectives on the Concept of Social Control." *Annual Review of Sociology* 8: 35–55.

Merton, Robert K. 1968. *Social Theory and Social Structure*. New York: Free Press.

Messner, Steven F. 1988. "Merton's 'Social Structure and Anomie': The Road Not Taken." *Deviant Behavior* 9: 33–53.

Messner, Steven F., and Richard Rosenfeld. 1997. *Crime and the American Dream*, 2nd ed. Belmont, CA: Wadsworth.

Mills, C. Wright. 1943. "The Professional Ideology of Social Pathologists." *American Journal of Sociology* 49: 165–180.

O'Malley, Pat. 1987. "Marxist Theory and Marxist Criminology." *Crime and Social Justice* 29: 70–87.

Osgood, D. Wayne, and Jeff M. Chambers. 2000. "Social Disorganization Outside the Metropolis: An Analysis of Rural Youth Violence." *Criminology* 38: 81–115.

Platt, Tony. 1974. "Prospects for a Radical Criminology in the United States." *Crime and Social Justice* 1: 2–10.

Quinney, Richard. 1979. *Criminology: Analysis and Critique of Crime in America*, 2nd ed. Boston: Little, Brown.

Quinney, Richard. 1980. *Class, State, and Crime*, 2nd ed. New York: Longman.

Reiman, Jeffery H. 1984. *The Rich Get Richer and the Poor Get Prison: Ideology, Class, and Criminal Justice*, 2nd ed. New York: John Wiley & Sons.

Sagarin, Edward. 1975. *Deviants and Deviance: An Introduction to the Study of Disvalued People and Behavior*. New York: Holt, Rinehart and Winston.

Schwendinger, Herman, and Julia Schwendinger. 1974. *Sociologists of the Chair*. New York: Basic Books.

Scull, Andrew. 1988. "Deviance and Social Control." Pp. 667–693 in *Handbook of Sociology*, Neil J. Smelser, ed. Beverly Hills, CA: Sage.

Simon, David, and D. Stanley Eitzen. 1987. *Elite Deviance*, 2nd ed. Boston: Allyn and Bacon.

Simon, William, and John Gagnon. 1976. "The Anomie of Affluence: A Post-Mertonian Conception." *American Journal of Sociology* 82: 356–378.

Spitzer, Steven. 1975. "Toward a Marxian Theory of Deviance." *Social Problems* 22: 638–651.

Takagi, Paul. 1974. "A Garrison State in a 'Democratic' Society." *Crime and Social Justice* 1: 27–33.

Taylor, Ian, Paul Walton, and Jock Young. 1973. *The New Criminology: For A Social Theory of Deviance*. London: Routledge and Kegan Paul.

Taylor, Ralph B., and Jeanette Covington. 1993. "Community Structural Change and Fear of Crime." *Social Problems* 40: 374–397.

Thomas, W. I., and Florian Znaniecki. 1918. *The Polish Peasant in Europe and America*. New York: Alfred A. Knopf.

Traub, Stuart H., and Craig B. Little, eds. 1994. *Theories of Deviance*, 4th ed. Itasca, IL: F. E. Peacock.

Turk, Austin T. 1969. *Criminality and Legal Order*. Chicago: Rand McNally.

Turk, Austin. 1984. "Political Crime." Pp. 119–135 in *Major Forms of Crime*, Robert F. Meier, ed. Beverly Hills, CA: Sage.

Vaillant, George E., Jane R. Bright, and Charles MacArthur. 1970. "Physicians' Use of Mood-Altering Drugs." *New England Journal of Medicine* 282: 365–370.

Vold, George B. 1958. *Theoretical Criminology*. New York: Oxford University Press.

Whyte, William F. 1943. *Street Corner Society*. Chicago: University of Chicago Press.

Wright, Erik Olin, Andrew Levine, and Elliott Sober. 1992. *Reconstructing Marxism: Essays on Explanation and the Theory of History*. London: Verso.

Young, Jock, and Roger Matthews, eds. 1992. *Rethinking Criminology: The Realist Debate*. Newbury Park, CA: Sage.

CHAPTER 6   Labeling, Control, and Learning Theories of Deviance

Akers, Ronald L. 1985. *Deviant Behavior: A Social Learning Perspective*, 3rd ed. Belmont, CA: Wadsworth.

Akers, Ronald L. 1991. "Self-Control as a General Theory of Crime." *Journal of Quantitative Criminology* 2: 201–211.

Akers, Ronald L. 1998. *Social Learning and Social Structure: A General Theory of Crime and Deviance*. Boston: Northeastern University Press.

Akers, Ronald L. 2000. *Criminological Theories: Introduction, Evaluation, and Application*, 3rd ed. Los Angeles: Roxbury.

Andrews, D. A. 1980. "Some Experimental Investigations of the Principles of Differential Association through Deliberate Manipulations of the Structure of Service Systems." *American Sociological Review* 45: 448–462.

Arnold, William R., and Terrance M. Brungardt. 1983. *Juvenile Misconduct and Delinquency*. Boston: Houghton Mifflin.

Becker, Howard S. 1973. *Outsiders: Studies in the Sociology of Deviance*, enlarged ed. New York: Free Press.

Bernard, Thomas J. 1987. "Structure and Control: Reconsidering Hirschi's Concept of Commitment." *Justice Quarterly* 4: 409–424.

Biernacki, Patricia. 1986. *Pathways from Heroin Addiction: Recovery without Treatment*. Philadelphia: Temple University Press.

Blumer, Herbert. 1969. *Symbolic Interactionism*. Englewood Cliffs, NJ: Prentice-Hall.

Box, Steven. 1981. *Deviance, Reality and Society*, 2nd ed. New York: Holt, Rinehart and Winston.

Brantingham, Paul, and Patricia Brantingham. 1984. *Patterns of Crime*. New York: Macmillan.

Coleman, James W. 1985. *The Criminal Elite: The Sociology of White Collar Crime*. New York: St. Martin's Press.

Cressey, Donald R. 1971. *Other People's Money*. Belmont, CA: Wadsworth; originally published 1953.

Dotter, Daniel L., and Julian B. Roebuck. 1988. "The Labeling Approach Re-examined: Interactionism and the Components of Deviance." *Deviant Behavior* 9: 19–32.

Downs, William R. 1987. "A Panel Study of Normative Structure, Adolescent Alcohol Use, and Peer Alcohol Use." *Journal of Studies on Alcohol* 48: 167–175.

Durkheim, Emile. 1933. *The Division of Labor in Society*. New York: Macmillan; originally published 1893.

Erikson, Kai T. 1962. "Notes on the Sociology of Deviance." *Social Problems* 9: 307–314.

Federal Bureau of Investigation. 1999. *Crime in the United States, 1998*. Washington, DC: Department of Justice, Government Printing Office.

Felson, Richard B., and James T. Tedeschi. 1993. "Grievances: Development and Reaction." Pp. 13–45 in *Aggression and Violence*, Richard B. Felson and James T. Tedeschi, eds. Washington, DC: American Psychological Association.

Fenton, Steve. 1984. *Durkheim and Modern Sociology.* Cambridge: Cambridge University Press.

Garfinkel, Harold. 1967. *Studies in Ethnomethodology.* Englewood Cliffs, NJ: Prentice-Hall.

Geis, Gilbert. 2000. "On the Absence of Self-Control as the Basis for a General Theory of Crime." *Theoretical Criminology* 4: 35–53.

Gibbons, Don C. 1994. *Talking about Crime and Criminals: Problems and Issues in Theory Development in Criminology.* Englewood Cliffs, NJ: Prentice-Hall.

Gibbs, Jack P. 1996. "Major Notions and Theories in the Sociology of Deviance." Pp. 64–83 in *Readings in Deviant Behavior,* Alex Thio and Thomas Calhoun, eds. New York: HarperCollins.

Giordano, Peggy C., and Sharon Mohler Rockwell. 2000. "Differential Association and Female Crime." Pp. 3–24 in *Of Crime and Criminality: The Use of Theory in Everyday Life,* Sally S. Simpson, ed. Thousand Oaks, CA: Pine Forge.

Goffman, Erving. 1963. *Asylums.* New York: Doubleday Anchor.

Gottfredson, Michael, and Travis Hirschi. 1990. *A General Theory of Crime.* Palo Alto, CA: Stanford University Press.

Hirschi, Travis. 1969. *Causes of Delinquency.* Berkeley: University of California Press.

Hirschi, Travis. 1984. "A Brief Commentary on Akers' 'Delinquent Behavior, Drugs, and Alcohol: What Is the Relationship?'" *Today's Delinquent* 3: 49–52.

Hirschi, Travis, and Michael Gottfredson, eds. 1993. *The Generality of Deviance.* New Brunswick, NJ: Transaction.

Hirschi, Travis, and Michael R. Gottfredson. 2000. "In Defense of Self-Control." *Theoretical Criminology* 4: 55–69.

Kaplan, Howard B., and Kelly R. Damphousse. 1997. "Negative Social Sanctions, Self-Derogation, and Deviant Behavior: Main and Interactive Effects in Longitudinal Perspective." *Deviant Behavior* 18: 1–26.

Kempf, Kimberly L. 1993. "The Empirical Status of Hirschi's Control Theory." Pp. 143–185 in *New Directions in Criminological Theory: Advances in Criminological Theory,* Vol. 4, Freda Adler and William S. Laufer, eds. New Brunswick, NJ: Transaction.

Kitsuse, John I. 1962. "Societal Reactions to Deviant Behavior: Problems of Theory and Method." *Social Problems* 9: 247–257.

Kitsuse, John I. 1980. "Coming Out All Over: Deviants and the Politics of Social Problems." *Social Problems* 28: 1–13.

Kornhauser, Ruth Rosner. 1978. *Social Sources of Delinquency: An Appraisal of Analytic Models.* Chicago: University of Chicago Press.

Langevin, Ron, ed. 1985. *Erotic Preference, Gender Identity, and Aggression in Men: New Research Studies.* Hillsdale, NJ: Lawrence Erlbaum Associates.

LeBanc, Marc, and Aaron Caplan. 1993. "Theoretical Formalization, a Necessity: The Example of Hirschi's Bonding Theory." Pp. 237–336 in *New Directions in Criminological Theory: Advances in Criminological Theory,* Vol. 4, Freda Adler and William S. Laufer, eds. New Brunswick, NJ: Transaction.

Lemert, Edwin M. 1951. *Social Pathology.* New York: McGraw-Hill.

Lemert, Edwin M. 1972. *Human Deviance, Social Problems and Social Control*, 2nd ed. Englewood Cliffs, NJ: Prentice-Hall.

Lester, David. 1987. *Suicide as a Learned Behavior.* Springfield, IL: Charles C. Thomas.

Liazos, Alexander. 1972. "The Poverty of the Sociology of Deviance: Nuts, Sluts, and Preverts." *Social Problems* 20: 103–120.

Mankoff, Milton. 1971. "Societal Reaction and Career Deviance: A Critical Analysis." *The Sociological Quarterly* 12: 204–218.

Margolin, Leslie. 1992. "Deviance on Record: Techniques for Labeling Child Abusers in Official Documents." *Social Problems* 39: 58–70.

Matsueda, Ross L. 1982. "Testing Control Theory and Differential Association: A Causal Modeling Approach." *American Sociological Review* 47: 489–504.

Matsueda, Ross L. 1988. "The Current State of Differential Association Theory." *Crime and Delinquency* 34: 277–306.

Matsueda, Ross L., and Karen Heimer. 1987. "Race, Family Structure, and Delinquency: A Test of Differential Association and Social Control Theories." *American Sociological Review* 52: 826–840.

McCaghy, Charles H., and Timothy A. Capron. 1994. *Deviant Behavior: Crime, Conflict, and Interest Groups*, 3rd ed. New York: Macmillan.

Meier, Robert F. 1983. "Shoplifting: Behavioral Aspects." Pp. 1,497–1,500 in *Encyclopedia of Crime and Justice*, Sanford H. Kadish, ed. New York: Free Press.

Moffitt, Terrie E., Robert F. Krueger, Avshalom Caspi, and Jeff Fagan. 2000. "Partner Abuse and General Crime: How Are They the Same? How Are They Different?" *Criminology* 38: 199–232.

Nye, F. Ivan. 1958. *Family Relationships and Delinquent Behavior.* New York: John Wiley & Sons.

Orcutt, James D. 1991. "The Social Integration of Beers and Peers: Situational Contingencies in Drinking and Intoxication." Pp. 198–215 in *Society, Culture, and Drinking Patterns Reexamined*, David J. Pittman and Helene Raskin White, eds. New Brunswick, NJ: Rutgers Center of Alcohol Studies.

Osgood, D. Wayne, Lloyd D. Johnston, Patrick M. O'Malley, and Jerald G. Bachman. 1988. "The Generality of Deviance in Late Adolescence and Early Adulthood." *American Sociological Review* 53: 81–93.

Plummer, Ken. 1979. "Misunderstanding Labeling Perspectives." Pp. 85–121 in *Deviant Interpretations*, David Downes and Paul Rock, eds. Oxford: Martin Robertson.

Randall, Susan, and Vicki McNickle Rose. 1984. "Forcible Rape." Pp. 47–72 in *Major Forms of Crime*, Robert F. Meier, ed. Beverly Hills, CA: Sage.

Reckless, Walter. 1973. *The Crime Problem*, 4th ed. New York: Appleton.

Rubington, Earl, and Martin S. Weinberg. 1996. *Deviance: The Interactionist Perspective*, 6th ed. Boston: Allyn and Bacon.

Scheff, Thomas J. 1984. *Being Mentally Ill*, rev. ed. New York: Aldine.

Schur, Edwin M. 1971. *Labeling Deviant Behavior.* New York: Harper and Row.

Schur, Edwin M. 1979. *Interpreting Deviance.* New York: Harper and Row.

Schur, Edwin M. 1980. *The Politics of Deviance: Stigma Contests and the Uses of Power.* Englewood Cliffs, NJ: Prentice-Hall.

Schur, Edwin M. 1984. *Labeling Women Deviant: Gender, Stigma, and Social Control.* Englewood Cliffs, NJ: Prentice-Hall.

Schutz, Alfred. 1967. *The Phenomenology of the Social World.* Evanston, IL: Northwestern University Press.

Seeman, Melvin, and Carolyn S. Anderson. 1983. "Alienation and Alcohol: The Role of Work Mastery and Community in Drinking Behavior." *American Sociological Review* 48: 60–77.

Shover, Neal. 1985. *Aging Criminals.* Beverly Hills, CA: Sage.

Simon, Rita J. 1975. *Women and Crime.* Lexington, MA: Lexington Books.

Stafford, Mark C., and Richard R. Scott. 1986. "Stigma, Deviance, and Social Control: Some Conceptual Issues." Pp. 77–91 in *The Dilemma of Difference: A Multidisciplinary View of Stigma,* Stephen C. Ainlay, Gaylene Becker, and Lerita M. Coleman, eds. New York: Plenum.

Sutherland, Edwin H. 1947. *Criminology,* 4th ed. Philadelphia: Lippincott.

Sutherland, Edwin H., Donald R. Cressey, and David F. Luckenbill. 1992. *Principles of Criminology,* 11th ed. Chicago: General Hall.

Sykes, Gresham M., and David Matza. 1957. "Techniques of Neutralization: A Theory of Delinquency." *American Sociological Review* 22: 664–670.

Tannenbaum, Charles. 1938. *Crime and the Community.* Boston: Ginn.

Thomas, W. I., and Florian Znaniecki. 1918. *The Polish Peasant in Europe and America.* New York: Alfred A. Knopf.

Thornberry, Terence P., Alan J. Lizotte, Marvin D. Krohn, Margaret Farnworth, and Sun Joon Jang. 1994. "Delinquent Peers, Beliefs, and Delinquent Behavior: A Longitudinal Test of Interactional Theory." *Criminology* 32: 47–83.

Toby, Jackson. 1957. "Social Disorganization and Stake in Conformity: Complementary Factors in the Predatory Behavior of Hoodlums." *Journal of Criminal Law, Criminology and Police Science* 48: 12–17.

Traub, Stuart H., and Craig B. Little, eds. 1999. *Theories of Deviance,* 5th ed. Itasca, IL: F. E. Peacock.

Triplett, Ruth. 2000. "The Dramatization of Evil: Reacting to Juvenile Delinquency in the 1990s." Pp. 121–140 in *Of Crime and Criminality: The Use of Theory in Everyday Life,* Sally S. Simpson, ed. Thousand Oaks, CA: Pine Forge.

Vaillant, George E., Jane R. Bright, and Charles MacArthur. 1970. "Physicians' Use of Mood-Altering Drugs." *New England Journal of Medicine* 282: 365–370.

Warr, Mark. 1993. "Age, Peers, and Delinquency." *Criminology* 31: 17–40.

Wellford, Charles F., and Ruth A. Triplett. 1993. "The Future of Labeling Theory." Pp. 1–22 in *New Directions in Criminological Theory: Advances in Criminological Theory,* Vol. 4, Freda Adler and William S. Laufer, eds. New Brunswick, NJ: Transaction.

Wiatrowski, Michael D., David B. Griswold, and Mary K. Roberts. 1981. "Social Control Theory and Delinquency." *American Sociological Review* 46: 525–541.

Wilsnack, Richard W., Sharon C. Wilsnack, and Albert D. Klassen. 1987. "Antecedents and Consequences of Drinking and Drinking Problems in Women: Patterns from a U.S. National Survey." Pp. 85–158 in *Alcohol and Addictive Behavior: Nebraska Symposium on Motivation, 1986,* P. Clayton Rivers, ed. Lincoln: University of Nebraska Press.

Wilson, James Q., and Richard J. Herrnstein. 1985. *Crime and Human Nature.* New York: Simon and Schuster.

Wrong, Dennis. 1961. "The Oversocialized Conception of Man in Modern Sociology." *American Sociological Review* 26: 183–193.

CHAPTER 7   Crimes of Interpersonal Violence

Amir, Menachem. 1971. *Patterns of Forcible Rape*. Chicago: University of Chicago Press.

Archer, Dane, and Rosemary Gartner. 1984. *Violence and Crime in Cross-National Perspective*. New Haven, CT: Yale University Press.

Athens, Lonnie H. 1992. *The Creation of Dangerous Violent Criminals*. London: Routledge.

Athens, Lonnie H. 1997. *Violent Criminal Acts and Actors Revisited*. Urbana: University of Illinois Press.

Bachman, R. 1994. *Violence against Women: A National Crime Victimization Survey Report*. Washington, DC: Bureau of Justice Statistics.

Ball-Rokeach, Sandra J. 1973. "Values and Violence: A Test of the Subculture of Violence Thesis." *American Sociological Review* 38: 736–749.

Baron, Larry, and Murray A. Straus. 1989. *Four Theories of Rape in American Society: A State-Level Analysis*. New Haven, CT: Yale University Press.

Baumgartner, M. P. 1993. "Violent Networks: The Origins and Management of Domestic Conflict." Pp. 209–231 in *Aggression and Violence*, Richard B. Felson and James T. Tedeschi, eds. Washington, DC: American Psychological Association.

Bensing, Robert C., and Oliver Schroeder, Jr. 1960. *Homicide in an Urban Community*. Springfield, IL: Charles C. Thomas.

Berger, R. J., W. L. Neuman, and P. Searles. 1994. "Impact of Rape Law Reform: An Aggregate Analysis of Police Reports and Arrests." *Criminal Justice Review* 19: 1–23.

Berk, Richard A., Alec Campbell, Ruth Klap, and Bruce Western. 1992. "The Deterrent Effect of Arrest in Incidents of Domestic Violence: A Bayesian Analysis of Four Field Experiments." *American Sociological Review* 57: 698–708.

Bing, Leon. 1991. *Do or Die*. New York: Harper and Row.

Blau, Judith R., and Peter M. Blau. 1982. "The Cost of Inequality: Metropolitan Structure and Violent Crime." *American Sociological Review* 47: 114–129.

Brantingham, Paul, and Patricia Brantingham. 1984. *Patterns in Crime*. New York: Macmillan.

Brownmiller, Susan. 1975. *Against Our Will: Men, Women, and Rape*. New York: Simon and Schuster.

Brownstein, Henry H. 2000. *The Social Reality of Violence and Violent Crime*. Boston: Allyn and Bacon.

Bullock, Henry Allen. 1955. "Urban Homicide in Theory and Fact." *Journal of Criminal Law, Criminology, and Police Science* 45: 565–575.

Bureau of Justice Statistics. 1987. *Criminal Victimization in the United States, 1985*. Washington, DC: Department of Justice, Government Printing Office.

Bursik, Robert J., Jr., and Harold Grasmick. 1993. *Neighborhoods and Crime: The Dimensions of Effective Community Control*. New York: Lexington.

Cao, Liqun, Anthony Adams, and Vickie J. Jensen. 1997. "A Test of the Black Subculture of Violence Thesis: A Research Note." *Criminology* 35: 367–379.

Clinard, Marshall B., and Daniel J. Abbott. 1973. *Crime in Developing Countries: A Comparative Perspective*. New York: John Wiley & Sons.

Clinard, Marshall B., Richard Quinney, and John Wildeman. 1994. *Criminal Behavior Systems: A Typology*, 3rd ed. Cincinnati, OH: Anderson.

Cohen, Murray L., Ralph Garofalo, Richard Boucher, and Theoharis Seghorn. 1975. "The Psychology of Rapists." Pp. 113–140 in *Violence and Victims*, Stefan A. Pasternack, ed. New York: Spectrum Publications.

Curtis, Lynn A. 1974. "Victim Precipitation and Violent Crime." *Social Problems* 21: 594–605.

Daly, Martin, and Margo Wilson. 1988. *Homicide*. Hawthorne, NY: Aldine de Gruyter.

Dawson, John M., and Barbara Boland. 1993. *Murder in Large Urban Counties, 1988*. Washington, DC: U.S. Department of Justice, Bureau of Justice Statistics.

Deming, Mary Beard, and Ali Eppy. 1981. "The Sociology of Rape." *Sociology and Social Research* 65: 357–380.

Dixon, Jo, and Alan J. Lizotte. 1987. "Gun Ownership and the 'Southern Subculture of Violence.'" *American Journal of Sociology* 93: 383–405.

Dunford, Franklyn W., David Huizinga, and Delbert S. Elliott. 1990. "The Role of Arrest in Domestic Assault: The Omaha Police Experiment." *Criminology* 28: 183–206.

Dworkin, Andrea. 1987. *Intercourse*. New York: Free Press.

Erlanger, Howard S. 1974. "The Empirical Status of the Subculture of Violence Thesis." *Social Problems* 22: 280–292.

Erlanger, Howard S. 1979. "Estrangement, Machismo, and Gang Violence." *Social Science Quarterly* 60: 235–248.

Farrington, David. 1993. "Understanding and Preventing Bullying." Pp. 381–458 in *Crime and Justice: A Review of Research*, Michael Tonry, ed. Chicago: University of Chicago Press.

Felson, Richard B. 1993. "Motives for Sexual Coercion." Pp. 233–253 in *Aggression and Violence*, Richard B. Felson and James T. Tedeschi, eds. Washington, DC: American Psychological Association.

Federal Bureau of Investigation. 1999. *Crime in the United States, 1998*. Washington, DC: Department of Justice, Government Printing Office.

Ferracuti, Franco, Renato Lazzari, and Marvin E. Wolfgang, eds. 1970. *Violence in Sardinia*. Rome: Mario Bulzoni.

Ferracuti, Franco, and Graeme Newman. 1974. "Assaultive Offenses." Pp. 194–195 in *Handbook of Criminology*, Daniel Glaser, ed. Chicago: Rand McNally.

Ferraro, Kenneth F. 1995. *Fear of Crime: Interpreting Victimization Risk*. Albany: State University of New York.

Finkelhor, David. 1982. "Sexual Abuse: A Sociological Perspective." *Child Abuse and Neglect* 6: 95–102.

Finkelhor, David. 1984. *Child Sexual Abuse: New Theory and Research*. New York: Free Press.

Finkelhor, David. 1994. "International Epidemiology of Child Sexual Abuse." *Child Abuse and Neglect* 18: 409–417.

Fleisher, Mark. 1995. *Beggars and Thieves: Lives of Urban Street Criminals*. Madison: University of Wisconsin Press.

Fox, James Alan. 1996. "The Calm Before the Juvenile Crime Storm." *Population Today* 24: 4–5.

Friedrichs, David O. 1996. *Trusted Criminals: White-Collar Crime in Contemporary Society*. Belmont, CA: Wadsworth.

Garbarino, James. 1989. "The Incidence and Prevalence of Child Maltreatment." Pp. 219–261 in *Family Violence*, Lloyd E. Ohlin and Michael Tonry, eds. Chicago: University of Chicago Press.

Gelles, Richard J. 1985. "Family Violence." *Annual Review of Sociology* 11. 347–367.

Gelles, Richard J., and Murray Straus. 1979a. "Violence in the American Family." *Journal of Social Issues* 35: 15–39.

Gelles, Richard J., and Murray Straus. 1979b. *Family Violence*. Beverly Hills, CA: Sage.

Gibson, Lorne, Rick Linden, and Stuart Johnson. 1980. "A Situational Theory of Rape." *Canadian Journal of Criminology* 25: 51–65.

Green, Edward, and Russell P. Wakefield. 1979. "Patterns of Middle and Upper Class Homicide." *Journal of Criminal Law and Criminology* 70: 172–181.

Greenfeld, Lawrence A., Michael R. Rand, Diane Craven, Patsy A. Klaus, Craig A. Perkins, Cheryl Ringel, Greg Warchol, Cathy Mason, and James Alan Fox. 1998. *Violence Among Intimates: Analysis of Data on Crimes by Current or Former Spouses, Boyfriends, and Girlfriends*. Washington, DC: U.S. Department of Justice.

Groth, A. Nicholas. 1979. *Men Who Rape: The Psychology of the Offender*. New York: Plenum Press.

Hartnagel, Timothy F. 1980. "Subculture of Violence: Further Evidence." *Pacific Sociological Review* 23: 217–242.

Helfer, R., and H. Kempe. 1974. *The Battered Child*. Chicago: University of Chicago Press.

Hepburn, John R., and Harwin L. Voss. 1973. "Violent Behavior in Interpersonal Relationships." *The Sociological Quarterly* 14: 419–429.

Hills, Stuart L. 1980. *Demystifying Social Deviance*. New York: McGraw-Hill.

Hoff Sommers, Christine. 1994. *Who Stole Feminism? How Women Have Betrayed Women*. New York: Simon and Schuster.

Holmstrom, Lynda Lytle, and Ann Wolbert Burgess. 1990. "Rapists' Talk: Linguisitic Strategies to Control the Victim." Pp. 556–576 in *Deviant Behavior: Readings in the Sociology of Deviant Behavior*, Clifton D. Bryant, ed. New York: Hemisphere.

Jacobs, Joanne. 1990. "Some Are Trying to Broaden Definition of Rape." *Des Moines Register*, December 18, p. 9A.

Jensen, Gary F., and Maryaltani Karpos. 1993. "Managing Rape: Exploratory Research on the Behavior of Rape Statistics." *Criminology* 31: 363–385.

Klein, Malcolm. 1995. *The American Street Gang*. New York: Oxford University Press.

Knowles, Gordon James. 1999. "Male Prison Rape: A Search for Causation and Prevention." *Howard Journal of Criminal Justice* 38: 267–282.

Lee, Matthew R., and William B. Bankston. 1999. "Political Structure, Economic Inequality, and Homicide: A Cross-National Analysis." *Deviant Behavior* 19: 27–55.

Lloyd, D. 1991. *What Do We Know About Child Sexual Abuse Today?* Washington, DC: National Center on Child Abuse and Neglect.

Luckenbill, David F. 1984. "Murder and Assault." Pp. 19–45 in *Major Forms of Crime*, Robert F. Meier, ed. Beverly Hills, CA: Sage.

Lundsgaarde, Henry P. 1977. *Murder in Space City: A Cultural Analysis of Houston Homicide Patterns*. New York: Oxford University Press.

Maine, Margo. 2000. *Body Wars: Making Peace with Women's Bodies*. Carlsbad, CA: Gurze.

McClintock, F. H. 1963. *Crimes of Violence*. London: Macmillan.

Messner, Steven F., and Reid M. Golden. 1992. "Racial Inequality and Racially Disaggregated Homicide Rates: An Assessment of Alternative Theoretical Explanations." *Criminology* 30: 421–447.

O'Brien, Robert M. 1991. "Sex Ratios and Rape Rates: A Power-Control Theory." *Criminology* 29: 99–114.

Ogle, Robbin S., Daniel Maier-Katkin, and Thomas J. Bernard. 1995. "A Theory of Homicidal Behavior among Women." *Criminology* 33: 173–193.

Ohlin, Lloyd E., and Michael Tonry, eds. 1989. *Family Violence.* Chicago: University of Chicago Press.

Oliver, William. 1994. *The Violent Social World of Black Men.* New York: Lexington.

Pagelow, Mildred Daley. 1989. "The Incidence and Prevalence of Criminal Abuse of Other Family Members." Pp. 263–313 in *Family Violence.* Edited by Lloyd E. Ohlin and Michael Tonry. Chicago: University of Chicago Press.

Parker, Robert Nash. 1993. "Alcohol and Theories of Homicide." Pp. 113–141 in *New Directions in Criminological Theory: Advances in Criminological Theory*, Vol. 4, Freda Adler and William S. Laufer, eds. New Brunswick, NJ: Transaction.

Pate, Antony M., and Edwin E. Hamilton. 1992. "Formal and Informal Deterrents to Domestic Violence: The Dade County Spouse Assault Experiment." *American Sociological Review* 57: 691–697.

Perkins, Craig, and Patsy Klaus. 1996. *Criminal Victimization, 1994.* Washington, DC: Bureau of Justice Statistics.

Phillips, David P. 1983. "The Impact of Mass Media Violence on U.S. Homicides." *American Sociological Review* 48: 560–568.

Pillemer, Karl A., and Rosalie S. Wolf, eds. 1986. *Elder Abuse: Conflict in the Family.* Dover, MA: Auburn House.

Pino, Nathan, and Robert F. Meier. 1999. "Gender Differences in Rape Reporting." *Sex Roles*, 40: 979–990.

Posner, Richard A., and Katharine B. Silbaugh. 1996. *A Guide to America's Sex Laws.* Chicago: University of Chicago Press.

Ptacek, James. 1999. *Battered Women in the Courtroom: The Power of Judicial Responses.* Boston: Northeastern University Press.

Randall, Susan, and Vicki McNickle Rose. 1984. "Forcible Rape." Pp. 47–72 in *Major Forms of Crime*, Robert F. Meier, ed. Beverly Hills, CA: Sage.

Reiss, Albert J., Jr., and Jeffery A. Roth, eds. 1993. *Understanding and Preventing Violence.* Washington, DC: National Academy Press.

Rennison, Callie Marie, and Sarah Welchans. 2000. *Intimate Partner Violence.* Washington, DC: Bureau of Justice Statistics.

Rhodes, Richard. 1999. *Why They Kill.* New York: Alfred A. Knopf.

Rodgers, K. 1994. "Wife Assault: The Findings of a National Survey." *Juristat Service Bulletin* 14: 1–21.

Sagatun, Inger J., and Leonard P. Edwards. 1995. *Child Abuse and the Legal System.* Chicago: Nelson Hall.

Sarafino, Edward P. 1979. "An Estimate of Nationwide Incidence of Sexual Offenses against Children." *Child Welfare* 58: 127–134.

Scacco, Anthony M., Jr. 1982. *Male Rape*. New York: AMS.

Schwartz, Martin. D. 1987. "Gender and Injury in Spousal Assault." *Social Forces* 20: 61–75.

Schwendinger, Julia R., and Herman Schwendinger. 1981. "Rape, the Law, and Private Property." *Crime and Delinquency* 28: 271–291.

Scully, Diana, and Joseph Marolla. 1984. "Convicted Rapists' Vocabulary of Motive: Excuses and Justifications." *Social Problems* 31: 530–544.

Secretary of Health and Human Services. 1990. *Alcohol and Health: Seventh Special Report to the U.S. Congress*. Rockville, MD: National Institute of Alcohol Abuse and Alcoholism, Government Printing Office.

Sherman, Lawrence W., and Richard A. Berk. 1984. "Deterrent Effects of Arrest for Domestic Violence." *American Sociological Review* 49: 261–272.

Sherman, Lawrence W., and Douglas A. Smith. 1992. "Crime, Punishment, and Stake in Conformity: Legal and Informal Control of Domestic Violence." *American Sociological Review* 57: 680–690.

Simon, Rita J., and Sandra Baxter. 1989. "Gender and Violent Crime." Pp. 171–197 in *Violent Crime, Violent Criminals*. Neil Alan Weiner and Marvin E. Wolfgang, eds. Beverly Hills, CA: Sage.

Spohn, Cassia, and Julie Horney. 1992. *Rape Law Reform: A Grassroots Revolution and Its Impact*. New York: Plenum.

Straus, Murray, Richard J. Gelles, and Suzanne Steinmetz. 1980. *Behind Closed Doors: Violence in the American Family*. Garden City, NY: Doubleday Anchor.

Tang, C. 1994. "Prevalence of Spouse Aggression in Hong Kong." *Journal of Family Violence* 9: 347–356.

Tittle, Charles R. 1995. *Control Balance: Toward a General Theory of Deviance*. Boulder, CO: Westview Press.

Voss, Harwin L., and John R. Hepburn. 1968. "Patterns of Criminal Homicide in Chicago." *Journal of Criminal Law, Criminology and Police Science* 59: 234–243.

Wallace, H. 1999. *Family Violence — Legal, Medical, and Social Perspectives*, 2nd ed. Boston: Allyn and Bacon.

Warr, Mark. 1985. "Fear of Rape among Urban Women." *Social Problems* 32: 238–250.

Weiner, Neil Alan. 1989. "Violent Criminal Careers and 'Violent Criminal Careers': An Overview of the Research Literature." Pp. 35–138 in *Violent Crime, Violent Criminals and Pathways to Criminal Violence*, Neil Alan Weiner and Marvin E. Wolfgang, eds. Beverly Hills, CA: Sage.

Weiner, Neil Alan, and Marvin E. Wolfgang, eds. 1989. *Violent Crime, Violent Criminals and Pathways to Criminal Violence*. Beverly Hills, CA: Sage.

Widom, Cathy Spatz. 1989. "The Intergenerational Transmission of Violence." Pp. 137–201 in *Violent Crime, Violent Criminals*, Neil Alan Weiner and Marvin E. Wolfgang, eds. Beverly Hills, CA: Sage.

Widom, Cathy Spatz. 1992. *The Cycle of Violence. A Research in Brief*. Washington, DC: Bureau of Justice Statistics.

Wolfgang, Marvin E. 1958. *Patterns of Criminal Homicide*. Philadelphia: University of Pennsylvania Press.

Wolfgang, Marvin E., and Franco Ferracuti. 1982. *The Subculture of Violence.* Beverly Hills, CA: Sage; originally published 1967.

Wolfgang, Marvin E., and Margaret A. Zahn. 1983. "Homicide: Behavioral Aspects." Pp. 849–855 in *Encyclopedia of Crime and Justice*, Vol. 2, Sanford H. Kadish, ed. New York: Free Press.

Wright, J. A. 1995. "Using the Female Perspective in Prosecuting Rape." *Prosecutor* 29: 19–25.

Zorza, J., and L. Woods. 1994. *Mandatory Arrest: Problems and Possibilities.* New York: National Center on Women and Family Law.

CHAPTER 8   Nonviolent and White-Collar Crime

Abadinsky, Howard. 1981. *Organized Crime.* Boston: Allyn and Bacon.

Albanese, Jay S. 1996. *Organized Crime in America*, 3rd ed. Cincinnati, OH: Anderson.

Anderson, Annelise Graebner. 1979. *The Business of Organized Crime: A Cosa Nostra Family.* Stanford, CA: Hoover Institution Press.

Anderson, Robert T. 1965. "From Mafia to Cosa Nostra." *American Journal of Sociology* 81: 302–310.

Attorney General's Commission on Pornography. 1986. *Final Report.* Washington, DC: Government Printing Office.

Bayh, Birch. 1975. *Our Nation's Schools: A Report Card—'A' in School Violence and Vandalism.* Washington, DC: Government Printing Office.

Blankenburg, Erhard. 1976. "The Selectivity of Legal Sanctions: An Empirical Investigation of Shoplifting." *Law and Society Review* 11: 110–130.

Block, Alan A., and William J. Chambliss. 1981. *Organizing Crime.* New York: Elsevier.

Cameron, Mary Owen. 1964. *The Booster and the Snitch: Department Store Shoplifting.* New York: Free Press.

Catanzaro, Raimondo. 1992. *Men of Respect: A Social History of the Sicilian Mafia.* Translated by Raymond Rosenthal. New York: Free Press.

Chambliss, Bill. 1972. *Box Man: A Professional Thief's Journey.* New York: Harper and Row.

Chambliss, William J. 1978. *On the Take: From Petty Crooks to Presidents.* Bloomington: Indiana University Press.

Clark, John P., and Richard C. Hollinger. 1983. *Theft by Employees in Work Organizations.* Lexington, MA: Lexington.

Clinard, Marshall B., Richard Quinney, and John Wildeman. 1994. *Criminal Behavior Systems: A Typology*, 3rd ed. Cincinnati, OH: Anderson.

Cohen, Albert K. 1977. "The Concept of Criminal Organization." *British Journal of Criminology* 17: 97–111.

Cressey, Donald R. 1969. *Theft of a Nation.* New York: Harper and Row.

Cressey, Donald R. 1972. *Criminal Organization.* New York: Harper and Row.

Cromwell, Paul, Lee Parker, and Shawna Mobley. 1999. *The Five-Finger Discount: An Analysis of Motivations for Shoplifting.* Los Angeles: Roxbury.

Decker, Scott, Richard Wright, and Robert Logie. 1993. "Perceptual Deterrence among Active Residential Burglars: A Research Note." *Criminology* 31: 135–147.

Douglas, Jack D. 1974. "Watergate: Harbinger or the American Prince." *Theory and Society* 1: 89–97.

Eisenstadter, Werner J. 1969. "The Social Organization of Armed Robbery." *Social Problems* 17: 67–68.

Fleming, Zachary. 1999. *The Thrill of It All: Youthful Offenders and Auto Theft*. Los Angeles: Roxbury.

Gasser, Robert Louis. 1963. "The Confidence Man." *Federal Probation* 27: 47–54.

Gibbons, Don C. 1965. *Changing the Lawbreaker*. Englewood Cliffs, NJ: Prentice-Hall.

Greenberg, Peter S. 1982. "Fun and Games with Credit Cards." Pp. 377–379 in *Contemporary Criminology*, Leonard D. Savitz and Norman Johnston, eds. New York: John Wiley & Sons.

Hagan, Frank E. 1997. *Political Crime: Ideology and Criminality*. Boston: Allyn and Bacon.

Haller, Mark H. 1990. "Illegal Enterprise: A Theoretical and Historical Interpretation." *Criminology* 28: 207–235.

Hayno, David M. 1977. "The Professional Poker Player: Career Identification and the Problem of Respectability." *Social Problems* 24: 556–565.

Hollinger, Richard C., and John P. Clark. 1982. "Formal and Informal Social Controls over Employee Deviance." *The Sociological Quarterly* 23: 333–343.

Holzman, Harold R. 1982. "The Serious Habitual Property Offender as 'Moonlighter': An Empirical Study of Labor Force Participation among Robbers and Burglars." *Journal of Criminal Law and Criminology* 73: 1,774–1,792.

Homer, Frederic D. 1974. *Guns and Garlic: Myths and Realities of Organized Crime*. West Lafayette, IN: Purdue University Press.

Ianni, Francis A. J. 1972. *A Family Business: Kinship and Social Control in Organized Crime*. New York: Russell Sage.

Ianni, Francis A. J. 1975. *Black Mafia: Ethnic Succession in Organized Crime*. New York: Simon and Schuster.

Ianni, Francis A. J., and Elizabeth Reuss-Ianni. 1983. "Organized Crime: Overview," Pp. 1,094–1,106 in *Encyclopedia of Crime and Justice*, Vol. 3, Sanford Kadish, ed. New York: Free Press.

Inciardi, James A. 1975. *Careers in Crime*. Chicago: Rand McNally.

Inciardi, James A. 1984. "Professional Theft." Pp. 221–243 in *Major Forms of Crime*, Robert F. Meier, ed. Beverly Hills, CA: Sage.

Inciardi, James A. 1990. *Criminal Justice*. Fort Worth, TX: Harcourt.

Inciardi, James A., Ruth Horowitz, and Anne E. Pottieger. 1993. *Street Kids, Street Drugs, Street Crime: An Examination of Drug Use and Serious Delinquency in Miami*. Belmont, CA: Wadsworth.

Irwin, John, and James Austin. 1997. *It's about Time: America's Imprisonment Binge*, 2nd ed. Belmont, CA: Wadsworth.

Jacobs, James B., and Lauryn P. Gouldin. 1999. "Cosa Nostra: The Final Chapter?" *Crime and Justice: A Review of Research* 25: 129–189.

Jaworski, Leon. 1977. *The Right and the Power: The Prosecution of Watergate*. New York: Pocket Books.

Kempf, Kimberly. 1987. "Specialization and the Criminal Career." *Criminology* 25: 399–420.

Kenney, Dennis J., and James O. Finckenauer. 1995. *Organized Crime in America.* Belmont, CA: Wadsworth.

Klein, Malcolm W. 1995. *The American Street Gang: Its Nature, Prevalence and Control.* New York: Oxford University Press.

Klemke, Lloyd W. 1992. *The Sociology of Shoplifting: Boosters and Snitches Today.* Westport, CT: Praeger.

Klockars, Carl B. 1974. *The Professional Fence.* New York: Free Press.

Knapp Commission. 1977. "Official Corruption and the Construction Industry." Pp. 225–232 in *Official Deviance: Readings in Malfeasance, Misfeasance, and Other Forms of Corruption,* Jack D. Douglas and John M. Johnson, eds. Philadelphia: Lippincott.

Kruissink, M. 1990. *The Halt Program: Diversion of Juvenile Vandals.* The Hague, Netherlands: Research and Documentation Center, Dutch Ministry of Justice.

Lemert, Edwin M. 1958. "The Behavior of the Systematic Check Forger." *Social Problems* 6: 141–149.

Lemert, Edwin M. 1972. *Human Deviance, Social Problems, and Social Control,* 2nd ed. Englewood Cliffs, NJ: Prentice-Hall, 1972.

Light, Ivan. 1977. "Numbers Gambling among Blacks: A Financial Institution." *American Sociological Review* 42: 892–904.

MacKenzie, Doris Layton, and James W. Shaw. 1993. "The Impact of Shock Incarceration on Technical Violations and New Criminal Activities." *Justice Quarterly* 10: 463–487.

MacKenzie, Doris Layton, and Claire C. Souryal. 1991. "Boot Camp Survey." *Corrections Today,* October, pp. 90–96.

Matsueda, Ross L., Rosemary Gartner, Irving Piliavin, and Michael Polakowski. 1992. "The Prestige of Criminal and Conventional Occupations: A Subcultural Model of Criminal Activity." *American Sociological Review* 57: 752–770.

Maurer, David W. 1949. *The Big Con.* New York: Pocket Books.

Maurer, David W. 1964. *Whiz Mob.* New Haven, CT: College and University Press.

Meier, Robert F. 1983. "Shoplifting: Behavioral Aspects." Pp. 1,497–1,500 in *Encyclopedia of Crime and Justice,* Sanford H. Kadish, ed. New York: Free Press.

Miethe, Terance D., and Richard McCorkle. 1998. *Crime Profiles: The Anatomy of Dangerous Persons, Places, and Situations.* Los Angeles: Roxbury.

Minor, William J. 1975. "Political Crime, Political Justice and Political Prisoners." *Criminology* 12: 385–398.

Morash, Merry. 1984. "Organized Crime." Pp. 191–220 in *Major Forms of Crime,* Robert F. Meier, ed. Beverly Hills, CA: Sage.

Morris, Norval, and Gordon Hawkins. 1971. *The Honest Politician's Guide to Crime Control.* Chicago: University of Chicago Press.

Morris, Norval, and Michael Tonry. 1990. *Between Prison and Parole: Intermediate Punishments in a Rational Sentencing System.* New York: Oxford University Press.

O'Kane, James M. 1992. *The Crooked Ladder: Gangsters, Ethnicity, and the American Dream.* New Brunswick, NJ: Transaction.

Plate, Thomas. 1975. *Crime Pays!* New York: Ballantine.

Polsky, Ned. 1964. "The Hustlers." *Social Problems* 12: 9–17.

Prus, R., and C. R. D. Sharper. 1979. *The Road Hustler: The Career Contingencies of Professional Card and Dice Hustlers.* Toronto: Gage.

Ramirez Amaya, Atilio, Miguel Angel Amaya, Carlos Alberto Avilez, Josefina Ramirez, and Miguel Angel Reyes. 1987. "Justice and the Penal System in El Salvador." *Crime and Social Justice* 30: 1–27.

Reuter, Peter. 1983. *Disorganized Crime: Illegal Markets and the Mafia.* Cambridge, MA: MIT Press.

Rhodes, Robert P. 1984. *Organized Crime: Crime Control vs. Civil Liberties.* New York: Random House.

Richards, Pamela. 1979. "Middle-Class Vandalism and Age-Status Conflicts." *Social Problems* 26: 482–497.

Robin, Gerald. 1974. "White-Collar Crime and Employee Theft." *Crime and Delinquency* 20: 251–262.

Roebuck, Julian B. 1983. "Professional Criminal: Professional Thief." Pp. 1,260–1,263 in *Encyclopedia of Crime and Justice*, Vol. 3, Sanford H. Kadish, ed. New York: Free Press.

Roebuck, Julian B., and Mervyn L. Cadwallader. 1961. "The Negro Armed Robber as a Criminal Type: The Construction and Application of a Typology." *Pacific Sociological Review* 4: 21–26.

Roebuck, Julian B., and Ronald C. Johnson. 1963. "The 'Short Con' Man." *Crime and Delinquency* 10: 235–248.

Rowan, Roy. 1986. "The Biggest Mafia Bosses." *Fortune*, November 10, pp. 24–38.

Sanchez Jankowski, Martin. 1991. *Islands in the Street: Gangs and American Urban Society.* Berkeley: University of California Press.

Schafer, Stephen. 1974. *The Political Criminal: The Problem of Morality and Crime.* New York: Free Press.

Shover, Neal, and David Honaker. 1996. "The Socially Bounded Decision Making of Persistent Property Thieves." Pp. 10–22 in *In Their Own Words: Criminals on Crime*, Paul Cromwell, ed. Los Angeles: Roxbury.

Simon, Carl P., and Ann D. Witte. 1982. *Beating the System: The Underground Economy.* Boston: Auburn House.

Smith, Dwight C. 1975. *The Mafia Mystique.* New York: Basic Books.

Steffensmeier, Darrell J. 1986. *The Fence: In the Shadow of Two Worlds.* Totowa, NJ: Rowan and Littlefield.

Stevenson, Robert J. 1998. *The Boiler Room and Other Telephone Sales Scams.* Champaign, IL: University of Illinois Press.

Sutherland, Edwin H. 1937. *The Professional Thief.* Chicago: University of Chicago Press.

Traub, Stuart H. 1997. "Battling Employee Crime: A Review of Corporate Strategies and Programs." *Crime and Delinquency* 42: 244–256.

Tunnell, Kenneth D. 1992. *Choosing Crime: The Criminal Calculus of Property Offenders.* Chicago: Nelson Hall.

Turk, Austin T. 1982. *Political Criminality: The Defiance and Defense of Authority.* Beverly Hills, CA: Sage.

Tyler, Gus. 1981. "The Crime Corporation." Pp. 273–290 in *Current Perspectives in Criminal Behavior*, 2nd ed., Abraham S. Blumberg, ed. New York: Alfred A. Knopf.

Vera Institute of Justice. 1981. *Felony Arrests*, rev. ed. New York: Longman.

Wade, Andrew L. 1967. "Social Processes in the Act of Juvenile Vandalism." Pp. 94–109 in *Criminal Behavior Systems: A Typology*, Marshall B. Clinard and Richard Quinney, eds. New York: Holt, Rinehart and Winston.

Walker, Andrew. 1981. "Sociology and Professional Crime." Pp. 153–178 in *Current Perspectives in Criminal Behavior*, 2nd ed., Abraham S. Blumberg, ed. New York: Alfred A. Knopf.

Walker, Leslie. 2000. "Taking a Whack at Hackers." *Washington Post*, January 13, p. E-1.

Wattenberg, William W., and James Balistieri. 1952. "Automobile Theft: A 'Favored-Group' Delinquency." *American Journal of Sociology* 57: 575–579.

Williams, Kristen M., and Judith Lucianovic. 1979. *Robbery and Burglary: A Study of the Characteristics of the Persons Arrested and the Handling of Their Cases in Court.* Washington, DC: Institute for Law and Social Research.

Wright, Richard T., and Scott H. Decker. 1996. "Choosing the Target." Pp. 34–46 in *In Their Own Words: Criminals on Crime*, Paul Cromwell, ed. Los Angeles: Roxbury.

CHAPTER 9   White-Collar and Corporate Crime

Adams, Stuart. 1989. "Ripping Off Books." Pp. 32–33 in *Degrees of Deviance: Student Accounts of Their Deviant Behavior*, Stuart Henry, ed. Brookfield, VT: Avebury.

Albanese, Jay S. 1995. *White Collar Crime in America*. Englewood Cliffs, NJ: Prentice Hall.

Beirne, Piers, and James Messerschmidt. 2000. *Criminology*, 3rd ed. Boulder, CO: Westview.

Benson, Michael L. 1996. "Deny the Guilty Mind: Accounting for Involvement in White-Collar Crime." Pp. 66–73 in *In Their Own Words: Criminals on Crime*, Paul Cromwell, ed. Los Angeles: Roxbury.

Blankenship, Michael B., ed. 1993. *Understanding Corporate Criminality*. New York: Garland.

Braithwaite, John. 1984. *Corporate Crime in the Pharmaceutical Industry*. London: Routledge and Kegan Paul.

Braithwaite, John. 1985. "White Collar Crime." *Annual Review of Sociology* 11: 1–25.

Brickey, Kathleen F. 1995. *Corporate and White-Collar Crime: Cases and Materials*, 2nd ed. Boston: Little, Brown.

Calavita, Kitty, Henry R. Pontell, and Robert H. Tillman. 1997. *Big Money Crime: Fraud and Politics in the Savings and Loan Crisis*. Berkeley: University of California Press.

Clinard, Marshall B. 1952. *The Black Market: A Study of White Collar Crime*. New York: Rinehart and Company.

Clinard, Marshall B. 1979. *Illegal Corporate Behavior*. Washington, DC: Government Printing Office.

Clinard, Marshall B. 1990. *Corporate Corruption: The Abuse of Power*. New York: Praeger.

Clinard, Marshall B., and Peter C. Yeager. 1980. *Corporate Crime*. New York: Free Press.

Cohen, Albert K. 1966. *Deviance and Control*. Englewood Cliffs, NJ: Prentice-Hall.

Coleman, James W. 1997. *The Criminal Elite: The Sociology of White Collar Crime*, 4th ed. New York: St. Martin's Press.

Coleman, James W., and Linda Ramos. 1998. "Subcultures and Deviant Behavior in the Organizational Context." *Research in the Sociology of Organizations* 15: 3–34.

Cressey, Donald R. 1953. *Other People's Money*. New York: Free Press.

Cullen, Francis T., William J. Maakestad, and Gray Cavender. 1987. *Corporate Crime Under Attack: The Ford Pinto Case and Beyond*. Cincinnati: Anderson.

Ermann, David and Richard Lundman, eds. 1982. *Corporate Deviance*. New York: Oxford University Press.

Fisse, Brent, and John Braithwaite. 1983. *The Impact of Publicity on Corporate Offenders*. Albany: State University of New York Press.

Friedrichs, David O. 1996. *Trusted Criminals in Contemporary Society*. Belmont, CA: Wadsworth.

Gandossey, Robert P. 1985. *Bad Business: The OPM Scandal and the Seduction of the Establishment*. New York: Basic Books.

Geis, Gilbert. 1967. "The Heavy Electrical Equipment Antitrust Cases of 1961." Pp. 139–150 in *Criminal Behavior Systems*, Marshall B. Clinard and Richard Quinney, eds. New York: Holt, Rinehart and Winston.

Geis, Gilbert. 1972. "Criminal Penalties for Corporate Criminals." *Criminal Law Bulletin*, 16: 380–381.

Geis, Gilbert. 1995. "A Review, Rebuttal, and Reconciliation of Cressey and Braithwaite and Fisse on Criminological Theory and Corporate Crime." Pp. 399–428 in *The Legacy of Anomie Theory*, Freda Adler and William S. Laufer, eds. New Brunswick, NJ: Transaction.

Geis, Gilbert, and Colin Goff. 1983. "Introduction." In *White Collar Crime: The Uncut Version*, Edwin H. Sutherland. New Haven, CT: Yale University Press.

Geis, Gilbert, and Robert F. Meier, eds. 1977. *White-Collar Crime*. New York: Free Press.

Geis, Gilbert, Robert F. Meier, and Lawrence S. Salinger, eds. 1995. *White-Collar Crime: Classic and Contemporary Views*. New York: Free Press.

Han, Valerie P. 2000. *Business on Trial: The Civil Jury and Corporate Responsibility*. New Haven, CT: Yale University Press.

Hartung, Frank E. 1950. "White-Collar Offenses in the Wholesale Meat Industry in Detroit." *American Journal of Sociology* 56: 25–35.

Helmkamp, James C., Kitty C. Townsend, and Jenny A. Sundra. 1997. "How Much Does White Collar Crime Cost?" Unpublished paper. Morgantown, West Virginia: National White Collar Crime Center.

Kaban, Elif. 1996. "Copyright Experts Worldwide Seek 'Basic Level of Protection' on Internet." *Des Moines Register*, December 3, p. 8S.

Kolko, Gabriel. 1965. *Railroads and Regulation, 1877–1916*. Princeton, NJ: Princeton University Press.

Lemert, Edwin M. 1972. *Human Deviance, Social Problems, and Social Control*, 2nd ed. Englewood Cliffs, NJ: Prentice Hall.

McKay, John P., Bennett D. Hill, and John Buckler. 1991. *A History of Western Society*. Boston: Houghton Mifflin.

Meier, Robert F., Leslie W. Kennedy, and Vincent F. Sacco. 2000. "Crime and the Criminal Event Perspective." Chapter 1 in *The Structure and Process of Crime*, Robert F. Meier, Leslie W. Kennedy, and Vincent F. Sacco, eds. New Brunswick, NJ: Transaction.

Meier, Robert F. and James F. Short, Jr. 1982. "The Consequences of White-Collar Crime." Pp. 23–49 in *White-Collar Crime: An Agenda for Research*, Herbert Edelhertz and Thomas A. Overcast, eds. Lexington, MA: Lexington Books.

Miethe, Terance D., and Robert F. Meier. 1994. *Crime and Its Social Context.* Albany: State University of New York Press.

Nettler, Gwynn. 1974. "Embezzlement Without Problems." *British Journal of Criminology* 14: 70–77.

Packer, Herbert L. 1968. *The Limits of the Criminal Sanction.* Stanford: Stanford University Press.

Quinney, Richard. 1964. "The Study of White-Collar Crime: Toward a Reorientation in Theory and Research." *Journal of Criminal Law, Criminology, and Police Science* 55: 208–214.

Rebovich, Don, and Jenny Layne. 2000. *The National Public Survey on White Collar Crime.* Morgantown, WV: National White Collar Crime Center.

Rosoff, Stephen M., Henry N. Pontell, and Robert Tillman. 1998. *Profit Without Honor: White-Collar Crime and the Looting of America.* Englewood Cliffs, NJ: Prentice Hall.

Schlegel, Kip, and David Weisburd, eds. 1992. *White-Collar Crime Reconsidered.* Boston: Northeastern University Press.

Schuessler, Karl. 1973. "Introduction." In Edwin H. Sutherland, *On Analyzing Crime.* Chicago: University of Chicago Press.

Shapiro, Susan P. 1984. *Wayward Capitalists: Target of the Securities and Exchange Commission.* New Haven, CT: Yale University Press.

Shapiro, Susan P. 1990. "Collaring the Crime, Not the Criminal: Reconsidering the Concept of White-Collar Crime." *American Sociological Review* 55: 346–365.

Shover, Neal, and Kevin M. Bryant. 1993. "Theoretical Explanations of Corporate Crime." Pp. 141–176 in *Understanding Corporate Crime,* Michael Blankenship, ed. New York: Garland.

Simon, David R., and Frank E. Hagan 1999. *White-Collar Deviance.* Boston: Allyn and Bacon.

Simpson, Sally S. 1987. "Cycles of Illegality: Antitrust Violations in Corporate America." *Social Forces* 65: 943–963.

Simpson, Sally S., M. Lyn Exum, and N. Craig Smith. 2000. "The Social Control of Corporate Criminals: Shame and Informal Sanction Threats." Pp. 141–158 in *Of Crime and Criminality: The Use of Theory in Everyday Life,* Sally S. Simpson, ed. Thousand Oaks, CA: Pine Forge.

Soble, Ronald L., and Robert E. Dallos. 1975. *The Impossible Dream: The Equity Funding Story: The Fraud of the Century.* New York: Putnam.

Sutherland, Edwin H. 1940. "White-Collar Criminality." *American Sociological Review* 5: 1–12.

Sutherland, Edwin. 1949. *White Collar Crime.* New York: Dryden.

Sutherland, Edwin H. 1983. *White-Collar Crime: The Uncut Version.* Introduction by Gilbert Geis and Colin Goff. New Haven, CT: Yale University Press.

Tien, James M., Thomas F. Rich, and Michael F. Cahn. 1986. *Electronic Fund Transfer Systems Fraud: Computer Crime. Public Systems Evaluation, Inc., 1985.* Washington, DC: Department of Justice, Government Printing Office.

Tonry, Michael, and Albert J. Reiss, Jr., eds. 1993. *Beyond the Law: Crime in Complex Organizations.* Chicago: University of Chicago Press.

Vaughan, Diane. 1982. "Transaction Systems and Unlawful Organizational Behavior." *Social Problems* 29: 373–379.

Vaughan, Diane. 1983. *Controlling Unlawful Organizational Behavior: Social Structure and Corporate Misconduct.* Chicago: University of Chicago Press.

Vaughan, Diane. 1996. *The Challenger Launch Decision*. Chicago: University of Chicago Press.

Vaughan, Diane. 1999. "The Dark Side of Organizations: Mistake, Misconduct, and Disaster." *Annual Review of Sociology* 25: 271–305.

Vaughan, Diane, and Giovanna Carlo. 1975. "The Appliance Repairman: A Study of Victim-Responsiveness." *Journal of Research in Crime and Delinquency* 12: 153–161.

Wheeler, Stanton, David Weisburd, and Nancy Bode. 1982. "Sentencing the White-Collar Offender: Rhetoric and Reality." *American Sociological Review* 47: 641–659.

Zuckerman, M. J. 1996. "FBI Takes on Security Fight in Cyberspace." *USA Today*, November 21, p. 4B.

CHAPTER 10   Drug Use and Addiction

Adler, Patricia A. 1993. *Wheeling and Dealing: An Ethnography of an Upper-Level Drug Dealing and Smuggling Community*, 2nd ed. New York: Columbia University Press.

Advisory Council on the Misuse of Drugs. 1982. *Treatment and Rehabilitation: Report of the Advisory Council on the Misuse of Drugs*. London: Her Majesty's Stationery Office.

Agar, Michael. 1973. *Ripping and Running: A Formal Ethnography of Urban Heroin Addicts*. New York: Seminar Press.

Agar, Michael, and Richard C. Stephens. 1975. "The Methadone Street Scene: The Addict's View." *Psychiatry* 38: 381–387.

Akers, Ronald L. 1992. *Drugs, Alcohol, and Society: Social Structure, Process, and Policy*. Belmont, CA: Wadsworth.

Anglin, M. Douglas, Yih-Ing Hser, and William McGlothlin. 1987. "Sex Differences in Addict Careers. 2. Becoming Addicted." *American Journal of Drug and Alcohol Abuse* 13: 59–71.

Ball, John C., and Carl D. Chambers, eds. 1970. *The Epidemiology of Opiate Addiction in the United States*. Springfield, IL: Charles C. Thomas.

Ball, John C., and M. P. Lau. 1966. "The Chinese Narcotic Addict in the United States." *Social Forces* 45: 68–72.

Ball, John C., Lawrence Rosen, John A. Flueck, and David N. Nurco. 1982. "Lifetime Criminality of Heroin Addicts in the United States." *Journal of Drug Issues* 12: 225–239.

Baum, Dan. 1996. *Smoke and Mirrors: The War on Drugs and the Politics of Failure*. Boston: Little, Brown.

Becker, Howard S. 1973. *Outsiders: Studies in the Sociology of Deviance*, enlarged ed. New York: Free Press.

Bennett, Trevor. 1986. "A Decision-Making Approach to Opioid Addiction." Pp. 83–102 in *The Reasoning Criminal: Rational Choice Perspectives on Offending*, Derek B. Cornish and Ronald V. Clarke, eds. New York: Springer-Verlag.

Biernacki, Patricia. 1986. *Pathways from Heroin Addiction: Recovery without Treatment*. Philadelphia: Temple University Press.

Blackwell, Judith Stephenson. 1983. "Drifting, Controlling, and Overcoming: Opiate Users Who Avoid Becoming Chronically Dependent." *Journal of Drug Issues* 13: 219–235.

Blum, Richard H., with Eva Blum and Emily Garfield. 1976. *Drug Education: Results and Recommendations*. Lexington, MA: Lexington Books.

Booth, Leo. 1991. *When God Becomes a Drug: Breaking the Chains of Religious Addiction and Abuse*. Los Angeles: Tarcher.

Chambers, Carl D., and Michael T. Harter. 1987. "The Epidemiology of Narcotic Abuse among Blacks in the United States, 1935–1980." Pp. 191–223 in *Chemical Dependencies: Patterns, Costs, and Consequences*, Carl D. Chambers, James A. Inciardi, David M. Peterson, Harvey A. Siegal, and O. Z. White, eds. Athens: Ohio University Press.

Chein, Isador, Donald L. Gerard, Robert S. Lee, and Eva Rosenfeld. 1964. *The Road to H.* New York: Basic Books.

Chitwood, Dale D., James E. Rivers, and James A. Inciardi. 1996. *The American Pipe Dream: Crack Cocaine and the Inner City.* Fort Worth, TX: Harcourt Brace.

Cohen, Sidney. 1984 (April–June). "Recent Developments in the Use of Cocaine." *Bulletin on Narcotics:* 9.

Cohen, Sidney. 1987. "Causes of the Cocaine Outbreak." Pp. 3–9 in *Cocaine: A Clinician's Handbook*, Arnold M. Washton and Mark S. Gold, eds. New York: Guilford Press.

Coombs, Robert H. 1981. "Drug Abuse as a Career." *Journal of Drug Issues* 4: 369–387.

Courtwright, David T. 1982. *Dark Paradise.* Cambridge, MA: Harvard University Press.

Cox, Terrance, Michael R. Jacobs, A. Eugene Leblanc, and Joan A. Marshman. 1983. *Drugs and Drug Abuse: A Reference Text.* Toronto: Addiction Research Foundation.

Cuskey, Walter R., and Richard B. Wathey. 1982. *Female Addiction: A Longitudinal Study.* Lexington, MA: Lexington Books.

Dembo, Richard, and Michael Miran. 1976. "Evaluation of Drug Prevention Programs by Youths in a Middle-Class Community." *The International Journal of Addictions* 11: 881–903.

Estroff, Todd Wilk. 1987. "Medical and Biological Consequences of Cocaine Abuse." Pp. 23–32 in *Cocaine: A Clinician's Handbook*, Arnold M. Washton and Mark S. Gold, eds. New York: Guilford Press.

Faupel, Charles E., and Carl B. Klockars. 1987. "Drugs-Crime Connections: Elaborations from the Life Histories of Hard-Core Heroin Addicts." *Social Problems* 34: 54–68.

Finestone, Harold. 1957. "Cats, Kicks, and Color." *Social Problems* 5: 3–13.

Galliher, John F., and Allyn Walker. 1977. "The Puzzle of the Social Origins of the Marihuana Tax Act of 1937." *Social Problems* 24: 366–376.

Gonzales, Laurence. 1987. "Addiction and Rehabilitation." *Playboy* 34: 149–152, 182–198.

Goode, Erich. 1999. *Drugs in American Society*, 5th ed. New York: Alfred A. Knopf.

Goode, Erich. 1997. *Between Politics and Reason: The Drug Legalization Debate.* New York: St. Martin's Press.

Goodstadt, Michael S., Margaret A. Sheppard, and Godwin C. Chan. 1982. "Relationships between Drug Education and Drug Use: Carts and Horses." *Journal of Drug Issues* 12: 431–441.

Grabowski, John, Maxine L. Stitzer, and Jack E. Henningfield. 1984. *Behavioral Intervention Techniques in Drug Abuse Treatment.* Washington, DC: National Institute of Drug Abuse.

Graevan, David B., and Kathleen A. Graevan. 1983. "Treated and Untreated Addicts: Factors Associated with Participation in Treatment and Cessation of Heroin Use." *Journal of Drug Issues* 13: 207–218.

Grinspoon, Lester, and James B. Bakalar. 1979. "Cocaine." Pp. 241–247 in *Handbook on Drug Abuse*, Robert I. Dupont, Avram Goldstein, and John O'Donnell, eds. Washington, DC: National Institute of Drug Abuse.

Haas, Ann Pollinger, and Herbert Hendin. 1987. "The Meaning of Chronic Marijuana Use among Adults: A Psychosocial Perspective." *Journal of Drug Issues* 17: 333–348.

Hanson, Bill, George Beschner, James M. Walters, and Elliott Bovelle. 1985. *Life with Heroin*. Lexington, MA: Lexington Books.

Hart, Don. 1989. "Drugs among Medical Students." Pp. 45–49 in *Degrees of Deviance: Student Accounts of Their Deviant Behavior*, Stuart Henry, ed. Brookfield, VT: Avebury.

Hser, Yih-Ing, M. Douglas Anglin, and William McGlothlin. 1987. "Addict Careers. 1. Initiation of Use." *American Journal of Drug and Alcohol Abuse* 13: 33–57.

Hunt, Leon Gibson, and Carl D. Chambers. 1976. *The Heroin Epidemics: A Study of Heroin Use in the United States, 1965–1975*. New York: Spectrum Publications.

Inciardi, James A. 1977. *Methadone Diversion: Experience and Issues*. Washington, DC: National Institute of Drug Abuse.

Inciardi, James A. 1986. *The War on Drugs*. Palo Alto, CA: Mayfield.

Inciardi, James A., ed. 1990. *Handbook of Drug Control in the United States*. New York: Greenwood Press.

Inciardi, James A., Ruth Horowitz, and Anne E. Pottieger. 1993. *Street Kids, Street Drugs, Street Crime: An Examination of Drug Use and Serious Delinquency in Miami*. Belmont, CA: Wadsworth.

Inciardi, James A., Dorothy Lockwood, and Anne Pottieger. 1993. *Women and Crack Cocaine*. New York: Macmillan.

Inciardi, James A., and Duane C. McBride. 1990 (September–October). "Legalizing Drugs: A Normless, Naive Idea." *The Criminologist* 15: 1, 3–4.

Inciardi, James A., and Anne E. Pottieger. 1986. "Drug Use and Crime among Two Cohorts of Women Narcotics Users: An Empirical Assessment." *Journal of Drug Issues* 16: 91–106.

Inciardi, James A., and Christine Saum. 1996. "Legalization Madness." *The Public Interest* 123: 72–82.

Inciardi, James A., Hilary L. Surratt, Dale D. Chitwood, and Clyde B. McCoy. 1996. "The Origins of Crack." Pp. 1–14 in *The American Pipe Dream: Crack Cocaine and the Inner City*, Dale D. Chitwood, James E. Rivers, and James A. Inciardi, eds. Fort Worth: Harcourt Brace.

Johnson, Bruce D., Paul J. Goldstein, Edward Preble, James Schmidler, Douglas S. Lipton, Barry Spunt, and Thomas Miller. 1985. *Taking Care of Business: The Economics of Crime by Heroin Abusers*. Lexington, MA: Lexington Books.

Johnston, Lloyd D., Patrick M. O'Malley, and Jerald G. Bachman. 1993a. *National Survey Results on Drug Use from the Monitoring the Future Study, 1975–1992: Volume 1, Secondary School Students*. Washington, DC: NIH Publication 93-3597, Government Printing Office.

Johnston, Lloyd D., Patrick M. O'Malley, and Jerald G. Bachman. 1993b. *National Survey Results on Drug Use from the Monitoring the Future Study, 1975–1992: Volume 2, College Students and Young Adults*. Washington, DC: NIH Publication 93-3598, Government Printing Office.

Johnston, Lloyd D., Patrick M. O'Malley, and Jerald G. Bachman. 1996. "Press Release on Findings from 1996 Survey of Drug Use." Ann Arbor, MI: News and Information Service, University of Michigan.

Johnston, Lloyd D., Patrick M. O'Malley, and Jerald G. Bachman. 1999. "Press Release on Findings from 1999 Survey of Drug Use." Ann Arbor, MI: News and Information Service, University of Michigan.

Jones, Paul. 1989. "Marijuana Smoking." Pp. 59–61 in *Degrees of Deviance: Student Accounts of Their Deviant Behavior*, Stuart Henry, ed. Brookfield, VT: Avebury.

Jorquez, James S. 1983. "The Retirement Phase of Heroin Using Careers." *Journal of Drug Issues* 13: 343–365.

Kaplan, John. 1983. *The Hardest Drugs: Heroin and Public Policy.* Chicago: University of Chicago Press.

Kaufman, Joel, Sidney Wolfe, and the Public Citizen Health Research Group. 1983. *Over the Counter Pills that Don't Work.* New York: Pantheon.

Koutroulis, Glenda Y. 1998. "Withdrawal from Injecting Heroin Use: Thematizing the Body." *Critical Public Health* 8: 207–224.

Kramer, John C., Vitezslav S. Fishman, and Don C. Littlefield. 1967. "Amphetamine Abuse Patterns and Effects of High Doses Taken Intravenously." *Journal of the American Medical Association* 201: 305–309.

Kreek, Mary Jeanne. 1979. "Methadone in Treatment: Physiological and Pharmacological Issues." Pp. 57–86 in *Handbook on Drug Abuse*, Robert I. Dupont, Avram Goldstein, and John O'Donnell, eds. Washington, DC: National Institute of Drug Abuse.

Levison, Peter K., Dean R. Gerstein, and Deborah R. Maloff. 1983. *Commonalities in Substance Abuse and Habitual Behavior.* Lexington, MA: Lexington Books.

Lindesmith, Alfred R. 1968. *Addiction and Opiates.* Chicago: Aldine.

Lindesmith, Alfred R. 1975. "A Reply to McAuliffe and Gordon's 'A Test of Lindesmith's Theory of Addiction.'" *American Journal of Sociology* 81: 149–150.

Lindesmith, Alfred R., and John H. Gagnon. 1964. "Anomie and Drug Addiction." Pp. 162–178 in *Anomie and Deviant Behavior: A Discussion and Critique*, Marshall B. Clinard, ed. New York: Free Press.

McAuliffe, William E. 1986. "Psychoactive Drug Use by Young and Future Physicians." *New England Journal of Medicine* 315: 805–810.

McAuliffe, William E., and Robert A. Gordon. 1974. "A Test of Lindesmith's Theory of Addiction: Frequency of Euphoria among Long-Term Addicts." *American Journal of Sociology* 79: 795–840.

McBride, Robert B. 1983. "Business as Usual: Heroin Distribution in the United States." *Journal of Drug Issues* 13: 147–166.

Meier, Robert F., and Gilbert Geis. 1997. *Victimless Crime? Prostitution, Homosexuality, Drugs, and Abortion.* Los Angeles: Roxbury.

Merton, Robert K. 1968. *Social Theory and Social Structure*, enlarged ed. New York: Free Press.

Milby, Jesse B. 1981. *Addictive Behavior and Its Treatment.* New York: Springer.

Morgan, H. Wayne. 1981. *Drugs in America: A Social History, 1800–1980.* Syracuse, NY: Syracuse University Press.

Musto, David F. 1973. *The American Disease: The Origins of Narcotics Control.* New Haven, CT: Yale University Press.

Nadelmann, Ethan A. 1991. "The Case for Legalization." Pp. 17–44 in *The Drug Legalization Debate*, James A. Inciardi, ed. Newbury Park, CA: Sage.

National Commission on Marihuana and Drug Abuse. 1972. *Marihuana: A Signal of Misunderstanding.* Washington, DC: Government Printing Office.

Newcomb, Michael D., and Peter M. Bentler. 1990. "Antecedents and Consequences of Cocaine Use: An Eight-Year Study from Early Adolescence to Young Adulthood."

Pp. 158–181 in *Straight and Deviant Pathways from Childhood to Adulthood*. Cambridge: Cambridge University Press.

Newman, Robert G., Sylvia Bashkow, and Margot Cates. 1973. "Arrest Histories before and after Admission to a Methadone Maintenance Program." *Contemporary Drug Problems* 2: 417–430.

Office of National Drug Control Policy. 1996. *National Drug Control Strategy*. Washington, DC: Government Printing Office.

Olin, William. 1980. *Escape from Utopia: My Ten Years in Synanon*. Santa Cruz, CA: Unity Press.

Orcutt, James D., and J. Blake Turner. 1993. "Shocking Numbers and Graphic Accounts: Quantified Images of Drug Problems in the Print Media." *Social Problems* 40: 190–206.

Peele, Stanton. 1990. *The Diseasing of America: Addiction Treatment Out of Control*. Lexington, MA: Lexington Books.

Petersen, Robert C. 1984. "Marijuana Overview." Pp. 1–17 in *Correlates and Consequences of Marijuana Use*, Meyer D. Glantz, ed. Rockville, MD: National Institute of Drug Abuse, U.S. Department of Health and Human Services.

Platt, Jerome J. 1986. *Heroin Addiction*. Melbourne, FL: Krieger.

Poundstone, William. 1983. *Big Secrets: The Uncensored Truth about All Sorts of Stuff You Are Never Supposed to Know*. New York: Quill.

Ratner, Mitchell S., ed. 1993. *Crack Pipe as Pimp: An Ethnographic Investigation of Sex-for-Crack Exchanges*. New York: Lexington.

Ray, Oakley. 1983. *Drugs, Society, and Human Nature*, 3rd ed. St. Louis, MO: C. V. Mosby.

Reznikov, Diane. 1987 (March). "States Spend $1.3 Billion on Substance Abuse Services, Treat 305,000 Drug Abuses in FY 1985." *NIDA (National Institute of Drug Abuse) Notes* 2: 11–12.

Rosecran, Jeffrey S., and Henry I. Spitz. 1987. "Cocaine Reconceptualized: Historical Overview." Pp. 5–16 in *Cocaine Abuse: New Directions in Treatment and Research*, Henry I. Spitz and Jeffrey S. Rosecran, eds. New York: Brunner/Mazel.

Rosenbaum, Marsha. 1981. *Women on Heroin*. New Brunswick, NJ: Rutgers University Press.

Rowe, Dennis. 1987. "UN Drug Meet Draws 138 Nations." *C. J. International* 3: 9.

Schaef, Ann Wilson. 1987. *When Society Becomes an Addict*. New York: Harper and Row.

Schlaadt, Richard G., and Peter T. Shannon. 1994. *Drugs*, 4th ed. Englewood Cliffs, NJ: Prentice-Hall.

Schur, Edwin M. 1983. *Labeling Women Deviant: Gender, Stigma, and Social Control*. New York: Random House.

Seidenberg, Robert. 1976. "Advertising and Drug Acculturation." Pp. 19–25 in *Socialization in Drug Use*, Robert H. Coombs, Lincoln J. Fry, and Patricia G. Lewis, eds. Cambridge, MA: Schenkman.

Shalala, Donna. 1999. *Press Release on Findings from 1998 National Household Survey on Drug Abuse*. Washington, DC: Substance Abuse and Mental Health Services Administration, Health and Human Services.

Siegal, Harvey Alan. 1987. "Current Patterns of Psychoactive Drug Use: Epidemiological Observations." Pp. 45–113 in *Chemical Dependencies: Patterns, Costs, and Consequences*, Carl D. Chambers, James A. Inciardi, David M. Peterson, Harvey A. Siegal, and O. Z. White, eds. Athens: Ohio University Press.

Skoll, Geoffrey R. 1992. *Walk the Walk and Talk the Talk: An Ethnography of a Drug Abuse Treatment Facility.* Philadelphia: Temple University Press.

Smart, Lisa. 1989. "College Kids as Coke Characters." Pp. 69–72 in *Degrees of Deviance: Student Accounts of Their Deviant Behavior,* Stuart Henry, ed. Brookfield, VT: Avebury.

Smith, David E., Donald R. Wesson, and Richard B. Seymour. 1979. "The Abuse of Barbiturates and Other Sedative-Hypnotics." Pp. 233–240 in *Handbook on Drug Abuse,* Robert I. Dupont, Avram Goldstein, and John O'Donnell, eds. Washington, DC: National Institute of Drug Abuse.

Stephens, Richard C. 1987. *Mind-Altering Drugs: Use, Abuse, and Treatment.* Beverly Hills, CA: Sage.

Stephens, Richard C. 1991. *The Street Addict Role: A Theory of Heroin Addiction.* Albany: State University of New York Press.

Stephens, Richard C., and Emily Cottrell. 1972. "A Follow-Up Study of 200 Narcotic Addicts Committed for Treatment under the Narcotic Addict Rehabilitation Act (NARA)." *British Journal of Addiction* 67: 45–53.

Sullivan, Barbara. 1990. "Society's Hang Ups Wear Other Labels Now." *Des Moines Register,* October 16, p. 1T.

Swan, Neil. 1993. "Despite Advances, Drug Craving Remains an Elusive Research Target." *NIDA (National Institute of Drug Abuse) Notes* 8: 1, 4–5.

Symposium on the British Drug System. 1983. *British Journal of Addictions* 78: entire issue.

Tennant, Forest. 1990. "Outcomes of Cocaine-Dependence Treatment." Pp. 314–415 in *Problems of Drug Dependence, 1989,* Louis S. Harris, ed. Rockville, MD: National Institute of Drug Abuse.

Trebach, Arnold S. 1982. *The Heroin Solution.* New Haven, CT: Yale University Press.

Trebach, Arnold S. 1987. *The Great Drug War.* New York: Macmillan.

Troyer, Ronald J., and Gerald E. Markle. 1983. *Cigarettes: The Battle over Smoking.* New Brunswick, NJ: Rutgers University Press.

Vaillant, George E., Jane R. Bright, and Charles MacArthur. 1970. "Physicians' Use of Mood-Altering Drugs." *New England Journal of Medicine* 282: 365–370.

Volkman, Rita, and Donald R. Cressey. 1963. "Differential Association and the Rehabilitation of Drug Addicts." *American Journal of Sociology* 69: 129–142.

Waldorf, Dan. 1973. *Careers in Dope.* Englewood Cliffs, NJ: Prentice-Hall.

Waldorf, Dan. 1983. "Natural Recovery from Opiate Addiction: Some Social-Psychological Processes of Untreated Recovery." *Journal of Drug Issues* 13: 237–280.

Waldorf, Dan, Craig Reinarman, and Sheigla Murphy. 1991. *Cocaine Changes: The Experience of Using and Quitting.* Philadelphia: Temple University Press.

Weil, Andrew. 1996. "Why People Take Drugs." Pp. 3–11 in *The American Drug Scene,* James A. Inciardi and Karen McElrath, eds. Los Angeles: Roxbury.

Welti, Charles V. 1987. "Fatal Reactions to Cocaine." Pp. 33–54 in *Cocaine: A Clinician's Handbook,* Arnold M. Washton and Mark S. Gold, eds. New York: Guilford Press.

Willie, Rolf. 1983. "Processes of Recovery from Heroin Dependence: Relationship to Treatment, Social Changes, and Drug Use." *Journal of Drug Issues* 13: 333–342.

Wilson, Lana. 1989. "The Bong and the Weed." Pp. 57–59 in *Degrees of Deviance: Student Accounts of Their Deviant Behavior,* Stuart Henry, ed. Brookfield, VT: Avebury.

Wilson, William Julius. 1996. *When Work Disappears: The World of the New Urban Poor.* New York: Alfred A. Knopf.

Winick, Charles. 1959–1960. "The Use of Drugs by Jazz Musicians." *Social Problems* 7: 240–254.

Winick, Charles. 1961. "Physician Narcotic Addicts." *Social Problems* 9: 174–186.

Yablonsky, Lewis. 1965. *The Tunnel Back: Synanon.* New York: Macmillan.

Zimring, Franklin E., and Gordon Hawkins. 1992. *The Search for Rational Drug Control.* New York: Cambridge University Press.

CHAPTER 11   Drunkenness and Alcoholism

Akers, Ronald L. 1992. *Drugs, Alcohol, and Society: Social Structure, Process, and Policy.* Belmont, CA: Wadsworth.

Alasuutari, Pertti. 1992. *Desire and Craving: A Cultural Theory of Alcoholism.* Albany: State University of New York Press.

Bahr, Howard M. 1973. *Skid Row: An Introduction to Dissaffiliation.* New York: Oxford University Press.

Barr, Kellie, Michael P. Farrell, Grace M. Barnes, and John Welte. 1993. "Race, Class, and Gender Differences in Substance Abuse: Evidence of Middle-Class/Underclass Polarization among Black Males." *Social Problems* 40: 314–327.

Beckman, Linda S. 1975. "Women Alcoholics: A Review of Social and Psychological Studies." *Journal of Studies on Alcohol* 36: 797–824.

Berreman, Gerald D. 1956. "Drinking Patterns of the Aleuts." *Quarterly Journal of Studies on Alcohol* 17: 503–514.

Biegel, Allan, and Stuart Chertner. 1977. "Toward a Social Model: An Assessment of Social Factors which Influence Problem Drinking and Its Treatment." Pp. 197–233 in *Treatment and Rehabilitation of the Chronic Alcoholic*, Benjamin Kassin and Genri Geglieter, eds. New York: Plenum Press.

Biernacki, Patricia. 1986. *Pathways from Heroin Addiction: Recovery without Treatment.* Philadelphia: Temple University Press.

Bogg, Richard A., and Janet M Ray. 1990. "Male Drinking and Drunkenness in Middletown." *Advances in Alcohol and Substance Abuse* 9: 13–29.

Bradstock, M. Kirsten, James S. Marks, Michelle R. Forman, Eileen M. Gentry, Gary C. Hogelin, Nancy J. Binkin, and Frederick L. Trowbridge. 1987. "Drinking-Driving and Health Lifestyle in the United States: Behavioral Risk Factor Surveys." *Journal of Studies on Alcohol* 48: 147–152.

Bromet, Evelyn, and Rudolf Moos. 1976. "Sex and Marital Status in Relation to the Characteristics of Alcoholics." *Journal of Studies on Alcohol* 37: 1,302–1,312.

Bureau of Justice Statistics. 1985. *Jail Inmates in 1985.* Washington, DC: Department of Justice, Government Printing Office.

Bureau of Justice Statistics. 1993. *Jail Inmates 1992.* Washington, DC: Department of Justice, Government Printing Office.

Burgess, Marilouise. 1998. *Alcohol Involvement in Fatal Crashes.* Washington, DC: National Highway Traffic Safety Administration.

Caetano, Raul. 1984. "Ethnicity and Drinking in Northern California: A Comparison among Whites, Blacks, and Hispanics." *Alcohol and Alcoholism* 19: 31–44.

Caetano, Raul. 1987. "Public Opinion about Alcoholism and Its Treatment." *Journal of Studies on Alcohol* 48: 153–160.

Caetano, Raul. 1989. "Drinking Patterns and Alcohol Problems in a National Sample of U.S. Hispanics." Pp. 147–162 in *The Epidemiology of Alcohol Use and Abuse among U.S. Minorities.* Rockville, MD: National Institute of Alcohol Abuse and Alcoholism, Government Printing Office.

Caetano, Raul, and Kaskutas, Lee Ann. 1995. "Changes in Drinking Patterns among Whites, Blacks and Hispanics, 1984–1992." *Journal of Studies on Alcohol* 56: 558–565.

Cahalan, Don. 1982. "Epidemiology: Alcohol Use in American Society." Pp. 96–118 in *Alcohol, Science and Society Revisited,* Edith Lisansky Gomberg, Helene Raskin White, and John A. Carpenter, eds. Ann Arbor: University of Michigan Press.

Carlson, Per, and Denny Vagero. 1998. "The Social Pattern of Heavy Drinking in Russia during Transition: Evidence from Taganrog 1993." *European Journal of Public Health*, 8: 280–285.

Carver, Raymond, ed. 1983. *Fire, Essays, Poems, Stories.* New York: Vantage Books.

Chafetz, Morris E. 1983. "Is Compulsory Treatment of the Alcoholic Effective?" Pp. 294–302 in *Alcoholism: Introduction to Theory and Treatment,* 2nd ed., David A. Ward, ed. Dubuque, IA: Kendall-Hunt.

Cherpitel, Cheryl, J. Stephens, and Haydee Rosovsky. 1990. "Alcohol Consumption and Casualties: A Comparison of Emergency Room Populations in the United States and Mexico." *Journal of Studies on Alcohol* 51: 319–326.

Clark, Walter B. 1981. "The Contemporary Tavern." Pp. 425–470 in *Research Advances in Alcohol and Drug Problems,* Vol. 6, Y. Israel, F. B. Glaser, H. Kalant, R. Popham, W. Schmidt, and R. Smart, eds. New York: Plenum Press.

Conrad, Peter, and Joseph W. Schnieder. 1980. *Deviance and Medicalization: From Badness to Sickness.* St. Louis, MO: C. V. Mosby.

Cospers, Ronald, and Florence Hughes. 1983. "So-Called Heavy Drinking Occupations: Two Empirical Tests." *Journal of Studies on Alcohol* 43: 110–118.

Cospers, Ronald L., Ishmael O. Okraku, and Brigette Neumann. 1987. "Tavern Going in Canada: A National Survey of Regulars at Public Drinking Establishments." *Journal of Studies on Alcohol* 48: 252–259.

Cressey, Donald R. 1955. "Changing Criminals: The Application of the Theory of Differential Association." *American Journal of Sociology* 61: 116–120.

Cross, G. M., C. W. Morgan, A. J. Mooney III, C. A. Martin, and J. A. Rafter. 1990. "Alcoholism Treatment: A Ten-Year Follow-Up Study." *Alcohol Clinical Experimental Research* 14: 169–173.

Dawson, Deborah A., Bridget F. Grant, S. Patricia Chou, and Roger P. Pickering. 1995. "Subgroup Variation in U.S. Drinking Patterns: Results of the 1992 National Longitudinal Alcohol Epidemiologic Study." *Journal of Substance Abuse* 7: 331–344.

Denzin, Norman K. 1993. *The Alcoholic Society: Addiction and Recovery of the Self.* New Brunswick, NJ: Transaction Publishers.

Donavan, John E., Richard Jessor, and Lee Jessor. 1983. "Problem Drinking in Adolescence and Young Adulthood: A Follow-Up Study." *Journal of Studies on Alcohol* 44: 109–115.

Downs, William R. 1987. "A Panel Study of Normative Structure, Adolescent Alcohol Use, and Peer Alcohol Use." *Journal of Studies on Alcohol* 48: 167–175.

Duncan, Terry E., Elizabeth Tildesley, Susan C. Duncan, and Hyman Hops. 1995. "The Consistency of Family and Peer Influences on the Development of Substance Use in Adolescence." *Addiction* 12: 1647–1660.

Eckhardt, Michael J., Thomas C. Harford, Charles T. Kaelber, Elizabeth S. Parker, Laura S. Rosenthal, Ralph S. Ryback, Gian C. Salmoiraghi, Ernestine Vanderveen, and Kenneth R. Warren. 1981. "Health Hazards Associated with Alcohol Consumption." *Journal of the American Medical Association* 246: 648–666.

Fagin, Ronald W., Jr., and Armand L. Mauss. 1978. "Padding the Revolving Door: An Initial Assessment of the Uniform Alcoholism and Intoxication Treatment Act in Practice." *Social Problems* 26: 232–246.

Federal Bureau of Investigation. 1996. *Crime in the United States, 1995.* Washington, DC: Department of Justice, Government Printing Office.

Field, Eugene. 1897. *The Colonial Tavern.* Providence, RI: Preston and Rounds.

Fingarette, Herbert. 1988. *Heavy Drinking: The Myth of Alcoholism as a Disease.* Berkeley: University of California Press.

Firebaugh, W. C. 1928. *The Inns of Greece and Rome.* Chicago: F. M. Morris.

General Accounting Office. 1985. *Homelessness: A Complex Problem and the Federal Response.* GAO/HRD-85-40. Gaithersburg, MD: Government Accounting Office, Government Printing Office.

Gilbert, F. S. 1991. "Development of a 'Steps Questionnaire.'" *Journal of Studies on Alcohol* 52: 353–360.

Glascow, Russell E., Robert C. Klesges, and Michael W. Vasey. 1983. "Controlled Smoking for Chronic Smokers: An Extension and Replication." *Addictive Behaviors* 8: 143–150.

Greenslade, Liam, Maggie Pearson, and Moss Madden. 1995. "A Good Man's Fault: Alcohol and Irish People at Home and Abroad." *Alcohol and Alcoholism* 30: 407–417.

Gusfield, Joseph R. 1963. *Symbolic Crusade: Status Politics and the American Temperance Movement.* Urbana: University of Illinois Press.

Hamill, Pete. 1994. *A Drinking Life: A Memoir.* Boston: Little Brown.

Harper, Frederick D., and Elaheh Saifnoorian. 1991. "Drinking Patterns among Black Americans." Pp. 327–338 in *Society, Culture and Drinking Patterns Reexamined,* David J. Pittman and Helene Raskin White, eds. New Brunswick, NJ: Rutgers Center of Alcohol Studies.

Hasin, Deborah S., Bridget Grant, and Thomas C. Harford. 1990. "Male and Female Differences in Liver Cirrhosis Mortality in the United States, 1961–1985." *Journal of Studies on Alcohol* 51: 123–129.

Herd, Denise. 1985. "Migration, Cultural Transformation, and the Rise of Black Liver Cirrhosis Mortality." *British Journal of Addiction* 80: 397–410.

Herd, Denise. 1989. "The Epidemiology of Drinking Patterns and Alcohol-Related Problems among U.S. Blacks." In *The Epidemiology of Alcohol Use and Abuse among U.S. Minorities.* Rockville, MD: National Institute of Alcohol Abuse and Alcoholism, Government Printing Office.

Herd, Denise. 1990. "Subgroup Differences in Drinking Patterns among Black and White Men: Results from a National Survey." *Journal of Studies on Alcohol* 51: 221–232.

Herd, Denise. 1994. "Predicting Drinking Problems among Black and White Men: Results from a National Survey." *Journal of Studies on Alcohol* 55: 61–71.

Hill, Shirley Y., Stuart R. Steinhauer, and Joseph Zubin. 1987. "Biological Markers for Alcoholism: A Vulnerability Model Conceptualization." Pp. 207–256 in *Alcohol and Addictive Behavior: Nebraska Symposium on Motivation, 1986,* P. Clayton Rivers, ed. Lincoln: University of Nebraska Press.

Hilton, Michael E. 1988. "Trends in U.S. Drinking Patterns: Further Evidence from the Past 20 Years." *British Journal of Addictions* 83: 269–278.

Hilton, Michael E., and Walter B. Clark. 1987. "Changes in American Drinking Patterns and Problems, 1967–1984." *Journal of Studies on Alcohol* 48: 515–522.

Hilton, Michael E., and B. M. Johnstone. 1988. "Symposium on International Trends in Alcohol Consumption." *Contemporary Drug Problems* 13: 145–154.

Hingson, Ralph, and Jonathan Howland. 1987. "Alcohol as a Factor for Injury or Death Resulting from Accidental Falls: A Review of the Literature." *Journal of Studies on Alcohol* 48: 212–219.

Hingson, Ralph, Tom Mangione, Allen Meyers, and Norman Scotch. 1982. "Seeking Help for Drinking Problems: A Study in the Boston Metropolitan Area." *Journal of Studies on Alcohol* 43: 273–288.

Horgan, M. M., M. D. Sparrow, and R. Brazeau. 1986. *Alcoholic Beverage Taxation and Control Policies*, 6th ed. Ottawa: Brewer's Association of Canada.

Indian Health Service. 1992. *Trends in Indian Health, 1991*. Rockville, MD: Indian Health Service.

Ivanets, N. N., I. P. Anokhina, V. F. Egorov, Y. V. Valentik, and S. B. Shesterneva. 1992. "Present State and Prospects of Treatment of Alcoholism in the Soviet Union." Pp. 9–21 in *Cure, Care, or Control: Alcoholism Treatment in Sixteen Countries*. Albany: State University of New York Press.

"J." 1996. *A Simple Program. A Contemporary Translation of the Book* Alcoholics Anonymous. New York: Hyperion.

J. W. C. 1998. "My Life with an Alcoholic." *AA Grapevine*, April.

Jacob, Theodore. 1987. "Alcoholism: A Family Interaction Perspective." Pp. 159–206 in *Alcohol and Addictive Behavior: Nebraska Symposium on Motivation, 1986*, P. Clayton Rivers, ed. Lincoln: University of Nebraska Press.

Jacobs, James B. 1989. *Drunk Driving: An American Dilemma*. Chicago: University of Chicago Press.

Jellinek, E. M. 1960. *The Disease Concept of Alcoholism*. Highland Park, NJ: Hillhouse Press.

Johnston, Lloyd D., Patrick M. O'Malley, and Jerald G. Bachman. 1996. *Press Release on Findings from 1996 Survey of Drug Use*. Ann Arbor, MI: News and Information Service, University of Michigan.

Johnston, Lloyd D., Patrick M. O'Malley, and Jerald G. Bachman. 1999. *Press Release on Findings from 1999 Survey of Drug Use*. Ann Arbor, MI: News and Information Service, University of Michigan.

Kandel, Denise B. 1980. "Drug and Drinking Behavior among Youth." *Annual Review of Sociology* 6: 235–285.

Keller, Mark. 1982. "On Defining Alcoholism: With Comment on Some Other Relevant Words." Pp. 119–133 in *Alcohol, Science and Society Revisited*, Edith Lisansky Gomberg, Helene Raskin White, and John A. Carpenter, eds. Ann Arbor, MI: University of Michigan Press.

Keller, Mark, and Vera Efron. 1955. "The Prevalence of Alcoholism." *Quarterly Journal of Studies on Alcohol* 16: 628–637.

Keller, Mark, and Vera Efron. 1961. *Selected Statistical Tables on Alcoholic Beverages, 1950–1960, and on Alcoholism, 1930–1960*. New Brunswick, NJ: Quarterly Journal of Studies on Alcohol.

Kitano, H. H. L., H. Hatanaka, W. T. Yeung, and S. Sue. 1985. "Japanese-American Drinking Patterns." Pp. 335–357 in *The American Experience with Alcohol: Contrasting Cultural Perspectives*, L. A. Bennett and G. M. Ames, ed. New York: Plenum.

Kjaerheim, Kristina, Reidar Mykletun, Olaf G. Aasland, Tor Haldorsen, and Aage Andersen. 1995. "Heavy Drinking in the Restaurant Business: The Role of Social Modelling and Structural Factors of the Work-Place." *Addiction* 90: 1487–1495.

Klatsky, A. L., A. B. Siegelaub, C. Landy, and G. D. Friedman. 1985. "Racial Patterns of Alcoholic Beverage Use." *Alcoholism: Clinical and Experimental Research* 7: 372–377.

Klein, Hugh. 1991. "Cultural Determinants of Alcohol Use in the United States." Pp. 114–134 in *Society, Culture and Drinking Patterns Reexamined*, David J. Pittman and Helene Raskin White, eds. New Brunswick, NJ: Rutgers Center of Alcohol Studies.

Kurtz, Ernest. 1982. "Why A.A. Works: The Intellectual Significance of Alcoholics Anonymous." *Journal of Studies on Alcohol* 43: 38–80.

Larkin, John R. 1965. *Alcohol and the Negro: Explosive Issues*. Zebulon, NC: Record Publishing Company.

LeMasters, Ersel E. 1975. *Blue-Collar Aristocrats: Life-Styles at a Working Class Tavern*. Madison: University of Wisconsin Press.

Lemert, Edwin M. 1954. *Alcohol and the Northwest Coast Indians*. University of California Publications on Culture and Society, Vol. 2, No. 6. Berkeley: University of California Press.

Lemert, Edwin M. 1972. "Alcohol, Values, and Social Control." Pp. 112–122 in *Human Deviance, Social Problems and Social Control*, 2nd ed., Edwin M. Lemert, ed. Englewood Cliffs, NJ: Prentice-Hall.

Lemert, Edwin M. 1982. "Drinking among American Indians." Pp. 80–95 in *Alcohol, Science and Society Revisited*, Edith Lisansky Gomberg, Helene Raskin White, and John A. Carpenter, eds. Ann Arbor, MI: University of Michigan Press.

Lender, Mark Edward, and James Kirby Martin. 1982. *Drinking in America: A History*. New York: Free Press.

Lex, B. W. 1985. "Alcohol Problems in Special Populations." Pp. 89–187 in *The Diagnosis and Treatment of Alcoholism*, 2nd ed., J. H. Mendelson and N. K. Mello, eds. New York: McGraw-Hill.

Linsky, Arnold S., John P. Colby, Jr., and Murray A. Straus. 1986. "Drinking Norms and Alcohol-Related Problems in the United States." *Journal of Studies on Alcohol* 47: 384–393.

Linsky, Arnold S., John P. Colby, Jr., and Murray A. Straus. 1987. "Social Stress, Normative Constraints, and Alcohol Problems in the United States." *Social Science and Medicine* 24: 875–883.

Lolli, Giorgio, Emilio Serianni, Grace M. Golder, and Peirpaolo Luzzatto-Fegis. 1958. *Alcohol in Italian Culture*. New York: Free Press.

Longclaws, Lyle, Gordon E. Barnes, Linda Grieve, and Ron Dumoff. 1980. "Alcohol and Drug Use among Broken Head Ojibwa." *Journal of Studies on Alcohol* 41: 21–36.

Lowman, C., T. C. Hardford, and C. T. Kaelber. 1983. "Alcohol Use among Black Senior High School Students." *Alcohol Health and Research World* 7: 37–46.

MacAndrew, Craig, and Robert B. Edgerton. 1969. *Drunken Comportment: A Social Explanation*. Chicago: Aldine.

Mangin, William. 1957. "Drinking among Andean Indians." *Quarterly Journal of Studies on Alcohol* 18: 55–66.

Marchant, Ward. 1990. "Secular Approach to Alcoholism: Groups Offer Alternatives to AA." *Santa Barbara News Press*, October 20, p. 6.

Martin, Casey, and Sally Casswell. 1987. "Types of Male Drinkers: A Multivariate Study." *Journal of Studies on Alcohol* 48: 109–118.

McCarthy, Dennis, Sherry Morrison, and Kenneth C. Mills. 1983. "Attitudes, Beliefs, and Alcohol Use." *Journal of Studies on Alcohol* 44: 328–341.

McCord, William, and Joan McCord. 1961. *Origins of Alcoholism.* Stanford, CA: Stanford University Press.

Meier, Robert F., and Gilbert Geis. 1997. *Victimless Crimes? Prostitution, Homosexuality, Drugs, and Abortion.* Los Angeles: Roxbury.

Meilman, Philip W., Janet E. Stone, Michael S. Gaylor, and John H. Turco. 1990. "Alcohol Consumption by College Undergraduates: Current Use and 10-year Trends." *Journal of Studies on Alcohol* 51: 385–395.

Miller, W. R., and R. K. Hester. 1980. "Treating the Problem Drinker: Modern Approaches." Pp. 11–14 in *The Addictive Behaviors: Treatment of Alcoholism, Drug Abuse, Smoking, and Obesity.* Oxford: Pergamon Press.

Montgomery, Henry A., William R. Miller, and J. Scott Tonigan. 1995. "Does Alcoholics Anonymous Involvement Predict Treatment Outcome?" *Journal of Substance Abuse Treatment* 12: 241–246.

Mosse, Philippe. 1992. "The Rise of Alcohology in France: A Monopolistic Competition." Pp. 205–221 in *Cure, Care, or Control: Alcoholism Treatment in Sixteen Countries.* Albany: State University of New York Press.

Mulford, Harold A. 1964. "Drinking and Deviant Behavior, USA." *Quarterly Journal of Studies on Alcohol* 25: 634–650.

Mulkern, V., and R. Spence. 1984. *Alcohol Abuse/Alcoholism among Homeless Persons: A Review of the Literature. Final Report.* Washington, DC: Government Printing Office.

National Highway Traffic Safety Administration. 1988. *Drunk Driving Facts.* Washington, DC: National Center for Statistics and Analysis, Government Printing Office.

Neff, James Alan. 1991. "Race, Ethnicity, and Drinking Patterns: The Role of Demographic Factors, Drinking Motives, and Expectancies." Pp. 339–356 in *Society, Culture and Drinking Patterns Reexamined*, David J. Pittman and Helene Raskin White, eds. New Brunswick, NJ: Rutgers Center of Alcohol Studies.

Nordstrom, Goran, and Mats Berglund. 1987. "A Prospective Study of Successful Long-Term Adjustment in Alcohol Dependence: Social Drinking versus Abstinence." *Journal of Studies on Alcohol* 48: 95–103.

Nusbaumer, Michael R., Armand L. Mauss, and David C. Pearson. 1982. "Draughts and Drunks: The Contributions of Taverns and Bars to Excessive Drinking in America." *Deviant Behavior* 3: 329–358.

Orcutt, James D. 1976. "Ideological Variations in the Structure of Deviant Types: A Multivariate Comparison of Alcoholism and Heroin Addiction." *Social Forces* 55: 419–437.

Parker, Robert Nash. 1993. "Alcohol and Theories of Homicide." Pp. 113–141 in *New Directions in Criminological Theory: Advances in Criminological Theory*, Vol. 4. New Brunswick, NJ: Transaction Books.

Parker, Robert Nash. 1995. "Bringing 'Booze' Back in: The Relationship between Alcohol and Homicide." *Journal of Research in Crime and Delinquency* 32: 3–38.

Patrick, Charles H. 1952. *Alcohol, Culture and Society.* Durham, NC: Duke University Press.

Peele, Stanton. 1990. *The Diseasing of America: Addiction Treatment Out of Control.* Lexington, MA: Lexington Books.

Pendery, Mary L., Irving M. Maltzman, and L. Jolyon West. 1982 (July). "Controlled Drinking by Alcoholics." *Science* 217: 169–175.

Pittman, David J. 1991. "Social Policy and Habitual Drunkenness Offenders." Pp. 219–233 in *Alcohol: The Development of Sociological Perspectives on Use and Abuse,* Paul Roman, ed. New Brunswick, NJ: Rutgers Center of Alcohol Studies.

Pittman, David J., and C. Wayne Gordon. 1958. *The Revolving Door.* New York: Free Press.

Plaut, Thomas F. A. 1967. *Alcohol Problems: A Report to the Nation by the Cooperative on the Study of Alcoholism.* New York: Oxford University Press.

Popham, Robert E. 1962. "The Urban Tavern: Some Preliminary Remarks." *Addictions* 9: 17–26.

Presley, Cheryl, Jami S. Leichliter, and Philip Meilman. 1998. *Alcohol and Drugs on American Campuses: A Report to College Presidents.* Carbondale, IL: The CORE Institute.

"Reactions to the Sobells." 1995. *Addictions* 90: 1,157–1,177.

Robinson, David. 1972. "The Alcohologist's Addiction: Some Implications of Having Lost Control over the Disease Concept of Alcoholism." *Quarterly Journal of Studies on Alcohol* 33: 1,028–1,042.

Roman, Paul M. 1981. "From Employee Alcoholics to Employee Assistance: Deemphasis on Prevention and Alcohol Problems in Work-Based Programs." *Journal of Studies on Alcohol* 42: 244–272.

Room, Robin. 1976. "Ambivalence as a Sociological Explanation: The Case of Cultural Explanations of Alcohol Problems." *American Sociological Review* 41: 1,047–1,065.

Room, Robin. 1983. "Alcohol and Crime: Behavioral Aspects." Pp. 34–44 in *Encyclopedia of Crime and Justice,* Vol. l, Sanford H. Kadish, ed. New York: Free Press.

Ross, H. Laurence. 1982. *Deterring the Drunk Driver: Legal Policy and Social Control.* Lexington, MA: Lexington Books.

Ross, H. Laurence. 1993. *Confronting Drunk Driving: Social Policy for Saving Lives.* New Haven, CT: Yale University Press.

Ross, H. L., S. Simon, J. Cleary, R. Lewis, and D. Storkamp. 1995. "Causes and Consequences of Implied Consent Test Refusal." *Alcohol, Drugs and Driving* 11: 57–72.

Rubington, Earl. 1991. "The Chronic Drunkenness Offender: Before and after Decriminalization." Pp. 733–752 in *Society, Culture and Drinking Patterns Reexamined,* David J. Pittman and Helene Raskin White, eds. New Brunswick, NJ: Rutgers Center of Alcohol Studies.

Sadoun, Roland, Giorgio Lolli, and Milton Silverman. 1965. *Drinking in French Culture.* New Brunswick, NJ: Rutgers Center of Alcohol Studies.

Secretary of Health and Human Services. 1981. *Alcohol and Health: Fourth Special Report to the U.S. Congress.* Rockville, MD: National Institute of Alcohol Abuse and Alcoholism, Government Printing Office.

Secretary of Health and Human Services. 1987. *Alcohol and Health: Sixth Special Report to the U.S. Congress.* Rockville, MD: National Institute of Alcohol Abuse and Alcoholism, Government Printing Office.

Secretary of Health and Human Services. 1990. *Alcohol and Health: Seventh Special Report to the U.S. Congress.* Rockville, MD: National Institute of Alcohol Abuse and Alcoholism, Government Printing Office.

Secretary of Health and Human Services. 1993. *Alcohol and Health: Eighth Special Report to the U.S. Congress.* Rockville, MD: National Institute of Alcohol Abuse and Alcoholism, Government Printing Office.

Secretary of Health and Human Services. 2000. *Alcohol and Health: Tenth Special Report to the U.S. Congress.* Rockville, MD: National Institute of Alcohol Abuse and Alcoholism, Government Printing Office.

Snyder, Charles R. 1978. *Alcohol and the Jews.* Carbondale, IL: Southern Illinois University Press; originally published 1958.

Sobell, Mark B., and Linda C. Sobell. 1973. "Alcoholics Treated by Individualized Behavior Therapy." *Behavior Research and Therapy* 11: 599–618.

Sobell, Mark B., and Linda C. Sobell. 1993. *Problem Drinkers: Guided Self-Change Treatment.* New York: Guilford.

Sobell, Mark B., and Linda C. Sobell. 1995. "Controlled Drinking after 25 Years: How Important Was the Great Debate?" *Addiction* 90:1149–1153.

Sonnenstuhl, William J., and Harrison M. Trice. 1991. "The Workplace as Locale for Risks and Interventions in Alcohol Abuse." Pp. 255–288 in *Alcohol: The Development of Sociological Perspectives on Use and Abuse*, Paul Roman, ed. New Brunswick, NJ: Rutgers Center of Alcohol Studies.

Spradley, James A. 1970. *You Owe Yourself a Drunk: An Ethnography of Urban Nomads.* Boston: Little, Brown.

Sterne, Muriel W. 1967. "Drinking Patterns and Alcoholism among American Negroes." Pp. 71–74 in *Alcoholism*, David J. Pittman, ed. New York: Harper and Row.

Stivers, Richard. 1976. *A Hair of the Dog: Irish Drinking and American Stereotype.* University Park, PA: Pennsylvania State University Press.

Straus, Robert. 1982. "The Social Costs of Alcohol." Pp. 137–147 in *Alcohol, Science and Society Revisited*, Edith Lisansky Gomberg, Helene Raskin White, and John A. Carpenter, eds. Ann Arbor: University of Michigan Press.

Sue, S., H. H. L. Kitano, H. Hatanaka, and W. T. Yeung. 1985. "Alcohol Consumption among Chinese in the United States." Pp. 359–371 in *The American Experience with Alcohol: Contrasting Cultural Perspectives*, L. A. Bennett and G. M. Ames, eds. New York: Plenum.

Taylor, John R., Terri Combs-Orme, and David A. Taylor. 1983. "Alcohol and Mortality: Diagnostic Considerations." *Journal of Studies on Alcohol* 44: 17–25.

Traml, Vladimir G. 1975. "Production and Consumption of Alcoholic Beverages in the USSR." *Journal of Studies on Alcohol* 36: 285–320.

Trice, Harrison M. 1959. "The Affiliation Motive and Readiness to Join Alcoholics Anonymous." *Quarterly Journal of Studies on Alcohol* 20: 313–321.

Trice, Harrison M. 1966. *Alcoholism in America.* New York: McGraw-Hill.

Trice, Harrison M., and Paul M. Roman. 1972. *Spirits and Demons at Work: Alcohol and Other Drugs on the Job.* Ithaca, NY: Industrial and Labor Relations Paperback.

Trice, Harrison M., and William J. Sonnenstuhl. 1990. "Alcohol and Mental Health Programs in the Workplace." *Research in Community and Mental Health* 6: 351–378.

Trice, Harrison M., and J. Richard Wahl. 1958. "A Rank Order Analysis of the Symptoms of Alcoholism." *Quarterly Journal of Studies on Alcohol* 19: 636–648.

Umanna, Ifekandu. 1967. "The Drinking Culture of a Nigerian Community: Onitsha." *Quarterly Journal of Studies on Alcohol* 28: 529–537.

Waldorf, Dan. 1983. "Natural Recovery from Opiate Addiction: Some Social-Psychological Processes of Untreated Recovery." *Journal of Drug Issues* 13: 237–280.

Wallack, Lawrence, Warren Breed, and John Cruz. 1987. "Alcohol on Prime-Time Television." *Journal of Studies on Alcohol* 48: 33–38.

Walsh, D. C., R. W. Hingson, D. M. Merrigan, S. M. Levenson, L. A. Supples, R. Heeren, G. Coffman, C. A. Becker, T. A. Barker, S. K. Hamilton, T. G. McGuire, and C. A. Kelly. 1991. "A Randomized Trial of Treatment Options for Alcohol-Abusing Workers." *New England Journal of Medicine* 325: 775–781.

Wanberg, Kenneth W., and John L. Horn. 1970. "Alcoholism Patterns in Men and Women." *Quarterly Journal of Studies on Alcoholism* 31: 40–61.

Ward, David A. 1990. "Conceptions of Alcoholism." Pp. 4–13 in *Alcoholism: Introduction to Theory and Treatment*, 3rd ed., David A. Ward, ed. Dubuque, IA: Kendall-Hunt.

Washburne, Chandler. 1961. *Primitive Drinking: A Study of the Uses and Functions of Alcohol in Preliterate Societies*. New Haven, CT: College and University Press.

Wechsler, Henry, Beth E. Molnar, Andrea E. Davenport, and John S. Baer. 1999. "College Alcohol Use: A Full or Empty Glass?" *Journal of American College Health* 47: 247–252.

Weed, Frank J. 1990. "The Victim-Activist Role in the Anti-Drunk Driving Movement." *Sociological Quarterly* 31: 459–473.

Welte, John W., and Grace M. Barnes. 1987. "Alcohol Use among Adolescent Minority Groups." *Journal of Studies on Alcohol* 48: 329–336.

Welte, John W., and Grace M. Barnes. 1992. "Drinking among Homeless and Marginally Housed Adults in New York State." *Journal of Studies on Alcohol* 53: 303–319.

White, Helene Raskin, Marsha E. Bates, and Valerie Johnson. 1991. "Learning to Drink: Familial, Peer, and Media Influences." Pp. 177–197 in *Society, Culture and Drinking Patterns Reexamined*, David J. Pittman and Helene Raskin White, eds. New Brunswick, NJ: Rutgers Center of Alcohol Studies.

Williams, G. D., D. Doernberg, F. Stinson, and J. Noble. 1986. "National, State, and Regional Trends in Apparent per Capita Consumption of Alcohol." *Alcohol Health and Research World* 10: 60–63.

Wilsnack, Richard W., Sharon C. Wilsnack, and Albert D. Klassen. 1987. "Antecedents and Consequences of Drinking and Drinking Problems in Women: Patterns from a U.S. National Survey." Pp. 85–158 in *Alcohol and Addictive Behavior: Nebraska Symposium on Motivation, 1986*, P. Clayton Rivers, ed. Lincoln: University of Nebraska Press.

Wiseman, Jacqueline P. 1981. "Sober Comportment: Patterns and Perspectives on Alcohol Addiction." *Journal of Studies on Alcohol* 42: 106–126.

Wolfgang, Marvin E. 1958. *Patterns of Criminal Homicide*. Philadelphia: University of Pennsylvania Press.

CHAPTER 12   Sexual Deviance

Abbott, Sharon A. 2000. "Motivations for Pursuing an Acting Career in Pornography." Pp. 17–34 in *Sex for Sale: Prostitution, Pornography and the Sex Industry*, Ronald Weitzer, ed. New York: Routledge.

Abramson, Paul R., and Haruo Hayashi. 1984. "Pornography in Japan: Cross-Cultural and

Theoretical Considerations." Pp. 173–183 in *Pornography and Sexual Aggression*. Edited by Neil M. Malamuth and Edward Donnerstein. New York: Academic Press.

Aday, David P., Jr. 1990. *Social Control at the Margins: Toward a General Understanding of Deviance*. Belmont, CA: Wadsworth.

Angel, Ronald, and Marta Tienda. 1982. "Determinants of Extended Household Structure: Cultural Pattern or Economic Necessity?" *American Journal of Sociology* 87: 1,360–1,383.

Attorney General's Commission on Pornography. 1986. *Final Report*. Washington, DC: Government Printing Office.

Baron, Larry, and Murray A. Straus. 1984. "Sexual Stratification, Pornography, and Rape in the United States." Pp. 186–209 in *Pornography and Sexual Aggression*, Neil M. Malamuth and Edward Donnerstein, eds. New York: Academic Press.

Barry, Kathleen. 1984. *Female Sexual Slavery*. New York: New York University Press.

Beller-Hann, Ildiko. 1995. "Prostitution and Its Effects in Northeast Turkey." *European Journal of Women's Studies* 2: 219–235.

Bellis, David J. 1990. "Fear of AIDS and Risk Reduction among Heroin-Addicted Female Street Prostitutes: Personal Interviews with 72 Southern California Subjects." *Journal of Alcohol and Drug Education* 35: 26–37.

Blumstein, Phillip, and Pepper Schwartz. 1983. *American Couples*. New York: Morrow.

Boeringer, Scot B. 1994. "Pornography and Sexual Aggression: Associations of Violent and Nonviolent Depictions with Rape and Rape Proclivity." *Deviant Behavior* 15: 289–304.

Bracey, Dorothy R. 1979. *"Baby Pros": Preliminary Profiles of Juvenile Prostitutes*. New York: John Jay Press.

Brock, Deborah R. 1998. *Making Work, Making Trouble: Prostitution as a Social Problem*. Toronto: University of Toronto Press.

Brosius, Jans-Bernd, James B. Weaver III, and Joachim F. Staab. 1993. "Exploring the Social and Sexual 'Reality' of Contemporary Pornography." *The Journal for Sex Research* 30: 161–170.

Bryan, James H. 1965. "Apprenticeships in Prostitution." *Social Problems* 12: 278–297.

Bryan, James H. 1966. "Occupational Ideologies and Individual Attitudes of Call Girls." *Social Problems* 13: 437–447.

Buunk, Bram P., and Barry van Driel. 1989. *Variant Lifestyles and Relationships*. Beverly Hills, CA: Sage.

Campbell, Carole A. 1991. "Prostitution, AIDS, and Preventive Health Behavior." *Social Science and Medicine* 32: 1,367–1,378.

Charkalis, Diana Mckeon. 1996. "Everything You Ever Wanted to Know about Cybersex." *Des Moines Register*, October 3, p. 3T.

Chattopadhyay, Molly, S. Bandyopadhyay, and C. Duttagupta. 1994. "Biosocial Factors Influencing Women to Become Prostitutes in India." *Social Biology* 41: 252–259.

Cohen, Bernard. 1980. *Deviant Street Networks: Prostitution in New York City*. Lexington, MA: Lexington Books.

Commission on Obscenity and Pornography. 1970. *Final Report*. New York: Bantam Books; originally published by Government Printing Office, 1970.

Davis, Kingsley. 1937. "The Sociology of Prostitution." *American Sociological Review* 2: 744–755.

Davis, Nanette J. 1981. "Prostitutes." Pp. 305–313 in *Deviance: The Interactionist Perspective*, 4th ed., Earl Rubington and Martin S. Weinberg, eds. New York: Macmillan.

Davis, Nanette J., ed. 1993. *Prostitution: An International Handbook on Trends, Problems, and Policies*. Westport, CT: Greenwood.

Davis, Nanette, J. 2000. "From Victims to Survivors: Working with Recovering Street Prostitutes." Pp. 139–155 in Sex *for Sale: Prostitution, Pornography and the Sex Industry*, Ronald Weitzer, ed. New York: Routledge.

D'Emilio, John, and Estelle B. Freedman. 1988. *Intimate Matters: A History of Sexuality in America*. New York: Harper and Row.

De Graaf, Ron, I. Vanwesenbeeck, G. Van Zessen, C. J. Straver, and J. H. Visser. 1994. "Male Prostitutes and Safe Sex: Different Settings, Different Risks." *AIDS Care* 6: 277–288.

Dekker, Rudolf. 1989. *The Tradition of Female Transvestism in Early Modern Europe*. New York: Macmillan.

DeLamater, John. 1981. "The Social Control of Sexuality." *Annual Review of Sociology* 7: 263–290.

Diana, Lewis. 1985. *The Prostitute and Her Clients*. Springfield, IL: Charles C. Thomas.

Donnerstein, Edward, Daniel Linz, and Steven Penrod. 1987. *The Question of Pornography: Research Findings and Policy Implications*. New York: Free Press.

Downs, Donald Alexander. 1989. *The New Politics of Pornography*. Chicago: University of Chicago Press.

Eysenck, Hans J. 1972. "Obscenity—Officially Speaking." *Penthouse* 3 (11): 95–102.

Federal Bureau of Investigation. 1999. *Crime in the United States, 1998*. Washington, DC: Department of Justice, Government Printing Office.

Ferguson, Frances. 1995. "Pornography: The Theory." *Critical Inquiry* 21: 670–695.

Fisher, William A., and Guy Grenier. 1994. "Violent Pornography, Antiwoman Thoughts, and Antiwoman Acts: In Search of Reliable Effects." *Journal of Sex Research* 31: 23–38.

Forsyth, Craig J., and Lee Fournet. 1987. "A Typology of Office Harlots: Mistresses, Party Girls, and Career Climbers." *Deviant Behavior* 8: 319–328.

Frost, Janet. 1989. "Affairs." Pp. 25–27 in *Degrees of Deviance: Student Accounts of Their Deviant Behavior*, Stuart Henry, ed. Brookfield, VT: Avebury.

Gagnon, John H., and William Simon. 1970a. *Sexual Conduct: The Social Sources of Human Sexuality*. Chicago: Aldine.

Gagnon, John H., and William Simon. 1970b. "Perspectives on the Sexual Scene." Pp. 1–12 in *The Sexual Scene*, John H. Gagnon and William Simon, eds. Chicago: Aldine.

Gebhard, Paul H., John H. Gagnon, Wardell B. Pomeroy, and Cornelia V. Christenson. 1965. *Sex Offenders*. New York: Harper and Row.

Goldstein, M. J., H. S. Kant, and J. J. Hartman. 1974. *Pornography and Sexual Deviance*. Berkeley: University of California Press.

Gray, Diana. 1973. "Turning Out: A Study of Teenage Prostitution." *Urban Life and Culture* 1: 401–425.

Hadaway, C. Kirk, Penny Long Marlet, and Mark Chaves. 1993. "What the Polls Don't Show: A Closer Look at U.S. Church Attendance." Paper presented at the annual meeting of the American Sociological Association, Miami, Florida.

Harris, Louis. 1987. *Inside America*. New York: Vintage.

Heyl, Barbara Sherman. 1979. *The Madam as Entrepreneur: Career Management in House Prostitution*. New Brunswick, NJ: Transaction Books.

Hobson, Barbara Meil. 1987. *Uneasy Virtue: The Politics of Prostitution and the American Reform Tradition*. New York: Basic Books.

Hodgson, Douglas. 1995. "Combating the Organized Sexual Exploitation of Asian Children: Recent Developments and Prospects." *International Journal of Law and the Family* 9: 23–53.

Holzman, Harold R., and Sharon Pines. 1982. "Buying Sex: The Phenomenology of Being a John." *Deviant Behavior* 4: 124–135.

Housebeck, Kathryn and Barbara G. Brents. 2000. "Inside Nevada's Brothel Industry." Pp. 217–243 in *Sex for Sale: Prostitution, Pornography and the Sex Industry*, Ronald Weitzer, ed. New York: Routledge.

Howitt, Dennis. 1995. "Pornography and the Paedophile: Is It Criminogenic?" *British Journal of Medical Psychology* 1: 15–27.

Hunt, Morton. 1974. *Sexual Behavior in the 1970s*. Chicago: Playboy Press.

Inciardi, James A. 1984. "Little Girls and Sex: A Glimpse at the World of 'Baby Pros.'" *Deviant Behavior* 5: 71–78.

Inciardi, James A., Dorothy Lockwood, and Anne Pottieger. 1993. *Women and Crack Cocaine*. New York: Macmillan

Itzin, Catherine, ed. 1992. *Pornography: Women, Violence, and Civil Liberties*. New York: Oxford University Press.

Jackman, Norman R., Richard O'Toole, and Gilbert Geis. 1963. "The Self-Image of the Prostitute." *The Sociological Quarterly* 4: 150–161.

Jackson, Emily. 1995. "The Problem with Pornography: A Critical Survey of the Current Debate." *Feminist Legal Studies* 3: 49–70.

James, Jennifer. 1977. "Prostitutes and Prostitution." Pp. 368–428 in *Deviants: Voluntary Actors in a Hostile World*, Edward Sagarin and Fred Montanino, eds. Morristown, NJ: General Learning Press.

Janus, Samuel S., and Cynthia L. Janus. 1993. *The Janus Report on Sexual Behavior*. New York: John Wiley & Sons.

Jenness, Valerie. 1993. *Making It Work: The Prostitutes' Rights Movement*. New York: Aldine DeGruyter.

Karlen, Arno. 1988. *Threesomes: Studies in Sex, Power, and Intimacy*. New York: William Morrow.

Kempadoo, Kamala, and Jo Doezema, eds. 1998. *Global Sex Workers: Rights, Resistance, and Redefinition*. New York: Routledge.

Kinsey, Alfred C., Ward B. Pomeroy, and Charles Martin. 1948. *Sexual Behavior in the Human Male*. Philadelphia: W. B. Saunders.

Kinsey, Alfred C., Ward B. Pomeroy, Charles Martin, and Paul H. Gebhard. 1953. *Sexual Behavior in the Human Female*. Philadelphia: W. B. Saunders.

Klassen, Albert D., Colin J. Williams, and Eugene E. Levitt. 1989. *Sex and Morality in the U.S.* Middletown, CT: Wesleyan University Press.

Kunstel, Marcia, and Joseph Albright. 1987. "Prostitution Thrives on Young Girls." *C. J. International* 3: 9–11.

Laumann, Edward O., John H. Gagnon, Robert T. Michael, and Stuart Michaels. 1994.

The Social Organization of Sexuality: Sexual Practices in the United States. Chicago: University of Chicago Press.

Leuchtag, Alice. 1995. "The Culture of Pornography." *Humanist* 5: 4–6.

Little, Craig B. 1983. *Understanding Deviance and Control: Theory, Research, and Control.* Itasca, IL: F. E. Peacock.

MacKinnon, Catharine. 1993. *Only Words.* Cambridge, MA: Harvard University Press.

Malamuth, Neil M., and Edward Donnerstein, eds. 1984. *Pornography and Sexual Aggression.* New York: Academic Press.

Matlock, Jann. 1993. "Masquerading Women, Pathologized Men: Cross-Dressing, Fetishism, and the Theory of Perversion, 1882–1935." Pp. 31–61 in *Fetishism as Cultural Discourse.* Edited by Emily Apther and William Pietz. Ithaca, NY: Cornell University Press.

McNamara, Robert P. 1994. *The Times Square Hustler: Male Prostitution in New York City.* New York: Praeger.

McWilliams, Peter. 1993. *Ain't Nobody's Business If You Do.* Los Angeles: Prelude Press.

Meier, Robert F., and Gilbert Geis. 1997. *Victimless Crime? Prostitution, Homosexuality, Drugs, and Abortion.* Los Angeles. Roxbury.

Michael, Robert T., John H. Gagnon, Edward O. Laumann, and Gina Kolata. 1994. *Sex in America: A Definitive Survey.* Boston: Little, Brown.

Miller, Eleanor M. 1986. *Street Woman.* Philadelphia: Temple University Press.

Miller, Jody. 1995. "Gender and Power on the Streets: Street Prostitution in the Era of Crack Cocaine." *Journal of Contemporary Ethnography* 23: 427–452.

Miller, Jody, and Martin D. Schwartz, 1995. "Rape Myths and Violence against Street Prostitutes." *Deviant Behavior* 16: 1–23.

Muekeking, George D. 1977. "Pornography and Society." Pp. 463–502 in *Deviants: Voluntary Actors in a Hostile World*, Edward Sagarin and Fred Montanino, eds. Morristown, NJ: General Learning Press.

Murphcy, Michacl. 1987. "She Sells Sex on the Dark Side of the Street." *Spokane Spokesman Review*, November 11, p. A7.

Ouellet, Lawrence J., W. Wayne Wiebel, Antonio D. Jimenez, and Wendell A. Johnson. "Crack Cocaine and the Transformation of Prostitution in Three Chicago Neighborhoods." Pp. 69–96 in *Crack Pipe as Pimp: An Ethnographic Investigation of Sex-for-Drugs Exchanges*, Mitchell S. Ratner, ed. New York: Lexington.

Plummer, Ken. 1982. "Symbolic Interactionism and Sexual Conduct: An Emergent Perspective." Pp. 223–241 in *Human Sexual Relations: Towards A Redefinition of Sexual Politics*, Mike Brake, ed. New York: Pantheon.

Polsky, Ned. 1967. *Hustlers, Beats, and Others.* Chicago: Aldine.

Potter, Gary W. 1989. "The Retail Pornography Industry and the Organization of Vice." *Deviant Behavior* 10: 233–251.

Ratner, Mitchell S., ed. 1993. *Crack Pipe as Pimp: An Ethnographic Investigation of Sex-for-Drugs Exchanges.* New York: Lexington.

Reiss, Ira L. 1967. *The Social Context of Premarital Sexual Permissiveness.* New York: Holt, Rinehart and Winston.

Serrill, Michael S. 1993. "Defiling the Children." *Time*, June 21, pp. 53–55.

Shifter, Jacobo. 1998. *Lila's House: Male Prostitution in Latin America.* New York: Haworth Press.

Shuger, Scott. 2000. "Hookers.com." *Slate Magazine*, January 30 (http://slate.msn.com).

Skolnick, Jerome H. 1975. *Justice without Trial: Law Enforcement in Democratic Society*, 2nd ed. New York: John Wiley & Sons.

Smith, M. Dwayne, and Carl Hand. 1987. "The Pornography/Aggression Linkage: Results from a Field Study." *Deviant Behavior* 8: 389–399.

Smith, Thomas W. 1990. "The Sexual Revolution?" *Public Opinion Quarterly* 54: 334–349.

Spaulding, Jay, and Stephanie Beswick. 1995. "Sex, Bondage, and the Market: The Emergence of Prostitution in Northern Sudan, 1750–1950." *Journal of the History of Sexuality* 5: 512–534.

Stack, Carol B. 1974. *All Our Kin: Strategies for Survival in a Black Community*. New York: Harper Colophon.

Steele, Valerie. 1988. *Paris Fashion*. New York: Oxford University Press.

Strossen, Nadine. 1995. "The Perils of Pornophobia." *Humanist* 55: 7–9.

Tewksbury, Richard. 1994. "Gender Construction and the Female Impersonator: The Process of Transforming 'He' to 'She.'" *Deviant Behavior* 15: 27–43.

Ullman, Sharon R. 1995. "'The Twentieth Century Way': Female Impersonation and Sexual Practice in Turn-of-the-Century America." *Journal of the History of Sexuality* 5: 573–600.

Valentine, Bettylou. 1978. *Hustling and Other Hard Work*. New York: Free Press.

Waddell, Charles. 1996. "HIV and the Social World of Female Commercial Sex Workers." *Medical Anthropological Quarterly* 10: 75–82.

Weaver, James. 1992. "The Social Science and Psychological Research Evidence: Perceptual and Behavioural Consequences of Exposure to Pornography." Pp. 284–309 in *Pornography: Women, Violence, and Civil Liberties*, Catherine Itzin, ed. New York: Oxford University Press.

Weisberg, D. Kelly. 1985. *Children of the Night: A Study of Adolescent Prostitution*. Lexington, MA: Lexington Books.

Weitzer, Ronald, ed. 2000. *Sex For Sale: Prostitution, Pornography, and the Sex Industry*. New York: Routledge.

Winick, Charles, and John T. Evans. 1994. "Is There a National Standard with Respect to Attitudes toward Sexually Explicit Media Material?" *Archives of Sexual Behavior* 23: 405–419.

Winick, Charles, and Paul M. Kinsie. 1971. *The Lively Commerce: Prostitution in the United States*. New York: Quadrangle.

Woodhouse, Annie. 1989. *Fantastic Women: Sex, Gender and Transvestism*. New York: Macmillan.

CHAPTER 13   Suicide

Akers, Ronald L. 1985. *Deviant Behavior: A Social Learning Approach*, 3rd ed. Belmont, CA: Wadsworth.

Ansel, Edward L., and Richard McGee. 1971. "Attitudes toward Suicide Attempters." *Bulletin of Suicidology* 8: 22–29.

Bearman, Peter S. 1991. "The Social Structure of Suicide." *Sociological Forum* 6: 501–524.

Berman, Alan L. 1991. *Adolescent Suicide: Assessment and Intervention*. Washington, DC: American Psychological Association.

Black, Stephen T., and David Lester. 1995. "Distinguishing Suicide Notes from Completed and Attempted Suicides." *Perceptual and Motor Skills* 81: 802.

Blackstone, William. 1765–1769. *Commentaries on the Laws of England*, Vol. IV.

Bould, Sally, Beverly Sanborn, and Laura Reif. 1989. *Eighty-Five Plus: The Oldest Old.* Belmont, CA: Wadsworth.

Brealt, Kevin D. 1986. "Suicide in America: A Test of Durkheim's Theory of Religious and Family Integration, 1933–1980." *American Journal of Sociology* 92: 628–656.

Brealt, Kevin D. 1988. "Beyond the Quick and Dirty: Problems Associated with Analyses Based on Small Samples or Large Ecological Aggregates: Reply to Girard." *American Journal of Sociology* 93: 1,479–1,486.

Breed, Warren. 1963. "Occupational Mobility and Suicide among White Males." *American Sociological Review* 28: 179–188.

Brent, David A., Joshua A. Perper, Grace Moritz, Marianne Baugher, Claudia Roth, Lisa Balach, and Joy Schweers. 1993. "Stressful Life Events, Psychopathology and Adolescent Suicide: A Case Control Study." *Suicide and Life-Threatening Behavior* 23: 179–187.

Bridge, T. Peter, Steven G. Potkin, William W. K. Zung, and Beth J. Soldo. 1977. "Suicide Prevention Centers: Ecological Study of Effectiveness." *The Journal of Nervous and Mental Disease* 164: 18–24.

Bureau of the Census. 1992. *Statistical Abstract of the United States, 1992.* Washington, DC: Government Printing Office.

Bureau of the Census. 1993. *Statistical Abstract of the United States, 1993.* Washington, DC: Government Printing Office.

Bureau of the Census. 1999. *Statistical Abstract of the United States, 1999.* Washington, DC: Government Printing Office.

Bush, James A. 1978. "Similarities and Differences in Precipitating Events Between Black and Anglo Suicide Attempts." *Suicide and Life-Threatening Behavior* 8: 243–249.

Canetto, Silvia Sara, and David Lester. 1995. "The Epidemiology of Women's Suicidal Behavior." Pp. 35–57 in *Women and Suicidal Behavior*, Silvia Sara Canetto and David Lester, eds. New York: Springer.

Canetto, Silvia Sara, and David Lester, eds. 1995. *Women and Suicidal Behavior.* New York: Springer.

Cole, Debra E., Howard O. Protinsky, and Lawrence H. Cross. 1992. "An Empirical Investigation of Adolescent Suicidal Ideation." *Adolesence* 27: 813–818.

Daly, Martin, and Margo Wilson. 1988. *Homicide.* Hawthorne, NY: Aldine de Gruyter.

Davis, Robert A. 1979. "Black Suicide in the Seventies: Current Trends." *Suicide and Life-Threatening Behavior* 9: 131–140.

Davis, Robert A. 1981. "Female Labor Force Participation, Status Integration, and Suicide, 1950–1969." *Suicide and Life-Threatening Behavior* 11: 111–123.

Davis, Robert A., and James F. Short, Jr. 1978. "Dimensions of Black Suicide: A Theoretical Model." *Suicide and Life-Threatening Behavior* 8: 161–173.

Denzin, Norman K. 1992. "The Suicide Machine." *Society* 12: 7–10.

Douglas, Jack D. 1967. *The Social Meanings of Suicide.* Princeton, NJ: Princeton University Press.

Dublin, Louis I. 1963. *Suicide: A Sociological and Statistical Study.* New York: Ronald Press.

Dublin, Louis I., and Bessie Bunzel. 1933. *To Be or Not to Be.* New York: Harrison Smith and Robert Haas.

Durkheim, Emile. 1951. *Suicide.* Translated by John A. Spaulding and George Simpson. New York: Free Press; originally published 1895.

Farber, Maurice L. 1968. *Theory of Suicide.* New York: Funk and Wagnalls.

Farber, Maurice L. 1979. "Suicide in France: Some Hypotheses." *Suicide and Life-Threatening Behavior* 9: 154–162.

Farberow, Norman L. 1977. "Suicide." Pp. 503–505 in *Deviants: Voluntary Actors in an Involuntary World*, Edward Sagarin and Fred Montanino, eds. Morristown, NJ: General Learning Press, 1977.

Farberow, Norman L., ed. 1980. *The Many Faces of Suicide: Indirect Self-Destructive Behavior.* New York: McGraw-Hill.

Faris, Robert E. L. 1948. *Social Disorganization.* New York: Ronald Press.

Ferracuti, Franco. 1957. "Suicide in a Catholic Country." Pp. 57–74 in *Clues to Suicide*, Edwin S. Schneidman and Norman L. Farberow, eds. New York: McGraw-Hill.

Folse, Kimberly A., and Dennis L. Peck. 1996. "A Phenomenological Analysis of Suicide." Pp. 405–409 in *Readings in Deviant Behavior*, Alex Thio and Thomas Calhoun, eds. New York: HarperCollins.

Gibbs, Jack P. 1971. "Suicide." Pp. 271–312 in *Contemporary Social Problems*, 3rd ed., Robert K. Merton and Robert Nisbet, eds. New York: Harcourt Brace Jovanovich.

Gibbs, Jack P. 1982. "Testing the Theory of Status Integration and Suicide Rates." *American Sociological Review* 47: 227–237.

Gibbs, Jack P., and Walter T. Martin. 1958. "Theory of Status Integration and Its Relationship to Suicide." *American Sociological Review* 23: 140–147.

Gibbs, Jack P., and Walter T. Martin. 1964. *Status Integration and Suicide: A Sociological Study.* Eugene: University of Oregon Books.

Girard, Chris. 1993. "Age, Gender, and Suicide: A Cross-National Analysis." *American Sociological Review* 58: 553–574.

Gould, Madelyn S., David Shaffer, and Mark Davies. 1990. "Truncated Pathways from Childhood to Adulthood: Attrition in Follow-Up Studies Due to Death." Pp. 3–9 in *Straight and Devious Pathways from Childhood to Adulthood*, Lee N. Robins and Michael Rutter, eds. Cambridge: Cambridge University Press.

Hacker, Andrew. 1983. *U/S: A Statistical Portrait of the American People.* Baltimore: Penguin Books.

Hankoff, L. D. 1979. "Judaic Origins of the Suicide Prohibition." Pp. 3–20 in *Suicide: Theory and Research*, L. D. Hankoff and Bernice Einsidler, eds. Littleton, MA: PSG Publishing.

Hawton, Keith. 1986. *Suicide and Attempted Suicide among Children and Adolescents.* Beverly Hills, CA: Sage.

Hawton, Keith, and Jose Catalan. 1982. *Attempted Suicide: A Practical Guide to Its Nature and Management.* Oxford: Oxford University Press.

Headley, Lee A., ed. 1983. *Suicide in Asia and the Near East.* Berkeley: University of California Press.

Hendin, Herbert. 1964. *Suicide in Scandinavia.* New York: Grune and Stratton.

Hendin, Herbert. 1969. *Black Suicide.* New York: Basic Books.

Henry, Andrew F., and James F. Short, Jr. 1954. *Suicide and Homicide*. New York: Free Press.

Hoelter, Jon W. 1979. "Religiosity, Fear of Death, and Suicide Acceptability." *Suicide and Life-Threatening Behavior* 9: 163–172.

Humphry, Derek. 1993. *Lawful Exit: The Limits of Freedom for Help in Dying*. Junction City, OR: Norris Lane Press.

Iga, Mamora. 1981. "Suicide of Japanese Youth." *Suicide and Life-Threatening Behavior* 11: 17–30.

Ingram, Ellen, and Jon B. Ellis. 1992. "Attitudes toward Suicidal Behavior: A Review of the Literature." *Death Studies* 16: 31–43.

Jacobs, Jerry. 1967. "A Phenomenological Study of Suicide Notes." *Social Problems* 15: 60–73.

Jacobs, Jerry. 1970. "The Use of Religion in Constructing the Moral Justification of Suicide." Pp. 229–252 in *Deviance and Respectability: The Social Construction of Moral Meanings*, Jack D. Douglas, ed. New York: Basic Books.

Jacobson, Gerald F., and Stephen H. Portuges. 1978. "Relation of Marital Separation and Divorce to Suicide: A Report." *Suicide and Life-Threatening Behavior* 8: 217–225.

Jobes, David A., Alan L. Berman, and Arnold R. Josselsen. 1986. "The Impact of Psychological Autopsies on Medical Examiners' Determination of Manner of Death." *Journal of Forensic Science* 31: 177–189.

Jobes, David A., Alan L. Berman, and Arnold R. Josselsen. 1987. "Improving the Validity and Reliability of Medical-Legal Certifications of Suicide." *Journal of Suicide and Life-Threatening Behavior* 17: 310–325.

Johnson, David, Starla D. Fitch, Jon P. Alston, and William Alex McIntosh. 1980. "Acceptance of Conditional Suicide and Euthanasia among Adult Americans." *Suicide and Life-Threatening Behavior* 10: 157–166.

Kirk, Alton R., and Robert A. Zucker. 1979. "Some Sociopsychological Factors in Attempted Suicide among Urban Black Males." *Suicide and Life-Threatening Behavior* 9: 76–86.

Kitagawa, Evelyn M., and Philip M. Hauser. 1973. *Differential Mortality in the United States: A Study in Socioeconomic Epidemiology*. Cambridge, MA: Harvard University Press.

Kobler, Arthur L., and Ezra Stotland. 1964. *The End of Hope: A Socio-Clinical Study of Suicide*. New York: Free Press.

Kosky, Robert, Sven Silburn, and Stephen R. Zubrick. 1990. "Are Children and Adolescents Who Have Suicidal Thoughts Different from Those Who Attempt Suicide?" *Journal of Nervous and Mental Disease* 178: 38–43.

Labovitz, Sanford. 1968. "Variations in Suicide Rates." Pp. 57–73 in *Suicide*, Jack P. Gibbs, ed. New York: Harper and Row.

Leenaars, Antoon A. 1988. *Suicide Notes: Predictive Clues and Patterns*. New York: Human Sciences Press.

Leenaars, Antoon A., and David Lester. 1991. "Myths about Suicide Notes." *Death Studies* 15: 303–308.

Leenaars, Antoon A., and David Lester. 1992. "Comparison of Rates and Patterns of Suicide in Canada and the United States, 1960–1988." *Death Studies* 16: 417–430.

Leenaars, Antoon A., David Lester, Susanne Wenckstern, Colleen McMullin, Donald Rudzinski, and Alison Brevard. 1992. "Comparison of Suicide Notes and Parasuicide Notes." *Death Studies* 16: 331–342.

Lester, David. 1992a. "Suicide in Police Officers: A Survey of Nations." *Police Studies* 15: 146–147.

Lester, David. 1992b. "A Test of Durkheim's Theroy of Suicide in Primitive Societies." *Suicide and Life-Threatening Behavior* 22: 388–395.

Lester, David. 1993. "The Effectiveness of Suicide Prevention Centers." *Suicide and Life-Threatening Behavior* 23: 263–267.

Lester, David, and Margot Tallmer, eds. 1994. *Now I Lay Me Down: Suicide in the Elderly.* Philadelphia: Charles Press.

Levi, Ken. 1982. "Homicide and Suicide: Structure and Process." *Deviant Behavior* 3: 91–115.

Li, Wen L. 1972. "Suicide and Educational Attainment in a Transitional Society." *The Sociological Quarterly* 13: 253–258.

Litman, Robert E. 1972. "Experiences in a Suicide Prevention Center." Pp. 217–230 in *Suicide and Attempted Suicide*, Jan Waldenstrom, Trage Larsson, and Nils Ljungstedt, eds. Stockholm: Nordiska Bokhandelns Forlag.

Litman, Robert E. 1987. "Mental Disorders and Suicidal Intention." *Suicide and Life-Threatening Behavior* 17: 85–92.

Males, Mike, and Kim Smith. 1991. "Teen Suicide and Changing Cause of Death Certification, 1958–1987." *Suicide and Life-Threatening Behavior* 21: 245–259.

Malinowski, Bronislaw. 1926. *Crime and Custom in Savage Society.* London: Routledge and Kegan Paul.

Maris, Ronald W. 1969. *Social Forces in Urban Suicide.* Homewood, IL: Dorsey Press.

Maris, Ronald W. 1981. *Pathways to Suicide: A Survey of Self-Destructive Behaviors.* Baltimore: Johns Hopkins University Press.

Markides, Kyriakos S. 1981. "Death-Related Attitudes and Behavior among Mexican Americans: A Review." *Suicide and Life-Threatening Behavior* 11: 75–85.

Marra, Realino, and Marco Orru. 1991. "Social Images of Suicide." *British Journal of Sociology* 42: 273–288.

Marshall, James. 1981. "Political Integration and the Effect of War on Suicide." *Social Forces* 59: 771–785.

McGinnis, J. Michael. 1987. "Suicide in America—Moving up the Public Health Agenda." *Suicide and Life-Threatening Behavior* 17: 18–32.

McIntosh, John L. 1992. "Older Adults: The Next Suicide Epidemic?" *Suicide and Life-Threatening Behavior* 22: 322–332.

Meneese, William B., and Barbara A. Yutrzenka. 1990. "Correlates of Suicidal Ideation among Rural Adolescents." *Suicide and Life-Threatening Behavior* 20: 206–212.

Miller, H. L., D. W. Coombs, J. D. Leeper, and S. N. Barton. 1984. "An Analysis of the Effects of Suicide Prevention Facilities on Suicide Rates in the United States." *American Journal of Public Health* 74: 340–343.

Nelson, Franklyn L., Norman L. Farberow, and Douglas R. MacKinnon. 1978. "The Certification of Suicide in Eleven Western States: An Inquiry into the Validity of Reported Suicide Rates." *Suicide and Life-Threatening Behavior* 8: 75–88.

Norstrom, Thor. 1995. "The Impact of Alcohol, Divorce, and Unemployment on Suicide: A Multilevel Analysis." *Social Forces* 74: 293–314.

Nuland, Sherwin B. 2000. "Editorial: Physician-Assisted Suicide and Euthanasia in Practice." *New England Journal of Medicine* 342, February 24.

Orbach, Israel, and Hanna Bar-Joseph. 1993. "The Impact of a Suicide Prevention Program for Adolescents on Suicidal Tendencies, Hopelessness, Ego, Identity, and Coping." *Suicide and Life-Threatening Behavior* 23: 120–129.

Osgood, Nancy J., and Barbara A Brant. 1990. "Suicide Behavior in Long-Term Care Facilities." *Suicide and Life-Threatening Behavior* 20: 113–122.

Paerregaard, Grethe. 1980. "Suicide in Denmark: A Statistical Review for the Past 150 Years." *Suicide and Life-Threatening Behavior* 10: 150–156.

Pescosolido, Bernice A. 1990. "The Social Context of Religious Integration and Suicide: Purusing the Network Explanation." *Sociological Quarterly* 31: 337–357.

Phillips, David P. 1974. "The Influence of Suggestion on Suicide: Substantive and Theoretical Implications of the Werther Effect." *American Sociological Review* 39: 340–354.

Phillips, David P. 1979. "Suicide, Motor Vehicle Fatalities, and the Mass Media: Evidence toward a Theory of Suggestion." *American Journal of Sociology* 84: 1,150-1,174.

Phillips, David P. 1980. "Airplane Accidents, Murder, and the Mass Media: Towards a Theory of Imitation and Suggestion." *Social Forces* 58: 1,001–1,024.

Phillips, David P., and John S. Wills. 1987. "A Drop in Suicides around Major National Holidays." *Suicide and Life-Threatening Behavior* 17: 1–12.

Pinquet, Maurice. 1993. *Voluntary Death in Japan.* Cambridge, England: Polity Press.

Pope, Whitney. 1976. *Durkheim's "Suicide": A Classic Analyzed.* Chicago: University of Chicago Press.

Rich, Alexander R., Joyce Kirkpatrick-Smith, Ronald L. Bonner, and Frank Jans. 1992. "Gender Differences in the Psychosocial Correlates of Suicidal Ideation among Adolescents." *Suicide and Life-Threatening Behavior* 22: 364–373.

Ringel, Erwin. 1977. "The Presuicidal Syndrome." *Suicide and Life-Threatening Behavior* 6: 131–149.

Roa, A. Venkoba. 1983. "India." Pp. 210–237 in *Suicide in Asia and the Near East*, Lee A. Headley, ed. Berkeley: University of California Press.

Robins, Lee N., Patricia A. West, and George E. Murphey. 1977. "The High Rate of Suicide in Older White Men: A Study Testing Ten Hypotheses." *Social Psychiatry* 12: 1–20.

Rudd, M. David. 1990. "An Integrative Model of Suicide Ideation." *Suicide and Life-Threatening Behavior* 20: 16–30.

Sainsbury, Peter. 1963. "Social and Epidemiological Aspects of Suicide with Special Reference to the Aged." Pp. 155–178 in *Processes of Aging*, Vol. II. Edited by Richard H. Williams, Clark Tibbitts, and Wilma Donahue. New York: Atherton Press.

Sakamaki, Sachiko. 1996. "Fates worse than death." *Far Eastern Economic Review* 159: 38–40.

Sanborn, Charlotte J. 1990. "Gender Socialization and Suicide: American Association of Suicidology Presidential Address, 1989." *Suicide and Life-Threatening Behavior* 20: 148–155.

Sawyer, John B., and Elizabeth M. Jameton. 1979. "Chronic Callers to a Suicide Prevention Center." *Suicide and Life-Threatening Behavior* 9: 97–104.

Schneidman, Edwin S., ed. 1976. *Suicidology: Contemporary Developments.* New York: Grune and Stratton.

Schneidman, Edwin S., and Norman L. Farberow, eds. 1957. *Clues to Suicide*. New York: McGraw-Hill.

Schneidman, Edwin S., and Normal L. Farberow. 1961. "Statistical Comparisons between Attempted and Committed Suicides." Pp. 19–47 in *The Cry for Help*, Norman L. Farberow and Edwin S. Schneidman, eds. New York: McGraw-Hill.

Secretary of Health and Human Services. 1993. *Alcohol and Health: Eighth Special Report to the U.S. Congress*. Rockville, MD: National Institute of Alcohol Abuse and Alcoholism, Government Printing Office.

Seiden, Richard H., and Raymond P. Freitas. 1980. "Shifting Patterns of Deadly Violence." *Suicide and Life-Threatening Behavior* 10: 195–209.

Spirito, Anthony, Larry Brown, James Overholser, Gregory Spitz, and Andrea Bond. 1991. "Use of the Risk-Rescue Rating Scale with Adolescent Suicide Attempters: A Cautionary Note." *Death Studies* 15: 269–280.

Stack, Steven. 1980. "Interstate Migration and the Rate of Suicide." *International Journal of Social Psychiatry* 26: 17–26.

Stack, Steven. 1982. "Suicide: A Decade Review of the Sociological Literature." *Deviant Behavior* 4: 41–66.

Stack, Steven. 1990. "The Effect of Divorce on Suicide in Denmark, 1951–1980." *Sociological Quarterly* 31: 359–370.

Stack, Steven, and Ira M. Wasserman. 1992. "The Effect of Religion on Suicide Ideology: An Analysis of the Networks Perspective." *Journal for the Scientific Study of Religion* 31: 457–466.

Stack, Steven, and Ira M. Wasserman. 1995. "Marital Status, Alcohol Abuse and Attempted Suicide: A Logit Model." *Journal of Addictive Diseases* 14: 43–51.

Stafford, Mark C., and Jack P. Gibbs. 1985. "A Major Problem with the Theory of Status Integration and Suicide." *Social Forces* 63: 643–660.

Stafford, Mark C., and Jack P. Gibbs. 1988. "Change in the Relation between Marital Integration and Suicide Rates." *Social Forces* 66: 1,060–1,079.

Stengel, Erwin. 1964. *Suicide and Attempted Suicide*. Baltimore: Penguin.

Stephans, B. Joyce. 1987. "Cheap Thrills and Humble Pie: The Adolescence of Female Suicide Attempters." *Suicide and Life-Threatening Behavior* 17: 107–118.

Sullivan, Amy D., Katrina Hedberg, and David W. Fleming. 2000. "Legalized Physician Assisted Suicide in Oregon: The Second Year." *New England Journal of Medicine* 342: 598–604.

Tatai, Kechinosuke. 1983. "Japan." Pp. 12–58 in *Suicide in Asia and the Near East*, Lee A. Headley, ed. Berkeley: University of California Press.

Trout, Deborah. 1980. "The Role of Social Isolation in Suicide." *Suicide and Life-Threatening Behavior* 10: 10–23.

United Nations. 1996. *Prevention of Suicide: Guidelines for the Formulation and Implementation of National Strategies*. New York: United Nations.

U.S. Department of Health and Human Services. 1985. *Suicide Surveillance Summary: 1970–1980*. Atlanta: Centers for Disease Control.

Violanti, John M. 1995. "Trends in Police Suicide." *Psychological Reports* 77: 688–690.

Weir, Robert F. 1992. "The Morality of Physician Assisted Suicide." *Law, Medicine, and Health Care* 20: 1–2.

West, Donald J. 1965. *Murder Followed by Suicide*. London: William Heinemann.

Westermark, Edward A. 1908. *Origin and Development of Moral Ideas*, Vol. II. London: Macmillan.

Wilson, Michele. 1981. "Suicide Behavior: Toward an Explanation of Differences in Female and Male Rates." *Suicide and Life-Threatening Behavior* 11: 131–140.

Wold, Carl I. 1970. "Characteristics of 26,000 Suicide Prevention Center Patients." *Bulletin of Suicidology* 7: 24–28.

Wolfgang, Marvin E. 1958. *Patterns of Criminal Homicide*. Philadelphia: University of Pennsylvania Press.

Yagla Mack, Kristin, Danny R. Hoyt, and Martin Miller. 1994 (March). "Adolescent Suicide Ideation." Paper presented at the annual meetings of the Midwest Sociology Society, St. Louis, Missouri.

Yang, Bijou. 1992. "The Economy and Suicide: A Time-Series Study of the U.S.A." *American Journal of Economics and Sociology* 51: 87–99.

CHAPTER 14   Physical Disabilities and Eating Disorders

Albrecht, Gary L. 1976. "Socialization and the Disability Process." Pp. 3–38 in *The Sociology of Physical Disability and Rehabilitation*, Gary L. Albrecht, ed. Pittsburgh: University of Pittsburgh Press.

Albrecht, Gary L. 1992. *The Disability Business: Rehabilitation in America*. Newbury Park, CA: Sage.

Bailey, Carol, and Stephen G. Gibbons. 1989. "Physical Victimization and Bulimia-Like Symptoms: Is There a Relationship?" *Deviant Behavior* 10: 335–352.

Baldwin, Marjorie L., and William G. Johnson. 1995. "Labor Market Discrimination against Women with Disabilities." *Industrial Relations* 34: 555–577.

Becker, Gaylene. 1980. *Growing Old in Silence*. Berkeley: University of California Press.

Becker, Howard S. 1973. *Outsiders: Studies in the Sociology of Deviance*, enlarged ed. New York: Free Press.

Blaska, Joan. 1993. "The Power of Language: Speak and Write Using 'Person First.'" Pp. 25–32 in *Perspectives on Disabilities*, 2nd ed., Mark Nagler, ed. Palo Alto, CA: Health Markets Research.

Bordo, Susan. 1993. *Unbearable Weight: Feminism, Western Culture and the Body*. Berkeley: University of California Press.

Cahill, Spencer E., and Robin Eggleston. 1995. "Reconsidering the Stigma of Physical Disability: Wheelchair Use and Public Kindness." *Sociological Quarterly* 36: 681–698.

Cahnman, Werner J. 1968. "The Stigma of Obesity." *Sociological Quarterly* 9: 283–299.

Colvin, Robert H., and Susan B. Olson. 1983. "A Descriptive Analysis of Men and Women Who Have Lost Significant Weight and Are Highly Successful at Maintaining the Loss." *Addictive Behaviors* 8: 287–295.

Connors, Mary E., and Craig L. Johnson. 1987. "Epidemiology of Bulimia and Bulimic Behaviors." *Addictive Behaviors* 12: 165–179.

Corkrean, Jennifer. 1994. "Anorexia Nervosa: A Puzzling Disease." *Iowa State Daily*, February 2, pp. 1, 5.

Cuthbert, David. 2000. "Reeve's 'Walk' Disturbs Doctors and Disabled." *Seattle Times*, February 14.

Davis, Fred. 1961. "Deviance Disavowal: The Management of Strained Interaction by the Visibly Handicapped." *Social Problems* 9: 120–132.

Davis, Fred. 1972. *Illness, Interaction and the Self.* Belmont, CA: Wadsworth.

Deegan, Mary Jo. 1985. "Multiple Minority Groups: A Case Study of Physically Disabled Women." Pp. 37–55 in *Women and Disability: The Double Handicap.* New Brunswick, NJ: Transaction Books.

DeJong, William. 1980. "The Stigma of Obesity: The Consequences of Naive Assumptions Concerning the Causes of Physical Deviance." *Journal of Health and Social Behavior* 21: 75–87.

Edgerton, Robert B. 1967. *The Cloak of Competence: Stigma in the Lives of the Mentally Retarded.* Berkeley: University of California Press.

Edgerton, Robert B. 1979. *Mental Retardation.* Cambridge, MA: Harvard University Press.

Eisenberg, M. G., C. Griggins, and R. J. Duval, eds. 1982. *Disabled People as Second-Class Citizens.* New York: Springer.

Elliott, Gregory C., Herbert L. Ziegler, Barbara M. Altman, and Deborah R. Scott. 1982. "Understanding Stigma: Dimensions of Deviance and Coping." *Deviant Behavior* 3: 275–300.

English, Cliff. 1993. "Gaining and Losing Weight: Identity Transformations." *Deviant Behavior* 14: 227–241.

Evans, Daryl Paul. 1983. *The Lives of Mentally Retarded People.* Boulder, CO: Westview Press.

Fairburn, Christopher G., and G. Terence Wilson, eds. 1993. *Binge Eating: Nature, Assessment, and Treatment.* New York: Guilford.

Fallon, Patricia, Melanie A. Katzman, and Susan C. Wooley, eds. 1993. *Feminist Perspectives on Eating Disorders.* New York: Guilford.

Fox, Renee. 1977. "The Medicalization and Demedicalization of American Society." *Daedalus* 106: 12–19.

Francis, Leslie Pickering, and Anita Silvers, eds. 2000. *Americans with Disabilities: Exploring Implications of the Law for Individuals and Institutions.* New York: Routledge.

Friedson, Eliot. 1965. "Disability as Social Deviance." Pp. 70–82 in *Sociology and Rehabilitation*, Marvin B. Sussman, ed. Washington, DC: American Sociological Association.

Furio, Joanne. 1996. "Global Beauty." *Marie Claire* 3 (October): 42–44.

Garrity, Thomas F. 1973. "Vocational Adjustment after First Myocardial Infarction: Comparative Assessment of Several Variables Suggested in the Literature." *Social Science and Medicine* 7: 705–717.

Goffman, Erving. 1963. *Stigma: Notes on the Management of Spoiled Identity.* Englewood Cliffs, NJ: Prentice-Hall.

Haber, Lawrence D., and Richard T. Smith. 1971. "Disability and Deviance: Normative Adaptations of Role Behavior." *American Sociological Review* 36: 87–97.

Habib, Lina Abu. 1995. "'Women and Disability Don't Mix!': Double Discrimination and Disabled Women's Rights." *Gender and Development* 3: 49–53.

Hainer, Cathy. 1996. "The Renaissance of Fat Pride." *USA Today*, November 21, pp. 1D–2D.

Hamli, Katherine A., James R. Falk, and Estelle Schwartz. 1981. "Binge-Eating and Vomiting: A Survey of a College Population." *Psychological Medicine* 11: 697–706.

Hawkins, R. C., and P. F. Clement. 1980. "Development and Construct Validation of a Self Report Measure of Binge Eating Tendencies." *Addictive Behaviors* 5: 219–226.

Hays, Diane, and Catherine E. Ross. 1987. "Concern with Appearance, Health Beliefs, and Eating Habits." *Journal of Health and Social Behavior* 28: 120–130.

Hesse-Biber, Sharlene. 1996. *Am I Thin Enough Yet? The Cult of Thinness and the Commercialization of Identity*. New York: Oxford University Press.

Higgins, Paul C. 1980. *Outsiders in a Hearing World*. Beverly Hills, CA: Sage.

Howards, Irving, Henry P. Brehm, and Saad Z. Nagi. 1980. *Disability: From Social Problem to Federal Program*. New York: Praeger.

Humphries, Laurie L., Sylvia Wrobel, and H. Thomas Wiegert. 1982. "Anorexia Nervosa." *American Family Physician* 26: 199–204.

Hyman, Marvin. 1971. "Disability and Patients' Perceptions of Preferential Treatment." *Journal of Chronic Diseases* 24: 329–342.

Kalb, Claudia. 1999. "Our Quest to be Perfect." *Newsweek*, August 9: 52–59.

Kaplan, H. I., and H. S. Kaplan. 1957. "The Psychosomatic Concept of Obesity." *Journal of Nervous and Mental Disease* 125: 181–201.

Kiernan, Louise. 1996. "International Incident: Miss Universe Gains 18 Pounds." *Des Moines Register*, September 2, p. 3T.

Klein, Richard. 1996. *Eat Fat*. New York: Pantheon.

Koestler, Frances A. 1976. *The Unseen Minority: A Social History of Blindness in the United States*. New York: David McKay.

Krause, Elliott A. 1976. "The Political Sociology of Rehabilitation." Pp. 201–221 in *The Sociology of Physical Disability and Rehabilitation*, Gary L. Albrecht, ed. Pittsburgh, PA: University of Pittsburgh Press.

Kübler-Ross, Elisabeth. 1969. *On Death and Dying*. New York: Macmillan.

Laslet, Barbara, and Carol A. B. Warren. 1975. "Losing Weight: The Organizational Promotion of Behavior Change." *Social Problems* 23: 69–80.

Lemert, Edwin M. 1951. *Social Pathology*. New York: McGraw-Hill.

Levitin, Teresa E. 1975. "Deviants as Active Participants in the Labeling Process: The Visibly Handicapped." *Social Problems* 22: 548–557.

Lord, M. G. 1994. *Forever Barbie: The Unauthorized Biography of a Real Doll*. New York: William Morrow.

Maddox, George L., Kurt W. Back, and Veronica R. Liederman. 1968. "Overweight as Social Deviance and Social Disability." *Journal of Health and Social Behavior* 9: 287–298.

Maine, Margo. 2000. *Body Wars: Making Peace with Women's Bodies*. Carlsbad, CA: Gurze.

Mayer, Jean. 1968. *Overweight: Causes, Cost, and Control*. Englewood Cliffs, NJ: Prentice-Hall.

McLorg, Penelope A., and Diane E. Taub. 1987. "Anexoria Nervosa and Bulimia: The Development of Deviant Identities." *Deviant Behavior* 8: 177–189.

McLorg, Penelope A., and Diane E. Taub. 1996. "Anorexics and Bulimics: Developing Deviant Identities." Pp. 392–397 in *Readings in Deviant Behavior*, Alex Thio and Thomas Calhoun, eds. New York: HarperCollins.

Mechanic, David. 1995. "Sociological Dimensions of Illness Behavior." *Social Science and Medicine* 41: 1,207–1,216.

Mercer, Jane R. 1973. *Labeling the Mentally Retarded*. Berkeley: University of California Press.

Millman, M. 1980. *Such a Pretty Face: Being Fat in America.* New York: W. W. Norton.

Mitchell, James E., Richard L. Pyle, Elke D. Eckert, Dorothy Hatsukami, and Elizabeth Soll. 1990. "Bulimia Nervosa in Overweight Individuals." *Journal of Nervous and Mental Disease* 178: 324–327.

Morgan, Carolyn Stout, Marilyn Affleck, and Orin Solloway. 1990. "Gender Role Attitudes, Religiosity, and Food Behavior: Dieting and Bulimia in College Women." *Social Science Quarterly* 71: 142–151.

Myerson, Lee. 1971. "Physical Disability as a Social Psychological Problem." Pp. 205–210 in *The Other Minorities*, Edward Sagarin, ed. Boston: Ginn.

Nagi, Saad Z. 1969. *Disability and Rehabilitation.* Columbus: Ohio State University Press.

Nagler, Mark. 1993a. "The Disabled: The Acquisition of Power." Pp. 33–36 in *Perspectives on Disabilities*, 2nd ed. Palo Alto, CA: Health Markets Research.

Nagler, Mark, ed. 1993b. *Perspectives on Disabilities*, 2nd ed. Palo Alto, CA: Health Markets Research.

Nonbeck, Michael E. 1973. *The Meaning of Blindness: Attitudes toward Blindness and Blind People.* Bloomington: Indiana University Press.

Parsons, Talcott. 1951. *The Social System.* New York: Free Press.

Pate, Jennifer, Andres J. Pumareiga, Colleen Hester, and David M. Garner. 1996. "Cross-Cultural Patterns in Eating Disorders: A Review." *American Academy of Child and Adolescent Psychiatry* 31: 802–809.

Powers, Pauline S., Richard G. Schulman, Alice A. Gleghorn, and Mark E. Prange. 1987. "Perceptual and Cognitive Abnormalities in Bulimia." *American Journal of Psychiatry* 144: 1,456–1,460.

Richardson, Stephen A., Norman Goodman, Albert H. Hastorf, and Sanford M. Dornbush. 1961. "Cultural Uniformity in Reaction to Physical Disabilities." *American Sociological Review* 26: 241–247.

Ross, Catherine E. 1994. "Overweight and Depression." *Journal of Health and Social Behavior* 35: 63–75.

Roth, William. 1981. *The Handicapped Speak.* Jefferson, NC: McFarland.

Rubinstein, Sharon, and Benjamin Caballero. 2000. "Is Miss America an Undernourished Role Model?" *Journal of the American Medical Association*, 282 (March 22/29).

Ruderman, Audrey J. 1983. "Obesity, Anxiety, and Food Consumption." *Addictive Behaviors* 8: 235–242.

Safilios-Rothschild, Constantina. 1970. *The Sociology and Social Psychology of Disability and Rehabilitation.* New York: Random House.

Safilios-Rothschild, Constantina. 1976. "Disabled Persons' Self-Definitions and Their Implications for Rehabilitation." Pp. 39–56 in *The Sociology of Physical Disability and Rehabilitation*, Gary L. Albrecht, ed. Pittsburgh, PA: University of Pittsburgh Press.

Sagarin, Edward. 1975. *Deviants and Deviance.* New York: Praeger.

Schur, Edwin M. 1984. *Labeling Women Deviant: Gender, Stigma, and Social Control.* New York: Random House.

Scott, Robert A. 1969. *The Making of Blind Men: A Study of Adult Socialization.* New York: Russell Sage.

Scott, Robert A. 1980. "Introduction." Pp. 2–10 in *Outsiders in a Hearing World*, Paul C. Higgins. Beverly Hills, CA: Sage.

Seid, Roberta Pollack. 1989. *Never Too Thin: Why Women Are at War with Their Bodies.* Englewood Cliffs, NJ: Prentice-Hall.

Shapiro, Deane H., Jr., Barton J. Blinder, Jennifer Hagman, and Steven Pituck. 1993. "A Psychological 'Sense-of-Control' Profile of Patients with Anorexia Nervosa and Bulimia Nervosa." *Psychological Reports* 73: 530–541.

Shorter, Edward. 1987. "The First Great Increase in Anorexia Nervosa." *Journal of Social History* 21: 80–89.

Sobal, Jeffery, and Donna Maurer, eds. 1999. *Weighty Issues: Fatness and Thinness as Social Problems.* Hawthorne, NY: Aldine de Gruyter.

Stafford, Mark C., and Richard R. Scott. 1986. "Stigma, Deviance, and Social Control: Some Conceptual Issues." Pp. 77–91 in *The Dilemma of Difference: A Multidisciplinary View of Stigma*, Stephen A. Ainlay, Gaylene Becker, and Lerita M. Coleman, eds. New York: Plenum Press.

Stone, Sharon Dale. 1995. "The Myth of Bodily Perfection." *Disability and Society* 10: 413–424.

Taub, Diane E., and Penelope A. McLorg. 1990. "The Sociocultural Context of Anorexia Nervosa and Bulimia Nervosa: A Review and Discussion." Unpublished paper, Department of Sociology, Southern Illinois University.

Thomas, David. 1982. *The Experience of Handicap.* London: Methuen.

Thompson, Becky W. 1994. *A Hunger So Wide and So Deep.* Minneapolis: University of Minnesota Press.

Topliss, Eda. 1982. *Social Responses to Handicap.* New York: Longman.

Turner, R. Jay, and Morton Beiser. 1990. "Major Depression and Depressive Symptomatology among the Physically Disabled." *Journal of Nervous and Mental Disease* 178: 343–350.

Vandereycken, Walter, and Eugene L. Lowenkopf. 1990. "Anorexia Nervosa in 19th Century America." *Journal of Nervous and Mental Disease* 178: 531–535.

Wang, Caroline. 1993. "Culture, Meaning, and Disability: Injury Prevention Campaigns and the Production of Stigma." Pp. 77–90 in *Perspectives on Disabilities*, 2nd ed., Mark Nagler, ed. Palo Alto, CA: Health Markets Research.

Warren, Carol A. B. 1974. "The Use of Stigmatized Labels in Conventionalizing Deviant Behavior." *Sociology and Social Research* 58: 303–311.

Wolinsky, Fredric D., and Sally R. Wolinsky. 1981. "Expecting Sick-Role Legitimation and Getting It." *Journal of Health and Social Behavior* 22: 229–242.

Wright, Beatrice A. 1960. *Physical Disability—A Psychological View.* New York: Harper and Row.

Zincand, H., R. J. Cadoret, and R. B. Widman. 1984. "Incidence and Detection of Bulimia in a Family Practice Population." *Journal of Family Practice* 18: 555–560.

CHAPTER 15   Homosexuality and Homophobia

Adam, Barry D. 1987. *The Rise of a Gay and Lesbian Movement.* Boston: Twayne.

Akers, Ronald L. 1985. *Deviant Behavior: A Social Learning Approach*, 3rd ed. Belmont, CA: Wadsworth.

Allen, Laura S., and Roger A Gorski. 1992. "Sexual Orientation and the Size of the Anterior Commissure." *Proceedings of the National Academy of Science* 89: 7,199–7,202.

American College Health Association. 1987. "AIDS: What Everyone Should Know." Rockville, MD: American College Health Association.

Ames, Lynda J., Alana B. Atchinson, and Rose D. Thomas. 1995. "Love, Lust, and Fear: Safe Sex Decision Making among Gay Men." *Journal of Homosexuality* 30: 53–73.

Appiah, K. Anthony. 1996. "The Marrying Kind." *New York Review of Books*, June 20, pp. 48–54.

Baum, Andrew, and Lydia Temoshok. 1990. "Psychosocial Aspects of Acquired Immunodeficiency Syndrome." Pp. 1–16 in *Psychosocial Perspectives on AIDS: Etiology, Prevention, and Treatment*, Lydia Temoshok and Andrew Baum, eds. Hillsdale, NJ: Laurence Erlbaum Associates.

Bauman, Laurie J., and Karolyn Siegel. 1990. "Misperception among Gay Men of the Risk of AIDS Associated with Their Sexual Behavior." Pp. 81–101 in *Psychosocial Perspectives on AIDS: Etiology, Prevention, and Treatment*, Lydia Temoshok and Andrew Baum, eds. Hillsdale, NJ: Laurence Erlbaum Associates.

Bawer, Bruce. 1996. *Beyond Queer: Challenging Gay Left Orthodoxy*. New York: Free Press.

Bell, Alan P., and Martin S. Weinberg. 1978. *Homosexualities: A Study of Diversity among Men and Women*. New York: Simon and Schuster.

Bell, Alan P., Martin S. Weinberg, and Sue Kiefer Hammersmith. 1981. *Sexual Preference: Its Development in Men and Women*. Bloomington: Indiana University Press.

Berger, Gregory, Lori Hank, Tom Ravzi, and Lawrence Simkins. 1987. "Detection of Sexual Orientation by Heterosexuals and Homosexuals." *Journal of Homosexuality* 13: 83–100.

Berger, Raymond M. 1986. "Gay Men." Pp. 162–180 in *Helping the Sexually Oppressed*, Harvey L. Gochros, Jean S. Gochros, and Joel Fischer, eds. Englewood Cliffs, NJ: Prentice-Hall.

Berger, Raymond M. 1990. "Men Together: Understanding the Gay Couple." *Journal of Homosexuality* 19: 31–49.

Bigner, Jerry J., and R. Brooke Jacobsen. 1992. "Adult Responses to Child Behavior and Attitudes toward Fathering: Gay and Nongay Fathers." *Journal of Homosexuality* 23: 99–112.

Bishop, William C. 1993. "Possible Genetic Link with Sexual Preference." *Chronicle of Higher Education* 43, July 21: B1.

Blumstein, Philip W., and Pepper Schwartz. 1974. "Lesbianism and Bisexuality." Pp. 278–295 in *Sexual Deviance and Sexual Deviants*, Erich Goode and Richard T. Troiden, eds. New York: William Morrow.

Blumstein, Philip W., and Pepper Schwartz. 1983. *American Couples*. New York: William Morrow.

Boston Lesbian Psychologies Collective, eds. 1987. *Lesbian Psychologies: Explorations and Challenges*. Urbana: University of Illinois Press.

Boswell, John. 1980. *Christianity, Social Tolerance, and Homosexuality*. Chicago: University of Chicago Press.

Brannock, JoAnn C., and Beata E. Chapman. 1990. "Negative Sexual Experiences with Men among Heterosexual Women and Lesbians." *Journal of Homosexuality* 19: 105–110.

Britton, Dana M. 1990. "Homophobia and Homosociality: An Analysis of Boundary Maintenance." *Sociological Quarterly* 31: 423–439.

Brody, Stuart. 1997. *Sex at Risk.* New Brunswick, NJ: Transaction.

Browning, Christine. 1987. "Therapeutic Issues and Intervention Strategies with Young Adult Lesbian Clients: A Developmental Approach." *Journal of Homosexuality* 14: 45–52.

Browning, Frank. 1994. *The Culture of Desire.* New York: Simon and Schuster.

Bryant, Anita. 1977. *The Anita Bryant Story: The Survival of Our Nation's Families and the Threat of Militant Homosexuality.* Old Tappan, NJ: Revell.

Card, Claudia. 1992. "Lesbianism and Choice." *Journal of Homosexuality* 23: 39–51.

Coates, Thomas J., Ron D. Stall, and Colleen C. Hoff. 1990. "Changes in Sexual Behavior among Gay and Bisexual Men Since the Beginning of the AIDS Epidemic." Pp. 103–137 in *Psychosocial Perspectives on AIDS: Etiology, Prevention, and Treatment*, Lydia Temoshok and Andrew Baum, eds. Hillsdale, NJ: Laurence Erlbaum Associates.

Coleman, Eli. 1981–1982. "Developmental Stages in the Coming Out Process." *Journal of Homosexuality* 7: 31–43.

Connell, R. W. 1992. "A Very Straight Gay: Masculinity, Homosexual Experience, and the Dynamics of Gender." *American Sociological Review* 57: 735–751.

Cronin, Denise M. 1974. "Coming Out among Lesbians." Pp. 265–277 in *Sexual Deviance and Sexual Deviants*, Erich Goode and Richard T. Troiden, eds. New York: William Morrow.

Cruikshank, Margaret. 1992. *The Gay and Lesbian Liberation Movement.* New York: Routledge.

Dank, Barry. 1971. "Coming Out in the Gay World." *Psychiatry* 34: 192–198.

Dank, Barry. 1972. "Why Homosexuals Marry Women." *Medical Aspects of Human Sexuality* 6: 14–23.

Davies, Christie. 1982. "Sexual Taboos and Social Boundaries." *American Journal of Sociology* 87: 1,032–1,063.

Dooley, Janne. 1986. "Lesbians." Pp. 181–190 in *Helping the Sexually Oppressed*, Harvey L. Gochros, Jean S. Gochros, and Joel Fischer, eds. Englewood Cliffs, NJ: Prentice-Hall.

Dover, Kenneth J. 1978. *Greek Homosexuality.* Cambridge, MA: Harvard University Press.

Dworkin, Ronald. 1977. *Taking Rights Seriously.* New York: Oxford University Press.

Endleman, Robert. 1990. *Deviance and Psychopathology: The Sociology and Psychology of Outsiders.* Malabar, FL: Robert Krieger Publishing.

Eskridge, William N. 1996. *The Case for Same-Sex Marriage: From Sexual Liberty to Civilized Commitment.* New York: Free Press.

Ficarrotto, Thomas J. 1990. "Racism, Sexism, and Erotophobia: Attitudes of Heterosexuals toward Homosexuals." *Journal of Homosexuality* 19: 111–116.

Fone, Bryne. 2000. *Homophobia: A History.* New York: Henry Holt.

Ford, Clellan S., and Frank A. Beach. 1951. *Patterns of Sexual Behavior.* New York: Harper and Row.

Fumento, Michael. 1990. *The Myth of Heterosexual AIDS.* New York: Basic Books.

Gallup Poll. 1986. "Sharp Decline Found in Support for Legalizing Gay Relations." *The Gallup Report* 254.

Gauthier, DeAnn K., and Craig J. Forsyth. 1999. "Bareback Sex, Bug Chasers, and the Gift of Death." *Deviant Behavior* 20: 85–100.

Geis, Gilbert. 1972. *Not the Law's Business?* Rockville, MD: National Institute of Mental Health.

Gilligan, Carol. 1982. *In a Different Voice: Psychological Theory and Women's Development.* Cambridge, MA: Harvard University Press.

Glassner, Barry, and Carol Owen. 1976. "Variations in Attitudes toward Homosexuality." *Cornell Journal of Social Relations* 11: 161–176.

Goffman, Erving. 1963. *Stigma: Notes on the Management of Spoiled Identity.* Englewood Cliffs, NJ: Prentice-Hall.

Goode, Erich, and Richard T. Troiden, eds. 1974. *Sexual Deviance and Sexual Deviants.* New York: William Morrow.

Green, Richard. 1987. *The "Sissy Boy Syndrome" and the Development of Homosexuality.* New Haven, CT: Yale University Press.

Greenberg, David F. 1988. *The Construction of Homosexuality.* Chicago: University of Chicago Press.

Hamer, Dean, and Peter Copeland. 1994. *The Science of Desire: The Search for the Gay Gene and the Biology of Behavior.* New York: Simon and Schuster.

Hammersmith, Sue Kiefer. 1987. "A Sociological Approach to Counseling Homosexual Clients and Their Families." *Journal of Homosexuality* 14: 173–190.

Harry, Joseph. 1982. *Gay Children Grown Up: Gender Culture, and Gender Deviance.* New York: Praeger.

Harry, Joseph. 1984. "Sexual Orientation as Destiny." *Journal of Homosexuality* 10: 111–124.

Harry, Joseph. 1990. "A Probability Sample of Gay Men." *Journal of Homosexuality* 19: 89–104.

Hedblom, Jack H. 1972. "The Female Homosexual: Social and Attitudinal Dimensions." Pp. 50–65 in *The Homosexual Dialectic,* James A. McCaffrey, ed. Englewood Cliffs, NJ: Prentice-Hall.

Henry, William A., III. 1993. "Born Gay?" Time, 142: (July 26): 36–39.

Herdt, Gilbert. 1981. *The Guardians of the Flute.* New York: McGraw-Hill.

Hetrick, Emery S., and A. Damien Martin. 1987. "Developmental Issues and Their Resolution for Gay and Lesbian Adolescents." *Journal of Homosexuality* 14: 25–43.

Humphreys, Laud. 1972. *Out of the Closets: The Sociology of Homosexual Liberation.* Englewood Cliffs, NJ: Prentice-Hall.

Humphreys, Laud. 1975. *Tearoom Trade: Impersonal Sex in Public Places,* enlarged edition. Chicago: Aldine.

Humphreys, Laud, and Brian Miller. 1980. "Identities in the Emerging Gay Culture." Pp. 142–156 in *Homosexual Behavior: A Modern Reappraisal,* Judd Marmor, ed. New York: Basic Books.

Jay, Karla, and Allen Young. 1979. *The Gay Report: Lesbians and Gay Men Speak Out about Sexual Experiences and Lifestyles.* New York: Summit Books.

Johnston, Jill. 1973. *Lesbian Nation: The Feminist Solution.* New York: Simon and Schuster.

Kantor, Martin. 1998. *Homophobia: Description, Development, and Dynamics of Gay Bashing.* Westport, CT: Praeger.

Katz, Jonathan, ed. 1976. *Gay American History: Lesbians and Gay Men in the U.S.A.* New York: Cromwell.

Katz, Jonathan. 1983. *Gay/Lesbian Almanac: A New Documentary.* New York: Harper and Row.

Kelly, Robert J. 1990. "AIDS and the Societal Reaction." Pp. 47–61 in *Perspectives on Deviance: Dominance, Degradation, and Denigration*, Robert J. Kelly and Donal E. J. MacNamara, eds. Cincinnati: Anderson.

Kinsey, Alfred C., Ward B. Pomeroy, and Charles E. Martin. 1948. *Sexual Behavior in the Human Male*. Philadelphia: W. B. Saunders.

Kinsey, Alfred C., Ward B. Pomeroy, Charles E. Martin, and Paul H. Gebhard. 1953. *Sexual Behavior in the Human Female*. Philadelphia: W. B. Saunders.

Klassen, Albert D., Colin J. Williams, and Eugene E. Levitt. 1989. *Sex and Morality in the U.S.* Middletown, CT: Wesleyan University Press.

Langevin, Ron. 1985. "Introduction." Pp. 1–13 in *Erotic Preference, Gender Identity, and Aggression in Men: New Research Studies*, Ron Langevin, ed. Hillsdale, NJ: Lawrence Erlbaum Associates.

Laumann, Edward O., John H. Gagnon, Robert T. Michael, and Stuart Michaels. 1994. *The Social Organization of Sexuality: Sexual Practices in the United States*. Chicago: University of Chicago Press.

LaVay, Simon. 1991. "A Difference in Hypothalamic Structure between Heterosexual and Homosexual Men." *Science* 253: 1,034–1,037.

Lewin, Ellen, and Terrie A. Lyons. 1982. "Everything in Its Place: The Coexistence of Lesbianism and Motherhood." Pp. 249–273 in *Homosexuality: Social, Psychological, and Biological Issues*, William Paul, James D. Weinrich, John C. Gonsiorek, and Mary E. Hotvedt, eds. Beverly Hills, CA: Sage.

Lindner, Robert. 1963. "Homosexuality and the Contemporary Scene." Pp. 60–82 in *The Problem of Homosexuality in Modern Society*, Hendrix Ruitenbeek, ed. New York: E. P. Dutton.

Luckenbill, David F. 1986. "Deviant Career Mobility: The Case of Male Prostitutes." *Social Problems* 33: 283–293.

Lynch, Frederick R. 1987. "Non-Ghetto Gays: A Sociological Study of Suburban Homosexuals." *Journal of Homosexuality* 13: 13–42.

Magee, Brian. 1966. *One in Twenty: A Study of Homosexuality in Men and Women*. New York: Stein and Day.

Marsiglio, William. 1993. "Attitudes toward Homosexual Activity and Gays as Friends: A National Survey of Heterosexual 15- to 19-year-old Males." *The Journal of Sex Research* 30: 12–17.

McCaghy, Charles H., and James K. Skipper, Jr. 1969. "Lesbian Behavior as an Adaptation to the Occupation of Stripping." *Social Problems* 17: 262–270.

McDonald, A. P. 1976. "Homophobia: Its Roots and Meanings." *Homosexual Counseling Journal* 3: 23–33.

McDonald, Gary J. 1982. "Individual Differences in the Coming Out Process for Gay Men: Implications for Theoretical Models." *Journal of Homosexuality* 8: 47–60.

McWhirter, David P., and Andrew M. Mattison. 1984. *The Male Couple: How Relationships Develop*. Englewood Cliffs, NJ: Prentice-Hall.

McWilliams, Peter. 1993. *Ain't Nobody's Business If You Do*. Los Angeles: Prelude Press.

Meier, Robert F., and Gilbert Geis. 1997. *Victimless Crime? Prostitution, Homosexuality, Drugs, and Abortion*. Los Angeles: Roxbury.

Michael, Robert T., John H. Gagnon, Edward O. Laumann, and Gina Kolata. 1994. *Sex in America: A Definitive Survey.* Boston: Little, Brown.

Miller, Neil. 1989. *In Search of Gay America.* New York: Atlantic Monthly Press.

Miller, Neil. 1992. *Out in the World: Gay and Lesbian Life from Buenos Aires to Bangkok.* New York: Random House.

Money, John. 1988. *Gay, Straight, and In-Between: The Sociology of Erotic Orientation.* New York: Oxford.

Muchmore, Wes, and William Hanson. 1989. "A Gay Man's Guide to Coming Out." Pp. 72–74 in *The Alyson Almanac.* Boston: Alyson Publications.

Murray, Stephen O. 2000. *Homosexualities.* Chicago: University of Chicago Press.

Paul, Jay P. 1990. Review of Richard Green, "The 'Sissy Boy Syndrome' and the Development of Homosexuality." *Journal of Homosexuality* 19: 140–147.

Paul, William, and James D. Weinrich. 1982. "Whom and What We Study: Definition and Scope of Sexual Orientation." Pp. 23–28 in *Homosexuality: Social, Psychological, and Biological Issues,* William Paul, James D. Weinrich, John C. Gonsiorek, and Mary E. Hotvedt, ed. Beverly Hills, CA: Sage.

Peplau, Letitia Anne, Christine Padesky, and Mykol Hamilton. 1982. "Satisfaction in Lesbian Relationships." *Journal of Homosexuality* 8: 23–35.

Plummer, Kenneth. 1975. *Sexual Stigma: An Interactionist Account.* London: Routledge and Kegan Paul.

Plummer, Kenneth, ed. 1981. *The Making of the Modern Homosexual.* London: Hutchinson.

Plummer, Kenneth. 1989. "Lesbian and Gay Youth in England." *Journal of Homosexuality* 17: 195–223.

Posner, Richard A. 1992. *Sex and Reason.* Cambridge, MA: Harvard University Press.

Price, Monroe. 1989. *Shattered Mirrors: Our Search for Identify and Community in the AIDS Era.* Cambridge, MA: Harvard University Press.

Quadland, Michael C., and William D. Shattis. 1987. "AIDS, Sexuality, and Sexual Control." *Journal of Homosexuality* 13: 13–42.

Quinn, D. Michael. 1996. *Same-Sex Dynamics among Nineteenth Century Americans: A Mormon Example.* Urbana: University of Illinois Press.

Rechy, John. 1963. *City of the Night.* New York: Grove Press.

Reiss, Albert J., Jr. 1961. "The Social Integration of Queers and Peers." *Social Problems* 9: 102–120.

Reiss, Ira L. 1986. *Journey into Sexuality: An Exploratory Voyage.* Englewood Cliffs, NJ: Prentice-Hall.

Reitzes, Donald C., and Juliette K. Diver. 1982. "Gay Bars as Deviant Community Organizations: The Management of Interactions with Outsiders." *Deviant Behavior* 4: 1–18.

Richards, David A. 1999. *Identity and the Case for Gay Rights: Race, Gender, Religion as Analogies.* Chicago: University of Chicago Press.

Roper, W. G. 1996. "The Etiology of Male Homosexuality." *Medical Hypotheses* 46: 85–88.

Ross, H. Laurence. 1971. "Modes of Adjustment of Married Homosexuals." *Social Problems* 18: 385–393.

Rosenzweig, Julie M., and Wendy C. Lebow. 1992. "Femme on the Streets, Butch in the Sheets? Lesbian Sex-Roles, Dyadic Adjustment, and Sexual Satisfaction." *Journal of Homosexuality* 23: 1–20.

Russell, Paul. 1996. *The Gay 100: A Ranking of the Most Influential Gay Men and Lesbians, Past and Present.* New York: Citadel Press.

Rust, Paula C. 1992. "The Politics of Sexual Identity: Sexual Attraction and Behavior among Lesbian and Bisexual Women." *Social Problems* 39: 366–386.

Saghir, Marcel T., and Eli Robins. 1973. *Male and Female Homosexuality: A Comparative Investigation.* Baltimore: Williams and Wilkins.

Saghir, Marcel T., and Eli Robins. 1980. "Clinical Aspects of Female Homosexuality." Pp. 286–315 in *Homosexual Behavior: A Modern Reappraisal,* Judd Marmor, ed. New York: Basic Books.

Savin-Williams, Ritch C. 1999. *. . . And Then I Became Gay.* New York: Routledge.

Schofield, Michael. 1965. *Sociological Aspects of Homosexuality: A Comparative Study of Three Types of Homosexuals.* Boston: Little, Brown.

Schur, Edwin M. 1984. *Labeling Women Deviant: Gender, Stigma, and Social Control.* New York: Random House.

Schwanberg, Sandra L. 1990. "Attitudes toward Homosexuality in American Health Care Literature, 1983–1987." *Journal of Homosexuality* 19: 117–136.

Shapiro, Joseph P. 1994. "Straight Talk about Gays." *U.S. News and World Report,* July 5, p. 47.

Shilts, Randy. 1987. *And the Band Played On: Politics, People and the AIDS Epidemic.* New York: St. Martin's Press.

Signorile, Michelangelo. 1993. *Queer in America: Sex, The Media, and the Closets of Power.* New York: Random House.

Silverstein, Charles. 1981. *Man to Man: Gay Couples in America.* New York: William Morrow.

Simon, William, and John H. Gagnon. 1967. "The Lesbians: A Preliminary Overview." Pp. 247–282 in *Sexual Deviance,* William Simon and John H. Gagnon, eds. New York: Harper and Row.

Simpson, Ruth. 1976. *From the Closet to the Courts: The Lesbian Tradition.* Baltimore: Penguin Books.

Soards, Marion. 1995. *Scripture and Homosexuality: Biblical Authority and the Church Today.* Louisville, KY: Westminster John Knox Press.

Spector, Malcolm. 1977. "Legitimizing Homosexuality." *Society* 14: 52–56.

Stephan, G. Edward, and Douglas R. McMullin. 1982. "Tolerance of Sexual Nonconformity: City Size as a Situational and Early Learning Determinant." *American Sociological Review* 47: 411–415.

St. Lawrence, Janet S., Brenda A. Husfeldt, Jeffrey A. Kelly, Harold V. Hood, and Steve Smith, Jr. 1990. "The Stigma of AIDS: Fear of Disease and Prejudice toward Gay Men." *Journal of Homosexuality* 19: 85–101.

Sullivan, Andrew. 1995. *Virtually Normal: An Argument about Homosexuality.* New York: Alfred A. Knopf.

Swaab, D. F., and M. A. Hofman. 1990. "An Enlarged Suprachiasmatic Nucleus in Homosexual Men." *Brain Research* 537: 141–148.

Troiden, Richard R. 1979. "Becoming Homosexual: A Model of Gay Identity Acquisition." *Psychiatry* 42: 362–373.

Troiden, Richard R. 1988. *Gay and Lesbian Identity: A Sociological Analysis.* New York: General Hall.

Troiden, Richard R. 1989. "The Formation of Homosexual Identities." *Journal of Homosexuality* 17: 43–73.

Valocchi, Steve. 1999. "The Class-Inflected Nature of Gay Identity." *Social Problems* 46: 207–224.

Warren, Carol A. B. 1972. *Identity and Community in the Gay World.* New York: John Wiley & Sons.

Weinberg, Martin S., and Colin J. Williams. 1974. *Male Homosexuals: Their Problems and Adaptations.* New York: Oxford University Press.

Weinberg, Martin S., and Colin J. Williams. 1975. "Gay Baths and the Social Organization of Impersonal Sex." *Social Problems* 23: 124–136.

Westwood, Gordon. 1960. *A Minority: A Report on the Life of the Male Homosexual in Great Britain.* London: Longmans Green.

Whitam, Frederick L., and Robin M. Mathy. 1985. *Male Homosexuality in Four Societies: Brazil, Guatemala, the Philippines, and the United States.* New York: Praeger.

Wheeler, David A. 1993. "Researchers Explore Genetic Link to Homosexuality." *Chronicle of Higher Education* 43, July 21: A1, A17.

World Health Organization. 1999. *AIDS Epidemic Update—December 1999.* New York: United Nations Press Release.

Zimet, Gregory D., Rina Lazebnik, Ralph J. DeClemente, Trina M. Anglin, Paul Williams, and Elise M. Ellick. 1993. "The Relationship of Magic Johnson's Announcement of HIV Infection to the AIDS Attitudes of Junior High School Students." *The Journal of Sex Research* 30: 129–134.

CHAPTER 16   Mental Disorders

Aday, David P., Jr. 1990. *Social Control at the Margins: Toward a General Understanding of Deviance.* Belmont, CA: Wadsworth.

American Psychiatric Association. 1994. *Diagnostic and Statistical Manual-IV,* 4th ed. Washington, DC: American Psychiatric Association.

Bassett, Anne S. 1991. "Linkage Analysis of Schizophrenia: Challenges and Promise." *Social Biology* 38: 189–196.

Bastide, Roger. 1972. *The Sociology of Mental Disorder.* Trans. by Jean McNeil. New York: David McKay.

Blazer, Dan, Dana Hughes, and Linda K. George. 1987. "Stressful Life Events and the Onset of a Generalized Anxiety Syndrome." *American Journal of Psychiatry* 114: 1,178–1,183.

Brown, George W., and Tirril Harris. 1978. *Social Origins of Depression: A Study of Psychiatric Disorder in Women.* New York: Free Press.

Cameron, Norman. 1947. *The Psychology of Behavior Disorders.* Boston: Houghton Mifflin.

Carstairs, G. M. 1959. "The Social Limits of Eccentricity: An English Study." Pp. 145–157 In *Culture and Mental Health,* Marvin K. Opler, ed. New York: Macmillan.

Carter, Mary, and Samuel Flesher. 1995. "The Neurosociology of Schizophrenia: Vulnerability and Functional Disability." *Psychiatry* 58: 209–224.

Cockerham, William C. 1995. *Sociology of Mental Disorder,* 3rd ed. Englewood Cliffs, NJ: Prentice-Hall.

Conger, Rand D., and Glen H. Elder, Jr., eds. 1994. *Families in Troubled Times: Adapting to Change in Rural America.* New York: Aldine De Gruyter.

Conrad, Peter, and Joseph W. Schneider. 1980. *Deviance and Medicalization: From Badness to Sickness.* St. Louis, MO: Mosby.

Dohrenwend, Bruce P., and Barbara Snell Dohrenwend. 1969. *Social Status and Psychological Disorder.* New York: Wiley-Interscience.

Dohrenwend, Bruce P., and Barbara Snell Dohrenwend. 1980. "Psychiatric Disorders and Susceptibility to Stress." Pp. 183–197 in *The Social Consequences of Psychiatric Illness,* Lee N. Robins, Paula J. Clayton, and John K. Wing, eds. New York: Brunner/Mazel.

Dohrenwend, Bruce P., Barbara Snell Dohrenwend, M. S. Gould, B. Link, R. Neugebauer, and R. Wunsch-Hitzig. 1980. *Mental Illness in the United States: Epidemiological Estimates.* New York: Praeger.

Dohrenwend, Bruce P., Itzhak Levav, Patrick E. Shrout, Sharon Schwartz, Guedalia Naveh, Bruce G. Link, Andrew E. Skodol, and Ann Stueve. 1992. "Socioeconomic Status and Psychiatric Disorders: The Causation-Selection Issue." *Science* 155: 946–952.

Eaton, William W., and Carles Muntaner. 1999. "Socioeconomic Stratification and Mental Disorders." Pp. 259–283 in *Handbook for the Study of Mental Health: Social Contexts, Theories, and Systems,* Allan V. Horwitz and Teresa L. Scheid, eds. New York: Cambridge University Press.

Flaum, M. 1992 (December 14–18). "DSM-IV: The Final Stage." Paper presented at the 31st annual meeting of the American College of Neuropsychopharmacology, San Juan, Puerto Rico.

Flor-Henry, P. 1990. "Influence of Gender in Schizophrenia as Related to Other Psychopathological Syndromes." *Schizophrenia Bulletin* 16: 211–227.

General Accounting Office. 1988. *Homeless Mentally Ill: Problems and Options in Estimating Numbers and Trends.* Washington, DC: Committee on Labor and Human Resources, U.S. Senate.

Goffman, Erving. 1959. *Presentation of Self in Everyday Life.* Garden City, NY: Anchor Doubleday.

Goffman, Erving. 1961. *Asylums.* Garden City, NY: Anchor Doubleday.

Goleman, D. 1985. "New Psychiatric Syndromes Spur Protest." *New York Times,* November 19, pp. C-1, C-16.

Gove, Walter R. 1970. "Societal Reaction as an Explanation of Mental Illness: An Evaluation." *American Sociological Review* 35: 873–884.

Gove, Walter R. 1972. "The Relationship between Sex Roles, Marital Status, and Mental Illness." *Social Forces* 51: 34–44.

Gove, Walter R. 1982. "The Current Status of the Labeling Theory of Mental Illness." Pp. 285–297 in *Deviance and Mental Illness,* Walter R. Gove, ed. Beverly Hills, CA: Sage.

Gove, Walter R., and Michael Hughes. 1989. "A Theory of Mental Illness: An Attempted Integration of Biological, Psychological, and Sociological Variables." Pp. 61–76 in *Theoretical Integration in the Study of Deviance and Crime: Problems and Prospects,* Steven F. Messner, Marvin D. Krohn, and Allen E. Liska, eds. Albany: State University of New York Press.

Greenley, James. 1972. "The Psychiatric Patient's Family and Length of Hospitalization." *Social Problems* 13: 25–37.

Grob, Gerald N. 1994. *The Mad among Us: A History of the Care of America's Mentally Ill.* New York: Free Press.

Grusky, Oscar, and Melvin Pollner, eds. 1981. *The Sociology of Mental Illness: Basic Studies.* New York: Holt, Rinehart and Winston.

Herman, Nancy J. 1987. "'Mixed Nutters' and 'Looney Tuners': The Emergence, Development, Nature, and Functions of Two Informal, Deviant Subcultures of Chronic, Ex-Psychiatric Patients." *Deviant Behavior* 8: 235–258.

Hollingshead, August B., and Frederick Redlich. 1958. *Social Class and Mental Illness.* New York: John Wiley & Sons.

Horney, Karen. 1937. *The Neurotic Personality in Our Time.* New York: W. W. Norton.

Isaac, Rael Jean, and Virginia C. Armat. 1990. *Madness in the Streets: How Psychiatry and the Law Abandoned the Mentally Ill.* New York: Free Press.

Johnson, Ann Braden. 1990. *Out of Bedlam: The Truth about Deinstitutionalization.* New York: Basic Books.

Joint Commission on Mental Illness and Health. 1961. *Action for Mental Health, Final Report.* New York: Basic Books.

Keith, Samuel J., Darrel A. Regier, and Donald S. Rae. 1991. "Schizophrenic Disorders." Pp. 33–52 in *Psychiatric Disorders in America: The Epidemiological Catchment Area Study*, Lee Robins and Darrel A. Regier, eds. New York: Free Press.

Kennedy, John F. 1967. "The Role of the Federal Government in the Prevention and Treatment of Mental Disorders." Pp. 297–300 in *The Sociology of Mental Disorders: Analyses and Readings in Psychiatric Sociology*, S. Kirson Weinberg, ed. Chicago: Aldine.

Kessler, Ronald C. 1982. "A Disaggregation of the Relationship between Socioeconomic Status and Psychological Distress." *American Sociological Review* 47: 752–764.

Kessler, Ronald C., Roger L. Brown, and Clifford L. Broman. 1981. "Sex Differences in Psychiatric Help-Seeking: Evidence from Four Large-Scale Surveys." *Journal of Health and Social Behavior* 22: 49–64.

Kessler, R. C., C. B. Nelson, K. A. McKinagle, M. J. Edlund, R. G. Frank, and P. J. Leaf. 1996. "The Epidemiology of Co-Occurring Addictive and Mental Disorders: Implications for Prevention and Service Utilization." *American Journal of Orthopsychiatry* 66: 17–31.

Kiesler, Charles A., and Amy E. Sibulkin. 1987. *Mental Hospitalization: Myths and Facts about a National Crisis.* Beverly Hills , CA: Sage.

Kirkpatrick, Brian, and Robert W. Buchanan. 1990. "The Neural Basis of the Deficit Syndrome of Schizophrenia." *Journal of Nervous and Mental Disease* 178: 545–555.

Leaf, Philip J., and Martha Livingston Bruce. 1987. "Gender Differences in the Use of Mental Health-Related Services: A Re-examination." *Journal of Health and Social Behavior* 28: 171–183.

Lemert, Edwin M. 1972. *Human Deviance, Social Problems and Social Control*, 2nd ed. Englewood Cliffs, NJ: Prentice-Hall.

Link, Bruce G., and Francis T. Cullen. 1990. "The Labeling Theory of Mental Disorder: A Review of the Evidence." *Research in Community and Mental Health* 6: 75–105.

Link, Bruce G., Mary Clare Lennon, and Bruce P. Dohrenwend. 1993. "Socioeconomic Status and Depression: The Role of Occupations Involving Direction, Control, and Planning." *American Journal of Sociology* 98: 1,351–1,387.

Link, Bruce G., Jerrold Mirotznik, and Francis T. Cullen. 1991. "The Effectiveness of Stigma Coping Orientations: Can Negative Consequences of Mental Illness Labeling Be Avoided? *Journal of Health and Social Behavior* 32: 302–320.

Mechanic, David. 1972. "Social Class and Schizophrenia: Some Requirements for a Plausible Theory of Social Influence." *Social Forces* 50: 305–309.

Mechanic, David. 1989. *Mental Health and Social Policy*, 3rd ed. Englewood Cliffs, NJ: Prentice-Hall.

Mechanic, David. 1999. "Mental Health and Mental Illness: Definitions and Perspectives." Pp. 12–28 in *Handbook for the Study of Mental Health: Social Contexts, Theories, and Systems*, Allan V. Horwitz and Teresa L. Scheid, eds. New York: Cambridge University Press.

Menninger, Karl. 1946. *The Human Mind*. New York: Alfred A. Knopf.

Murphey, Jane M. 1982. "Cultural Shaping and Mental Disorders." In *Deviance and Mental Illness*, Walter R. Gove, ed. Beverly Hills, CA: Sage Publications.

Myers, Jerome K., Jacob J. Lindenthal, and Max P. Pepper. 1974. "Social Class, Life Events, and Psychiatric Symptoms: A Longitudinal Study." Pp. 191–205 in *Stressful Life Events: Their Nature and Effects*, Barbara Snell Dohrenwend and Bruce P. Dohrenwend, eds. New York: John Wiley & Sons.

Paykel, E. S. 1974. "Life Stress and Psychiatric Disorder: Applications of the Clinical Approach." Pp. 135–149 in *Stressful Life Events: Their Nature and Effects*, Barbara Snell Dohrenwend and Bruce P. Dohrenwend, eds. New York: John Wiley & Sons.

Pearlin, Leonard L. 1999. "Stress and Mental Health: A Conceptual Overview." Pp. 161–175 in *Handbook for the Study of Mental Health: Social Contexts, Theories, and Systems*, Allan V. Horwitz and Teresa L. Scheid, eds. New York: Cambridge University Press.

Penn, D. L., and J. Martin. 1998. "The Stigma of Severe Mental Illness: Some Potential Solutions for a Recalcitrant Problem." *Psychiatric Quarterly* 69: 235–247.

Perucci, Robert. 1974. *Circle of Madness: On Being Insane and Institutionalized in America*. Englewood Cliffs, NJ: Prentice-Hall.

Prior, Lindsay. 1993. *The Social Organization of Mental Illness*. London: Sage.

Raine, Adrian. 1993. *The Psychopathology of Crime: Criminal Behavior as a Clinical Disorder*. New York: Academic Press.

Redlich, Frederick C. 1957. "The Concept of Health in Psychiatry." Pp. 145–146 in *Explorations in Social Psychiatry*, Alexander H. Leighton, John A. Clausen, and Robert N. Wilson, eds. New York: Basic Books.

Robins, Lee, and Darrel A. Regier, eds. 1991. *Psychiatric Disorders in America: The Epidemiological Catchment Area Study*. New York: Free Press.

Rogler, Lloyd R., and August B. Hollingshead. 1965. *Trapped*. New York: John Wiley & Sons.

Rosenfield, Sarah. 1999. "Gender and Mental Health: Do Women Have More Psychopathology, Men More, or Both the Same (and Why)?" Pp. 348–360 in *Handbook for the Study of Mental Health: Social Contexts, Theories, and Systems*, Allan V. Horwitz and Teresa L. Scheid, eds. New York: Cambridge University Press.

Rosenhan, David L. 1973. "On Being Sane in Insane Places." *Science* 179: 250–258.

Rotenberg, Mordechai. 1975. "Self-Labeling Theory: Preliminary Findings among Mental Patients." *British Journal of Criminology* 15: 360–375.

Roth, Martin, and Jerome Kroll. 1986. *The Reality of Mental Illness.* Cambridge: Cambridge University Press.

Rothman, David. 1971. *The Discovery of the Asylum: Social Order and Disorder in the New Republic.* Boston: Little, Brown.

Scheff, Thomas J. 1974. "The Labeling Theory of Mental Illness." *American Sociological Review* 39: 444–452.

Scheff, Thomas J. 1975. *Labeling Madness.* Englewood Cliffs, NJ: Prentice-Hall.

Scheff, Thomas J. 1999. *Being Mentally Ill: A Sociological Theory*, 3rd ed. Hawthorne, NY: Aldine de Gruyters.

Scull, Andrew. 1977. *Decarceration: Community Treatment and the Deviant—A Radical View.* Englewood Cliffs, NJ: Prentice-Hall.

Shibutani, Tamotsu. 1986. *Social Processes.* Berkeley: University of California Press.

Srole, Leo, Thomas S. Langner, Stanley T. Michael, Price Kirkpatrick, Marvin K. Opler, and Thomas A. C. Rennie. 1977. *Mental Health in the Metropolis, Book Two*, rev. ed., Leo Srole and Anita K. Fisher, eds. New York: Harper and Row.

Srole, Leo, Thomas S. Langner, Stanley T. Michael, Marvin K. Opler, and Thomas A. C. Rennie. 1962. *Mental Health in the Metropolis: The Midtown Manhattan Study.* New York: McGraw-Hill.

Stanton, Alfred H., and Morris S. Schwartz. 1954. *The Mental Hospital.* New York: Basic Books.

Stone, Alan A. 1982. "Psychiatric Abuse and Legal Reform: Two Ways to Make a Bad Situation Worse." *International Journal of Law and Psychiatry* 5: 9–28.

Szasz, Thomas S. 1974. *The Myth of Mental Illness*, rev. ed. New York: Harper and Row.

Szasz, Thomas S. 1987. *Insanity: The Idea and Its Consequences.* New York: John Wiley & Sons.

Thoits, Peggy A. 1987. "Gender and Marital Status Differences in Control and Distress: Common Stress versus Unique Stress Explanations." *Journal of Health and Social Behavior* 28: 7–22.

Thoits, Peggy A. 1994. "Stressors and Problem-solving: The Individual as Psychological Activist." *Journal of Health and Social Behavior* 35: 143–160.

Torrey, E. Fuller. 1974. *The Death of Psychiatry.* Radnor, PA: Chilton Books.

Torrey, E. Fuller. 1988. *Nowhere to Go: The Tragic Odyssey of the Homeless Mentally Ill.* New York: Harper and Row.

Torrey, E. Fuller, with Ann E. Bowler, Edward H. Taylor, and Irving I. Gottesman. 1994. *Schizophrenia and Manic-Depressive Disorder: The Biological Roots of Mental Illness as Revealed by the Landmark Study of Identical Twins.* New York: Basic Books.

Townsend, J. Marshall. 1975. "Cultural Conceptions, Mental Disorders, and Social Roles: A Comparison of Germany and America." *American Sociological Review* 40: 739–752.

Turner, R. Jay. 1999. "Social Support and Coping." Pp. 198–210 in *Handbook for the Study of Mental Health: Social Contexts, Theories, and Systems*, Allan V. Horwitz and Teresa L. Scheid, eds. New York: Cambridge University Press.

U.S. Surgeon General. 1999. *Mental Health: A Report of the Surgeon General, 1999.* Washington, DC: U.S. Public Health Service, Department of Health and Human Services.

Van Praag, Herman M. 1990. "The DSM-IV (Depression) Classification: To Be or Not to Be?" *Journal of Nervous and Mental Disease* 178: 147–149.

Weinberg, S. Kirson, and H. Warren Dunham. 1960. *The Culture of the State Mental Hospital*. Detroit, MI: Wayne State University Press.

Wheaton, Blair. 1983. "Stress, Personal Coping Resources, and Psychiatric Symptoms." *Journal of Health and Social Behavior* 24: 208–229.

Zigler, Edward, and Marion Glick. 1986. *A Development Approach to Adult Psychopathology*. New York: John Wiley & Sons.

Zimmerman, M. 1988. "Why Are We Rushing to Publish DSM-IV?" *Archives of General Psychiatry* 45: 1,135–1,138.

NAME INDEX